National Apprenticeship Training Program for Cooks

A Training Guide for the Culinary Industry

atp
AMERICAN TECHNICAL PUBLISHERS
ORLAND PARK, ILLINOIS 60467-5756

ACF
American Culinary Federation
The Standard of Excellence for Chefs

Cover Photos Courtesy of: Sullivan University; Vulcan-Hart, a division of the ITW Food Equipment Group LLC; Kolpak; Bobby McRee Watson; NSF International

American Technical Publishers, Inc., Editorial Staff

Editor in Chief:
Jonathan F. Gosse
Vice President—Production:
Peter A. Zurlis
Director of Product Development:
Cathy A. Scruggs
Art Manager:
James M. Clarke
Multimedia Manager:
Carl R. Hansen
Technical Editors:
Cathy A. Scruggs
Larry E. Pierce
Copy Editor:
Catherine A. Mini

Cover Design:
Jennifer M. Hines
Illustration/Layout:
Jennifer M. Hines
Melanie G. Doornbos
Samuel T. Tucker
CD-ROM Development:
Gretje Dahl
Nicole S. Polak
Daniel Kundrat

1 2 3 4 5 6 7 8 9 – 12 – 9 8 7 6 5 4 3 2

Printed in the United States of America

ISBN 978-0-8269-4196-1

This book is printed on recycled paper.

Acknowledgments

The American Culinary Federation Education Foundation (ACFEF) apprenticeship program and all its success are the result of the efforts of many individuals who work for the betterment of our apprentices and our organization as a whole. First and foremost, we thank the apprentices. Without their dedication and passion for becoming the next generation of professional chefs, we would have no goal to strive for. They work long hours in tough environments to become better culinarians. Their determination does not go unnoticed and is often what fuels the rest of us.

Just as importantly, we acknowledge the steadfast commitment of our many supervising chefs and program coordinators. They dedicate countless hours of time and effort to enriching the education of numerous apprentices, often without additional compensation or recognition. They tirelessly lead a new generation of culinarians to professional enlightenment.

No acknowledgment would be complete without mention of the many volunteers who make up the ACFEF National Apprenticeship Committee. These volunteers were invaluable in crafting and honing the current ACFEF apprenticeship program. Their time commitment was exceeded only by the value of their combined expertise. Without them, none of this would be possible.

American Technical Publishers and the American Culinary Federation would also like to acknowledge the experts who contributed content, reviewed content, and provided images and technical assistance during the development of this training guide for the culinary industry.

Content Development

Michael J. McGreal, CEC, CCE, CHE, FMP, CHA, MCFE
Department Chair, Culinary Arts & Hospitality Management
Joliet Junior College

Thomas J. Hickey, CEC, CCE, CFE, HOGT
Chef Ambassador
Sullivan University

Daniel Pliska, CEC
Executive Chef, University Club and University Catering
University of Missouri

Christopher Allen Tanner, CEC, WCC, CHE
Chef-Instructor
Schenectady County Community College
President, ACF Capital District–Central New York

Tina Powers, CEC, CCE, CEPC
Program Coordinator
Oregon Coast Culinary Institute

Acknowledgments

Technical Review

CW4 Russell D. Campbell, CEC
Advanced Food Service Training Division
U.S. Army

Robert C. Cawley, CEC, CEPC, AAC
Executive Chef
Trenholm State Technical College

Kevin C. Clarke, CCE, J.D.
Professor of Culinary Arts
Director of Culinary Education
Colorado Mountain College Culinary Institute

SFC Gabriel Delagarza
TWI Apprenticeship Representative
U.S. Army

Elizabeth Dickson, CEPC, CCI
Culinary Arts Instructor
Bossier Parish Community College

Marvis Hinson, CCE
Formerly with Savannah Technical Institute

Brian D. Knecht, CEC, CCA, ACE, AAC
Department Chair
Sarasota County School Board

Cindy A. Komarinski, CCC, CCE, HAAC, MBA
Associate Professor, Culinary Arts & Hospitality
Westmoreland County Community College

Steven F. Learned
Owner
Adventures in Catering

Donald R. Miller, CEC, CCE, AAC
Executive Chef
University of Notre Dame Food Services

Daniel Pliska, CEC
Executive Chef, University Club and University Catering
University of Missouri

Michael W. Riggs, PhD, CEC, CCA, FMP
Associate Professor of Culinary Arts
Bowling Green Technical College

Michael T. Roddey, MS Ed, CCE, CCC, CDM, CFPP, FMP
Department Chair/Assistant Professor, Culinary Arts & Hospitality
UAF Community & Technical College, Hutchison Institute of Technology

SFC Ronald Stafford
Enlisted Aide Instructor
Advance Culinary Instructor
92G Apprenticeship Program Coordinator

Mark W. Warren
Sergeant Major, USA
Joint Culinary Center of Excellence
U.S. Army

David W. Weir, MBA, CEC, CCE
Assistant Professor/Chef Instructor
Daytona State College
School of College of Hospitality and Culinary Management

Images and Technical Assistance

2008 ACF Culinary Team USA
ACP, Inc.
Agricultural Research Service, USDA
All-Clad Metalcrafters
Alpha Baking Co., Inc.
American Egg Board
American Lamb Board
American Metalcraft, Inc.
BakeMax
Barilla America, Inc.
Barker Company
Basic American Foods
Beef Checkoff
Blodgett Oven Company
Bobby McRee Watson
Bridgford Foods Corporation
Browne Foodservice
California Strawberry Commission
Canada Beef Inc.
Cape Cod Cranberry Growers' Association
Carlisle FoodService Products
Charlie Trotter's
Chef Daniel Pliska
Chef Eric LeVine
Chef Gui Alinat
Chef's Choice® by EdgeCraft Corporation
Cintas Corporation
Cooper-Atkins Corporation
Cres Cor
Czimer's Game & Seafoods, Inc.
Daniel NYC
D'Artagnan, Photography by Doug Adams Studio
Daymark® Safety Systems
Detecto, A Division of Cardinal Scale Manufacturing Co.
Dexter-Russell, Inc.
Edlund Co.
Edward Don & Company
Eloma Combi Ovens
Entourage
Florida Department of Agriculture and Consumer Services
Florida Department of Agriculture and Consumer Services, Bureau of Seafood and Aquaculture Marketing
Florida Department of Citrus
Florida Tomato Committee
Fluke Corporation
Fortune Fish Company
Frieda's Specialty Produce
Harbor Seafood, Inc.
Henny Penny Corporation
HerbThyme Farms
Hobart
House Foods
Idaho Potato Commission
Indian Harvest Specialtifoods, Inc./ Rob Yuretich
In-Sink-Erator
InterMetro Industries Corporation
Irinox USA
Kolpak
Kyocera Advanced Ceramics
Lauren Frisch
Lincoln Foodservice
L. Isaacson and Stein Fish Company
MacArthur Place Hotel, Sonoma
MacFarlane Pheasants, Inc.
ManYee DeSandies/Pics4Learning, Inc.
Matfer Bourgeat USA
McCain Foods USA
Melissa's Produce
Mercer Cutlery
Merrychef
Messermeister
Mushroom Council
National Cancer Institute
National Cherry Growers and Industries Foundation
National Chicken Council
National Garden Bureau Inc.
National Honey Board
National Oceanic and Atmospheric Administration/Department of Commerce
National Onion Association
The National Pork Board
National Pork Producers Council
National Turkey Federation
National Watermelon Promotion Board
New Zealand Greenshell™ Mussels
NSF International
Oregon Raspberry & Blackberry Commission
Paderno World Cuisine
Pear Bureau Northwest
Pepper-Passion Inc.
Perdue Foodservice, Perdue Farms Incorporated
Planet Hollywood International, Inc.
Plitt Seafood
Rishi Tea
Robot Coupe USA
San Jamar
Shenandoah Growers
Southern Pride
The Spice House
Strauss Free Raised
Sullivan University
Tanimura & Antle®
Trails End Chestnuts
Tru, Chicago
True FoodService Equipment, Inc.
Tyco/ANSUL
United States Department of Agriculture
United States Potato Board
U.S. Apple® Association
USA Rice Federation
U.S. Army
USDA National Nutrient Database
U.S. Fish & Wildlife Service
U.S. Navy Photo by MC3 Shawn Stewart
U.S. Range
U.S. Wellness Meats
Viking Commercial
Vita-Mix® Corporation
The Vollrath Company, LLC
Vulcan-Hart, a division of the ITW Food Equipment Group LLC
Wisconsin Milk Marketing Board, Inc.

Technical Support at Video Locations

Joliet Junior College
Sullivan University

Contents

STEWARD STATION

BREAKFAST STATION

VEGETABLE, STARCH & PASTA STATION

Contents

FABRICATION STATION

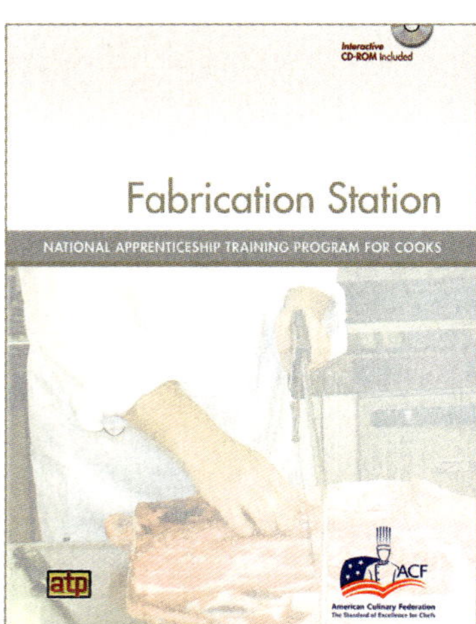

Contents

BROILER/CHAR-GRILL STATION

SOUP & SAUCE STATION

Contents

PANTRY & COLD FOODS STATION

COOKING TECHNIQUES STATION

BAKING & PASTRY STATION

Contents

SUPERVISORY DEVELOPMENT STATION

INTERACTIVE DVD CONTENTS

- Using This Interactive DVD
- Quick Quizzes®
- Illustrated Glossary
- Flash Cards
- Media Clips
- Checkpoints
- Culinary Math Applications
- Certification Exam Preparation
- Apprenticeship Online Portal
- ATPeResources.com

Procedures Index

BROILER/CHAR-GRILL STATION

SOUP & SAUCE STATION

PANTRY & COLD FOODS STATION

Procedures Index

COOKING TECHNIQUES STATION

BAKING & PASTRY STATION

Introduction

Welcome to the American Culinary Federation Education Foundation (ACFEF) apprenticeship program. Whether you are registered as an ACFEF apprentice or have a thirst for learning more, the culinary techniques outlined in this book have been validated by the culinary industry and sanctioned by the U.S. Department of Labor and the ACFEF National Apprenticeship Committee.

The ACFEF apprenticeship program focuses on hands-on training that allows apprentices to learn while being mentored by leading chefs in the industry. The program is rigorous and requires apprentices to complete 445 hours of classroom instruction with a minimum of 4,000 working hours in the kitchen. From the first to last day of the apprenticeship program, ACFEF apprentices are fully immersed in the smells, tastes, and textures of their culinary creations.

Several different options for the apprenticeship program are offered to meet the specific needs of apprentices. Program options range from two to three years and include the required classroom hours. Training is organized into 10 different stations, allowing apprentices to work at their own pace. It is an "earn while you learn" approach to training. Graduates from the ACFEF apprenticeship program earn recognition as an apprentice from the U.S. Department of Labor and the American Culinary Federation (ACF) and are given the opportunity to test for industry-recognized certification.

The ACFEF apprenticeship program also addresses sustainability. The quality of the food a chef serves is directly related to the health of the ecosystem. The program encompasses a multitude of topics from recycling and composting to purchasing locally and organically.

The journey to becoming a chef can be challenging and at times may seem difficult. But there is no substitute for valuable experience. Best of luck to you as you embark on your way to becoming a chef.

The ACFEF Apprenticeship Team

American Culinary Federation
Education Foundation

Book Features

The *National Apprenticeship Training Program for Cooks* includes several features to make learning easier.

Introductions focus attention on the concepts to be addressed.

Production Tips highlight ways to work efficiently in the professional kitchen.

Nutrition Notes highlight key nutrition information about specific foods.

Procedures list the steps required to prepare a specific food. Complex procedures often include illustrations.

Media Clip icons indicate that an applicable media clip can be referenced on the Interactive DVD.

QR codes link to Internet resources and are directly tied to the topic by which they appear.

Checkpoints appear throughout the book, and each one serves as a comprehension check of a specific portion of the book content.

Sustainability Corner addresses content related to topics such as recycling, composting, energy conservation, and product repurposing. A QR code is included with this feature.

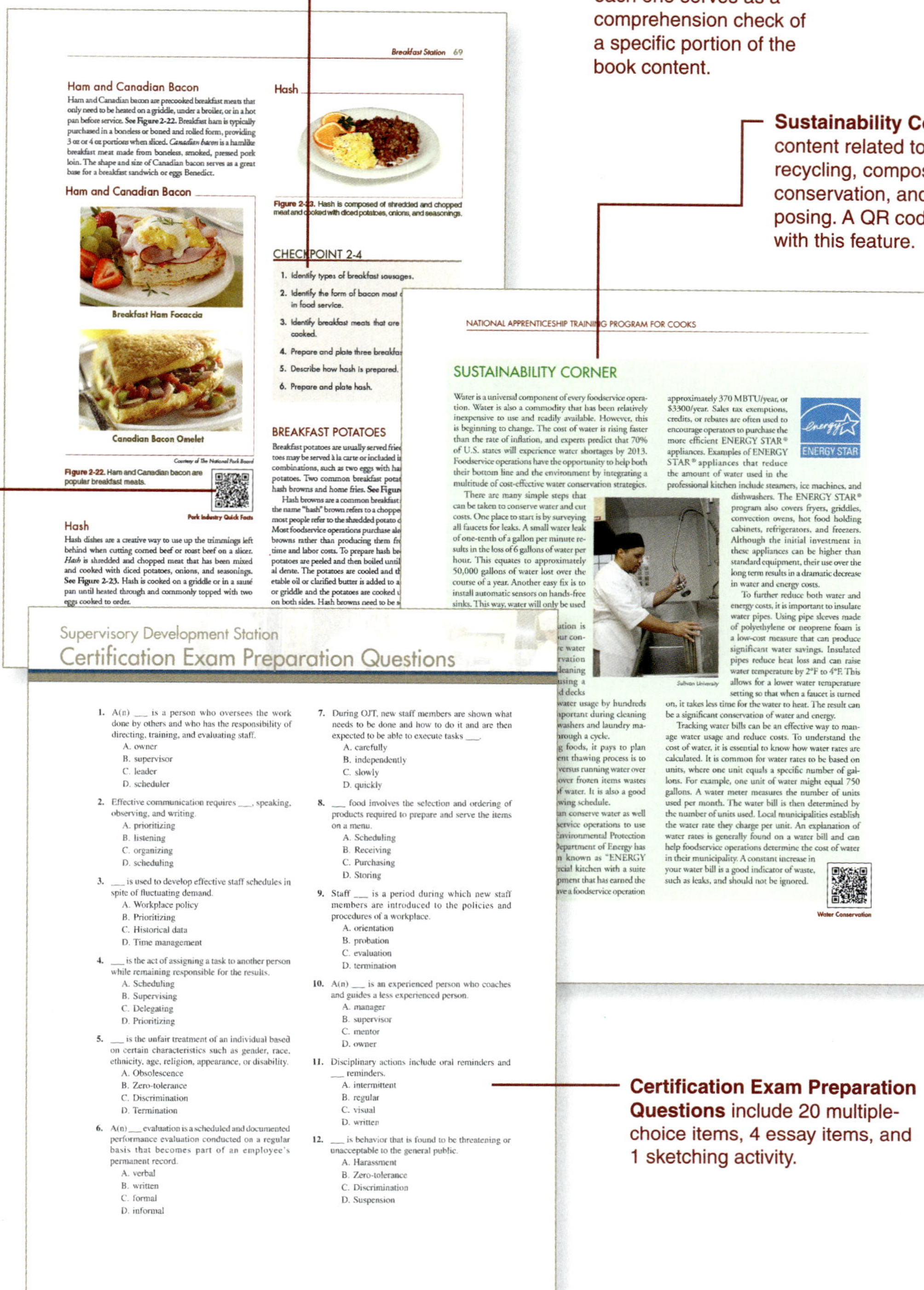

Certification Exam Preparation Questions include 20 multiple-choice items, 4 essay items, and 1 sketching activity.

Digital Features

Quick Response (QR) codes located throughout the book offer easy access to related information on the Internet using smartphone technology. To access content using a QR code, follow these steps:

1. Download a QR code reader application to the smartphone.
2. Open the application and scan the QR code on the book page with the smartphone.
3. View the displayed content or click on the hyperlink to access the related content.

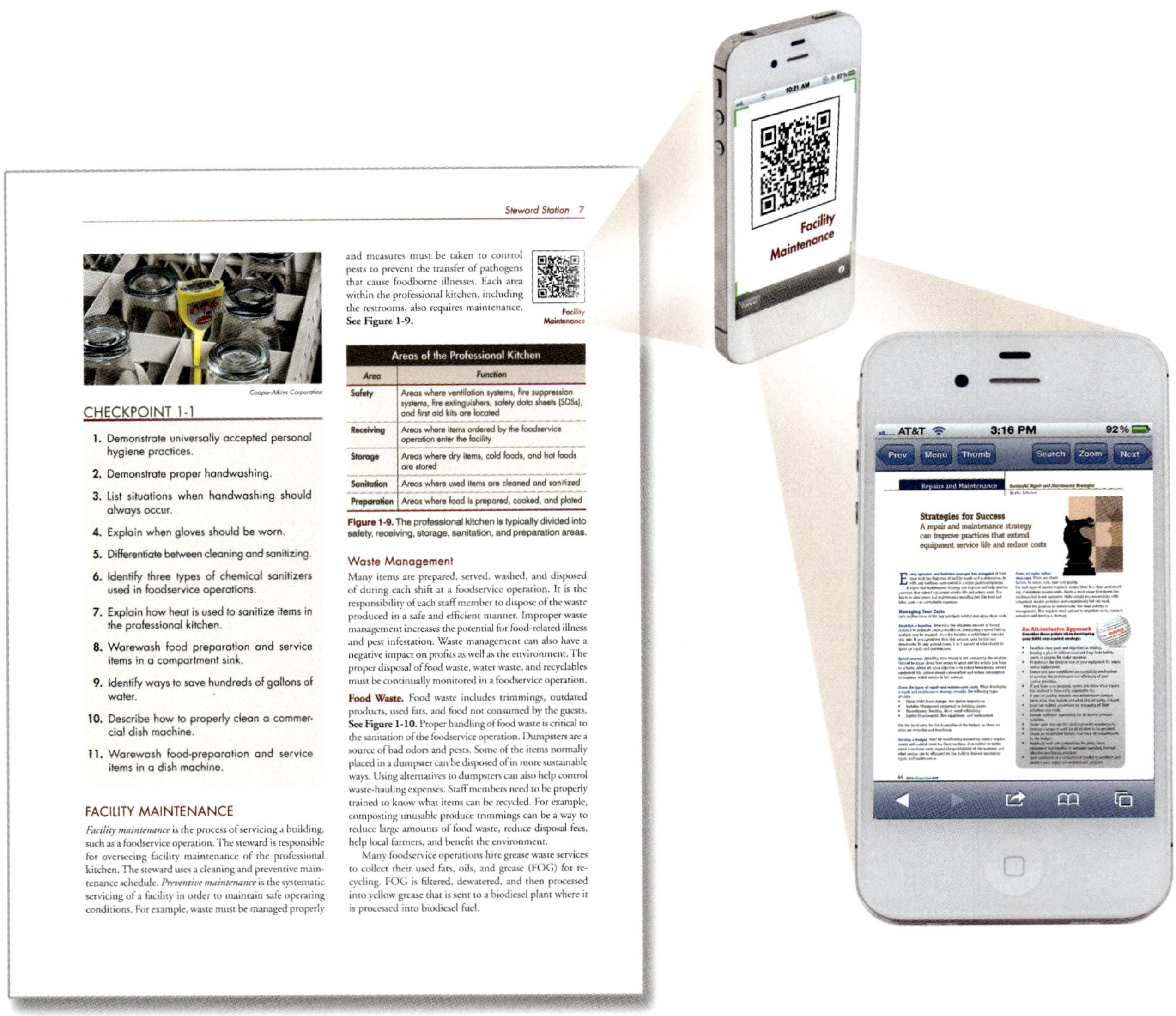

The *National Apprenticeship Training Program for Cooks* Interactive DVD is a self-study aid that reinforces content. This DVD is both Windows® and Mac OS® compatible.

Quick Quizzes® provide 25 interactive questions with embedded links to highlighted content within the book and to the Illustrated Glossary.

Illustrated Glossary provides a reference to commonly used terms. Selected terms are linked to interactive illustrations or media clips.

Flash Cards display terms and definitions, units of measure, measurement equivalents, knives, smallwares, and foodservice equipment.

Media Clips provide animations and video clips that reinforce and expand upon book content.

Checkpoints provide editable PDF files of the checkpoint review questions.

Culinary Math Applications offer animations and video clips that review and reinforce math skills used in the professional kitchen.

Certification Exam Preparation provides 25 multiple-choice questions in certification exam format.

Apprenticeship Online Portal allows apprentices to track training hours, make logbook entries, review culinary resources, and build online portfolios. Supervising chefs and program coordinators can use the portal to evaluate apprentice training, grade logbooks, and communicate with apprentices.

ATPeResources.com links to online reference materials that support continued learning.

Steward Station

Introduction

In the first station in the ACFEF apprenticeship program, the steward station, you will learn what makes a professional kitchen run smoothly in terms of equipment and supplies, while also learning about warewashing, waste management, and storeroom practices. In this station, you will learn how to be responsible for the overall cleanliness and maintenance of the kitchen at large as well as the equipment, tools, and supplies.

The steward station will teach you about kitchen equipment, from brooms to walk-in freezers to dry storage shelving units. Simple as it may sound, understanding the storeroom and how products should be stored is a foundational learning component of becoming successful in food service. This station will also teach you how to receive and distribute goods, take inventory, and prepare daily food orders.

Kitchens produce tons of waste. You will learn waste management methods, including how to properly dispose of food waste and maintain the garbage collection area. Maintaining cleanliness in each work area and with equipment is also critical. In addition, identifying pest infestations and where they originate from is a necessary component in prepping the kitchen for proper pest control practices. Learning how to perform an unofficial health inspection will be a skill of great value throughout your career.

You will also learn sustainable practices for purchasing and procurement by identifying environmentally friendly cleaning products. Understanding what products may be replaced with biodegradable items in a cost-effective manner will help you further sustainable best practices in the kitchen.

Guests expect a foodservice operation to be clean and to serve food that is safe to eat. The function of a steward is to focus on the safety, sanitation, and maintenance practices used in the professional kitchen. A steward oversees how goods are received, stored, and handled throughout the flow of food. A steward also oversees the safe use and maintenance of the equipment used to prepare and serve food as well as the physical kitchen facility. The proper handling of every item that enters the operation and is stored, used, and disposed of creates an efficient and productive workplace.

SANITATION PRACTICES

The foodservice staff is the most important part of any food safety and sanitation program. Sanitary work conditions reduce the threat of contaminants and foodborne illness. Following good personal hygiene standards, adhering to handwashing procedures, properly cleaning and sanitizing the work environment, and managing pests help promote sanitary work conditions.

Personal Hygiene

Personal hygiene is the physical care maintained by an individual. Foodservice staff must always practice good personal hygiene to help eliminate the spread of pathogens. **See Figure 1-1.** An appropriate head covering, a clean and pressed jacket, and a clean apron are always required. Gloves are worn when working with foods that will not be cooked prior to service. The following are universally accepted personal hygiene practices for staff in foodservice operations:

- Shower or bathe daily.
- Keep hair washed, combed, and covered as required.
- Keep facial hair clean, trimmed, and covered as required.
- Wear proper and clean clothing for the job.
- Do not wear nail polish or artificial fingernails.
- Except for a plain wedding band, do not wear jewelry.
- Keep hands and fingernails clean.
- Handle food only as required and always with clean or gloved hands.
- Do not allow dirty utensils or equipment to touch food.
- Eat and smoke only in designated areas.
- Bandage all cuts and sores, and wear gloves over injured hands.
- Do not cough, spit, or sneeze near food or in food preparation areas. Always cover your mouth with your arm when coughing and sneeze into a handkerchief or tissue. Wash hands immediately after coughing or using a handkerchief or tissue.
- Notify a supervisor and stay home when experiencing vomiting, diarrhea, jaundice, or a sore throat and fever.

Personal Hygiene

Sullivan University

Figure 1-1. Foodservice staff must always practice good personal hygiene to help eliminate the spread of pathogens.

Handwashing

Every member of the foodservice staff must follow FDA-approved handwashing procedures. The FDA Food Code § 2-301.12, Personal Cleanliness—Hands and Arms, specifies the following handwashing procedure for all foodservice employees. Handwashing stations must be located in preparation, service, and dishwashing areas. Handwashing stations must be equipped with soap and an approved means for drying hands. Hand sanitizers must be FDA-approved for contact with food and food preparation utensils.

Procedure for Handwashing

1. Wet hands and arms with hot water (at least 100°F).

2. Lather fingers, fingertips, areas between the fingers, hands, and lower arms with soap.

3. Scrub vigorously for at least 20 seconds.

4. Clean under the fingernails and between fingers.

5. Rinse hands and arms thoroughly with warm water.

6. Dry hands with disposable paper towels, a heated-air hand drying device, or a drying device that delivers high-velocity pressurized air.

Handwashing
Media Clip

Restrooms used by foodservice employees must display signs notifying employees that they must wash their hands before returning to work. Foodservice staff must also wash their hands immediately before beginning any food preparation task and after the following:

- handling raw poultry, seafood, or meat
- handling money
- using the restroom
- touching the hair, face, or body
- coughing, sneezing, or using a tissue
- eating, drinking, chewing gum, and using tobacco products
- handling chemicals
- disposing of garbage

Gloves

Gloves are worn when handling and serving ready-to-eat foods such as salads and sandwiches. However, gloves are not a substitute for handwashing. Before applying gloves, hands should be washed and dried using the proper handwashing procedure. Gloves need to be changed when an employee leaves a station, after handling raw foods, and when gloves are torn. **See Figure 1-2.** Each time gloves are removed, hands should be washed again.

Gloves

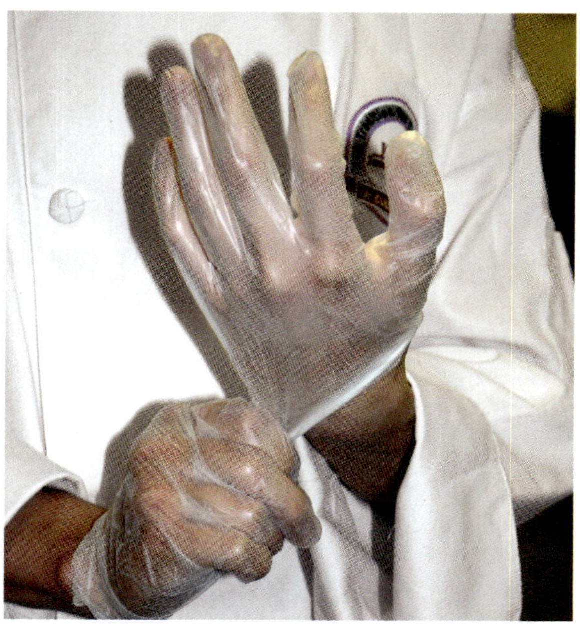

Figure 1-2. Gloves need to be changed when an employee leaves a station, after handling raw food, and when gloves are torn.

Cleaning and Sanitizing

All food contact surfaces must be cleaned and sanitized after each use, before working with any food, or a minimum of every 4 hours. *Cleaning* is the process of removing food and residue from a surface. **See Figure 1-3.** Different types of cleaning agents are used depending on the item to be cleaned. For example, detergents remove residue from surfaces, and solvent cleaners cut through grease. An item must be cleaned before it can be sanitized.

Cleaning Surfaces

Cintas Corporation

Figure 1-3. Cleaning is the process of removing food and residue from a surface.

Sanitizing is the process of using chemicals or heat to destroy or reduce harmful microorganisms to a safe level. **See Figure 1-4.** *Chemical sanitizing* is the process of using a chemical solution to reduce harmful pathogens to a safe level. Quaternary ammonium compounds (quats), chlorine, and iodine are types of chemical sanitizers used in the professional kitchen. It is important to know which sanitizers work best with local water sources. *Heat sanitizing* is the process of using dry heat, very hot water, or steam to reduce harmful pathogens to a safe level. An example of heat sanitizing is immersing an object in water that is at least 171°F for a minimum of 30 seconds.

Health Inspections

Common Sanitizers			
Sanitizing Agent	*Concentration Level*	*Recommended Temperature*	*Minimum Contact Time*
Quaternary ammonium	200–400* ppm	75°F	30 seconds
Chlorine	50 ppm	100°F	10 seconds
	100 ppm	55°F	
Iodine	12.5 ppm	75°F	30 seconds
Steam	NA	200°F	5 minutes
Dry heat	NA	180°F	20 minutes
Water	NA	170°F	5–20 minutes

* *varies based on water hardness*

Figure 1-4. An object can be sanitized using chemicals or heat.

Proper cleaning and sanitation of equipment is essential for safe food handling. The presence of bacteria on kitchen equipment can create cross-contamination and spread foodborne illnesses. Even though a surface may be visibly clean, it may not be sanitary. All surfaces in a foodservice operation need to be clean and any surface that can potentially come in contact with food needs to be sanitized. Cleaning tools, sanitizing tools, handwashing sinks, prep sinks, warewashing sinks, and dish machines are used to clean and sanitize the professional kitchen.

Cleaning Tools. Brushes, scrub pads, brooms, mops, buckets, and cleaning towels are all used in foodservice operations. A different set of cleaning tools should be used for food contact surfaces, surfaces that do not come into contact with food, and restroom surfaces to prevent cross-contamination. Many foodservice operations use color coding to designate cleaning tools for specific areas.

Cleaning tools need to be cleaned and sanitized to prevent cross-contamination. Once washed and sanitized, cleaning tools should be stored away from food preparation areas. Mops and brooms should be hung on hooks to prevent them from touching the floor and to allow them to dry properly. **See Figure 1-5.** Mop buckets and general purpose cleaning buckets should never be dumped into food preparation sinks. A designated utility sink should be used to dispose of wastewater.

Cleansers used in a foodservice operation include detergents, degreasers, delimers, and abrasive cleaners. Detergents contain cleaning agents called surfactants, which loosen dirt and cooked-on items, making cleaning surfaces easier. Degreasers contain detergents that dissolve cooked-on grease. Delimers are used to dissolve mineral deposits, tarnish, and rust. Abrasive cleaners are used to scour off cooked-on foods.

Mop and Broom Storage

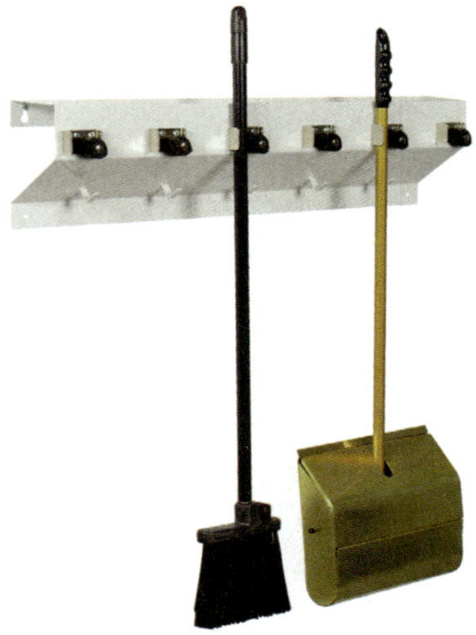

Edward Don & Company

Figure 1-5. Mops and brooms should be hung on hooks to prevent them from touching the floor and allow them to dry properly.

Sanitizing Tools. Sanitizing tools are used to remove harmful microorganisms from or prevent their growth on food contact surfaces. Sanitizing tools should be sanitized between uses and stored separately from other cleaning tools. Tools used for sanitizing include sanitizing buckets, sanitizing sinks, sanitizer test kits, and towels.

Sanitizing buckets are used to hold mild sanitizing solutions. **See Figure 1-6.** A cloth is dipped into the sanitizing solution and used to wipe down items after they have been washed to reduce the possibility of cross-contamination. Sanitizing solutions can be used to soak, spray, or rinse items.

Cleaning and Sanitizing Tools

Edward Don & Company

Figure 1-6. Sanitizing buckets are used to hold mild sanitizing solutions and sanitizing towels.

Handwashing Sinks. Small sinks designated for handwashing are located in the preparation, service, dishwashing, and restroom areas. Handwashing stations are equipped with approved soaps and hand-drying materials or equipment. **See Figure 1-7.** Notices reminding staff members to wash their hands prior to returning to their work stations must be posted near all handwashing sinks. Handwashing sinks must be cleaned and sanitized and the soap and towels replenished during each shift.

Handwashing Stations

Edward Don & Company

Figure 1-7. Handwashing stations are equipped with approved soaps and hand-drying materials or equipment.

Prep Sinks. A *prep sink* is a sink that is used in the preparation of food, such as rinsing whole produce. Prep sinks can have a single basin or several basins. They should be equipped with an approved chemical sanitizer for washing field produce such as potatoes and carrots. Prior to each use, prep sinks should be cleaned and sanitized. Prep sinks should not be used for handwashing or warewashing.

Production Tip

Prep sinks should not be used to thaw foods. The use of a thawing schedule that allows foods to thaw in a refrigerator saves gallons of water.

Warewashing Sinks. *Warewashing* is the process of cleaning and sanitizing all items used to prepare and serve food. Warewashing can be accomplished using a three- or four-compartment sink or a dish machine. Each compartment of a warewashing sink, commonly known as a pot sink, should be properly cleaned and sanitized before washing items. A warewashing sink is often used to wash and sanitize large items and items that are difficult to clean due to adhered foods. During heavy use periods, the water should be drained and the sink cleaned before refilling it. Warewashing sinks are used as a backup if the dish machine is not functioning properly.

Procedure for Warewashing Items in Three-Compartment Sinks

1. Scrape and rinse off each item.
2. In the first compartment, wash each item in a detergent and water solution that is at least 110°F.
3. In the second compartment, rinse the items by immersing them under water.
4. In the third compartment, immerse each item for 30 seconds in a chemical sanitizing solution mixed, per manufacturer recommendations, with 75°F to 120°F water.
5. Allow items to air dry.

Production Tip

Check all faucets to make sure they are turned off completely and not dripping. Hundreds of gallons of water are wasted from dripping faucets.

Dish Machines. A dish machine is an efficient way to wash and sanitize items that do not require special attention. Each dish is scraped, rinsed, and then placed in a rack so that all sides of the dish will be exposed to the spray action of the dish machine. Solid food material is scraped into a compost bin and then the dish is rinsed prior to being loaded into a dish machine. Dishes are often rinsed over a food waste disposer to eliminate smaller bits of food that would otherwise be caught in the warewashing sink drain. A *food waste disposer* is a food grinder mounted beneath a sink that is used to grind and eliminate solid food material.

Both low-temperature and high-temperature dish machines are used in the professional kitchen. In a low-temperature dish machine, the chemical sanitizer needs to reach at least 120°F and instructions for the chemical being used need to be followed carefully. The water temperature in a high-temperature dish machine should reach 180°F to properly sanitize items.

A single-tank dish machine has a door that is raised up to load racks of scraped dirty dishes and glassware. After the rack is loaded, the door is closed and the washing cycle automatically begins. A multitank dish machine takes racks by conveyor through the prewash, wash, rinse, sanitize, and dry cycle. **See Figure 1-8.** The racks emerge from the other end of the dish machine where they can be removed and stored. After washing, items are checked to ensure that they are clean. Items that come out of the dish machine soiled are run through the machine a second time.

Dish machines must be powered down and thoroughly cleaned at the end of a shift. During heavy use periods a dish machine may need to be cleaned every two hours. Properly cleaning dish machines reduces the potential for pest infestation and helps ensure that dishes are cleaned and sanitized properly.

Procedure for Cleaning a Dish Machine

1. Use a squeegee to pull water and food particles from the dish tables into the dish machine.
2. Thoroughly clean the dish tables with an approved cleaner, taking care to clean the corners as well as the top, sides, and bottoms of the tables.
3. Remove the scrap screens from inside the dish machine and wash them in a warewashing sink.
4. Use an approved nonabrasive cleaner to clean the interior and exterior of the dish machine.
5. Use the prerinse sprayer to remove the cleaner and rinse the dish machine.
6. Replace the screens and leave the dish machine doors open to allow the machine to air dry.

Production Tip

Only run a dish machine when it is full and turn it off at closing every day to save energy and money.

Commercial Dish Machines

Hobart

Figure 1-8. A multitank dish machine takes racks by conveyor through the prewash, wash, rinse, sanitize, and dry cycle.

Cooper-Atkins Corporation

CHECKPOINT 1-1

1. Demonstrate universally accepted personal hygiene practices.
2. Demonstrate proper handwashing.
3. List situations when handwashing should always occur.
4. Explain when gloves should be worn.
5. Differentiate between cleaning and sanitizing.
6. Identify three types of chemical sanitizers used in foodservice operations.
7. Explain how heat is used to sanitize items in the professional kitchen.
8. Warewash food preparation and service items in a compartment sink.
9. Identify ways to save hundreds of gallons of water.
10. Describe how to properly clean a commercial dish machine.
11. Warewash food-preparation and service items in a dish machine.

FACILITY MAINTENANCE

Facility maintenance is the process of servicing a building, such as a foodservice operation. The steward is responsible for overseeing facility maintenance of the professional kitchen. The steward uses a cleaning and preventive maintenance schedule. *Preventive maintenance* is the systematic servicing of a facility in order to maintain safe operating conditions. For example, waste must be managed properly and measures must be taken to control pests to prevent the transfer of pathogens that cause foodborne illnesses. Each area within the professional kitchen, including the restrooms, also requires maintenance. **See Figure 1-9.**

Facility Maintenance

Areas of the Professional Kitchen	
Area	***Function***
Safety	Areas where ventilation systems, fire suppression systems, fire extinguishers, safety data sheets (SDSs), and first aid kits are located
Receiving	Areas where items ordered by the foodservice operation enter the facility
Storage	Areas where dry items, cold foods, and hot foods are stored
Sanitation	Areas where used items are cleaned and sanitized
Preparation	Areas where food is prepared, cooked, and plated

Figure 1-9. The professional kitchen is typically divided into safety, receiving, storage, sanitation, and preparation areas.

Waste Management

Many items are prepared, served, washed, and disposed of during each shift at a foodservice operation. It is the responsibility of each staff member to dispose of the waste produced in a safe and efficient manner. Improper waste management increases the potential for food-related illness and pest infestation. Waste management can also have a negative impact on profits as well the environment. The proper disposal of food waste, water waste, and recyclables must be continually monitored in a foodservice operation.

Food Waste. Food waste includes trimmings, outdated products, used fats, and food not consumed by the guests. **See Figure 1-10.** Proper handling of food waste is critical to the sanitation of the foodservice operation. Dumpsters are a source of bad odors and pests. Some of the items normally placed in a dumpster can be disposed of in more sustainable ways. Using alternatives to dumpsters can also help control waste-hauling expenses. Staff members need to be properly trained to know what items can be recycled. For example, composting unusable produce trimmings can be a way to reduce large amounts of food waste, reduce disposal fees, help local farmers, and benefit the environment.

Many foodservice operations hire grease waste services to collect their used fats, oils, and grease (FOG) for recycling. FOG is filtered, dewatered, and then processed into yellow grease that is sent to a biodiesel plant where it is processed into biodiesel fuel.

Food Waste

Barilla America, Inc.

Figure 1-10. Food waste includes food not consumed by guests.

Water Waste. Water is used in almost every facet of a foodservice operation. Water is used to wash produce, steam foods, and clean the facility. *Wastewater* is water that has been used to prepare food or beverages or for cleaning. The proper and immediate disposal of wastewater helps prevent harmful pathogens, dirt, and chemicals from coming into contact with food or equipment used to produce food and beverages.

All foodservice operations have floor drains and custodial sinks that are used to dispose of wastewater. Hoses with high-pressure nozzles are often used to clean floors, floor mats, and even walls. **See Figure 1-11.** Water brooms are a water-efficient alternative to hoses that use half as much water. After using a hose or a water broom, a caution sign should be placed on the wet area to warn others of the danger of the wet surface. A caution sign should also be used any time a spill occurs.

Floor Mats

Cintas Corporation

Figure 1-11. Hoses with high-pressure nozzles are often used to clean floors, floor mats, and even walls.

Recyclables. Foodservice operations generate a large amount of waste on a daily basis. By separating items such as cardboard, glass, and produce trimmings, foodservice staff members can reduce the amount of waste in the environment. Many recycling companies provide bins for recyclables to help with the sorting of common recyclable items. Recycle containers should be cleaned thoroughly each day and allowed to dry upside down.

Pest Control

Insects and rodents are pests that can damage supplies, contaminate food, and transmit disease. Foodservice operations follow a regular pest maintenance schedule to control pests. Upon receipt, all food and supplies are checked for any indication of pests. Suspect items are rejected.

Since pests need food and water to survive, garbage areas, food waste disposal areas, and dish areas are prime places for them to thrive. **See Figure 1-12.** Garbage dumpsters and recycling collection areas should be located as far as possible from the exterior doors of the facility, and garbage and recycling should not be allowed to accumulate. Performing the following tasks during each shift can help control pests:

- Dispose of food scraps immediately.
- Clean up spills immediately.
- Wash soiled items immediately.
- Do not leave open doors unattended.
- Do not allow trash to accumulate.
- Empty compost bins.
- Clean restrooms regularly.

Production Tip

Pests can enter facilities through the smallest of openings. To prevent pests from entering a facility, all doors, windows, air vent screens, and other openings need to be securely sealed. Checking seals and gaskets on doors, windows, and equipment creates a more energy efficient kitchen.

Most foodservice operations contract a pest control company to inspect the premises and apply pest control chemicals on a regular basis. A pest control technician will review the inspection report with the supervisor on duty. The report will contain any observations made during the inspection and possible solutions, as well as list the chemicals used and the locations where they were applied. Most health departments review these reports during health department inspections.

Pest Magnets

Garbage Cans

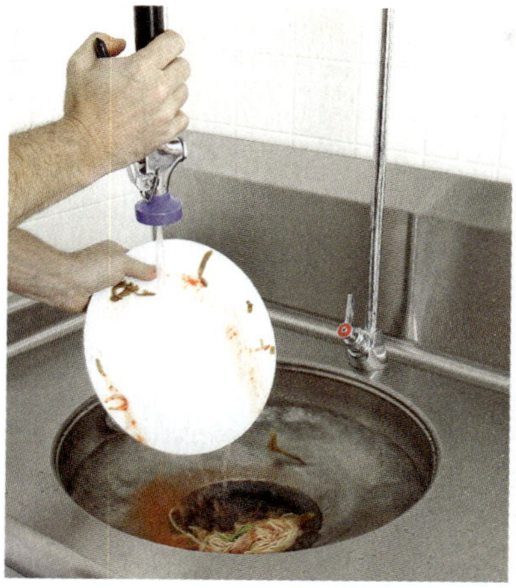

In-Sink-Erator

Disposals

Dish Areas

Figure 1-12. Since pests need food and water to survive, garbage areas, food waste disposal areas, and dish areas are prime places for them to thrive.

Restrooms

Restrooms should always be well-maintained. If restrooms are unclean, guests may wonder if the kitchen is unclean. Restrooms must be clean and maintained during each shift. Toilets and sinks must be cleaned regularly and their handles sanitized. Floors must be swept and mopped. Mirrors should be cleaned and the trash emptied during each shift.

Sanitation Areas

Sanitation areas include locations in the kitchen and dining room where items are cleaned and sanitized, such as a ware-washing station and a custodial closet where cleaning supplies and equipment are stored. Most municipalities require sanitation supplies and equipment to be stored behind closed doors.

Receiving Areas

The first steps for safe food handling in a professional kitchen occur the moment food enters the building through the receiving area. The *receiving area* is the area of the professional kitchen where all delivered items are checked for freshness, proper amount, correct temperature, and accurate price. If a problem is detected with an item at the time it is received, the item is immediately returned to the supplier.

Storage Areas

Dry storage areas are used to hold dry or packaged items that do not require refrigeration. Dry storage rooms are kept at temperatures between 50°F and 70°F. All dry storage items must be kept at a minimum of 6 inches above the floor. **See Figure 1-13.** Dry storage shelving units are typically cleaned with soap and water.

Dry Storage

Sullivan University

Figure 1-13. Dry storage items must be kept at a minimum of 6 inches above the floor.

Cold storage areas include refrigeration and freezer units that may be as small as an under-counter refrigerated cabinet or as large as a refrigerated warehouse. Refrigerator temperature settings must be cold enough to keep the internal temperature of food at 41°F or below and freezers set at −10°F to keep products frozen. Aluminum dunnage racks, vinyl-coated wire racks, and heavy-duty polymer racks are used in cold storage areas because they are made from materials that do not rust. Shelving units need to be cleaned with soap and water. Some shelves are treated with an antimicrobial compound, but this does not replace the need for regular cleaning.

Ventilation Systems

A *ventilation system* is a large exhaust system that draws heat, smoke, and fumes out of the kitchen and into the outside air. A properly working ventilation system is essential in any foodservice operation to remove excess heat, smoke, and fumes that can build up in the kitchen. **See Figure 1-14.** A large fan located at the end of the ventilation system ductwork creates a vacuum that draws exhaust air out from the building. As hot, greasy air is drawn up into the hood of a ventilation system, grease tends to build up. This grease must be cleaned off of the hood and the fire-suppression components regularly to prevent potential fires.

Ventilation Systems

Photo Courtesy of Tyco/ANSUL

Figure 1-14. A properly working ventilation system is essential in any foodservice operation to remove excess heat, smoke, and fumes that can build up in the kitchen.

Production Tip

Grease needs to be disposed of properly and not poured down the drain. Grease will clog drains and contaminate groundwater, which adversely affects the environment.

The grease reservoir at the edge of the exhaust hood must also be emptied regularly and run through a dish machine. The interior and exterior surfaces of the exhaust hood must be cleaned daily with a mild grease-cleaning detergent and polished with a towel.

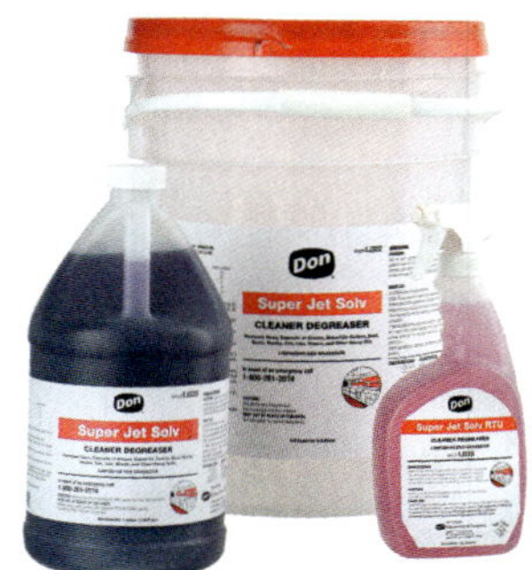

Edward Don & Company

The ventilation filters must be removed and cleaned, or replaced, according to a set schedule, such as once per week. Ventilation filters can be thoroughly washed in a warewashing sink or run through a dish machine to remove grease and dirt. The area behind the filters must also be inspected for grease buildup and cleaned periodically.

Depending on the overall volume of the kitchen, the exhaust hood should have a detailed cleaning monthly, quarterly, or annually based on the recommendation of the National Fire Protection Association (NFPA®) or local fire inspector. The detailed cleaning should be performed by a licensed and bonded professional who uses a steam-driven power washer.

Preparation Areas

A *work section* is an area within the professional kitchen that is dedicated to a specific set of tasks. For example, a baking work section is dedicated to the production of baked goods. This area is not used for other purposes such as meat fabrication. Common work sections within the professional kitchen include sanitation, hot foods, short order, cold foods, baking and pastry, and beverage sections. **See Figure 1-15.**

Work sections are further divided into work stations. A *work station* is an area within a work section that is relatively self-sufficient and provides the necessary tools and equipment to perform specific tasks within the section. For example, a broiler station would be equipped with a broiler, hand tools such as tongs, cold storage to hold uncooked food items, seasonings, and plates to hold the prepared food.

Preparation in the Professional Kitchen		
Work Sections	*Work Stations*	*Common Equipment*
Sanitation	• Warewashing • Dish machine • Maintenance	Warewashing equipment, dish racks, dish machines, cleaning and sanitizing supplies and equipment
Hot Foods	• Sauté/Sauce • Vegetable • Fry	Ranges, ovens, steamers, fryers, holding equipment, refrigeration units, work tables
Short Order	• Broiler • Grill/Griddle	Broiler, grill, griddle, refrigeration units, work tables
Cold Foods	• Salad • Sandwich • Plated dessert	Slicers, mixers, blenders, refrigeration units, work tables
Baking and Pastry	• Baking • Pastry	Scales, mixers, proofing cabinets, sheeters, ovens, refrigeration units, bakeshop work tables
Beverages	• Hot beverage • Cold beverage • Alcoholic beverage	Coffee and tea brewers, ice bins, soft drink dispensers, drink blenders, bar sinks, refrigeration units

Figure 1-15. Common work sections within the professional kitchen include sanitation, hot foods, short order, cold foods, baking and pastry, and beverage sections.

CHECKPOINT 1-2

1. Define facility maintenance.
2. Identify three types of waste that must be continually managed.
3. Identify three ways foodservice operations can reduce waste management costs.
4. Explain how FOG can be recycled.
5. Identify six tasks that can help control pests in a foodservice operation.
6. Identify six areas of the professional kitchen that need to be maintained.
7. Explain how dry and cold storage areas are kept clean.
8. Explain how a ventilation system is cleaned.
9. Differentiate between work sections and work stations.
10. Identify common work sections and the equipment used in each.

EQUIPMENT MAINTENANCE

Prior to operating equipment in the professional kitchen, safety guidelines must be understood to protect against personal injury and damage to equipment. Unsafe or careless operation of equipment can lead to serious injury. The following safety guidelines are recommended for operating and maintaining professional kitchen equipment:

- Prior to use, read manufacturer instructions for safe equipment operation.
- Install and use all available safety features, such as guards, on equipment.
- Securely station or anchor equipment to prevent slipping or falling.
- Turn off and unplug all equipment before cleaning or disassembling.
- Check cords on electrical appliances to verify they are in safe operating condition.
- Never use extension cords to operate commercial equipment.
- Sanitize equipment after cleaning to prevent foodborne illness.
- Turn off and unplug equipment before reassembling and after cleaning or service.
- Notify a supervisor immediately if equipment malfunctions or damage is detected.

- Post a caution notice on malfunctioning or damaged equipment.
- Never use abrasive cleaners or steel scrubbing pads to clean equipment as they will damage the equipment surfaces.

The equipment used in a professional kitchen must withstand wear and tear and be able to be easily cleaned. Some pieces of equipment are designed to reduce preparation and cooking times and increase consistency, while other items simply provide a safe and durable work surface. NSF-certified products bear the NSF mark, indicating the product has passed rigorous inspection, can be easily maintained, and can withstand the daily wear and tear of a professional kitchen. **See Figure 1-16.**

The NSF mark indicates that the product has a smooth, nonporous, nontoxic, corrosion-resistant surface. The NSF mark also means that the internal corners of the item are sealed and smooth and that the external edges are rounded and smooth. An NSF-certified item also can be easily cleaned and easily taken apart for routine cleaning and maintenance.

Equipment Standards

NSF Marks

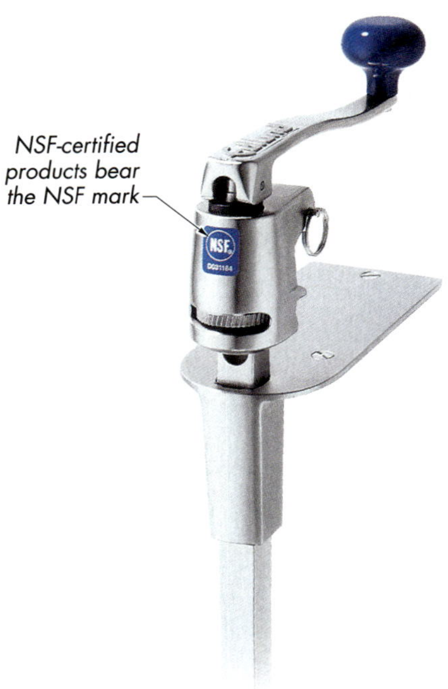

Edlund Co.

Figure 1-16. NSF-certified products bear the NSF logo, indicating the product is able to withstand the daily wear and tear of a professional kitchen.

Cold Storage Equipment

Refrigeration and freezer units are used to store cold items in the professional kitchen. All refrigeration units must be kept at 41°F or below and all freezers must be kept at 0°F or below. This often requires setting refrigeration units a few degrees lower in order to compensate for the doors being opened frequently. **See Figure 1-17.** To keep freezer temperatures in an acceptable range, items should be placed below the freezer load line, and the freezer doors should only be opened when necessary.

Refrigerated Storage

Kolpak

Figure 1-17. Refrigerators are set a few degrees below 41°F to compensate for the doors being opened frequently.

Refrigerator and freezer temperatures should be checked regularly. Refrigeration units can be easily checked by looking at door readout panels that show the internal temperature of the unit. Periodic verification that the refrigeration unit thermometer is accurate can be accomplished by placing a refrigerator thermometer inside the warmest area of the unit and comparing it to the unit thermometer.

The interior walls, shelves, and racks of refrigeration and freezer units should be cleaned regularly with a solution of baking soda and water. Four major types of refrigeration and freezer units used in the professional kitchen are walk-in units, roll-in units, reach-in units, and blast chillers. **See Figure 1-18.**

Cold Storage Equipment

Figure 1-18. Walk-in units, roll-in units, reach-in units, lowboys, and blast chillers are used for cold storage.

Walk-in Units. A *walk-in unit* is a room-size insulated storage unit used to store bulk quantities of cold or frozen food. Walk-in units are often outfitted with adjustable shelving. The shelving units used in refrigerators and freezers are either stainless steel or metal that has been dipped into a plastic-like resin to prevent rust or corrosion. Speed racks can be wheeled directly into walk-in units. Walk-in units can be built in any size or shape.

Roll-in Units. A *roll-in unit* is a refrigeration unit that allows speed racks to be rolled in and out of the unit through a door opening that is just above floor height. Roll-in units are similar to reach-in units in size and dimensions, but have a small ramp at the door opening to allow speed racks to be easily rolled in or out.

Reach-in Units. A *reach-in unit* is a temperature-controlled cabinet for storing cold or frozen food items. Reach-in units can be individual units or may have two or three doors, each with shelves the size of a standard sheet pan. Reach-in units are usually located throughout the kitchen, creating convenient refrigerated or frozen storage near where items are needed. A *lowboy* is a reach-in refrigerated unit located beneath a work surface. This allows the foodservice worker to prepare foods on the work surface and have adequate refrigerated storage beneath.

Blast Chillers. A *blast chiller* is a specialized cooling unit that rapidly reduces the temperature of foods, rendering them safe for immediate storage. A blast chiller can reduce the temperature of a cooked food to below 41°F very quickly. Items can also be quickly frozen in a blast chiller, resulting in a high-quality frozen product.

Preparation Equipment

Much of food preparation requires the use of preparation equipment such as slicers, blenders, juicers, food processors, buffalo choppers, vertical cutter/mixers, and bench and floor mixers.

Slicers. A *slicer* is an appliance that is used to uniformly slice foods such as meat and cheese. It has a regulator providing a wide range of slice thicknesses and a feed grip that firmly holds the top of the food item or serves as a pusher plate for slicing small end pieces. A slicer has a circular blade that rotates at a high speed, slicing items as they move across it. It is extremely important to ensure that the safety guards remain in place when using a slicer.

A slicer should be cleaned and sanitized after cutting each type of product. Failure to clean and sanitize slicers before slicing a different type of product can result in cross-contamination. **See Figure 1-19.**

Slicers

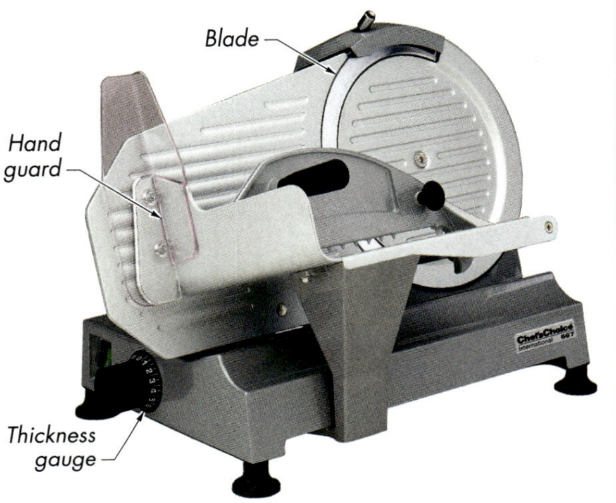

Chef's Choice® by EdgeCraft Corporation

Figure 1-19. Failure to clean and sanitize slicers before slicing a different type of product can result in cross-contamination.

Procedure for Cleaning a Slicer

1. Unplug the slicer and then remove the food carriage and the top and bottom blade covers.
2. Clean and sanitize the removed parts in a warewashing sink.
3. Carefully insert a folded paper towel between the knife blade and the surrounding edge guard. **CAUTION:** Slowly rotate the blade past the paper towel to remove any food residue along the edge of the blade. Remove the paper towel and discard.
4. Wash the surface of the slicer and the blade using a towel and a mild soap-and-water solution.
5. Rinse the surface of the slicer and the blade with a wet towel.
6. Wipe the surface of the slicer with a mild sanitizing solution.
7. Allow the slicer to air dry and then reassemble.

Blenders. Different types of blenders are used to process a variety of foods quickly and evenly. **See Figure 1-20.** A *blender* is a tall appliance with a slender canister that is used to chop, blend, purée, or liquefy food. It is designed for puréeing soups, soft foods, and beverages and can also be used to crush ice. An *immersion blender*, also known as a stick mixer, is a narrow, handheld blender with a rotary blade that is used to purée a product in the container in which it is being prepared. For example, an immersion blender can be inserted into a saucepot to purée a soup.

Blenders must be cleaned thoroughly between uses to prevent cross-contamination. Simply rinsing a blender will not remove food residue on or beneath the cutting blade.

Blenders

Figure 1-20. Different types of blenders are used to process a variety of foods quickly and evenly.

Procedure for Cleaning a Blender

1. Unplug the blender and then remove the canister from the base unit that houses the motor.
2. Unscrew the bottom of the canister to remove the threaded base, rubber gasket, and cutting blade.
3. Clean and then sanitize the canister, threaded base, rubber gasket, and cutting blade in a warewashing sink and allow them to air dry.
4. Wipe down the base with soap and water. Do not submerge the base.
5. Sanitize the base unit and allow it to air dry.
6. Reassemble the blender.

Immersion blenders are cleaned by carefully removing the cutting blade and then washing and sanitizing the blade and the bottom shaft of the blender. Do not submerge the motor of an immersion blender. Simply wipe it down with soap and water. Allow the equipment to air dry and then reassemble the immersion blender.

Juicers. A *juicer* is a device used to extract juice from fruits and vegetables. **See Figure 1-21.** A *juice extractor* is an electric machine that creates juice by liquefying raw vegetables and fruits and separating the fiber or pulp from the juice. Juice extractors are often referred to as juicers in the professional kitchen.

Juicers

Robot Coupe USA

Figure 1-21. A juicer is a device used to extract juice from fruits and vegetables.

Procedure for Cleaning a Juicer

1. Unplug the juicer and remove the canister, strainer basket, cutting blade, and all other parts from the base.
2. Clean and then sanitize all of the parts, except the motor, in a warewashing sink and allow them to air dry.
3. Wipe down the base of the juicer with soap and water. Do not submerge the base in water.
4. Use a wet towel to rinse the base.
5. Sanitize the base and allow it to air dry.
6. Reassemble the juicer.

Food Processors. A *food processor* is an appliance with an S-shaped blade and a removable bowl and lid that can be used to quickly chop, purée, blend, or emulsify foods. **See Figure 1-22.** A food processor also comes with blades for shredding, grating, slicing, and julienning foods.

Food Processors

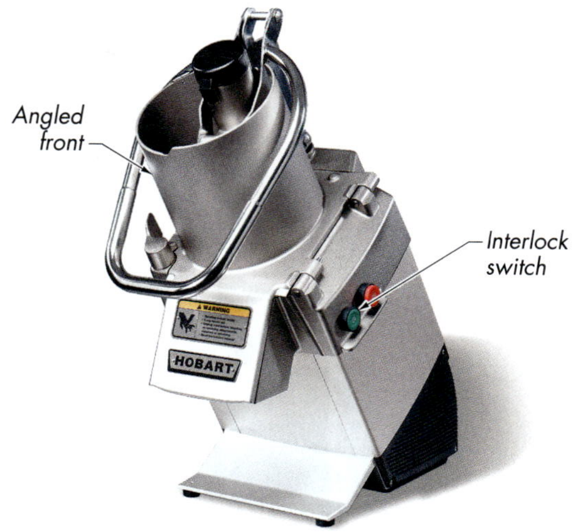

Hobart

Figure 1-22. A food processor is an appliance used to quickly chop, purée, blend, or emulsify foods.

Procedure for Cleaning a Food Processor

1. Unplug the food processor and remove the lid, cutting blade, and canister.
2. Clean and then sanitize all of the components, except the motor, in a warewashing sink and allow them to air dry.
3. Wipe down the base of the food processor with soap and water. Do not submerge the base.
4. Use a wet towel to rinse the base.
5. Sanitize the base and allow it to air dry.
6. Reassemble the food processor.

Robot Coupe USA

Buffalo Choppers. A *buffalo chopper* is an appliance used to process larger amounts of a product into roughly equal-size pieces. **See Figure 1-23.** Food passes under a hood, which houses a large, S-shaped blade. The coarseness of the cut is dependent on how long the food is left in the machine. The more times food passes under the blade of a buffalo chopper, the finer the cut. There is a built-in safety switch that prevents the machine from operating if the hood is open.

Buffalo Choppers

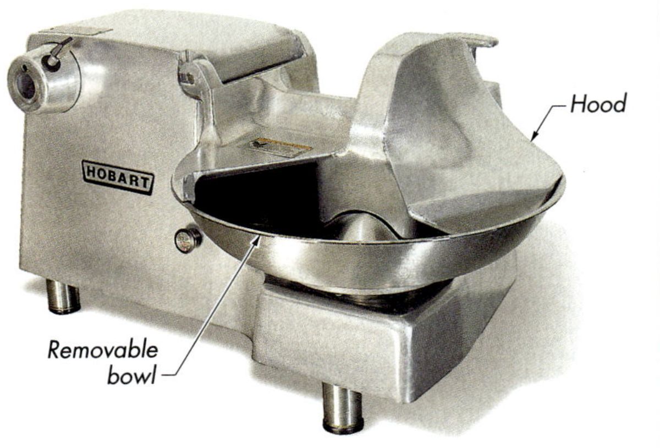

Hobart

Figure 1-23. A buffalo chopper is used to process larger amounts of a product into roughly equal-size pieces.

Procedure for Cleaning a Buffalo Chopper

1. Unplug the chopper and then unlock the latch that secures the bowl cover.
2. Remove the cover, blade comb, blade securing knob, cutting blade, and bowl.
3. Clean and then sanitize each disassembled part in a warewashing sink and allow the parts to air dry.
4. Wipe down the base of the buffalo chopper with soap and water. **CAUTION:** Be careful when handling the sharp blade of the chopper.
5. Use a wet towel to rinse the chopper.
6. Sanitize the chopper and allow it to air dry.
7. Reassemble the buffalo chopper.

Vertical Cutter/Mixers. A *vertical cutter/mixer (VCM)* is an appliance used to cut and mix foods simultaneously using high-speed blades and a mixing baffle, which is used to manually move the product into the blades. **See Figure 1-24.** A VCM is usually floor mounted with a 15–80 qt capacity bowl and has a built-in safety switch that prevents the machine from operating if the lid is open.

Vertical Cutters/Mixers

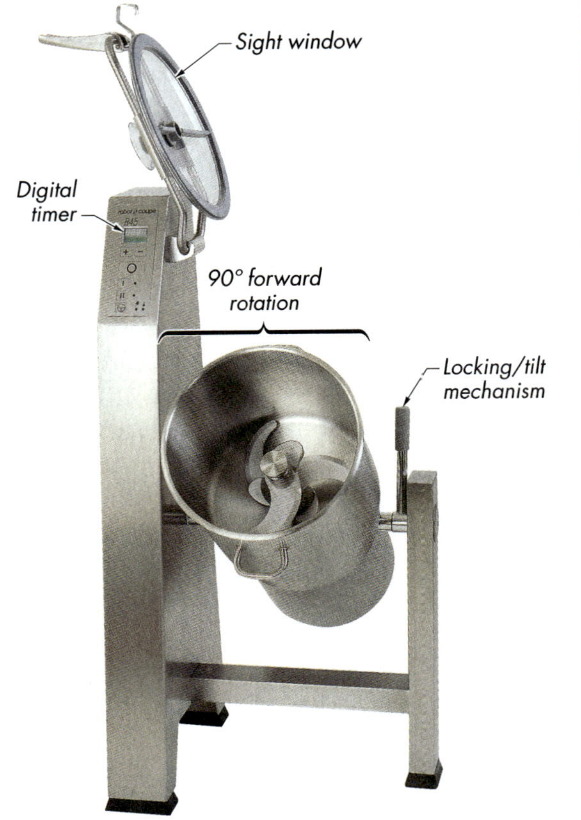

Robot Coupe USA

Figure 1-24. A VCM is used to cut and mix foods simultaneously using high-speed blades and a mixing baffle.

Procedure for Cleaning a Vertical Cutter/Mixer (VCM)

1. Unplug the VCM and then open the lid completely and remove the blade from the unit.
2. Using a mild soap-and-water solution, saturate the interior of the VCM and the inside of the lid. Allow them to soak for a few minutes.
3. Clean and then sanitize all of the attachments in a warewashing sink.
4. Allow the attachments to air dry and then reassemble. **CAUTION:** Be careful when handling the sharp blade of the VCM unit.
5. Clean the interior and exterior surfaces of the VCM unit using a mild soap-and-water solution to remove all visible food particles.
6. Rinse the VCM unit with clean rinse water.
7. Sanitize the VCM unit and allow it to air dry.
8. Reassemble the VCM.

Bench and Floor Mixers. A *mixer* is a versatile electric appliance with U-shaped arms that secure a stainless steel mixing bowl under a rotating head that accommodates various attachments. **See Figure 1-25.** Mixers can be operated at different speeds for light or aggressive mixing and come in both bench and floor models. Bench mixers range in size from 4½–20 qt capacities. Floor mixers range in size from 30–140 qt capacities.

A paddle attachment, also known as a beater, is used for mixing and creaming. A whip attachment is used for whipping volume into products. A hook attachment is used for kneading dough. A pastry knife attachment is used to cut fat into flour when making pastry dough. A chopper/grinder attachment is used to chop or grind meat, and a slicer/shredder attachment is used to slice or shred vegetables such as cabbage for coleslaw.

Mixer and Attachments

Hobart

Figure 1-25. A mixer is a versatile electric appliance that accommodates various attachments.

Procedure for Cleaning a Mixer

1. Unplug the mixer and remove the safety cage, the attachment, and the mixing bowl.
2. Clean and sanitize the safety cage, any attachments, and mixing bowl in a warewashing sink and then allow to air dry.
3. Clean the entire surface of the mixer with a mild soap-and-water solution to remove all visible food particles.
4. Use a wet towel to rinse the mixer.
5. Sanitize the mixer and then let it dry.
6. Reattach the mixing bowl, attachment, and safety cage.

Cooking Equipment

Cooking equipment is usually placed in a central location in the professional kitchen. When choosing cooking equipment for a professional kitchen, the menu should be considered, as the types of items frequently cooked will help to determine what equipment is needed. For example, if the menu is for a steakhouse, there is a greater need for broilers and grills than for griddles and open-flame ranges. Common cooking equipment includes ranges, induction cooktops, griddles, grills, broilers, steamers, steam-jacketed kettles, tilt skillets, fryers, and a variety of ovens.

Ranges. A *range* is a large appliance with surface burners. A variety of open-burner ranges and flat-top ranges are used in the professional kitchen. Open-burner ranges have open-flame burners that allow easy regulation of the intensity of heat. Flat-top ranges have a flat top that covers the burners, providing even heat across a larger cooking surface. Although a flat-top range takes longer to initially heat, it allows more cooking vessels to be heated at one time. Some ranges have both open burners and a flat top within the same unit.

Ranges should be spot-cleaned throughout each shift and thoroughly cleaned at the end of each day. **See Figure 1-26.** To make cleaning easier, the drip trays should be lined with foil to catch spills. Soiled aluminum foil should be discarded as required and replaced with clean foil.

Equipment Specifications

Procedure for Cleaning a Range

1. Check to make sure the range is off and then remove the burner grates and place them in a warewashing sink for a few minutes to loosen the baked-on food particles.
2. Clean under and around the burners with a mild soap-and-water solution.
3. Wash and sanitize the burner grates in the warewashing sink.
4. Remove the drip tray beneath the burners and clean it. Reline the tray with foil.
5. Clean the entire surface of the range, including the knobs, front, top, and sides, using a mild soap-and-water solution.
6. Use a wet towel to rinse the surface of the range.
7. Sanitize the surface of the range.
8. Reassemble the dry burner grates and the foil-lined drip tray.
9. Fire up the range to burn off any remaining moisture to prevent rusting.
10. Polish the exterior of the range using stainless steel cleaner if desired.

Ranges

Vulcan-Hart, a division of the ITW Food Equipment Group LLC

Open-Burner Range

U.S. Range

Flat-Top Range

Figure 1-26. Ranges should be spot-cleaned throughout each shift and thoroughly cleaned at the end of each day.

Induction Cooktops. An *induction cooktop* is an electromagnetic unit that uses a magnetic coil below a flat surface to heat food rapidly. Induction cooktops interact with magnetic cookware, quickly creating heat within the cookware itself. **See Figure 1-27.** The surface of an induction cooktop does not heat up like a traditional electric burner.

Induction Cooktops

Edward Don & Company

Figure 1-27. Induction cooktops interact with magnetic cookware, quickly creating heat within the actual cookware.

Procedure for Cleaning an Induction Cooktop

1. Unplug the induction cooktop.
2. Clean the entire surface of the unit using a mild soap-and-water solution to remove food particles. *Note:* If the ceramic surface is stained, use a gentle paste cleanser.
3. Use a wet towel to rinse the unit. Do not submerge the unit.
4. Sanitize the unit and allow it to air dry.

Griddles. A *griddle* is a solid cooking surface made of metal on which foods are cooked. Griddles are commonly used to cook items such as pancakes, eggs, and hamburgers. The surface temperature of a griddle is controlled with a thermostat.

Griddles need to be thoroughly cleaned at the end of each shift, and they may also need to be cleaned during a shift if heavily used. Cleaning involves removing all cooked-on food particles and carbon residue in order to get back to a shiny surface. A griddle brick is often used to clean the surface of a griddle. **See Figure 1-28.** The brick should only be used in the direction of the grain of the metal and never in a circular motion. A handle may be attached to the griddle brick for added leverage.

Cleaning Griddles

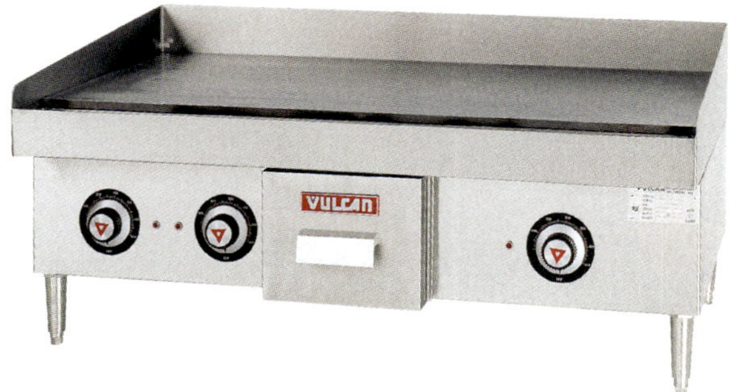

Vulcan-Hart, a division of the ITW Food Equipment Group LLC

Carlisle FoodService Products

Griddle Brick

Carlisle FoodService Products

Brick Handle

Figure 1-28. A griddle brick is often used to clean the surface of a griddle.

Procedure for Cleaning a Griddle

1. Heat the griddle to 200°F and pour a small amount of water onto the surface to loosen the cooked-on food and carbon residue. *Note:* If preferred, vinegar or lemon juice can be used.
2. Use a griddle brick to loosen any cooked-on food from the metal surface.
3. Use a griddle scraper to scrape the loosened food and carbon into the grease drawer.
4. Pour a small amount of water on the surface to rinse it. Use a clean towel to wipe the surface thoroughly clean.
5. Remove the grease drawer and discard its contents.
6. Wash and sanitize the grease drawer in a warewashing sink. After air drying, return the grease drawer to its place.
7. Clean and sanitize the splash guards and the griddle controls with a wet towel.
8. Use a towel to apply a very thin coat of oil to the surface of the griddle to keep it seasoned.

Production Tip

A cleaning gel may be spread over the surface of a griddle and allowed to sit for 10 minutes to loosen burnt oils and carbonized foods. The surface should be thoroughly rinsed prior to use.

Grills. A *grill* is a cooking unit consisting of a large metal grate placed over a heat source. The heat source may be gas or another burning fuel such as charcoal or wood. Grills are commonly used to cook meats, seafood, poultry, and some fruits and vegetables. **See Figure 1-29.** Food is placed on the preheated metal grates and turned over halfway through the grilling process to cook the other side.

Grills

Vulcan-Hart, a division of the ITW Food Equipment Group LLC

Figure 1-29. Grills are commonly used to cook meats, seafood, poultry, and some fruits and vegetables.

A grill is cleaned by being preheated and then using a wire brush to scrape the hot grates clean. The grates are then wiped with a towel lightly coated in vegetable oil to prevent food from sticking.

Procedure for Cleaning a Grill

1. Preheat the grill. Use a wire brush to scrape clean the grates of the hot grill.

2. Wipe the grates with a towel that is lightly coated in vegetable oil.

Broilers. A *broiler* is a large piece of cooking equipment in which the heat source is located above or behind the food instead of below it. Broilers can be stand-alone or combined with an oven. A broiler is cleaned in the same manner as a grill.

The three basic types of broilers include the standard broiler, rotisserie, and salamander. A *rotisserie* is a sideways broiler in which foods are placed on a steel rod or spit that revolves past the heat source to ensure even heating. **See Figure 1-30.** Rotisseries are most often used to cook poultry and meats. A *salamander* is a small overhead broiler that is usually attached to a range. A salamander is primarily used to brown, glaze, melt, or finish cooking foods.

Broilers

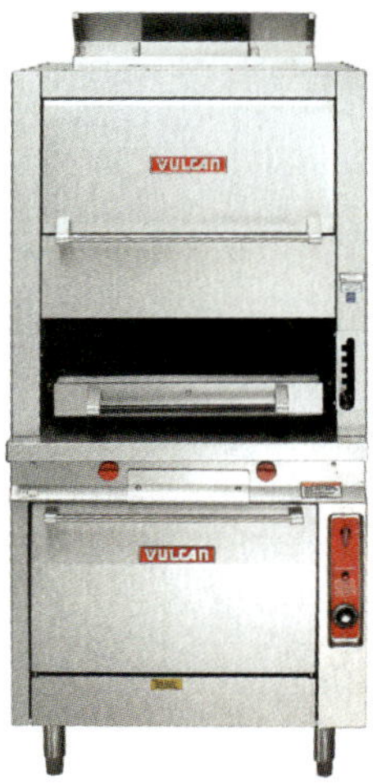

Vulcan-Hart, a division of the ITW Food Equipment Group LLC

Standard Broiler

Henny Penny Corporation

Rotisserie

Figure 1-30. A rotisserie is a sideways broiler in which food on a spit revolves past the heat source to ensure even heating.

Production Tip

Griddles are more energy efficient than broilers, and ovens are more energy efficient than rotisseries.

Steamers. Steamers are used to cook foods very quickly. **See Figure 1-31.** A *convection steamer* is a steamer that generates steam using an internal boiler, which circulates around the food to cook it rapidly. A *pressure steamer* is a steamer that uses water heated within a pressure-controlled, sealed cabinet to cook foods much quicker than a convection steamer. To prevent potential steam burns, the door gaskets on the steamer should be checked prior to each use.

Steamers

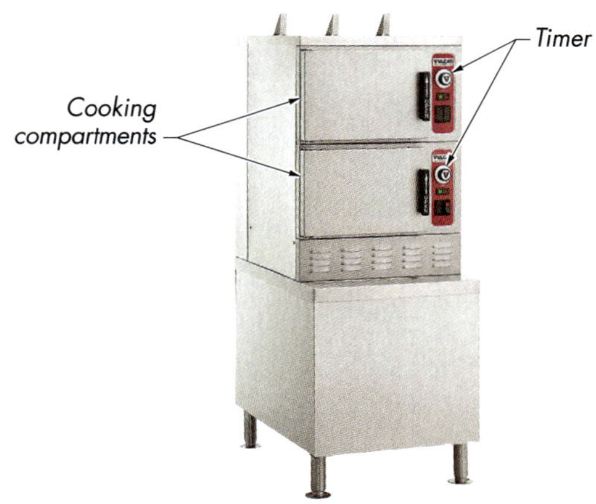

Convection Steamer

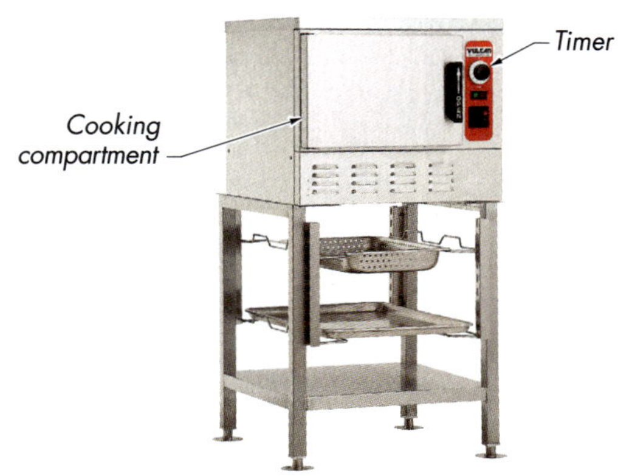

Pressure Steamer

Vulcan-Hart, a division of the ITW Food Equipment Group LLC

Figure 1-31. Both convection steamers and pressure steamers are used in the professional kitchen.

A mild soap-and-water solution is used to remove visible food particles from the interior and exterior of a steamer. Once cleaned, the steamer is rinsed with a wet towel and then sanitized. Many manufacturers also recommend running a descaling chemical through the steamer monthly to remove any hard water buildup. The exterior of the steamer can be polished using stainless steel cleaner.

Steam-Jacketed Kettles. A *steam-jacketed kettle,* also known as a steam kettle, is a large cooking kettle that has a hollow lining, known as a jacket, into which steam is injected to rapidly and uniformly cook foods. It cooks foods quickly and evenly while reducing the chance of scorching foods by heating the food without allowing the steam to touch it. A *trunnion kettle* is a small steam-jacketed kettle that is tilted by pulling a lever or turning a wheel to empty the kettle. Steam-jacketed kettles can vary greatly in size and style. **See Figure 1-32.**

Steam-Jacketed Kettles

Vulcan-Hart, a division of the ITW Food Equipment Group LLC

Figure 1-32. Steam-jacketed kettles and trunnion kettles cook foods quickly and evenly without allowing steam to touch the food.

Production Tip

The cleaning water from steam-jacketed kettles can be reused to mop floors.

A steam-jacketed kettle is cleaned by pouring off or draining any water while the kettle is still warm but with the power turned off. The kettle is then filled with clean, warm water and a mild detergent. The inside of the kettle is then scrubbed.

After being cleaned, the kettle is sanitized and allowed to air dry. The outside of the unit is wiped down with a mild soap-and-water solution, followed by a sanitizer, and then allowed to air dry. The exterior also can be polished using stainless steel cleaner.

Sullivan University

Tilt Skillets. A *tilt skillet,* also known as an electric braiser, is a versatile piece of cooking equipment with a large-capacity pan, a thermostat, a tilting mechanism, and a cover. The pan of a tilt skillet can hold 30–40 gal. and can be used as an oversize skillet, bain-marie, stockpot, kettle, or evenly heated cooktop. **See Figure 1-33.** Although a tilt skillet can be used to pan-fry or sauté, it should not be used for deep-frying. It is best to clean the tilt skillet while it is still warm as cooked-on foods can be removed easily.

Tilt Skillets

Blodgett Oven Company

Figure 1-33. A tilt skillet can be used as an oversized skillet, bain-marie, stockpot, kettle, or evenly heated cooktop.

Procedure for Cleaning a Tilt Skillet

1. Place approximately 2 gal. of mild detergent and water in the tilt skillet.
2. Use a brush to gently remove heavy food particles from under the cover, sides, strainer, pour spout, and interior of the skillet. **CAUTION:** Do not get water on the controls or thermostat.
3. Pour off the soapy water and rinse the skillet well to remove all food particles and soap residue.
4. Remove the food-receiving tray and clean and sanitize it in a warewashing sink.
5. Sanitize the skillet and then gently rinse the sanitizing solution from the surface.
6. Allow the skillet to air dry.
7. Polish the exterior of the tilt skillet using stainless steel cleaner if desired.

Fryers. A *fryer* is a cooking unit used to cook foods in hot fat. It operates by heating fat to a temperature between 300°F and 375°F. A thermostat regulates the temperature. Fryers are sized by the number of pounds of fat they can hold. For example, a 20 lb fryer holds 20 lb of fat.

Daily fryer cleaning and filtering are essential to maintaining clean fryer fat because particles of cooked food and food coatings will continue to cook in the hot fat, break down, and thereby shorten the life of the fat. Fryer fat quality should be checked per manufacturer specifications. If the fryer fat is no longer usable, it should be transferred to a grease recycling bin. Only metal containers or filtering machines should be used to filter hot fat. A shortening shuttle is often used to transfer the waste fats, oils, and grease (FOG) to a grease recycling bin. **See Figure 1-34.**

Shortening Shuttles

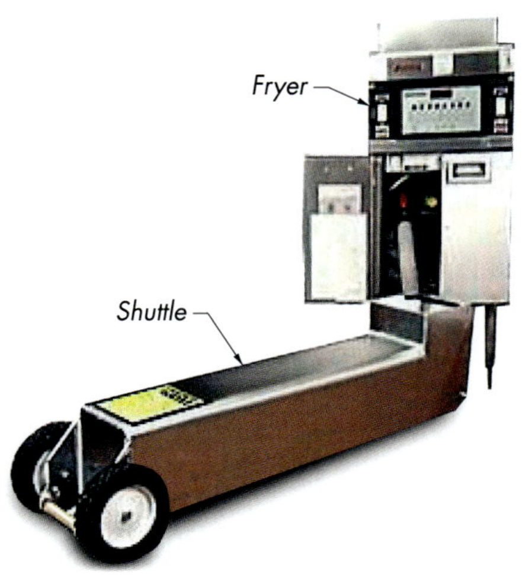

Edward Don & Company

Figure 1-34. A shortening shuttle is often used to transfer FOG to a grease recycling bin.

If fryer oil can be reused, the oil should be properly filtered. Because fryer oil is hot during the filtering process, extreme caution must be taken to avoid injury. Filter holders, grease filters, a drain extension pipe (if required), protective eyewear, and heat-resistant protective gloves may also be needed to filter fryer fat.

Procedure for Cleaning a Fryer

1. With the fryer turned off but the fat still hot, set a filter cone in a filter cone holder and place it on a shortening shuttle.
2. If required, screw the drain extension pipe into the fryer drain.
3. Position the shortening shuttle so that the drain extension pipe fits directly into the opening.
4. If the fryer is an electric model, raise the electric coil unit out of the hot fat and secure it in place.
5. Open the drain and allow the hot fat to pass through the filter cone into the shuttle container.
6. Once all the fat has been removed, thoroughly wipe the interior of the fryer clean of any crumbs using a towel.
7. Close the fryer drain and remove the drain extension.
8. Confirm that the drain has been closed, and then return the fat to the fryer. *Note:* Additional fat may be needed to reach the fill line.
9. Clean the exterior surface of the fryer with a mild detergent solution. **CAUTION:** Do not let any water make contact with the hot fat.
10. Polish the exterior of the fryer using stainless steel cleaner if desired.

Production Tip

Self-cleaning fryers have a pump that circulates the fat over a filtering media. Manufacturer instructions should be closely followed when cleaning these fryers.

Convection Ovens. A *convection oven* is a gas or electric oven with an interior fan that circulates dry, hot air throughout the cabinet. **See Figure 1-35.** This fan creates intense and even heating as it keeps the air moving while items are cooking. Cooking temperatures are 50°F lower in a convection oven than in a conventional oven. The airflow leads to more efficient cooking, reduced shrinkage, and cooks more uniformly than a conventional oven.

Convection Ovens

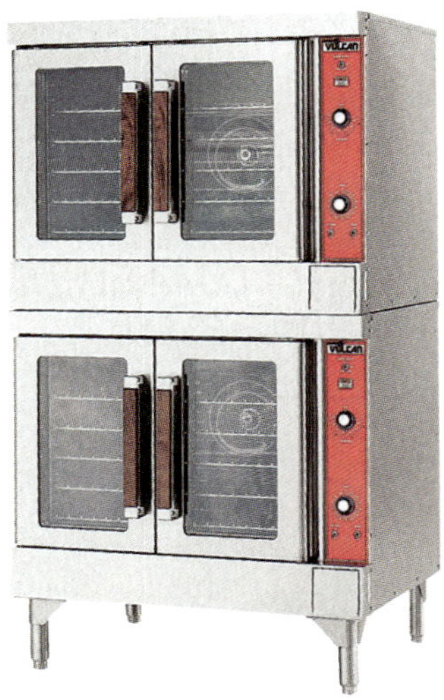

Vulcan-Hart, a division of the ITW Food Equipment Group LLC

Figure 1-35. A convection oven is a gas or electric oven with an interior fan that circulates dry, hot air throughout the cabinet.

Food and grease on the interior of a convection oven can affect airflow, especially if grease builds up on the surface of the fan. Grease can also accumulate on the fan and cause it to stop working. A convection oven can be cleaned daily by wiping down the interior and exterior of the oven using a moist towel, paying special attention to the fan.

If the convection oven is in need of deep cleaning, a commercial oven cleaner can be used to clean the interior. The racks, rack supports, and baffle should be removed from the oven and soaked in a mild detergent in a warewashing sink. **CAUTION:** A commercial oven cleaner should not be used on the blower wheel because it can damage the surface of the wheel. Gloves and protective eyewear should always be worn when using an oven cleaner. The door seal of the oven should also be checked for debris and the debris removed. The exterior of the oven can be polished using stainless steel cleaner.

Production Tip

It is important to calibrate all ovens every few months to ensure that they bake foods at the proper temperature.

Combi Ovens. A *combi oven,* also known as a combination oven, is an oven that has both convection and steaming capabilities. Combi ovens can steam and circulate hot air at the same time, decrease cooking time, save energy, and enable even roasting and baking. **See Figure 1-36.** At the end of each shift, the combi cavity, hinges, racks, and exterior of the oven should be cleaned using a mild detergent solution. If easily accessible, the fan should be wiped clean to ensure proper airflow. Most combi ovens are equipped with a spray nozzle that can be used to spray-rinse the interior of the oven compartment for light cleaning.

Combi Ovens

Hobart

Figure 1-36. Combi ovens can steam and circulate hot air at the same time, decrease cooking time, save energy, and enable even roasting and baking.

Combi ovens often have an automatic cleaning cycle feature for general cleaning. One type of automatic cleaning cycle requires the operator to dissolve a cleaning tablet into a spray bottle containing warm water. The interior of the oven is then thoroughly sprayed with the cleaning solution, the door is closed, and the steam-clean cycle is initiated. Once the cleaning cycle is finished, the oven is rinsed, sanitized, and allowed to air dry. The filter located on the bottom of most combi ovens needs to be cleared after the cleaning process is complete.

A completely automatic cleaning model has bottles installed inside the unit that dispense cleaning and sanitizing chemicals. When the cleaning cycle is initiated, the chemicals are automatically released. A standby signal indicates the cleaning cycle is complete and the oven is ready for use. Combi ovens also need to be delimed regularly depending on the amount of use and the hardness of the local water supply. The exterior of the oven can be polished using stainless steel cleaner.

Cook-and-Hold Ovens. A *cook-and-hold oven,* also known as a retherm oven, is an oven with two separately controlled compartments within one stainless steel cabinet that can be used to cook, roast, reheat, and hold a variety of foods. **See Figure 1-37.** Cook-and-hold ovens use halo heat, a uniform and controlled heat source that gently surrounds foods. Thermostats allow cooking up to 325°F and holding up to 200°F. Some cook-and-hold ovens have a built-in smoking tray that can be filled with wood chips for hot and cold smoking applications.

Cook-and-Hold Ovens

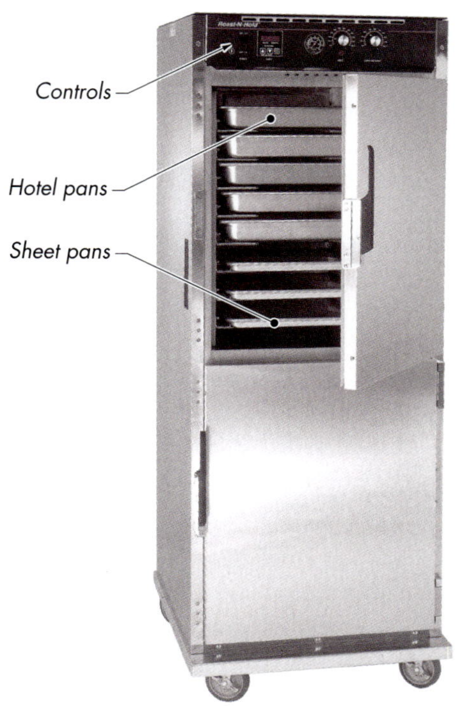

Cres Cor

Figure 1-37. A cook-and-hold oven can be used to cook, roast, reheat, and hold a variety of foods.

To clean a cook-and-hold oven, the wire shelves, side racks, and drip pans are removed and cleaned in a warewashing sink. The air tunnel from the back of the unit is removed and cleaned of any debris. The interior cavity and exterior of the oven are wiped clean using a mild detergent solution. A water-soluble degreaser may be used on heavily soiled areas. After cleaning, the oven is rinsed off using a clean damp towel and then sanitized. The controls should not come into contact with water because water could severely damage the electrical connections. Once the oven is clean, the racks and air tunnel can be reassembled.

The food probes, cables, and brackets need to be cleaned after every use and sanitized with disposable alcohol pads or a food-safe sanitizing solution. The vents need to be inspected regularly and cleaned as needed. The exterior of the oven can be polished using stainless steel cleaner.

Deck Ovens. A *deck oven* is a drawer-like oven that is commonly stacked one on top of another, providing multiple-temperature baking shelves. Food items are placed directly on the floor of a deck oven. Deck ovens are commonly used to bake pizzas and breads. **See Figure 1-38.**

Deck Ovens

Blodgett Oven Company

Figure 1-38. Deck ovens are commonly used to bake pizzas and breads.

Deck ovens need to be swept clean using a steel-bristle oven brush to remove crumbs and spilled foods. While the deck oven is still hot, the brush can be used to scrape the surface clean. The doors and handles of the deck oven are cleaned with a mild detergent, rinsed with a moistened towel, and then sanitized. The exterior of the oven can be polished using stainless steel cleaner.

Rotating Rack Ovens. A *rotating rack oven,* also known as a reel oven, is a large oven that rotates 10–80 pans of food as it cooks. **See Figure 1-39.** Some rotating rack ovens are available with a steam kettle, hood, and trough used for cooking bagels. Rotating rack ovens should be cleaned at the end of each shift.

Rotating Rack Ovens

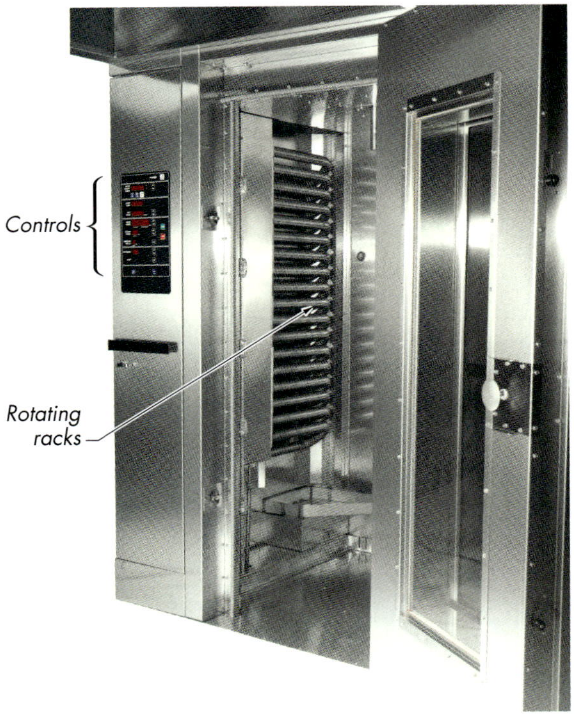

Hobart

Figure 1-39. A rotating rack oven is a large oven that rotates 10–80 pans of food as it cooks.

A rotating rack oven must be cool before it can be cleaned. The oven cavity, hinges, rotating hooks in the ceiling of the oven, racks, and exterior should be wiped clean using a mild detergent solution. When a more thorough cleaning is necessary, the oven rack should be removed and washed in a warewashing sink. The rack is removed from the oven by removing a spring-loaded pin located at the bottom of the rack. A special cleaning product is dissolved in water, added to a spray bottle, and then sprayed over the entire interior of the oven and allowed to stand for 10–20 minutes. Gloves and protective eyewear must be worn when wiping the interior to remove grease and baked-on residue. Once the surface has been cleaned, it is rinsed with a damp towel and then sanitized. The exterior of the rotating rack oven can be polished using stainless steel cleaner.

Impinger Conveyor Ovens. An *impinger conveyor oven* is an oven that directs heat from both above and below a food item as it moves along a conveyor belt. Sandwich shops use impinger conveyor ovens to toast bread, melt cheese on bread as it toasts, and to cook pizzas. **See Figure 1-40.** Be sure to keep all utensils away from the conveyor belt as they can become lodged in the belt and cause damage to the conveyor.

Impinger Conveyor Ovens

Lincoln Foodservice

Figure 1-40. Sandwich shops use impinger conveyor ovens to toast bread, melt cheese on bread as it toasts, and to cook pizzas.

To clean an impinger conveyor oven, the interior and exterior surfaces should be wiped clean using a mild detergent solution. The surfaces can be rinsed using a damp towel and then sanitized. Care should be taken to avoid getting the unit too wet. The conveyor drive chain should be checked periodically to make sure it remains snug. A loose conveyor chain can damage the drive motor.

The conveyor chain should not be cleaned while the impinger is in the ON position. If the conveyor chain requires deep cleaning, it can be removed and soaked in a degreasing solution overnight. The conveyor chain can then be rinsed, sanitized, and reinstalled. The crumb tray also needs to be removed, washed, rinsed, and sanitized in a warewashing sink. The exterior of the oven can be polished using stainless steel cleaner.

Smoker Ovens. A *smoker oven* is a gas or electric oven that generates wood smoke and is most often used to smoke or barbeque meats and poultry. Meat can be placed on racks or hung inside a smoker. As the meat cooks, the smoke covers and adds flavor to the meat. As the temperature inside the meat increases, the seasoning deposited on the outside of the meat is absorbed into the meat, giving it a smoky barbeque flavor. **See Figure 1-41.**

Smoker Ovens

Southern Pride

Figure 1-41. As the temperature inside the meat increases, the seasoning deposited on the outside of the meat is absorbed into the meat, giving it a smoky barbeque flavor.

Smoker ovens should be cleaned often to prevent carbon build up. Once carbon is allowed to build up on the interior, much effort is required to clean the oven. In addition, the intense smoke released from the smoking wood and the grease released from the food items as they smoke build up on the interior walls and produce a dark, thick residue. Some smoker ovens have self-contained smoking cabinets that can be hosed out.

A mild detergent will clean most of the surfaces. In areas where the grease cannot be removed with soap and water, a commercial oven cleaner may be used. Abrasives should never be used on a smoker. Once the grease is removed, the oven should be rinsed with clean water and sanitized using a food-safe sanitizer. The exterior of the oven can be polished using stainless steel cleaner.

Infrared and Microwave Ovens. An *infrared oven* is an oven that uses infrared radiation to evenly and efficiently bake flat foods such as pizza. A *flashbake oven* is an oven that uses both infrared radiation and light waves to cook foods quickly and evenly from above and below. Because the heat is so intense, there is no loss of flavor or moisture in the foods cooked in an infrared oven. A *microwave oven* is a cooking unit that uses microwaves to heat the water molecules within foods. A microwave oven requires no preheating. Metal containers and utensils cannot be put in a microwave because they will damage the oven.

Infrared and microwave ovens should be cleaned during each shift to remove any spills or crumbs. **See Figure 1-42.** The interior of infrared and microwave ovens can be cleaned with a mild detergent solution. At the end of each shift, the spatter shield should be cleaned to remove any accumulated food spatter and then rinsed and sanitized. The air intake filter located at the lower front of the unit should be removed and cleaned once a week. The filter can be cleaned with a mild detergent and hot water. The oven should not be used without a filter. The exterior of the oven can be polished using stainless steel cleaner.

Infrared and Microwave Ovens

Vulcan-Hart, a division of the ITW Food Equipment Group LLC

Infrared Oven

ACP, Inc.

Microwave Oven

Figure 1-42. Infrared and microwave ovens should be cleaned during each shift to remove any spills or crumbs.

Holding and Serving Equipment

Storing hot foods requires holding equipment that provides a safe environment for hot foods during service and transport. Hot foods must be held at temperatures above 140°F to prevent bacteria growth. Overhead food warmers, holding cabinets, proofing cabinets, steam tables, chafing dishes, and insulated carriers are used to hold hot foods.

Overhead Warmers. An *overhead warmer,* also known as a heat lamp, is a heat source located above a prepared food that keeps the food hot for service. **See Figure 1-43.** Overhead warmers contain electric or infrared rod-style elements or bulbs that radiate intense heat downward, keeping the food beneath them hot.

To clean an overhead warmer, the electric heating element should be allowed to cool and the unit should be unplugged. Once cooled, the unit can be cleaned with a mild detergent solution, rinsed, and then sanitized. Care should be taken to avoid applying too much water due to the risk of electrical shock. Abrasive cleaners and steel scrubbing pads should never be used to clean overhead warmers because they could damage the surface of the warmer as well as the heating element. The exterior of the warmer can be polished using stainless steel cleaner.

Holding and Proofing Cabinets. A *holding cabinet,* also known as a hot box, is a tall and narrow stainless steel box on wheels that accommodates standard sheet pans and contains temperature controls. A *proofing cabinet,* also known as a proofer, is a holding cabinet that contains both temperature and humidity controls. A proofing cabinet is used for proofing dough and for holding hot food without drying it out. **See Figure 1-44.**

Overhead Warmers

Carlisle FoodService Products

French Fry Station

Cres Cor

Carving Station

Figure 1-43. An overhead warmer keeps prepared food hot for service.

Holding/Proofing Cabinets

InterMetro Industries Corporation

Figure 1-44. A proofing cabinet is used for proofing dough and for holding hot food without drying it out.

Holding and proofing cabinets should be cleaned and sanitized at least once per day. To clean a holding or proofing cabinet, the cabinet interior and exterior are wiped with a mild detergent solution to remove visible food particles and dirt. The unit is then rinsed and sanitized. The exterior of the cabinet can be polished using stainless steel cleaner.

Production Tip

Adding a small amount of vinegar can minimize the amount of mineral deposits that may form in a proofing cabinet.

Edward Don & Company

Steam Tables. A *steam table* is an open-top table with heated wells that are filled with water. Foods are placed in hotel pans and the pans are placed into the wells of the steam table. **See Figure 1-45.** The heated wells are controlled by a thermostat so temperature can be adjusted as needed. Foods are kept hot by the heat from the hot water in each well of the steam table. Foods in a steam table must be kept covered to prevent heat loss.

To clean a steam table, the hotel pans need to be removed and all of the water needs to be drained from the well. The hotel pans should be washed in a warewashing sink or a dish machine. The well and the entire surface of the steam table should be cleaned using a mild detergent solution. Once clean, the unit needs to be rinsed and sanitized. The exterior of the steam table can be polished using stainless steel cleaner.

Steam Tables

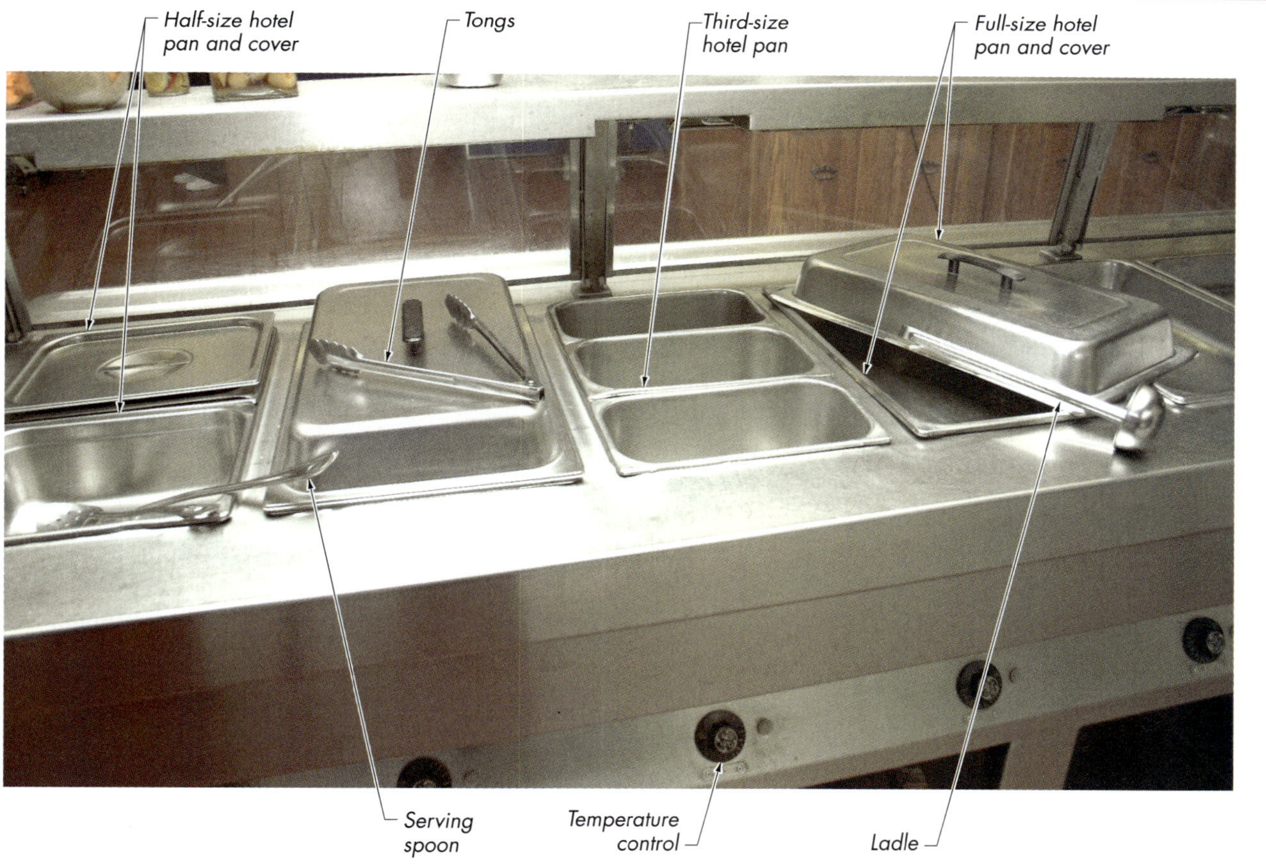

Figure 1-45. A steam table has temperature-controlled heated wells that are filled with water in order to keep hotel pans of food warm.

Chafing Dishes. A *chafing dish* is a hotel pan inside of a stand with a water reservoir and a portable heat source, such as canned fuel, underneath. **See Figure 1-46.** *Canned fuel* is a flammable gel that provides several hours of heat once it is lit.

Chafing Dishes

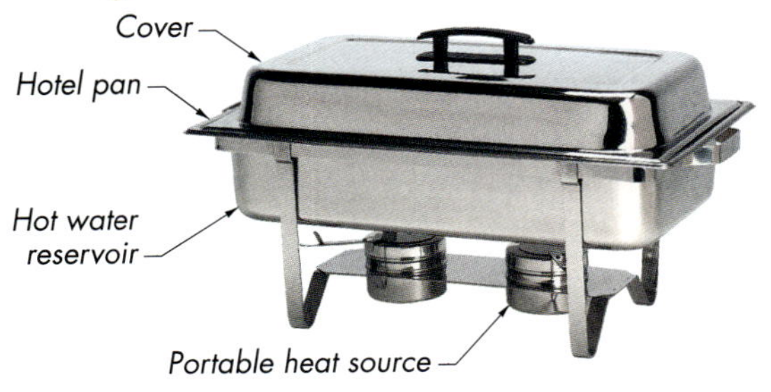

American Metalcraft, Inc.

Figure 1-46. A chafing dish is a hotel pan inside of a stand with a water reservoir and a portable heat source, such as canned fuel, underneath.

Cleaning a chafing dish is similar to cleaning a steam table. The water pan is emptied of water and washed in a warewashing sink or dishwasher. Insert pans can be washed in a warewashing sink or dish machine. The chafing dish unit itself should be cleaned using a mild detergent solution and then rinsed and sanitized. Many chafing dishes are highly polished and care should be taken not to dull the surface.

Insulated Carriers. An *insulated carrier* is a container made of heavy polyurethane that is designed to hold pans of hot or cold foods at an appropriate temperature during transport. **See Figure 1-47.** Insulated carrier lids and doors have insulated gaskets that seal and lock securely in place. Many insulated carriers come with an encased stone, which can be heated to 350°F and placed in the carrier to keep foods hot for extended periods of time. Insulated carriers can be cleaned by wiping the interior and exterior surfaces with a mild detergent solution. Once clean, the surfaces of the insulated containers should be rinsed and then sanitized.

Insulated Carriers

Carlisle FoodService Products

Figure 1-47. An insulated carrier is designed to hold hotel pans of hot or cold foods during transport.

CHECKPOINT 1-3

1. List 11 safety guidelines for operating and maintaining foodservice equipment.
2. List the specifications that must be met in order for equipment to be NSF certified.
3. Describe four basic types of cold storage units.
4. Clean and sanitize a slicer.
5. Clean and sanitize a blender, juicer, or food processor.
6. Clean and sanitize a buffalo chopper.
7. Clean and sanitize a VCM.
8. Clean and sanitize a mixer.
9. Clean and sanitize a range.
10. Clean and sanitize an induction cooktop.
11. Clean and sanitize a griddle.
12. Clean and sanitize a grill or a broiler.
13. Clean and sanitize a steamer.
14. Clean and sanitize a steam-jacketed kettle and reuse the cleaning water for mopping.
15. Clean and sanitize a tilt skillet.
16. Clean and sanitize a fryer.
17. Clean and sanitize a convection oven, a combi oven, or a cook-and-hold oven.
18. Clean and sanitize a deck oven or a rotating rack oven.
19. Clean and sanitize an impinger conveyor oven or a smoker oven.
20. Clean and sanitize an infrared or a microwave oven.
21. Clean and sanitize an overhead warmer.
22. Clean and sanitize a holding cabinet or a proofing cabinet.
23. Clean and sanitize a steam table.
24. Clean and sanitize an insulated carrier.

FOOD SAFETY

Food safety standards are established and regulated by agencies, such as the Food and Drug Administration (FDA), and organizations, such as NSF International. **See Figure 1-48.** Food safety standards are enforced by local health departments and other authorities having jurisdiction. Local agencies that oversee the enforcement of health codes will perform periodic on-site health inspections. To verify that the foodservice operation is using good food safety practices, a trained manager should complete a food safety audit or unofficial health inspection.

Most states also require foodservice managers to complete ServSafe® manager certification training and pass the certification exam. ServSafe® certification is valid for five years. However, some states and employers have even more stringent guidelines regarding certification renewal.

Foodborne Illnesses

A *foodborne illness* is an illness that is carried or transmitted to people through contact with or consumption of contaminated food. Foodborne illnesses can come on rapidly or over the course of many days. Symptoms may include vomiting, cramping, headache, sweats, chills, diarrhea, and fever. In some cases, a foodborne illness can be life threatening. Infants, young children, pregnant women, the elderly, people with compromised immune systems, and those who have life-threatening allergies are considered highly susceptible to foodborne illness.

Production Tip

Menu items that contain raw or undercooked eggs, fish, or meat must be marked clearly on menus so that these items can be avoided by guests who are highly susceptible to foodborne illness.

Foodborne illnesses are most often caused by pathogens. A *pathogen* is a microorganism that can cause disease. There are many types of pathogens that can cause foodborne illnesses. **See Figure 1-49.** Pathogens can be transferred through a variety of sources including raw foods, contact surfaces, garbage, pests, and people. Pathogens can cause foodborne illness shortly after contaminated food is eaten or up to several days later. Pathogens that cause foodborne illness are often categorized as physical, chemical, or biological contaminants.

Regulation of Food Safety Standards	
Agency/Organization	*Mission*
U.S. Food and Drug Administration (FDA)	Government agency that ensures the safety of all food except meat, poultry, and egg products; publishes the *Food Code*, a document that recommends licensing, inspection, and enforcement regulations for the foodservice industry
USDA Food Safety and Inspection Service (FSIS)	Government agency that ensures the safety of meat, poultry, and egg products
USDA Animal and Plant Health Inspection Service (APHIS)	Government agency that ensures protection of animals and plants from diseases or pests
United States Environmental Protection Agency (EPA)	Government agency that establishes the levels of pesticide residue that can be tolerated by humans
Centers for Disease Control (CDC)	Government agency that investigates foodborne illnesses and plays a key role in supporting state and local health departments
NSF International	Not-for-profit organization that sets safety and sanitation standards for tools and equipment used in the foodservice industry; publishes safety standards for food, water, and consumer goods
Underwriters Laboratories (UL)	Independent safety certification organization that develops standards for various tools and equipment used by the foodservice industry

Figure 1-48. Food standards are established and regulated by agencies, such as the Food and Drug Administration (FDA), and organizations, such as NSF International.

Common Pathogens that Cause Foodborne Illnesses		
Pathogen	*Methods of Transmission*	*Common Symptoms*
Amisakis simplex	Herring; cod; halibut; mackerel; Pacific salmon; certain shellfish	*Noninvasive type:* throat tingle and coughing up worms *Invasive type:* stomach pain, diarrhea, nausea, and vomiting
Bacillus cereus	Contaminated meats; cooked corn; cooked vegetables; baked potatoes; cooked rice	*Diarrhea type:* watery diarrhea and abdominal cramps *Vomiting type:* nausea and vomiting
Campylobacter jejuni	Contaminated water; poultry	Fever, headache, and abdominal cramps followed by diarrhea
Clostridium botulinum	Improperly canned foods; untreated garlic and oil mixtures; reduced-oxygen-packaged (ROP) food; baked potatoes	Vomiting and nausea; double vision, trouble speaking and swallowing, and weakness
Clostridium perfringens	Food left for long periods in the temperature danger zone; poultry; meats; stews; gravies	Diarrhea and abdominal pains
Cryptosporidium parvum	Food contaminated by food handlers or herd animals; contaminated water; contaminated produce	Severe watery diarrhea, nausea, and weight loss
Cyclospora cayetanensis	Contaminated water; contaminated produce; food contaminated by infected food handlers	Low fever, nausea, abdominal cramping, and diarrhea that alternates with constipation
E. coli O157:H7	Raw or rare ground beef; contaminated fruits and vegetables	Diarrhea or bloody diarrhea, abdominal cramps; can cause hemolytic uremic syndrome (HUS)
Giardia duodenalis	Contaminated water; food prepared by infected food handlers	Fever, diarrhea, abdominal cramps, and nausea
Hepatitis A virus	Contaminated water; food contaminated by food handlers; fruits; vegetables; shellfish; salads; ready-to-eat food	Fever, nausea, abdominal pain, and weakness; jaundice
Listeria monocytogenes	Ready-to-eat food; raw meats; soft cheeses; unpasteurized milk	Spontaneous abortion in third trimester of pregnancy; sepsis, pneumonia, and meningitis in newborns
Noroviruses (Norwalk and Norwalk-like viruses)	Contaminated water; contaminated shellfish; food contaminated by infected food handlers; ready-to-eat food	Nausea, vomiting, diarrhea, and abdominal cramps
Salmonella spp.	Raw or undercooked eggs, poultry, and meat; unpasteurized milk and dairy products; food contaminated by infected food handlers	Abdominal pain, diarrhea, vomiting, and fever
Shigella spp.	Fecal contamination of food and water; food contacted by food handlers with poor personal hygiene; salads that contain potentially hazardous foods such as eggs	Diarrhea containing blood, fever, and abdominal cramps
Staphylococcus aureus	Improperly handled food; salads that contain potentially hazardous foods; deli meat	Severe nausea, abdominal cramps, vomiting
Vibrio parahaemolyticus	Raw or partially cooked oysters	Low fever, chills, nausea, vomiting, diarrhea, abdominal cramps
Vibrio vulnificus	Raw or partially cooked oysters	*Septicemia:* diarrhea, nausea, vomiting, skin lesions, fever, and sudden chills *Gastroenteritis:* diarrhea and abdominal cramps

Figure 1-49. There are many types of pathogens that can cause foodborne illnesses.

Physical Contaminants. A *physical contaminant* is any object that can be hazardous if it is inhaled, swallowed, or otherwise absorbed into the body. Examples of physical contaminants found in food include bone fragments, bits of metal or glass, and tiny stones. Most physical contaminants in food are accidental. Physical contamination can be minimized by carefully inspecting food items and storing them in safe, well-maintained locations.

Chemical Contaminants. A *chemical contaminant* is any chemical substance that can be hazardous if it is inhaled, swallowed, or otherwise absorbed into the body. Chemical contaminants include cleaning supplies, pesticides, and other chemical-based compounds used in a foodservice operation. **See Figure 1-50.** These items should be stored separately from food products and in properly labeled containers. Containers used for chemicals cannot be recycled for any food preparation task or used for future storage.

Chemical Contaminants

Hobart

Figure 1-50. Chemical contaminants include cleaning supplies, pesticides, and other chemical-based compounds used in a foodservice operation.

Biological Contaminants. A *biological contaminant* is a living microorganism that can be hazardous if it is inhaled, swallowed, or otherwise absorbed. Biological contaminants that can be found in food include parasites, viruses, and bacteria.

- A *parasite* is a pathogen that relies on a host for survival in a way that benefits the organism and harms the host. Parasites inhabit the stomachs and intestinal tracts of animals and are passed on to people when contaminated food is consumed. Anisakis simplex is an example of a parasite caused by a roundworm that is found in contaminated fish and squid.
- A *virus* is a pathogen that grows inside the cells of a host. Human beings are often hosts to viruses. Sneezing or coughing on food is one way that viruses can spread. Viruses can also survive on the surface of food or production equipment. Viruses can withstand freezing temperatures. However, heat can destroy viruses. Ready-to-use foods, such as deli meats, do not require heating, making them more susceptible to viruses.
- *Bacteria* are pathogens that live in soil, water, or organic matter such as the bodies of plants and animals. Bacteria, such as E. coli, divide every 12–20 minutes and can multiply into millions in a matter of hours. **See Figure 1-51.**

E. Coli Bacteria

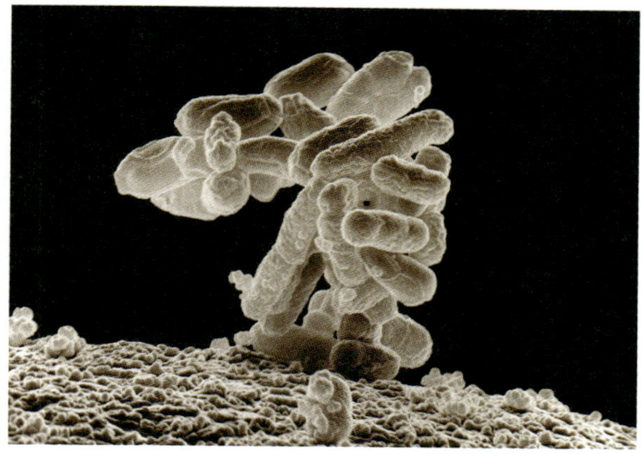

Agricultural Research Service, USDA

Figure 1-51. Bacteria, such as E. coli, divide every 12–20 minutes and can multiply into millions in a matter of hours.

Some biological contaminants can cause detectable changes in food as food begins to spoil. For example, bacteria decompose food and cause changes that can be seen, smelled, or tasted. A *food spoilage indicator* is a condition that signifies that food is deteriorating and is no longer safe for consumption. If food changes in color, odor, or texture it should be discarded to prevent foodborne illnesses.

Bacterial Contamination

Bacterial contamination is the leading cause of foodborne illness. FAT TOM is the acronym that identifies the six factors that contribute to bacterial growth. FAT TOM stands for food, acidity, temperature, time, oxygen, and moisture.

Food. Bacteria thrive on protein-rich foods for the energy they need to reproduce. High-protein foods such as poultry, meats, seafood, and dairy products easily support bacterial growth.

Acidity. Every food falls between 0 and 14 on the pH scale, where 0 is the highest in acidity and 14 is the highest in alkalinity. **See Figure 1-52.** Most bacteria growth is inhibited by very acidic conditions, below a 4.6 pH. Acidic ingredients, such as lemon juice, do not provide a favorable climate for bacteria.

Food Safety and Quality

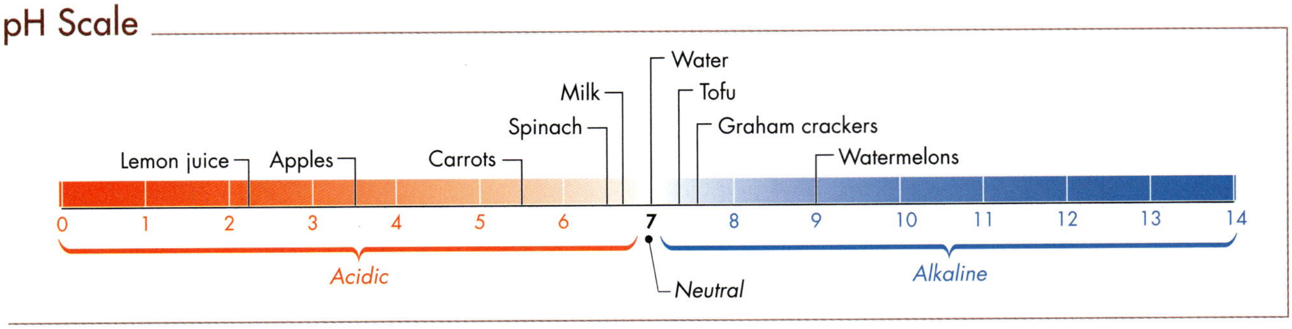

Figure 1-52. Every food falls between 0 and 14 on the pH scale, where 0 is the highest in acidity and 14 is the highest in alkalinity.

Temperature. The *temperature danger zone* is a range of temperature, between 41°F and 135°F, in which bacteria thrive. **See Figure 1-53.** Most bacteria levels are safe at temperatures of 165°F or above. Bacteria multiply very slowly at temperatures below 41°F and stop completely at temperatures of 0°F and below. However, the bacteria do not die. When frozen bacteria are returned to temperatures between 41°F and 135°F they begin to grow again.

Time. The less time that foods are left in the temperature danger zone, the less opportunity there is for bacteria to multiply and reach dangerous levels.

Oxygen. Almost all foodborne pathogens require oxygen to grow. For example, campylobacter and E. coli require oxygen to grow. Clostridium botulinum, the pathogen that causes botulism, does not require oxygen to grow.

Moisture. Foods such as poultry, seafood, meats, soft cheeses, and produce contain high amounts of water and are therefore more susceptible to fast-growing bacteria. If moisture is extracted from food, the food can be stored with less chance of bacterial growth. However, when dried or dehydrated foods are rehydrated, bacteria can multiply.

Food Code

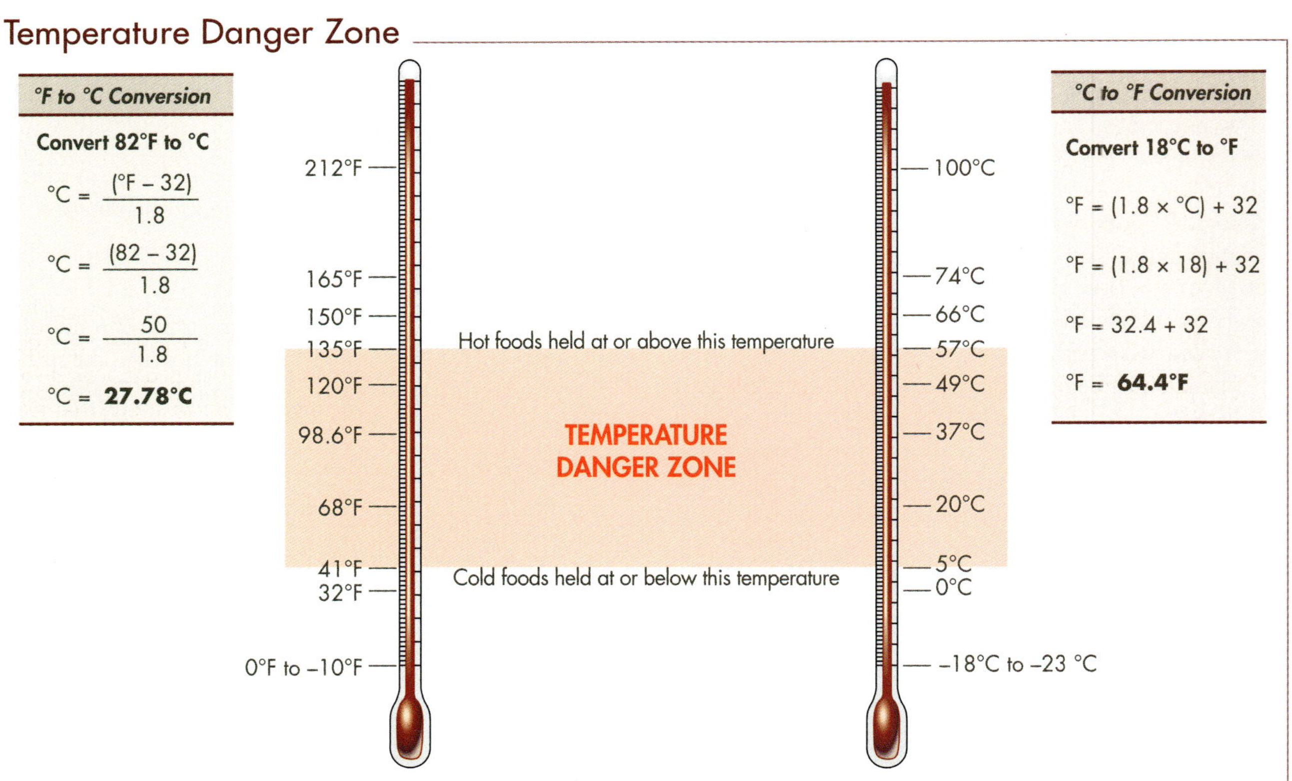

Figure 1-53. The temperature danger zone is a range of temperature, between 41°F and 135°F, in which bacteria thrive.

CHECKPOINT 1-4

1. Identify seven agencies and organizations responsible for regulating food safety standards.
2. Define foodborne illness.
3. Explain how pathogens cause foodborne illnesses.
4. Identify three types of physical contaminants that can be found in food.
5. Identify three types of chemical contaminants that can be found in foodservice operations.
6. Describe three types of biological contaminants that can be found in food.
7. Define food spoilage indicator.
8. Name the acronym that describes the six factors that promote bacterial growth.
9. Describe how food contributes to bacterial growth.
10. Describe how acidity contributes to bacterial growth.
11. Define the temperature danger zone.
12. Describe how time contributes to bacterial growth.
13. Describe how oxygen contributes to bacterial growth.
14. Describe how moisture contributes to bacterial growth.

THE FLOW OF FOOD

Preventing foodborne illness starts long before a meal is served. Contaminants can enter at any stage in the flow of food. The flow of food is the path food takes in a foodservice operation as it moves from purchasing to service. The flow of food includes purchasing, receiving, storing, preparing, cooking, cooling, holding, reheating, and serving. **See Figure 1-54.**

Food Safety

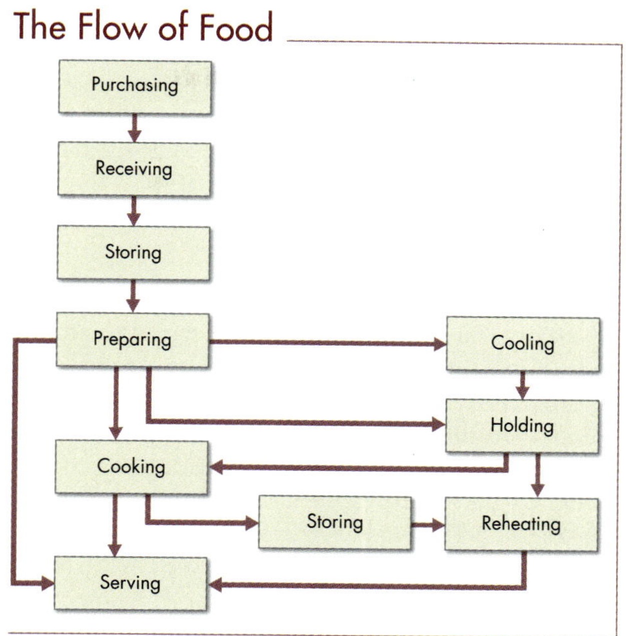

Figure 1-54. The flow of food includes purchasing, receiving, storing, preparing, cooking, cooling, holding, reheating, and serving.

Purchasing Food

When choosing a supplier, it is critical to purchase goods and products from reputable suppliers that follow food safety guidelines. This includes confirming that suppliers are getting products from approved sources, have staff trained in food safety practices, and follow a good manufacturing practice (GMP) program that addresses the minimum sanitation and processing requirements set forth by the FDA to help ensure the safe production of food.

Receiving Food

Once food is received, inspections are necessary to uphold quality and safety standards. Items should be rejected if they are damaged, past their expiration date, or do not meet temperature requirements. Frozen foods that show signs of thawing or refreezing, such as ice crystals or freezer burn, should be rejected. Unless otherwise specified by the manufacturer, ready-to-eat foods should be received at 41°F or less with packaging intact. This pertains to foods such as precut meats, salads containing potentially hazardous foods such as eggs or cold cuts, and refrigerated entrees that require heating before service.

Special attention needs to be given to potentially hazardous foods. A *potentially hazardous food* is a food that requires temperature control in order to keep it safe for consumption. Potentially hazardous foods have a neutral or slightly acidic pH level and contain moisture and protein.

Potentially hazardous foods must be maintained at a temperature at or below 41°F. An exception to this is shell eggs, which are allowed to maintain a minimum temperature of 45°F. Other exceptions include milk and shellfish, which may have alternate minimum temperatures specified by state laws.

Storing Food

Once food has been properly inspected, it needs to be stored in a safe and sanitary manner. All food should be properly covered, labeled, and dated. To promote quality and avoid hazards, stock rotation is important when storing deliveries. *First-in, first-out (FIFO)* is the process of dating new items as they are placed into inventory and placing them behind or below older items to ensure that older items are used first. **See Figure 1-55.** Goods used in foodservice operations fall into one of three storage categories: refrigerated storage, frozen storage, or dry storage.

Refrigerated Storage. Refrigerator temperature settings must be cold enough to keep the internal temperature of food at 41°F or below. This often requires setting the refrigeration unit a few degrees lower in order to compensate for the doors being opened frequently.

First-in, First-out (FIFO)

Sullivan University

Figure 1-55. The FIFO method requires new items to be stored behind or below older items to ensure that older items are used first.

First-In, First-Out (FIFO)
Media Clip

Production Tip

In top-to-bottom refrigerated storage, cooked and ready-to-use items are stored on the top shelf of the refrigerator above raw seafood, meat, and poultry. This is done to prevent the juices of raw food from potentially dripping on cooked and ready-to-use items.

Frozen Storage. Freezers should be set at −10°F to keep all products frozen. To keep freezer temperatures in an acceptable range, items should be placed below the freezer load line, and the freezer doors should only be opened when necessary. Freezer temperature should be checked regularly. In addition to temperature, it is essential to monitor dates on labels of frozen food to maintain safety and quality.

Raw food that has been frozen requires special care when thawing. Proper thawing helps prevent contamination and the growth of bacteria. The following are ways to thaw foods properly:

- Food can be thawed in the refrigerator as long as the food maintains a temperature of 41°F or below once it thaws.
- Food can be submerged under clean, running water that is 70°F or below. No portion of the food should reach 41°F or above. An overflow drain must be used to catch any particles from the food, and the work area must be cleaned and sanitized immediately after thawing.
- Food can thaw during the cooking process as long as the minimum internal temperature for the cooked food is met and held.

Dry Storage. Dry storage facilities should be kept dry, clean, and between 50°F and 70°F. All dry storage items must be kept at a minimum of 6 inches above the floor. The condition of dry storage items and the dates on labels need to be checked regularly. Items should be discarded if they are damaged.

Preparing Food

While preparing food, it is important to keep food out of the temperature danger zone (41°F to 135°F) as much as possible. Preparing food efficiently and using proper time and temperature controls can limit the amount of time food spends in the temperature danger zone. **See Figure 1-56.** Time and temperature controls are vital during the processes of cooking, cooling, reheating, and holding food.

Preparing Food

Figure 1-56. Preparing food efficiently and using proper time and temperature controls can limit the amount of time food spends in the temperature danger zone.

Cooking Food. The presence of harmful pathogens can be reduced to a safe level by thoroughly cooking foods. The look, smell, or taste of a food is not always enough to determine if it is thoroughly cooked. A thermometer inserted into the thickest part of the food is the best way to determine its temperature. Food needs to reach a minimum internal cooking temperature and hold that temperature for a specific amount of time in order to ensure food is safe to eat. **See Figure 1-57.** Different foods have different internal temperature and time requirements.

Minimum Internal Cooked Temperatures and Times		
Food	**Temperature**	**Time**
Eggs	145°F	15 seconds
Poultry	165°F	15 seconds
Fish and shellfish	145°F	15 seconds
Beef, lamb, and veal:		
Steaks and chops	145°F	15 seconds
Roasts	145°F	4 minutes
Ground	155°F	15 seconds
Pork:		
Chops and cutlets	145°F	15 seconds
Roasts	145°F	4 minutes
Ground	160°F	15 seconds
Stuffed poultry, fish, shellfish, meat, and pasta	165°F	15 seconds
Reheated foods	165°F	15 seconds

Figure 1-57. Food needs to reach a minimum internal cooking temperature and hold that temperature for a specific amount of time in order to ensure food is safe to eat.

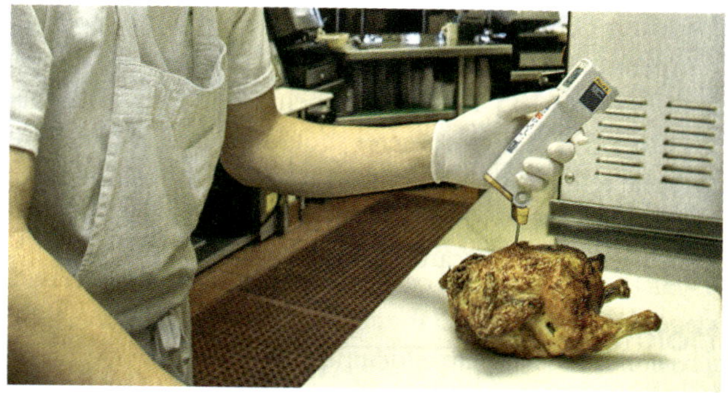

Fluke Corporation

Cooling Food. When cooling a hot food, it must be cooled to 70°F within 2 hours. Over the next 4 hours, the food must be cooled to a minimum of 41°F. The following methods are used to cool foods:

- Foods such as soups, stocks, sauces, and stews can be poured from a hot pot into a chilled container. A cooling paddle, also known as a cooling wand, can be used to stir the food until it cools to 70°F. **See Figure 1-58.** The container is then covered and refrigerated.
- Food can be placed in a blast chiller. When the food has cooled, the container is then covered and refrigerated.
- Food can be placed in shallow pans, loosely covered, and placed on an upper shelf in the refrigerator. When the food has cooled, the pan is tightly covered and returned to the refrigerator.
- An ice bath can be created by filling a larger pan or a sink with ice and then placing the pan of hot food in the ice bath. The food is stirred every 10–15 minutes and the ice is replaced as it melts. The cooled food is then covered and refrigerated.

Cooling Food

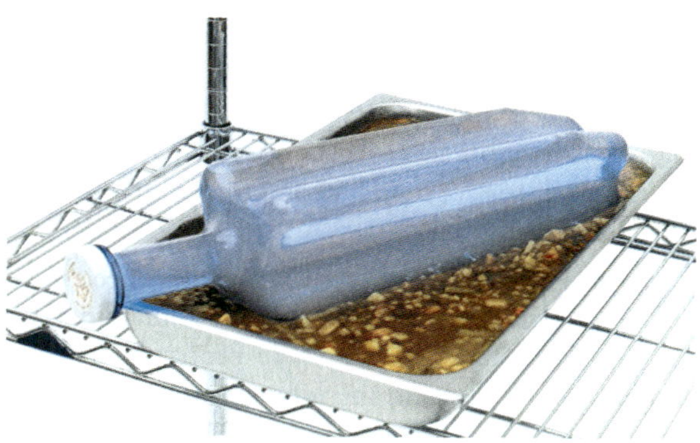

San Jamar

Figure 1-58. A cooling paddle can be used to cool foods such as soups.

Critical Limits. A *critical limit* is the point in a HACCP plan where a minimum or maximum value is established for a CCP in order to prevent, eliminate, or reduce a hazard to a safe level. A critical limit needs to be established for each CCP. These limits must be met to eliminate that hazard or reduce it to a safe limit. An example of a critical limit would be meeting minimum internal temperature and holding times when cooking food. **See Figure 1-63.**

Monitoring Procedures. A HACCP plan must identify what must be monitored, how those items are monitored, the times and frequency at which those items should be monitored, and the individual responsible for the monitoring. Monitoring procedures should address each CCP. Monitoring procedures might also require determining the individual responsible for checking cooking temperatures and times and how they are checked.

Corrective Actions. A *corrective action* is the point in a HACCP plan that identifies the steps that must be taken when food does not meet a critical limit. These corrective actions must be identified in advance and implemented if critical limits are not met. For example, a corrective action could be disposing of food if the minimum cooking temperature is not met. Corrective actions must be clearly defined and easy to implement. Corrective actions are documented to allow modifications that can result in the elimination of the problem.

Edward Don & Company

Critical Control Point	Preventive Measure	Critical Limit
Receiving	Receive foods at proper temperatures	41°F or lower for potentially hazardous foods
	Put perishable foods in cold storage quickly	4 hour maximum time in temperature danger zone
	Obtain supplies from approved sources	Suppliers must be regulated and inspected by proper authorities
Storage	Maintain temperature control	41°F for perishable foods 0°F for frozen foods
	Prevent cross-contamination	Ready-to-eat foods do not contact raw meat, seafood, or eggs
Preparation	Minimize bacterial growth	4 hour maximum time in the temperature danger zone
	Minimize contamination from staff and equipment	Ready-to-eat foods do not contact raw meat, seafood, or eggs Employees wash hands when changing tasks and at least once every 4 hours
Cooking	Cook foods to their proper temperatures	145°F for eggs, fish, and meat 165°F for poultry 135°F for fruits and vegetables
Cooling	Use rapid chill refrigeration equipment	Cool to 70°F in 2 hours following cooking
	Stir hot foods in an ice water bath	Cool to 41°F in 6 hours following cooking
Holding	Keep hot foods at temperatures above the temperature danger zone	Hold at temperature of 135°F or above
	Keep cold foods at temperatures below the temperature danger zone	Hold at temperature of 41°F or below

Figure 1-63. An example of a critical limit would be meeting minimum internal temperatures and holding times when cooking food.

System Verification. A HACCP plan needs to be verified to determine if it is preventing, reducing, or eliminating the identified hazards. Verification is conducted by someone other than the person directly responsible for performing the activities specified in the food safety management system. The person conducting the verification checks receiving, cooling, handwashing, and cooking logbooks to see that the required entries were made by each shift.

Documentation. Documents should be kept and maintained in order for a HACCP plan to be the most effective. Logs, graphs, charts, receipts, and notes are examples of these documents. These documents can pertain to suppliers, equipment, CCPs, monitoring activities, and any additional component essential to the operation's HACCP plan.

A relatively new documentation source for foodservice operations comes in the form of radio frequency identification (RFID) technology. *Radio frequency identification (RFID)* is a form of technology that uses electronic tags to store data that can be monitored from remote distances. This means that foods can be RFID tagged and monitored for temperature requirements during transportation as well as during storage. **See Figure 1-64.**

HACCP Documentation Forms

RFID Tags

Figure 1-64. Foods with an RFID tag can be monitored during transportation as well as during storage.

RFID tags promote compliance with the Food Safety Modernization Act (FSMA) of 2011, which focuses on tracing high-risk foods, such as raw fruits and vegetables. RFID tags also can alert foodservice operators if a food has been recalled and can be used to manage inventory from remote locations.

CHECKPOINT 1-5

1. Describe the flow of food.
2. Outline the flow of food used to prepare a cold dish and a hot dish.
3. Define potentially hazardous food.
4. Define first-in, first-out (FIFO).
5. Explain why a minimum internal cooking temperature is important to food safety.
6. Identify an energy-efficient way to cool hot foods.
7. Cool a hot food such as soup or stock.
8. Monitor the temperature of a hot food that is being held for 3 hours prior to service.
9. Test the temperature of items on a salad bar to determine if they are in the temperature danger zone.
10. Explain the purpose of a HACCP plan.
11. List the seven Hazard Analysis and Critical Control Point (HACCP) principles.
12. Define a critical control point (CCP) and a critical limit.
13. Describe a corrective action.
14. Identify at least five types of documentation that can be used in a HACCP plan.
15. Explain how foodservice operations can use radio frequency identification (RFID).
16. Create a HACCP plan for a high-risk menu item. Identify critical control points (CCPs) for each food process.

FACILITY SAFETY

It is important to be informed of all safety procedures upon employment. Employers are responsible for safety training and ensuring that employees follow Occupational Safety and Health Administration (OSHA) standards. Foodservice staff members are responsible for reporting safety hazards, accidents, and injuries to a supervisor. The following lifting safety, fire safety, chemical safety, and security measures promote a safe work environment.

Lifting Safety

A lot of heavy items are used in a foodservice operation during each shift. Dollies and carts are often used to move these heavy items from one location to another. However, lifting items onto dollies and carts can be dangerous when done incorrectly. Heavy objects need to be lifted with the legs, not with the back. Using proper lifting guidelines can greatly reduce the risk of injury.

Procedure for Lifting

1. Stand close to the item to gain a safe lifting position. Keep a wide stance with the feet turned out and heels down.
2. Squat by bending at the hips and the knees so that the shoulders and hips nearly form a straight, vertical line.
3. Pull the load close to the body and grasp the object firmly.
4. Rise from the squatting position, using the legs to power the lift. *Note:* Do not bend over at the neck, shoulders, or waist while lifting.
5. To unload, face the chosen location and lower the item slowly using the legs and not the back.
6. Bend at the knees to lower the load at the same time as the body lowers. *Note:* Keep the back comfortably straight.

Production Tip

Traffic paths should always be clear of items that could cause a dolly or cart to overturn or cause someone to trip.

Fire Safety

Foodservice staff should be trained in how to prevent fires and what to do in the event of a fire. OSHA standards require employers to provide a fire protection plan, fire exits, fire extinguishers, a fire-suppression system, and an emergency action plan.

Fire Protection Plan. A fire protection plan typically includes housekeeping procedures. Housekeeping procedures include proper storage of flammable materials and the proper maintenance of heat producing equipment such as ovens, steamers, and fryers.

Fire Exits. Each foodservice operation facility must have at least two properly marked fire exits that are clear of obstructions. To keep the fire exits clear, all items should be stored in designated areas.

Fire Extinguishers. Fire extinguishers must be stationed at various areas throughout the kitchen and the entire foodservice operation. Employees must know where fire extinguishers are located and how to use them properly. Class A, Class C, and Class K fire extinguishers are commonly used in foodservice operations. **See Figure 1-65.**

- Class A fire extinguishers are for fires involving common combustible materials such as trash, wood, or paper.
- Class C fire extinguishers are used to extinguish electrical fires.
- Class K fire extinguishers are used specifically for grease fires. It is important to check fire extinguishers frequently to make sure they are adequately charged in case of an emergency.

Facility Safety

Fire Extinguishers

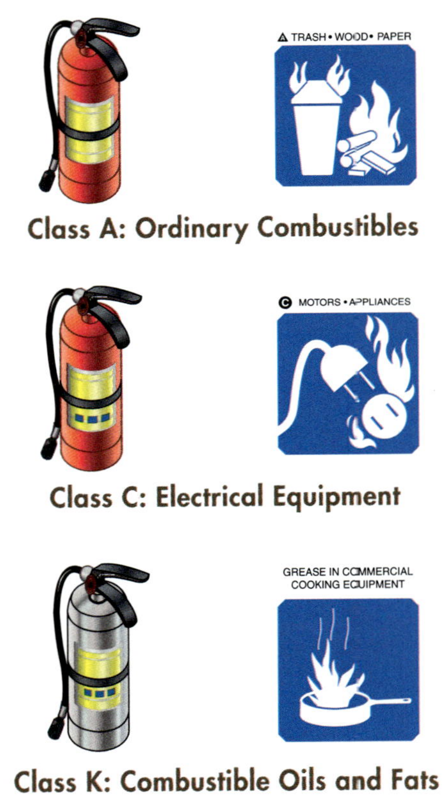

Figure 1-65. Class A, Class C, and Class K fire extinguishers are commonly used in foodservice operations.

Production Tip

Checking fire extinguishers at the end of each month is a safe practice to follow.

The proper procedure for using a fire extinguisher should be followed. The acronym "PASS" is used to explain the simple four step process:

P–Pull the pin at the top of the fire extinguisher.

A–Aim the hose at the base of the fire.

S–Squeeze the fire extinguisher handle.

S–Sweep the hose from side to side to smother the fire.

Fire-Suppression Systems. A *fire-suppression system* is an automatic fire extinguishing system that is activated by the intense heat generated by a fire. A fire-suppression system is designed to provide fire protection in areas with open flames, such as ranges, grills, broilers, and fryers. **See Figure 1-66.** Foodservice fire-suppression systems extinguish fires by spraying special chemicals from overhead nozzles when the system is activated. Fire codes vary by locale, and some municipalities have stricter codes for fire-suppression systems.

Emergency Action Plans. An *emergency action plan* is a written plan intended to organize employees during an emergency situation. In the event of a fire, an emergency action plan includes procedures for reporting fires, evacuation and escape routes, how to account for all employees after an evacuation, personnel that are trained for medical duties to be performed, and contact information for emergency services, security, building management, and utilities.

Emergency plans need to be in place to provide staff members with direction when an emergency occurs. A foodservice operation may have separate emergency plans for fire, natural disasters, utility outages, chemical spills, and burglary.

Chemical Safety

A *hazardous material* is a chemical present in the workplace that is capable of causing harm. Improper handling of chemicals can have harmful effects such as respiratory tract irritation, burns, and eye damage. Examples of hazardous materials include cleaning supplies, sanitizers, and pesticides. Hazardous materials such as chemicals must be stored in an area that will not contaminate food. **See Figure 1-67.** Chemicals placed in a separate container from the original must be labeled with the name of the chemical in the container, the hazardous warnings for that chemical, and the name and address of the manufacturer.

Fire-Suppression Systems

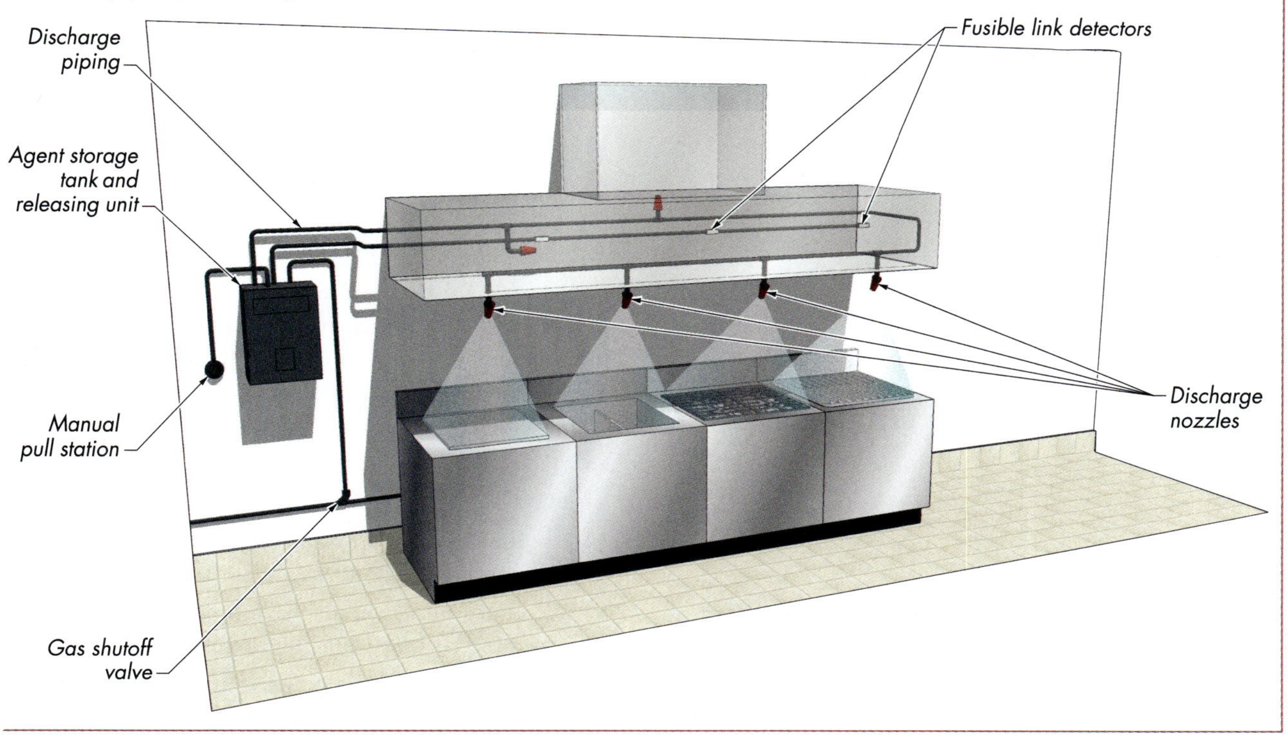

Figure 1-66. A fire-suppression system is an automatic fire-extinguishing system with discharge nozzles located above each piece of cooking equipment.

Storing Chemicals

Sullivan University

Figure 1-67. Hazardous materials such as chemicals must be stored in an area that will not contaminate food.

Each hazardous material is also required to have a safety data sheet that is accessible for review. A *safety data sheet (SDS)*, formerly known as a material safety data sheet, is a document that provides detailed information describing a chemical, instructions for its safe use, the potential hazards, and appropriate first-aid measures. SDSs and a first aid kit should always be located in an easily accessible area. **See Figure 1-68.**

Safety Data Sheets

First Aid Kits

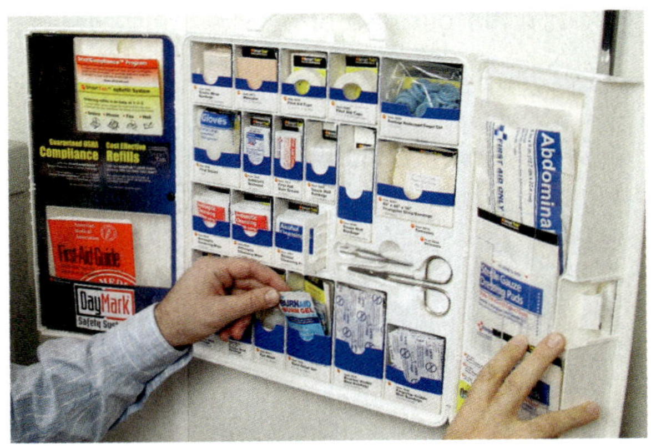

Daymark® Safety Systems

Figure 1-68. Safety data sheets and a first aid kit should always be located in an easily accessible area.

Security Measures

Threats of violence and acts of terrorism have caused foodservice operations to focus security efforts on protecting staff, as well as protecting inventory against theft. Locked doors, good lighting, an alarm system, and an aware staff are the best preventive security measures. Several steps can be taken to increase the security of a foodservice operation:

- All exterior lights should be turned on so that the building is well illuminated.
- All exterior doors and windows should be secured and locked once guests have left the building.
- All video cameras should be kept operational as they are helpful when security is breached.
- All service doors should have a peephole to prevent unwanted persons from entering the facility.
- Money should never be counted in front of windows or doors with windows that can be seen from the outside.
- Security measures should be reviewed with all foodservice staff members on a regular basis.

CHECKPOINT 1-6

1. Explain how to lift heavy items properly.
2. Identify five components of fire safety.
3. Identify three types of fire extinguishers used in the professional kitchen.
4. Differentiate between ventilation systems and fire-suppression systems.
5. Define hazardous material.
6. Explain the purpose of a safety data sheet (SDS).
7. Fill out a safety data sheet (SDS) for a hazardous material used in the professional kitchen.
8. Describe the importance of staff members understanding security measures.

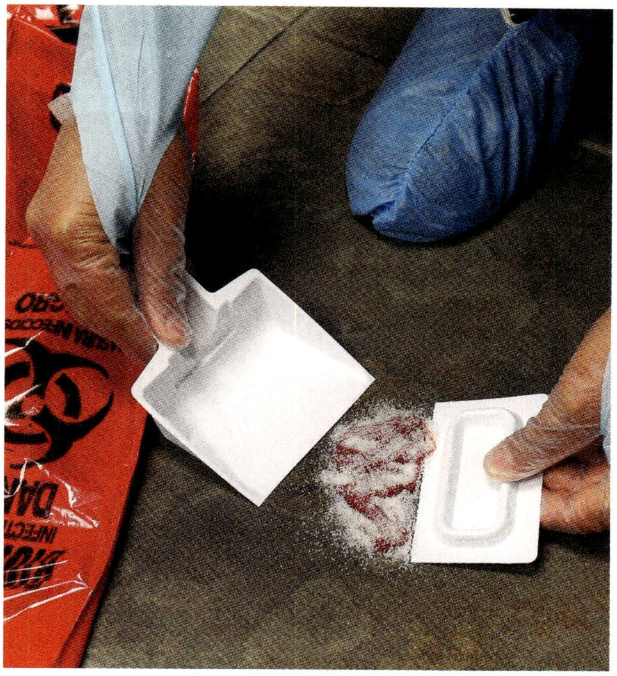

Daymark® Safety Systems

Refer to DVD for **Quick Quiz®** questions

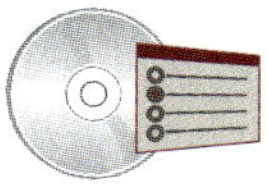

SUSTAINABILITY CORNER

The foodservice industry relies on many nonfood products that can easily be replaced with recyclable, reusable, or biodegradable items. By analyzing the options, foodservice operators can make purchasing decisions that encourage sustainability while reducing costs. For example, washable plates, cups, and utensils are reusable and generally the best option for dine-in customers. Reusable tableware preserves natural resources and reduces waste pick-up fees.

When to-go containers are necessary, biodegradable products suitable for composting can be used. However, biodegradable containers that are thrown into traditional trash bins end up in landfills and take decades to decompose. By contrast, composting piles rely on the earth's organic matter and heat to break down biodegradable materials over a few months. The result is a nutrient-rich soil ideal for planting an organic garden. When customers are encouraged to compost to-go containers, sustainable practices can be even more far-reaching by creating the opportunity for others to help sustain and promote the health of the planet.

Promoting sustainability when purchasing nonfood items, however, involves more than buying eco-friendly, take-out containers. Cleaning products can be toxic to the environment and a costly investment. Learning which products are safe for the environment and which can be replaced economically also helps promote sustainable practices.

Reading labels on commercial cleaning products is an important first step for a better understanding of a product's impact on the environment. Labels often contain the words "toxic" or "danger," indicating the possibility of harm. However, labels stating "ecologically friendly" or "natural" do not always mean a product is nontoxic. To ensure that a cleaning product meets environmentally friendly and/or nontoxic criteria, only items that have been approved by a credible third party, such as Green Seal or the EPA, should be purchased.

Often, chemical cleaning products do not have scents, so artificial scents are added to indicate to individuals that they are using, for example, a gallon of bleach and not a gallon of water. Artificial scents are just one type of additive found in commercial cleaners that may not be safe for the environment. A cost-saving way to clean with natural products is to use items like vinegar or hydrogen peroxide as a degreaser. For areas that do not need disinfecting, an all-purpose spray cleaner can be created from baking soda, dish soap, and hot water.

Charlie Trotter's

Table linens and work towels are another example of nonfood items used by foodservice operations that can be made sustainable. Table linens can be replaced with reusable natural-fiber placemats that do not require harsh chemicals for cleaning. If linens are used, colored linens may be an option. Colored linens can be cleaned without the harmful bleaches and starches needed to clean white linens. Using unbleached work towels made of 100% natural fibers is also a way to facilitate sustainable practices. Procuring additional items such as kitchen aprons and uniforms made of colored or natural fibers can be an additional way to promote sustainability.

Eco-Friendly Cleaning Products

Steward Station

Certification Exam Preparation Questions

1. ___ is the process of removing food and residue from a surface.
   A. Cleaning
   B. Sanitizing
   C. Handwashing
   D. Warewashing

2. Quats, chlorine, and iodine are types of ___ sanitizers used in the professional kitchen.
   A. cooling
   B. heat
   C. chemical
   D. biological

3. A(n) ___ is an appliance used to process larger amounts of a product into roughly equal-size pieces.
   A. buffalo chopper
   B. immersion blender
   C. vertical cutter/mixer
   D. bench mixer

4. A ___ is a pathogen that relies on a host for survival in a way that benefits the organism and harms the host.
   A. bacteria
   B. quat
   C. virus
   D. parasite

5. ___ maintenance is the systematic servicing of a facility in order to maintain safe operating conditions.
   A. Facility
   B. Preventive
   C. Regular
   D. Equipment

6. A ___ system is a large exhaust system that draws heat, smoke, and fumes out of the kitchen.
   A. ventilation
   B. waste management
   C. fire safety
   D. maintenance

7. A(n) ___ is an appliance with an S-shaped blade and a removable bowl and lid that can be used to quickly chop, purée, blend, or emulsify foods.
   A. blender
   B. juicer
   C. food processor
   D. immersion blender

8. ___ is the process of using chemicals or heat to destroy or reduce harmful microorganisms to a safe level.
   A. Cleaning
   B. Sanitizing
   C. Handwashing
   D. Hygiene

9. A ___ is a large cooking kettle that has a hollow lining into which steam is injected to rapidly and uniformly cook food.
   A. steam-jacketed kettle
   B. trunnion kettle
   C. pressure steamer
   D. convection steamer

10. A combi oven is an oven that has both convection and ___ capabilities.
    A. microwave
    B. steaming
    C. smoking
    D. infrared

11. Most states require foodservice managers to complete ___ manager certification training and pass a certification exam.
    A. NSF
    B. ServSafe®
    C. FDA
    D. USDA

12. A ___ is an illness that is carried or transmitted to people through contact with or consumption of contaminated food.
    A. pathogen
    B. bacteria
    C. foodborne illness
    D. physical illness

13. ___ is the process of cleaning and sanitizing all items used to prepare and serve food.
A. Warewashing
B. Handwashing
C. Sanitation
D. Hygiene

14. A ___ plan is a written document detailing what policies and procedures will be followed to help ensure the safety of food.
A. USDA
B. HACCP
C. cleaning
D. sanitation

15. Class A, Class C, and Class ___ fire extinguishers are commonly used in foodservice operations.
A. E
B. F
C. H
D. K

16. FAT TOM stands for food, acidity, temperature, time, oxygen, and ___.
A. moisture
B. mold
C. mildew
D. microorganisms

17. ___ is the process of dating new items as they are placed into inventory and placing them behind or below older items to ensure that older items are used first.
A. Storing
B. Receiving
C. First-in, first-out (FIFO)
D. The flow of food

18. A ___ is the point where a hazard can be prevented, eliminated, or reduced.
A. quat
B. prep
C. CCP
D. HACCP

19. The proper procedure for using a fire extinguisher is known by the acronym "___."
A. LEAN
B. LAST
C. PASS
D. PAST

20. A ___ material is a chemical present in the workplace that is capable of causing harm.
A. flammable
B. hazardous
C. waste
D. preventive

21. Describe the procedure for cleaning a griddle.

____

____

____

____

____

22. Describe the procedure for cleaning a fryer.

____

____

____

____

____

____

____

____

23. Identify common pathogens that cause foodborne illness.

____

____

____

____

____

____

24. Describe three types of waste that need to be monitored in a foodservice operation.

____

____

____

SKETCHING EXERCISE

25. Sketch the flow of food.

Breakfast Station

Introduction

The second and perhaps most important station in the ACFEF apprenticeship program is the breakfast station. Multiple skills and techniques are learned while preparing for breakfast service. For an apprentice, breakfast can be a unique opportunity to experience the pressure of producing a high volume of made-to-order items. For these reasons and many more, breakfast is considered a favorite station by many apprentices.

When you have learned what items to prepare and how to prepare them safely, you can begin actual production. As part of this station, you will make fried eggs and poached eggs as well as complex egg dishes such as soufflé omelets and Benedicts. You will also prepare a wide variety of breakfast meats, including a few favorites such as bacon, sausage, and ham steak.

Breakfast batters will also be part of your production training. From standard waffles and pancakes to more refined crêpes, you will prepare a multitude of batter-based breakfast items, as well as the many toppings and fillings that accompany them. Next will be breakfast potatoes, such as hash browns and home fries, as well as hot cereals, toast, oatmeal, creamed wheat, and English muffins. In addition, you will learn to make quality coffee, tea, and fresh-squeezed juices.

Lastly, you will learn about several diverse sustainability issues pertaining to the culinary industry and specifically to breakfast service. Breakfast service produces a large amount of compostable material. You will need to identify and separate these materials for proper processing. You will also research the money-saving benefits of recycling and learn how to control waste in the kitchen. Most importantly, you will take this newfound knowledge and seek out composting and recycling agencies in your area to learn their policies.

Edward Don & Company

Breakfast foods require attention to detail because many of these menu items are made to order. The griddle and/or sauté station is a busy area during breakfast service. Griddles, also known as flat-top grills, are used to prepare many egg dishes, pancakes, French toast, and a variety of breakfast meats. The ingredients and tools used at this station must be organized to ensure speed and accuracy of prepared orders. Station set-up depends on the menu items offered, the volume of orders, and the physical size of the station. Mastering efficient breakfast preparation can lead to success in a variety of foodservice careers.

EGGS

Eggs can be tempting menu items at any time of day. Eggs are the most popular breakfast food due to their versatility, availability, and reasonable cost. Simple preparations such as scrambled eggs are relatively easy, while more complex preparations, such as eggs Benedict, require more skill. Knowledge of egg composition and proper cooking methods is essential to creating egg-based preparations.

Egg Composition

An egg is composed of four main parts: the shell, shell membrane, yolk, and albumen (white). **See Figure 2-1.** Due to the porous shell, eggs should be stored away from strong smelling foods such as garlic and onions. The porous shell also allows the moisture inside the egg to evaporate over time.

In addition to the outer shell of the egg, the fragile yolk and white are also protected by a shell membrane. The *shell membrane* is a thin, skinlike material located directly under the eggshell. The *egg yolk* is the yellow portion of the egg. Although the yolk is only about one-third the weight of the egg, it contains more than three-fourths of the calories and all of the cholesterol in the egg.

The *albumen,* also known as the egg white, is the clear portion of the raw egg, which makes up two-thirds of the egg, consists mostly of ovalbumin protein, and provides a cushion from the egg shell. The albumen contains no cholesterol, a trace amount of fat, and only about 17 calories per large egg. When cooked, the albumen turns white, which is why it is commonly referred to as the egg "white."

Another component found in the albumen are chalazae. The *chalazae* are twisted cordlike strands of egg white that anchor the yolk to the center of the egg white.

Egg Composition

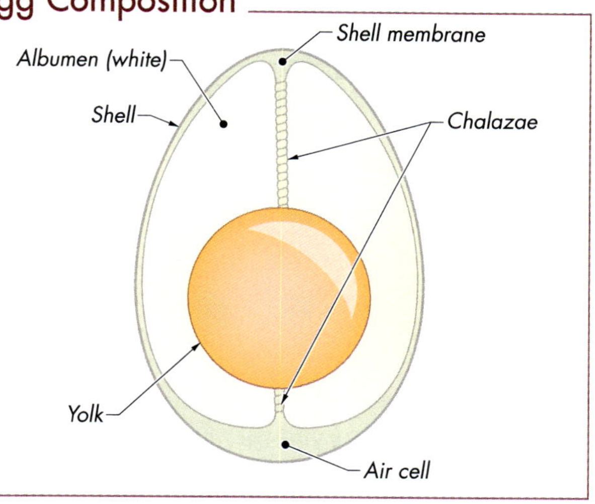

Figure 2-1. An egg is composed of four main parts: the porous shell, shell membrane, yolk, and albumen (white).

Egg Nutrition

Eggs are a nutrient-dense food, which means they have a high proportion of nutrients to calories. **See Figure 2-2.** Eggs are an affordable source of high-quality protein that includes all nine essential amino acids and two antioxidants, lutein and zeaxanthin, which help prevent common causes of age-related blindness.

Nutritional Data for a Large (Raw) Egg

Total Calories	72
From Fat (63%)	46.03
From Protein (35%)	25.89
From Carbohydrates (2%)	1.58
Total Fat	4.75 g
Saturated Fat	1.50 g
Monounsaturated Fat	1.91 g
Polyunsaturated Fat	0.68 g
Cholesterol	211.50 mg
Sodium	71 mg
Potassium	67 mg
Total Carbohydrates	0.36 g
Fiber	0 g
Sugar	0.18 g
Protein	6.28 g

Vitamins	*Amount*
Vitamin A	270 IU
Vitamin D	41 IU
Vitamin E	0.53 mg
Vitamin K	0.10 mcg
Vitamin C	0 IU
Thiamin (Vitamin B1)	0.02 mg
Riboflavin (Vitamin B2)	0.23 mg
Niacin (Vitamin B3)	0.04 mg
Pantothenic Acid (Vitamin B5)	0.77 mg
Vitamin B6	0.08 mg
Choline	125.50 mg
Folate (Vitamin B9)	24 mcg
Vitamin B12	0.45 mcg

Minerals	*Amount*
Calcium	28 mg
Copper	0.04 mg
Fluoride	0.6 mcg
Iodine	23.70 mcg
Iron	0.88 mg
Manganese	0.01 mg
Magnesium	6 mg
Phosphorus	99 mg
Potassium	69 mg
Selenium	15.30 mcg
Zinc	0.65 mg

Source: USDA National Nutrient Database, SR 23

Figure 2-2. Eggs are a nutrient-dense food, which means they have a high proportion of nutrients to calories.

Egg yolks do contain a lot of cholesterol, but they are also packed with vitamins, minerals, and essential fatty acids. Although the yolk is only one third of the egg, it is the nutrition powerhouse. In addition to cholesterol and fat, the yolk contains the majority of the thiamin, pantothenic acid, B6, folate, B12, calcium, iron, phosphorous, and zinc. The egg white contains most of the sodium, magnesium, and niacin.

Egg Substitutes

Eggs (more specifically egg yolks) are naturally high in fat and cholesterol, which causes concern for people with high cholesterol, heart disease, and related health concerns. Egg whites, egg substitutes, and eggless substitutes each offer alternatives to whole eggs and egg yolks. For example, egg whites do not contain any cholesterol but can be used to make a fluffy omelet.

An *egg substitute* is an egg product made from eggs that have been separated to remove the yolks. Egg substitutes are produced from pasteurized egg whites and a combination of ingredients such as vegetable gums, milk solids, vegetable oil, and flavorings that mimic the texture and color of whole beaten eggs. One-fourth cup of liquid egg substitute contains only 1 mg of cholesterol as compared to 211.5 mg in one large egg yolk. Egg substitutes are used to prepare omelets, scrambled eggs, quiches, and custards. Egg substitutes are commonly sold in cartons and usually frozen for a longer shelf life. **See Figure 2-3.**

An *eggless egg substitute* is a yellow-colored liquid composed of soy, vegetable gums, and starches derived from corn or flour. Eggless egg substitutes do not contain any egg product. They have the appearance of beaten eggs but not the flavor. For this reason, eggless egg substitutes are not used to prepare egg dishes such as omelets, scrambled eggs, or custards. Instead, they are almost exclusively used to produce cholesterol free or lower cholesterol baked goods.

Egg Substitutes

American Egg Board

Figure 2-3. Egg substitutes are sold in cartons and usually frozen for a longer shelf life.

Egg Allergies

Some customers have an allergic reaction when they eat eggs or egg-derived ingredients. As one of the eight most common food allergens, current FDA labeling regulations require the presence of eggs be marked on food labels in bold print. It is important to note that eggs and egg-derived ingredients may be listed as the following:

- eggs, powdered eggs, egg whites, or egg yolks
- any ingredient with the prefix ovo- or ova- (meaning "egg" in Latin)
- albumin, globulin, livetin, or lysozyme (which are all egg proteins)
- Simplesse® (a fat substitute made from egg protein)
- mayonnaise

Egg Allergy

Egg Sizes and Grades

Large eggs are the most common size of egg used in foodservice operations, and most recipes use large eggs as the standard. Eggs are classified according to size based on the minimum weight per dozen eggs. Each named size weighs about 3 oz per dozen more than the previous size. **See Figure 2-4.**

The size does not refer to the dimensions of an egg or how big it appears. For example, some eggs may be slightly smaller or larger than others in the same carton. Size refers to the total weight of the dozen.

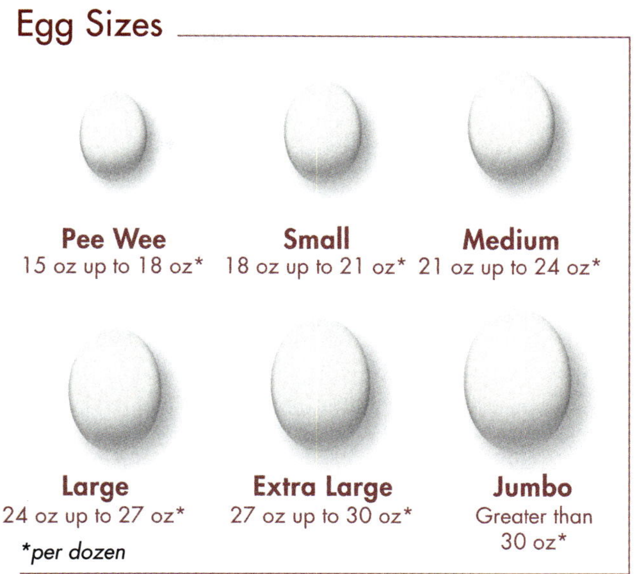

Figure 2-4. Eggs are classified according to size based on the minimum weight per dozen eggs.

The size of an egg, the color of its shell, and its grade have no bearing on its nutritional value. Eggs are graded as AA, A, or B. The highest quality egg, Grade AA, has a firm yolk and white that both stand tall when the egg is broken onto a flat surface. As an egg ages or if it is of a lower grade, the yolk and the white will lie flatter and spread out farther on a flat surface. **See Figure 2-5.** The white will also be looser and more watery, and the air pocket inside the egg will be larger.

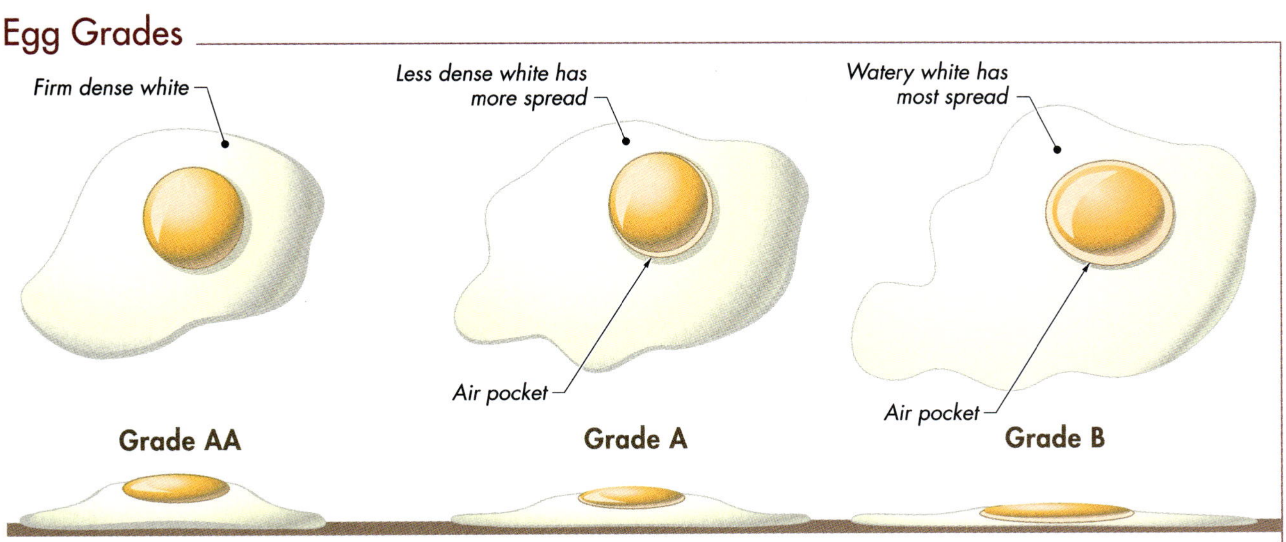

Figure 2-5. Grade AA eggs have a firm yolk and white that both stand tall, whereas the yolk and white of lower grade eggs lie flatter and spread out farther.

Eggs may also be purchased in liquid, frozen, and dried forms. These alternative forms of eggs are convenient in volume food preparation settings, especially for preparing baked products. Dried eggs are not often used in foodservice establishments.

Egg Safety

Fresh eggs are used more than any other form of egg in the professional kitchen. Like all protein-rich foods, eggs are potentially hazardous when not handled and stored properly. Salmonella enteritidis is the bacteria found inside eggs that have become contaminated from infected laying hens. **See Figure 2-6.** The potential presence of salmonella enteritidis is why it is best to thoroughly cook eggs prior to eating them.

Salmonella Enteritidis

Agricultural Research Service, USDA

Figure 2-6. Eggs are tested for salmonella enteritidis, the bacteria found inside eggs that are contaminated from infected laying hens.

Egg Quality and Composition

To avoid the contamination issue, some foodservice operations use pasteurized eggs and egg products. A *pasteurized egg* is an egg that has been heated to a specific temperature for a specific period of time to kill bacteria that can cause foodborne illnesses. The pasteurization process takes place before the egg is packaged, either in the shell or as a liquid egg product. The use of pasteurized eggs and egg products reduces the potential of a foodborne illness being contracted from eating foods that contain uncooked or partially cooked eggs. Pasteurized eggs are a good option for preparing egg dishes that are not fully cooked, such as sunny-side up eggs and over-easy eggs.

Egg Storage

Shell eggs should always be refrigerated at 40°F or below and may be kept refrigerated for about a month past the Julian date. The Julian date stamped on the egg carton indicates the day the eggs were packed. Julian dates begin with January 1 as day 001 and number successively through December 31 as day 365 to represent the days of the calendar year.

Opened liquid egg substitutes can be stored for three days. Frozen egg products should always be thawed in the refrigerator and then stirred thoroughly before use. Dried eggs should always be stored in a cool, dry place and then refrigerated once they have been reconstituted.

Production Tip

Leftover egg whites and egg yolks should be stored in a tightly covered container before being refrigerated and should be used within 2–4 days.

CHECKPOINT 2-1

1. Identify the four main parts of the egg.
2. Identify the part of the egg that contains most of the calories and all of the cholesterol.
3. Explain why eggs are considered a nutrient-dense food.
4. Describe the purpose of egg substitutes and eggless egg substitutes.
5. Explain why it is important to know if a dish contains eggs or egg-derived ingredients.
6. Describe the six classifications of eggs according to size.
7. Explain how eggs are graded.
8. Explain the advantage of using pasteurized eggs.
9. Describe the storage requirements for eggs.

EGG PREPARATION

Eggs can be prepared many different ways using dry-heat and moist-heat cooking methods. Eggs can be fried, scrambled, made into omelets and frittatas, baked, and poached.

The most common mistake made when cooking eggs is overcooking or undercooking them. Most griddles have thermostats that control the temperature of the griddle plate or manual controls that can be set on low, medium, or high. Regardless of the type of griddle being used, it is necessary to verify the temperature using either an infrared, manual, or digital contact thermometer. Using very high heat or too long a cooking time will cause eggs to become rubbery. It can also change their color and flavor. For example, overcooked scrambled eggs that sit in a steam table too long often develop a green cast.

U.S. Navy Photo by MC3 Shawn Stewart

Fried Eggs

Sunny-side up, basted, over-easy, over-medium, and over-hard eggs are cooked to order. Although they are all referred to as fried eggs, these egg dishes are technically sautéed. **See Figure 2-7.** The main difference among each dish is the variation in cooking time and technique.

Fried eggs should always be cooked to order. The fresher the egg, the more the yolk stands up and the less runny the white will be. If fried eggs are cooked in butter, only clarified butter should be used due to it having a higher smoke point than whole butter, which burns easily. Many foodservice operations use nonstick cooking spray instead of butter.

Fried Eggs

Figure 2-7. Fried eggs are technically sautéed.

Sunny-Side Up Eggs. A *sunny-side up egg* is a lightly fried egg with an unbroken yolk that is not flipped over to cook the other side. The yolk of a sunny-side up egg is still liquid inside and bright yellow on top.

Basted Eggs. A *basted egg* is a fried egg with an unbroken yolk that is cooked like a sunny-side up egg, yet the tops are slightly cooked by tilting the pan and basting the top of the egg with hot butter. The yolk is still liquid inside, but is covered with a thin white film.

Over-Easy Eggs. An *over-easy egg* is a fried egg with an unbroken yolk that is cooked until the egg white has gone from translucent to white on the bottom and then flipped over and cooked on the other side just until the white is no longer translucent. The yolk of an over-easy egg is still liquid inside and covered with a white film.

Over-Medium Eggs. An *over-medium egg* is a fried egg with a completely cooked white and a yolk that is cooked nearly all the way through. Over-medium eggs are simply cooked a little longer than over-easy eggs. The yolk of an over-medium egg is fairly set around the edges, has an almost fully cooked appearance, and yet is soft in the very center.

Over-Hard Eggs. An *over-hard egg* is a fried egg with a bright, firm white that is not rubbery and a pale yellow yolk that looks almost fluffy. Over-hard eggs are cooked a little longer than over-medium eggs.

Prior to cooking, whole eggs should be cracked into a nonreactive bowl so that the raw eggs can be inspected for quality, as well as to help ensure that the yolks do not break when added to the oiled and preheated cooking surface. Then, the eggs are gently poured onto the hot surface, immediately solidifying the whites. Fried eggs are cooked to the desired doneness.

Procedure for Frying (Sautéing) Eggs

1. Heat a griddle to 325°F and add a small amount of fat.
2. Crack the eggs into a bowl without breaking the yolks.
3. Gently slide the eggs onto the cooking surface and reduce the heat.
4. Cook eggs until the whites have set.
5. Carefully turn the eggs with a smooth motion using an offset spatula. When turning the eggs, a portion of the white should remain in contact with the griddle to minimize the chance of breakage.
6. Cook each egg to order: over-easy, over-medium, or over-hard.

Photos: Sullivan University

Tips for Frying Eggs

- Use a conditioned pan or griddle. A poorly conditioned pan or griddle can cause eggs to break and burn.
- Use the correct amount of fat. Too much fat results in greasy eggs that may also burn. Too little fat results in eggs that stick to the pan and break when trying to remove them from the pan.
- Use the correct temperature. Too high a temperature can cause eggs to burn and overcooking may result in a tough exterior and a runny interior. Too low a temperature can cause egg whites to spread too rapidly.

Scrambled Eggs

Scrambled eggs are a popular egg dish that is often served plated with breakfast meats or as an entrée choice on buffets. **See Figure 2-8.** Scrambling is the easiest method to use when preparing eggs in quantity. First, the eggs are broken into a nonreactive bowl and seasoned with salt and pepper. An aluminum bowl should never be used because it will discolor the eggs as they are whisked. The eggs are lightly beaten with a whisk or kitchen fork. If desired, about 1 tsp of milk, cream, or water per egg may be added. Too much added liquid will cause the eggs to weep (give off water) after they are cooked.

Breakfast Serving Line

U.S. Army

Figure 2-8. Scrambled eggs are a popular item on a breakfast service line.

Production Tip

When scrambled eggs are to be held longer than 5 minutes, a small amount of béchamel sauce or heavy whipping cream can be added to extend the holding time and prevent the eggs from drying out and discoloring.

Procedure for Scrambling Eggs

1. Break the eggs into a nonreactive bowl and beat the eggs slightly with a whisk.
2. Pour the beaten eggs onto a 325°F preheated tilt skillet or griddle coated with a small amount of fat. *Note:* Scrambled eggs can also be prepared in a steam-jacketed kettle, steamer, or sauté pan.
3. Reduce heat and lift the eggs carefully from the bottom and stir gently so the uncooked portion will settle to the bottom and cook.
4. When the eggs are fluffy and light-golden in color, serve immediately. If scrambled eggs are to be held for service, undercook the eggs slightly.

Photos: Sullivan University

Tips for Scrambling Eggs

- Use the correct amount of fat. Too much fat yields greasy eggs. Too little fat can cause eggs to stick, become tough and dry, or burn.
- Cook eggs at the correct temperature. Too high a temperature yields overcooked, dry, brown eggs that may also burn.
- Do not stir eggs excessively. Excessive stirring breaks the egg into very fine particles, resulting in a poor appearance once they are cooked.
- Do not hold scrambled eggs too long. Holding eggs too long results in an unappetizing texture and a green cast.

Then the eggs are sautéed in a steam-jacketed kettle, in a tilt skillet, on a griddle, in a double boiler, or in a sauté pan to produce fluffy curds. Scrambled eggs should always be slightly undercooked, because they become firm as they are held for service. Scrambled eggs are properly cooked when they are fluffy, moist, and not runny. **See Figure 2-9.**

Scrambled Egg Wrap

Courtesy of The National Pork Board

Figure 2-9. Scrambled eggs can be served in wraps that are plated or as a to-go breakfast option.

Omelets

An *omelet* is an egg dish made with beaten eggs and cooked into a solid form. Omelets are one of the most commonly ordered egg dishes in restaurants. Omelets may be filled with a variety of cheeses, vegetables, and meats. **See Figure 2-10.** A filled omelet is often named for the filling ingredient, such as a bacon omelet or a mushroom omelet.

Omelet Fillings

Carlisle FoodService Products

Figure 2-10. Omelets may be filled with a variety of cheeses, vegetables, and meats.

Omelets are prepared to order and usually consist of two or more eggs per order. If omelets are held for even a short period of time, they lose their fluffiness and become rubbery. There are four common varieties of omelets: rolled (French) omelets, folded omelets, soufflé omelets, and frittatas.

Rolled (French) Omelets. A *rolled omelet,* also known as a French omelet, is an omelet that is cooked and then rolled onto a plate and cooked filling ingredients are added through a slit cut into the top. To prepare a rolled omelet, cook an omelet until it is slightly set but still very moist. Then, tilt the pan about 60° and use a high-heat spatula to roll the omelet. The eggs should be pale yellow and browned as little as possible or not at all. Once the egg portion of the omelet is cooked, roll the omelet tightly on the plate and add cooked ingredients and cheeses through a slit cut into the top of the omelet.

Production Tip

Almost any food can be used to fill an omelet. Filling ingredients can be savory or sweet and should total about ⅓ to ½ cup per omelet. Fillings should be prepared in advance and held at safe temperatures. Meats should be fully cooked and cut into bite-sized pieces before being added to an omelet.

Procedure for Preparing Rolled (French) Omelets

1. Add fat to a heated omelet pan over medium heat.
2. Whisk eggs in a nonreactive bowl. If desired, season with salt and pepper.
3. Pour the egg mixture in the pan and gently shake the pan while gently stirring with a heat-resistant spatula.

4. When the mixture is almost set, gently lift one side of the omelet to allow any liquid to run underneath and cook. Then, sweep the sides of the pan with the spatula to round out the omelet.
5. Fold one side of the omelet one-third of the way toward the center.

6. Place the omelet pan over the serving plate with the unfolded end of the omelet nearest the plate and gently roll the omelet onto the plate, seam side down. Serve immediately.

Preparing Rolled (French) Omelets
Media Clip

Tips for Preparing Omelets

- Use a conditioned cooking surface. A poorly conditioned cooking surface can cause omelets to stick as well as break when rolled or folded.
- Use the correct amount of fat. Too much fat yields greasy eggs. Too little fat can cause eggs to stick, become tough and dry, or burn.
- Cook eggs at the correct temperature. Too high a temperature yields brown omelets that may also crack when rolled or folded.
- Do not stir the eggs once the omelet begins to set or the end product will resemble scrambled eggs.
- Do not hold omelets. Held omelets are tough and rubbery.

Folded Omelets. A *folded omelet* is an omelet that is cooked until nearly done and then folded before serving. Any vegetables or meats need to be precooked, or at least heated, prior to adding them to the beaten eggs. The exception to the rule is cheese, which is added just before the omelet is folded near the end of the cooking process.

Soufflé Omelets. A *soufflé omelet,* also known as a fluffy omelet, is a lighter variation of a folded omelet. To make a soufflé omelet, the eggs are beaten to incorporate air, which increases the volume. Air can be incorporated into a soufflé omelet two different ways. One method is to separate the yolks from the whites and then whisk each into a soft foam. Next, the whisked whites and the whisked yolks are folded together and then poured into a hot, oiled pan and cooked in the same manner as a folded omelet. An immersion blender can also be used to whip air into the egg mixture. When this technique is used, the egg mixture will cook up even lighter than an egg mixture that was whisked by hand.

The omelet is then placed under a broiler for 10–15 seconds to crisp. A soufflé omelet will puff slightly due to the air incorporated from using the blender and the addition of whipped egg whites.

Frittatas. A *frittata* is an omelet that is served open-faced after being browned in a broiler or hot oven. The ingredients in a frittata are all precooked and mixed with the eggs, but instead of folding the omelet, it is left open and cheese is often added on top. The frittata is then placed under a broiler or in a very hot oven to melt and brown the cheese. **See Figure 2-11.**

Frittatas

Figure 2-11. A frittata is basically a traditional folded omelet served open-faced after being browned in a broiler or hot oven.

Procedure for Preparing Folded Omelets

1. Place shelled eggs into a nonreactive bowl. Beat the eggs just enough to blend the yolks and whites.
2. Add salt and pepper, mixing thoroughly.
3. Using a 4 oz ladle for individual omelets, pour egg mixture onto a lightly oiled and preheated 325°F griddle.

4. Cook until the bottom of the omelet is lightly browned. If necessary, gently lift the cooked portion with a spatula to permit the uncooked mixture to flow underneath. Continue cooking until the eggs are set.
5. Add desired fillings to each omelet.

6. Fold one-third of the omelet to the middle.
7. Then, fold the opposite side of the omelet to the middle to enclose the fillings, making a long oval-shaped omelet. Serve immediately or hold in a hot steam table pan.

Photos: Sullivan University

Procedure for Preparing Frittatas

1. Add fat to a heated omelet pan over medium heat.
2. Add fillings to the omelet pan and cook until done.
3. Whisk eggs in a nonreactive bowl. If desired, season with salt and pepper.
4. Pour beaten eggs over the cooked fillings and stir with a heat-resistant spatula until the eggs are somewhat set.
5. When the mixture is almost set, gently lift one side of the frittata to allow any liquid to run underneath and cook.
6. Top the frittata with cheese, if desired.
7. Place the frittata under a broiler or in a hot oven to melt and brown the cheese. Serve immediately.

Egg Sandwiches

A fried egg, scrambled eggs, or a mini omelet can be added to a hot biscuit, bagel, English muffin, tortilla, or croissant to create a delicious handheld breakfast. **See Figure 2-12.** Cheese, vegetables, or meats may also be added to enhance the flavor and the size of the sandwich.

Breakfast Sandwiches

American Egg Board

Courtesy of The National Pork Board

Figure 2-12. A handheld breakfast can be created by adding a fried egg, scrambled eggs, or a mini omelet to a hot biscuit, bagel, English muffin, tortilla, or croissant.

Preparing Frittatas
Media Clip

Quiches

A *quiche* is a baked egg dish composed of a savory custard baked in a precooked piecrust. **See Figure 2-13.** A savory custard consists of beaten eggs mixed with cream, milk, and seasonings. The cream and milk added to the beaten eggs cause the egg protein in the custard to be more delicate and tender when cooked. To prepare a quiche, a precooked piecrust can be filled with a variety of cooked vegetables, meat, and cheeses. Then, the savory custard mixture is poured on top of the chosen ingredients and the quiche is baked until set. It is important to note that the custard should not be cooked to an internal temperature above 185°F or the egg proteins will begin to curdle and separate from the water molecules in the egg.

Quiches

Courtesy of The National Pork Board

Figure 2-13. A quiche is a baked egg dish composed of a savory custard baked in a precooked piecrust.

Procedure for Preparing Quiche

1. Roll 8 oz of pie dough into a 10 inch circle.
2. Place the dough into an 8 inch pie shell and press gently to set. Place in refrigerator until well chilled.
3. Once the crust has chilled completely, place a second pie tin inside the first pie tin and bake upside down in a 450°F preheated oven for 10 minutes to prebake the crust.
4. While the crust is prebaking, prepare the filling ingredients (e.g., diced bacon or shredded cheese).
5. Remove the piecrust from the oven and add the prepared filling ingredients.
6. Whisk milk, eggs, cream, salt, and pepper together to create the custard.
7. Pour the custard on top of the prepared filling ingredients.
8. Place the quiche on a sheet pan to catch any spillover during baking.
9. Bake quiche in a 375°F oven for 35–45 minutes or until the custard is firm. *Note:* When a toothpick inserted into the center of the quiche comes out clean, the quiche is done.

Poached Eggs

Poached eggs are very popular on breakfast menus for dishes such as eggs à la Florentine and eggs Benedict. Poached eggs are removed from the shell prior to cooking. Vinegar is used to set the white firmly around the yolk when the egg is placed in the water. The acid from the vinegar actually toughens the protein in the egg white and does not affect the flavor of the egg.

Although poached eggs are very easy to prepare, the challenge is in determining doneness as well as transporting eggs from the poaching water to the plate without breaking the yolks. Eggs cooked at too high a temperature are usually overcooked and tough. Eggs cooked at too low a temperature are too tender and very difficult to handle when they are served.

About 8 eggs can be poached at a time in a gallon of water. The water may be used for three different batches before being discarded. Eggs that are poached in quantity are slightly undercooked and then immediately placed in cold water to stop further cooking and held until service. To serve, the precooked eggs are reheated in hot salted water. This quantity method is often used for banquet or buffet preparations. For individual breakfast preparations, eggs should be poached to order. **See Figure 2-14.**

Procedure for Poaching Eggs

1. Fill a pan with enough water to cover the eggs.
2. Add 1 tsp of salt and 1 tbsp of distilled white vinegar per quart of water.
3. Bring the water to a boil and then lower the heat to approximately 180°F.
4. Break each egg into a nonreactive bowl and then gently slide the eggs into the simmering liquid so the yolk remains in the center of the white and does not break.
5. Cook to desired doneness (3–5 minutes).
6. Remove poached eggs with a skimmer or slotted spoon and drain well. Serve immediately.

Poached Eggs

National Turkey Federation

Figure 2-14. Eggs should be poached to order for individual breakfast preparations.

Eggs Benedict

Traditional eggs Benedict, known as the most lavish of egg dishes, consists of two toasted English muffin halves, each topped with a slice of Canadian bacon and a poached egg. The poached eggs are then topped with a coating of hollandaise sauce.

Procedure for Preparing Eggs Benedict

1. Prepare hollandaise sauce and set aside until needed.
2. Prepare poaching liquid for eggs and hold between 160°F and 180°F.
3. Heat 2 slices of Canadian bacon on a griddle or in a sauté pan until warmed through. Set aside until needed.
4. Crack two eggs into a small bowl and pour them gently into the poaching water.
5. Split an English muffin and toast in toaster until done. Set on serving plate until eggs are ready.
6. Top each muffin half with one slice of heated Canadian bacon.
7. Using a slotted or perforated spoon, remove eggs from poaching water at desired degree of doneness (soft, medium, or hard cooked). Place the spoon with eggs on a folded paper towel to draw any excess water from the eggs.
8. Place well drained egg on top of each piece of Canadian bacon and muffin half.
9. Ladle an ample portion of hollandaise sauce over each poached egg and serve immediately.

Nutrition Note

Since poached eggs are cooked in water, no additional fat is added to the eggs. However, dishes such as eggs Benedict and eggs Florentine usually include heavy sauces that contain added fat and calories.

The richness of the poached eggs and hollandaise sauce complement the saltiness of the Canadian bacon and the crunch of a toasted English muffin. Variations of traditional eggs Benedict are offered by many foodservice operations. For example, avocado and turkey can be placed under the egg and hollandaise sauce and atop an English muffin to offer a unique flavor combination. **See Figure 2-15.**

Eggs Benedict

National Turkey Federation

Figure 2-15. Eggs Benedict is a lavish egg dish showcasing poached eggs atop an English muffin with many popular variations.

CHECKPOINT 2-2

1. Identify five types of fried eggs.
2. Prepare and plate five types of fried eggs.
3. Identify the type of egg preparation most often used when preparing eggs in quantity.
4. Prepare and plate scrambled eggs.
5. Contrast the preparation of a rolled (French) omelet, a folded omelet, a soufflé omelet, and a frittata.
6. Prepare and plate a rolled (French) omelet, a folded omelet, a soufflé omelet, and a frittata.
7. Identify five bread products often used to create egg sandwiches.
8. Prepare and plate two different types of egg sandwiches.
9. Explain how to prepare a quiche.
10. Prepare a quiche.
11. Explain how to poach eggs.
12. Explain how to prepare traditional eggs Benedict.
13. Prepare and plate eggs Benedict.

Variations of Eggs Benedict

Variation	*Description*
Country Benedict	Replace English muffin and Canadian bacon with a buttermilk biscuit and a sausage patty; then top eggs with country-style white gravy
Eggs Blackstone	Replace Canadian bacon with a slice of regular smoked ham topped with a tomato slice
Eggs Bombay	Serve poached eggs on a bed of rice pilaf and top with a creamy curry sauce
Eggs Florentine	Replace Canadian bacon with sautéed spinach and top eggs with hollandaise or Mornay sauce
Eggs Sardou	Replace English muffins with cooked artichoke bottoms and the Canadian bacon with two crossed anchovies; then top eggs with a slice of truffle
Irish Benedict	Replace Canadian bacon with corned beef hash
Lobster Benedict	Replace Canadian bacon with slices of sautéed lobster tail or claw meat

BREAKFAST BATTERS

Batters are used to create several breakfast items that may be served alone or as a complement to egg dishes or breakfast meats. Pancakes, waffles, crêpes, blintzes, and French toast are made from breakfast batters.

Pancakes and Waffles

Pancakes and waffles can be served with a variety of different toppings. Pancakes and waffles are always served piping hot. Berries, apples, bananas, or pecans are often added to pancake batter or served on top of waffles.

Pancakes are also known as hotcakes, griddle cakes, or flapjacks. Pancakes are often plated separately from other breakfast items served to the same guest. **See Figure 2-16.** The basic recipe for pancakes consists of flour, sugar, salt, baking powder, baking soda, milk or buttermilk, eggs, and butter or oil. The dry ingredients are sifted together, and the milk, eggs, and butter are whisked together in a separate bowl. Then the liquid mixture is slowly combined with the dry mixture and the batter is ladled on a hot oiled griddle. Pancakes are ready to be flipped over when the surface is nearly covered in bubbles. When the second side turns golden brown, the pancakes are ready to be plated.

The procedure for making waffles is very similar to pancakes except that a waffle iron is used in place of a griddle. Waffle batter is poured onto a well-oiled or nonstick waffle iron and the lid is closed. The batter is simultaneously cooked for a few minutes on both sides. **See Figure 2-17.** Waffles are most often served with a fruit topping and whipped cream.

Pancakes

Figure 2-16. Pancakes are often plated separately from other breakfast foods.

Waffles

Chef's Choice® by EdgeCraft Corporation

Figure 2-17. A waffle iron is used to simultaneously cook both sides of a waffle.

Procedure for Preparing Pancakes

1. Use a ladle to place pancake batter onto a lightly oiled, preheated 375°F griddle. Repeat this process quickly for larger orders of pancakes.

Sullivan University

2. Cook batter on one side until the top is covered with bubbles and the underside has browned.
3. Flip pancakes and cook the other side.

Sullivan University

4. Pancakes that are ready to be removed from the griddle will appear to have a gap between the two sides.
5. Shingle prepared pancakes by slightly overlapping them in a 2 inch hotel pan that has been dusted with granulated sugar.

Pancakes and waffles may be prepared in advance, although the quality is not nearly as good as when they are freshly made. Both are completely cooked and then shingled on sheet pans. The pancakes and waffles are then warmed in the oven before plating.

Crêpes and Blintzes

A *crêpe* is a French pancake that is light and very thin. Crêpe batter does not contain leavening agents and therefore remains very thin when cooked. Crêpes are cooked either in a very hot crêpe pan, in a small sauté pan, or on a crêpe maker. The surface of a crêpe is cooked to a light golden brown color. Typically, crêpes are filled with fruit or cheese, rolled up, and topped with confectioners' sugar. Crêpes may be served for breakfast, as a dessert, or filled with savory items for lunch or dinner. Although crêpes are delicate, they can be frozen.

A *blintz* is a filled crêpe that is only cooked on one side and not flipped over to cook the other side. Like crêpes, blintzes can be filled with sweet or savory fillings. Sweet blintzes are often filled with a sweetened ricotta or a farmer's cheese mixture and then rolled or folded into small pillow-shaped parcels. **See Figure 2-18.** Then, the blintz is put back into a hot sauté pan with clarified butter and browned on the top and bottom. Blintzes are garnished with any variety of fruit compote, warmed fruit preserves, or sour cream. A savory blintz can be filled with any variety of sautéed vegetables, but the most common filling is potato and onion garnished with sour cream.

Procedure for Preparing Crêpes

1. Add fat to an omelet pan over medium heat.
2. Add just enough batter to lightly coat the pan.

3. Spread the batter across the diameter of the pan and cook over medium-low heat.

4. Before the crêpe begins to turn golden, flip with a rubber spatula.

5. When the batter is set, add filling if desired.

6. Fold an edge of the crêpe toward the center and then position the pan over the plate.

7. Roll the crêpe onto the plate and serve immediately.

Preparing Crêpes
Media Clip

Blintzes

Figure 2-18. Blintzes are often filled with a sweetened ricotta or a farmer's cheese mixture and then rolled or folded into small pillow-shaped parcels.

French Toast

French toast is prepared by dipping slices of bread into a batter made from eggs, milk, sugar, and vanilla. The batter-dipped bread is then placed on a lightly oiled, preheated griddle or pan and cooked on each side until golden brown. **See Figure 2-19.**

French Toast

National Honey Board

Figure 2-19. French toast is batter-dipped bread that is cooked on a lightly oiled griddle or in a lightly oiled pan until each side is golden brown.

Crunchy French toast is prepared the same way as traditional French toast except that the batter-dipped bread is also dipped in a crunchy coating such as crumbled Corn Flakes™ or sliced nuts. The toast is then cooked on a griddle or in a pan until it is golden on both sides. It is then placed on a sheet pan and placed in a 350°F oven until crispy and cooked through. Both traditional and crunchy French toast can be served with a fruit topping or syrup.

CHECKPOINT 2-3

1. Describe the preparation of pancakes and waffles.
2. Prepare and plate pancakes and waffles.
3. Contrast the preparation of crêpes and blintzes.
4. Prepare and plate crêpes and blintzes.
5. Describe the preparation of French toast.
6. Prepare and plate French toast.

BREAKFAST MEATS

Breakfast meats commonly include a variety of sausages, bacon, ham and Canadian bacon, and hash. Most breakfast meats may be precooked and then reheated for service when the breakfast demand is large. When the demand is smaller, breakfast meats are usually cooked to order.

Sausages

Breakfast pork sausages often have a strong flavor of sage and crushed red pepper flakes. Turkey sausages offer a lower fat and calorie alternative to pork sausage. However, turkey sausage contains more cholesterol than pork sausage. **See Figure 2-20.**

Nutrition Note

A cooked turkey sausage patty contains 55 calories, 3 g of fat, 26 mg of cholesterol, and 7 g of protein. In contrast, a fresh cooked pork sausage patty contains 95 calories, 8 g of fat, 24 mg of cholesterol, and 5 g of protein.

Turkey Sausages

Photo Courtesy of Perdue Foodservice, Perdue Farms Incorporated

Figure 2-20. Turkey sausages offer a lower fat and calorie alternative to pork sausage.

Sausages come in many flavors, which often connect to their cuisine of origin. For example, chorizo con huevos mixes chopped and fried chorizo sausage with scrambled eggs. *Chorizo* is a Mexican pork sausage flavored with paprika that can be sweet or spicy. Likewise, *kielbasa* is a Polish pork or beef sausage flavored with garlic, pimento, and cloves. *Boudin* is a highly seasoned Creole link sausage made of pork, pork liver, and rice. In the midwestern United States, bratwurst is a common accompaniment for eggs.

Breakfast sausages are served in the form of patties or links. Sausage patties may be purchased in bulk or prepared in the professional kitchen. Patties are usually portioned into 1 oz or 2 oz servings. Sausage links typically range in size from about 1 oz each (16 per lb) to 2 oz each (8 per lb). The portion served for breakfast is generally three or four links if they are smaller and one or two links if they are larger. After cooking, sausage links are placed in a hotel pan and held for service.

Production Tip

Sausage links should always be cooked in small batches as any leftovers will be too dry to use the next day.

Bacon

Different types of bacon are served at breakfast. Pork bacon comes from the belly of a pig and is the most common breakfast bacon. Hickory smoked, applewood smoked, peppered, and turkey bacon frequently appear on breakfast menus. Italian breakfasts often feature strips of pancetta. *Pancetta* is unsmoked pork belly that has been cured in salt and spices, such as pepper and nutmeg, and then dried for a few months. Unlike bacon, pancetta is not traditionally smoked.

Sliced bacon is packaged in a few different forms. Slab-packed bacon is presliced but the user has to pull the slices apart. Shingled bacon consists of slightly overlapped sliced strips of bacon. Layer-packed bacon is packaged with each slice separated and laid out on sheets of parchment paper. **See Figure 2-21.** Layer-packed bacon is the type commonly used in foodservice operations, because it does not require someone to spend time taking the slices apart.

Layer-Packed Bacon

Figure 2-21. Layer-packed bacon is packaged with each slice separated and laid out on sheets of parchment paper.

The most common thickness of sliced bacon is an 18–22 count, meaning there will be between 18 and 22 slices per pound. Bacon slices are typically placed side by side, but not touching, on a sheet pan and baked at 350°F until they are three-quarters done. The bacon is then removed from the oven and placed in a perforated hotel pan to drain the excess grease. The bacon is finished just before service. This method of preparation also reduces shrinkage and curling. Bacon slices may also be cooked in a broiler or on a griddle.

Procedure for Preparing Bacon

1. Place bacon slices side by side, but not touching, on a baking sheet in a 350°F oven.
2. Cook until bacon is three-quarters done.
3. Remove bacon and place in a perforated hotel pan.

Ham and Canadian Bacon

Ham and Canadian bacon are precooked breakfast meats that only need to be heated on a griddle, under a broiler, or in a hot pan before service. **See Figure 2-22.** Breakfast ham is typically purchased in a boneless or boned and rolled form, providing 3 oz or 4 oz portions when sliced. *Canadian bacon* is a hamlike breakfast meat made from boneless, smoked, pressed pork loin. The shape and size of Canadian bacon serves as a great base for a breakfast sandwich or eggs Benedict.

Ham and Canadian Bacon

Breakfast Ham Focaccia

Canadian Bacon Omelet

Courtesy of The National Pork Board

Figure 2-22. Ham and Canadian bacon are popular breakfast meats.

Pork Industry Quick Facts

Hash

Hash dishes are a creative way to use up the trimmings left behind when cutting corned beef or roast beef on a slicer. *Hash* is shredded and chopped meat that has been mixed and cooked with diced potatoes, onions, and seasonings. **See Figure 2-23.** Hash is cooked on a griddle or in a sauté pan until heated through and commonly topped with two eggs cooked to order.

Hash

Figure 2-23. Hash is composed of shredded and chopped meat and cooked with diced potatoes, onions, and seasonings.

CHECKPOINT 2-4

1. Identify types of breakfast sausages.
2. Identify the form of bacon most commonly used in food service.
3. Identify breakfast meats that are purchased precooked.
4. Prepare and plate three breakfast meats.
5. Describe how hash is prepared.
6. Prepare and plate hash.

BREAKFAST POTATOES

Breakfast potatoes are usually served fried or griddled. Potatoes may be served à la carte or included in featured breakfast combinations, such as two eggs with ham and hash brown potatoes. Two common breakfast potato preparations are hash browns and home fries. **See Figure 2-24.**

Hash browns are a common breakfast side dish. Although the name "hash" brown refers to a chopped, recooked potato, most people refer to the shredded potato dish as hash browns. Most foodservice operations purchase already prepared hash browns rather than producing them from scratch to save time and labor costs. To prepare hash browns from scratch, potatoes are peeled and then boiled until tender, yet slightly al dente. The potatoes are cooled and then shredded. Vegetable oil or clarified butter is added to a very hot sauté pan or griddle and the potatoes are cooked until golden brown on both sides. Hash browns need to be served immediately.

Breakfast Potatoes

Basic American Foods

Hash Browns

Idaho Potato Commission

Home Fries

Figure 2-24. Two popular breakfast potato preparations are hash browns and home fries.

Home fries are another simple breakfast potato dish that may or may not be made from scratch, depending on the foodservice operation. Home fries are made by peeling and boiling red mealy potatoes until tender, yet al dente. The potatoes are then cooled and diced or sliced and browned on all sides in a hot sauté pan with butter. They are seasoned with salt and pepper and served immediately.

CHECKPOINT 2-5

1. Differentiate between hash browns and home fries.
2. Prepare and plate hash browns.
3. Prepare and plate home fries.

BREAKFAST BREADS

Common breakfast breads include toast, biscuits, bagels, and English muffins. **See Figure 2-25.** Breakfast breads are often accompanied by butter, honey, jellies, jams, or preserves.

Toast is made from a variety of loaf breads. These breads include, but are not limited to, wheat, white, rye, multigrain, and Texas toast and are toasted immediately before being served.

Breakfast Breads

Biscuits

Bagels

Photo Courtesy of Perdue Foodservice, Perdue Farms Incorporated

Figure 2-25. Biscuits and bagels are two common breakfast breads.

A *biscuit* is a quick bread made by mixing solid fat, baking powder or baking soda, salt, and milk with flour. At breakfast, biscuits are often topped with sausage gravy or served as a side for egg dishes and with grits. Biscuits are also used as a base for breakfast sandwiches.

Bagels are usually sliced in half horizontally and then toasted. Typical condiments include butter, plain or flavored cream cheese, jams, preserves, and peanut butter. Bagels also are used as a base for breakfast sandwiches.

English muffins are often served toasted and also may be used as the base for breakfast sandwiches. English muffins are split horizontally, heated, and buttered before service.

CHECKPOINT 2-6

1. Identify four common breakfast breads.
2. Prepare and plate breakfast breads.

FRUITS

Any type of fruit may be served for breakfast. Fruit that is in season is most often served as an accompaniment. A combination of melons, such as honeydew and cantaloupe, are often sliced or served in chunks. Fresh berries are washed and served whole or as a garnish on breakfast plates. **See Figure 2-26.** Oranges are often cut into slices or wedges to add color and freshness to a hot breakfast plate. Frozen or canned fruit is typically used to fill crêpes and blintzes and top pancakes, waffles, or French toast.

Fresh Fruits

Wisconsin Milk Marketing Board, Inc.

Figure 2-26. Fruit that is in season is most often served as an accompaniment.

Strawberry Handling

CHECKPOINT 2-7

1. Explain various ways that fruit is served for breakfast.
2. Prepare and plate four different presentations of breakfast fruits.

BREAKFAST CEREALS

Foodservice operations commonly serve hot cereals and granolas for breakfast. Hot cereals such as oatmeal, grits, and farina are made of whole, cracked, ground, or flaked grains. Hot cereals need to be cooked on the stovetop or in a steam-jacketed kettle and may be served with milk, cream, nuts, fruit, cinnamon, brown sugar, butter, or cheese.

Hot cereals are often served with butter or cream and brown sugar. Oatmeal, farina, and grits are popular hot breakfast cereals made from ground grains. **See Figure 2-27.**

Hot Cereals

Figure 2-27. Oatmeal and grits are popular hot breakfast cereals.

Oatmeal can be made from crushed, rolled, steel cut, or coarsely ground oats that are slowly simmered. Oatmeal may be served with milk or cream, brown sugar, nuts, fruit, or butter.

Farina is made from a ground soft wheat called semolina and commonly sold under the brand name of Cream of Wheat®. Farina is boiled until the semolina becomes tender and creamy. Farina is often served alone or with a side of toast or fruit.

Grits are made from ground corn called hominy. Grits are simmered in water or milk until the grain becomes tender and creamy. Both white and yellow grits have an earthy flavor and are often served with eggs and toast or biscuits. These hearty cereals are often served only during the morning hours due to how quickly they coagulate when held too long.

Granola is a baked mixture of rolled oats, nuts, dried fruit, and honey. Granolas may also contain other grains, cinnamon, coconut, chocolate bits, and other tasty items. Granolas and muesli are often served with yogurts, milks, fresh fruit, or porridge. **See Figure 2-28.** *Muesli* is an unbaked mixture of rolled oats, wheat flakes, oat bran, raisins, dates, sunflower seeds, hazelnuts, and wheat germ. Fresh cut fruit, such as bananas, may be added to granola or muesli along with milk.

Granolas

Alpha Baking Co., Inc.

Figure 2-28. Granolas are often served with yogurts, milks, fresh fruit, or porridge.

CHECKPOINT 2-8

1. Identify three common hot breakfast cereals.
2. Prepare and plate a hot breakfast cereal.
3. Contrast granola and muesli.
4. Prepare and plate a granola.

BREAKFAST BEVERAGES

Almost as important as the breakfast menu items are the breakfast beverages that accompany the meal. A wide variety of coffees, teas, juices, milks, and smoothies may be served.

Coffees

Coffee is prepared from the fruit of a coffee tree, which is found in tropical and subtropical climates around the world. The small red fruit, known as a cherry, contains two small green seeds called coffee beans. Ripened coffee cherries are harvested, fermented, and hulled to remove the two green coffee beans inside each cherry. During processing, the pulp, parchment, and silverskin are removed. **See Figure 2-29.** The green coffee beans are then dried and roasted to produce a rich, intensely flavored, and aromatic beverage called coffee.

Coffee Beans

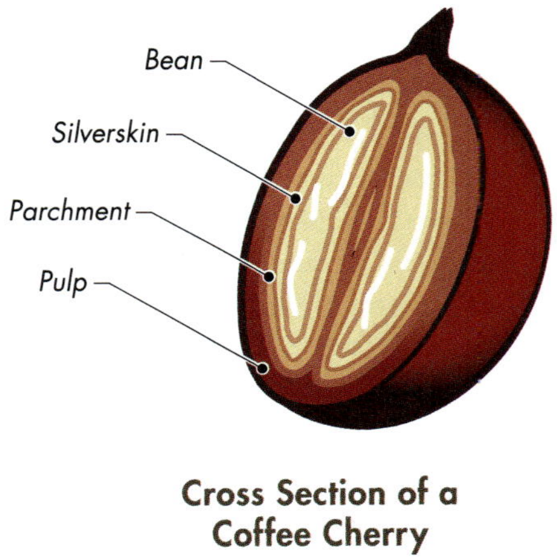

Figure 2-29. A coffee cherry contains two small green seeds called coffee beans. The pulp, parchment, and silverskin are removed during processing. The bean is then roasted and ground before brewing.

Coffee beans are roasted to varying levels of darkness to produce different levels of flavor and aromas. The lighter the beans and the shorter the time period that they are roasted, the lighter the flavor and less intense the aroma. The darker the beans and the longer time period that they are roasted, the darker the flavor and the more intense the aroma. Most coffees are roasted to at least a medium roast because most coffee drinkers prefer a robust flavor.

Types of Coffee Roasts		
Roast	**Common Name**	**Description**
Medium	City roast	Coffee beans are roasted to a medium-brown color, yielding a medium body and medium-flavored coffee
Medium dark	Viennese roast	Coffee beans are roasted to a rich, dark-brown color, yielding a rich and intensely flavored coffee
Dark	French roast	Coffee beans are roasted to a very dark-brown color, yielding a very intensely flavored coffee with a strong aroma and a slightly bitter taste
Espresso	Italian roast	Coffee beans are roasted until they almost reach a burnt state, yielding a bitter, full-flavored, and intensely strong aromatic coffee

Although there are numerous species of coffee beans grown around the world, arabica beans and canephora beans are used in approximately 90% of coffee production. Arabica beans are typically used to produce higher-quality and more expensive coffees and are the best beans for brewing. Canephora beans, commonly known as robusta beans, produce a less flavorful coffee but are often blended with arabica beans to produce a less bitter and smoother tasting coffee.

Espresso. *Espresso* is an intensely flavored coffee made from beans that have been roasted to the very dark or espresso-roasted stage. The almost burnt beans are then ground to a very fine powder, allowing a high extraction of flavor from the dark beans. Espresso is used as the base for many specialty coffee drinks such as cappuccinos, lattes, and mochas.

Common Coffee Drinks	
Coffee Drink	**Description**
Americano	Equal ratio of espresso to very hot water
Cappuccino	Equal ratio of espresso to frothy steamed milk
Frappuccino™	Iced coffee blended to make a coffee milkshake
Iced coffee	Espresso with milk and sugar served over ice
Latte	Espresso with double the ratio of steamed milk to espresso
Macchiato	Espresso served with a small amount of frothy steamed milk
Mocha	Espresso mixed with cocoa or hot chocolate and served with a dollop of whipped cream

Decaffeinated Coffees. Decaffeinated coffee undergoes a process that removes most of the caffeine from the coffee beans. Removing the caffeine allows a coffee drinker to consume coffee without the effects of the stimulant.

Coffee Beans

American Metalcraft, Inc.

Teas

Tea is the primary breakfast drink for a majority of the world's population. In some locations this beverage is also enjoyed with an afternoon snack or after an evening meal. **See Figure 2-30.** Tea can be served hot or cold and comes in many varieties and flavors, depending on the growing region and length of processing. Common teas include green, black, and oolong. Tea-like beverages such as chamomile are often listed on menus as teas.

Green Teas. Green teas are dried without fermentation. It is a slightly bitter flavored tea and is most commonly served hot, without the addition of milk, lemon, or any sweetener. One cup of green tea contains approximately 20–22 mg of caffeine. Sencha is the most common variety of green tea. Green tea loses it flavor more quickly than black tea and should not be stored for extended periods of time.

Teas

Rishi Tea

Figure 2-30. Tea is the primary breakfast drink for a majority of the world's population.

Black Teas. Black tea is a strongly flavored tea resulting from the fermentation of tea leaves. Black tea varieties vary from bold and smoky to light and fruity. To produce black tea, tea leaves are left to dry and then cracked to begin the oxidation process. When oxidation is complete, the once green leaves are left either deep brown or black. Unlike green tea, black tea does not lose its flavor for several years. The types of black tea derive their names from the places where they are produced.

Black tea is graded according to the shape and size of the tea leaves. The smaller the tea leaf, the shorter the brewing time needed to release the maximum amount of flavor. Likewise, tea leaves rolled one way adopt a particular flavor that differs from the same tea leaves rolled a different way. For example, orange pekoe is a tea industry term that refers to leaf size, not a type of tea.

It is important to note that quality is determined by the taste of the tea, not its grade. The purpose of grading tea leaves is to create teas that have a uniform leaf density and appearance and that will extract at the same rate when brewed. The more broken the leaf, the more full-bodied and pungent the tea and the shorter the brewing time. Black teas are often served hot and cold with milk, lemon, and sweetener. A cup of black tea contains approximately 40–42 mg of caffeine.

Oolong Teas. Oolong tea is partially fermented, meaning that it goes through a short period of fermentation that turns the tea leaves a red-brown color. Oolong tea is pale yellow and has a fruity, floral quality with a hint of smoke. Oolong is commonly served hot and without milk, lemon, or sweeteners.

Common Tea Grades

Tea Grade	*Grade Translation*	*Description*
S.	Souchong	Whole twisting leaf that is often light in color; China is the main producer
F.O.P.	Flowery orange pekoe	Long whole leaf with a crushed flower appearance; India is the main producer
O.P.	Orange pekoe	Thin, wiry leaf with a tighter roll than F.O.P.
T. and G.	Tippy and golden	Modifiers used to describe both whole leaf and broken grades; indicates colorful tips on the leaf
P.	Pekoe	Curly, large broken leaf without a visible tip; Sri Lanka is the main producer
B.O.P.	Broken orange pekoe	Small, squarish broken leaf with good body and strength; India is the main producer
F.	Fannings	Smaller than B.O.P. with less keeping quality; used for commercial tea bags
D.	Dust	Smallest grade produced; quick liquoring

Herbal Beverages. Tea is only made from the leaves of an evergreen plant named "camellia sinensis" or one of its subspecies. Tea-like beverages are called tisanes or herbal infusions. A *tisane* is an herbal beverage created by steeping herbs, spices, flowers, dried fruits, or roots in boiling water. The act of steeping creates an infusion. Common tisanes or herbal infusions include chamomile, rose hip, peppermint, and lemon verbena.

Juices

Both fruit and vegetable juices are common breakfast menu items. Juices may be purchased fresh, frozen, or canned. Breakfast juices commonly include orange, grapefruit, cranberry, pineapple, tomato, prune, and mixed or blended juices such as cranapple. **See Figure 2-31.** A standard serving of juice is a 4 fl oz glass.

Fruit Juices

Florida Department of Citrus

Figure 2-31. Grapefruit juice is just one of the many fruit juices commonly offered on breakfast menus.

Milks

Milk is a nutritional beverage from the mammary glands of cows, goats, sheep, or water buffalo. Milk is available in whole, low-fat, and skim and can be flavored, such as with chocolate and strawberry milk. Milk is served as a stand-alone beverage as well as an accompaniment to coffee, hot tea, and porridge. Milk is extremely perishable, so proper storage and handling procedures must always be followed.

Milks are classified according to the percentage of milk fat that remains after processing. Each milk product contains a different percentage of milk fat depending on the amount of milk fat removed. Milk classifications are determined based on the following characteristics:

- Whole milk must contain at least 3.5% milk fat in order to be labeled as whole milk.
- Low-fat milk is commonly labeled as 1% or 2% because the majority of the milk fat has been removed. The percentage of milk fat in low-fat milk can vary between 0.5% and 2%.
- Skim milk must contain less than 0.5% milk fat in order to be labeled as skim milk. Nonfat milk is another name for skim milk.
- Flavored milks contain flavoring ingredients such as chocolate or strawberry syrup plus added sugars or sweeteners. Flavored milk is typically higher in calories than plain milk.

Milklike beverages made from plants such as soy, rice, almond, hemp, and oat milk are also available. However, these products are not typically served in commercial foodservice operations.

Smoothies

A *smoothie* is a blended drink made of fruit, such as berries or bananas, yogurt, and/or milklike beverages. **See Figure 2-32.** All of the ingredients are blended together to the desired thickness and served in a glass with a straw. Using frozen fruit will make the consistency of the smoothie similar to that of a milkshake. The use of frozen fruit also means the smoothie will have less ice crystals than a smoothie made with ice cubes.

Smoothies

Frieda's Specialty Produce

Figure 2-32. Smoothies are blended drinks made of fruit, such as berries or bananas, yogurt, and/or milklike beverages.

Beverages

CHECKPOINT 2-9

1. Identify four common coffee roasts.
2. Identify the type of coffee that is used to create common coffee drinks such as cappuccino.
3. Explain the benefit of drinking decaffeinated coffee.
4. Name three common types of tea.
5. Explain how black teas are graded.
6. Describe how oolong tea is served.
7. Describe a tisane.
8. Name two different types of breakfast juices.
9. Identify three forms in which juices are sold.
10. Explain how milks are classified.
11. Explain how a smoothie is made.
12. Prepare and present a hot and a cold breakfast beverage.

PLATING BREAKFAST

Common plating of breakfast includes the use of round or oval plates for the entrée and placement of the main protein as the focal point of the dish. Accompaniments such as toast are commonly served on side plates that are about half the size of the entrée plate. However, many restaurants and hotels serve pancakes on larger side plates. Simple garnishes add color and freshness to breakfast plates. Common breakfast garnishes include sliced or fanned strawberries and orange slices. **See Figure 2-33.**

Beverage cups and glasses are chosen based on the guest-to-waitstaff ratio. If a restaurant has a large number of guests per waitstaff, the staff may not be able to keep up with demand if smaller cups and glasses are used. Slightly larger cups and glasses mean fewer trips to the table for refills. However, it is important to not serve hot coffee or hot tea in too large of a cup as these beverages may become cold prior to the guest consuming them.

Simple Garnishes

Idaho Potato Commission

Figure 2-33. Common breakfast garnishes include sliced or fanned strawberries and orange slices.

When planning table layout for breakfast service, it is important to consider table size in proportion to the number of breakfast plates used in an average order. Guests perceive more value if their meal is provided on multiple plates, but the table must be able to accommodate all of the plates without overcrowding the guests' eating space.

CHECKPOINT 2-10

1. Explain how to plate a variety of breakfast dishes.
2. Garnish a breakfast plate.

BREAKFAST BUFFETS

A successful breakfast buffet offers a variety of foods that are well prepared and arranged in a manner that stimulates the appetite. **See Figure 2-34.** Breakfast buffets allow guests to eat at their own pace and take pressure off the kitchen staff to quickly turn orders. Breakfast buffets also allow a foodservice establishment to charge a higher price for breakfast. Most people perceive buffets as a better value because there are more choices and no limit to how much a person can consume. However, it is important to note that most people do not eat that much more at a buffet than they eat when ordering an à la carte breakfast.

Breakfast Buffets

MacArthur Place Hotel, Sonoma

Figure 2-34. A successful breakfast buffet offers a variety of foods that are well prepared and arranged to showcase a wide variety of colors.

The success of a breakfast buffet is dependent not only on the quality of the food but the overall design of the buffet. The appearance of the buffet should be inviting and pleasant. Decorative pieces should be used as a way of accenting the theme of the buffet or foodservice establishment. Using various heights and textures in the display will highlight the food items. When the guests' eyes are drawn to a variety of items, the whole buffet can be appreciated. Also, many foodservice establishments display their awareness of health and wellness by placing tags near their high-protein, low-fat, low-carbohydrate, or gluten-free breakfast items.

The use of negative space is also important because the presentation should not be cluttered. Every type of item, including glassware and plates, should be allowed at least one linear foot of space. If there are 20 items on the buffet, the buffet should be at least 20 feet long. **See Figure 2-35.**

The buffet should be laid out so that all of the items are easily accessible. Having items that are positioned behind another element of the display can slow the movement of the guests through the line. For example, plates should always be placed at the very beginning of the buffet. Also, it is important to provide an ample supply of plates to ensure that customers use a clean plate each time they return to the buffet table. The reuse of a dirty plate can cause cross-contamination.

Buffet Layout

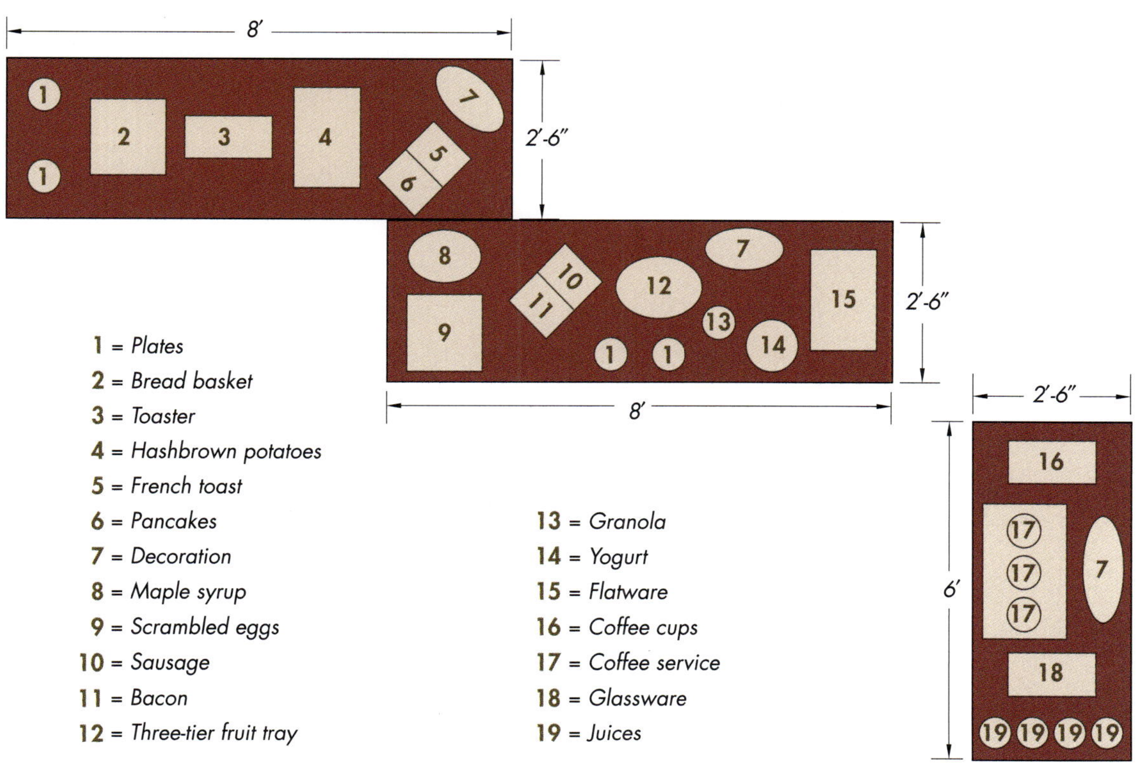

Figure 2-35. Every type of item on a breakfast buffet should be allowed at least one linear foot of space.

Buffet Table Setting

Many buffets require the use of chafing dishes to hold hot foods. **See Figure 2-36.** The heat source can be canned heat or an electric heating element. To prevent scorching, food is placed in a shallow pan that is then set inside a larger pan containing heated water. Double-panning allows for a more even heat distribution. Occasionally, a chafing dish is used for holding cold items. In this case, the heating element is not used, and the water in the pan is replaced with ice water.

The order in which food is placed on the buffet must be logical and appealing to the guests. Heated items are usually grouped together, and chilled items are usually grouped together. Starches such as potatoes, breakfast batter items, breads, and hot cereals are commonly placed at the beginning of the buffet. Following the starches, egg dishes and breakfast meats are presented. The last items placed include fruits, yogurts, and condiments. Coffee, juices, and utensils are usually offered at the end of the buffet.

Although buffets are commonly self-service, some buffets have action stations where chefs prepare, carve, or serve individual items to guests. Action stations give a buffet an upscale appearance and create the potential for greater customer service through increased interaction between guests and staff.

Chafing Dishes

Figure 2-36. A chafing dish has a portable heat source underneath to keep hot foods at serving temperature.

CHECKPOINT 2-11

1. Describe how to design a breakfast buffet.
2. Set up a breakfast buffet.

Refer to DVD for **Quick Quiz®** questions

SUSTAINABILITY CORNER

The foodservice industry is responsible for preparing and serving large amounts of food, which can result in a great deal of waste. A successful waste control system that includes composting and recycling may help counteract these losses and create positive benefits for the environment as well.

Compostable items are items that are biodegradable. This means that the products will break down into carbon dioxide, water, and biomass within the natural environment. When products break down, they release valuable nutrients into the soil, which in turn aids the growth of plants.

Foodservice establishments can contribute a vast array of items for composting. Appropriate items include, but are not limited to, leftover foods such as fruit and vegetable trimmings, coffee grounds and filters, tea bags, wooden stir sticks, and paper products. In addition, some products that were traditionally disposable, such as plates, cups, and cutlery, are now being produced with materials that are biodegradable, making them ideal for composting.

Recyclable products refer to goods that can be collected and reprocessed to make new products. Many items suitable for recycling are marked with the universal recycling symbol. Examples of recyclable materials include paper, glass, aluminum, and plastic.

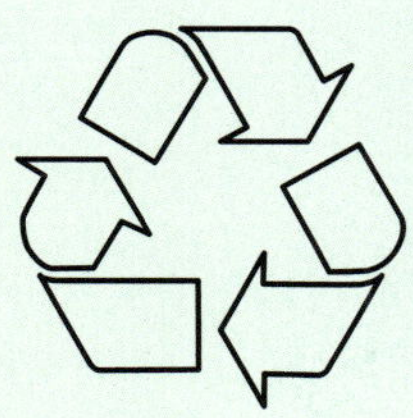

It is important to separate products to be composted from products to be recycled because compostable items will contaminate the recycling process. The simplest way to do this is to use the containers provided by the local waste collection services. Some food establishments, however, maintain a compost area on site to help minimize costs and promote sustainability, or if composting facilities are unavailable. Composting and recycling services can be researched by contacting local waste management municipalities or by visiting the Environmental Protection Agency (EPA) food waste web site (www.epa.gov/epawaste/conserve/materials/organics/food/fd-live.htm).

In addition to products suitable for composting and recycling, it is vital to minimize waste within the professional kitchen by monitoring inventory levels, storage procedures, food preparation, and reuse of products. Close attention should be paid to "use by" dates and careful consideration given to food items that may be used in other recipes. For example, parts left over from deboning a chicken may be used to make stock, or stale breads may be turned into croutons. Other in-house efforts to help reduce waste include reusing nonfood items, such as flatware and towels.

United States Department of Agriculture

An effective waste control system, including composting and recycling, can be a positive way to reduce a large amount of waste, diminish hauling fees, and help farmers grow and sustain crops. Some local governments even offer monetary incentives for participating in composting and recycling programs. In addition, customers may be drawn to environmentally conscious foodservice establishments. Ultimately, a successful waste management program can mean daily savings that contribute substantially to annual profits, while making the planet a better place.

Foodservice Waste Management

Certification Exam Preparation Questions

1. A ___ is an omelet that is cooked until nearly done and then folded before service.
   A. folded omelet
   B. rolled omelet
   C. frittata
   D. soufflé omelet

2. Grade ___ eggs are of the highest quality and have a firm yolk and white that both stand tall when broken onto a flat surface.
   A. AAA
   B. AA
   C. A
   D. B

3. An omelet loses its ___ and becomes tough and rubbery if it is prepared too far in advance.
   A. bright color
   B. flavor
   C. fluffiness
   D. aroma

4. The basic recipe for ___ and waffles consists of flour, sugar, salt, baking powder, baking soda, milk or buttermilk, eggs, and butter or oil.
   A. crêpes
   B. French toast
   C. pancakes
   D. blintzes

5. Using too much fat when preparing omelets ___.
   A. causes eggs to stick
   B. causes eggs to burn
   C. yields greasy eggs
   D. yields tough eggs

6. Eggs whites spread too rapidly when cooked at ___ temperature.
   A. too low of a
   B. too high of a
   C. medium
   D. high

7. Precooked breakfast meats include ham and ___.
   A. sausage
   B. bacon
   C. Canadian bacon
   D. all of the above

8. ___ eggs are the most common size of egg used in foodservice establishments and most recipes use this size as the standard.
   A. Pee wee
   B. Medium
   C. Large
   D. Extra large

9. A griddle is often used to prepare breakfast ___.
   A. meats
   B. potatoes
   C. eggs
   D. all of the above

10. Eggs Benedict is traditionally served on a(n) ___.
    A. biscuit
    B. bagel
    C. English muffin
    D. all of the above

11. ___ can be served fresh, frozen, or canned.
    A. Fruits
    B. Cereals
    C. Egg substitutes
    D. Eggless egg substitutes

12. ___ is a hot breakfast cereal.
    A. Quiche
    B. Granola
    C. Muesli
    D. Oatmeal

13. Common types of coffee roasts include French, Italian, city, and ___.
    A. Viennese
    B. Arabica
    C. Columbian
    D. Spanish

14. ___ and bagels can be used to make breakfast sandwiches.
    A. Croissants
    B. Biscuits
    C. Tortillas
    D. all of the above

15. ___ is used to set the egg white around the yolk when an egg is poached.
    A. Sugar
    B. Vinegar
    C. Butter
    D. Oil

16. A quiche is composed of a ___ custard baked in a precooked piecrust.
    A. sweet
    B. savory
    C. salty
    D. syrupy

17. A(n) ___ is a crêpe that can be filled with sweet or savory fillings.
    A. frittata
    B. blintz
    C. quiche
    D. omelet

18. Black ___ is graded according to the shape and size of its leaves.
    A. coffee
    B. tea
    C. milk
    D. juice

19. Simple ___ add color and freshness to breakfast plates.
    A. garnishes
    B. accompaniments
    C. decorations
    D. none of the above

20. A successful breakfast ___ is dependent on the overall design as well as the food.
    A. plate
    B. buffet
    C. layout
    D. order

21. Given a breakfast plate consisting of scrambled eggs, bacon, hash browns, and buttered toast, list a substitute for each plated item that is lower in calories and fat.

    ____
    ____
    ____
    ____

22. Describe the procedure for preparing a rolled (French) omelet.

    ____
    ____
    ____
    ____
    ____
    ____
    ____

23. Describe the procedure for preparing bacon.

    ____
    ____
    ____
    ____

24. List some examples of how breakfast items can be composted and recycled.

    ____
    ____
    ____
    ____
    ____
    ____
    ____

SKETCHING EXERCISE

25. Sketch a breakfast buffet layout and identify each item on the display.

Notes

Vegetable, Starch & Pasta Station

Vulcan-Hart, a division of the ITW Food Equipment Group LLC

Introduction

The third station in the ACFEF apprenticeship program encompasses a wide variety of techniques related to the preparation of vegetable, starch, and pasta products. A major learning component of this station is the identification of various types of vegetables, sea vegetables, mushrooms, fruits, potatoes, grains, and pastas. Another important learning component of this station is knife skills.

Upon completion of this station, you will be able to identify products as the edible root, bulb, tuber, stem, leaf, flower, or seed of a plant. You will also be able to identify different types of starches and varieties of pasta. Another critical component of this station is knife safety and developing a proficient use of knives and special cutting tools.

Once you are well-versed and knowledgeable about each type of food product and basic cutting techniques, you will learn how to use appropriate cooking techniques to prepare vegetables, fruits, starches, and pastas in the professional kitchen and how to properly store these products. You will also learn how to plate these foods to help create balance and nutrition.

Finally, you will learn many sustainable practices at this station that will put you one step closer to completing the ACFEF apprenticeship program. Sustainable practices include using a power-up/power-down schedule for high-energy equipment and replacing incandescent light bulbs with fluorescent bulbs. Also, using ENERGY STAR® rated equipment in the professional kitchen will conserve energy and result in substantial savings.

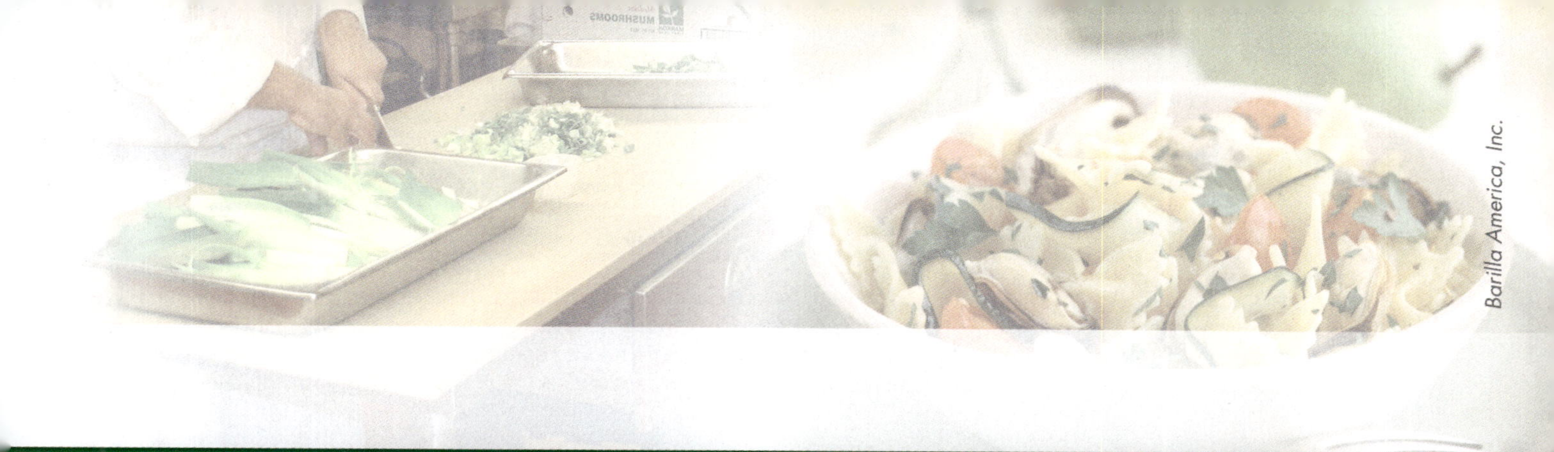

Barilla America, Inc.

Vegetables, fruits, potatoes, grains, and pastas are versatile foods, and it is important to be able to identify their many varieties. Knowledge of the market forms of these foods and how they are stored in the professional kitchen is also necessary. Equally important is the ability to correctly use large and small knives and cutting tools while preparing these foods. Many are prepped before the day of service and then finished using a variety of cooking methods. When knowledge of these foods and proficient knife skills have been acquired, an apprentice will be well prepared to tackle a multitude of recipes, from simple to complex.

VEGETABLE CLASSIFICATIONS

A *vegetable* is an edible root, bulb, tuber, stem, leaf, flower, or seed of a nonwoody plant. **See Figure 3-1.** Each vegetable offers culinary professionals unique opportunities to enhance menus with healthy, earthy flavors. Most vegetables can be served raw or cooked. Vegetables are an excellent source of vitamins, minerals, and fiber. They are also low in fat and calories. Vegetables have the best flavor and color when purchased in season.

Foodservice professionals should be able to identify common root, bulb, tuber, stem, leaf, flower, and seed vegetables. Likewise, foodservice professionals should be able to identify common fruit-vegetables, sea vegetables, and mushrooms that are prepared and served like vegetables.

Nutrition Note

Eating five servings per day of vegetables and fruits is optimal for weight management and chronic disease prevention. Providing a wide variety of vegetables on the menu can increase guest consumption of these nutritious foods.

Vegetables

Barilla America, Inc.

Figure 3-1. A vegetable is an edible root, bulb, tuber, stem, leaf, flower, or seed of a nonwoody plant.

Root Vegetables

A *root vegetable* is an earthy-flavored vegetable that grows underground and has leaves that extend above ground. Root vegetables include beets, carrots, celeriac, jicamas, lotus roots, parsnips, radishes, rutabagas, salsify, and turnips. **See Figure 3-2.** *Note:* Ginger and horseradish are also edible roots, but they are not classified as vegetables.

Root Vegetables	
Beets  *Melissa's Produce*	A *beet* is a round root vegetable that is a deep reddish purple or gold color. Beets are rich in vitamins A and C and potassium. Quality beets are firm with a smooth skin and no spots or bruising. Beets are served raw, pickled, sautéed, or roasted.
Carrots 	A *carrot* is an elongated root vegetable that is rich in vitamin A. Carrots come in many colors and are sold with and without their green tops. Carrots can be served raw, blanched, steamed, sautéed, or broiled.
Celeriac  *Frieda's Specialty Produce*	*Celeriac,* also known as celery root, is a knobby, brown root vegetable cultivated from a type of celery grown for its root rather than its stalk. Celery root measures about 4–5 inches in diameter and can weigh 2–4 lb. Celeriac should be washed thoroughly and peeled before use. Since it oxidizes quickly, lemon juice or vinegar should be added to the water prior to cooking to avoid discoloration. Celeriac can be served raw, boiled, braised, sautéed, puréed, or baked.
Jicama  *Melissa's Produce*	A *jicama* is a large, brown root vegetable that ranges in size from 4 oz to 6 lb. Jicama is native to Central America and a good source of vitamin C and potassium. The thin brown skin must be removed before use. The crisp, white flesh has a delicate, sweet flavor. Small jicamas are more flavorful than larger ones. Jicama can be served raw, steamed, boiled, puréed, baked, or fried.
Lotus Roots  *Melissa's Produce*	A *lotus root* is the underwater root vegetable of an Asian water lily that looks like a solid-link chain about 3 inches in diameter and up to 4 feet in length. Removal of the reddish-brown skin reveals a creamy white flesh that tastes like fresh coconut and has the texture of a raw potato. Lotus root can be served as a vegetable and as a dessert. Water chestnuts and sunchokes may be substituted for lotus roots.
Parsnips  *Frieda's Specialty Produce*	A *parsnip* is an off-white root vegetable 5–10 inches in length. Parsnips harvested in late fall and winter are sweeter. Parsnips can be served raw, blanched, steamed, broiled, or roasted.

Figure 3-2. (continued on next page)

Root Vegetables	
Radishes 	A *radish* is a root vegetable with a white flesh and a peppery taste that comes in many colors and shapes. Radishes are a good source of vitamin C, potassium, and folic acid. The pungent flavor of radishes is found in the skin, which is commonly removed prior to cooking. A red radish is about 1 inch in diameter with a red exterior and a crisp white flesh that has a sharp, peppery flavor. A daikon radish is usually 8–12 inches in length with an elongated shape similar to a carrot. Daikons are milder in flavor than red radishes. Radishes can be served raw or added to stir-fries and soups.
Rutabagas  *Melissa's Produce*	A *rutabaga* is a round root vegetable with a yellow-tinted flesh that is the result of a cross between a Savoy cabbage and a turnip. Rutabagas are often confused with turnips, but rutabagas are longer and rounder by comparison. Rutabagas are high in potassium and vitamin C and have a more distinct flavor than turnips. Rutabagas can be roasted, boiled, steamed, mashed, stewed, or stir-fried.
Salsify  *Melissa's Produce*	*Salsify* is a white or black root vegetable that can grow up to 12 inches in length. The thin inedible skin is usually removed after the root is boiled. Peeled salsify should be immediately immersed in water mixed with vinegar or lemon juice to prevent discoloration. When cooked, white salsify tastes like an artichoke heart. Black salsify has a savory fish flavor. Salsify is often cooked with other vegetables or battered and fried.
Turnips  *Melissa's Produce*	A *turnip* is a round, fleshy root vegetable that is purple and white in color. Turnips have a peppery flavor, similar to a radish, and are a good source of vitamin C and potassium. Turnips need to be washed and peeled prior to being blanched, sautéed, puréed, or roasted.

Figure 3-2. Root vegetables grow underground and have leaves that extend above ground.

Nutrition Note

Radishes are high in fiber and water content and low in calories. Radish greens contain more vitamin C, calcium, and iron than the radish root.

Bulb Vegetables

A *bulb vegetable* is a strongly flavored vegetable that grows underground and consists of a short stem base with one or more buds that are enclosed in overlapping membranes or leaves. Bulb vegetables are very fragrant and are used for their aromatic qualities as well as their flavors. Examples of bulb vegetables include garlic, leeks, onions, ramps, scallions, and shallots. **See Figure 3-3.**

Tubers

A *tuber* is a short, fleshy vegetable that grows underground and bears buds capable of producing new plants. Examples of tubers include ocas, potatoes, sunchokes, sweet potatoes, water chestnuts, and yams. **See Figure 3-4.**

Bulb Vegetables	
Garlic  *Eloma Combi Ovens*	*Garlic* is a bulb vegetable made up of several small cloves that are enclosed in a thin, husklike skin. White garlic is the most common variety and the most pungent in flavor. Pink garlic is named for its pinkish outer covering. Elephant garlic contains much larger cloves and is milder in flavor than smaller varieties. The flavor of garlic is released when a clove is cut, crushed, or minced and increases the more finely the clove is cut. Cut garlic must be refrigerated. Garlic is used to flavor a wide variety of dishes.
Leeks  *Tanimura & Antle®*	A *leek* is a long, white bulb vegetable, with long, wide, flat leaves. Leeks are similar in appearance to scallions, but are much larger, milder, and sweeter than onions. The green leaves are often used to flavor soups and stocks, whereas the white portion of the leek is used in a variety of recipes.
Onions  *Melissa's Produce*	An *onion* is a bulb vegetable made up of many concentric layers of fleshy leaves. Onion varieties include white, yellow, red, and pearl onions. The variety of the onion and the climate where it was grown determine how strong its flavor is. White onions have a slightly sweet flavor. Yellow onions, also known as Spanish onions, are very mild in flavor. Red onions are the sweetest variety and are commonly added to salads and sandwiches for color. Pearl onions are very small, white, and mild in flavor. Onions can be served raw, sautéed, grilled, roasted, stir-fried, or deep-fried.
Ramps  *Melissa's Produce*	A *ramp* is a wild leek with a flavor similar to scallions, yet with more zing. Ramps are often diced for use in salads or on sandwiches. They can also be sautéed for use in egg or potato dishes.
Scallions  *Tanimura & Antle®*	A *scallion,* also known as a green onion, is a small bulb vegetable with a slightly swollen base and long, slender, green leaves that are hollow. Scallions are mildly flavored compared to onions. Scallions are often added to salads or used as a garnish.
Shallots  *Melissa's Produce*	A *shallot* is a very small bulb vegetable similar in shape to garlic and has two or three cloves inside. The outer covering can be bronze-colored, rose-colored, or pale gray. Shallots have a pink-tinged ivory flesh and a more subtle flavor than onions. When purchasing shallots, it is important to choose those that are firm and dry-skinned and to avoid any that are sprouting.

Figure 3-3. Bulb vegetables are used for their aromatic qualities as well as their flavors.

Tubers	
Ocas  *Melissa's Produce*	An *oca,* also known as a New Zealand yam, is a small, knobby tuber that has a potato-like flesh and ranges in flavor from very sweet to slightly acidic. Ocas are white, pink, or red in color and are a good source of carbohydrates, calcium, dietary fiber, and iron. Ocas must be kept in a cool, dark place at room temperature or refrigerated in a crisper. Ocas can be served raw in salads, cooked like potatoes, or pickled.
Potatoes  *Frieda's Specialty Produce*	A *potato* is a round, oval, or elongated tuber that is the only edible part of the potato plant. The color of potato skin differs among varieties and can be brown, red, yellow, white, orange, blue, or purple. Potato flesh can be creamy white to yellow-gold or purple in color. When purchasing potatoes, firm, undamaged potatoes with no signs of sprouting should be chosen. Potatoes must be stored in a dry, cool, dark place that allows them to breathe. If potatoes do not have adequate ventilation they quickly rot. Potatoes can be baked, sautéed, broiled, grilled, or fried.
Sunchokes  *Frieda's Specialty Produce*	A *sunchoke,* also known as a Jerusalem artichoke, is a tuber with thin, brown, knobby-looking skin. The skin is edible but is often removed before cooking. The white flesh is crisp and sweet. Sunchokes can be blanched, steamed, puréed, or used to flavor soups.
Sweet Potatoes  *Melissa's Produce*	A *sweet potato* is a tuber that has a paper-thin skin and flesh that ranges in color from ivory to dark orange. Sweet potatoes are an excellent source of vitamin A and potassium. The skin is edible, although it is often removed before cooking. Peeled or cut sweet potatoes oxidize, so it is important to place them in cold water until they are used. Sweet potatoes can be prepared in the same manner as potatoes. They also can be incorporated into breads and desserts.
Water Chestnuts  *Frieda's Specialty Produce*	A *water chestnut,* also known as a water caltrop, is a small tuber with brownish-black skin and white flesh. A water chestnut is crunchy and juicy and resembles a chestnut in exterior color and shape. Water chestnuts have a mild, sweet flavor. Fresh water chestnuts must be peeled before use and may be refrigerated for up to one week if they are tightly wrapped. Canned water chestnuts are also available, but inferior in quality.
Yams  *Melissa's Produce*	A *yam* is a large tuber that has thick, barklike skin and a flesh that varies in color from ivory to purple. The skin is inedible. Yams are commonly confused with sweet potatoes because they are often labeled as sweet potatoes in the United States, but they are different vegetables. Common varieties of yams include the tropical yam, garnet yam, and jewel yam. Yams are low in fat and a good source of carbohydrates, protein, and vitamins A and C. Yams can be prepared in the same manner as sweet potatoes.

Figure 3-4. Tubers grow underground and bear buds capable of producing new plants.

Stem Vegetables

A *stem vegetable* is the main trunk of a plant that develops buds and shoots instead of roots. Stem vegetables contain a lot of cellulose and become tougher as they continue to develop. Therefore, stems are usually harvested while tender. Examples of stem vegetables include asparagus, celery, fennel, hearts of palm, kohlrabi, and rhubarb. **See Figure 3-5.**

Stem Vegetables	
Asparagus 	*Asparagus* is a green, white, or purple edible stem vegetable that is referred to as a spear. Green asparagus is the most common variety of asparagus. White asparagus is grown covered in soil to prevent photosynthesis from taking place and harvested as soon as the spears begin to emerge. White asparagus is more tender, but less flavorful, than green asparagus. Purple asparagus is sweeter than green asparagus because it has a higher sugar content. A cancer-fighting phytochemical called anthocyanin gives purple asparagus its purple hue. Raw asparagus is a popular ingredient in salads, omelets, quiches, and pasta dishes. Asparagus can be served raw, broiled, grilled, steamed, or puréed.
Celery  *Tanimura & Antle®*	*Celery* is a green stem vegetable that has multiple stalks measuring 12–20 inches in length. The inner stalks, or hearts, are sweeter and more tender than the outer stalks. Celery should be shiny, firm, and crisp. Celery can be served raw, sautéed, stir-fried, or roasted.
Fennel  *Melissa's Produce*	*Fennel* is a celery-like stem vegetable with overlapping leaves that grow out of a large bulb at its base. Fennel has a mild, sweet flavor that is often associated with licorice or anise. Fennel can be served raw, sautéed, broiled, blanched, or steamed.
Hearts of Palm  *Melissa's Produce*	A *heart of palm* is a slender, white, stem vegetable that is surrounded by a tough husk. Hearts of palm are about 4 inches long and can be up to 1½ inches thick, although most are very thin. They are good sources of fiber and do not contain any cholesterol. Once the husks have been removed, hearts of palm can be served raw, steamed, or fried.
Kohlrabi  *Frieda's Specialty Produce*	*Kohlrabi* is a sweet, crisp, stem vegetable that has a pale-green or purple, bulbous stem and dark-green leaves. Kohlrabi is created by crossbreeding a cabbage and a turnip. Although the entire kohlrabi is edible, the bulbous stem is the portion primarily used in cooking. The inner part of the stem base may be removed to produce a cavity that can be stuffed. Kohlrabi can be eaten raw, blanched, sautéed, or stir-fried.
Rhubarb  *Frieda's Specialty Produce*	*Rhubarb* is a tart stem vegetable that ranges in color from pink to red and is usually prepared like a fruit. It may be peeled or left with the skin intact, depending on the use. Rhubarb leaves are never used because they contain a poisonous toxin called oxalate. Rhubarb can be sweetened and stewed to make sauces, but it is most commonly used to make pies, tarts, and other desserts.

Figure 3-5. Stem vegetables contain a lot of cellulose and become tougher as they continue to develop.

Leaf Vegetables

Leaf vegetables, also known as greens, are plant leaves that are often accompanied by edible leafstalks and shoots. Although edible leaves or greens can be eaten raw, they are often cooked to decrease their bitterness and increase their palatability. Leaf vegetables include beet greens, Belgian endive, bok choy, Brussels sprouts, butterhead lettuces, chard, chicory, collards, crisphead lettuces, dandelion greens, fiddlehead ferns, frisée, head cabbages, kale, looseleaf lettuces, mâche, mesclun greens, microgreens, mustard greens, Napa cabbages, nopales, radicchio, romaine lettuces, Savoy cabbages, sorrel, spinach, tatsoi, turnip greens, and watercress. **See Figure 3-6.**

Leaf Vegetables	
Beet Greens  *Tanimura & Antle®*	A *beet green* is the green, edible leaf that grows out of the top of the beet root vegetable. Beet greens are full of vitamins and minerals and contain more iron than spinach. They have a bitter taste that is similar to chard. Beet greens are often sautéed.
Belgian Endive  *Tanimura & Antle®*	*Belgian endive* is a leaf vegetable with a slender, tightly packed, elongated head that forms a point. It is the sprouting head of a chicory root. The leaves are creamy white in the center and yellow, slightly green, or purple along the edges. Belgian endive can be split in half lengthwise and then braised or grilled.
Bok Choy  *Frieda's Specialty Produce*	*Bok choy* is a leaf vegetable that has tender white ribs, bright-green leaves, and a more subtle flavor than head cabbage. There are many varieties of bok choy, some having short ribs and others having long ribs. Bok choy can be served raw, sautéed, or stir-fried.
Brussels Sprouts  *Melissa's Produce*	A *Brussels sprout* is a very small round head of tightly packed leaves that looks like a tiny cabbage. Brussels sprouts grow along an upright stalk and are ready to be harvested when they reach a diameter of about 1 inch. The best sprouts are bright green and have no yellowing leaves. Brussels sprouts can be steamed, broiled, grilled, or sautéed.
Butterhead Lettuces 	A *butterhead lettuce* is a lettuce with pale-green leaves and a sweet, buttery flavor, tender texture, and delicate structure. The natural shape of the leaf forms a shallow bowl that can be used as an edible serving dish for portion-size salads or single-portion foods. Butterhead lettuce can also be used as a wrap in place of a tortilla. Before use, the leaves must be thoroughly washed. Boston lettuce and Bibb lettuce are two varieties of butterhead lettuce that are similar in appearance and flavor, but they have differently sized leaves.

Figure 3-6. (continued on next page)

Leaf Vegetables	
Chard  *Frieda's Specialty Produce*	*Chard* is a large, dark-green, leaf vegetable with white or reddish stalks. Swiss chard is grown for its silvery stalks and crinkly leaves. Ruby chard has deep-red leaves tinged with green and bright-red stalks. The flavor of ruby chard is milder than Swiss chard. Chard is a good source of vitamins A and C and iron. Chard can be prepared in the same manner as spinach.
Chicory  *Tanimura & Antle®*	*Chicory,* also known as escarole, is a curly leaf vegetable with a slightly bitter-tasting flavor. Chicory comes in various shades including green, yellow, and red. Frisée, Belgian endive, and radicchio are varieties of chicory. Chicory can be used in salads or served as cooked greens.
Collards  *Melissa's Produce*	A *collard,* also known as a collard green, is a large, dark-green, edible leaf with a thick, white vein. The flavor of collards is a cross between kale and cabbage. Collards are a variety of cabbage that does not form a head, but grows in clusters at the top of a tall stem. Collards are an excellent source of vitamins A and C, calcium, and iron. Collards must be thoroughly washed before being prepared. Collards are prepared in the same manner as cabbage or spinach.
Crisphead Lettuces  *Tanimura & Antle®*	A *crisphead lettuce* is a lettuce with a large, round, tightly packed head of pale-green leaves. This lettuce is crisp in texture, very mild in flavor, and wilt-resistant. Crisphead lettuce remains crisp even after being processed. It only requires a light washing because the compact head shields the leaves from dirt. The outer leaves and the core are removed before use. Iceberg is the most common type of crisphead lettuce. Crisphead lettuce is often used in salads and on sandwiches.
Dandelion Greens  *Frieda's Specialty Produce*	A *dandelion green* is the dark-green leaf of the dandelion plant. Dandelion greens are quite bitter. Dandelion greens are often added to salads or also sautéed.
Fiddlehead Ferns  *Melissa's Produce*	A *fiddlehead fern* is the curled tip of an ostrich fern frond that has a nutty and slightly bitter flavor similar to asparagus and artichokes. Fiddlehead ferns are only available for a few weeks in spring. They are a great source of vitamin A, vitamin C, and antioxidants. Before use, fiddlehead ferns should be rubbed between the hands to remove the brown scales and silk. Fiddlehead ferns can be boiled, grilled, sautéed, or roasted.

Figure 3-6. (continued on next page)

Leaf Vegetables	
Frisée  *Tanimura & Antle®*	*Frisée*, also known as curly endive, is a leaf vegetable with twisted, thin leaves that grow in a loose bunch. The leaves vary in color from dark green outer leaves to pale green or white inner leaves. Frisée has a bitter flavor and pairs well with strongly flavored cheeses and acidic vinaigrettes. It can also be used as a garnish.
Head Cabbages 	*Head cabbage* is a tightly packed, round head of overlapping leaves that can be green, purple, red, or white in color. Head cabbage usually ranges from 2–8 pounds and from 4–10 inches in diameter. The base of the head where the leaves attach to the stalk is known as the heart. The inedible heart is removed during preparation. The best heads are heavy and compact. Head cabbage can be eaten raw, steamed, braised, roasted, or stir-fried.
Kale  *Melissa's Produce*	*Kale* is a large, frilly, leaf vegetable that varies in color from green and white to shades of purple. Although all varieties of kale are edible, green varieties are better for cooking. Because of its bitterness, kale is rarely eaten raw. The center stalk is often removed before kale is cooked. Kale is an ample source of vitamins A and C, folic acid, calcium, and iron. Kale may be prepared in the same manner as spinach.
Looseleaf Lettuces  *Tanimura & Antle®*	A *looseleaf lettuce* is a mild flavored, rich-colored lettuce with a cascade of leaves held loosely together at the root. Some varieties have thick leaves, and others have thin leaves. Some leaves are flat, while others are frilled or curled. Looseleaf lettuces are commonly used in salads and on sandwiches.
Mâche 	*Mâche*, also called lamb's lettuce, is a small, leafy green with a velvety texture. Mâche can have dark-green, scoop-shaped leaves or elongated, pale-green leaves. It is tender and mild in flavor and pairs well with vinaigrettes.
Mesclun Greens 	*Mesclun greens* are a mix of young greens that range in color, texture, and flavor. Mesclun greens can be tender and sweet or crisp and peppery. Mesclun is most often a mix of 10–12 varieties of greens, often including romaine, radicchio, endive, baby spinach, red oak, leaf lettuce, arugula, frisée, and tatsoi. Some mesclun mixes contain as many as 30 different plants, including flowers and herbs.

Figure 3-6. (continued on next page)

Leaf Vegetables	
Microgreens 	*Microgreens* are the first sprouting leaves of an edible plant. In addition to being used in salads, microgreens are commonly used to garnish hors d'oeuvres. Common varieties of microgreens include beet greens, turnip greens, spinach, and kale.
Mustard Greens 	A *mustard green* is a large, dark-green leaf vegetable from the mustard plant that has a strong peppery flavor. Mustard greens are an excellent source of vitamins A and C, thiamin, and riboflavin. Mustard greens must be thoroughly washed before being cooked. Mustard greens may be steamed, braised, sautéed, or stir-fried.
Napa Cabbages  *Frieda's Specialty Produce*	*Napa cabbage,* also known as celery cabbage, is an elongated head of crinkly and overlapping edible leaves that are a pale yellow-green color with a white vein. Napa cabbage is often referred to as Chinese cabbage. Its leaves are more tender than those of head cabbage and it has a very delicate flavor. Napa cabbage is most often served raw or stir-fried.
Nopales  *Melissa's Produce*	A *nopal* is the green, edible leaf of the prickly pear cactus. Nopales measure about 5 inches in length and 3–4 inches in width. They are crunchy, slippery, and slightly tart. Nopales are a great source of calcium and vitamins A and C. When preparing nopales the eyes, prickles, and all dry areas are removed. Nopales may be served raw, steamed, or sautéed.
Radicchio  *Tanimura & Antle®*	*Radicchio* is a small, compact head of red leaves, similar to a small head of red cabbage. Radicchio leaves form a bowl shape and can hold individually sized salads. Radicchio is also commonly sautéed or braised and served as a side.
Romaine Lettuces  *Tanimura & Antle®*	A *romaine lettuce* is a lettuce with long, green leaves that grow in a loosely packed, elongated head on crisp center ribs. The outer edges of the leaves are darker in color and lighten to a pale celadon near the rib. Romaine lettuce does not bruise easily when cut and is high in vitamin A, vitamin C, vitamin K, and folate. Dirt collects in the ridges of the loose leaves, so romaine lettuce must be washed thoroughly. Romaine lettuce has a mild, sweet flavor and blends well with other greens.

Figure 3-6. (continued on next page)

Leaf Vegetables	
Savoy Cabbages  *Melissa's Produce*	*Savoy cabbage* is a conical-shaped head of tender, crinkly, edible leaves that are blue-green on the exterior and pale green on the interior. The leaves are very pliable and have a distinct sweet flavor. Savoy cabbages lack the sulfur-like odor often associated with cooking other cabbage varieties. Savoy cabbage can be stir-fried, stuffed, or used raw in salads.
Sorrel  *Melissa's Produce*	*Sorrel* is a large, green, leaf vegetable that ranges in color from pale green to dark green and from 2–12 inches in length. Sorrel is quite acidic in flavor. Strongly flavored sorrel is called sour dock or sour grass. The acidic flavor comes from the presence of oxalic acid. Sorrel is a good source of vitamin A. Like spinach, sorrel can be served raw or cooked.
Spinach  *Tanimura & Antle®*	*Spinach* is a dark-green leaf vegetable with a slightly bitter flavor that may have flat or curly leaves, depending on the variety. Fresh spinach is rich in vitamins A and C, folate, potassium, iron, and magnesium. Spinach is usually very gritty and must be thoroughly rinsed. The tough stems are usually removed before the spinach is cooked. Spinach can be served raw, steamed, or sautéed.
Tatsoi 	*Tatsoi* is a spoon-shaped, emerald-colored leaf vegetable native to Japan. Tatsoi has a mild flavor and is a good source of vitamins, minerals, and antioxidants. Tatsoi may be served raw in salads, steamed, sautéed, or boiled.
Turnip Greens 	A *turnip green* is a dark-green leaf vegetable that grows out of the top of the turnip root vegetable. Young turnip greens have a sweet flavor. As the plant ages, the leaves become bitter. Turnip greens are an excellent source of vitamins A and C and a good source of riboflavin, calcium, and iron. Turnip greens must be thoroughly washed before being cooked. The removal of the ribs from the leaves will yield a more tender batch of greens. Turnip greens may be steamed, braised, sautéed, or stir-fried.
Watercress  *Tanimura & Antle®*	*Watercress* is a small, crisp, dark-green, leaf vegetable with a pungent, yet slightly peppery flavor. It is typically sold in bouquets and can be refrigerated for up to five days if the stems are kept in water. Watercress is often used to garnish sandwiches, salads, soups, and stir-fries.

Figure 3-6. Leaf vegetables, also known as greens, are plant leaves that are often accompanied by edible leafstalks and shoots.

Edible Flowers

Edible flowers are the flowers of nonwoody plants that are prepared as vegetables. Edible flowers can be eaten raw or cooked. Examples of edible flowers include artichokes, broccoli, cauliflower, and squash blossoms. **See Figure 3-7.**

Production Tip

Edible flowers can be used to add color and flavor. For example, roses or lavender can be used to add color and sweetness to salads and desserts. Nasturtiums have a peppery flavor similar to watercress, and their buds can be substituted for capers.

Edible Flowers	
Artichokes  *Melissa's Produce*	An *artichoke* is the edible flower bud of a large, thistle-family plant that comes in many varieties. Fresh artichokes should have a tight leaf formation, be deep-green in color, and are heavy for their size. In general, the smaller the artichoke, the more tender it will be. Artichokes must be washed before cooking and prepared in nonreactive bowls and pans to prevent off-flavors. The petals of an artichoke, called leaves, are the edible part of the bud. The dense area where the leaves attach to the stem is known as the heart. Artichokes are often steamed, grilled, or filled with a mixture of lemon, garlic, and bread crumbs and baked.
Broccoli  *Tanimura & Antle®*	*Broccoli* is an edible flower that is a member of the cabbage family and has tight clusters of dark-green florets on top of a pale-green stalk with dark-green leaves. Fresh broccoli should be firm and evenly colored. Broccoli can be eaten raw or can be steamed, blanched, broiled, sautéed, or stir-fried.
Cauliflower  *Tanimura & Antle®*	*Cauliflower* is an edible flower that is a member of the cabbage family and has tightly packed white florets on a short, white-green stalk with large, pale-green leaves. Some varieties of cauliflower have a purple or greenish tinge to the florets. Cauliflower grows covered with numerous layers of leaves attached to the stalk and surrounding the head. Fresh cauliflower should be firm and compact. Cauliflower can be served raw, steamed, blanched, sautéed, stir-fried, or broiled.
Squash Blossoms 	A *squash blossom* is the edible flower of a summer or a winter squash. They come in varying shades of yellow and orange and often taste a bit like the parent squash. Ideal squash blossoms are closed buds, extremely perishable, and cannot be refrigerated more than a day. Squash blossoms are used to flavor salads and soups. They also can be lightly sautéed, stuffed with soft cheese and baked, or batter dipped and fried.

Figure 3-7. Edible flowers are the flowers of nonwoody plants that are prepared as vegetables.

Edible Seeds

Edible seeds include some of the oldest recorded forms of food. Many edible seeds can be eaten raw, and all of them can be cooked. Examples of edible seeds include all varieties of legumes and sprouts. A *legume* is the edible seed of a nonwoody plant that grows in multiples within a pod. In some cases, the pods are eaten along with the seeds. Legumes are rich in fiber and protein and contain little or no fat. There are thousands of varieties of legumes, including beans, peas, pulses, and lentils.

Beans. A *bean* is the oval or kidney-shaped edible seed of various plants in the legume family. Beans can be purchased fresh, canned, frozen, or dried. Popular varieties of beans include limas, cannellinis, anasazis, peruanos, calypsos, flageolets, pintos, kidney beans, great northern beans, and black beans. **See Figure 3-8.** Beans are often used to make hearty soups and are also served as sides. Some beans, such as pintos, can be puréed to make refried beans. Lima beans can be used to make succotash.

Most fresh beans can be eaten raw, steamed, sautéed, grilled, or fried. Some fresh bean varieties are called edible pods, meaning that both the exterior skin and the interior seeds are edible. For example, fresh green beans and fresh wax beans are actually immature beans with underdeveloped pods that are therefore edible. **See Figure 3-9.**

Beans with Edible Pods

Melissa's Produce

Figure 3-9. Green beans and wax beans are beans with edible pods.

Beans

Figure 3-8. Popular varieties of beans include limas, cannellinis, anasazis, peruanos, calypsos, flageolets, pintos, kidney beans, great northern beans, and black beans.

Other varieties of fresh beans, such as edamame, do not have edible pods. *Edamame* are green soybeans housed within a fibrous, inedible pod. **See Figure 3-10.**

Edamame

Figure 3-10. Edamame are green soybeans housed within a fibrous, inedible pod.

Peas. A *pea* is the round edible seed of various plants in the legume family. Some peas, called split peas, are harvested fully mature, left to dry, and then split. Some varieties of peas have edible pods. For example, snow peas have a flat pod that is entirely edible. Sugar snap peas have a more rounded pod, but are still tender enough to eat. **See Figure 3-11.** These edible pods can be eaten raw, steamed, sautéed, or fried.

Peas with Edible Pods

Melissa's Produce

Figure 3-11. Edible pods can be eaten raw, steamed, sautéed, or fried.

Pulses. A *pulse* is a dried seed of a legume. Dried beans and peas, such as cannellini beans and black-eyed peas, are shelled and then left to dry until they become rock hard. Pulses must be rehydrated by soaking them overnight or by using a quick-soaking process. Rehydration decreases the total cooking time and results in an even texture throughout. **See Figure 3-12.**

Pulses

Indian Harvest Specialtifoods, Inc./Rob Yuretich

Figure 3-12. Pulses are the dried seeds of legumes and must be rehydrated before being cooked.

Lentils. A *lentil* is a very small, dried pulse that has been split in half. There are many varieties of lentils, with colors ranging from white to green. **See Figure 3-13.** Unlike dried beans and peas, lentils do not have to be soaked. However, lentils must be thoroughly washed before cooking because they often contain small stones. Lentils are used to make soups, added to salads, combined and served with other vegetables, and served as sides. Lentils turn mushy when overcooked.

Lentils

Figure 3-13. Varieties of lentils range in color from white to green.

Lentils

Nutrition Note

Lentils contain calcium and vitamins A and B. They are also a good source of iron and phosphorus.

Fruit-Vegetables

A *fruit-vegetable* is a botanical fruit that is sold, prepared, and served as a vegetable. Fruit-vegetables are typically more tart than sweet. Fruit-vegetables used in the professional kitchen include bell peppers, chiles, cucumbers, eggplants, okra, pumpkins, summer squashes, sweet corn, tomatillos, tomatoes, and winter squashes. **See Figure 3-14.**

Fruit-Vegetables	
Bell Peppers  Barilla America, Inc.	A *bell pepper* is a fruit-vegetable with three or more lobes of crisp flesh that surround hundreds of seeds in an inner cavity. Bell peppers turn yellow and ultimately red if left to ripen on the vine. The longer the pepper stays on the vine to ripen, the sweeter it becomes. Red peppers are the sweetest peppers because they are the ripest. Once the core and seeds have been removed, bell peppers are usually julienned before use. Bell peppers can be served raw or cooked.
Chiles  Frieda's Specialty Produce	A *chile* is a brightly colored fruit-vegetable pod with distinct mild to hot flavors. Chiles come in many colors, shapes, and sizes. There are more than 200 varieties, including the jalapeño, habanero, poblano, and serrano. Chiles range in color from yellow to green and fire red to black, and from ¼ – 4 inches in size. The seeds and veins, or membranes, located inside the chile pod contain capsaicin. *Capsaicin* is a potent compound that gives chiles their hot flavor. Smaller chiles typically contain more seeds and veins and are therefore hotter. Chiles are used to flavor sauces and soups. Some chiles, like the poblano, can be stuffed.
Cucumbers  Melissa's Produce	A *cucumber* is a green, cylindrical fruit-vegetable that has an edible skin, edible seeds, and a moist flesh. Common varieties of cucumbers include English (burpless), Japanese, Mediterranean, and dosakai, which are yellow, round cucumbers. Pickling cucumbers have a bumpy, light-green skin and are smaller and thicker than cucumber varieties that are eaten fresh. Cucumbers are used in salads and soups and to garnish sandwiches. Cucumbers are often pickled.
Eggplants  Frieda's Specialty Produce	An *eggplant* is a deep-purple, white, or variegated fruit-vegetable with edible skin and a yellow to white, spongy flesh that contains small, brown, edible seeds. There are many varieties of eggplant. An eggplant begins to discolor as soon as it is cut, so it is important either to cook the eggplant immediately after it is sliced or to sprinkle it with lemon juice. Sliced, raw eggplant can be lightly salted and left on paper towels to drain some of the moisture from the eggplant before cooking.
Okra 	*Okra* is a green fruit-vegetable pod that contains small, round, white seeds and a gelatinous liquid. When preparing okra, the stem end is trimmed, and the pod is thoroughly rinsed in cold water before cooking. Okra should be cooked only in stainless steel cookware to prevent it from turning dark. Okra may be blanched, steamed, boiled, fried, or pickled.
Pumpkins  National Garden Bureau Inc.	A *pumpkin* is a round fruit-vegetable with a hard orange skin and a firm flesh that surrounds a cavity filled with seeds. Pumpkins vary in size and weight. The flesh has a mild, sweet flavor and can be prepared in the same manner as winter squash. Pumpkin seeds are often roasted for use as a garnish or to add flavor to salads.

Figure 3-14. (continued on next page)

Fruit-Vegetables	
Summer Squashes 	A *summer squash* is a fruit-vegetable that grows on a vine and has edible skin, flesh, and seeds. The tender flesh of a summer squash has a high water content and mild flavor. Summer squashes come in many shapes, sizes, and colors. *Zucchini* is an elongated summer squash that resembles a cucumber and is available in green or yellow varieties. *Yellow squash* is a summer squash that resembles a bowling pin with either a straight or crooked neck. *Pattypan* is a round, shallow summer squash with scalloped edges that is harvested when it is no larger than 2–3 inches in diameter. Summer squashes can be served raw, sautéed, grilled, stir-fried, baked, or deep-fried.
Sweet Corn 	*Sweet corn* is a fruit-vegetable that has edible seeds called kernels that grow in rows on a spongy cob encased by thin leaves (husks), forming what is referred to as an ear of corn. Beneath the husks, the kernels are covered with fine, hairlike material called silk. An ear of corn can contain 200–400 kernels and be 4–10 inches in length. Sweet corn can be grilled, roasted, steamed, or boiled as corn on the cob or as kernels that have been cut off the cob.
Tomatillos  *Melissa's Produce*	A *tomatillo,* also known as a Mexican husk tomato, is a small fruit-vegetable with a thin, papery husk covering a pale-green skin that encases a pale-green flesh. The husk is always removed before cooking. Green tomatillos are tart in flavor, but become sweeter as they ripen and turn yellow. Tomatillos are used to make salsa verde.
Tomatoes  *Florida Tomato Committee*	A *tomato* is a juicy fruit-vegetable that contains edible seeds. Beefsteak, cherry, yellow, pear, and plum tomatoes are just a few of the thousand plus varieties. Tomatoes are available in many different colors, sizes, and shapes. Most tomatoes are red, but some tomatoes are yellow, orange, pink, purple, green, black, white, multicolored, or striped. Tomatoes can be eaten raw, added to salads or sandwiches, and made into soups, sauces, and juice.
Winter Squashes  *Melissa's Produce*	A *winter squash* is a fruit-vegetable that grows on a vine and has a thick, hard, inedible skin and firm flesh surrounding a cavity filled with seeds. The firm flesh of a winter squash is deep yellow to orange in color, and varies in color and taste by variety. *Acorn squash* is a winter squash that looks like a large, dark-green acorn. *Butternut squash* is a large, bottom-heavy, tan-colored winter squash. *Spaghetti squash* is a dark-yellow winter squash with pale-yellow flesh that can be separated into spaghetti-like strands after it is cooked. Spaghetti squash can be used as a substitute for spaghetti noodles. Winter squash requires longer cooking times than summer squash. Most winter squashes are baked, roasted, made into soups, or puréed and served as a side dish.

Figure 3-14. Fruit-vegetables are botanical fruits that are sold, prepared, and served as vegetables.

Sea Vegetables

Sea vegetables are edible saltwater plants that contain high amounts of dietary fiber, vitamins, and minerals. People living in coastal Asian countries, New Zealand, the Pacific Islands, coastal South American countries, Ireland, Scotland, Norway, and Iceland have been harvesting and eating sea vegetables for thousands of years.

Sea vegetables lend a salty flavor to food because of the minerals they absorb from the ocean. When adding sea vegetables to a dish, no salt should be added. Many sea vegetables also contain alginic acid, which is used as a stabilizer and a thickener when making processed foods such as ice creams, puddings, and pie fillings. Sea vegetables can be roasted along with other vegetables or crumbled and added to soups, sauces, salads, pastas, and rice dishes. Common varieties of sea vegetables used in the professional kitchen include dulse, kombu, nori, and wakame. **See Figure 3-15.**

Sea Vegetables	
Dulse 	*Dulse* is a stringy, reddish-brown sea vegetable with a fishy odor that is rich in iron, iodine, potassium, and vitamin A. Dulse is sold in whole, crushed, flaked, and powdered form. Dulse powder is often used in soups and stews as a salt substitute. Whole dulse can be pan-fried until it is crispy to add crunch to sandwiches and salads. Whole dulse has a chewy texture and a savory flavor similar to bacon.
Kombu 	*Kombu* is a long, dark-brown to purple sea vegetable that is used to flavor dashi stock. It is sold in dried strips or sheets. Quality kombu is almost black with a white residue on its surface that is quite flavorful. It contains iodine, calcium, magnesium, and iron. When kombu is added to the cooking liquid for rice, soups, and beans, extra liquid is required because it soaks up liquid and doubles in volume. Kombu may be substituted for monosodium glutamate (MSG).
Nori 	*Nori* is a thin, purple-black sea vegetable that turns green when it is toasted. Nori contains iodine and vitamin C and is often used as a wrapper for sushi. Nori is often toasted and crushed before being used as a condiment for soups, salads, and grain dishes.
Wakame 	*Wakame* is a long, tender, grayish-green sea vegetable that expands to seven times its size when soaked in water. Wakame is high in dietary fiber and potassium. Wakame can be eaten raw, added to soups and stir-fries, or roasted and sprinkled on salads.

Figure 3-15. Sea vegetables contain high amounts of dietary fiber, vitamins, and minerals.

Mushrooms

Although mushrooms are not vegetables, they are prepared and used in the same manner as vegetables. A *mushroom* is the fleshy, spore-bearing body of an edible fungus that grows above the ground. Many mushrooms are commercially raised, but others are harvested from the wild. Both fresh and dried mushrooms are used in the professional kitchen.

Fresh mushrooms should be firm and not spotted or slimy. Fresh mushrooms must be stored in a cool, dry place. They also require air circulation to stay fresh, so storing them in paper bags works best. Fresh mushrooms should be cleaned with a damp towel and lightly rinsed, if necessary, before being used in a recipe. Dried mushrooms must be rehydrated before they are used in a recipe.

Courtesy of Chef Eric LeVine

Mushrooms are used to add flavor to many dishes and can be served raw, sautéed, stir-fried, or deep-fried. Common varieties of mushrooms used in the professional kitchen include button, chanterelle, enokitake, morel, oyster, porcini, portobello, shiitake, and wood ear mushrooms. **See Figure 3-16.**

Mushrooms	
Button Mushrooms 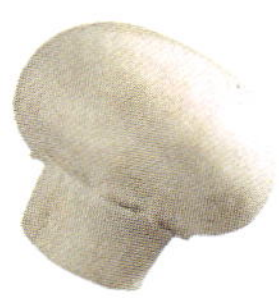 Mushroom Council	A *button mushroom,* also known as a white mushroom, is a cultivated mushroom with a very smooth, rounded cap and completely closed gills atop a short stem. Button mushrooms are one of the most widely consumed mushrooms. Button mushrooms have a mild flavor that complements most foods.
Chanterelle Mushrooms  Mushroom Council	A *chanterelle mushroom* is a trumpet-shaped mushroom that ranges in color from bright yellow to orange and has a nutty flavor and a chewy texture. Chanterelles tend to toughen when overcooked, so it is best to add them to a dish near the end of the cooking time.
Enokitake Mushrooms  Mushroom Council	An *enokitake mushroom,* also known as an enoki or a snow puff mushroom, is a crisp, delicate mushroom that has spaghetti-like stems topped with white caps. Enokitake mushrooms have a crunchy texture and an almost fruity flavor. Fresh enokitake mushrooms should be wrapped in a paper towel and then a plastic bag before being refrigerated.
Morel Mushrooms  Melissa's Produce	A *morel mushroom* is an uncultivated mushroom with a cone-shaped cap that ranges in height from 2–4 inches and in color from tan to very dark brown. Morels belong to the same fungus species as the truffle and are favored for their earthy, nutty flavor. Darker-colored morels tend to have a stronger flavor. Dried morels have a more intense and smoky flavor. Morel mushrooms are typically sautéed in butter.

Figure 3-16. (continued on next page)

Mushrooms

Mushrooms	
Oyster Mushrooms 	An *oyster mushroom* is a broad, fanlike or oyster-shaped mushroom that varies in color from white to gray or tan to dark brown. Oyster mushrooms span 2–10 inches. Their flavor is mild and they have a slight odor that resembles anise. The firm white flesh varies in thickness. The gills of the mushroom are white to cream and descend toward the stalk, which is often not present at the point of sale. Oyster mushrooms may be sautéed or stir-fried.
Porcini Mushrooms 	A *porcini mushroom,* also known as a cèpe, is an uncultivated, pale-brown mushroom with a smooth, meaty texture and a pungent flavor. Porcini mushrooms can weigh anywhere from 1 oz to 1 lb and have a cap that ranges from 1–10 inches in diameter. Dried porcini mushrooms must be softened in hot water for about 20 minutes before use. Porcinis can be substituted for cultivated mushrooms in most recipes.
Portobello Mushrooms  *Mushroom Council*	A *portobello mushroom* is a very large and mature, brown cremini mushroom that has a flat cap measuring up to 6 inches in diameter. The gills of the portobello mushroom are fully exposed, leaving the mushroom without much moisture and creating a dense, meaty texture. Their woody stems are removed and used to flavor stocks and soups. The caps are typically used whole, but can be diced for use in a wide variety of dishes. Portobello mushrooms have a meaty, savory flavor and are often grilled for sandwiches or sliced and added to salads.
Shiitake Mushrooms 	A *shiitake mushroom,* also known as a forest mushroom, is an amber, tan, brown, or dark-brown mushroom with an umbrella shape and curled edges. Shiitake mushroom caps range in size from 3–10 inches in diameter. Cooked shiitakes release a pinelike aroma and have a rich, earthy, savory flavor. The tough stems are usually removed and used to flavor stocks and soups. Shiitake mushrooms can be sautéed, broiled, or baked.
Wood Ear Mushroom 	A *wood ear mushroom,* also known as a cloud ear or a tree ear mushroom, is a brownish-black, ear-shaped mushroom that has a slightly crunchy texture. When dried wood ear mushrooms are reconstituted they increase 5–6 times in size. The albino variety of wood ear mushrooms is white in color. Wood ear mushrooms have a delicate flavor and often absorb the taste of other ingredients in a dish.

Figure 3-16. An edible mushroom is the fleshy, spore-bearing body of an edible fungus that grows above the ground.

Purchasing Fresh Vegetables

Fresh vegetables are packed in cartons, lugs, flats, crates, or bushels and sold by weight or count. The weight or count of a packed container depends on the size and type of vegetable. Because most fresh vegetables have a short shelf life, it is important to know how to maximize their use. **See Figure 3-17.**

Fresh Vegetables

Vegetables

Barilla America, Inc.

Figure 3-17. Fresh vegetables are packed in cartons, lugs, flats, crates, or bushels and sold by weight or count.

The USDA voluntary grading system for fresh vegetables includes U.S. Extra Fancy, U.S. Fancy, U.S. Extra No. 1, U.S. No. 1, U.S. No. 2, and U.S. No. 3. The grade of vegetable to purchase depends on how the vegetable will be used. Recipes using fresh or slightly cooked vegetables require premium ingredients, while lesser grades are acceptable in soup recipes. Fresh vegetables are less expensive during their peak season.

Most vegetables should be stored in a produce cooler at a temperature of 41°F or below. Vegetables should always be stored away from fruits that emit ethylene gas, such as apples and bananas, as the gas can cause the vegetables to overripen and spoil. It is also important to store vegetables away from poultry, meat, seafood, and dairy products.

Vegetables such as potatoes, onions, garlic, and squash should be stored in a cool, dry location that is between 60°F and 70°F. Storing these vegetables in a refrigerator causes their starches to convert to sugars, which negatively alters their texture and flavor.

Canned Vegetables

Canned vegetables are a staple in the professional kitchen. Canned vegetables have already been cleaned, cut, peeled, cooked, and treated with heat to kill any harmful microorganisms. However, the canning process often softens vegetables and can sometimes cause nutrient loss. Canned vegetables are USDA graded as U.S. Grade A or U.S. Fancy, U.S. Grade B, and U.S. Grade C or U.S. Standard. Canned vegetables are packed by weight and sold in standard commercial sizes. **See Figure 3-18.**

Vegetable Can Sizes		
Can Size	*Weight*	*Cans Per Case*
No. 300	14–15 oz	36
No. 303	16–17 oz	36
No. 2	20 oz	24
No. 2½	28 oz	24
No. 5	46–51 oz	12
No. 10	6 lb 10 oz	6

Figure 3-18. Canned vegetables are a staple in the professional kitchen and come in standard sizes.

Canned vegetables can be stored for long periods if they are kept in a cool, dry place. Dented or bulging cans should be discarded as they may contain harmful bacteria. After opening canned vegetables, the unused portion should be placed in an airtight storage container, labeled, dated, and refrigerated.

Frozen Vegetables

Frozen vegetables offer the same convenience as canned vegetables, with an additional advantage. Frozen vegetables retain their color and nutrients better than canned vegetables. **See Figure 3-19.** The USDA grading system used for canned vegetables also applies to frozen vegetables. Frozen vegetables are usually packed in 1 lb or 2 lb bags. Some vegetables are individually quick-frozen to preserve their texture and appearance. Some vegetables are blanched before being frozen, which reduces overall cooking time. Other frozen vegetables are already fully cooked and need only to be heated for service.

Frozen Vegetables

Figure 3-19. Frozen vegetables often retain their color and nutrients better than canned vegetables.

Dried Vegetables

Dried vegetables have had most of their moisture removed by a food dehydrator or the freeze drying process. **See Figure 3-20.** *Freeze drying* is the process of removing the water content from a food and replacing it with a gas. Freeze dried vegetables retain more color, texture, and shape than dehydrated vegetables.

Dried Vegetables

Frieda's Specialty Produce

Figure 3-20. Dried vegetables, such as sun-dried tomatoes, have had most of their moisture removed by the dehydration process.

Dried onions are commonly used for convenience. For example, 1 lb of dried onions can be used instead of 8 lb of fresh onions. For maximum shelf life, dried vegetables should be stored in an airtight container in a cool, dry place.

CHECKPOINT 3-1

1. Describe common root vegetables used in the professional kitchen.
2. Describe common bulb vegetables used in the professional kitchen.
3. Describe common tubers used in the professional kitchen.
4. Describe common stem vegetables used in the professional kitchen.
5. Describe common leaf vegetables used in the professional kitchen.
6. Describe edible flowers used in the professional kitchen.
7. Describe common seed vegetables used in the professional kitchen.
8. Describe common fruit-vegetables used in the professional kitchen.
9. Describe common sea vegetables used in the professional kitchen.
10. Describe common mushrooms used in the professional kitchen.
11. Describe factors to consider when purchasing fresh vegetables.
12. Explain the role of canned vegetables in the professional kitchen.
13. Explain the role of frozen vegetables in the professional kitchen.
14. Explain the role of dried vegetables in the professional kitchen.

COOKING VEGETABLES

Common methods for cooking vegetables include steaming, blanching, grilling, broiling, baking, roasting, sautéing, and frying. Vegetables should be cooked until they are just tender enough to be easily digested. At this stage of cooking, most vegetables retain the majority of their nutritional value, flavor, and color. Overcooked vegetables often lose their bright colors, may become mushy in texture, and lose nutrients because vitamins and minerals are destroyed by excess heat.

The addition of acidic or alkaline ingredients when cooking vegetables causes chemical reactions that affect the color and texture of the vegetables. Acids, such as lemon juice or vinegar, are often added to a cooking liquid to contribute flavor to the dish. An alkali such as baking soda can be added to tough vegetables to speed up the softening process. Adding acids or alkalis also alters the natural pigment present in a vegetable. **See Figure 3-21.** The types of natural pigments in vegetables include chlorophyll, carotenoids, and flavonoids.

- *Chlorophyll* is an organic pigment found in green vegetables. When an acid is added to the cooking liquid, green vegetables turn a drab olive color, but retain their naturally firm texture. If an alkali is added to the cooking liquid, green vegetables become brighter in color but mushy in texture.
- A *carotenoid* is an organic pigment found in orange or yellow vegetables. Acids have little to no effect on carotenoids. Alkalis do not affect the color of carotenoids, but do cause orange and yellow vegetables to become mushy.
- A *flavonoid* is an organic pigment found in purple, dark-red, and white vegetables. Acids cause purple and dark-red vegetables to turn bright red. In contrast, alkaline ingredients cause purple and dark-red vegetables to turn blue and white vegetables to turn yellow. Alkalis also cause purple, dark-red, and white vegetables to have a mushy texture.

Steaming Vegetables

Steaming is one of the quickest ways to cook vegetables and minimizes nutrient loss. Steamed vegetables also retain much of their color and most of their firm texture. Large quantities of fresh or frozen vegetables can be steamed in a pressurized or convection steamer. **See Figure 3-22.** Steamed vegetables may also be sautéed.

Steaming Vegetables

Figure 3-22. Large quantities of fresh or frozen vegetables can be steamed in a pressurized or convection steamer.

Acid and Alkali Reactions

Pigment	Cooked Vegetables*	Acid Added	Alkali Added
Chlorophyll	Broccoli 	Color loss 	Mushy texture 
Carotenoids	Carrots 	Little or no effect 	Mushy texture 
Flavonoids	Beets 	Brighter red 	Turns blue; mushy texture 

** No acidic or alkaline ingredients used*

Figure 3-21. The addition of acidic or alkaline ingredients when cooking vegetables causes chemical reactions that affect the color and texture of the vegetables.

Irinox USA

Blanching Vegetables

Blanching is a moist-heat cooking technique in which food is briefly parcooked and then shocked by placing it in ice-cold water to stop the cooking process. Fresh vegetables are often blanched to highlight their bright color before they are finished with another cooking technique. For example, vegetables such as asparagus are often blanched and then finished in a broiler or on a grill at time of service. Other vegetables, such as tomatoes, may be blanched to make it easier to remove their skin.

Nutrition Note

Blanching not only brightens the colors of some fruits and vegetables, but also allows foods to retain their vitamins and minerals.

Grilling and Broiling Vegetables

Grilling and broiling are both fast and easy ways to prepare vegetables. Fresh vegetables are seasoned as desired, drizzled with a little oil, and then placed under the broiler or directly on the grill. The size of the vegetables determines how long they need to cook. Grilling and broiling caramelize the sugars in vegetables, which results in a sweeter flavor. **See Figure 3-23.**

Grilled Vegetables

Tanimura & Antle®

Figure 3-23. Grilling vegetables caramelizes their sugars, giving them a sweeter flavor.

Procedure for Blanching

1. Bring a pot of water to a boil and prepare a separate ice bath. *Note:* Adding salt to the boiling water when blanching green vegetables intensifies their color.

2. Place the cleaned and prepared items in the rapidly boiling water until the desired result is achieved.

3. Remove the items and immediately submerge them in the ice bath to stop the cooking process.

Baking and Roasting Vegetables

Baking and roasting are excellent ways to prepare vegetables. Dishes such as a broccoli and cheese casserole are typically baked. Many root vegetables such as onions, carrots, turnips, parsnips, and various types of potatoes are often roasted together or placed alongside a large cut of meat. When baking and roasting vegetables, it is important to cut vegetables into uniform sizes to ensure doneness. Some vegetables, such as peppers, can be fire-roasted. Fire-roasting a vegetable over an open flame allows the item to be cooked whole.

Sautéing Vegetables

Vegetables prepared for sautéing or stir-frying are usually diced small or thinly sliced. Fresh vegetables are sautéed and stir-fried very quickly in a hot pan with a small amount of oil. **See Figure 3-24.** The finished vegetables are firm.

Sautéed and Stir-Fried Vegetables

National Honey Board

Sautéed

U.S. Apple® Association

Stir-Fried

Figure 3-24. Vegetables are sautéed and stir-fried very quickly in a hot pan with a small amount of oil.

Frying Vegetables

Deep-fried vegetables such as onions, mushrooms, cauliflower, zucchini, and eggplant are popular appetizers. **See Figure 3-25.** These vegetables are batter-coated and fried until crisp. French fries are also a popular fried vegetable.

Fried Vegetables

Photo Courtesy of McCain Foods USA

Figure 3-25. Deep-fried vegetables such as onions rings are popular appetizers.

Nutrition Note

When frying vegetables, it is difficult to calculate the nutritional value of the frying oil. The nutritional value varies depending on cook time and temperature, ingredient density, and type of oil used. To compensate, a 10% retention rate is typically used to calculate the nutritional value of the oil.

Plating Vegetables

Vegetables add color and texture to a plate. The conventional clock system of placing vegetables at 2 o'clock, proteins at 6 o'clock, and starches at 11 o'clock is a good method to follow when plating. Vegetables should cover about half of the plate, while the protein should cover a fourth and the starch the other fourth. Green vegetables should be complemented with vegetables of other colors, such as carrots, bell peppers, or beets. Also, it is important to avoid clustering the vegetables. This is so that the contrasting colors can be seen. Different shapes and cuts of vegetables also add visual interest. Contrasting hard with soft vegetables and smooth with coarse vegetables will enhance the presentation. **See Figure 3-26.**

Plating Vegetables

Daniel NYC / T. Schauer

Figure 3-26. Contrasting hard with soft vegetables and smooth with coarse vegetables will enhance the plate presentation.

CHECKPOINT 3-2

1. Explain how acidic and alkaline ingredients affect cooked vegetables.
2. Identify eight methods that can be used to cook vegetables.
3. Steam or blanch a vegetable and evaluate the quality of the prepared dish.
4. Grill or broil a vegetable and evaluate the quality of the prepared dish.
5. Bake or roast a vegetable and evaluate the quality of the prepared dish.
6. Explain how to fire-roast peppers.
7. Fire-roast a vegetable and evaluate the quality of the prepared dish.
8. Sauté a vegetable and evaluate the quality of the prepared dish.
9. Stir-fry a vegetable and evaluate the quality of the prepared dish.
10. Deep-fry a vegetable and evaluate the quality of the prepared dish.

FRUIT CLASSIFICATIONS

A *fruit* is the edible, ripened ovary of a flowering plant that usually contains one or more seeds. The vast assortment of available fruits is primarily due to the large number of fruit varieties and hybrid fruits created by fruit growers over the last 2000 years. **See Figure 3-27.** A *variety fruit* is a fruit that is the result of breeding two or more fruits of the same species that have different characteristics. For example, a Jonagold apple is a variety of apple created by breeding a Jonathan apple with a Golden Delicious apple. A *hybrid fruit* is a fruit that is the result of crossbreeding two or more fruits of different species to obtain a completely new fruit. For example, the loganberry was created by crossbreeding a raspberry and a blackberry.

Fruit Varieties

Figure 3-27. The vast assortment of available fruits is primarily due to the large number of fruit varieties and hybrid fruits created by fruit growers over the last 2000 years.

Fruits are nutritious because they are high in water, dietary fiber, vitamins, fructose, and antioxidants. Fruit adds color, texture, and flavor to a meal. Fruit is classified into several major categories, including berries, grapes, pomes, drupes, melons, citrus fruits, tropical fruits, and exotic fruits.

Berries

A *berry* is a type of fruit that is small and has many tiny, edible seeds. Quality berries are sweet and evenly colored. Some berries are actually aggregate fruits. An *aggregate fruit* is a cluster of very tiny fruits. Blackberries, boysenberries, loganberries, and raspberries are aggregate fruits. Berries are harvested ripe because they do not continue to ripen after harvest. Berries commonly used in the professional kitchen include blackberries, blueberries, boysenberries, cranberries, currants, gooseberries, loganberries, raspberries, and strawberries. **See Figure 3-28.**

Berries	
Blackberries 	A *blackberry* is a sweet, dark-purple to black, aggregate fruit that grows on a bramble bush. Blackberries have a shiny outer layer and are used in many sauces and desserts.
Blueberries 	A *blueberry* is a small, dark-blue berry that grows on a shrub. Blueberries are often used in salads, breads, and desserts.
Boysenberries  *Oregon Raspberry & Blackberry Commission*	A *boysenberry* is a deep-maroon, hybrid berry made by crossbreeding a raspberry, a blackberry, and a loganberry. Boysenberries are similar in size and shape to blackberries. This aggregate fruit can be made into syrup and wine.
Cranberries  *Melissa's Produce*	A *cranberry* is a small, red, round berry that has a tart flavor. The skin of a cranberry is white when young and turns a deep red when ripe. Because of their strong, sour flavor, cranberries are most often used in desserts, breads, sauces, and jellies. Fresh or frozen cranberries are typically cooked with sugar or simple syrup before being added to a dish.
Currants  *Frieda's Specialty Produce*	A *currant* is a small red, black, or golden-white berry that grows in grape-like clusters. Currants are used in sauces and some savory dishes.
Gooseberries  *Frieda's Specialty Produce*	A *gooseberry* is a smooth-skinned berry that can be green, golden, red, purple, or white and has many tiny seeds on the inside. Gooseberries are used to make a classic English dessert called a fool, which is composed of puréed gooseberries mixed with sweetened whipped cream. They are also used to make chutneys.

Figure 3-28. (continued on next page)

Berries	
Loganberries  Oregon Raspberry & Blackberry Commission	A *loganberry* is a red-purple hybrid berry made by crossbreeding a raspberry and a blackberry. Loganberries are similar in size and shape to blackberries but are redder in color. Loganberries can be made into a syrup.
Raspberries 	A *raspberry* is a slightly tart, red aggregate fruit that grows in clusters. The velvety soft texture of raspberries makes them one of the most fragile fruits. Raspberries are used in a variety of desserts and in sweet sauces to complement desserts, as well as in glazes for savory dishes such as roasted meats or poultry.
Strawberries  California Strawberry Commission	A *strawberry* is a bright-red, heart-shaped berry covered with tiny black seeds. A ripe strawberry should be evenly red and free of brown or soft spots. Strawberries are widely used in desserts, sauces, and salads.

Figure 3-28. A berry is a small fruit with many tiny, edible seeds.

Grapes

A *grape* is an oval fruit that has a smooth skin and grows on woody vines in large clusters. Grapes are the most widely grown fruit because of their use in winemaking. Table grapes are the varieties suitable for eating, as opposed to varieties grown specifically for winemaking. The two main classifications of table grapes are white grapes, which are usually green in color, and black grapes, which are usually red to dark blue in color. Most of the flavor of a grape comes from its skin. It is important to choose grapes that are firm with no discoloration. Table grapes frequently used in the professional kitchen include Thompson, red flame, and Concord grapes. **See Figure 3-29.** Dried grapes are called raisins.

Thompson Grapes. A *Thompson grape* is a seedless grape that is pale to light green in color. Thompson grapes may be used in salads or served with a cheese platter. Dried seedless grapes are called sultanas.

Red Flame Grapes. A *red flame grape* is a seedless grape that ranges from a light purple-red color to a dark-purple color. Red flame grapes are typically sweeter and crisper in texture than Thompson grapes. Red flame grapes are often used as an accompaniment to salads, cheese trays, or platters.

Grapes

Figure 3-29. Common varieties of table grapes include Concord, red flame, and Thompson grapes.

Concord Grapes. A *Concord grape* is a seeded grape with a deep black color. It is sometimes called a slipskin grape, as its skin is very easily separated from the fruit. Concord grapes are available in the market as a table grape but, because they have seeds, they are used less often than the Thompson or red flame varieties.

Pomes

A *pome* is a fleshy fruit that contains a core of seeds and has an edible skin. Pomes have thin skin, grow on trees or bushes, and are an excellent source of antioxidants and dietary fiber. Quality pomes are free of blemishes and bruises and have no soft spots. Pomes often used in the professional kitchen include apples, pears, and quinces. **See Figure 3-30.**

Pomes

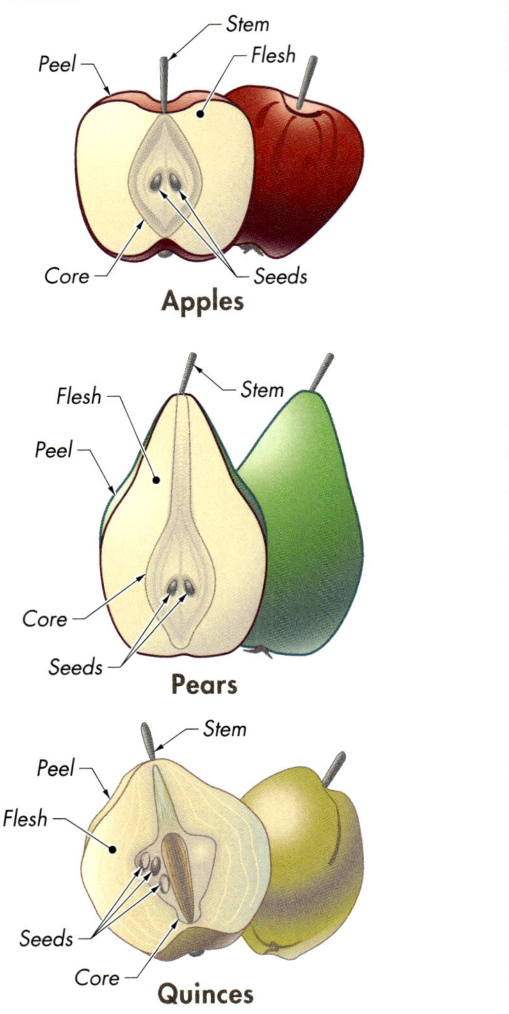

Figure 3-30. A pome is a fleshy fruit that contains a core of seeds and an edible peel. Apples, pears, and quinces are pomes.

Apples. An *apple* is a hard, round pome that can range in flavor from sweet to tart and in color from pale yellow to dark red. There are more than 7500 cultivated varieties of apples. Some varieties are not suitable for cooking purposes, and others become bitter when baked. In general, tart apples, or apples that have a high acid content, are desirable for cooking because of their intense flavor. **See Figure 3-31.**

Common Apple Varieties

Name	Description	Availability
Cortland	Sweet, hint of tartness, juicy, tender, white flesh	Sept. to Apr.
Crispin	Sweet, juicy, crisp	Oct. to Sept.
Empire	Sweet, tart, juicy, creamy-white flesh	Sept. to July
Fuji	Spicy, sweet, juicy, firm cream-colored flesh, tender skin	Oct. to June
Gala	Yellow to red, sweet, juicy, crisp yellow flesh	Sept. to June
Golden Delicious	Sweet, crisp, light yellow flesh	Sept. to June
Granny Smith	Tart, crisp, juicy	Sept. to June
Idared	Sweet, tart, juicy, firm, pale yellow-green or rosy pink flesh	Oct. to Aug.
Jonagold	Tangy, sweet	Oct. to May
McIntosh	Sweet, tangy, juicy, tender, white flesh	Sept. to June
Rome	Mildly tart, firm, greenish-white flesh	Oct. to Sept.

U.S. Apple® Association

Figure 3-31. Common varieties of apples are used to prepare a vast array of dishes, including salads, sauces, purées, desserts, and savory dishes.

Pears. A *pear* is a bell-shaped pome with a thin peel and sweet flesh. Pears should be harvested before they are ripe, but if picked too early they will not develop their full flavor. If left to fully ripen on the tree, pears develop concentrations of cellulose, resulting in a grainy texture. There are thousands of different kinds of pears. Common varieties of pears include Anjou, Asian, Bartlett, Bosc, Comice, and Seckel pears. **See Figure 3-32.**

Pears	
Anjou Pears  Pear Bureau Northwest	An *Anjou pear* is a plump, lopsided pear that is green in color. Anjou pears have a juicy, firm, and smooth flesh that is not as sweet as some other varieties. Ripe Anjou pears are best used in salads. Anjou pears that are not fully ripe can be baked, poached, or roasted.
Asian Pears  Frieda's Specialty Produce	An *Asian pear,* also known as an apple-pear, is a round pear with the texture of an apple and yellow-colored skin. Asian pears are ready to eat when harvested instead of being allowed to ripen and soften off the tree like other pears. Asian pears are used in salads and slaws.
Bartlett Pears  Pear Bureau Northwest	A *Bartlett pear* is a large, golden pear with a bell shape. Bartlett pears have a creamy, juicy flesh and a musky flavor and aroma. Bartlett pears can be used in sauces, sorbets, pies, and tarts.
Bosc Pears  Pear Bureau Northwest	A *Bosc pear* is a pear with a gourd-like shape and a brown-colored to bronze-colored peel. The peel of a quality Bosc pear has a yellow rather than a green undertone. The flesh of a Bosc pear is rich, syrupy, and sweet. Because of their firm texture, Bosc pears hold their shape when cooked.
Comice Pears  Pear Bureau Northwest	A *Comice pear* is a fairly large pear with a rotund body and a very short, well-defined neck. They are most often green-yellow in color with a red blush covering part of the surface, although some varieties are almost entirely red in color. Comice pears are very sweet and juicy, making them a poor choice for cooking.
Seckel Pears  Pear Bureau Northwest	A *Seckel pear* is a small pear that is sometimes called a honey pear or sugar pear because of its syrupy, fine-grained flesh and complex sweetness. Seckel pears are superb for use in desserts.

Figure 3-32. Common varieties of pears include Anjou, Asian, Bartlett, Bosc, Comice, and Seckel pears.

Quinces. A *quince* is a hard yellow pome that grows in warm climates. **See Figure 3-33.** Quinces are not eaten raw because they have a bitter taste. This characteristic disappears during cooking. Quinces are typically cooked in sugar syrup, which turns the fruit slightly darker with a pinkish tint.

Drupes

A *drupe,* also known as a stone fruit, is a type of fruit that contains one hard seed or pit. Drupes grow on shrubs and trees and are usually harvested before they are ripe. High-quality drupes are free of blemishes or bruises. Apricots, avocados, cherries, dates, nectarines, olives, peaches, and plums are all drupes that are used in the professional kitchen. **See Figure 3-34.**

Quinces

Frieda's Specialty Produce

Figure 3-33. Quinces are hard yellow pomes that grow in warm climates.

Drupes	
Apricots 	An *apricot* is a drupe that has pale orange-yellow skin with a fine, downy texture and a sweet and aromatic flesh. Apricots are quite delicate and may be harvested before they are ripe to avoid damage in shipping. Apricots are a popular choice for fruit tarts, dessert sauces, pastry fillings, and savory sauces for meats or poultry.
Avocados 	An *avocado,* also known as an alligator pear, is a pear-shaped drupe with a rough green skin and a large pit surrounded by yellow-green flesh. The inedible skin and pit must be removed before the avocado can be eaten. The color of the skin varies depending on the variety. The skin of a Hass avocado turns black when ripe. Other varieties remain green when ripe. Avocados are full of vitamins and minerals and contain both omega 3 and 6 fatty acids. The majority of the fat in avocados is monounsaturated. Avocados also are a good source of protein and do not contain cholesterol. Avocados are often made into guacamole, used in salads, as a sandwich spread, and as an accompaniment to savory dishes.
Cherries  National Cherry Growers and Industries Foundation	A *cherry* is a small, smooth-skinned drupe that grows in a cluster on a cherry tree. Cherries have a long, thin stem that holds them on the tree. The skin of cherries typically ranges in color from a bright red to a deep red that is nearly black. There are also golden-skinned varieties. The flesh of a cherry is pulpy and juicy and ranges in color from a dark yellow-orange to a deep reddish black. Cherries are classified as sweet or sour. Sweet cherries include Bing cherries, Gean cherries, and Rainier cherries. Sour cherries include Montmorency cherries and Morello cherries. Cherries are often used to make sauces, chutneys, pies, and a variety of desserts.

Figure 3-34. (continued on next page)

Drupes	
Dates  *Frieda's Specialty Produce*	A *date* is a plump, juicy, and meaty drupe that grows on a date palm tree. Dates are high in dietary fiber, carbohydrates, and potassium. They are also low in fat, cholesterol, and sodium. Dates may be dried or stuffed with savory or sweet fillings after the pit has been removed. Dates can also be candied.
Nectarines 	A *nectarine* is a sweet, slightly tart, orange to yellow drupe with a firm, yellow flesh and a large oval pit. Nectarines share many characteristics with peaches. Like peaches, nectarines need to ripen at room temperature to prevent them from becoming mealy. Virtually any recipe that calls for peaches can also be made with nectarines.
Olives 	An *olive* is a small, green or black drupe that is grown for both the fruit and its oil. Unripe olives are green. Olives that are tree-ripened naturally turn dark brown to black. Fresh olives are bitter and their flavor reflects how ripe they were at harvest and how they were processed. Most olives are brined or salt cured and then packed in olive oil or vinegar. Black olives, also known as Mission olives, get their color and flavor from lye curing and oxygenation. Greek Kalamata olives and French Niçoise olives are also popular varieties. Pitted and unpitted olives are used in both raw and cooked dishes.
Peaches 	A *peach* is a sweet, orange to yellow drupe with downy skin. The flesh of a peach is juicy, yet firm enough to hold its shape. The skin is edible, but the large oval pit inside the peach is not. Peaches should be ripened at room temperature and then refrigerated to prevent them from becoming mealy in texture. Peaches are used in salsas, chutneys, pies, and pastries.
Plums 	A *plum* is an oval-shaped drupe that grows on trees in warm climates and comes in a variety of colors such as blue-purple, red, yellow, or green. There are more than 2000 varieties of plums. Some types of plums are sweet, juicy, and fragrant, while others can be sour, crisp, or mealy. The flesh can be red-orange, yellow, or green-yellow. Plums can be baked in cobblers and tarts. Dried plums are known as prunes.

Figure 3-34. A drupe, also known as a stone fruit, contains one hard seed or pit.

Melons

A *melon* is a type of fruit that has a hard outer rind (skin) and a soft inner flesh that contains many seeds. The hard outer rind can be netted, ribbed, or smooth in texture. Most melons are picked just before they are ripe. Ripe melons are firm and have a good aroma. Melons are a good source of vitamin C and potassium and are relatively low in calories.

With the exception of watermelon, the seeds are typically removed from fresh melons before service. The rind may or may not be removed depending on the desired presentation. Melons can be served as sides, used as garnishes, or made into cold soups, sorbets, ice creams, or parfaits. Canary melons, cantaloupes, casaba melons, Crenshaw melons, honeydew melons, muskmelons, and watermelons are all used in the professional kitchen. **See Figure 3-35.**

Melons	
Canary Melons  Melissa's Produce	A *canary melon* is a fairly large, bright yellow melon with a smooth rind that is slightly waxy when ripe. Canary melons are served raw.
Cantaloupes  Frieda's Specialty Produce	A *cantaloupe* is an orange-fleshed melon with a rough, deeply grooved rind. The grooves run lengthwise from the stem. The most widely cultivated variety of the cantaloupe is the Charentais cantaloupe, which has a pale-green, lightly ridged rind and is native to Europe. American cantaloupe is actually a type of muskmelon.
Casaba Melons  Melissa's Produce	A *casaba melon* is a teardrop-shaped melon with a thick, bright-yellow, ridged rind and white flesh. Casaba melons are unique in that they do not have an aroma so the quality must be judged by looking for a deep yellow color and an absence of blemishes. Virtually any recipe that calls for cantaloupe can also be made with casaba melon.
Crenshaw Melons  Melissa's Produce	A *Crenshaw melon* is a large, pear-shaped melon with a yellow-green, slightly ribbed rind and an orange or salmon-colored flesh. Some Crenshaw melons weigh as much as 10 pounds. Crenshaw melons have a sweet and spicy flavor and are used like cantaloupes.
Honeydew Melons 	A *honeydew melon* is a melon with a smooth outer rind that changes from a pale-green color to a creamy-yellow color as it ripens. Honeydews have a mild, sweet flavor. Honeydew melons are used in the same manner as cantaloupes.

Figure 3-35. (continued on next page)

Melons	
Muskmelons 	A *muskmelon* is a round, orange-fleshed melon with a beige or brown, netted rind. The inside flesh can range in color from salmon to an orange-yellow color. There are many hybrids derived from muskmelon, such as the American cantaloupe. A *Santa Claus melon* is a large, mottled yellow and green variety of muskmelon that has a slightly waxy skin and soft stem end when ripe. A *Persian melon,* also known as a patelquat, is a green muskmelon with a finely textured net on the rind. The green coloring lightens and the netting turns brown as the melon ripens. Muskmelons are typically sliced and served raw.
Watermelons  *National Watermelon Promotion Board*	A *watermelon* is a sweet, extremely juicy melon that is round or oblong in shape, with pink, red, or golden flesh and green skin. The watermelon is named for its high water content—watermelons are over 90% water. The weight of a watermelon varies by variety, but can weigh up to 30 pounds. The thick rind ranges from light green to dark green in color and is often striped or solid. Some watermelons are seedless while others contain a lot of seeds. Watermelon seeds thicken and turn black when mature. Immature watermelon seeds are thin and white. Watermelon is most often enjoyed fresh or puréed into sauces and cold soups.

Figure 3-35. Melons have a hard outer rind and a soft inner flesh that contains many seeds.

Citrus Fruits

A *citrus fruit* is a type of fruit with a brightly colored, thick rind and pulpy, segmented flesh that grows on trees in warm climates. The peel and the pith are both removed during preparation because of their bitter taste. The *peel* is the thick outer rind of a citrus fruit. The *pith* is the white layer just beneath the peel of a citrus fruit. Citrus fruits are harvested fully ripe, as they do not continue to ripen after they are picked. Fresh citrus should be kept refrigerated to extend storage life. Citrus fruits are an excellent source of vitamin C and are quite acidic. Common citrus fruits used in the professional kitchen include grapefruits, lemons, limes, mandarins, oranges, tangerines, and ugli fruit. **See Figure 3-36.**

Citrus Fruits	
Grapefruits 	A *grapefruit* is a round citrus fruit with a thick, yellow outer rind and tart flesh. Common varieties of grapefruit include white grapefruit, pink grapefruit, and ruby red grapefruit. White grapefruit has pale-yellow flesh and pink grapefruit has pink-colored flesh. Ruby red grapefruit has red-colored flesh. Fresh grapefruit and grapefruit juice are commonly offered on breakfast menus.
Lemons 	A *lemon* is a tart yellow citrus fruit with high acidity levels. The juice and rind of lemons are used in many dishes. Lemon juice is used in desserts, as well as many types of sauces that flavor poultry, fish, and shellfish. Lemon juice can also be used as a salad dressing or as a flavoring in marinades and beverages. A *Meyer lemon* is cross between a lemon and an orange. Meyer lemons are round and have a smooth, dark-yellow peel. The flesh has a yellow-orange color and the juice is sweeter and less acidic than a regular lemon.

Figure 3-36. (continued on next page)

Citrus Fruits	
Limes 	A *lime* is a small citrus fruit that can range in color from dark green to yellow-green. A *key lime* is a variety of lime that is smaller, more acidic, and more strongly flavored than other limes. Key limes have a juicy green-yellow flesh. Limes are available year-round but are at peak of season in the summer. Lime juice adds a fresh, crisp flavor to many marinades, sauces, ethnic dishes, desserts, and beverages.
Mandarins  *Frieda's Specialty Produce*	A *mandarin* is a small, intensely sweet citrus fruit that is closely related to the orange, but is more fragrant. The peel of a mandarin is thinner than that of an orange and is more easily separated from the flesh. Mandarins are usually eaten raw because of their intense sweetness. A *Satsuma* is a small, seedless variety of mandarin. Mandarins and tangerines are close relatives of oranges. The peels of mandarins and tangerines are thinner than that of an orange and are more easily separated from the flesh.
Oranges 	An *orange* is a round, orange-colored citrus fruit. Valencia and navel oranges are two popular varieties of sweet oranges and are available year-round, but are at peak of season from December to April. Blood oranges are another variety of sweet orange. Blood oranges are smaller than the Valencia and navel varieties and have a deep-red flesh. They are often used to make beautifully colored sauces and to add color to salads. Some recipes call for bitter oranges, also known as Seville oranges, in which both the flesh and the peel are sour tasting. Oranges are used in sauces, juices, jams, jellies, and desserts.
Tangerines  *Florida Department of Citrus*	A *tangerine* is a small citrus fruit with a slightly red-orange peel that can be easily removed without a knife. Tangerines are a hybrid of a mandarin and a bitter variety of orange. A *clementine* is similar to a tangerine, but with a rougher skin. A *tangor* is a hybrid of a tangerine and a sweet orange. A *tangelo* is a hybrid of a tangerine and a grapefruit. Tangerines are typically eaten raw.
Ugli Fruit  *Frieda's Specialty Produce*	An *ugli fruit,* also known as a uniq fruit, is a large, teardrop-shaped, seedless citrus fruit. Ugli fruit is similar in size to a large grapefruit but is sweeter. The peel is lumpy with large pores and ranges from green to orange when fully ripe. The juicy flesh is yellow-orange and tastes like grapefruit with a hint of sweet orange. Ugli fruit is a hybrid made by crossbreeding a grapefruit and a tangerine and is available from December through April. Ugli fruit is eaten fresh and the juice can be used in salad dressings and marinades.

Figure 3-36. A citrus fruit has a brightly colored, thick rind and pulpy, segmented flesh.

Many recipes call for the zest of a citrus fruit, such as a lemon or a lime. *Zest* is the colored, outermost layer of the peel of a citrus fruit, which contains a high concentration of oil. Zest is used to add flavor to a dish. Fine shreds of zest are made by gently rubbing the fruit against a zester or a rasp grater. **See Figure 3-37.** Thin strips of zest can be obtained by running a five-hole zester along the surface of the rind. Large strips of zest can be obtained by using a vegetable peeler to peel off the colored surface of the rind and then be julienned. Zest may also be candied for use as a confection or a decoration.

Zesting

Browne Foodservice

Figure 3-37. Zesting is the process of removing the thin, colored peel of a citrus fruit.

Tropical Fruits

A *tropical fruit* is a type of fruit that comes from a hot, humid location but is readily available. Tropical fruits range in flavor from sweet to tangy and in texture from soft to crisp. Bananas, coconuts, figs, guavas, kiwifruit, mangoes, papayas, persimmons, pineapples, plantains, pomegranates, and prickly pears are tropical fruits that are commonly used in the professional kitchen. **See Figure 3-38.**

Frieda's Specialty Produce

Tropical Fruits	
Bananas 	A *banana* is a yellow, elongated tropical fruit that grows in hanging bunches on a banana plant. Banana bunches grow in tiers called hands, with each hand having up to 20 bananas. There are typically 5–20 hands of bananas in a bunch. If bananas are purchased green and allowed to ripen naturally, their flavor is notably richer. Bananas at peak ripeness have a bright yellow skin that is speckled with brown spots. Bananas can be eaten raw, grilled, fried, or baked.
Coconuts  Frieda's Specialty Produce	A *coconut* is a tropical drupe with a white meat that is housed within a hard, fibrous brown husk. Coconut meat contains less sugar and more protein than apples, bananas, and oranges. It is also a good source of iron, phosphorus, and zinc. However, 90% of the fat in a coconut is saturated. Coconuts are used in desserts, salads, soups, and sauces.
Figs  Melissa's Produce	A *fig* is the small pear-shaped fruit of the fig tree. There are more than 150 varieties, including black figs, green figs, white figs, and purple figs. Figs have a sweet, rich flavor and a gritty texture due to the massive amount of tiny seeds inside. Fresh figs are an excellent accompaniment to a cheese platter. Figs can also be stuffed or used in compotes and pastry fillings.

Figure 3-38. (continued on next page)

Tropical Fruits	
Guavas  *Florida Department of Agriculture and Consumer Services*	A *guava* is a small oval-shaped tropical fruit, usually 2–3 inches in diameter, with thin edible skin that can be yellow, red, or green. The flesh can be white, yellow, or pink in color. Guava juice is often blended into tropical fruit drinks. Guavas are also used in chutneys.
Kiwifruit 	A *kiwifruit* is a small, barrel-shaped tropical fruit with a thin, brown, fuzzy skin. It has bright green flesh and a white core that is surrounded by hundreds of tiny black seeds. It is approximately 3 inches long and weighs between 2–4 ounces. Kiwifruit contains twice as much vitamin C as an orange. It makes a colorful garnish and is often served raw.
Mangoes 	A *mango* is an oval or kidney-shaped tropical drupe with orange to orange-yellow flesh. The thin outer skin of the mango can include shades of light yellow, red, green, and pink. There are over 1000 varieties of mangoes grown around the world. The flesh of a mango clings to a large flat stone in the center of the fruit. The peel and stone must be removed before the mango can be used in a recipe. Mangoes can be used in sauces, salads, salsas, desserts, and beverages.
Papayas 	A *papaya* is a pear- or cylinder-shaped tropical fruit weighing 1–2 pounds with flesh that ranges in color from orange to red-yellow. A papaya is filled with numerous edible seeds that resemble peppercorns and have a sharp taste similar to pepper. Green papayas can be pickled or cooked for use in savory applications such as sauces and baked desserts. Ripe papayas should be added to fruit salads at the last minute so the enzymes will not soften the other fruits. Papayas are often used in chutneys and salsas. Papayas contain the enzyme papain, which is used as a meat tenderizer.
Persimmons  *Frieda's Specialty Produce*	A *persimmon* is a bright-orange tropical fruit that grows on trees and is similar in shape to a tomato. Ripe persimmons are very sweet. Unripe persimmons are tannic in flavor and have a chalky aftertaste. Persimmons are eaten raw or are baked into breads and desserts.
Pineapples 	A *pineapple* is a sweet, acidic tropical fruit with a prickly, pinecone-like exterior and juicy, yellow flesh. Pineapples take 18–20 months to fully mature, and must be fully ripe when harvested because they will not ripen after harvesting. Before pineapple can be used, the spiny outer covering and tough inner core must be removed. Pineapples are served raw, used in salads and desserts, grilled as an accompaniment, or made into juice.

Figure 3-38. (continued on next page)

Tropical Fruits	
Plantains 	A *plantain* is a tropical fruit that is a close relative of the banana, but is larger and has a dark brown skin when ripening. When extremely ripe, the skin of a plantain turns black and the flesh is a soft, deep yellow. Unripe plantains are firm and starchy, like potatoes. Plantains are usually fried and are sometimes dried and ground into banana meal.
Pomegranates 	The *pomegranate* is a round, bright-red tropical fruit with a hard, thick outer skin. Pomegranates measure about 3 inches in diameter and contain a thick white membrane that encloses hundreds of red seeds that are the edible portion of the pomegranate. The sweet and juicy seeds of the pomegranate are used in salads, soups, sauces, and beverages.
Prickly Pears  *Frieda's Specialty Produce*	The *prickly pear* is a pear-shaped tropical fruit with protruding prickly fibers that is a member of the cactus family. The prickly pear is 2–4 inches long and has a very thick, coarse outer skin that can be green, yellow, orange, red, or deep purple, depending on the variety. The flesh is sweet and juicy and contains sweet, crisp, edible seeds. Peeled and sliced prickly pears are typically served cold.

Figure 3-38. Tropical fruits range in flavor from sweet to tangy and in texture from soft to crisp.

Exotic Fruits

An *exotic fruit* is a type of fruit that comes from a hot, humid location but is not as readily available as a tropical fruit. Many exotic fruits are used in the professional kitchen. Breadfruit, dragon fruit, durians, jackfruit, kiwanos, kumquats, lychees, mangosteens, passion fruit, rambutans, and star fruit are considered exotic fruits due to their limited availability. **See Figure 3-39.**

Exotic Fruits	
Breadfruit  *Frieda's Specialty Produce*	A *breadfruit* is an exotic fruit that is native to Polynesia and has bumpy green skin and a white starchy flesh. Breadfruit is about the size of a small cantaloupe. When unripe, the flesh is cooked and used much like a potato and has a flavor similar to fresh bread. As the fruit ripens, the flesh becomes softer and slightly sweeter and can be filled with butter and brown sugar and roasted.
Dragon Fruit  *Frieda's Specialty Produce*	A *dragon fruit* is an exotic fruit of a cactus with an inedible pink skin, green scales, and white or red flesh speckled with small crunchy black seeds. The flesh is mildly sweet and similar in texture to kiwifruit. The flesh makes an interesting addition to a tropical fruit salad and can be juiced or blended into beverages.

Figure 3-39. (continued on next page)

Exotic Fruits

Exotic Fruits	
Durians  *Frieda's Specialty Produce*	A *durian* is an exotic fruit that contains several pods of sweet, yellow flesh and has a custard-like texture. Durians are about the size of a large cantaloupe, have a thick thorny skin that is golden to green in color, and can weigh up to 10 pounds. This Malaysian fruit is known for having a strong, unpleasant odor but a very appealing taste.
Jackfruit   *Florida Department of Citrus* *Melissa's Produce*	A *jackfruit* is an enormous, spiny, oval exotic fruit with yellow flesh that tastes like a banana and has seeds that can be boiled or roasted and then eaten. Jackfruit is the largest tree-borne fruit in the world. A jackfruit can weigh up to 100 pounds and be up to 36 inches long and approximately 20 inches in diameter. There are 100–500 edible seeds in a single jackfruit. Immature jackfruit is similar in texture to chicken, making it a vegetarian substitute known as "vegetable meat." The flesh can be eaten raw or preserved in syrup. Like pineapple, jackfruit is a multiple fruit derived from the convergence of many individual flowers and a fleshy stem axis.
Kiwanos  *Frieda's Specialty Produce*	A *kiwano,* also known as horned melon, is an exotic fruit with jagged peaks rising from an orange and red-ringed rind that is native to Africa. Its flesh is a brilliant lime-green color with cucumber-like seeds and a tart, refreshing kiwi-cucumber flavor. Kiwano melons can be cut in half lengthwise and eaten right out of the rind or sliced and added to fruit plates. The flesh can be strained to make a tangy-sweet sauce, salad dressing, or a refreshing beverage.
Kumquats  *Frieda's Specialty Produce*	A *kumquat* is a small, golden, oval-shaped exotic fruit with a thin, sweet peel and tart center. A kumquat is similar in appearance to a small citrus fruit but is eaten peel and all. Kumquats are available from November to April. Kumquats are used to make jellies, marmalades, and chutneys. They are also an attractive garnish, especially when displayed with their leaves.
Lychees  *United States Department of Agriculture*	A *lychee* is an exotic drupe covered with a thin, red, inedible shell and has a light-pink to white flesh that is refreshing, juicy, and sweet. The lychee also contains an inedible seed. Lychees must be harvested ripe, as they do not continue to ripen after being harvested. With the shell and seed removed, lychees can be eaten raw or in salads and desserts.

Figure 3-39. (continued on next page)

Exotic Fruits	
Mangosteens  Frieda's Specialty Produce	A *mangosteen* is a round, sweet and juicy exotic fruit about the size of an orange with a hard, thick, dark-purple rind that is inedible. Underneath the hard rind is an inedible red membrane that encloses the juicy, cream-colored flesh. The sweet-tart flesh is segmented similarly to a tangerine. Mangosteens must be ripe before being harvested as they spoil quickly. This prized fruit from Asia is eaten raw.
Passion Fruit  Florida Department of Agriculture and Consumer Services	A *passion fruit* is a small, oval-shaped exotic fruit that typically weighs 2–3 ounces and has firm, inedible skin that can be either yellow or purple. As the fruit ripens, the skin becomes thinner and wrinkles. The orange-yellow or pink-green pulp of the passion fruit is sweet and juicy, with a gelatin-like texture and a citrus flavor. Its small black seeds are edible. Passion fruit may be used in jams, jellies, ice creams, and a variety of desserts.
Rambutans  Frieda's Specialty Produce	A *rambutan* is a fragrant and sweet exotic fruit covered on the outside with soft, hair-like spikes. The rambutan is native to Malaysia and grows in clusters on an evergreen tree. Rambutans are easily opened revealing flesh, similar to that of a lychee, that surrounds a single, inedible seed.
Star Fruit  Frieda's Specialty Produce	A *star fruit,* also known as a carambola, is an exotic fruit that is shaped like a star when cut perpendicular to the stem. It has a thin, edible skin enclosing translucent flesh that is crisp and juicy. Star fruit is native to Asia and ranges from 2–5 inches in length, 1–1½ inches in diameter, and 3–4 ounces in weight. Its flavor ranges from very tart and acidic to somewhat sweet. Star fruit is ripe when the tips of the "star" begin to turn brown. Star fruit may be added to salads or used as a garnish.

Figure 3-39. Exotic fruits come from hot, humid locations but are not as readily available as tropical fruits.

Purchasing Fresh Fruit

Fresh fruit is packed in cartons, lugs, flats, crates, or bushels and is sold by weight or count. **See Figure 3-40.** Lugs can hold between 25–50 pounds of fruit, whereas flats are usually used to ship pint-size or quart-size containers of fruits such as blueberries or strawberries. Individual states have diverse regulations on the weight the containers in a flat can hold. Fruit size may also need to be specified when ordering from a vendor. For example, a 25 pound case of apples could contain 90, 110, or 125 apples, depending on the size of the fruit. Some fruits can also be purchased in a prepared form—cleaned, peeled, and cut.

Fresh Fruit

Figure 3-40. Fresh fruit is packed in cartons, lugs, flats, crates, or bushels and is sold by weight or count.

Ripening Fresh Fruit. It is important to know when fresh fruit should be purchased in order to purchase fruit with the best flavor and quality. As fruit begins to ripen, it changes in color and size and its flesh becomes soft and succulent. Left on the plant, fruit continues to ripen, breaking down in texture and flavor and eventually spoiling. Some fruits continue to ripen and mature after harvest, while other fruits do not. Fruits that continue to ripen after being harvested, such as bananas and pears, are often purchased before they are fully ripe. Other fruits, such as pineapples, have to be harvested fully ripe because they do not continue to ripen after they are picked.

The ripening process can be accelerated by storing fruit at room temperature or with other fruits that emit a large amount of ethylene gas. Apples, melons, and bananas give off ethylene gas and should be stored away from delicate fruits that could quickly ripen and spoil. Ripening can be delayed by chilling.

Fresh Fruit Grades. The USDA has a voluntary grading program for fresh fruit based on a variety of characteristics including uniformity of shape, size, color, texture, and the absence of defects. The USDA grades for fresh fruit include U.S. Fancy, U.S. No. 1, U.S. No. 2, and U.S. No. 3. Some fruit varieties have additional grades, such as U.S. Extra Fancy, U.S. Utility, or U.S. Commercial, that are specific to that particular variety. Most fruit used in restaurants is either U.S. Fancy or U.S. No. 1.

Some fruits are irradiated prior to being sold. *Irradiation* is the process of exposing food to low doses of gamma rays in order to destroy deadly organisms such as E. coli O157:H7, campylobacter, and salmonella. Irradiation also reduces spoilage bacteria, reduces insects and parasites such as fruit flies and the mango seed weevil, and in certain fruits and vegetables it inhibits sprouting and delays ripening. A statement indicating that the fruit was "treated with irradiation" and a radura, the international symbol of irradiation, appear on the label of irradiated foods. **See Figure 3-41.**

Radura Symbol

Figure 3-41. The radura symbol is required to be on the label of any food that has been irradiated.

Canned Fruit

Almost any type of fruit can be canned. Peaches, pears, and pineapples are commonly canned fruits. Fruit is canned at its peak. Depending on how long after harvest fresh fruit is eaten, canned fruit can have similar nutritional value. While the heating process does destroy some vitamins, canned fruits are adequate substitutes. Canned fruits that are packed in their own juice and those that have low amounts of added sugar and salt are the best choices.

Fruits are canned in water, light syrup, medium syrup, or heavy syrup. The packing method should be considered when using canned fruit in recipes, as the canning liquid can affect the flavor, texture, and nutritional value of the fruit.

Canned fruit can be stored for indefinite periods as long as it is kept in a cool, dry place. **See Figure 3-42.** Cans that are dented or bulging should be disposed of as they may contain harmful bacteria. After opening canned fruit, any unused portions should be transferred to an airtight storage container, dated and labeled, and then refrigerated.

Canned Fruit

InterMetro Industries Corporation

Figure 3-42. Canned fruits can be stored for indefinite periods when kept in a cool, dry place.

Frozen Fruit

Fruits, such as berries, are frozen whole, while other fruits, such as peaches, are cleaned, sliced, and frozen. **See Figure 3-43.** Freezing fruit inhibits the growth of microorganisms without affecting the nutritional value of the fruit. Individually quick-frozen fruits are convenient because the entire content need not be thawed to use a small portion. Some frozen fruits packaged in heavy syrup are also sold as purées for use in making ice creams, sorbets, and some sauces. The USDA grades for frozen fruit include U.S. Grade A or U.S. Fancy, U.S. Grade B or U.S. Choice, and U.S. Grade C or U.S. Standard.

Frozen Fruit

Figure 3-43. Frozen fruits, such as berries, are frozen whole and are convenient because the entire package need not be thawed in order to use a small portion.

Dried Fruit

Dried fruit is fruit that has had most of the moisture removed either naturally or through the use of a machine, such as a food dehydrator. Dried fruit has a much sweeter taste than fresh fruit due to the sugars being more concentrated. **See Figure 3-44.** Dried fruit is widely used in chutneys and compotes. It can also be added to breads and desserts. For maximum shelf life, dried fruit is stored in an airtight container in a cool, dry place.

Dried Fruit

Frieda's Specialty Produce

Figure 3-44. Dried fruits, such as cranberries, have a much sweeter taste than fresh fruit due to the sugars being more concentrated.

CHECKPOINT 3-3

1. Differentiate between variety fruits and hybrid fruits.
2. List the nutritional benefits of eating fruit.
3. Identify common berries used in the professional kitchen.
4. Identify three type of grapes used in the professional kitchen.
5. Identify common pomes used in the professional kitchen.
6. Identify common pears used in the professional kitchen.
7. Describe how quinces are typically cooked.
8. Identify common drupes (stone fruits) used in the professional kitchen.
9. Name the drupe that is harvested both for its fruit and its oil.
10. Identify common melons used in the professional kitchen.
11. Identify common citrus fruits used in the professional kitchen.
12. Identify common tropical fruits used in the professional kitchen.
13. Identify common exotic fruits used in the professional kitchen.
14. Describe factors to consider when purchasing fresh fruits.
15. Explain the role of canned fruits in the professional kitchen.
16. Explain the role of frozen fruits in the professional kitchen.
17. Explain the role of dried fruits in the professional kitchen.

COOKING FRUIT

Although fruit is often served raw, fruit can also be simmered, poached, grilled, broiled, baked, roasted, sautéed, and fried. Regardless of the cooking method used, it is important to remember that fruit is delicate and can become soft or mushy very quickly. Adding sugar or an acid such as lemon juice to fruit can help prevent it from becoming mushy in the cooking process. The sugar or acid is absorbed by the cells of the fruit, helping the fruit to plump up and stay firm. **See Figure 3-45.** However, adding an alkali such as baking soda will quickly break down the fruit, turning it to mush.

Cooking Fruit

Figure 3-45. Adding sugar to fruit helps the fruit stay plump.

Another characteristic of fruit to understand is the varying levels of pectin that different fruits contain. *Pectin* is a chemical present in all fruits that acts as a thickening agent when it is cooked in the presence of sugar and an acid. For example, when cranberries are simmered with sugar, they break down during the cooking process and pectin is released. As the mixture cools it thickens to a jelly-like consistency. Fruits that are high in pectin include apples, blackberries, cranberries, quinces, gooseberries, grapes, and plums. The peels of citrus fruits also contain a lot of pectin.

U.S Apple® Association

Simmering and Poaching Fruit

The simmering method is often used to make fruit compotes and stewed fruit. Fresh, frozen, canned, or dried fruit can be simmered. Simmering tenderizes and sweetens fruit. Simmered fruit can be served hot or cold and can accompany a dessert or entrée. Apples, pears, peaches, and plums may be poached in water, liquor, wine, or syrup. **See Figure 3-46.** Poaching is done at 185°F to ensure the fruit will retain its shape while being cooked.

Poached Fruit

Daniel NYC / Battman

Figure 3-46. Fruits such as pears may be poached in water, simple syrup, wine, or liquor.

Grilling and Broiling Fruit

Grilling or broiling fruit causes the sugars in the fruit to caramelize quickly because these two cooking methods require very high temperatures. Fruits that are good to grill or broil include pineapples, peaches, grapefruits, bananas, and pears. These fruits can be cut into slices or chunks and soaked in liquor or coated with sugar, honey, or liqueur for extra flavor before cooking. Fruit can be placed directly on the grill or cooked on skewers to make fruit kabobs. **See Figure 3-47.** When broiling fruit, the fruit should be placed on a sheet pan lined with parchment paper. Grilled or broiled fruit can be eaten alone or as an accompaniment.

Grilling Fruit

National Chicken Council

Figure 3-47. Fruit can be added to skewers and grilled as kabobs.

Baking and Roasting Fruit

Most berries, pomes, and drupes are well suited for baking. Pies, tarts, cobblers, strudels, and turnovers are typically filled with fruits such as apples, blueberries, cherries, or peaches before they are baked. **See Figure 3-48.** The inner cavity of an apple, fig, or pear can also be stuffed with a flavorful filling. Fruit can also be added to meats that are being roasted. For example, ham is often covered with pineapple rings to add extra sweetness to the meat. Placing peach halves atop pieces of chicken during the final stages of roasting adds flavor and beauty to the dish.

Sautéing and Frying Fruit

Fruit is often sautéed in butter, sugar, spices, or liquor. The fruit develops a sweet, rich flavor and a syrupy, caramelized glaze. Sautéed fruit can be used in dessert dishes such as in crêpes or as toppings for ice cream. It can also be incorporated into savory mixtures that include garlic, onions, or shallots. Savory fruit mixtures pair well with entrées such as pork and poultry. **See Figure 3-49.**

Cooking Fruit

Baked Fruit Pies and Tarts

National Cherry Growers and Industries Foundation

Figure 3-48. Baked fruit pies and tarts are typically filled with fruits such as cherries, apples, blueberries, or peaches.

Nutrition Note

Baking with canned fruit produces similar nutritional values to baking with fresh fruit. Canned fruits may contain less vitamin C, but the canning process stabilizes other nutrients.

Sautéed Fruit

Courtesy of The National Pork Board

Figure 3-49. Sautéed fruits pair well with entrées such as pork.

Apples, bananas, pears, and peaches are suitable fruits for frying because they do not break down when exposed to very high temperatures. Before frying, the fruit is sliced into uniformly sized pieces so that it cooks evenly. The fruit is then patted dry with a paper towel to help the batter adhere to the fruit. Next, the fruit is dipped in the batter and fried in fat. When the batter turns golden brown, the fruit is removed from the hot fat and cooled on a rack while the excess fat drains off. Fried fruits may be garnished with confectioners' sugar or melted chocolate.

Plating Fruit

In some cases, fruit will be added to the plate instead of vegetables. In other cases, fruit may be plated with the protein, as in sautéed apples with roast pork. Fruit may also be served on a separate plate. **See Figure 3-50.** Fruit may be used in a complementary sauce, salsa, or chutney, such as a pineapple-mango salsa atop a sea bass fillet or a blueberry gastric served over lamb chops. As with all plating, attention should be paid to contrasting colors and textures. Cooked fruit should be caramelized for added color and flavor. Breakfast is an ideal time to add fruit to a plate.

Plated Fruit

National Honey Board

Figure 3-50. Fruit may be served on a separate plate or on the same plate as the entrée.

CHECKPOINT 3-4

1. Explain how the pectin level in a fruit affects the cooking process.
2. List common methods of cooking fruit.
3. Explain why fruit is poached at 185°F.
4. Simmer or poach a fruit and evaluate the quality of the prepared fruit.
5. Grill or broil a fruit and evaluate the quality of the prepared fruit.
6. Bake or roast a fruit and evaluate the quality of the prepared fruit.
7. Sauté or fry a fruit and evaluate the quality of the prepared fruit.

KNIVES

Knives are the most fundamental tool used in the professional kitchen. The use of a sharp knife in skilled hands can accomplish a wide variety of cutting tasks with great efficiency. Well-constructed knives are comfortable and balanced in the hand. Each part of a knife has a specific function. **See Figure 3-51.**

Knife Blades

Most knife blades used in the professional kitchen are made of high-carbon stainless steel or ceramic material. High-carbon stainless steel produces a blade that is easy to keep sharp, does not change color, and does not transfer a metallic taste to foods. Ceramic blades provide a sharper edge for a longer period than any other material, do not react with acidic foods, and are very easy to keep clean. Ceramic blades are made from zirconium oxide, a less flexible material than stainless steel, and may chip or break if they strike hard surfaces, such as large bones, or are dropped on tile floors.

Kyocera Advanced Ceramics

Parts of a Knife

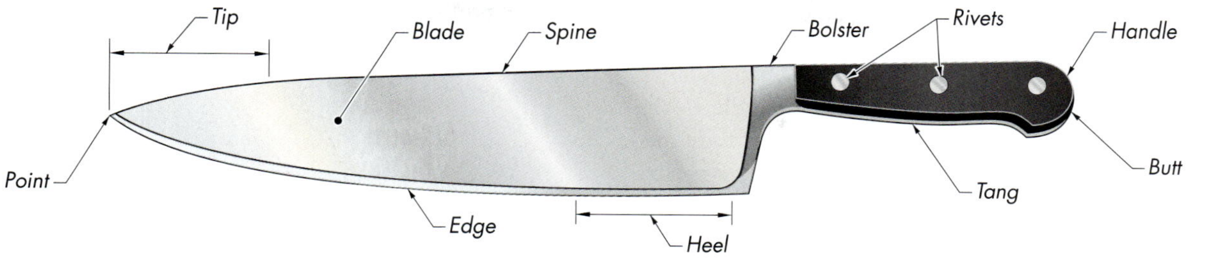

Figure 3-51. Each part of a knife, including all parts of the blade, as well as the tang, handle, bolster, and rivets, has a specific function.

A knife blade consists of five parts: the heel, tip, point, spine, and edge.

- The *heel* is the rear portion of the knife blade and is most often used to cut thick items where more force is required.
- The *tip* is the front quarter of the knife blade. Most cutting is accomplished with the section of the blade between the tip and the heel.
- The point of the blade is used as a piercing tool.
- The *spine* is the unsharpened top part of the knife blade that is opposite the edge.
- The *edge* is the sharpened part of the knife blade that extends from the heel to the tip. A knife with a sharp edge is safer than a knife with a dull edge because it requires less pressure to use. There are four basic types of knife blade edges: straight, serrated, granton, and hollow ground. **See Figure 3-52.**

Knife Blade Edges	
Straight 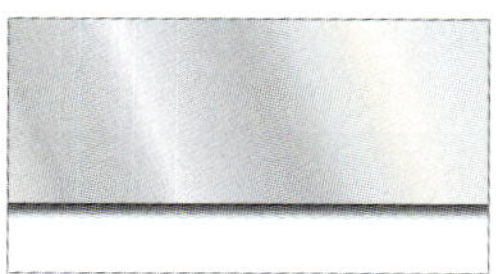	Straight edge blades are the most common type of knife blade.
Serrated 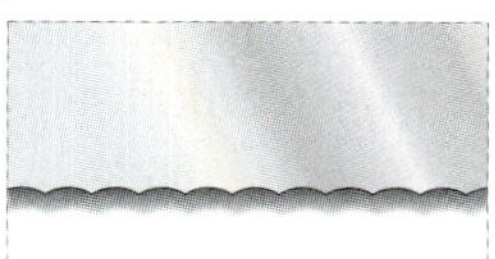	Serrated edge blades have scallop-shaped teeth that easily penetrate tough outer crusts or skins of food products such as breads and fruits.
Granton 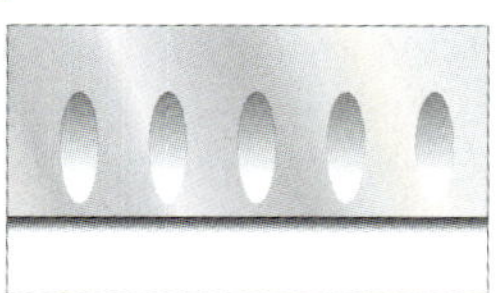	Granton edge blades have hollowed out grooves running along both sides that reduce the amount of friction as the edge of the blade cuts the food, allowing maximum contact. Granton edge blades are often used to cut meats and poultry.
Hollow Ground 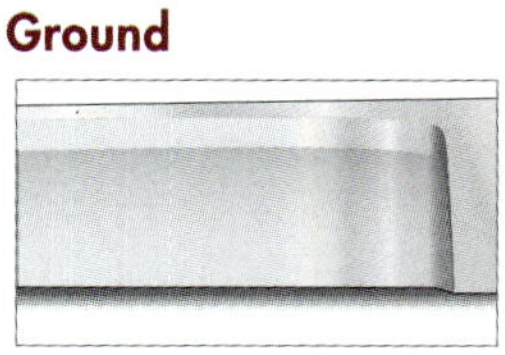	Hollow ground edge blades have been ground just below the midpoint of the blade to form a very thin cutting edge that is easily dulled. Hollow ground edge blades are ideal for skinning fish, peeling fruits, and preparing sushi.

Figure 3-52. Each of the four basic types of knife blade edges offers an advantage when cutting specific foods.

Tangs

The *tang* is the unsharpened tail of a knife blade that extends into the handle. The highest-quality knives have a tang that extends all the way to the end of the handle. The tang contains holes for securing the handle to the blade. A *partial tang* is a shorter tail of a knife blade that has fewer rivets than a full tang. Partial tang knives are less durable than full-tang knives, but may be acceptable for infrequent or light use. A *rat-tail tang* is a narrow rod of metal that runs the length of the knife handle but is not as wide as the handle. Rat-tail tangs are fully enclosed in the handle and are less durable than full or partial tangs.

Knife Handles

The handle of a knife can be made from wood, stainless steel, or synthetic materials such as plastic, nylon, styrene, resin, or polypropylene. Wood handles are becoming less common due to their lack of durability and how easily they trap bacteria. Stainless steel handles are virtually maintenance free, however, they become slippery when wet. Synthetic handles are easy to clean, last longer than wood, and are easier to grip when wet than stainless steel. However, synthetic materials crack over time and when they are exposed to extreme temperature changes. The end of a knife handle is referred to as the butt of the knife.

Production Tip

Knives should always be washed by hand. Knives should never be left in standing water or placed in a commercial dishwasher because this can cause the handles to crack or warp.

Bolsters and Rivets

A *bolster* is a thick band of metal located where the blade of a knife joins the handle. The purpose of the bolster is to provide strength to the blade and prevent food from entering the seam between the blade and the handle. A *rivet* is a metal fastener used to attach the tang of a knife to the handle. Some knives do not have bolsters and rivets.

Large Knives

Many different types of knives are used in the professional kitchen. Knowing which knife to use in a given application makes working with knives safer and more efficient. Large knives are used to make precise cuts in large areas or to separate food items such as meats. Large knives used in the professional kitchen include boning knives, bread knives, butcher's knives, chef's knives, cleavers, santoku knives, scimitars, slicers, and utility knives. **See Figure 3-53.**

Large Knives	
Boning Knives  *Mercer Cutlery*	A *boning knife* is a thin knife with a pointed 6 – 8 inch blade used to separate meat from bones with minimal waste. The blade may be either stiff (curved) or flexible (straight). Boning knives with stiff blades are used on larger cuts of meat. Those with flexible blades are used for filleting fish.
Bread Knives  *Messermeister*	A *bread knife* is a knife with a serrated blade 8 –12 inches long and is used to cut through the crusts of breads without crushing the soft interior. Serrated knives are hard to sharpen and chefs may opt to replace bread knives rather than have them sharpened. For this reason, the bread knives used may be of lesser quality than the other knives used in the professional kitchen.
Butcher's Knives 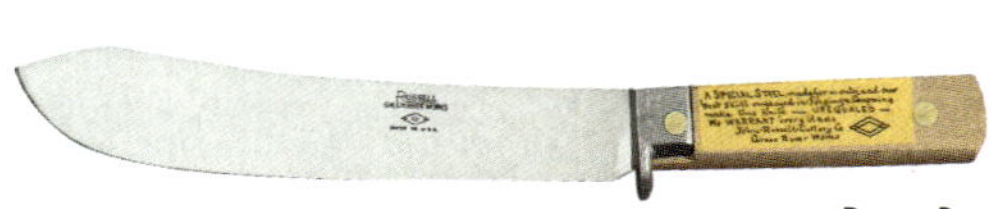 *Dexter-Russell, Inc.*	A *butcher's knife* is a heavy knife with a curved tip and a blade that is 7–14 inches in length. The tip of the blade curves upward about 25°. A butcher's knife is used to cut, section, and portion raw meats.

Figure 3-53. (continued on next page)

Large Knives	
Chef's Knives  *Mercer Cutlery*	A *chef's knife,* also known as a French knife, is a large and very versatile knife with a tapering blade used for slicing, dicing, and mincing. The heel of the blade is wide and tapers to a point. The weight of a chef's knife should be evenly balanced between the blade and the handle to prevent hand and wrist fatigue. The most popular blade lengths are 8, 10, and 12 inches.
Cleavers  *Mercer Cutlery*	A *cleaver* is a heavy, rectangular-bladed knife that is used to cut through bones and thick meat.
Santoku Knives  *Mercer Cutlery*	A *santoku knife* is a knife with a razor-sharp edge and a heel that is perpendicular to the spine. A santoku knife resembles a small cleaver with a pointed tip. It usually has a granton edge blade 5–8 inches in length, which prevents food from sticking to the blade.
Scimitars  *Mercer Cutlery*	A *scimitar* is a long knife with an upward curved tip that is used to cut steaks and primal cuts of meat. The shape of the scimitar blade resembles a boning knife, yet is much larger.
Slicers  *Mercer Cutlery*	A *slicer,* also known as a carving knife, is a knife with a narrow blade 10–14 inches long that is used to slice roasted meats. Slicers are available with a straight, serrated, or granton edge. The blade may be stiff or flexible. Slicers with stiff blades often have a rounded, blunt tip and are used to slice hot meats such as roasts. Flexible-blade slicers are better suited for cutting cold meats such as ham.
Utility Knives  *Mercer Cutlery*	A *utility knife* is a multipurpose knife with a stiff 6–10 inch blade that is similar in shape to a chef's knife but much narrower at the heel. The blade edge may be straight or serrated. This knife is a cross between a chef's knife and a paring knife.

Figure 3-53. Large knives are used to make precise cuts in large areas or to separate foods such as meats.

Small Knives

Small knives are used to make precise cuts in small areas or to open food items such as shellfish. Small knives commonly used in the professional kitchen include clam knives, oyster knives, paring knives, and tourné knives. **See Figure 3-54.**

Special Cutting Tools

In addition to knives, special cutting tools are used to cut food items for specific applications. Although there are many special cutting tools, those commonly used in the professional kitchen include channel knives, mandolines, parisienne scoops, peelers, and zesters. **See Figure 3-55.**

Small Knives	
Clam Knives 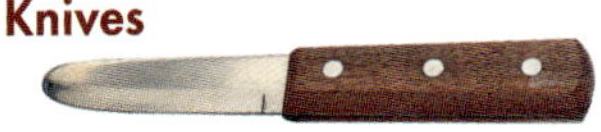 American Metalcraft, Inc.	A *clam knife* is a small knife with a short, flat, round-tipped sharp blade that is used to open clams. The proper use of a clam knife makes the task of opening clams an efficient and safe process.
Oyster Knives 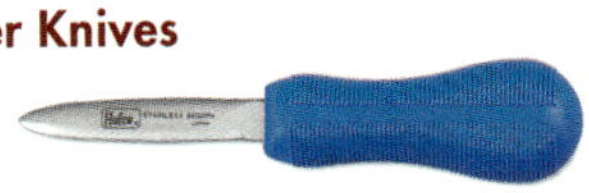 Browne Foodservice	An *oyster knife* is a small knife with a short, dull-edged blade with a tapered point that is used to open oysters. The proper use of an oyster knife makes the task of opening oysters an efficient and safe process.
Paring Knives  Mercer Cutlery	A *paring knife* is a short knife with a stiff 2– 4 inch blade used to trim and peel fruits and vegetables. A paring knife is often used in conjunction with a chef's knife to remove stems from produce.
Tourné Knives  Mercer Cutlery	A *tourné knife,* also known as a bird's beak knife, is a short knife with a curved blade that is primarily used to carve vegetables into a specific shape called a tourné, which is a seven-sided football shape with flat ends.

Figure 3-54. Small knives are used to make precise cuts in small areas or to open food items such as shellfish.

Special Cutting Tools	
Channel Knives  Dexter-Russell, Inc.	A *channel knife* is a special cutting tool with a thin metal blade within a raised channel that is used to remove a large string from the surface of a food item. A channel knife leaves a decorative pattern on the surface of an item, such as a cucumber.
Mandolines  Paderno World Cuisine	A *mandoline* is a special cutting tool with adjustable steel blades used to cut food into consistently thin slices. A mandoline can cut foods paper thin and also produce julienne cuts and waffle cuts. A hand guard needs to be in place when using a mandoline.
Parisienne Scoops  Paderno World Cuisine	A *parisienne scoop* is a special cutting tool that has a half-ball cup with a blade edge attached to a handle and is used to cut fruits and vegetables into uniform spheres.
Peelers  Carlisle FoodService Products	A *peeler* is a special cutting tool with a swiveling, double-edged blade that is attached to a handle and is used to remove the skin or peel from fruits and vegetables. The double-edged blade contours to the shape of the fruit or vegetable.
Zesters  Browne Foodservice	A *zester* is a special cutting tool with tiny blades inside of five or six sharpened holes that are attached to a handle. To use a zester, the cutting holes are drawn across the peel of a citrus fruit to yield small strings or "zest" that can be added to foods as a natural flavoring.

Figure 3-55. Special cutting tools are used to cut food items for specific applications.

Knife Safety

Improper use of knives can lead to injury. The following safety precautions should be taken when holding, using, carrying, washing, and storing knives:

- Always grip a knife properly to ensure safety and control. When using a knife, the more pressure that is applied the higher the risk of the knife slipping and of personal injury occurring.
- Always position the guiding hand properly when using, sharpening, and honing a knife.
- Always cut food items on a nonporous cutting board because the nonporous surface greatly reduces the risk of cross-contamination. Color-coded cutting boards may be used for specific types of foods. **See Figure 3-56.**
- Always pass a knife to a person by laying it on a table and sliding it forward.
- When walking with a knife, always keep the knife pointing down and hold it along the side of the body.
- Use only clean, sanitized knives on a whetstone or sharpening steel to avoid cross-contamination.
- Always wipe a blade after using a whetstone or sharpening steel to remove any metal residue.
- Always keep knives sharp. Injury is more likely to occur with a dull knife than a sharp one. Hone knives after each use to maintain a smooth, sharp edge.
- Always wipe a knife blade with the edge facing away from the hand.
- Always clean and sanitize knives before storing them.
- Store knives in sleeves, guards, or knife holders to avoid injury.
- Never leave knives in a sink as someone could reach in and be injured.
- Never wash knives in a commercial dish machine because the heat and chemicals can cause the handles to crack or warp.
- Never use a knife to pry a lid off of any type of container.
- Never attempt to catch a falling knife.

Knife Grip and Positioning

There are different acceptable methods for gripping a knife, but there is a common method used by culinary professionals that provides control and stability. To begin, the knife is held by the handle while resting the side of the index finger against one side of the blade and placing the thumb on the other side of the blade. The hand not holding the knife is referred to as the guiding hand. The guiding hand is responsible for guiding the item to be cut into the knife. To correctly position the fingers of the guiding hand, imitate the shape of a spider on the table. The fingertips should all be slightly tucked, yet touch the surface of the table. This guiding hand position is used to safely hold the food next to the blade of the knife.

Carlisle FoodService Products

Figure 3-56. Color-coded, nonporous cutting boards may be used for specific types of foods to reduce the risk of cross-contamination.

Using the proper knife grip, the tip of the knife is placed on the cutting board. The guiding hand is placed next to the knife blade in the proper position, with fingertips slightly tucked under near the back half of the blade. The side of the blade should rest against the knuckle of the middle finger of the guiding hand. This position reduces the chances of cutting fingers. **See Figure 3-57.**

Procedure for Using Proper Cutting Techniques

1. Using the proper cutting grip, place the tip of the knife on the cutting board and press down on the knife handle.

2. Continue pressing down on the handle and slide the blade forward, following the curve of the blade.

3. With the heel of the blade on the cutting surface, slide the blade backward and then raise the handle slightly to position the knife for the next slice.

Knife Grip and Positioning

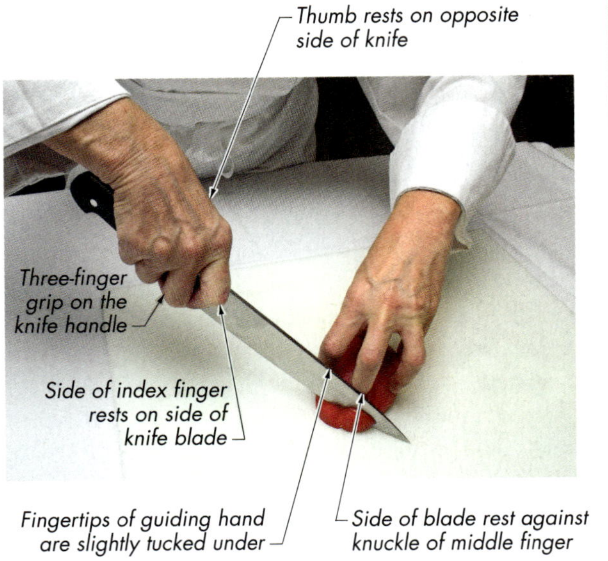

Figure 3-57. Using the proper knife grip with the knife hand, and with fingertips slightly tucked under with the guiding hand, the side of the blade should rest against the knuckle of the middle finger of the guiding hand.

With the proper knife grip and hand position, a rocking motion is used to cut with a chef's knife. Using the wrist as a fulcrum, the handle is brought down as the tip of the knife slides forward. Likewise, the handle is raised up as the tip of the knife slides backward. This rocking movement, coupled with the correct position of the guiding hand, creates a controlled motion that can be used to efficiently cut through food.

Production Tip

When cutting using a rocking motion, continually rest the blade of the knife against the knuckle of the middle finger of the guiding hand.

Sharpening Knives

Always check the edge of a knife to make sure it is sharp and properly maintained before using it. A sharp knife is much safer than a dull knife. Having a sharp knife helps to prevent injury because less pressure is required to use a sharp knife as compared to a dull knife.

Although hand-held or electric sharpeners can be used, a whetstone is typically used to sharpen professional knives. A *whetstone* is a stone used to grind the edge of a blade to the proper angle for sharpness. A three-sided whetstone has a coarse-grit, a medium-grit, and a fine-grit side. Two-sided whetstones have a medium-grit side and a fine-grit side.

To sharpen a knife, the blade of the knife is held at a specific angle to the stone. To achieve this angle, the knife blade is held at a 90° angle straight above the whetstone as if it were cutting the stone in half. Then, the knife is tilted halfway toward the stone, at a 45° angle, and then halfway again to find the perfect sharpening angle between 20° and 25°. **See Figure 3-58.**

Sharpening Angle

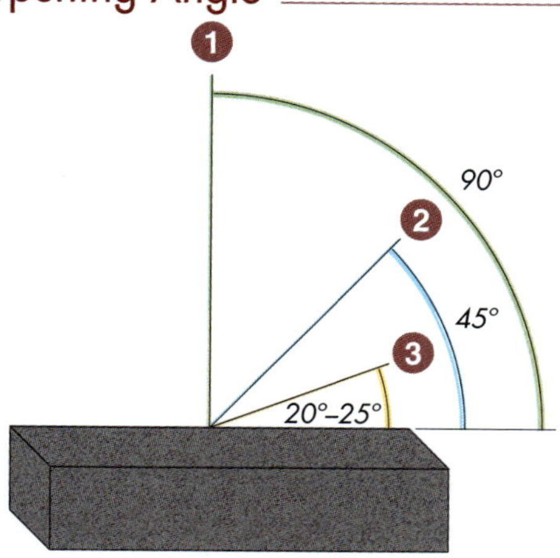

Figure 3-58. To sharpen a knife, the blade is held at a 20–25° angle against the whetstone.

After the proper angle is achieved, the knife blade is then slowly dragged across the stone from tip to heel while applying light pressure. **See Figure 3-59.**

Sharpening Knives

Figure 3-59. After establishing the correct angle, the blade is dragged across the whetstone from tip to heel while applying light pressure.

Honing Knives

After using a whetstone, it is important to "hone" or align the edge of a knife blade. *Honing,* also known as truing, is the process of aligning a blade's edge and removing any burrs or rough spots on the blade. A *steel,* also known as a butcher's steel, is a steel rod approximately 18 inches long attached to a handle and is used to align the edge of knife blades.

Sharpening Knives
Media Clip

The 20–25° angle used to sharpen knives is also used to hone knives. To achieve this angle, hold the steel perpendicular or pointed toward the floor with the guiding hand and hold the knife blade at a 20–25° angle in relation to the steel. This can be done by first holding the blade at a 90° angle to the steel, then adjusting it to about half that angle (45°), and finally adjusting it about half of that angle again to a 20–25° angle. **See Figure 3-60.** A steel should always be used to hone the blade of a knife after sharpening as well as between sharpenings to maintain a sharp, smooth edge.

Positioning Knife for Honing

1 *Hold the blade at a 90° angle to the steel.*

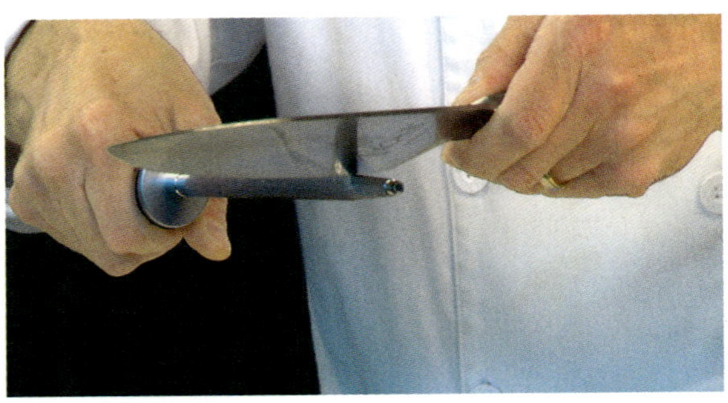

2 *Adjust the blade to half that angle (45°).*

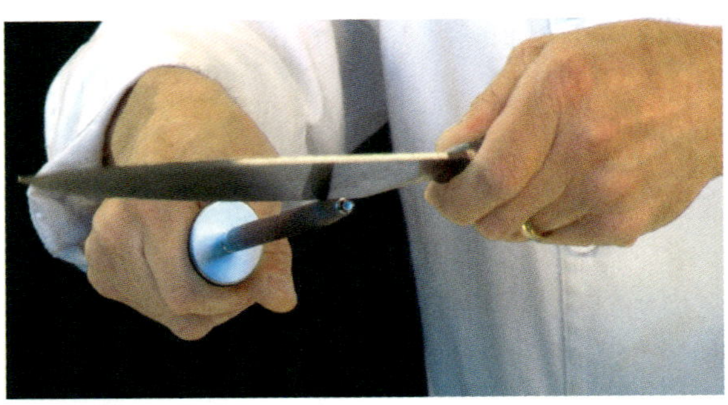

3 *Adjust the blade about halfway again to reach the correct 20–25° angle.*

Figure 3-60. To hone a knife, the blade is held at a 20–25° angle against the steel.

CHECKPOINT 3-5

1. Identify the main parts of a knife and explain the function of each.
2. Contrast high-carbon stainless steel blades and ceramic blades.
3. Explain why a knife with a sharp edge is safer to use than a knife with a dull edge.
4. Explain the advantage of using a knife with a granton edge blade.
5. Differentiate between a partial tang and a rat-tail tang.
6. Describe eight types of large knives used in the professional kitchen.
7. Describe four types of small knives used in the professional kitchen.
8. Describe five special cutting tools used in the professional kitchen.
9. Explain why nonporous cutting boards are used in the professional kitchen.
10. Describe how to safely pass a knife to another person.
11. Describe how to safely carry a knife when walking.
12. Describe how knives should be stored.
13. Provide four examples of what should never be done with a knife.
14. Demonstrate how to properly grip and position a knife.
15. Identify the term used to describe the hand that is not used to hold the knife.
16. Identify the angle at which a knife blade is held against a whetstone when being sharpened.
17. Explain the process of honing a knife.
18. Sharpen and hone a chef's knife.

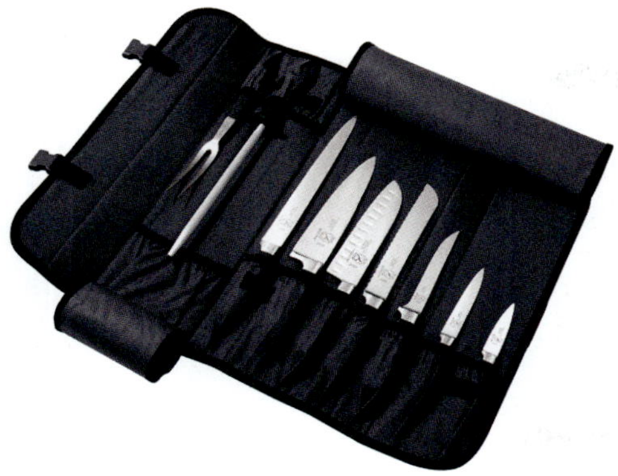

BASIC KNIFE CUTS

Every foodservice professional must know the dimensions of the basic knife cuts and be able to execute them accurately and efficiently. Uniform knife cuts ensure that items cook evenly and look appealing in the finished product. Common knife cuts used in the professional kitchen can be grouped into slicing cuts, stick cuts, dice cuts, chopping, mincing, fluted cuts, and tourné cuts.

Slicing Cuts

Slicing involves passing the blade of the knife slowly through an item to make long, thin pieces. In slicing, the knife is pulled backward or slid forward through the item. Slicing cuts include the rondelle, diagonal, oblique, and chiffonade.

Rondelle Cuts. A *rondelle cut,* also known as a round cut, is a slicing cut that produces disks. Rondelle cuts are produced from slicing cylindrical vegetables such as cucumbers and carrots straight through. To make a rondelle cut, the cylindrical vegetable is placed perpendicular to the blade and then sliced to create ¼, ⅛, or 1/16 inch disks. **See Figure 3-61.**

Rondelle Cuts

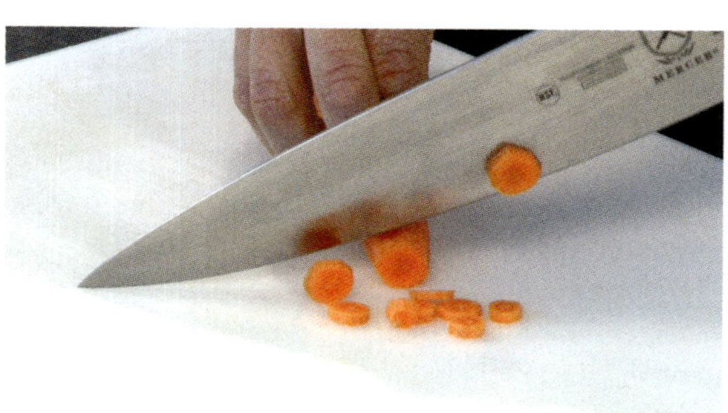

Figure 3-61. A rondelle cut, also known as a round cut, is a slicing cut that produces disks.

Diagonal Cuts. A *diagonal cut* is a slicing cut that produces flat-sided, oval slices. Diagonal cuts are made from cylindrical vegetables that are cut on the bias. To make a diagonal cut, place the item at a 45° angle to the knife blade. Then, guide the item toward the blade as each cut is made. **See Figure 3-62.**

Diagonal Cuts

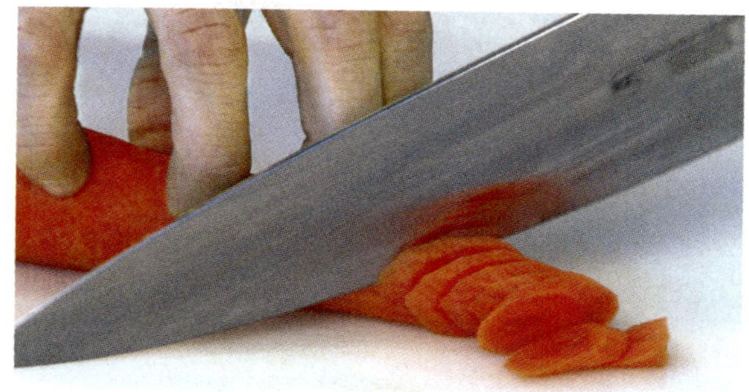

Figure 3-62. A diagonal cut is a slicing cut that produces flat-sided, oval slices.

Oblique Cuts. An *oblique cut,* also known as a rolled cut, is a slicing cut that produces wedge-shaped pieces with two angled sides. The sides are neither parallel nor perpendicular. The oblique cut is similar to the diagonal cut in that 45° angle slices are also made on a cylindrical item. However, the oblique cut produces larger pieces that are more wedge-shaped.

Chiffonade Cuts. A *chiffonade cut* is a slicing cut that produces thin shreds of leafy greens or herbs. Chiffonade-cut items can be used as ingredients or as a base under displayed foods. To make a chiffonade cut, first wash the leafy items, stack the leaves on top of one another, and roll the stack lengthwise like a cigar. Place the cigar-shaped roll on the cutting board perpendicular to the knife blade. Use a rocking motion to thinly slice the roll as it is fed with the guiding hand into the knife blade. The result is finely shredded leaves or herbs.

Procedure for Making Oblique Cuts

1. With fingers of the guiding hand tucked, guide the cylindrical vegetable toward the blade at a 45° angle and slice. Reserve this first cut for later use.
2. Roll the vegetable 180° and slice again, keeping the blade at a 45° angle. *Note:* This slice produces the first oblique cut.

3. Roll the item back 180° to the original position and slice again.

4. Continue rolling the item 180° between cuts to make wedge-shaped oblique cuts.

Procedure for Making Chiffonade Cuts

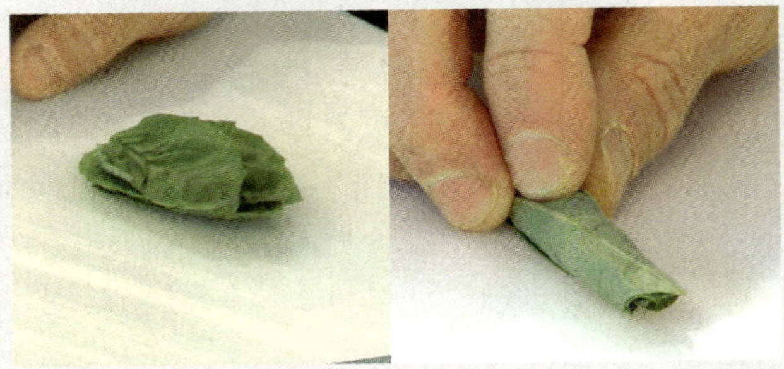

1. Place washed, dry leaves in a neat stack.
2. Roll the stack into a tight cylinder and place the cylinder perpendicular to the knife blade.

3. With the fingers of the guiding hand tucked, use a rocking motion to thinly slice the leaves.

Finished basil chiffonades are very finely cut.

Stick Cuts

Many fruits and vegetables are cut into sticklike shapes to create a uniform appearance and to ensure even cooking. Common stick cuts include batonnet, julienne, and fine julienne cuts.

- A *batonnet cut* is a stick cut that produces a stick-shaped item $\frac{1}{4} \times \frac{1}{4} \times 2$ inches long.
- A *julienne cut* is a stick cut that produces a stick-shaped item $\frac{1}{8} \times \frac{1}{8} \times 2$ inches long.
- A *fine julienne cut* is a stick cut that produces a stick-shaped item $\frac{1}{16} \times \frac{1}{16} \times 2$ inches long.

Oblique Cuts Media Clip

Chiffonade Cuts Media Clip

Julienne Cuts Media Clip

Procedure for Making Batonnet Cuts

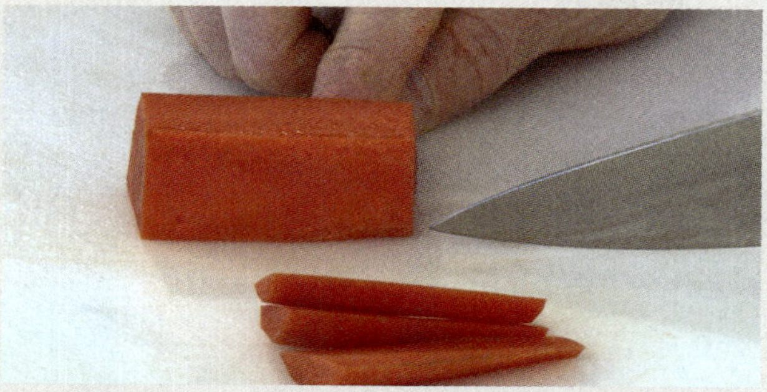

1. Cut washed and peeled vegetables, such as carrots, into pieces 2 inches in length.
2. Carefully square off three sides, leaving the fourth side rounded. Save scraps for later use.

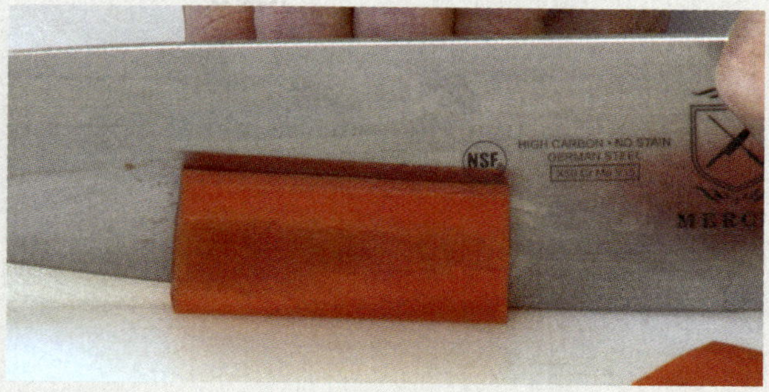

3. With the rounded side farthest away from the knife, cut even, ¼ inch thick planks. Save scraps.

4. Stack a few ¼ inch planks.
5. Carefully slice again into ¼ inch sticks.

Finished batonnets measure ¼ × ¼ × 2 inches.

All stick cuts begin by squaring off the item to be cut. There are industry-accepted dimensions for each stick cut. However, the length of stick cuts may vary depending on the desired result for a specific dish.

Dice Cuts

Dice cuts are precise cubes cut from uniform stick cuts. To produce a dice cut, a stick cut of the appropriate dimension is cut into cubes. **See Figure 3-63.** Common dice cuts include large dice, medium dice, small dice, brunoise, and fine brunoise. Items such as onions consist of many layers and require a modified procedure for dicing.

- A large dice is ¾ × ¾ × ¾ inch cubes cut from ¾ × ¾ × 2 inch sticks.
- A medium dice is ½ × ½ × ½ inch cubes cut from ½ × ½ × 2 inch sticks.
- A small dice is ¼ × ¼ × ¼ inch cubes cut from ¼ × ¼ × 2 inch sticks, or batonnets.

Dice Cuts

Figure 3-63. Large, medium, and small dice cuts are precise cubes cut from uniform stick cuts.

Batonnet Cuts
Media Clip

Procedure for Dicing Onions

1. Using a chef's knife, cut off the stem end and lightly trim the root end of an onion. *Note:* Do not cut the root end off completely as it holds the layers of the onion in place, preventing it from falling apart.

2. Cut the onion in half from the stem end to root end.

3. Make a thin slice from root end to stem end through the outer peel only.
4. Use the tip of the paring knife to pull off the top layer of the peel.

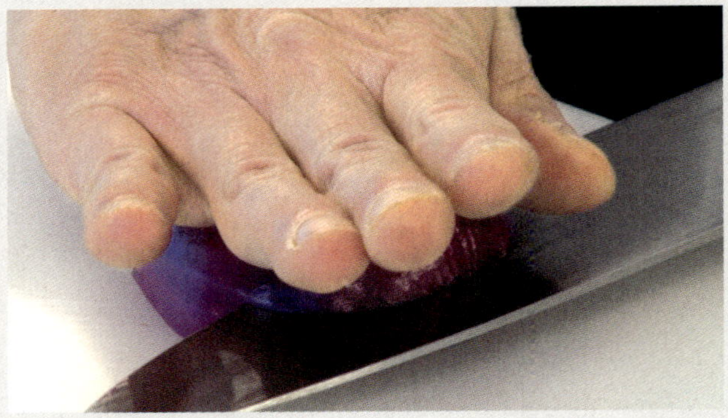

5. Position onion half on the cutting board with the flat side down. Use the chef's knife to make two or three horizontal cuts through the onion, leaving the root end intact.

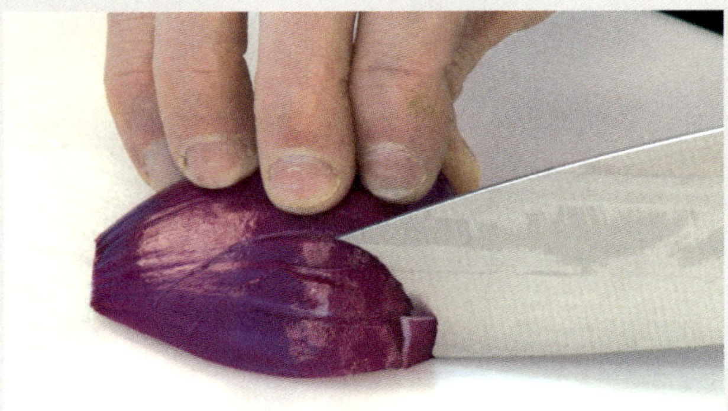

6. Make vertical slices through the onion from stem end to root end, again leaving the root end intact. *Note:* The closer together the slices, the smaller the finished dice.

7. Turn the onion a quarter turn and make cuts the thickness of the desired dice, slicing all the way through from stem end to root end. Repeat the dicing process on the other half of the onion.

Dicing Onions
Media Clip

A *brunoise cut* is a dice cut that produces a cube-shaped item with six equal sides measuring ⅛ inch each. A brunoise is cut from a julienne stick. A *fine brunoise cut* is a dice cut that produces a cube-shaped item with six equal sides measuring 1⁄16 inch. A fine brunoise is cut from a fine julienne stick. **See Figure 3-64.**

Procedure for Making Brunoise Cuts

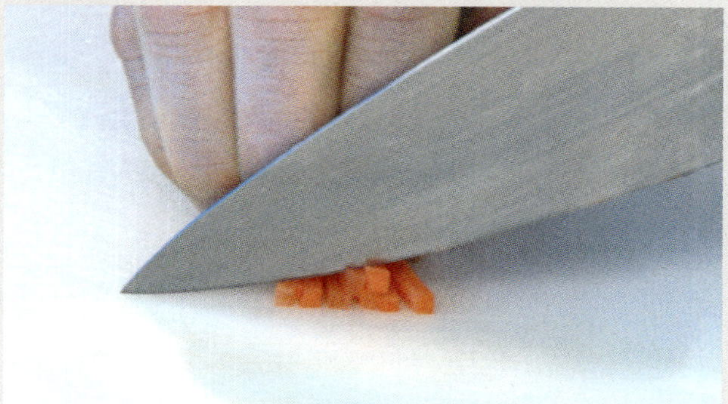

1. Place a small bundle of julienne sticks perpendicular to the knife blade.

2. With a slicing motion, cut through the bundle at equally spaced intervals to produce six-sided cubes.

Finished brunoise cuts measure ⅛ × ⅛ × ⅛ inches.

Brunoise Cuts
Media Clip

Fine Brunoise Cuts

Figure 3-64. A fine brunoise cut is a dice cut that uses a fine julienne stick to produce a cube-shaped item with six equal sides measuring 1⁄16 inch each.

Chopping

Chopping is rough-cutting an item so that there are relatively small pieces throughout, although there is no uniformity in shape or size. Parsley, hard-cooked eggs, and a rough-cut mix of vegetables called mirepoix are often chopped because a uniform shape is not required.

Mincing

Mincing and chopping have fewer applications in the professional kitchen than the other knife cuts. *Mincing* is finely chopping an item to yield a very small cut, yet not entirely uniform, product. Shallots, garlic, and fresh herbs are commonly minced.

Procedure for Mincing

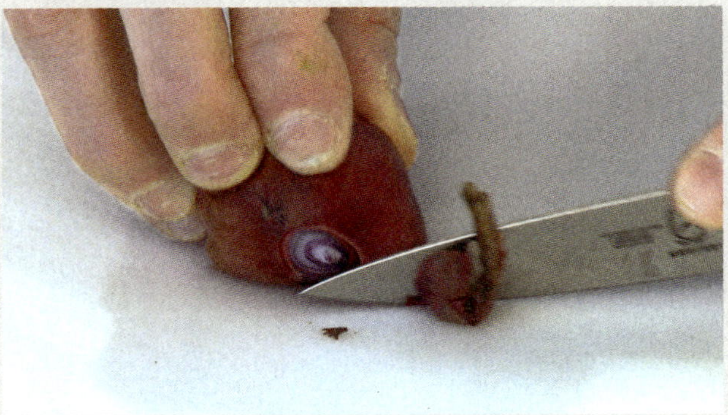

1. Using a paring knife, cut off the stem end and lightly trim the root end of a vegetable, such as a shallot. *Note:* Do not cut the root end off completely as it holds the layers of the shallot in place, preventing it from falling apart.

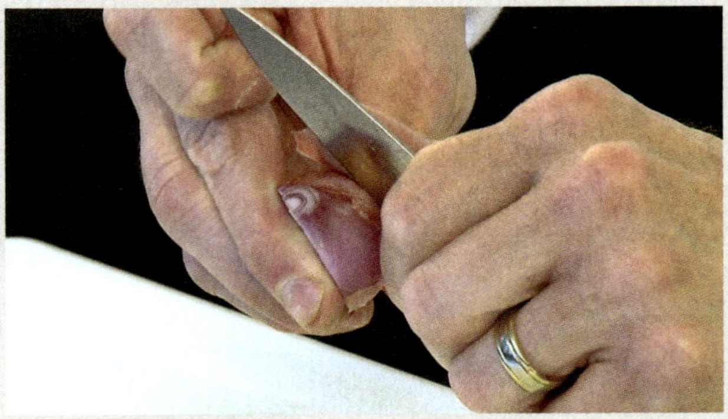

2. Make a thin slice from root end to stem end through the outer peel only.
3. Use the tip of the paring knife to pull off the top layer of the peel.

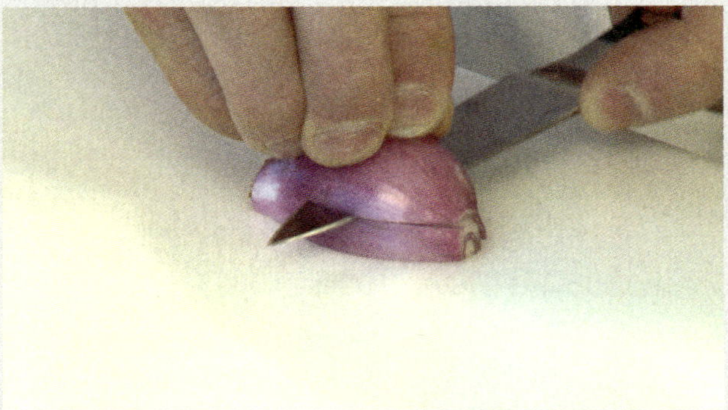

4. Cut the shallot in half lengthwise from stem end to root end.
5. Lay half of the shallot on the cutting board with the flat side down. Use the tip of the knife to make two or three horizontal cuts through the shallot, leaving the root end intact.

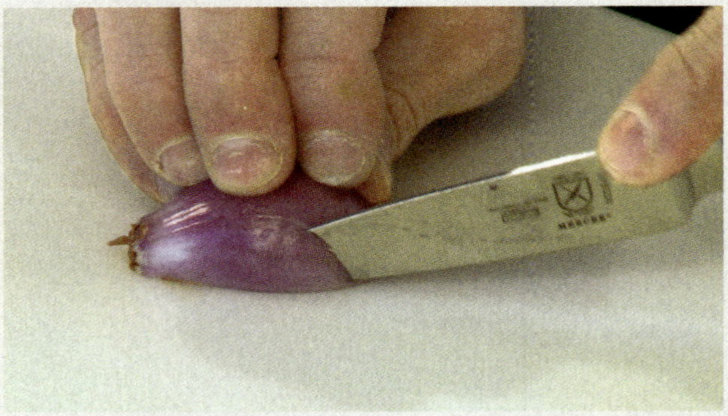

6. Make vertical slices through the shallot from stem end to root end, again leaving the root end intact. *Note:* The closer together the slices, the smaller the finished dice.

7. Using a chef's knife, turn the shallot and make cuts all the way through from stem end to root end until only the root remains.

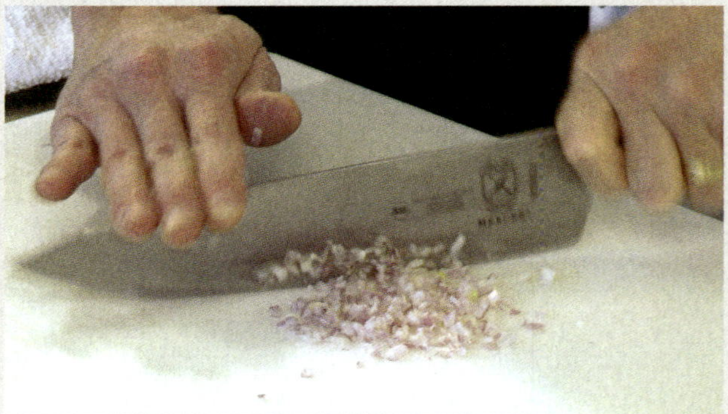

8. Place the guiding hand, opened and flat, on the top of the blade to help pivot the knife back and forth.

Mincing
Media Clip

Fluted Cuts

A *fluted cut* is a specialty cut that leaves a spiral pattern on the surface of an item by removing only a sliver with each cut. Button mushrooms are often fluted. Making fluted cuts requires good hand-eye coordination as each sliver is cut away and then the mushroom is turned slightly clockwise or counter clockwise before making the next cut from the same central point at the top of the mushroom.

Tourné Cuts

Tourné is a French word from the verb "to turn." A *tourné cut* is a carved, football-shaped cut with seven sides and flat ends. Turnips, carrots, and beets are often displayed as tournés on elegant plate presentations. This challenging specialty cut requires a lot of practice to master. Tournés are viewed as a demonstration of masterful knife skills.

Procedure for Cutting Flutes

1. Wipe mushroom with a damp towel. *Note:* Always start with a cold button mushroom.
2. Hold the mushroom with the index finger and thumb.
3. While holding a paring knife, place the thumb on the top of the blade, the index finger on the bottom of the blade, and the middle finger underneath the knife to hold it steady.

4. Hold the mushroom level and place the front third of knife blade on the center point of the top of the mushroom facing 12 o'clock.

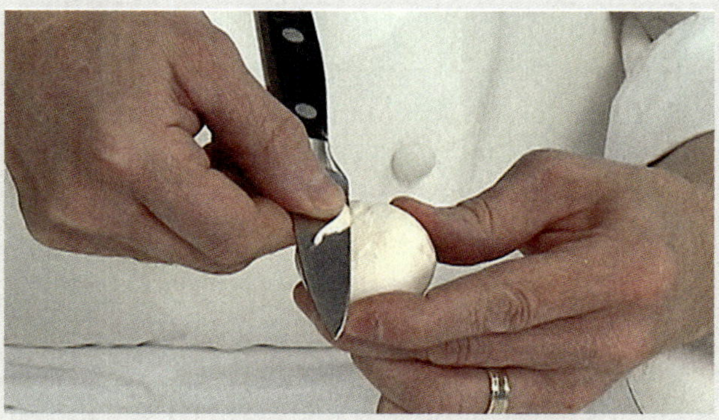

5. Turn the knife to 7 o'clock (right-handed) or 5 o'clock (left-handed), keeping the knife in contact with the mushroom at all times. *Note:* A thin strip of mushroom will be removed with each cut, leaving a fluted edge.

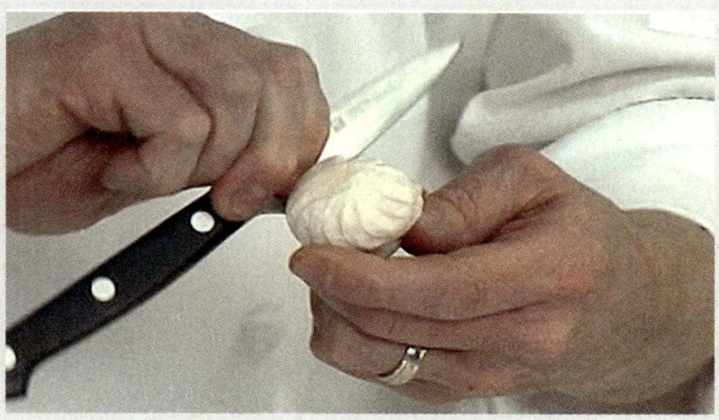

6. Rotate the mushroom clockwise (right-handed) or counter clockwise (left-handed) and continue cutting flutes until they intersect at the top center of the mushroom.

Flutes
Media Clip

Procedure for Cutting Tournés

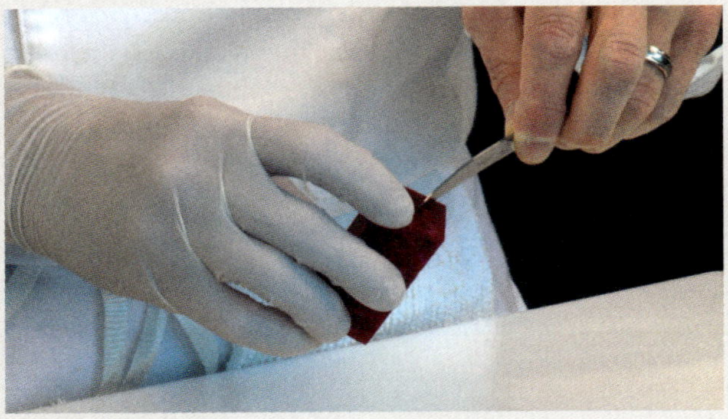

1. Cut a washed and peeled root vegetable, such as a beet or a carrot, into 2 inch long pieces. *Note:* Wider vegetables, such as potatoes, may need to be cut into 4–6 sections before being cut into 2 inch long pieces.
2. Cut each 2 inch piece into 1 inch widths.

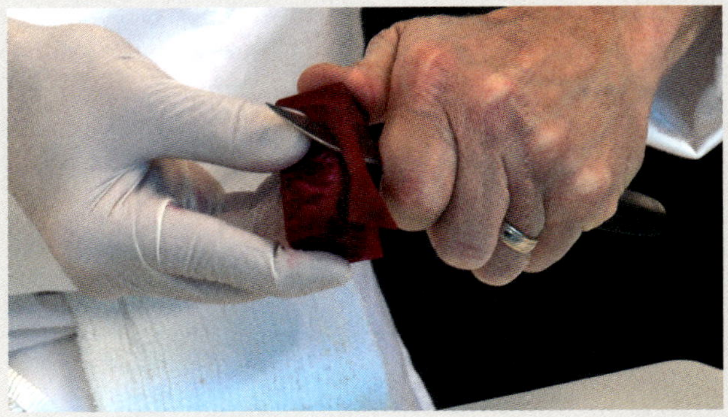

3. Holding the item in the guiding hand, place the index finger on one end and the thumb on the opposite end.
4. Using a tourné knife, slowly carve the item from one end to the other in a smooth, continuous stroke that creates a slightly rounded surface.

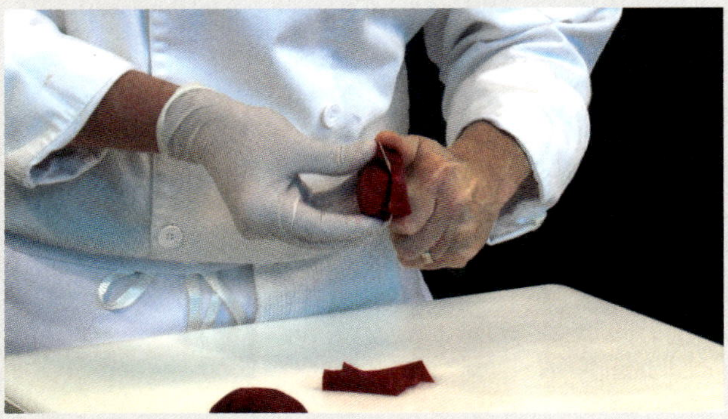

5. Turn the vegetable slightly and continue to carve the item in smooth, curved strokes until a seven-sided football shape is achieved. *Note:* As each slice is carved, the ends will become narrower than the middle.

A finished tourné has seven sides and is flat on each end.

CHECKPOINT 3-6

1. Explain why uniform knife cuts are used in the professional kitchen.
2. Describe the difference between a diagonal cut and an oblique cut.
3. Slice a root vegetable into rondelles, diagonals, and obliques.
4. Chiffonade a bunch of fresh greens.
5. Cut a root vegetable into batonnet, julienne, and fine julienne stick cuts.
6. Cut a whole fruit into large dice, medium dice, and small dice.
7. Dice an onion.
8. Cut julienne sticks into brunoise cubes and fine julienne sticks into fine brunoise cubes.
9. Differentiate between chopping and mincing.
10. Mince a shallot.
11. Flute a button mushroom.
12. Tourné a turnip, a beet, or a carrot.

Tournés
Media Clip

POTATO CLASSIFICATIONS

Potatoes are often referred to as starches in the professional kitchen. A *potato* is a round, oval, or elongated tuber that is the only edible part of the potato plant. The color of potato skin differs among varieties and can be brown, white, purple, red, or yellow. Depending on the variety, potato flesh can be creamy white, yellow-gold, or purple in color. A potato eaten with the skin on provides dietary fiber, vitamin C, vitamin B6, and potassium. Mealy potatoes, waxy potatoes, and new potatoes, as well as sweet potatoes and yams are used in the professional kitchen.

Mealy Potatoes

A *mealy potato* is a type of potato that is higher in starch and lower in moisture than other types of potatoes. Mealy potatoes are the preferred type of potato for baking, frying, mashing, puréeing, and making casseroles. After cooking, these potatoes become light and fluffy inside. Mealy potatoes include purple potatoes, russet potatoes, and white potatoes. **See Figure 3-65.**

Waxy Potatoes

A *waxy potato* is a type of potato with a thin skin and slightly waxy flesh that is lower in starch and higher in moisture than mealy potatoes. Compared to mealy potatoes, waxy potatoes stay much firmer in the center when fully cooked and also retain their shape better. Waxy potatoes can be roasted, sautéed, steamed, or simmered. Waxy potatoes are the common choice for cooking except when baking or deep-frying. Fingerling potatoes, red potatoes, and yellow potatoes are waxy potatoes. **See Figure 3-66.**

New Potatoes

A *new potato,* also known as an early crop potato, refers to any variety of potato that is harvested before the sugar is converted to starch. Because they are harvested so early, new potatoes are small and relatively uniform in size. **See Figure 3-67.** They have a thin, delicate skin and tender flesh. New potatoes hold their shape after cooking. They are often roasted, steamed, or simmered. Due to their high moisture content, new potatoes spoil more quickly than other types of potatoes.

Mealy Potatoes	
Purple Potatoes  *United States Potato Board*	A *purple potato,* also known as a blue potato, is a mealy potato with smooth, thin, bluishpurple skin and purple flesh. Purple potatoes have a fine texture and a mild, nutty flavor. They are typically roasted, baked, mashed, fried, or steamed. Purple potatoes can be used to make mashed potatoes and potato salad.
Russet Potatoes  *United States Potato Board*	A *russet potato* is a mealy potato with thin brown skin, an elongated shape, and shallow eyes. Russets have white flesh that is high in starch. They are commonly baked, mashed, or fried. Russet potatoes are sometimes referred to as bakers or baking potatoes. Common varieties of russets include Idaho and Burbank potatoes.
White Potatoes  *United States Potato Board*	A *white potato* is an oblong mealy potato with a thin, white or light-brown skin and tender, white flesh. White potatoes contain less starch than russet potatoes and can be prepared using almost any cooking method. They are often used to make scalloped and au gratin potatoes.

Figure 3-65. Mealy potatoes are higher in starch and lower in moisture than other types of potatoes.

Waxy Potatoes	
Fingerling Potatoes  *United States Potato Board*	A *fingerling potato* is a small, tapered, waxy potato with butter-colored flesh and tan, red, or purple skin. Popular varieties of fingerlings include Russian banana fingerlings and French fingerlings. Russian banana fingerlings have tan or yellow skin. French fingerlings have light-purple skin. Fingerlings can be roasted, baked, or steamed. They are often cooked whole or halved lengthwise and used to add visual interest to salads and entrées.
Red Potatoes  *Idaho Potato Commission*	A *red potato* is a round, waxy, red-skinned potato with white flesh. Red potatoes are often grilled, roasted, or simmered. Popular varieties of red potatoes include red bliss, chieftain, and Norland red potatoes.
Yellow Potatoes  *Idaho Potato Commission*	A *yellow potato* is an oval, waxy potato with thin, yellowish skin and flesh and pink eyes. The flesh has a buttery, nutty flavor and remains a yellowish color after cooking. Popular varieties include Yukon gold and yellow finn potatoes. Yellow potatoes are typically roasted, mashed, or simmered. Their sweet flavor pairs well with citrus, parsnips, and cauliflower.

Figure 3-66. Waxy potatoes have a slightly waxy flesh and are lower in starch and higher in moisture than mealy potatoes.

New Potatoes

United States Potato Board

Figure 3-67. Because they are harvested so early, new potatoes are small and relatively uniform in size.

Sweet Potatoes and Yams

Sweet potatoes and yams have a similar appearance but differ from each other. **See Figure 3-68.** A *sweet potato* is a tuber that grows on a vine and has a paper-thin skin and flesh that ranges in color from ivory to dark orange. Different varieties can be yellow, red, or brown in skin color with yellow to orange-red flesh. The edible skin is often removed before cooking. Sweet potatoes are often incorporated into breads and desserts. Sweet potatoes can be baked, roasted, sautéed, puréed, or fried. Baking or roasting caramelizes the sugars in sweet potatoes and releases a sweet flavor. Sweet potatoes are an excellent source of vitamin A and potassium.

Potatoes and Yams

Melissa's Produce

Figure 3-68. Sweet potatoes and yams have a similar outer appearance but differ from each other.

A *yam* is a large tuber that has thick, barklike skin and flesh that varies in color from ivory to purple. The flavor of a yam is somewhat dry and starchier than a sweet potato. Common varieties include tropical yams, garnet yams, and jewel yams. Yams are low in fat and a good source of carbohydrates, protein, vitamin A, and vitamin C.

Market Forms of Potatoes

Potatoes can be purchased fresh, frozen, canned, dehydrated, or as processed items such as potato chips. **See Figure 3-69.** Various market forms of potatoes can be used to reduce time and labor costs. For example, frozen precooked French fries, canned cooked whole potatoes, and dehydrated mashed potatoes are often used in the professional kitchen.

Market Forms of Potatoes

United States Department of Agriculture

Figure 3-69. Market forms of potatoes include processed, frozen, dehydrated, and fresh potatoes. Potatoes are also available canned.

Storing Fresh Potatoes

Fresh potatoes are generally sold in 50 lb cases and are packed according to the size of the potato. They may be packaged as 80 count, 90 count, or 100 count. Potatoes must be inspected upon receipt to ensure they are firm and undamaged and show no signs of sprouting or green patches. Potatoes that are purchased cleaned have a shorter storage life because cleaning removes a protective outer coating that deters bacteria. USDA grading of potatoes is voluntary. U.S. No. 1, U.S. Commercial, and U.S. No. 2 grade potatoes are used in the professional kitchen.

Fresh potatoes must be kept in a dry, cool, dark place that allows them to breathe. Fresh potatoes are best stored in cardboard boxes or mesh bags to allow adequate ventilation. **See Figure 3-70.** The temperature of the storage area should be between 45°F and 55°F. Higher temperatures will shorten the lifespan of the potatoes because potatoes begin to sprout and dehydrate under higher temperatures. Refrigeration is not recommended, as colder temperatures increase the conversion of starch to sugar.

Potato Packaging

Idaho Potato Commission

Figure 3-70. Potatoes are best stored in cardboard boxes or mesh bags to allow adequate ventilation.

Production Tip

Potatoes that are exposed to too much light will turn green and begin to sprout. If potatoes do not have adequate ventilation, they quickly rot.

CHECKPOINT 3-7

1. Describe the four major classifications of potatoes.
2. Identify five market forms of potatoes.
3. Describe the guidelines for receiving and storing potatoes.

COOKING POTATOES

Potatoes can be cooked using a variety of cooking methods and are served for breakfast, lunch, and dinner. Potatoes are often added to soups, braised dishes, and stewed dishes for additional texture, flavor, and nutrients. Potatoes also help thicken sauces by releasing starch during the cooking process. A shorter cooking time requires smaller pieces of potatoes than a dish that cooks for a longer period. If a dish with a longer cooking time calls for small-dice potatoes, the potatoes should be added late in the cooking process to avoid overcooking. Potatoes can be grilled, roasted, baked, sautéed, fried, or simmered.

Production Tip

Uncooked potatoes begin to oxidize immediately when exposed to air, causing them to discolor. To prevent discoloration, potatoes should be placed in cold water as soon as they are peeled or cut.

Grilling Potatoes

Grilled potatoes have a golden color and smoky flavor. **See Figure 3-71.** When grilling potatoes, they are sliced into ¼ inch thick slices or ½ inch wedges, coated with oil and seasonings, and then placed on a preheated grill. Grilled potatoes should be cooked al dente. Overcooking potatoes makes them soft and difficult to remove from the grill without breaking.

Grilled Potatoes

United States Potato Board

Figure 3-71. Grilled potatoes have a golden color and smoky flavor.

Roasting Potatoes

Potatoes can be roasted or baked. Roasted potatoes are usually cut up, tossed with oil and other seasonings, and then cooked in an oven. Waxy potatoes, such as red, yellow, and fingerling potatoes, can be tossed lightly in oil, seasoned, and then roasted until golden brown on the outside and soft and creamy on the inside. **See Figure 3-72.** Mealy potatoes can also be roasted in this way, but will have a slightly tough exterior. Roasted potatoes are done when a fork can be inserted into the potato with little resistance. Roasted potatoes are typically cooked between 350°F and 400°F for 20 – 45 minutes.

Roasted Potatoes

United States Potato Board

Figure 3-72. Waxy potatoes can be tossed lightly in oil, seasoned, and then roasted until golden brown.

Cooking Potatoes

Baking Potatoes

Baked potatoes are cooked whole. Russet potatoes make the best baked potatoes. When baked, these low-moisture potatoes produce fluffy and light flesh that readily absorbs butter, sour cream, and a wide variety of other toppings. **See Figure 3-73.** Baked potatoes are typically cooked between 350°F and 400°F until fork tender.

Baked potatoes should be served as quickly as possible because their quality deteriorates if they are held too long. They can be baked directly on the oven rack or can be placed on sheet pans. The skin of a potato is washed thoroughly and pricked with a fork prior to baking. The tiny holes left by the fork allow moisture to escape during the baking process.

Baked Potatoes

United States Potato Board

Plain Baked Potato

Idaho Potato Commission

Toppings

Figure 3-73. When baked, low-moisture potatoes produce fluffy and light flesh that readily absorbs butter, sour cream, and a wide variety of other toppings.

Potato casseroles are also baked. Mealy potatoes are the best choice for potato casseroles because they are low in moisture and can absorb the most flavor from a sauce. A potato casserole, such as scalloped potatoes or potatoes au gratin, can be baked with or without toppings, such as bread crumbs or cheese. **See Figure 3-74.** *Gratinée* is the process of topping a dish with a thick sauce, cheese, or bread crumbs and then browning it in a broiler or high-temperature oven. *Gratin* is any dish prepared using the gratinée method.

Potatoes used in casseroles may be sliced raw or partially cooked prior to slicing. The sauce used for a casserole can range from a cream sauce to a stock. Using parcooked (partially cooked) potatoes or a warmed sauce will decrease the required baking time.

Potato Casseroles

Scalloped

Au Gratin

Basic American Foods

Figure 3-74. A potato casserole, such as scalloped potatoes or potatoes au gratin, can be baked with or without toppings, such as bread crumbs or cheese.

Procedure for Preparing Potato Casseroles

1. Wash and peel the potatoes. Place them in cold water to avoid browning.

2. Slice potatoes to a uniform thickness. If precooked potatoes are desired, steam or simmer them until they are half done. Drain the cooking liquid from the potatoes and allow them to cool slightly.

3. Lay a single layer of the potatoes in a casserole dish.

4. Cover the potatoes with a warm, well-seasoned sauce.

5. Add another layer of potatoes and cover them with sauce. Continue this process until the potatoes and sauce are used.

6. Top with cheese or bread crumbs and cover the pan with foil.
7. Bake at 350°F until potatoes are almost tender. Uncover and allow the potatoes to brown.

Sautéing Potatoes

Waxy potatoes are best for sautéing. Large waxy potatoes are sliced for sautéing and small potatoes are usually cut in half. Sautéed potatoes are typically steamed or boiled first and then finished in a sauté pan to give them a golden-brown exterior. Potatoes that are steamed or boiled first should be removed from the cooking liquid and allowed to steam dry before they are sautéed. Never rinse steamed potatoes to cool them because they absorb even more water and will not brown properly.

Preparing Potato Casseroles
Media Clip

Frying Potatoes

Deep-fried potatoes, such as French fries and potato chips, are a popular side. Fried potatoes can be prepared in a variety of shapes including waffle-cut fries, shoestring fries, steak fries, cottage fries, and hash browns. **See Figure 3-75.** Low-moisture, mealy potatoes, such as russets, are the best choice for deep-frying. They crisp nicely on the outside and remain light and fluffy on the inside. Low-moisture potatoes also spatter less and stay crisp longer.

Fried Potatoes

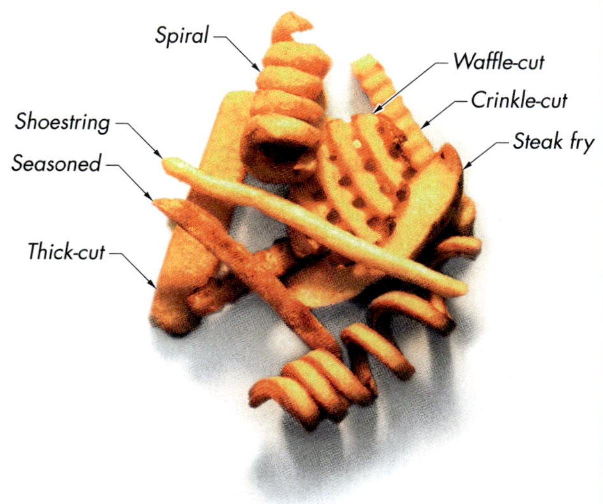

United States Potato Board

Figure 3-75. Fried potatoes can be prepared in a variety of shapes including waffle-cut fries, shoestring fries, steak fries, cottage fries, and hash browns.

Thinly cut potatoes, such as potato chips, can be deep-fried in a single step, usually in hot oil between 350°F and 375°F. French fries need to be blanched and cooled before being deep-fried. Blanching potatoes in 275°F oil removes some of the moisture, causing them to cook more evenly when they are deep-fried. Potatoes that are cooked in 375°F oil without being blanched first instantly sear, preventing their interior moisture from escaping. The trapped internal moisture causes the potatoes to become limp and soggy soon after frying.

Simmering Potatoes

Any type of potato can be simmered with good results. If the simmered potatoes are to be mashed, mealy potatoes are the best choice. Waxy potatoes are used when preparing whole, quartered, or sliced potatoes or potato salad because they retain their shape and do not fall apart as easily. **See Figure 3-76.** Herbs and aromatics can add subtle flavors to simmered potatoes. To add even more flavor when simmering potatoes, stocks or juices can be used as the cooking liquid instead of water.

Simmered Potatoes

United States Potato Board

Mashed

Idaho Potato Commission

Sliced

Figure 3-76. If simmered potatoes are to be mashed, mealy potatoes are best. Waxy potatoes are used for whole, quartered, or sliced potatoes and potato salad because they do not fall apart as easily.

If simmered potatoes are to be served whole, they should be placed on a sheet pan in a single layer to dry. Simmered potatoes have sufficiently dried and are ready to garnish and serve when steam no longer rises from the potatoes. Simmered potatoes are typically seasoned with butter and chopped fresh parsley.

Simmered potatoes can also be puréed before service. To purée simmered potatoes, the drained potatoes are passed through a ricer or food mill until a fine, smooth texture is achieved. Milk, butter, roasted garlic, fresh herbs, grated cheeses, and pesto can be added to simmered potatoes.

Determining Doneness of Potatoes

Fully cooked potatoes are fluffy and have little resistance when pierced by a fork. If a potato is hard in the center, it needs to be cooked longer. Roasted and fried potatoes should be crunchy on the outside and soft on the inside. Cutting open these types of cooked potatoes is the only way to determine if they are done. Lightly squeezing a baked potato will indicate whether it is done. The skin of a fully roasted potato will look slightly shriveled, and the skin of a simmered potato will begin to separate from the flesh when the potato is done.

Plating Potatoes

The potato is not always presented as elegantly as it could be. Too many times cooks only offer baked, mashed, or fried potatoes. The potato of today is a very versatile starch. The addition of blue cheese and fresh chives to whipped sweet potatoes adds variety, flavor, and contrast to the plate. Russet potatoes and sweet potatoes can be boiled separately and then piped simultaneously to create a swirl of color on the plate. Yukon gold potatoes offer a deep creamy color with great flavor. Mixing new potatoes and Peruvian purple potatoes can add intrigue and loads of flavor. Shape is another important element to keep in mind when choosing potatoes, such as fingerlings, to complement a protein or another vegetable on the plate. **See Figure 3-77.** Texture variations can be achieved by grilling, baking, roasting, sautéing, frying, or simmering potatoes.

CHECKPOINT 3-8

1. Describe six different methods of cooking potatoes.
2. Grill potatoes and evaluate the results.
3. Roast potatoes and evaluate the results.
4. Bake potatoes and evaluate the results.
5. Sauté potatoes and evaluate the results.
6. Fry potatoes and evaluate the results.
7. Simmer potatoes and evaluate the results.
8. Explain how to determine the doneness of potatoes.

United States Potato Board

Plating Potatoes

Idaho Potato Commission

French Fries

United States Potato Board

Steak Fries

Figure 3-77. Shape is an important element to keep in mind when choosing potatoes as a complement.

GRAIN CLASSIFICATIONS

A *grain* is the edible fruit, in the form of a kernel or seed, of a grass. The name of the grass is often the same name given to the grain from that plant. For instance, wheat grass produces the grain called wheat. Grains are often referred to as starches. Common grains include rice, wheat, corn, and barley, but there are many other types of edible grains used in the professional kitchen.

Grain Composition

A kernel of grain is composed of a husk, bran, endosperm, and germ. **See Figure 3-78.** The *husk,* also known as the hull, is the inedible, protective outer covering of grain. The *bran* is the tough outer layer of grain that covers the endosperm. While it is often removed during processing, bran provides necessary fiber, complex carbohydrates, vitamins, and minerals. Studies indicate that bran can help lower cholesterol.

Grain Composition

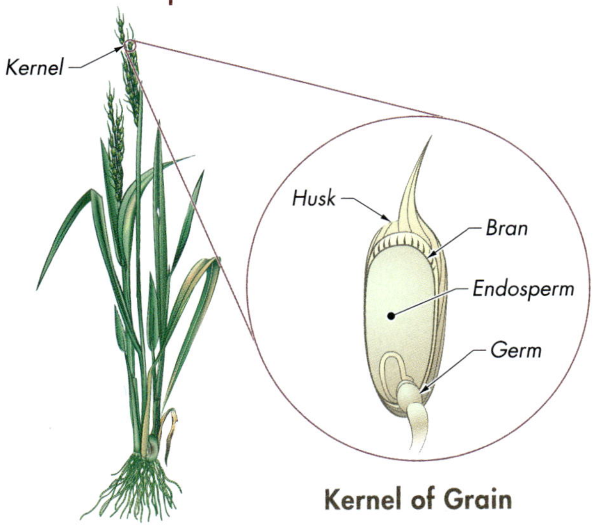

Figure 3-78. A kernel of grain is composed of a husk, bran, endosperm, and germ.

Grains

The *endosperm* is the largest component of a grain kernel and consists of carbohydrates and a small amount of protein. It is milled to produce flours and other grain products. Endosperm consists of carbohydrates (in the form of starches) and a small amount of protein. The *germ* is the smallest part of a grain kernel and contains a small amount of natural oils as well as vitamins and minerals.

Grains come in many shapes and sizes. Most grains are hard and all have a husk that is indigestible in its natural form. It is necessary to process grains to some extent to make them easier to digest. However, the amount of processing directly affects the nutritional values of the grain.

Whole Grains and Refined Grains

A *whole grain* is a grain that only has the husk removed. Because the rest of the grain is intact, whole grains require more time to cook than processed grains. Examples of whole grains include brown rice, wild rice, wheat berries, bulgur, oats, barley, rye, quinoa, millet, and spelt. A *cracked grain* is a whole kernel of grain that has been cracked by being placed between rollers. Bulgur wheat is a cracked grain. Some whole grains are parcooked to reduce the required amount of cooking time.

A *refined grain* is a grain that has been processed to remove the germ, bran, or both. **See Figure 3-79.** Refined grains are easier to digest, have a longer shelf life, can have their color changed, and take less time to cook. Processing often goes beyond removing the husk and can include removing the nutrient-rich bran and germ. The bran is removed from common all-purpose flour to make the product white in color instead of brown. Removing the germ can help preserve the product so that it does not spoil quickly. However, grains lose vitamins, minerals, and fiber during processing.

Refined Grains

United States Department of Agriculture

Figure 3-79. Refined grains have been processed to remove the germ, bran, or both.

Refined grains have been milled, pearled, or flaked. A *milled grain* is a refined grain that has been ground into a fine meal or powder. Meal, such as cornmeal, and all varieties of flour are milled grains. A *pearled grain* is a refined grain with a pearl-like appearance that results from having been scrubbed and tumbled to remove the bran. A *flaked grain,* also known as a rolled grain, is a refined grain that has been rolled to produce a flake. Oatmeal is a flaked grain.

Rice

Rice is a staple food source for two-thirds of the world's population. There are over 40,000 varieties of rice. Rice can be cooked whole or made into flour, noodles, paper, milk, wine, vinegar, and a wide variety of processed foods. Rice is a complex carbohydrate that contains only a trace amount of fat. Rice is cholesterol, sodium, and gluten free and a rich source of thiamine, niacin, iron, phosphorus, and potassium. The high starch content of rice allows it to absorb flavors from ingredients that are cooked with it.

USA Rice Federation

All varieties of rice are milled whole and then rolled to remove the husk. Many varieties are sold as either brown or white rice, depending upon how they were milled. *Brown rice* is rice that has had only the husk removed. Brown rice is chewier and nuttier in flavor than white rice and contains more vitamins and fiber. White rice has had the husk, bran, and germ removed and is more delicate than brown rice. Regardless of its color, rice is classified by the size of the grain. Grain sizes of rice include short-grain, medium-grain, and long-grain. **See Figure 3-80.** The volume of rice triples when it is cooked.

Rice	
Short-Grain Rice  Indian Harvest Specialtifoods, Inc./Rob Yuretich	*Short-grain rice* is rice that is almost round in shape and has moist grains that stick together when cooked. It is high in starch and commonly used to make risotto and rice pudding. Arborio rice is a short-grain rice used to make risotto, pilaf, and paella. Sweet rice, also known as sticky rice, is a short-grain rice used to make desserts. It is not used in sushi. Sweet rice is soaked for several hours and then steamed. It is sometimes ground into flour to make dumplings, pastries, or rice puddings.
Medium-Grain Rice  Indian Harvest Specialtifoods, Inc./Rob Yuretich	*Medium-grain rice* is a rice that contains slightly less starch than short-grain rice but is still glossy and slightly sticky when cooked. It has a similar appearance to short-grain varieties, with the grains being only slightly longer and plumper. Medium-grain rice is widely used to make sushi. Forbidden rice, also known as emperor's rice, is a black medium-grain rice that turns indigo when cooked. Thai sticky rice is a medium-grain rice that turns a deep-purple color when cooked. Each grain is purple and brown interspersed with flecks of white.
Long-Grain Rice  Indian Harvest Specialtifoods, Inc./Rob Yuretich	*Long-grain rice* is a rice that is long and slender and remains light and fluffy after cooking. *Basmati rice* is a long-grain rice that only expands lengthwise when it is cooked. It has a sweet, nutty flavor and is high in fiber. The dry nature of basmati makes it pair well with sauces. Jasmine rice is a long-grain rice that releases a floral aroma when cooked. Himalayan red rice has a rose-colored bran, firm texture, and nutty flavor. Wild rice is not true rice, but its long, black grains are high in nutrients and have a rich, nutty flavor. Wild rice comes from an aquatic grass and is prepared in the same manner as rice.

Figure 3-80. Regardless of its color, rice is classified by the size of the grain.

Rice

Corn

Corn, also known as maize, is a cereal grain cultivated from an annual grass that bears kernels on large woody cobs called ears. **See Figure 3-81.** It is prepared as a grain and as a vegetable. Tiny ears of immature corn, also known as baby corn, are often added to stir-fries and salads or eaten whole. Corn is used to make cornstarch, corn oil, corn syrup, and whiskey. Corn is also used to make cornmeal, grits, and hominy.

- Cornmeal is coarsely ground corn. It is commonly used to make cornbread and as a coating for fried foods.
- *Grits* are a coarse type of meal made from ground corn or hominy. Grits are traditionally served with butter, salt, and pepper. Cheese, bacon, or hot sauce may also be added.
- *Hominy* is the hulled kernels of corn that have been stripped of their bran and germ and then dried. White hominy is made from white corn kernels, and yellow hominy is made from yellow corn kernels. Hominy can be ground into flour and used to make tortillas or tamales. Hominy is boiled whole or ground into grits. Hominy is also used to make posole.

Corn

United States Department of Agriculture

Figure 3-81. Corn, also known as maize, is a cereal grain cultivated from an annual grass that bears kernels on large woody cobs called ears.

Wheat

Wheat is a light yellow cereal grain cultivated from an annual grass that yields the flour used in many pastas and baked goods. The most common form of wheat used in cooking is milled flour. Durum wheat, bulgur wheat, and wheat berries are derivatives of wheat that are frequently used in the professional kitchen. **See Figure 3-82.**

Durum Wheat By-Products

Moroccan Couscous

Wheat Berries

Indian Harvest Specialtifoods, Inc./Rob Yuretich

Figure 3-82. Derivatives of wheat are often used in the professional kitchen.

Durum Wheat. *Durum wheat* is the hardest type of wheat. Its high protein content and gluten strength make durum the wheat of choice for making pasta dough. Durum kernels are amber colored and larger than other wheat kernels. Its yellow endosperm gives pasta its golden hue. Semolina and couscous are made from durum wheat.

- *Semolina* is the granular product that results from milling the endosperm of durum wheat. When semolina is mixed with water it forms a stiff dough that can be forced through metal dies to create different pasta shapes.
- *Couscous* is a tiny, round pellet made from durum wheat that has had both the bran and germ removed. It is very fine in texture and similar in size to cornmeal. Israeli couscous is a larger variety.

Bulgur Wheat. Bulgur wheat is golden-brown, nutty-tasting wheat kernels. The husks and bran are removed and it is steamed, dried, and ground. Bulgur comes whole or is cracked into fine, medium, or coarse grains. Bulgur wheat is commonly simmered and seasoned with herbs, spices, and vegetables and cooks in less time than wheat berries. Tabbouleh is made from cooked, chilled bulgur wheat mixed with mint, lemon, olive oil, and parsley.

Wheat Berries. A wheat berry is a chewy wheat kernel with only the husk removed. It contains both the bran and germ. Wheat berries are simmered using a procedure similar to the procedure used for rice. After simmering for an extended period, they can be finished by sautéing them in a pan with various herbs and spices.

Other Grain Varieties

In addition to rice, corn, and wheat there are many other grains used in the professional kitchen. Barley, buckwheat, farro, millet, oats, quinoa, rye, and spelt are only a few of the grains that can be used to add texture, flavor, contrast, and aroma to many dishes. **See Figure 3-83.**

Grains	
Barley  *Canada Beef Inc.*	*Barley* is an ancient grain that resembles brown rice in shape, yet takes longer to cook. Barley contains high levels of soluble fiber and has a chewy texture. Pearled barley is polished barley with the bran removed. It is often used in salads and pilafs. Barley is often added to soups and stews for its earthy flavor and because it is a natural thickener.
Buckwheat 	*Buckwheat* is a dark, three-cornered seed of a plant unrelated to wheat that has a nutty, earthy flavor. Buckwheat is commonly ground into a gritty flour and used to make everything from pancakes to soba noodles. Buckwheat flour lacks gluten and is loaded with nutrients. A *buckwheat groat* is a crushed, coarse piece of whole-grain buckwheat that can be prepared like rice. *Kasha* is roasted buckwheat.
Farro  *Indian Harvest Specialtifoods, Inc./Rob Yuretich*	*Farro* is a hearty ancient grain that tastes similar to wheat, yet resembles brown rice. It is native to Italy. Farro is low in gluten and high in fiber and vitamins. Farro can be purchased as a whole grain or flour. Like barley, farro can be used to make risottos and is often pearled.
Millet 	*Millet* is a small, round, butter-colored grain that is gluten-free. It is high in iron and B vitamins. A native of Asia, millet resembles couscous, yet is prepared like rice. Millet is a quick-cooking grain with a mild flavor that can be toasted to bring forth a nuttier flavor and deep-yellow and light-brown coloring. Millet pairs nicely with chives, green onions, and garlic and is often used in salads, casseroles, and stuffing.
Oats 	*Oats* are grains derived from the berries of oat grass. Oats can be purchased in many different forms. An *oat groat* is an oat grain that only has the husk removed. *Steel-cut oats* are oat groats that have been toasted and cut into small pieces. *Rolled oats,* also known as old-fashioned oats, are oats that have been steamed and flattened into small flakes. Rolled oats require less cooking time than steel-cut oats. Oats are packed with cholesterol-fighting soluble fiber and are often served as a hot cereal or used to make breads and desserts.

Figure 3-83. (continued on next page)

Grains	
Quinoa  *Indian Harvest Specialtifoods, Inc./Rob Yuretich*	*Quinoa* is a small, round, gluten-free grain that is classified as a complete protein. Quinoa is one of the oldest known grains and is available in ivory, red, pink, brown, and black varieties. Quinoa contains fiber, protein, vitamins, and minerals. It is a rich source of the amino acid lysine, which promotes tissue growth and repair and supports the immune system. Quinoa must be rinsed before cooking to remove its bitter coating. It cooks quickly, has a mild flavor, and has a slightly crunchy texture. Quinoa is often used in salads, stuffing, quick breads, and as a side dish.
Rye 	*Rye* is a hearty grain with dark-brown kernels that are longer and thinner than wheat. Rye has a distinctive flavor. Rye is used to make flour, breads, crackers, and whiskey. Rye flour is heavier and darker in color than wheat flours. Soaked and cooked rye berries are sometimes added to breads for extra texture or used to make pilafs or hot breakfast cereals. Triticale is a hybrid grain made by crossing rye and wheat. It has a sweet, nutty flavor and contains more protein and less gluten than wheat. Like rye, triticale makes heavy, hearty loaves of bread. Triticale is sold as whole berries, flaked, and flour.
Spelt 	*Spelt* is an ancient grain with a nutty flavor and high protein content that is also a good source of riboflavin, zinc, and dietary fiber. Spelt is commonly mistaken for farro because of their similar appearance. Spelt can be used as a hot breakfast cereal or as a substitute for wheat in many dishes. Though it contains gluten, spelt is often tolerated by people with wheat allergies.

Figure 3-83. Barley, buckwheat, farro, millet, oats, quinoa, rye, and spelt can be used to add texture, flavor, contrast, and aroma to many dishes.

Nutrition Note

Grains do not contain any saturated fat or cholesterol. They are a rich source of protein, but the type of protein found in most grains lacks some of the essential amino acids. Carbohydrates account for 65–90% of the calorie content of grains. The bran and germ of whole grains are loaded with vitamins, minerals, and fiber.

Storing Grains

Grains should be stored in a cool, dry place in an airtight container to keep out moisture and prevent insects from getting in the product. **See Figure 3-84.** Some grains can absorb strong aromas, so they should be stored away from foods such as garlic and onions. Grains that still contain germ should be used quickly or kept in a refrigerator or freezer to prevent the germ from becoming rancid. Milled grains will keep indefinitely in sealed containers that are stored in a cool, dry place. Brown rice will last for several months.

Storing Grains

Carlisle FoodService Products

Figure 3-84. Grains should be stored in a cool, dry place in an airtight container to keep out moisture and prevent insects from getting into the product.

Shaped Pastas	
Campanelle  *Barilla America, Inc.*	Campanelle, also known as gigli, are shaped pasta that have been rolled into a fluted-cone shape.
Conchiglie (Large Shells)  *Barilla America, Inc.*	Conchiglie, also known as large shells, are shell-shaped pasta.
Farfalle  *Barilla America, Inc.*	Farfalle, also known as bow tie pasta, are flat squares of shaped pasta that are pinched in the center in the shape of bow ties.
Fiori 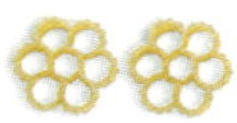 *Barilla America, Inc.*	Fiori are flower-shaped pasta.
Fusilli  *Barilla America, Inc.*	Fusilli are corkscrew-shaped pasta.
Gemelli  *Barilla America, Inc.*	Gemelli is formed by twisting two strands of pasta together.
Jumbo Shells  *Barilla America, Inc.*	Jumbo shells are large, shell-shaped pasta.
Orecchiette  *Barilla America, Inc.*	Orecchiette are small, ridged, bowl-shaped pasta.
Orzo  *Barilla America, Inc.*	Orzo are small oval-shaped pasta with an appearance similar to that of a grain.

Figure 3-90. (continued on next page)

Pastas

Shaped Pastas	
Radiatori  Barilla America, Inc.	Radiatori are deeply-ridged, curly-shaped pasta.
Rotini 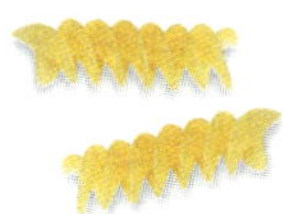 Barilla America, Inc.	Rotini are 2 inch long, twist-shaped pasta.
Stelline  Barilla America, Inc.	Stelline are tiny, star-shaped pasta.

Figure 3-90. Shaped pastas are made into complex shapes, such as corkscrews, shells, and stars, using an extruder.

Tube Pastas

A *tube pasta* is a pasta that has been pushed through an extruder and then fed through a cutter that cuts the tubes to desired length. **See Figure 3-91.** Tube pastas are often stuffed with cheese or meat and are often used in casseroles. Short tubes are often used in soups. Cellentani, ditalini, elbows, manicotti, penne, pipettes, rigatoni, and ziti are common types of tube pasta.

Photo Courtesy of Perdue Foodservice, Perdue Farms Incorporated

Tube Pastas	
Cellentani  Barilla America, Inc.	Cellentani, also known as cavatappi, are twisted, hollow tubes of pasta with a ribbed surface.
Ditalini  Barilla America, Inc.	Ditalini are short, hollow tubes of pasta with a smooth or ridged surface.
Elbows 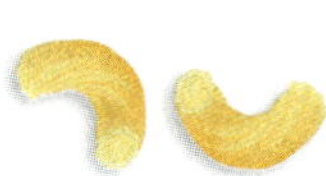 Barilla America, Inc.	Elbows are relatively short, slightly curved, hollow tubes of pasta.
Manicotti  Barilla America, Inc.	Manicotti, also known as cannelloni, are large, round tubes of pasta approximately 4 inches long and 1 inch in diameter. They can be straight cut or diagonal cut.

Figure 3-91. (continued on next page)

Tube Pastas	
Penne  *Barilla America, Inc.*	Penne are hollow, diagonally cut tubes of pasta approximately 1½–2 inches in length with a smooth or ribbed surface.
Pipettes  *Barilla America, Inc.*	Pipettes are curved, hollow, ridged tubes of pasta with one open end and one pinched end.
Rigatoni  *Barilla America, Inc.*	Rigatoni are wide, hollow, ridged tubes of pasta.
Ziti  *Barilla America, Inc.*	Ziti are straight, round tubes of pasta approximately ¼ inch in diameter of various lengths with a smooth or ribbed surface.

Figure 3-91. Tube pastas are pushed through a tube-shaped pasta extruder and then fed through a tube cutter that cuts the tubes to desired length.

Ribbon Pastas

A *ribbon pasta* is a thin, round strand or flat, ribbonlike strand of pasta. **See Figure 3-92.** To form ribbon pasta, the dough is rolled out and cut to the desired width by hand or using a pasta machine. Round strands of pasta are formed by passing the dough through small, round dies. Ribbon pastas are often dressed in sauces. Common types of ribbon pasta include capellini, egg noodles, fettuccine, lasagna, linguine, and spaghetti.

Ribbon Pastas	
Capellini *Barilla America, Inc.*	Capellini, also known as angel hair, are very fine, round, strandlike ribbons of pasta approximately 1⁄64 inch in diameter.
Egg Noodles 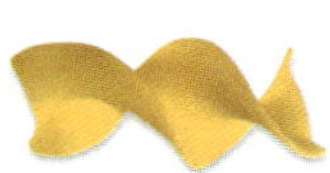 *Barilla America, Inc.*	Egg noodles are flat ribbons of pasta that can be cut long or short and thin, medium, or wide. To be labeled egg noodles, the pasta must contain at least 5.5% egg solids.
Fettuccine  *Barilla America, Inc.*	Fettuccine are long, thin, flat strips of ribbon pasta approximately ¼ inch wide.

Figure 3-92. (continued on next page)

Ribbon Pastas	
Lasagna  Barilla America, Inc.	Lasagna are flat, ripple-edged ribbons of pasta, approximately 2 inches wide.
Linguine  Barilla America, Inc.	Linguine are long, thin, flat strips of ribbon pasta, about 1/8 inch wide.
Spaghetti  Barilla America, Inc.	Spaghetti are long, round rods of ribbon pasta approximately 3/32 inch in diameter. Very thin strands of spaghetti are known as spaghettini.

Figure 3-92. Ribbon pasta is a thin, round strand or a flat, ribbonlike strand of pasta.

Stuffed Pastas

A *stuffed pasta* is a pasta that has been formed by hand or machine to hold fillings. **See Figure 3-93.** Individual portions of filling are added to a sheet of pasta and then a wash is applied around each filling. Another sheet of pasta is laid over the first sheet and the mounds of filling, the two sheets are sealed with the mounds of filling inside, and then the pasta is cut into individual portions. Stuffed pastas can be filled with savory and sweet cheeses, puréed meats, poultry, seafood, or vegetables, or a combination of ingredients made into a paste. Common types of stuffed pasta include ravioli, tortellini, and tortelloni.

Stuffed Pastas	
Ravioli  Barilla America, Inc.	Ravioli are stuffed pasta formed from equal size squares, or other shapes, of flat pasta and are filled.
Tortellini  Barilla America, Inc.	Tortellini are stuffed pasta formed by wrapping filled half circles of dough around a finger and pressing the ends together.
Tortelloni  Barilla America, Inc.	Tortelloni are stuffed pasta that resemble pot stickers, are larger than tortellini, and will hold more filling than tortellini.

Figure 3-93. Ravioli, tortellini, and tortelloni are stuffed pastas.

Asian Noodles

Some Asian noodles are wheat-based, yet many are made from vegetable starch, eggs, rice flour, or buckwheat flour. **See Figure 3-94.** Asian noodles are soaked briefly in hot water rather than cooked. Asian noodles may also be steamed, stir-fried, and deep-fried. Some Asian noodles are served cold as well as hot.

Asian Noodles

Asian Noodles	
Cellophane Noodles 	Cellophane noodles, also known as glass noodles, are made from mung bean starch. They are fairly transparent and are commonly rehydrated in hot water. Cellophane noodles can be deep-fried to produce a crispy and crunchy delicate noodle. If deep-frying, they do not need to be presoaked. Cellophane noodles are used hot or cold, in soups or stir-fries, or as fillings, especially in vegetarian dishes. Similar noodles are available made from sweet potato starch.
Hokkien Noodles 	Hokkien noodles are yellow egg noodles that resemble spaghetti. They are sold fresh, dried, or vacuum-sealed. Hokkien noodles work well in soups or stir-fries or served with a sauce.
La Mian Noodles 	La mian noodles are Chinese wheat noodles that are sold in round strands or ribbons of various widths. They are available fresh or dried. La mian noodles are popular in a wide range of dishes including soups, stir-fries, and chow mein. La mian noodles that have been deep-fried and then dehydrated are sold as ramen noodles.
Lo Mein Noodles 	Lo mein noodles, also known as Cantonese noodles, are egg noodles in varying widths that can be purchased fresh, dried, or frozen. Lo mein is a popular stir-fry dish named for the noodle itself.

Figure 3-94. (continued on next page)

Asian Noodles	
Rice Noodles 	Rice noodles are made from ground rice flour and are available dried in various shapes, including sheets, flat sticks of various widths, or vermicelli. They should be soaked in hot water before they are cooked and thoroughly rinsed after cooking to remove excess starch. Thin rice noodles can be deep-fried to produce a crispy, delicate noodle that can be used as a base for an entrée, in salads, or in desserts.
Shirataki Noodles 	Shirataki noodles are long, slender noodles made from Japanese yams. Shirataki noodles have no calories, carbohydrates, gluten, or fat. They have a distinct earthy odor before they are cooked.
Soba Noodles 	Soba noodles, also known as buckwheat noodles, are brown-gray noodles made from buckwheat flour. They can be used hot or cold, in soups, in main dishes, and as side dishes. Soba noodles can be substituted for spaghetti if desired.
Somen Noodles 	Somen noodles are long, thin wheat noodles that are white in color and are sold as bundled rods. Varieties of somen noodles include egg yolk and green tea. Somen noodles are used in stir-fries and soups or served cold in noodle dishes.
Udon Noodles 	Udon noodles are thick, white wheat noodles. They are available fresh or dried and as round or square strands. Udon noodles have a chewy, slippery texture and are served in soups, stews, or stir-fries, with meat dishes, or cold.

Figure 3-94. Some Asian noodles are wheat-based, yet many are made from eggs, rice flour, or buckwheat flour.

Storing Pasta

Pasta is often purchased ready to use. Purchased pasta should be placed in an airtight container and stored in a cool, dry place to keep moisture out and prevent insects from getting in the product.

CHECKPOINT 3-11

1. Identify three ways pasta can be purchased.
2. Describe shaped pastas.
3. Describe tube pastas.
4. Describe ribbon pastas.
5. Describe stuffed pastas.
6. Describe nine types of Asian noodles.

PREPARING AND COOKING PASTAS

Most pastas are boiled or baked. Pasta at least doubles in volume when cooked. For example, a pound of fresh pasta yields approximately 2–2½ lb of cooked pasta. Fresh pasta requires less cooking time than dried pasta. Cooking time varies depending on the shape, size, and quality of the pasta. **See Figure 3-95.** Pasta should always be cooked uncovered. After the minimum cooking time is reached, the pasta should be tested frequently to ensure the proper doneness. Fresh pasta cooks in approximately 3–5 minutes, while dried pasta cooks in approximately 3–13 minutes.

Approximate Cooking Times for Dried Pasta

Pasta	Minutes	Pasta	Minutes
Capellini	3–5	Spaghetti	9–10
Egg Noodles	7–8	Penne and Mostaccioli	9–11
Elbow Macaroni	7–8	Tortellini	10–11
Conchiglie	8–9	Farfalle	11–12
Lasagna	8–9	Fettuccini	12–13
Manicotti and Connelloni	8–9	Fusilli	12–13
Linguine	9–10	Jumbo shells	12–13
Orzo	9–10	Vermicelli	12–13

Figure 3-95. Cooking time varies depending on the shape, size, and quality of the pasta.

Preparing Pasta Doughs

Some chefs prefer to make their own pasta because fresh pasta is more tender and can be flavored and colored as desired. For example, squid ink can provide a salty, metallic flavor that will also color the pasta. Herbs, spices, and vegetables may also be added to enhance the flavor and appearance of pasta. Fresh pasta dough can be processed in a mixer or kneaded by hand.

Procedure for Preparing Pasta Dough

1. On a clean work surface, place flour in a mound and form a well in the center. Add eggs, oil, and salt to the well.
2. Slowly work the flour into the well with the egg mixture until all ingredients are mixed.

3. Knead the mixture until a smooth, dry ball of dough has formed. Cover the dough and allow it to rest for 20 minutes.

4. Roll the dough to the appropriate thickness and cut into desired shapes.

Preparing Pasta Dough
Media Clip

Preparing Ravioli. To make ravioli, one sheet of pasta is topped with a small amount of filling, and a second sheet of pasta is laid on top. The top sheet is pressed down around the filling, and individual ravioli are formed by cutting around the mounds of filling with a pastry wheel, ravioli cutter, or knife.

Procedure for Preparing Ravioli

1. Roll out a 12 inch, square sheet of pasta dough as thin as possible.

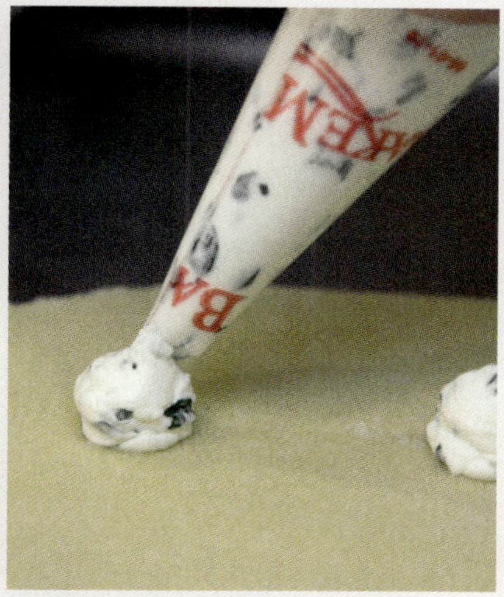

2. Use a pastry bag to deposit ¼ oz portions of filling on the dough, approximately 2 inches apart.

3. Use a pastry brush to egg wash the pasta around each portion of filling.

4. Prepare a second sheet of dough the same size as the first and place it on top of the first sheet.
5. Press down around each portion of filling to seal all edges of the two sheets of dough together.

6. Use a pastry wheel or knife to cut between the filled mounds.

7. Separate the ravioli and place on a sheet pan. Cook or cover and refrigerate or freeze for future use.

Preparing Ravioli
Media Clip

Preparing Tortellini. When making tortellini, thin pasta dough is cut into circles and a filling is placed in the center of each circle. The circle is then folded in half and the edges are pressed together to hold the filling in place. The half circle is then formed into a ring by wrapping it around a finger and pressing the two ends together.

Procedure for Preparing Tortellini

1. Roll out pasta dough as thin as possible.

2. Use a 2 inch cutter to cut the dough into rounds.

3. Place a small portion of filling in the center of each round.

4. Moisten the edge of each round with water or egg wash.

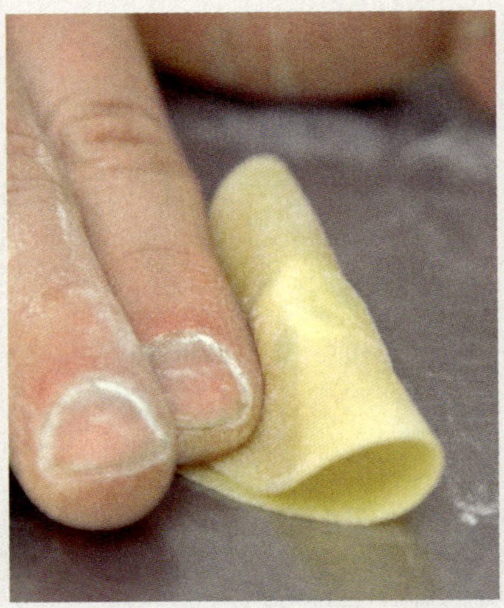

5. Fold each circle in half and then press the edges together until they are tightly sealed.

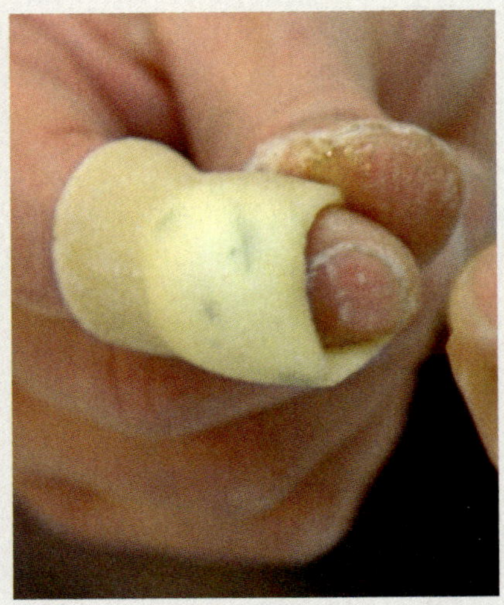

6. Slightly stretch the tips of each half circle to form a ring of dough around a finger. Press the tips together until they are tightly sealed. Cook or cover and refrigerate or freeze.

Determining Doneness of Pastas

Pasta should be cooked al dente, meaning "to the tooth." When pasta is cooked al dente, there should be a slight resistance in the center of the pasta when it is chewed. Pasta should not be undercooked and brittle nor overdone and mushy. For example, rigatoni tubes should not collapse and rotini should not unwind.

Plating Pastas

The ideal pasta plate has a wide, flat bottom and shallow sides that keeps the pasta warm and unruly noodles in their place while supporting the stacking of pasta for visual appeal. Pasta should only be tossed with enough sauce to coat, moisten, and flavor each piece. The pasta should not drown in the sauce. A pasta fork or spaghetti tongs are used to grab the pasta without crushing it and transfer it to the warmed plate. One to two generous tablespoons of sauce are then spooned over the pasta before it is topped with freshly grated cheese or a splash of extra virgin olive oil. **See Figure 3-96.** Pasta is often garnished with chopped parsley.

Plated Pasta

Barilla America, Inc.

Figure 3-96. Pasta is often topped with grated cheese.

Reheating Pastas

In the professional kitchen, pasta is often cooked in advance. When pasta is cooked in advance, it is rinsed in cold water to stop the cooking process, then it is tossed with a small amount of oil to prevent sticking, and stored. Parcooked pasta can be rinsed in warm water or dropped in simmering water to bring it to the temperature needed for service.

CHECKPOINT 3-12

1. Explain how pasta reacts to the cooking process.
2. Prepare fresh pasta dough and evaluate the results.
3. Prepare a shaped pasta dish and evaluate the results.
4. Prepare a tube pasta dish and evaluate the results.
5. Prepare a ribbon pasta dish and evaluate the results.
6. Prepare a stuffed pasta dish and evaluate the results.
7. Prepare a dish using vegetable noodles and evaluate the results.
8. Prepare a dish using egg noodles and evaluate the results.
9. Prepare a dish using rice noodles and evaluate the results.
10. Prepare a dish using buckwheat noodles and evaluate the results.
11. Prepare a dish using wheat noodles and evaluate the results.
12. Explain how to prepare ravioli.
13. Explain how to prepare tortellini.
14. Explain how to determine if pasta is cooked al dente.
15. Reheat parcooked pasta and evaluate the results.

Barilla America, Inc.

Refer to DVD for **Quick Quiz®** questions

SUSTAINABILITY CORNER

The restaurant business is one of the largest consumers of energy, with most of that energy being used by cooking equipment, food storage equipment, and dishwashers. The average professional kitchen uses five times more energy than the rest of the building. The cost of energy is a factor that affects the budget and profit margins of virtually all foodservice operations. Fortunately, there are a number of ways to provide high-quality food and service while still being energy efficient, conserving natural resources, and improving profits.

One simple yet effective strategy for reducing energy costs is to develop a power up/power down schedule. For example, broilers do not have a thermostatic control and can run at full power all day. However, according to the Food Service Technology Center, broilers operating 12 hours per day without an effective power up/power down schedule can add an extra $7344 per year to energy costs.

Also, lighting in a professional kitchen can be extensive, expensive, and heat producing, especially in an already hot kitchen. Traditional light bulbs use only 10% of their energy to produce light, with the remaining 90% burnt off as heat. Replacing traditional incandescent light bulbs with compact fluorescent light bulbs, or CFLs, is one of the easiest ways to shrink a power bill and stay cooler in the kitchen. Replacing 24-hour EXIT signs with CFLs and equipping the signs with off/on timers is one small change that can make a substantial difference.

A major strategy for saving energy and reducing costs is to invest in energy-efficient equipment and appliances. Using energy-efficient ovens, fryers, griddles, hot food holding cabinets, refrigerators, freezers, ice machines, and dishwashers in the professional kitchen can result in substantial savings. For example, an energy-rated commercial dishwasher can potentially reduce energy bills by $900 per year and water bills by $200 per year.

Another piece of kitchen equipment that is a huge consumer of energy is the ice machine. If a commercial ice machine has earned the ENERGY STAR® rating, savings can average 15% on energy and 10% on the amount of water used. Installing a timer and shifting ice production to nighttime off-peak hours will make the ice machine even more efficient.

Merrychef

The type of ovens used can also significantly impact energy use. Convection ovens conserve more energy than conventional ovens because they cook food faster at lower temperatures. Minimizing oven preheating times will also conserve energy. Fifteen minutes is generally enough time to preheat an oven. Efficient use of ovens also includes scheduling cooking times so the oven is fully loaded, keeping the oven door closed as much as possible, and turning ovens off when not in use.

Energy Efficient Kitchens

Vegetable, Starch & Pasta Station
Certification Exam Preparation Questions

1. ___ vegetables include beets, carrots, celeriac, jicamas, lotus roots, parsnips, radishes, rutabagas, salsify, and turnips.
   A. Leaf
   B. Stem
   C. Root
   D. Bulb

2. The USDA voluntary grading system for ___ vegetables includes U.S. Extra Fancy, U.S. Fancy, U.S. Extra No. 1, U.S. No. 1, U.S. No. 2, and U.S. No. 3.
   A. fresh
   B. canned
   C. frozen
   D. dried

3. The ___ is the unsharpened top part of a knife blade that is opposite the edge.
   A. tip
   B. heel
   C. spine
   D. bolster

4. A ___, also known as a stone fruit, is a type of fruit that contains one hard seed or pit.
   A. melon
   B. berry
   C. drupe
   D. pome

5. The USDA has a voluntary grading program for fresh ___ based on uniformity of shape, size, color, texture, and the absence of defects.
   A. potatoes
   B. fruit
   C. grains
   D. vegetables

6. Adding ___ or an acid such as lemon juice to fruit can help prevent it from becoming mushy in the cooking process.
   A. salt
   B. pepper
   C. flour
   D. sugar

7. The conventional clock system of placing vegetables at ___ o'clock, proteins at 6 o'clock, and starches at 11 o'clock is a good method to follow when plating.
   A. 2
   B. 4
   C. 8
   D. 12

8. When properly positioning a knife, the side of the blade should rest against the knuckle of the ___ finger of the guiding hand.
   A. index
   B. middle
   C. ring
   D. pinkie

9. A ___ cut is a stick cut that produces a stick-shaped item ⅛ × ⅛ × 2 inches long.
   A. julienne
   B. fine julienne
   C. batonnet
   D. fine batonnet

10. When making ___, thin pasta dough is cut into circles and a filling is placed in the center of each circle.
    A. ravioli
    B. tortellini
    C. capellini
    D. manicotti

11. ___ potatoes are the preferred type of potato for baking, frying, mashing, puréeing, and making casseroles.
    A. Mealy
    B. Waxy
    C. Red
    D. Purple

12. Fresh potatoes must be kept in a ___ place that allows them to breathe.
    A. dry, warm, dark
    B. moist, warm, dark
    C. dry, cool, dark
    D. moist, cool, dark

13. ___ is a small, round, gluten-free grain that is classified as a complete protein.
- A. Barley
- B. Buckwheat
- C. Quinoa
- D. Spelt

14. The ___ is the largest component of a grain kernel and consists of carbohydrates and a small amount of protein.
- A. husk
- B. bran
- C. germ
- D. endosperm

15. ___ rice is rice that has had only the husk removed.
- A. White
- B. Brown
- C. Short-grain
- D. Long-grain

16. Cutting open ___ potatoes is the only way to determine if they are done.
- A. baked
- B. fried
- C. grilled
- D. simmered

17. ___ is the most common method of cooking grains.
- A. Simmering
- B. Boiling
- C. Sautéing
- D. Stir-frying

18. Cellentani, ditalini, elbows, manicotti, penne, pipettes, rigatoni, and ziti are common types of ___ pasta.
- A. shaped
- B. ribbon
- C. tube
- D. stuffed

19. A tourné cut is a carved, football-shaped cut with ___ sides and flat ends.
- A. five
- B. six
- C. seven
- D. eight

20. When pasta is cooked al dente, there should be ___ resistance in the center of the pasta when it is chewed.
- A. no
- B. a slight
- C. a lot of
- D. complete

21. Describe the procedure for blanching vegetables.

____

____

____

____

____

22. Describe proper knife grip and positioning.

____

____

____

____

____

23. Describe the procedure for preparing risotto.

____

____

____

____

____

24. Describe the procedure for preparing tortellini.

____

____

____

____

____

____

____

____

SKETCHING EXERCISE

25. Sketch a chef's knife and label each part.

Fabrication Station

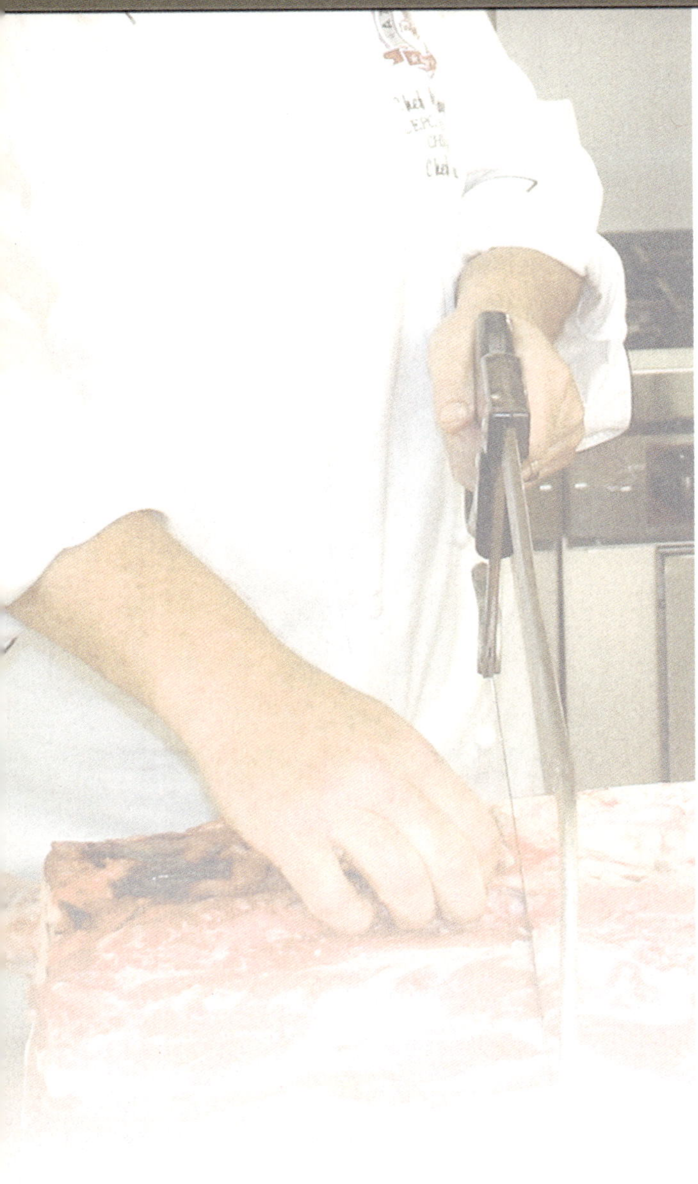

Introduction

In the fourth ACFEF apprenticeship station, you will learn fabrication techniques for poultry, fish, shellfish, and various meats. As an apprentice, you will learn to identify the different varieties and market forms of these proteins. Product identification is essential to learning the step-by-step fabrication of poultry, fish, shellfish, beef, veal, pork, and lamb. Fabricating beef will include techniques for turning a fillet of beef into Chateaubriand, filet mignons, and tournedos. Proteins such as veal, pork, and lamb will be broken down from primal cuts into scallopini, chops, and racks. You will also learn how to fabricate a rabbit.

In this station, you will practice knife-sharpening techniques using a steel grinder, stone grinder, and electric grinder. You will also learn proper handling, safety, sanitation, and storage methods for fabrication equipment, including bone saws and meat grinders.

As you advance your knowledge of fabrication, you will also learn about "food miles" and other sustainable practices, such as identifying protein products that can be purchased locally and procuring a vendor for these products. Understanding the concept of sustainable proteins and identifying the benefits of purchasing locally will further increase your use of best practices in food production.

Fabrication is the skill of breaking down poultry, fish, shellfish, and meats into standardized cuts. Standardized cuts enable a chef to maximize the yield from each animal. The ability to identify each cut is useful for ordering, receiving, and storing. Also, understanding the various market forms available and the terminology used is important for efficient purchasing. Knowledge of grading also ensures that the product ordered will satisfy the chef as well as the guests.

POULTRY

Poultry is the collective term for various kinds of birds that are raised for human consumption. The kinds of poultry recognized by the United States Department of Agriculture (USDA) that are commonly served in restaurants include chickens, turkeys, ducks, and geese. Poultry is inspected by the USDA for wholesomeness. However, the grading of poultry for quality is voluntary and paid for by poultry producers and/or processors. Each kind of poultry is subdivided into classes based on age and/or gender.

Chickens

Chicken is the most common kind of poultry raised for consumption. Chickens are classified by age and sometimes gender. Common chicken classifications include Cornish game hens, broilers/fryers, roasters, capons, and stewers (stewing hens). **See Figure 4-1.**

Cornish Game Hens. A *Cornish game hen,* also known as a Rock Cornish game hen, is either a female or male chicken that is less than five weeks old. Cornish game hens typically weigh 1½ lb or less. The term "hen" is commonly reserved for female birds, but the Cornish game hen is the exception. The flesh of a Cornish game hen is very tender and mildly flavored. Cornish game hens are most often stuffed and roasted whole.

Safe Poultry Handling

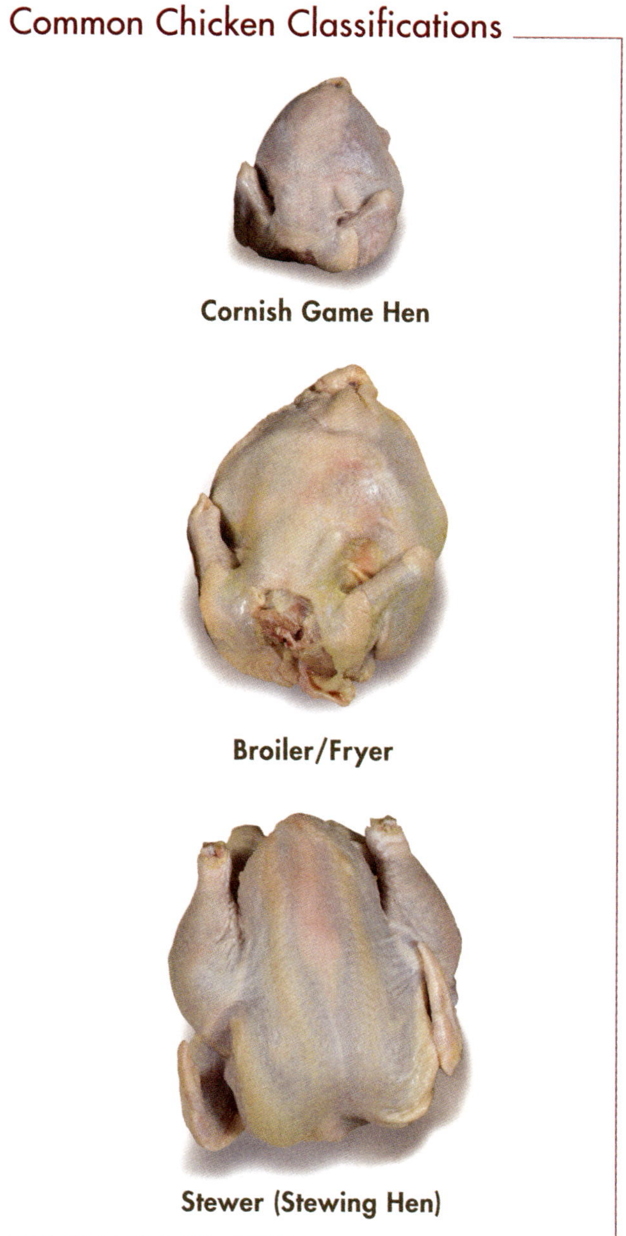

Figure 4-1. Chicken classifications include Cornish game hens, broilers/fryers, and stewers. Roasters and capons are also available.

Broilers/Fryers. A *broiler/fryer* is a young male or female chicken less than five months old. Broiler/fryers typically weigh 1½–3½ lb. Broilers/fryers have very tender flesh and smooth skin. They contain a slightly higher percentage of fat than Cornish hens. Any cooking method can be used to prepare broilers/fryers.

Roasters. A *roaster* is a young male or female chicken that is from 2–3 months old and has a ready-to-cook carcass weight of 5 lb or more. Roasters have tender flesh and smooth skin. Roasters can be prepared using any cooking method.

Capons. A *capon* is a surgically castrated male chicken that is less than four months old. Capons weigh approximately 4–7 lb. Capons are castrated to produce large, well-formed breasts with flesh that is more tender than that of broilers/fryers. Capons are most often roasted.

Stewers. A *stewer,* also known as a stewing hen, is a female chicken that is more than 10 months old. Stewers weigh approximately 3–8 lb. The flesh and skin of a stewer is tough but flavorful due to the age of the bird. A stewer requires slow moist-heat cooking methods, such as stewing or braising.

Turkeys

Turkeys are also classified by age. Classifications of turkeys include fryer/roaster turkeys, young turkeys, yearling turkeys, and mature turkeys.

Fryer/Roaster Turkeys. A *fryer/roaster turkey* is a male or female turkey that is less than three months old. Fryer/roaster turkeys weigh approximately 4–9 lb. Fryer/roaster turkeys are very tender and have soft, flexible skin. Male fryer/roaster turkeys are commonly called "toms." Female turkeys are commonly called "hens." Common preparation methods include roasting, sautéing, and pan-frying.

Young Turkeys. A *young turkey* is a male or female turkey that is less than eight months old. Young turkeys weigh approximately 8–22 lb. Young turkeys have tender flesh, smooth skin, and a firm breastbone. Young turkeys are commonly roasted or stewed.

Yearling Turkeys. A *yearling turkey* is a mature turkey that is less than 15 months old. Yearling turkeys weigh approximately 10–30 lb. The flesh of a yearling turkey is still tender. **See Figure 4-2.** Yearlings are commonly roasted or stewed.

Mature Turkeys. A *mature turkey* is a turkey that is more than 15 months old. Like yearlings, mature turkeys range in weight from 10–30 lb. The flesh and skin of a mature turkey is tough but flavorful due to the age of the bird. Mature turkeys are often roasted or stewed.

Yearling Turkeys

Photo Courtesy of D'Artagnan, Photography by Doug Adams Studio

Figure 4-2. The flesh of a yearling turkey is still tender.

Ducks

Ducks are classified by age. Classifications of duck commonly served in restaurants include broiler/fryer ducklings, roaster ducklings, and mature ducks.

Broiler/Fryer Ducklings. A *broiler/fryer duckling* is a duck that is less than two months old. Broiler/fryer ducklings weigh approximately 3–6½ lb. Broiler/fryer ducklings have very tender flesh and a soft windpipe. They are commonly roasted or stewed.

Roaster Ducklings. A *roaster duckling* is a duck that is less than four months old. Roaster ducklings weigh approximately 4–7½ lb. The flesh of a roaster duckling is tender, and the windpipe is just starting to harden. Roaster ducklings are most often roasted.

Mature Ducks. A *mature duck* is a duck that is more than six months old. The flesh of a mature duck is fairly tough and the windpipe is hardened. The flesh of mature ducks is typically used in processed duck products. Mature ducks are typically braised.

White Pekin ducks, Muscovy, Moulard, and Mallard ducks are sold for consumption. The Mallard duck is the breed from which all domestic ducks, except the Muscovy, are descended. Farm-raised Mallards are available on a limited basis. Each breed of duck is known for one or more products or uses. **See Figure 4-3.**

- The white Pekin duck, also known as Long Island duckling, is the most popular breed of duck consumed in the United States. The white Pekin is typically used for roast duck. This breed of duck should not be confused with the dish named Peking duck, which consists of duck served with pancakes, scallions, and hoisin sauce.
- The Muscovy duck, also known as the Barbary or Barbarie duck, has thinner skin and much less fat than

other breeds of duck. Male Muscovy ducks are meatier than the females. Muscovy flesh is redder and has a more pronounced flavor than other ducks. Each side of a Muscovy breast looks like a "duck steak" in that it weighs approximately 8 oz. The liver of the Muscovy is also sold, but it is not as preferred as Moulard liver.

- The Moulard duck is a sterile hybrid of a male Pekin duck and a female Muscovy duck. The Moulard is larger and considerably fatter than the white Pekin duck and the flesh has a strong gamey flavor. Moulards are bred for their exceptional livers and breast meat. The breast of a Moulard is called a magret and must be served rare in order to be tender.
- The Mallard duck is smaller in size and its flesh is tougher than the white Pekin duck. The flesh of a Mallard has very little fat, yet it can taste greasy and gamey.

Ducks have less flesh in proportion to bone and fat than most other kinds of poultry. For example, a duck yields half as much flesh as a chicken of the same size.

The majority of fat in a duck is located in and just beneath the skin. With the fatty skin removed, duck is lean in comparison to other kinds of poultry. When cooking duck, the fatty skin must be rendered slowly or it becomes nearly inedible.

Duck breasts are often seared in a sauté pan, finished in an oven, and served medium-rare. Duck legs are most often braised or roasted until well-done and tender. Whole ducks are usually roasted on a rack to allow the fat to render.

Duck legs and thighs may be prepared as confit. *Confit* is a French term for meat that has been cooked and preserved in its own fat. To make a duck confit, pieces of duck leg and thigh are salted and then simmered in rendered duck fat until tender. The confit can be served whole, or the flesh can be pulled from the bone and used as a flavorful ingredient in salads or other dishes. A *cassoulet* is a dish that consists of white beans stewed with duck fat, fresh sausage, and duck confit.

Foie gras is the fattened liver of a duck or goose. **See Figure 4-4.** Foie gras is a delicacy that is considered a staple of classic fine dining. A duck or goose that is used to make foie gras is fed a rich diet until the liver becomes almost solid fat. The fattened liver is removed and sold as the most expensive part of the duck or goose.

Foie Gras

Photo Courtesy of D'Artagnan, Photography by Doug Adams Studio

Figure 4-4. Foie gras is the fattened liver of a duck or goose.

Foie gras is graded based on the size, appearance, and texture of the liver. To receive a grade of A, the liver must weigh at least 1 lb 8 oz and be round and firm with no blemishes. Grade A foie gras is used for terrines and pâtés. Grade B foie gras weighs between 1 lb and 1 lb 7 oz, has a good texture, but is less round in shape than Grade A foie gras. Grade B is a good choice for poaching or sautéing. Foie gras that weighs less than 1 lb, is slightly flattened, and has some visual imperfections and soft spots is grade C. Grade C foie gras is used primarily for mousses and pâtés.

Whole Ducks and Duck Breasts

Pekin Duck

Muscovy Duck

Duck Breast

Photos Courtesy of D'Artagnan, Photography by Doug Adams Studio

Figure 4-3. Common market forms of duck include whole ducks and duck breasts.

Because foie gras is almost all fat, it is typically seared in a very hot sauté pan and served immediately. Foie gras may also be poached, cooled, and puréed to make a pâté de foie gras.

Geese

Geese have dark flesh and a large amount of fat in both the skin and the flesh. A goose is classified as either young or mature. Only young geese are used in foodservice operations.

A *young goose* is a goose that is usually less than six months of age and weighs approximately 4–10 lb. **See Figure 4-5.** Young geese have tender flesh and a windpipe that is easily dented. Their flesh has a rich taste due to the high fat content. The large flaps of thick, fatty skin around the neck and tail should be removed with a knife or kitchen shears prior to cooking. A young goose is commonly roasted at very-high temperatures to aid in rendering some of the fat from the skin and the flesh.

Goose

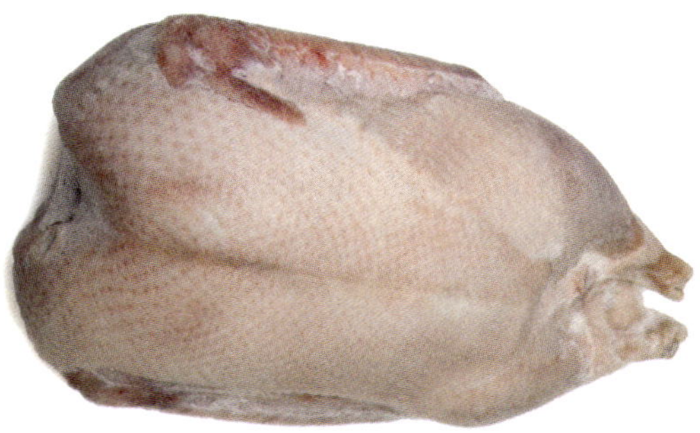

Czimer's Game & Seafoods, Inc.

Figure 4-5. A young goose weighs approximately 4–10 lb.

CHECKPOINT 4-1

1. Identify two traits that are used by the USDA to classify poultry.
2. Describe five classifications of chicken.
3. Describe four classifications of turkey.
4. Describe three classifications of duck.
5. Differentiate between white Pekin, Muscovy, Moulard, and Mallard ducks.
6. Contrast the three grades of foie gras.

MARKET FORMS OF POULTRY

Poultry is sold after the head, neck, feet, and feathers have been removed from the bird. Poultry can be sold fresh or frozen, whole or cut-up, boneless or bone in, ground, or processed into a prepared form such as chicken tenders. Foodservice operations purchase all market forms of poultry.

Some poultry has both light and dark flesh. Other poultry has only dark flesh. Leg and thigh flesh is always dark, but breasts and wings can be either light or dark. For example, chickens spend most of their time on their feet and therefore have dark flesh in their legs and thighs. Ducks have dark flesh throughout their entire bodies, because all of their muscle groups are used for slow, sustained movement as they fly. Because the breast and wing muscles get more exercise, more blood flows through them. Blood contains the protein myoglobin, which causes the darkening of muscle tissue. The more a muscle is used, the darker and more flavorful the muscle becomes.

The fat in poultry is found just beneath the skin, around the tail, and on the abdomen. **See Figure 4-6.** Because poultry does not have much intramuscular fat, it can become very dry if it is overcooked even slightly. As a bird ages, its breastbone and keel becomes less flexible and the flesh and skin toughen, intensifying the overall flavor. Therefore, younger birds are often desired for their tenderness and mild flavor.

Poultry Fat

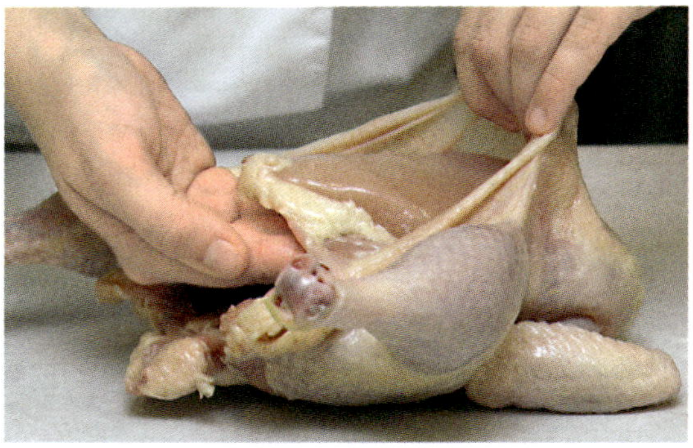

Figure 4-6. The fat in poultry is found just beneath the skin, around the tail, and on the abdomen.

Whole Poultry and Fabricated Cuts

Poultry is available in a variety of market forms, including whole birds and fabricated cuts. Knowledge of these market forms is necessary for accurate ordering. Purchasing whole poultry allows a chef to be more creative with the menu. It is also less expensive to purchase whole poultry and debone it in-house.

Fabricated cuts of poultry include breasts, tenders, tenderloins, wings, legs, leg quarters, breast quarters, halves, and ground poultry. **See Figure 4-7.** Cutlets and sausages are also available.

Whole and Fabricated Poultry Cuts

National Turkey Federation

Figure 4-7. Poultry is sold whole and in common fabricated cuts, including ground, wings, thighs, breasts, and legs.

- A *breast* is the top front portion of the flesh above the rib cage. The breast consists of white flesh in birds that only fly in quick, short bursts and dark flesh in birds that have the ability to fly long distances.
- An *airline breast* is a boneless skin-on chicken breast with the first wing section (bone-in) attached.
- A *chicken cutlet* is a boneless, skinless section of a chicken breast that has been tenderized.
- A *tender* is a small strip of a breast.
- A *poultry tenderloin* is the inner pectoral muscle that runs alongside the breastbone of a bird.
- A *wing* consists of a tip, paddle, and drummette. A *wing tip* is the outermost section of a wing. It is often used to make stock as it contains little flesh. A *wing paddle,* also known as a wing flat, is the second section of a wing located between the two wing joints. A *drummette* is the innermost section of a wing located between the first wing joint and the shoulder.
- A *poultry leg* consists of a drumstick and thigh. A *drumstick* is the lower portion of the leg located below the hip and above the knee joint. The *thigh* is the upper section of the leg located below the hip and above the knee joint.
- A *leg quarter* is a thigh, a drumstick, and a portion of the back.
- A *breast quarter* is half of a breast, a wing, and a portion of the back.
- A *poultry half* is a full half-length of a bird split down the breast and spine.
- *Ground poultry* is ground fabricated cuts of poultry.

To help ensure that foodservice operations and suppliers communicate efficiently, the USDA publishes the Institutional Meat Purchase specifications (IMPS) for commonly purchased meats, poultry, seafood, and game products. All cuts are numbered by category.

Giblets

When purchasing whole poultry, it is common to find a small bag containing the giblets inside the cavity of the bird. *Giblets* is the name for the grouping of the neck, heart, gizzard, and liver of a bird. **See Figure 4-8.** When ordering whole poultry it is common to specify whether the giblets are desired. When purchasing poultry without giblets, the acronym WOG is used by suppliers.

Giblets

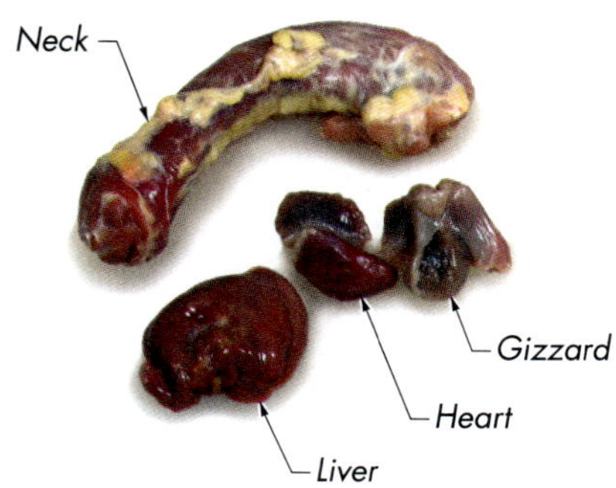

Figure 4-8. Giblets include the neck, heart, gizzard, and liver of a bird.

The neck, heart, and gizzard are often used to make giblet gravy. The neck contains an abundant amount of gelatin, which makes a rich and flavorful stock. Turkey hearts are served with a béchamel sauce in a classic preparation called creamed hearts. Gizzards are often trimmed of connective tissue, breaded or battered, and then fried. Livers are often breaded then fried or sautéed and served with caramelized onions. Livers are also used to make liver pâtés.

Receiving and Storing Poultry

Poultry must be stored at an internal temperature of 41°F or below. It should always be stored beneath other foods to prevent the juices from contaminating the other foods. Fresh poultry should be used within two to three days or should be frozen immediately to prevent loss of quality and potential spoilage.

Poultry spoils rapidly. Although it develops an odor as it spoils, it may be unsafe for consumption prior to developing offensive odors. Many foodservice operations specify that fresh poultry be packed in vacuum-sealed bags and sometimes gas flushed. Gas flushed bagging increases the shelf life of the poultry by replacing oxygen with nitrogen, which inhibits bacteria growth.

Frozen poultry should be frozen in its original packaging at 0°F or below. Frozen poultry should be moved to a refrigeration unit a day prior to use in order to thaw safely. Whole turkeys may need an additional one to two days to thaw because of their size. Poultry should never be refrozen once it has thawed.

Frozen convenience poultry products do not need to be thawed. Instead, frozen convenience poultry products such as chicken tenders should remain in the freezer until they are ready to be cooked. **See Figure 4-9.**

Frozen Poultry Products

Photo Courtesy of Perdue Foodservice, Perdue Farms Incorporated

Figure 4-9. Frozen convenience poultry products should remain in the freezer until they are ready to be cooked.

CHECKPOINT 4-2

1. Explain why some poultry has both light and dark flesh and some poultry has only dark flesh.
2. Explain the advantage of purchasing whole poultry.
3. Identify common fabricated cuts of poultry.
4. Name the items in the bag located inside whole poultry.
5. Identify special precautions to follow when receiving and storing poultry.
6. Explain how frozen poultry should be thawed.

FABRICATING POULTRY

There are many ways that poultry can be cut into portions. Typical fabrication methods for poultry include cutting them into halves, quarters, and eighths. Poultry may also be fabricated to produce boneless breasts and airline breasts. Fabricating techniques can be applied to almost any type of poultry because all types have similar bodies and bone structures. Chicken is the most economical bird to fabricate. Understanding the skeletal structure of a chicken can aid the fabrication process. **See Figure 4-10.**

Trussing Whole Poultry

When roasting a whole bird, the bird is usually trussed using butcher's twine. *Trussing* is the process of tying the legs and wings of a bird tightly to the body to keep a compact shape. Trussing helps the bird cook evenly and retain moisture. Trussing also gives the bird a pleasing finished appearance. Prior to trussing, any excess fat should be trimmed from around the neck area and tail portion of the bird. The skin is then pulled tightly and evenly across the breast to cover any exposed flesh and prevent the breast from drying out during cooking. If desired, the wing tips can be removed, as they have a tendency to burn during roasting. If the wings are left intact, the first joint is tucked behind the second joint for a neat appearance.

Skeletal Structure of Chickens

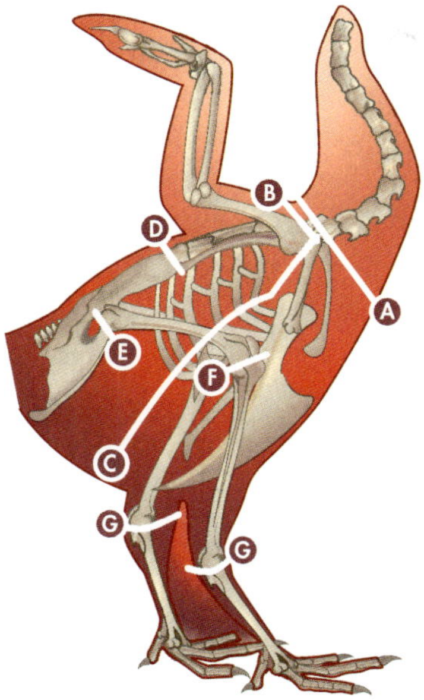

Cut Points	
A	Neck from breast/back
B	Wing from breast
C	Breast from back
D	Front half of back from back half of back
E	Leg from back
F	Thigh from drumstick
G	Drumstick from foot

Figure 4-10. Understanding the skeletal structure of a chicken can aid the fabrication process.

Procedure for Trussing Whole Poultry

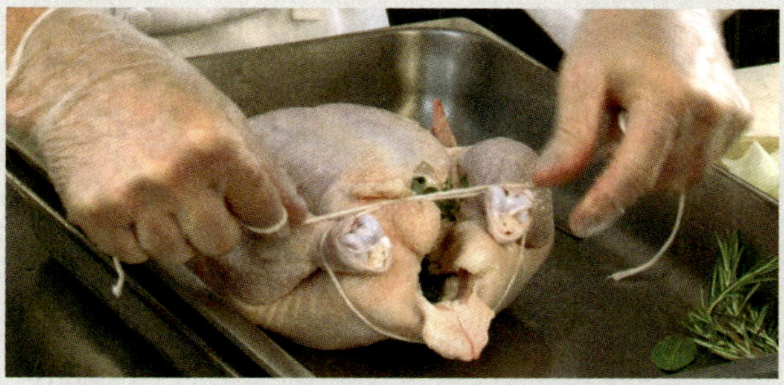

1. Cut a length of butcher's twine approximately three times the length of the bird to be trussed.
2. With the breast up, place the center of the twine beneath the bird, about 1 inch under the tail.

3. Bring the twine up around the legs and cross the ends, creating an "X" between the legs.
4. Pass the ends of the twine under the legs and pull tight.

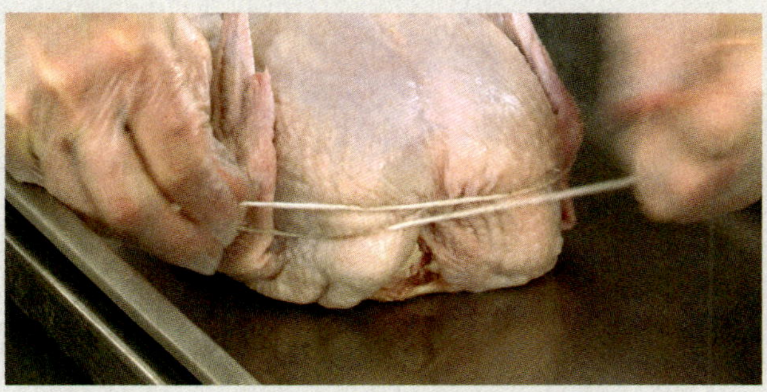

5. Turn the bird around, and pull the twine across the wings and tie a square knot at the neck to secure the truss.

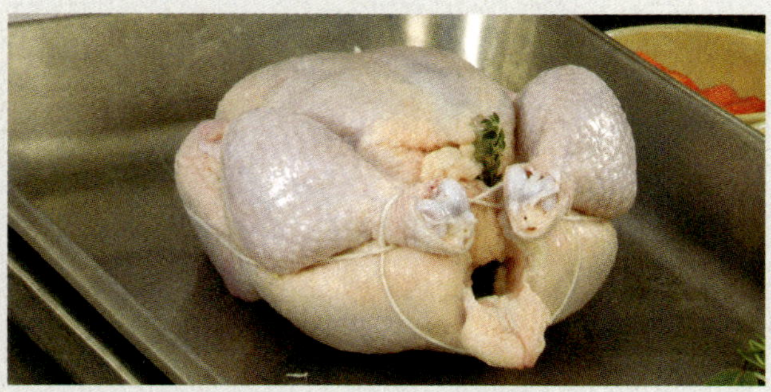

Finished truss.

Trussing Whole Poultry
Media Clip

Cutting Poultry into Halves

Poultry is commonly cut into halves. The bird is split from top to bottom between the breasts and along the backbone to the tail. This results in two equal portions.

Production Tip

Never use the same cutting boards or tools that were used with raw poultry for other products until they have been washed and sanitized.

Cutting Poultry into Quarters and Eighths

Poultry is often cut into quarters for grilling, broiling, or roasting. First, the bird is divided into leg and thigh sections and wing and breast sections. There are two in each section, yielding four quarters.

Poultry is commonly cut into eighths for pan-frying, deep-frying, grilling, broiling, and roasting. To cut a bird into eighths, quarters are cut into two breasts, two wings, two thighs, and two legs.

Procedure for Cutting Poultry into Halves

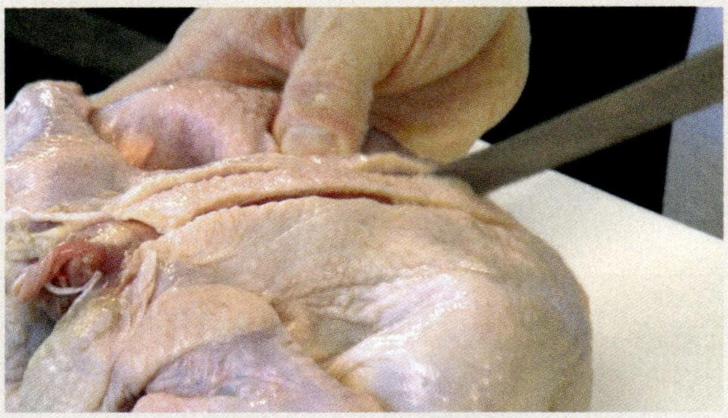

1. Square the bird by firmly squeezing the legs and wings toward the body.
2. With the breast side down, use a stiff (curved) boning knife to split the bird along both sides of the backbone from the neck to the tail.

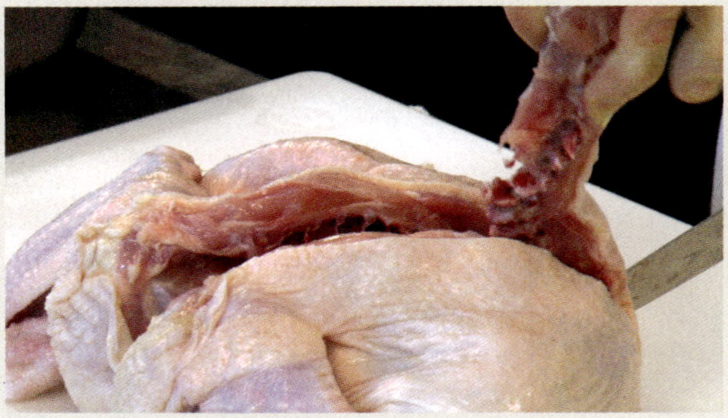

3. Remove the backbone from the carcass.

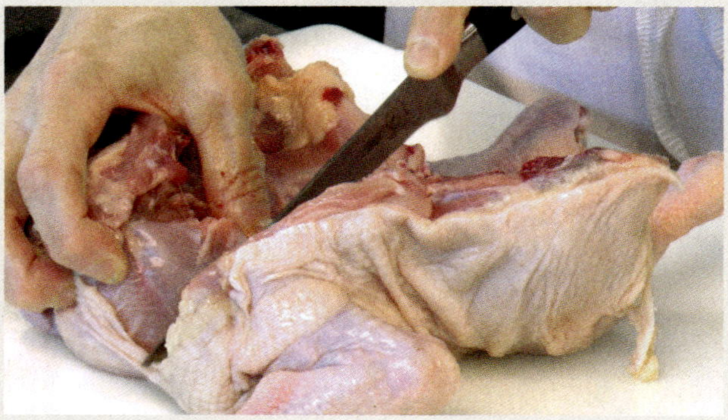

4. Open both sides of the bird to reveal the keel bone (breastbone). Cut through the keel bone and wishbone lengthwise from neck to tail. If necessary, hit the spine of the blade with the heel of the hand.
5. Cut through the flesh and skin behind the breastbone to separate the bird into halves.

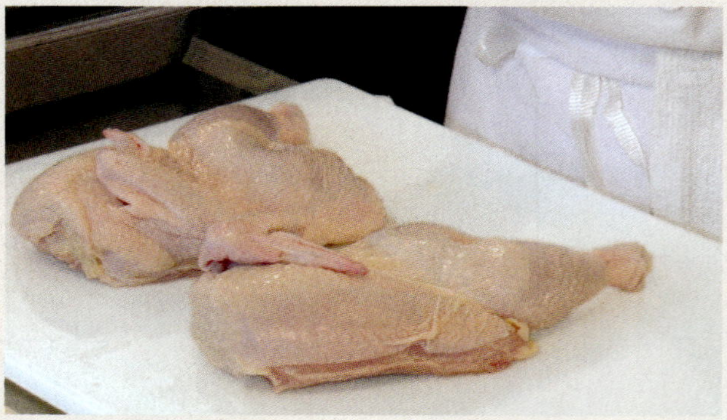

Poultry cut into halves.

Procedure for Cutting Poultry into Quarters and then Eighths

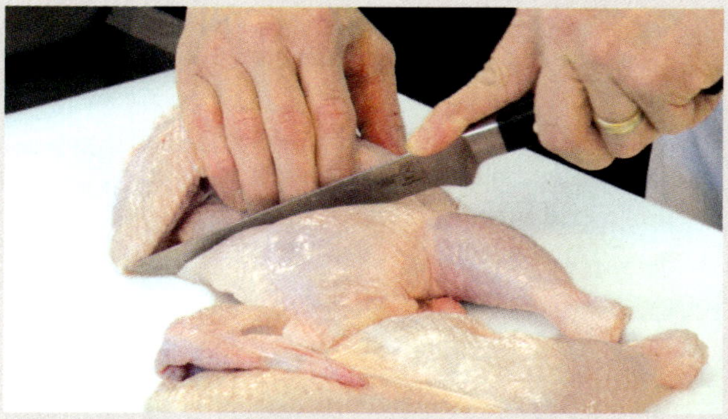

1. Cut the bird into two halves.
2. Cut through the flap of skin between the breast and thigh and pull the thigh away from the breast to expose the joint. Then, cut the joint to separate the thigh from the breast.

Poultry cut into quarters.

3. Cut the joint between the breast and the thigh to separate the breast and thigh from one half.

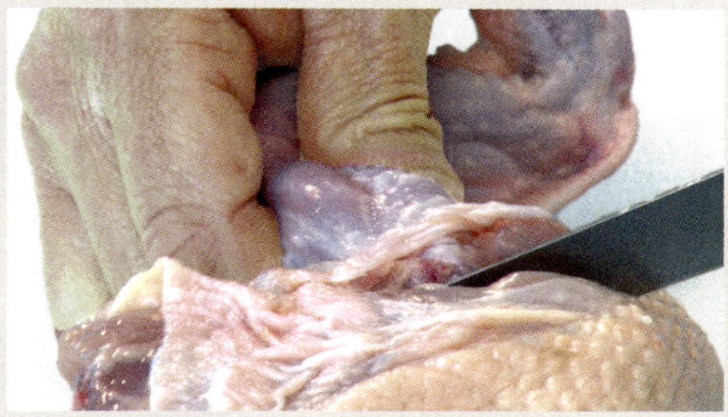

4. Hold the wing away from the breast, and cut the wing from the breast at the wing joint.

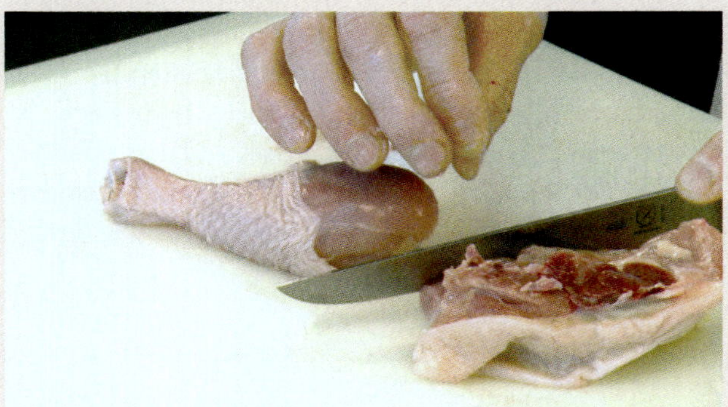

5. Hold the leg so that the inside thighbone is visible and locate the thin line of fat that separates the leg and thigh muscles. Cut along this line to separate the leg and thigh joint.

Poultry cut into eighths.

Cutting Poultry into Eighths
Media Clip

Cutting Boneless Breasts

Boneless breasts are the most popular poultry cut because of their versatility. They can be grilled, broiled, roasted, sautéed, pan-fried, poached, or stuffed. Larger breasts, such as those from a turkey, can be roasted whole or sliced into medallions and sautéed. A boneless breast is often flattened before being sautéed or stuffed. The breast is pounded flat with a meat mallet until the desired thickness is achieved. Boneless breasts are often fabricated in-house.

Procedure for Cutting Boneless Breasts

1. With the breast side down, use a stiff (curved) boning knife to split the bird along both sides of the backbone from the neck to the tail. Remove the backbone from the carcass.
2. Pull the leg and thigh away from the breast. Locate the thigh joint and cut the flesh down to the joint.
3. Twist the leg to break the thigh joint. Cut through the joint and separate the leg from the carcass.
4. Open the bird to reveal the keel bone. Cut through the keel bone and wishbone lengthwise from the neck to the tail. If necessary, hit the spine of the blade with the heel of the hand.
5. Cut through the flesh and skin behind the keel bone to separate the bird into halves.
6. Lay the bird skin-side down and cut the breast away from the rib bones. *Note:* Use care when cutting around the rib cage to keep the tenderloin attached to the breast.

Cutting Airline Breasts

Airline breasts, also known as suprêmes, are similar to boneless breasts but contain one drummette bone. **See Figure 4-11.** An airline breast makes an elegant presentation due to the exposed drummette bone.

Airline Breasts

Cutting Airline Breasts

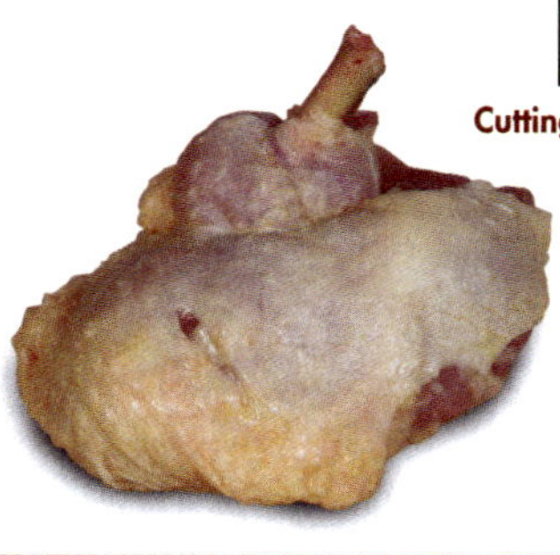

Figure 4-11. Airline breasts are similar to boneless breasts but contain one drummette bone.

Procedure for Cutting Airline Breasts

1. With the breast side down, use a stiff (curved) boning knife to split the bird along both sides of the backbone from the neck to the tail. Remove the backbone from the carcass.
2. Open the bird to reveal the keel bone. Cut through the keel bone and wishbone lengthwise from the neck to the tail. If necessary, hit the spine of the blade with the heel of the hand.
3. Cut through the flesh and skin behind the keel bone to separate the bird into halves.
4. Cut through the skin between the breast and the thigh. Pull the thigh away from the breast to expose the joint.
5. Cut the joint to separate the breast from the thigh.
6. Cut along one side of the breastbone, following the curve of the ribs, to separate the flesh from the bone.
7. Separate the wing from the rib cage by cutting the joint. Keep the wing attached to the breast.
8. Cut the breast meat free from the carcass.
9. Make a cut on the back of the joint between the drummette and the paddle bones.
10. Break the joint and pull back the flesh and skin to expose the drummette bone. Trim the end of the drummette bone to remove the cartilage. *Note:* If desired, the skin can be removed.

Boning Legs and Thighs

Legs and thighs have more flavor than breasts because they have dark flesh and additional fat. Boneless legs and thighs are often stuffed and roasted. When stuffing and roasting legs and thighs, the bones are removed but the flesh is left intact.

Boneless legs and thighs may be cut into smaller pieces for use in stir-fries and soups. They can also be flattened prior to being sautéed or stuffed. To flatten a boneless leg or thigh, the boneless leg or thigh is wrapped in plastic and then pounded flat with a meat mallet until the desired thickness or diameter is achieved. Partially boned legs and thighs are often desired to create a more elegant presentation.

Procedure for Boning Legs and Thighs

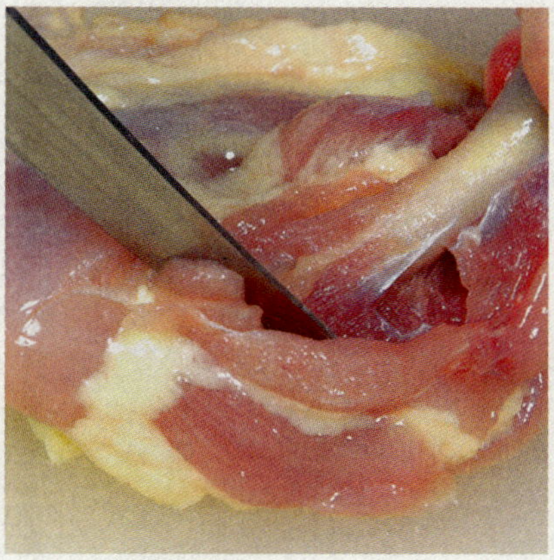

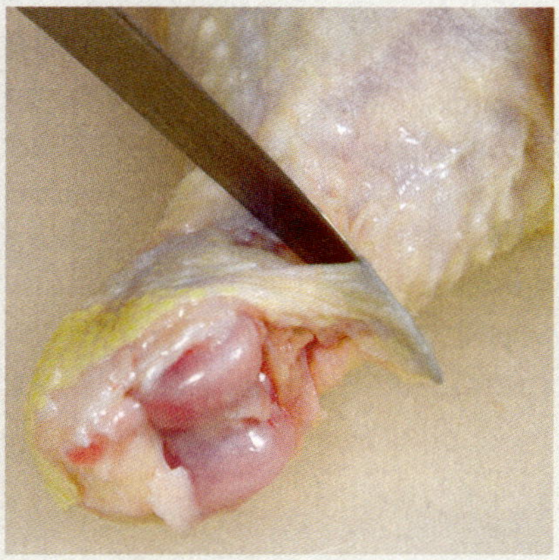

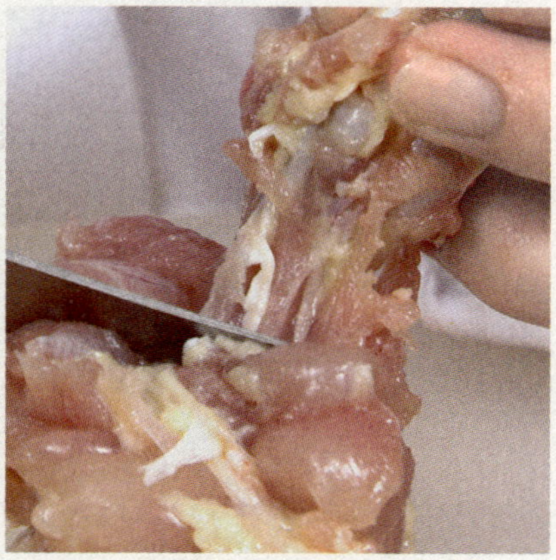

1. Place a leg quarter on a cutting board with the inside of the thigh facing upward.
2. Use a stiff (curved) boning knife to cut down the length of the thighbone and leg bone on each side and around the cartilage of the joint to free the bones from the flesh.
3. Pull the flesh away from the leg bone and cut the flesh off where it connects to the end joint.
4. With smooth, even strokes cut around the joint at the end of the leg until the L-shaped leg, thighbones, and cartilage are free from the flesh. Reserve the bones for stock.
5. Repeat with the other leg quarter. If desired, remove the skin.

Procedure for Partially Boning Legs and Thighs

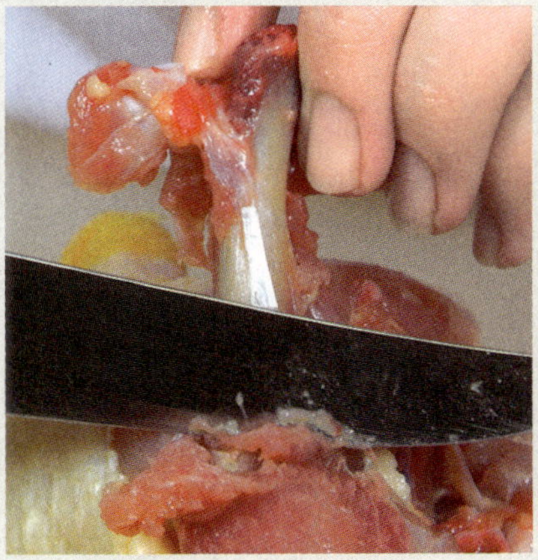

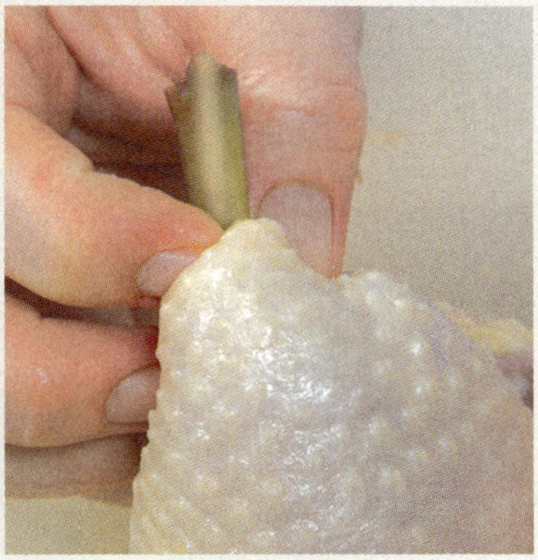

1. Use a stiff (curved) boning knife to cut down the length of the thighbone of a leg quarter.
2. Scrape the thigh flesh off the thigh-bone down to the joint.
3. Cut through the joint that connects the leg to the thighbone to remove the thighbone.
4. Use a chef's knife to chop off the joint.
5. Push the flesh away from the joint for a finished presentation.

Fully boned and partially boned legs and thighs may be stuffed. They are typically trussed or wrapped in caul fat before cooking to help to maintain the shape of the flesh.

Caul fat is a meshlike fatty membrane that surrounds sheep or pig intestines. Caul fat adds additional moisture and maintains the shape of the poultry being roasted.

Caul fat melts almost completely away during the cooking process, but not before it helps set the shape of the item being cooked. Stuffed legs and thighs can be seared in a sauté pan and finished in the oven, or they can be roasted or braised.

Boning Whole Poultry

Boning a whole bird is one of the more difficult fabrication techniques performed in the professional kitchen. In this technique the bones of the bird are removed while keeping the flesh intact. Small birds, such as quail, squab, and poussins, can be boned or partially boned for a variety of preparations and presentations. Larger birds, such as chicken, are boned for charcuterie preparations. A deboned chicken is often called a butterflied chicken because it can be laid flat on a grill and will cook evenly.

Procedure for Boning Whole Poultry

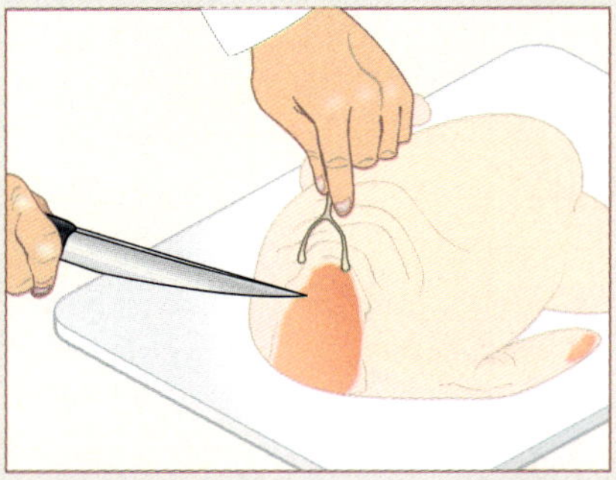

1. Place the bird breast-side up and stretch each wing flat against the cutting board by pulling on the tip. Use a stiff (curved) boning knife to cut off the wing tip and the next joint, leaving the drummette still attached.
2. Pull the skin of the neck area out of the way, and slide the knife along the underside of the wishbone. Cut around and under the wishbone until it is free and can be pulled out by hand.

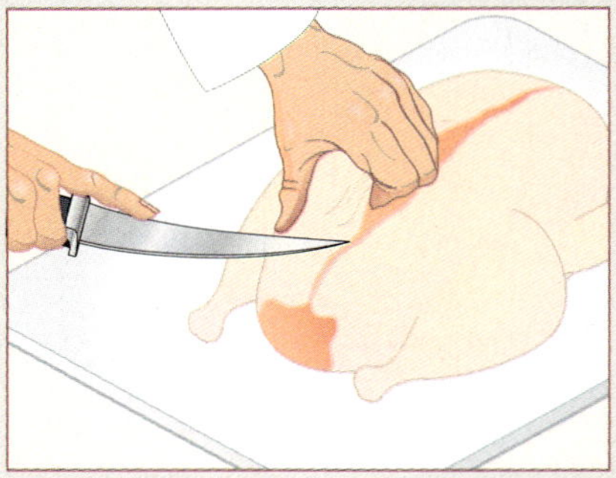

3. Turn the bird breast-side down. Use short strokes to cut along the backbone from the neck to the tail, keeping the knife close to the bones. Carefully pull the flesh away from the carcass as the flesh is cut.

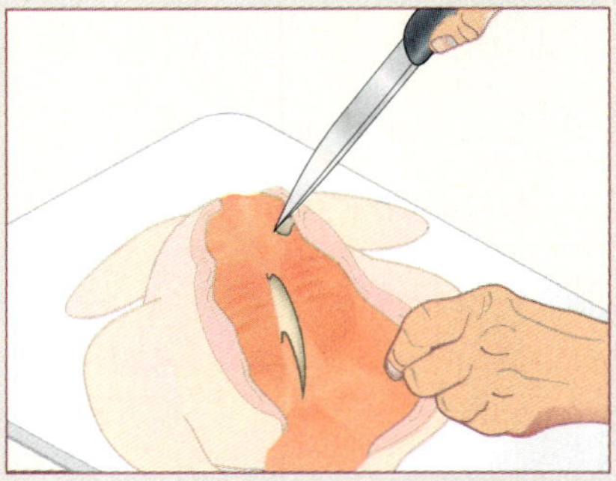

4. Cut through the ball-and-socket joints connecting each wing and thighbone to the carcass to separate them from the carcass while leaving them attached to the skin. *Note:* When this step is complete on both sides, the flesh will only be attached along the ridge of the breastbone.
5. Gently pull to separate the breastbone and carcass from the flesh.

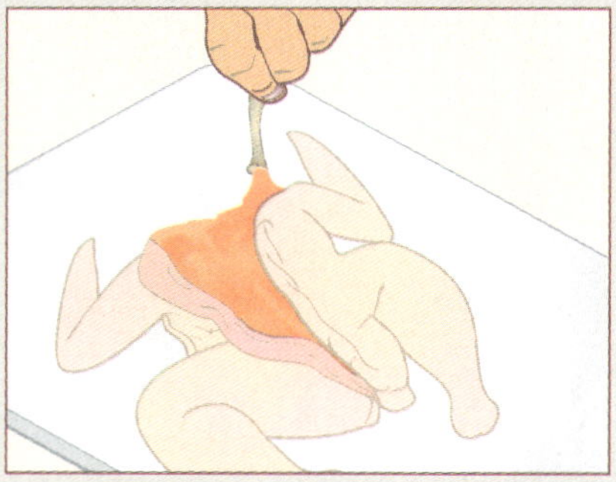

6. Cut the flesh from the curved bone near each wing to remove the bone.
7. Cut down the length of the thighbone and scrape the flesh down to the joint.
8. Cut between the leg and thighbones at the joint to remove the thighbone from each side.

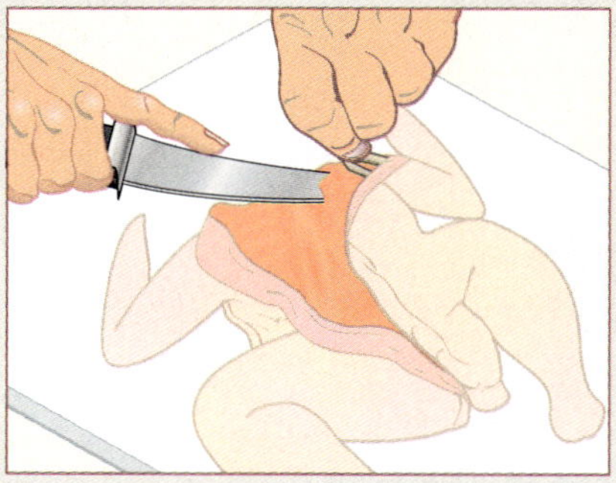

9. While holding the exposed portion of the wing bone, cut through the tendons and scrape the meat from the bones.
10. Pull out the wing bone and repeat on the other side.

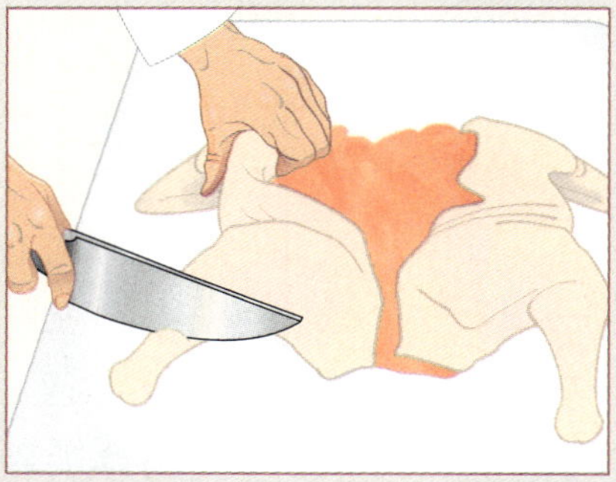

11. Use a chef's knife to cut off the end of the leg bone.

CHECKPOINT 4-3

1. Describe how to truss whole poultry.
2. Truss whole poultry.
3. Describe how to cut poultry into halves.
4. Cut poultry into halves.
5. Describe how to cut poultry into quarters.
6. Cut poultry into quarters.
7. Describe how to cut poultry into eighths.
8. Cut poultry into eighths.
9. Describe how to cut a boneless breast.
10. Describe how to cut an airline breast.
11. Cut a boneless breast and an airline breast.
12. Describe how to bone legs and thighs.
13. Bone a leg and a thigh.
14. Describe how to partially bone legs and thighs.
15. Partially bone a leg and a thigh.
16. Describe how to bone whole poultry.
17. Bone whole poultry.

FISH

A *fish* is any of a classification of animal that has fins, gills, and an internal skeleton made of bones or cartilage. Fish have edible flesh that contains protein, carbohydrates, fat, water, and trace amounts of vitamins and minerals. Unlike land animals, fish have delicate connective tissue. Because the connective tissue is very thin, fish flesh naturally flakes and comes apart easily. Fish are considered lean or fatty based on the amount of fat they contain. **See Figure 4-12.**

Production Tip

Avoiding the use of seafood that is endangered or being overharvested is key to sustainability. The Monterey Bay Aquarium's Seafood Watch program provides lists of seafood to avoid as well as substitution options.

Approximate Fat Content of Common Fish

Lean Fish		Fatty Fish	
Species	**Fat Content***	**Species**	**Fat Content***
Cod	1 g	Farmed catfish	6 g
Flounder	1 g	Rainbow trout	4 g
Halibut	2 g	Coho salmon	6 g
Striped bass	2 g	Shad	10 g

* per 3 oz serving

Figure 4-12. Fish are considered lean or fatty based on the amount of fat they contain.

The fat content of a fish can affect both flavor and the cooking method required. Lean fish contain very little fat and are often prepared using moist-heat cooking methods. Pacific cod is an example of a lean fish. Fatty fish contain more fat and are rich in omega-3 fatty acids, vitamin A, and vitamin D. Fatty fish, such as salmon, are best prepared using dry-heat cooking methods. The fat content of a particular type of fish can vary slightly by season.

Fish are most often grouped by freshwater or saltwater habitat and by their external shape and structure. Fish are classified as roundfish, flatfish, or cartilaginous fish, based on their shape and structure.

U.S. Fish & Wildlife Service

Roundfish

A *roundfish* is any fish with a cylindrical body, an eye located on each side of the head, and a backbone that runs from head to tail in the center of the body. **See Figure 4-13.** Roundfish are the most common type of fish and are found in freshwater lakes and streams as well as in saltwater. Trout and salmon are examples of roundfish.

Roundfish Shape and Structure

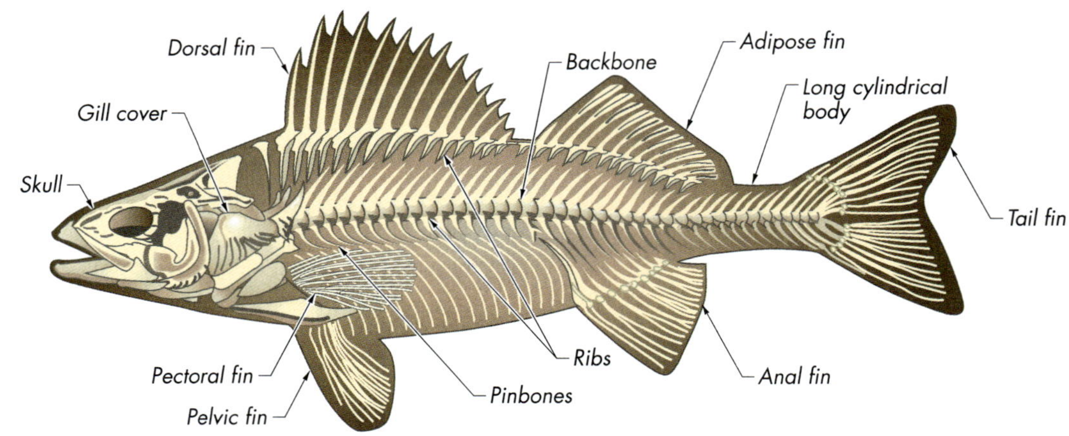

Figure 4-13. A roundfish is any fish with a cylindrical body, an eye located on each side of the head, and a backbone that runs from head to tail in the center of the body.

Flatfish

A *flatfish* is any thin, wide fish with both eyes located on one side of the head and a backbone that runs from head to tail through the lateral line of the body. **See Figure 4-14.** Flatfish are born round like other fish. However, as they grow, their eyes move to one side of the body so they can see all around when lying flat on the ocean or sea floor. Flatfish swim parallel to the surface of the water with both eyes facing toward the surface. The skin on the top side of a flatfish is typically dark and may change color to blend in with its environment. The bottom side of a flatfish is typically light in color. Flounder, halibut, and sole are examples of flatfish.

Cartilaginous Fish

A *cartilaginous fish* is any fish that has a skeleton composed of cartilage instead of bones. **See Figure 4-15.** Cartilaginous fish often have a smooth, tough outer skin without scales. Sharks, skates, and stingrays are examples of cartilaginous fish.

Flatfish Shape and Structure

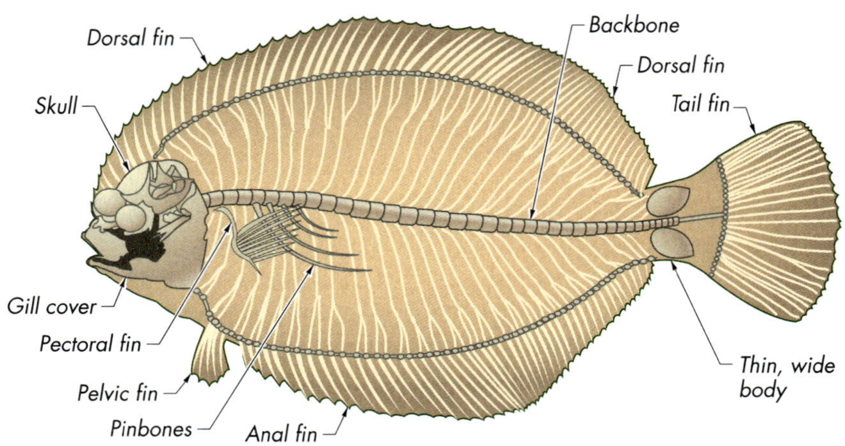

Figure 4-14. A flatfish is any thin, wide fish with both eyes located on one side of the head and a backbone that runs from head to tail through the lateral line of the body.

Cartilaginous Fish Structure

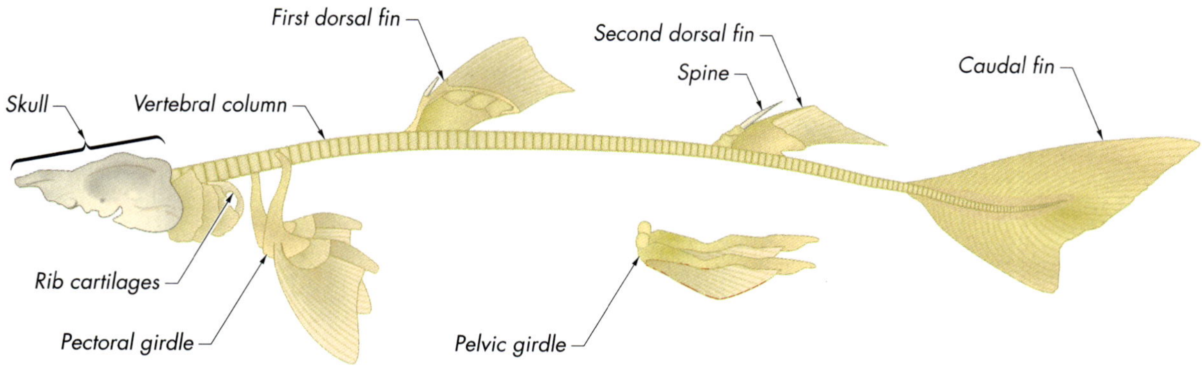

Figure 4-15. A cartilaginous fish is any fish that has a skeleton composed of cartilage instead of bones.

CHECKPOINT 4-4

1. Differentiate between lean and fatty fish.
2. Describe three classifications of fish based on external shape and structure.

MARKET FORMS OF FISH

Fish may be purchased fresh, frozen, or processed. The more that is done to a fish prior to delivery, the higher the cost per pound. Fish are commonly sold whole, drawn, pan-dressed, as steaks, as fillets, as butterflied fillets, as loins, and as wheels. **See Figure 4-16.**

Market Forms of Fish

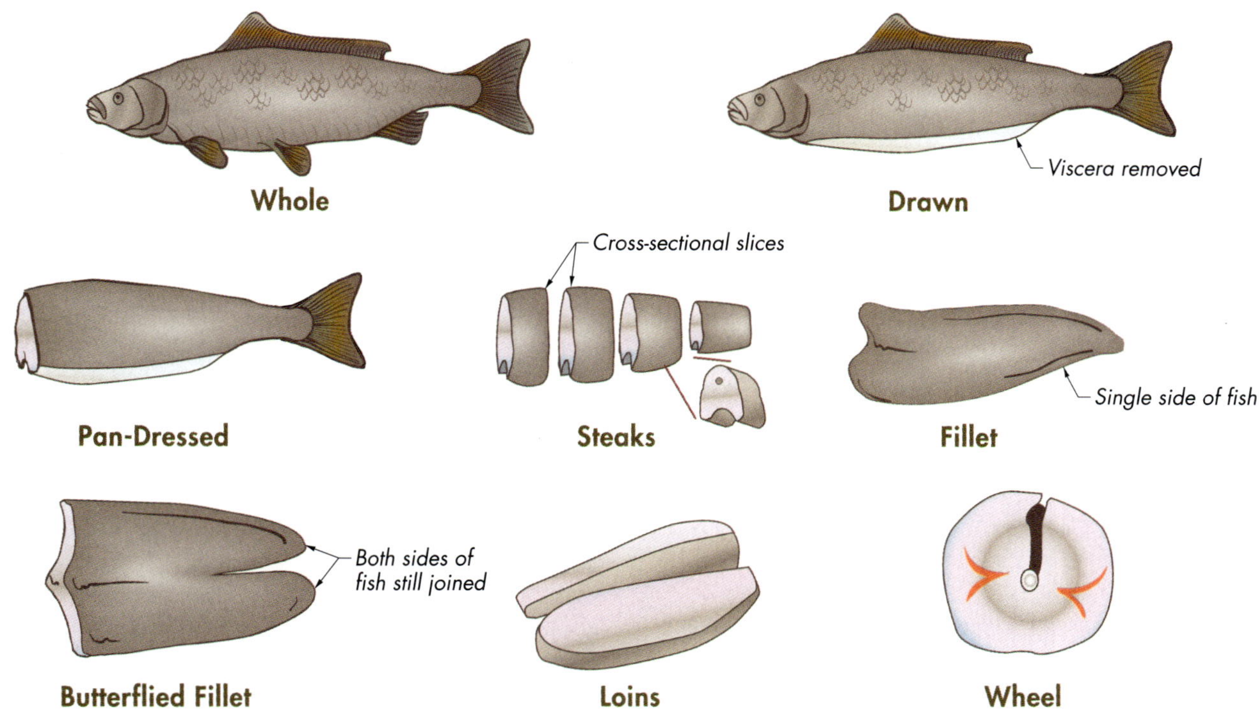

Figure 4-16. Fish are commonly sold whole, drawn, pan-dressed, as steaks, as fillets, as butterflied fillets, as loins, and as wheels.

Whole Fish

A *whole fish* is the market form of a fish that is taken from the water and sold as is. Nothing has been done to process it. Whole fish have the shortest shelf life of any market form because all the internal organs (viscera) are still present. Fish purchased whole cost less per pound, require more preparation, and yield more waste than any other market form.

Drawn and Dressed Fish

A *drawn fish* is a fish that has had only the viscera removed. Drawn fish can be prepared whole. A *dressed fish* is a fish that has been scaled and has had the viscera, gills, and fins removed. A *pan-dressed fish,* also known as a headed and gutted (H&G) fish, is a dressed fish that has had its head removed.

Fish Steaks

A *fish steak* is a cross section of a dressed fish. Fish steaks from large fish are fabricated from a wheel. Steaks are ready to cook when purchased. Generally, the only bone present in a steak is a small section of the backbone. Steaks from very large fish, such as swordfish, are boneless. Tuna, salmon, swordfish, and sharks are commonly sold as steaks.

Fish Fillets

A *fish fillet* is the lengthwise piece of flesh cut away from the backbone. Roundfish have two fillets, one on each side. Flatfish have four fillets, two on each side. Fillets can be purchased with or without belly bones and pin bones. Fillets can also be purchased with or without skin. Fillets with the skin left on are sold scaled. A *butterflied fillet* is two single fillets from a dressed fish that are held together by the uncut back or belly of the fish.

Fish may also be sold as loins, wheels, roasts, or portions. A fish loin is cut lengthwise from either side of the backbone of large roundfish. A *wheel* is the round center cut of a large fish from which steaks are cut. A fish portion is cut from fillets into smaller pieces that are sold individually.

Frozen Fish

Using frozen fish allows a foodservice operation to serve a wider variety of fish year-round. Frozen fish cannot be purchased whole. For example, a large percentage of salmon is sold frozen. Frozen salmon is held at 32°F until it is flash-frozen. It is protected from dehydration by glazing. *Glazing* is the process of covering an item with water to form a protective coating of ice before the item is frozen. Proper glazing of frozen salmon results in a product that tastes like fresh salmon.

Processed Fish

Processed fish are available canned, smoked, salted, or pickled. Sardines, anchovies, and tuna are the most common varieties of canned fish. Canned fish should be checked for signs of damage or bulging before they are used. Damaged cans should be discarded.

Popular smoked and cured fish include cod, haddock, salmon, sturgeon, and herring. **See Figure 4-17.** Cod is often salted, whereas salmon and herring are often pickled.

Processed Fish

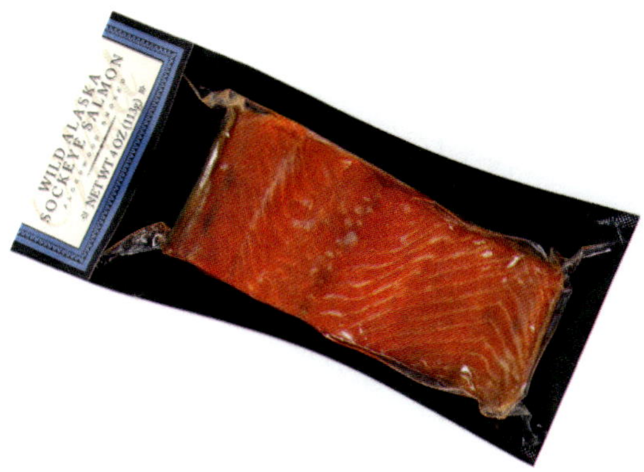

Figure 4-17. Salmon is often smoked.

Receiving and Storing Fresh Fish

Fresh fish spoils rapidly so it is essential to check the smell, appearance, and flesh upon receipt. **See Figure 4-18.** The smell of a fish is one of the easiest ways to determine freshness, but it can also be misleading. Fresh fish have a light, fishy smell. Strong fishy smells or ammonia odors are signs of deterioration. Fish that give off such smells should not be accepted.

The eyes and gills of fresh fish indicate age. The eyes should be round, slightly bulging, and clear. Cloudy or sunken eyes are signs of age and the fish should be rejected. The gills of a fresh fish should be bright red or reddish pink. If the gills are brown or missing, the fish should be rejected.

If fish has been cut into fillets, they should be moist and solid. Gently pressing on the fillet should not leave an indentation. The flesh should be firm and spring back when touched. A fillet should not be slimy, and the flakes of flesh should not separate when the fillet is bent slightly. The flesh should be of the proper color for the particular variety of fish. There also should not be any bruises or other signs of visible damage. If a fish shows any of these signs, it should be rejected.

Consideration for Purchasing Fresh Fish		
	Acceptable	*Unacceptable*
Smell	• Slight smell of seaweed or the ocean	• Strong fishy smell
	• Slight smell of fish	• Ammonia odor
Appearance	• Round, clear, bulging eyes	• Cloudy or sunken eyes
	• Bright-red or reddish-pink gills	• Brown gills or missing gills
	• Firmly attached scales	• Loose scales
	• Properly covered flesh	• Bruised, discolored, or damaged flesh
	• Moist and solid fillets	• Flesh of fillets separates when slightly bent
Touch	• Wet, slightly slippery exterior surface	• Slimy internal cavity
	• Smooth scales lying flat against body	• Rough scales
	• Moist, intact fins	• Dry fins
	• Firm flesh	• Mushy flesh

Figure 4-18. Fresh fish spoils rapidly so it is essential to check the smell, appearance, and flesh upon receipt.

The external surface of a fresh fish should be wet and a little slippery. Scales should be smooth and lie flat against the body of the fish. Scales falling off the fish are signs of age, and these fish should be rejected. Fins should be moist and intact. If the fins are dried out or removed, the fish may be older and the flesh may be starting to dry out. Fish with dry fins should be rejected.

Even if the fish being delivered is from an inspected facility, it may not have been held at proper temperatures since leaving the plant. Fresh fish should be packed in flaked ice to maintain a safe temperature. Fish should have an internal temperature of 41°F or lower to be accepted at delivery. Fish above 41°F should be rejected. Once accepted, fish must be stored at 41°F or below and used as quickly as possible to maintain quality and freshness.

Fresh fish can be stored a maximum of two days. If a fresh fish is not cooked upon receipt, it should be removed from the ice it was packaged in, placed in a self-draining container, covered with plastic wrap, and then covered with a thin layer of crushed ice. **See Figure 4-19.** Then the container should be placed in the coldest part of the refrigerator. The plastic wrap helps prevent the flesh of the fish from developing freezer burn from the ice and from absorbing water as the ice melts. The ice helps hold the proper temperature and reduces deterioration. The ice should be replaced daily. Fish must be stored away from other food in the refrigerator to prevent the odors from affecting the smell or taste of other foods.

Storing Fresh Fish

Figure 4-19. If a fresh fish is not cooked upon receipt, it should be removed from the ice it was packaged in, placed in a self-draining container, covered with plastic wrap, and then covered with a thin layer of crushed ice.

Receiving and Storing Frozen Fish

When receiving frozen fish, it should be received frozen solid. If the fish is even slightly thawed on the edges, it should be rejected. Fish that has been refrozen will have a poor texture. If there are dry-looking spots on the fish, it has slightly thawed and been refrozen and should be rejected. Signs of moisture on the outside of the box may also indicate that the product has thawed and been refrozen. Ice crystals are another sign that fish thawed or was not stored at or below 0°F. Frozen fish quickly deteriorate at temperatures above 0°F, so fish displaying ice crystals should be rejected.

Fish stored in the freezer must be wrapped properly with moistureproof wrapping to prevent freezer burn. Fatty fish should not be stored in the freezer for more than two months because they deteriorate quickly. Lean fish can be stored frozen for up to six months. Stock should be rotated so that the oldest fish is used first. Frozen fish must be thawed in the refrigerator as close as possible to the time the fish is needed. Once frozen fish has thawed, it must be used or discarded. Some seafood vendors offer a slacking or refreshing service. Slacking involves thawing the previously frozen seafood under refrigeration to ensure proper temperature control at the vendor's facility. The same quality indicators should be used as when receiving fresh fish.

Frozen fish can be individually quick-frozen (IQF), layer packed, cello-packed, or block frozen. *Individually quick-frozen (IQF)* is a designation for products preserved using a method in which each item is glazed with a thin layer of water and frozen individually. **See Figure 4-20.** IQF portions can be packaged together without sticking to other portions. This allows as much of an item to be used as needed without having to thaw an entire package. IQF products are packaged according to an average size, such as 2–3 oz fillets. IQF packaging makes it easy to remove the exact number of items needed.

Layer packs, also known as shatter packs, consist of high-quality, graded fish fillets layered on polyethylene sheets. The edges slightly overlap so that entire layers can be removed when desired.

Cello-packs contain ungraded fish fillets that are frozen in packets, typically one to three fillets per packet, wrapped in cellophane, frozen, and packaged six packets per box. Fillets packaged in this manner may be inconsistent in size and are relatively inexpensive.

Block-frozen fish is placed in a block-shaped form between two hollow stainless steel plates that have refrigerant flowing through them. The plates freeze the fish into a solid block within 2–4 hours.

Inspection and Grades of Fish

Fresh fish are not subject to a mandatory federal inspection. Instead, optional inspections are carried out by the National Marine Fisheries Service, which is part of the United States Department of Commerce (USDC). There are three types of optional inspections. These types of inspections include Type 1, Type 2, and Type 3.

A Type 1 inspection guarantees the fish or shellfish product is safe and wholesome for human consumption, is accurately labeled, has a good odor, and was processed in a sanitary, inspected facility. After being processed under a Type 1 inspection, a PUFI (processed under federal inspection) mark is affixed to the packing carton. **See Figure 4-21.** Type 1 inspection involves continual inspection of the fresh product from the time it arrives at the processing plant to the moment it is packaged for sale.

PUFI Marks

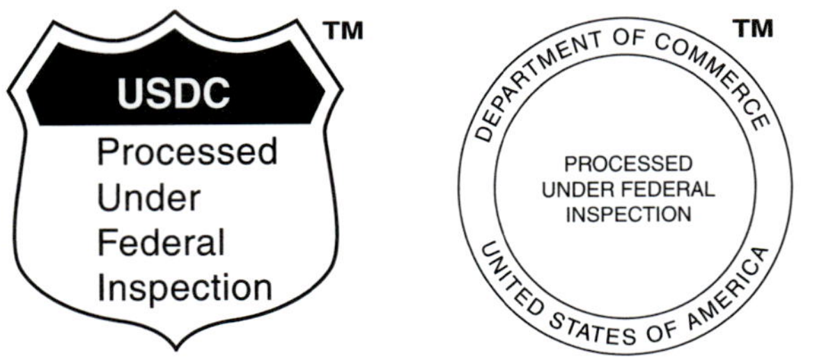

Figure 4-21. After being processed under a Type 1 inspection, a PUFI (processed under federal inspection) mark is affixed to the packing carton.

Individually Quick-Frozen (IQF) Fish

Czimer's Game & Seafoods, Inc.

Figure 4-20. Individually quick-frozen (IQF) is a designation for products preserved using a method in which each item is glazed with a thin layer of water and frozen individually.

A Type 2 inspection takes place in a warehouse or cold storage facility. The product is randomly inspected to ensure it meets the product specifications listed on a specification sheet.

A Type 3 inspection involves the examination of the fishing boats and processing plants to ensure that they are adhering to sanitation guidelines when handling and processing the product.

Only fish that is processed under a voluntary Type 1 inspection is eligible for grading. Fish may be graded A, B, or C. Because there are so many varieties of fish, the USDC only sets grade standards for the most common varieties. Grade A fish are of the best quality and do not have any visible defects. Grade B and Grade C are typically used for processed products.

Premium quality fish are used to make sashimi, which is masterfully presented small pieces of raw fish. Sashimi-grade fish must either be supercooled or kept alive until it is cut and served. *Supercooling* is the process of cooling an item below the freezing point without the item becoming solid or crystallizing. Sashimi-grade fish are typically supercooled to –76°F in order to maintain their texture and quality. The extremely low temperature also kills any parasites that could cause foodborne illnesses.

Production Tip

The practice of treating sashimi-grade fish with carbon monoxide to keep the flesh red for a longer time in storage can make spoiled fish appear fresh. It is best to purchase supercooled or live fresh fish for use as sashimi.

CHECKPOINT 4-5

1. Describe the various market forms of fish.
2. Explain how fresh fish are received and stored.
3. Explain how frozen fish are received and stored.
4. Name the government organization in charge of voluntary fish inspections.
5. Describe what happens during a Type 1 inspection.
6. Explain the importance of supercooling sashimi-grade fish if not using live fish.

FABRICATING FISH

Fish may be purchased whole and fabricated in-house or may be purchased in a portion-controlled or processed form. Knowing and using proper fabrication techniques for fish is an important skill to have in the professional kitchen.

Scaling Fish

A fish scaler is used to remove the scales from fish prior to preparation. **See Figure 4-22.** However, care should be taken to ensure the flesh is not damaged as the scales are removed.

Fish Scalers

Paderno World Cuisine

Figure 4-22. A variety of fish scalers can be used to remove the scales from fish.

Procedure for Scaling Fish

1. In a sink, firmly hold the tail of the fish with the guiding hand. Use a fish scaler to scrape against the scales down toward the head. *Note:* Use caution not to damage the flesh of the fish by applying too much pressure.
2. When finished scaling one side, turn the fish over and repeat the process.
3. Rinse the fish under cold running water to remove any loose scales.

A fish steak is a popular cut for many large roundfish varieties. For example, tuna is often purchased in loins. The loins are commonly cut into steaks. Belly loins commonly have a higher fat content than back loins.

A *blood line* is a dark-red, almost black muscle that runs along the lateral line of a fish. The blood line should be removed prior to preparation. There may also be a vein running along the loin that should be removed because it has a strong, undesirable flavor.

Fabricating Fish

Procedure for Cutting Roundfish into Steaks

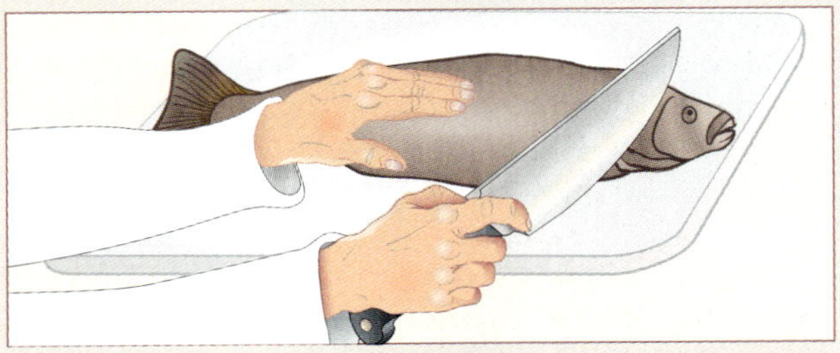

1. Use a boning knife to remove the scales, fins, viscera, head, and tail from the fish.

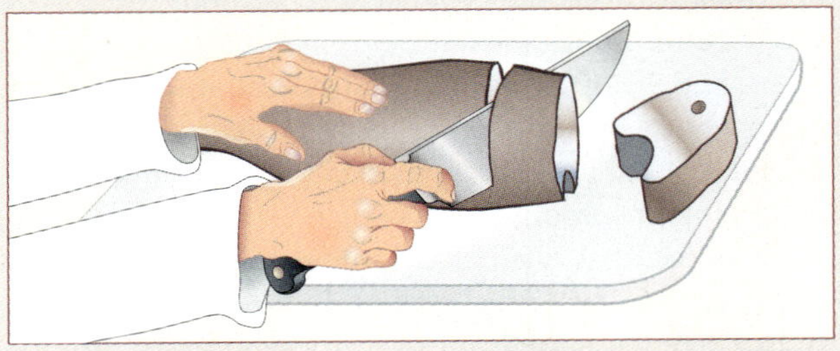

2. Slice the fish horizontally into equal sized pieces.

Filleting Roundfish

A fillet is the most common serving portion from a fish. It is important to remove all belly bones and pin bones from a fillet before cooking.

Skinning Fillets

A fillet can be prepared with or without the skin intact. A particular presentation or dish may require removing the skin from a fillet.

Procedure for Filleting Roundfish

1. Use a boning knife to make a cut about ½ inch behind the gills and down to, but not through, the backbone.
2. Make a second cut along the backbone from just behind the head all the way to the tail. Do not cut through the backbone.

3. Starting at the tail, carefully slice toward the head to cut the flesh away from the backbone.

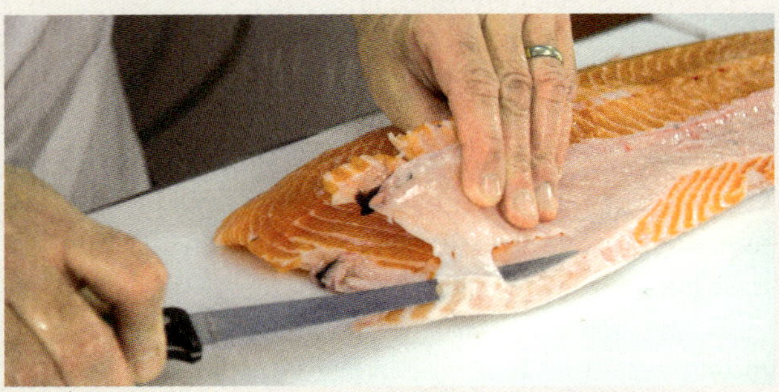

4. Carefully lift the fillet and cut away any rib bones that are still attached to the fillet. Trim any belly fat from the fillet.

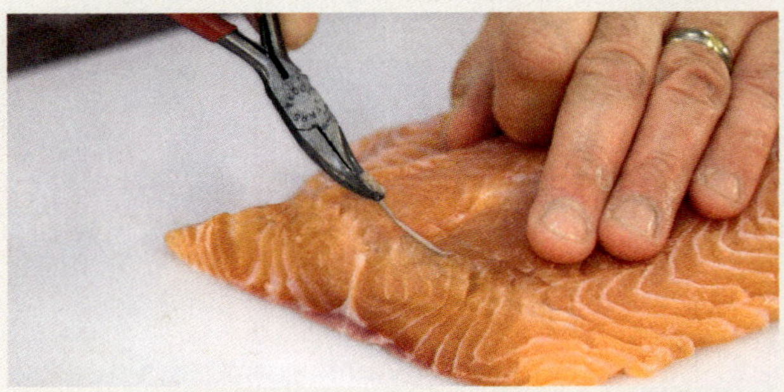

5. Run fingers gently along the surface of the fillet to raise the ends of any pinbones that may remain. Use needle-nose pliers to remove the pinbones.
6. Turn the fish over and repeat the entire process on the other side.

Filleting Roundfish
Media Clip

Procedure for Skinning Fillets

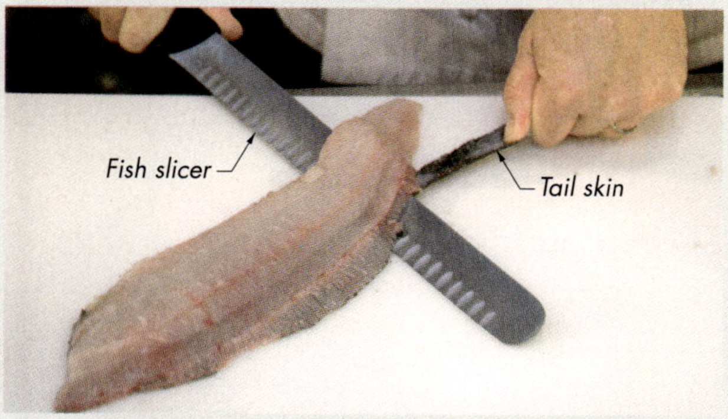

1. Place the fillet skin-side down. Starting at the tail, use a fish slicer or boning knife to carefully cut down through the flesh to the skin while using the guiding hand to firmly hold the tail skin.
2. Angle the edge of the blade slightly downward and cut between the skin and the flesh while moving the blade toward the head end.

Skinned fish fillet.

Filleting Flatfish

The backbone of a flatfish runs through the midline of the fish. As a result, flatfish yield four fillets, two on each side of the fish.

Production Tip

Using a pair of kitchen shears to remove the fins from both sides of a flatfish will make filleting the fish easier.

Procedure for Filleting Flatfish

1. Use a boning knife to cut along the backbone from the head to the tail.

2. Insert the tip of the blade near the head of the fish and carefully slice the flesh away from the bones on one side of the backbone.

3. Remove the second fillet using the same process.
4. Turn the fish over and repeat steps 1–3 to remove the other two fillets.

Filleting Flatfish
Media Clip

CHECKPOINT 4-6

1. Explain how to scale a fish.
2. Scale a fish.
3. Explain how to cut a roundfish into steaks.
4. Cut a roundfish into steaks.
5. Explain how to fillet a roundfish.
6. Fillet a roundfish.
7. Explain how to skin a roundfish fillet.
8. Skin a roundfish fillet.
9. Explain how to fillet a flatfish.
10. Fillet a flatfish.

SHELLFISH

Shellfish is the classification of aquatic invertebrates that may or may not have a hard, external shell. An external shell functions as a skeleton and is called an exoskeleton. Shellfish are commonly categorized as crustaceans or mollusks. **See Figure 4-23.**

Crustaceans

A *crustacean* is a shellfish that has a hard, segmented shell that protects soft flesh and does not have an internal bone structure. Crustaceans live in both freshwater and saltwater. Unlike fish, crustaceans can live out of water for a few days if they are kept moist. Crustaceans include shrimp, prawns, lobsters, crayfish, and crabs.

Shrimp and Prawns. Shrimp and prawns are different species of crustaceans that both have a tender white flesh. However, the terms shrimp and prawns are often used interchangeably. In the United Kingdom and Australia, the term prawn is used for both shrimp and prawns. In North America, larger shrimp are commonly marketed as prawns. **See Figure 4-24.**

Shrimp and Prawns

Figure 4-24. In North America, larger shrimp are commonly marketed as prawns.

Four common types of shrimp used in the professional kitchen are white, brown, pink, and black tiger. Although these four types of shrimp vary in color when caught, they differ very little in appearance when cooked. They also have similar flavor and nutritional value. Grilling, broiling, sautéing, and frying are popular methods of preparing shrimp. Cooked shrimp also can be chilled and served cold as appetizers, salads, or entrées.

Shellfish

Agricultural Research Service, USDA

Crustaceans

Florida Department of Agriculture and Consumer Services, Bureau of Seafood and Aquaculture Marketing

Mollusks with Shells (Univalves)

National Oceanic and Atmospheric Administration / Department of Commerce

Mollusks without Shells (Cephalopods)

Figure 4-23. Shellfish are commonly categorized as crustaceans or mollusks.

Lobsters. A *lobster* is a saltwater crustacean with a brown to bluish-black external shell and two large claws. **See Figure 4-25.**

Lobsters are most often steamed or poached. The two primary types of lobster are clawed lobsters and spiny lobsters. The two types are similar in size but vary in color. Clawed lobsters contain more meat than spiny lobsters. Clawed lobsters are sold live whereas spiny lobsters are harvested for their tails. Common varieties of lobster include Maine lobsters, spiny lobsters, and langoustines.

- A *Maine lobster* is a large clawed lobster with a dark bluish-green shell, two heavy claws, and eight slender legs. Maine lobsters are harvested from the cold North Atlantic waters.
- A *spiny lobster,* also known as a rock lobster, is a lobster that has spines covering its body and five slender legs on each side. Spiny lobsters have smaller claws than Maine lobsters. The tail section of a spiny lobster is typically eaten. Spiny lobsters are typically harvested from warm waters.
- A *langoustine,* also known as a Norway lobster or a Dublin Bay prawn, is a small lobster that resembles a very large shrimp, except it has very long front arms with long, thin claws. Only the tail section of the langoustine is eaten. Langoustines are harvested from both cold and warm waters such as the northeastern Atlantic Ocean, the North Sea, and the Mediterranean Sea.

Lobsters that are sold live must be kept alive until cooking. A *sleeper* is a lobster that is dying. Sleepers are sold at a reduced price and must be cooked immediately. Cooking and eating dead lobsters can be harmful.

Almost all lobster flesh is edible except for a small section near the eye area. Most of the flesh is found in the tail and the claws. The soft green substance found in the body cavity of a lobster is the liver and pancreas, which are collectively called the tomalley. The tomalley is considered a delicacy. It may be eaten alone or used to thicken sauces. Lobster paste, also known as lobster pâté, is a mixture of tomalley and lobster roe (coral).

Crayfish. A *crayfish,* also known as crawfish or crawdad, is a freshwater crustacean that resembles a tiny lobster. **See Figure 4-26.** Crayfish are dark brown to black in color and can range from 3–7 inches in length. Most of the crayfish harvested come from Louisiana and the Pacific Northwest. Crayfish have a flavor similar to that of shrimp, but a slightly tougher texture. They are commonly used in Creole, Cajun, and French cuisine.

Crayfish

Figure 4-26. A crayfish, also known as crawfish or crawdad, is a freshwater crustacean that resembles a tiny lobster.

Crabs. Crabs are a popular variety of shellfish with tender, sweet-tasting flesh that can be used in many menu items. Crabmeat is available fresh, frozen, and canned. When using fresh crabs, only live crabs should be used. Fresh crabs that have died should be discarded.

Anatomy of a Lobster

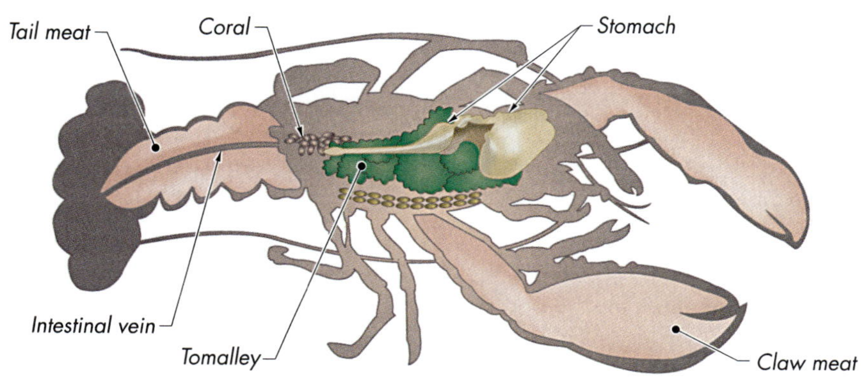

Figure 4-25. A lobster is a saltwater crustacean with a brown to bluish-black external shell and two large claws.

Most edible crabs can be distinguished from inedible varieties by counting the pairs of legs. Most edible crabs have five pairs of legs, four for walking and one pair that serve as arms. Most inedible crabs have only four pairs of legs. King crabs are an exception, as they have only four pairs of legs but are still edible. King crabs, blue crabs, Dungeness crabs, snow crabs, and stone crabs are frequently used in the professional kitchen. **See Figure 4-27.**

- *King crab* is the largest-sized variety of crab, typically weighing between 6–20 lb and measuring as much as 10 feet from the tip of one leg to the tip of the opposite leg. King crabs are one of the few edible crabs that have four pairs of legs. The raw flesh has a pinkish tinge that becomes snowy white when cooked. King crab is available in the shell as cooked frozen legs, as legs and claws, as leg and body flesh, or as shredded flesh. Canned or frozen king crabs are also available shucked or pulled (removed from the shell). King crab is not typically sold live because the legs are removed at catch and flash-frozen to preserve their quality.
- *Blue crab* is a North American crab with blue claws and a dark blue-green, oval shell. Blue crabs measure approximately 5 inches across and weigh approximately 5 oz. Hard-shell blue crab is sold live, cooked and frozen, or cooked, pickled, and canned. A *soft-shell crab* is a blue crab that has been harvested within 6 hours of molting, or shedding its shell in order to grow a larger shell. Molting season is from mid-May to the beginning of September. Soft-shell crab is only sold live during molting season. Soft-shell crabs must be handled with special care to ensure that they arrive at their destination alive. They must be kept alive until they are cooked or cleaned and immediately quick-frozen.
- *Dungeness crab* is a Pacific crab with a sweet-tasting flesh. Approximately 25% of the body weight of a Dungeness crab is flesh, which is the highest percentage found in any variety of crab. Most of the edible flesh in a Dungeness crab is body flesh, not leg flesh. Dungeness crabs are larger than blue crabs, weighing approximately 1¾–4 lb each. Dungeness crab is commonly available frozen or canned but can be purchased live.
- A *snow crab* is a crab that is similar to a king crab but is smaller and available in greater supply. Also known as a spider crab, snow crab is typically sold as cooked, frozen leg clusters. A snow crab has only a fraction of the flesh that a king crab has. Frozen leg clusters are commonly steamed or poached just enough to warm the flesh because they are already fully cooked. Snow crab can be steamed and served whole or served cold in a crab cocktail.

Crab Varieties

National Oceanic and Atmospheric Administration / Department of Commerce

King

Florida Department of Agriculture and Consumer Services, Bureau of Seafood and Aquaculture Marketing

Blue

U.S. Fish & Wildlife Service

Dungeness

National Oceanic and Atmospheric Administration / Department of Commerce

Snow

Figure 4-27. Crab varieties include king crab, blue crab, Dungeness crab, and snow crab.

- A *stone crab* is an Atlantic crab with a brownish-red shell and large claws of unequal size. Unlike other varieties of crab, when a stone crab is caught one claw is removed and the crab is placed back in the water. The claws are the only parts of the stone crab that are harvested. In approximately 12–24 months, the stone crab generates a new claw in place of the one that was removed. Each time a claw is removed it grows back slightly larger than before. Stone crab claws range from 2–5 oz and are typically served cold with most of the shell removed. They can also be served hot. Stone crabs have a similar taste and texture to lobster.

Mollusks

A *mollusk* is a shellfish with a soft, nonsegmented body. Some mollusks, such as clams, have a hard, external shell. Other types of mollusks, such as squid, do not have an external shell. Like crustaceans, mollusks do not have an internal skeleton. Mollusks should always be alive prior to cooking, unless they are frozen or canned. If a mollusk dies prior to cooking, it should be discarded. The three classifications that edible mollusks are divided into are univalves, bivalves, and cephalopods.

Univalves. A *univalve*, also known as a gastropod, is a mollusk that has a single solid shell and a single foot. A univalve uses its foot to move along the surface of underwater structures or on land. Univalve varieties include abalone and conch. **See Figure 4-28.**

- An *abalone* is a univalve contained in a brown, bowl-shaped shell with an iridescent, multicolored interior. The California abalone variety is considered the highest in quality, but all varieties of abalone have a sweet, slightly salty flesh. The texture of an abalone is tender like lobster, but has a creamier taste. It is very tough when overcooked. Aquafarmed abalones are raised in saltwater pens. Abalones are also imported from the coastal waters of Japan, New Zealand, and Mexico and are sold in canned and frozen form.
- A *conch* is a univalve that has a pinkish-orange shell with an interior that resembles a large snail. Conchs are found in the warm waters off the Florida Keys and throughout the Caribbean. Conch flesh is rubbery and is typically sliced thin and tenderized before it is cooked. It has a sweet flavor similar to that of a clam. Fresh conch is often eaten raw with lime juice and hot sauce.

Bivalve. A *bivalve* is a mollusk with a top shell and a bottom shell connected by a central hinge that can close for protection. Live bivalves should close tightly when gently tapped. Bivalves that are broken or do not close when tapped are dead and should be discarded. A bivalve that is noticeably heavier than others is probably dead and should be discarded. Common types of bivalves include clams, cockles, mussels, oysters, and scallops.

Nutrition Note

Mollusks are a good source of protein, vitamin B12, vitamin C, iron, zinc, and copper. Mollusks are also very high in cholesterol.

Univalves

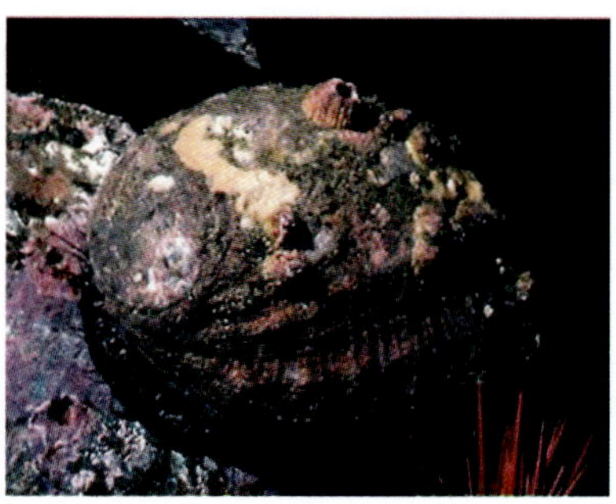

National Oceanic and Atmospheric Administration / Department of Commerce

Abalone

U.S. Fish & Wildlife Service

Conch

Figure 4-28. Univalve varieties include abalone and conch.

Clams. A *clam* is a bivalve found in both freshwater and saltwater. Clams can be fried or steamed with excellent results and are the key ingredient in clam chowders. There are many varieties of ocean clams found in different oceans. **See Figure 4-29.**

Ocean Clams

Figure 4-29. There are many varieties of ocean clams.

The Atlantic ocean produces soft-shell, hard-shell, and surf clams. A *soft-shell clam,* also known as a long-neck or steamer clam, is an Atlantic clam with a thin, brittle shell that breaks easily. Soft-shell clams have a protruding siphon that prevents the shell from closing completely, which causes them to dry out more quickly than hard-shell clams once removed from the water. A *siphon* is a tubular organ that is used to draw in or eject fluids. Soft-shell clams also have a tendency to be gritty due to excess sand settling inside the shell. They need to be soaked in a solution of salted water and cornmeal to remove the sand. Soft-shell clams have a tender, sweet-tasting flesh and are commonly steamed or fried.

A *hard-shell clam,* also known as a quahog, is an Atlantic clam with a blue-grey shell that contains a chewy flesh. These clams are rarely sold by the name hard-shell clam or quahog. Instead, they are sold by classifications that indicate their size. These classifications include littleneck, cherrystone, topneck, and chowder clams.

- Littlenecks are 1–2 inches in size. These bite-size clams are served raw on the half shell or steamed.
- Cherrystones are 2–3 inches in size. They are commonly steamed or served raw on the half shell.
- Topnecks are 3 inches in size and are often stuffed and baked.
- Chowder clams are larger and typically cut into strips or minced for making soups and chowders.

A *surf clam* is a large species of Atlantic clam that can grow to 8 inches in size. Half of a shucked surf clam is the siphon, which is typically cut into strips and then breaded and fried. The other half of the shucked surf clam is the adductor muscle. The *adductor muscle* is a muscle that opens and closes the shell of a bivalve. It is commonly chopped or ground for use in chowders. Surf clams have a sweet, mild flavor.

Pacific clams have a tougher texture than Atlantic clams. Pacific clams include Manila clams, razor clams, butter clams, and geoducks.

Manila clams are the most common variety of Pacific clam. The shell is covered with slight ridges from the lip to the hinge. Manila clams can be steamed or served raw and have a sweet and salty flavor.

Razor clams are named for their narrow, oval-shaped shell. Razor clams have a sweet flavor. Butter clams have a mild, buttery flavor.

A *geoduck* is a very large Pacific clam with a meaty siphon that protrudes from its shell. Geoducks can weigh up to 3 lb. The siphon of a geoduck is typically split in half lengthwise and then cut into very thin slices and served carpaccio-style. It can also be quickly sautéed, steamed, or poached. Geoducks become very tough if overcooked.

Cockles. A *cockle* is a 1 inch wide bivalve with a shell that has deep ridges. **See Figure 4-30.** The small, fleshy cockles are often served shucked with a squeeze of lemon juice. Cockles are often used to make paella and Southeast Asian and Thai dishes.

Cockles

Fortune Fish Company

Figure 4-30. A cockle is a 1 inch wide bivalve with a shell that has deep ridges.

Mussels. A *mussel* is a freshwater or saltwater bivalve with whisker-like threads that extend outside the shell to allow the animal to attach to items for protection. These threads are referred to as a "beard." In the wild, mussels are commonly found attached by their beards to rocks. Aquafarmed mussels attach their beards to ropes. Common types of mussels include blue mussels and greenlip mussels.

- Blue mussels are the most common variety of edible mussel. **See Figure 4-31.** Aquafarmed blue mussels have a thinner, blue-black shell, while wild-caught blue mussels have a thicker, silver-blue shell. Blue mussels have tender, sweet flesh that is bright orange in color. They vary in size, but typically there are 10–20 mussels per pound. Blue mussels can be steamed and served either hot or cold or shucked and then sautéed or simmered in a sauce.

Blue Mussels

New Zealand Greenshell™ Mussels

Figure 4-31. Blue mussels are the most common variety of edible mussel.

- Greenlip mussels have a distinctive green-edged shell and are larger than blue mussels. They average 8–12 mussels per pound. They have a sweet, plump, and tender flesh. Greenlip mussels are often steamed with white wine, lemon, and herbs or cooked and served cold with cocktail sauce and lemon wedges.

Oysters. An *oyster* is a saltwater bivalve with a very rough shell that is coated with calcium deposits. Although oysters are not found in freshwater, they can be found in brackish water (a mixture of freshwater and saltwater). Most oysters are cultivated from oyster beds that require care and attention if they are to continue to produce. Common varieties include Atlantic oysters, Pacific oysters, and European oysters. The variety of oyster and where it comes from impacts the flavor profile.

- An *Atlantic oyster,* also known as an Eastern oyster, is a variety of oyster that has a fairly flat shell and a distinctive, salty-flavored, plump, and tender flesh. **See Figure 4-32.** Atlantic oysters account for roughly 70% of all oyster production. Common varieties of Atlantic oysters include Blue Point oysters, Chesapeake Bay oysters, Long Island oysters, and Prince Edward Island (PEI) oysters. PEI is known for high-quality, fresh seafood that can be delivered to the customer within hours of harvest.

Atlantic Oysters

Florida Department of Agriculture and Consumer Services, Bureau of Seafood and Aquaculture Marketing

Figure 4-32. Atlantic oysters have a fairly flat shell and a distinctive, salty-flavored, plump, and tender flesh.

- A *Pacific oyster,* also known as a Japanese oyster, is a variety of large oyster that has fragile, curvy shells and a briny, sweet, and mild-tasting flesh. The plump, moist flesh is silver, gold, or white in color. Common varieties include Olympia, Penn Cove, and Kumamoto oysters.
- A *European oyster* is a variety of oyster with a relatively flat, cup-shaped shell and salty-sweet flavored, creamy-textured flesh. European oysters are commonly served on the half shell. Belon oysters are a rare variety of European oysters.

Oysters are available year-round, but most are at peak quality from September to April. Oysters may be served raw, roasted, baked, breaded and fried, or poached. The key to preparing oysters is to apply just enough heat to heat them through, leaving them plump and tender. A common preparation is to top an oyster with a savory filling. Raw oysters need to be held on ice during service to keep them within a safe temperature range for consumption.

Scallops. A *scallop* is a bivalve with a fan-shaped shell and a cream-colored adductor muscle with a sweet, delicate flavor. The well-developed adductor muscle is the lean and juicy edible portion of the scallop. The rest of the scallop body is made up of white or red roe called coral, which is often removed prior to sale. Color is the best way to judge the quality of scallops. High-quality scallops have a creamy, almost translucent color. If scallops are white, they have been packed wet and the flavor and texture have been impaired. Rose-colored scallops indicate the scallops were in the process of producing roe. If scallops are a brownish color, they are old and should not be used.

Some chefs use scallop shells as serving dishes when featuring seafood appetizers or entrées. Bay scallops and sea scallops are different varieties of scallops that vary in size and flavor. **See Figure 4-33.**

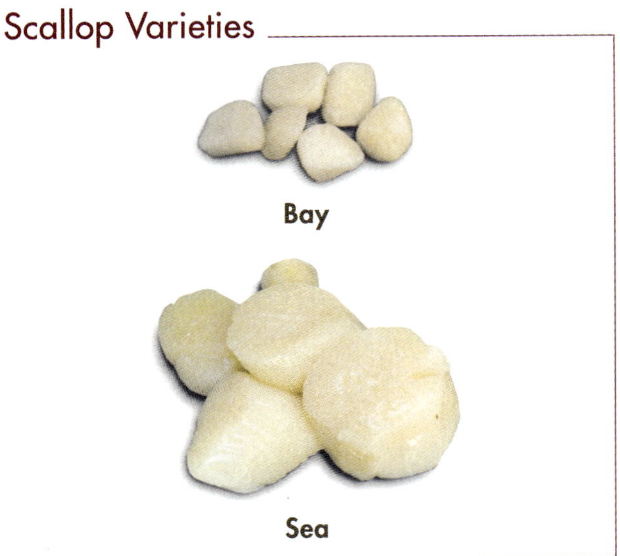

Figure 4-33. Bay scallops and sea scallops are different varieties of scallops.

A *bay scallop* is a fairly small scallop harvested from shallow saltwater. Bay scallops average 100 scallops per pound. Bay scallops are typically cleaned prior to sale and packed wet (soaked in a preservative that whitens the scallop and helps prevent spoiling), dry (untreated, without any preservatives), or individually quick-frozen. The most common preservative used with scallops is sodium tripolyphosphate, also known as STP. Bay scallops are commonly sautéed in butter, battered and fried, or marinated in lemon juice and served seviche-style.

A *sea scallop* is a large scallop with a coarse texture that is harvested from deep saltwater. Sea scallops are typically 2–5 times larger than bay scallops and average 30 scallops per pound. Sea scallops have a sweet, somewhat briny taste.

Fortune Fish Company

Cephalopods

A *cephalopod* is any of a variety of mollusks that do not have an external shell. Some cephalopods have an internal bone called a cuttlebone. Cephalopods have keen vision, a highly-developed nervous system, and a birdlike beak that is used to crack the shells of shrimp, crabs, lobsters, and other prey. Squid, octopuses, and cuttlefish are cephalopods. **See Figure 4-34.**

Squid. A *squid* is a translucent, head-footed cephalopod that has two tentacles, eight sucker-equipped arms, two lateral fins, and a flat, internal cuttlebone. The arms are attached near the eyes at the bottom of the head. The two tentacles are longer than the arms. Squid can change their skin color at will and expel a dark cloud of ink to confuse their prey. The ink is often used to color and flavor grain and pasta dishes. Most whole squid are packed 8–10 per pound. Larger squid are cut into steaks and sold frozen. Squid is often called by its Italian name, calamari. Squid is commonly sautéed, breaded and fried, or used to make stews. The flesh becomes tough if overcooked.

Octopuses. An *octopus* is a gray cephalopod with eight sucker-equipped arms, a birdlike beak, well-developed vision, and no internal or external shell. Octopuses can be small or quite large. However, they do not grow as large as giant squid. Octopuses are usually sold whole by the pound and are available fresh and frozen. When cooked, the gray skin turns deep purple or reddish-purple. Octopus flesh is white, firm, and sweet. The flesh becomes tough if overcooked. Cold, cooked octopus is used to make pulpo salad.

Cuttlefish. A *cuttlefish* is a translucent cephalopod with two tentacles, eight sucker-equipped arms, a hard internal cuttlebone, and large eyes at the base of its head. They can change their color at will. Cuttlefish expel a dark-brown cloud of ink to confuse their prey. The tentacles, arms, and ink are the only parts of a cuttlefish that are eaten. The ink is commonly used to make cuttlefish risotto, which is often dark in color. The tentacles and arms are typically cut into rings and sautéed, breaded and fried, or cooked in soups or stews. The flesh becomes tough if overcooked.

Cephalopods

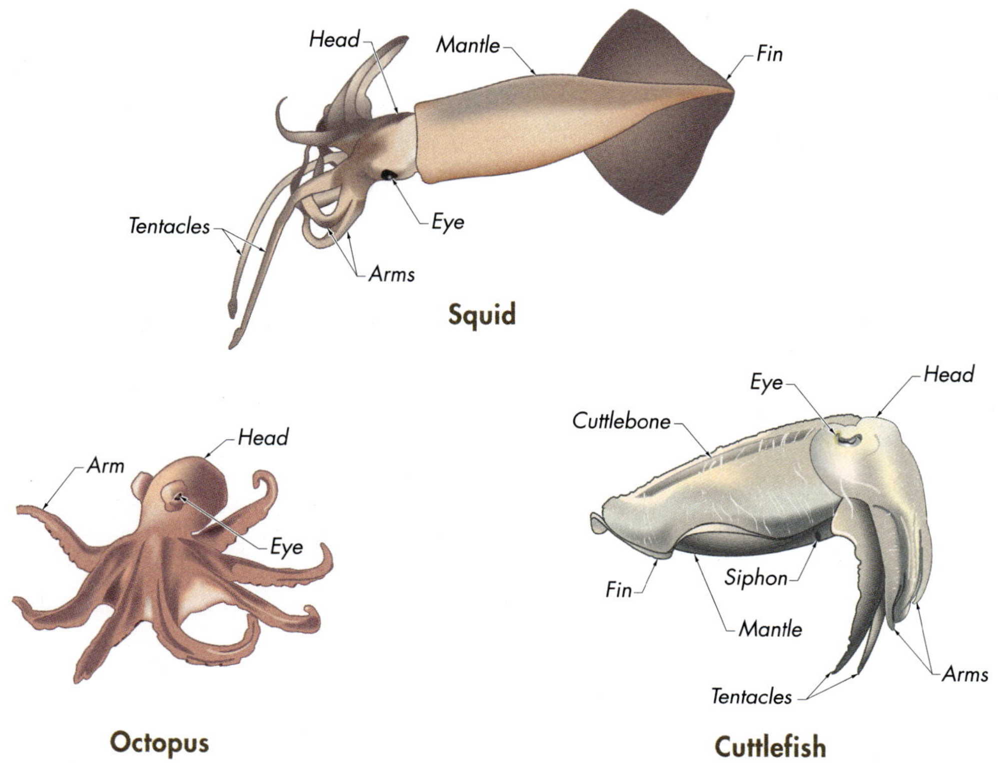

Figure 4-34. Squid, octopuses, and cuttlefish are cephalopods.

CHECKPOINT 4-7

1. Describe four types of crustaceans.
2. Describe two types of univalves.
3. Describe five types of bivalves.
4. Describe three types of cephalopods.

MARKET FORMS OF SHELLFISH

Each market form of shellfish has advantages and disadvantages in terms of cost, convenience, and labor. Shellfish are available live, shucked, or frozen. Shucked and cooked shellfish are also available canned.

Live Shellfish

Lobsters are usually purchased live and come in various sizes. On the coast both hard-shell and soft-shell crabs are sold live. Away from the coast only soft-shell crabs are available live.

Live clams, oysters, and scallops can be purchased by the gallon or the pound. They are usually packed by count. A higher count indicates a smaller size. If a container of oysters is marked 350 per gallon and another container is marked 225 per gallon, the 350 count container holds more oysters that are smaller. Live oysters and clams have tightly closed shells. If a shell is open or does not close fully when lightly tapped, the animal is dead and must be discarded. Oyster and clam shells that are unusually heavy probably contain mud and should also be discarded.

Shucked Shellfish

Clams, oysters, and scallops are often sold with the shell removed and are available both fresh and frozen. Shrimp may also be purchased cooked and canned. Cooked shrimp may be purchased by the pound either in the shell or peeled. Crabmeat can be cooked (fresh or frozen) and canned. Crabmeat also may be purchased fresh or frozen in the shell in a variety of cooked forms. **See Figure 4-35.** Cooked crabmeat should be kept packed in ice and refrigerated until it is used.

Forms of Cooked Crab	
Forms	**Descriptions**
Jumbo lump	Large pieces of white body flesh attached to the swimming legs
Back fin, lump, or special	Small pieces of white flesh from the body
Claws and fingers	Brownish flesh from the claws and legs
Cocktail claws	Claw flesh with a portion of the shell remaining as a handle for eating

Figure 4-35. Crabmeat also may be purchased fresh or frozen in the shell in a variety of cooked forms.

Imitation crabmeat, known as surimi, is also available. *Surimi* is a fish product made from a mixture of fish and/or shellfish and other ingredients. Surimi looks, cooks, and tastes like crabmeat. **See Figure 4-36.** It is high in protein and low in calories, sodium, fat, and cholesterol. Surimi is precooked and frozen. It can be purchased as legs, chunk meat, or flake meat.

Surimi

Harbor Seafood, Inc.

Figure 4-36. Surimi looks, cooks, and tastes like crabmeat.

Frozen Shellfish

Prepared lobster tails may be purchased frozen. The head and thorax (body) of shrimp and prawns are typically removed, and the tails are frozen while still at sea to ensure maximum freshness. Frozen shellfish is commonly packaged by count. **See Figure 4-37.** For example, shrimp packed and labeled 21/25 count indicates that there is an average of 21–25 shrimp per 1 lb package. Frozen shrimp can be purchased green (uncooked), peeled or unpeeled, cooked and peeled, or peeled, deveined, and breaded. Frozen shrimp are usually sold in 5 lb blocks.

Packaging by Count

Sizing Shrimp	
Count	**Size**
25 or fewer per lb	Jumbo
25 to 30 per lb	Large
30 to 42 per lb	Medium
42 or more per lb	Small

Frozen Shrimp

Fortune Fish Company

King Crab Leg Clusters

Figure 4-37. Frozen shellfish is commonly packaged by count.

Receiving and Storing Shellfish

To be acceptable upon delivery, live shellfish should be delivered with an air temperature of 45°F and shucked shellfish must have an internal temperature of 45°F or below. Once accepted, it is imperative to store shellfish between 30°F and 34°F or below and to use it as quickly as possible to maintain quality. If the shellfish will not be used within two days of purchase, it should be wrapped in moistureproof freezer paper and foil, to protect it from air and moisture, and then frozen. Shrimp, scallops, and crabmeat are often block frozen.

Live Shellfish. Lobsters, crabs, clams, oysters, and mussels are often purchased live in the shell. Live crustaceans should be covered with wet seaweed or damp newspaper to keep them from drying out. Crustaceans may also be stored in a saltwater tank. In the absence of a tank, live crustaceans live two to four days. When received, the lobsters must be carefully inspected. Live lobsters have a tightly curled tail. Live lobsters and crabs should show leg movement when received and must be kept alive until they are cooked. If a shellfish is dead, it should be rejected.

Production Tip

To pick up a lobster, grab it by the back just above where the tail connects to the body. Stay clear of the claws to avoid injury, even when the claws are bound with rubber bands.

Fresh shellfish are not subject to a mandatory federal inspection. However, all mollusks must be delivered with a shellstock tag. Shellstock tags are waterproof, tear-resistant tags attached to containers of mollusks. **See Figure 4-38.** The tags must remain attached to the bag of shellfish until the bag is empty. Shellstock tags list the dealer's contact information and identification number, the original harvester (if different than the dealer), the harvest date and general area, the type and quantity of shellfish, a 90 day retention notice, and a consumer advisory. Shellstock tags are kept on file for 90 days after the mollusks are harvested in case of foodborne illness. Mollusks delivered without shellstock tags should be rejected.

Shellstock Tags

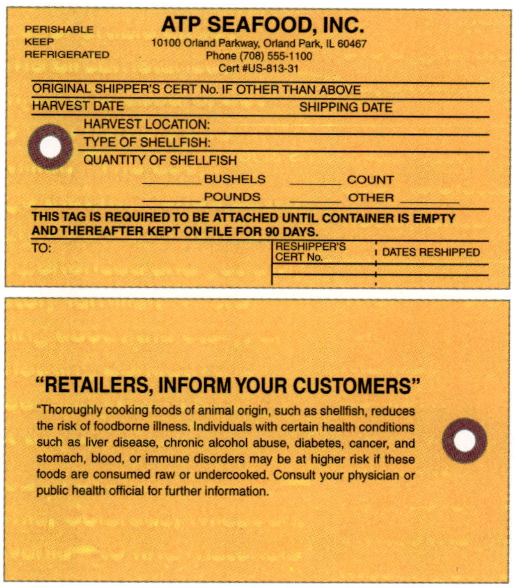

Figure 4-38. Shellstock tags are waterproof, tear-resistant tags attached to containers of mollusks.

Live bivalves should be refrigerated in the original box or netted bag and placed in a pan to prevent drips from contaminating other foods. Live mollusks should not be stored in a sealed container or plastic bag because they will die from a lack of oxygen. Under ideal conditions, fresh mollusks can live in refrigerated storage for up to a week. **See Figure 4-39.** Live oysters and clams have tightly closed shells. If a shell is open and does not close when handled, the oyster or clam is dead and should be discarded.

Storing Live Mollusks

Figure 4-39. Under ideal conditions, fresh mollusks can live in refrigerated storage for up to a week.

Shucked oysters are packed in metal containers and must be kept refrigerated and packed in ice at all times to prevent spoilage. If handled in the proper manner, oysters remain fresh for up to a week.

Frozen Shellfish. When receiving frozen shellfish, it is important to verify that the product is frozen and has not been thawed and refrozen. **See Figure 4-40.** If shellfish is thawed on the edges, it should be rejected. Shellfish that has been refrozen will have a poor texture. Dry spots on the shellfish indicate that it has thawed slightly and been refrozen. Ice crystals are another sign that shellfish has been thawed or not stored at appropriate temperatures. Shellfish displaying dry spots or ice crystals should be rejected. Maximum shelf life is obtained by storing frozen shellfish at 0°F or below.

Frozen Shellfish

Fortune Fish Company

Figure 4-40. When receiving frozen shellfish, verify the product is frozen and has not been thawed and refrozen.

CHECKPOINT 4-8

1. Describe various market forms of shellfish.
2. Explain how live shellfish are received and stored.
3. Describe the purpose of a shellstock tag.
4. Explain how frozen shellfish are received and stored.

FABRICATING SHELLFISH

Shellfish are often purchased whole and fabricated in-house. Common shellfish fabrication techniques include deveining shrimp, debearding mussels, splitting lobster tails, cleaning soft-shell crabs, shucking oysters, cleaning squid, and shucking clams.

Deveining Shrimp

Shrimp contain a sand vein, or intestinal tract, that must be removed prior to further preparation. The shells are also removed.

Debearding Mussels

Mussels have hairlike threads called beards that need to be removed prior to preparation. If the beards are removed too soon, the mussels will die.

Procedure for Debearding Mussels

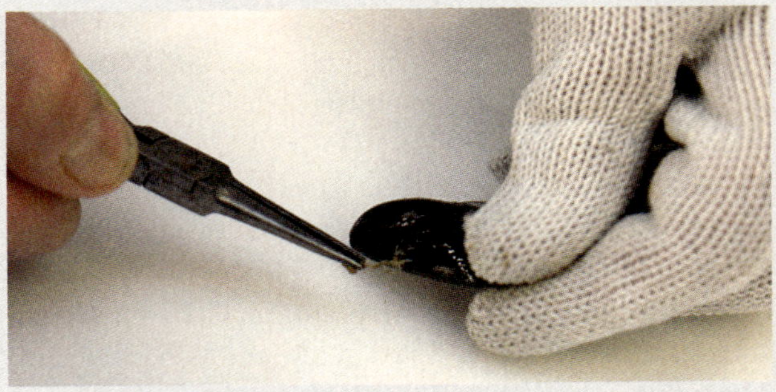

1. Brush the mussel under running water to remove excess mud, sand, and debris.
2. Use a pair of needle-nose pliers to grab the beard and gently pull it free from the mussel.

Splitting Lobster Tails

Broiled lobster tail is a preparation that requires the shell of the tail to be split open. The tail meat is pulled up from, but left inside, the shell for service.

Procedure for Deveining Shrimp

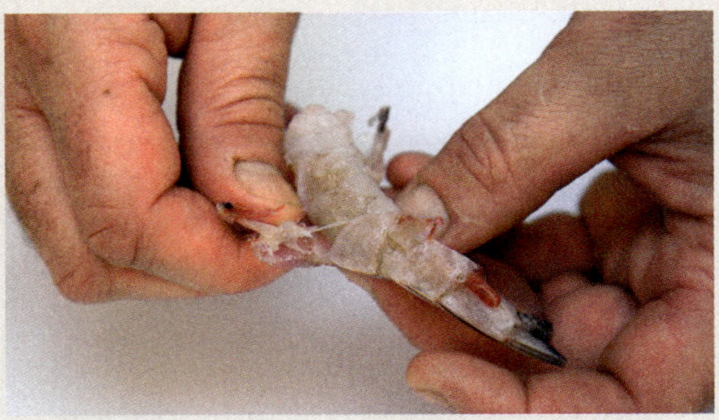

1. Hold the tail fins in one hand with the underside of the shrimp facing up. Pinch the legs between the thumb and index finger of the other hand and gently pull to remove.
2. Grasp the tail between the thumb and index finger of the guiding hand. Use the other hand to gently grasp the shell and twist it slightly to separate the shell from the body.

3. Use a paring knife to make a shallow slice along the back of the shrimp to expose the sand vein. *Note:* If the shrimp is to be butterflied, make a deeper slice to open the flesh so that it lays flat.
4. Gently pull the sand vein to remove it. Discard the vein. *Note:* Holding the shrimp under cold running water may facilitate the removal of the sand vein.

Deveining Shrimp
Media Clip

Procedure for Splitting Lobster Tails

1. With the underside of the lobster facing down, insert the tip of a chef's knife into the center of the tail just above the bottom fins.

2. Use the guiding hand to push the handle of the knife down to slice through the top portion of the shell lengthwise.

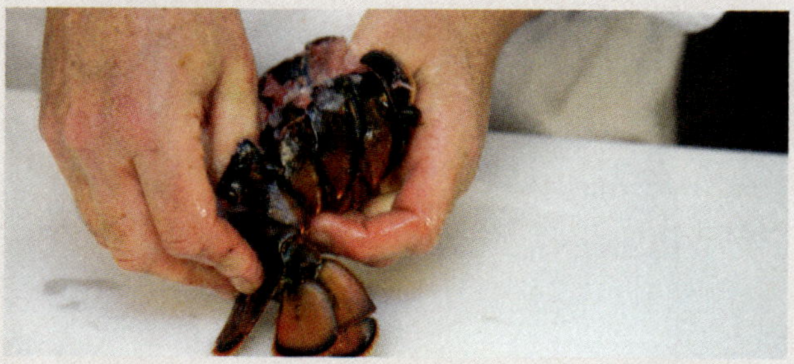

3. Spread the cut shell all the way open to reveal the tail flesh.

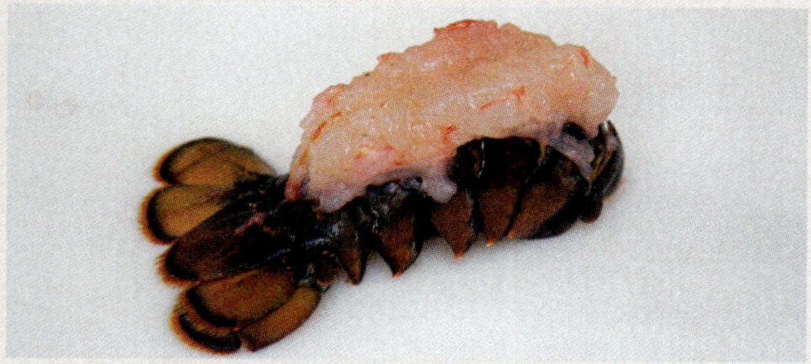

4. Gently pull the tail flesh up through the cut shell for presentation.
5. Make a few shallow slices along the underside of the tail flesh.

Cleaning Soft-Shell Crabs

Cleaning soft-shell crabs involves removing the inedible portions, such as the eyes and the apron. **See Figure 4-41.** The apron is tucked under the body and used to carry and conceal eggs.

Soft-Shell Crabs

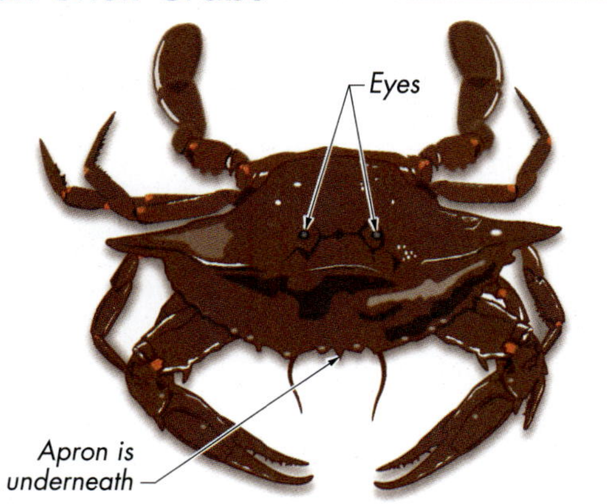

Figure 4-41. Cleaning soft-shell crabs involves removing the inedible portions, such as the eyes and the apron.

Nutrition Note

Crabs are a good source of protein, vitamin B12, zinc, and copper. King crabs are also high in vitamin C. However, crabs are very high in cholesterol and sodium.

Procedure for Cleaning Soft-Shell Crabs

1. Peel back the pointed top shell of the crab to reveal the stringlike gills. Use a boning knife to carefully scrape away the gills.
2. Locate the eyes on the top of the head. Use kitchen shears to remove the mouth and head by cutting just behind the eyes.
3. Gently squeeze the body just behind where the head was removed to release a green bubble of fluid. Rinse the green fluid away.
4. Turn the crab over and locate the apron. Firmly twist and then pull the apron to remove it. Discard the apron. *Note:* The intestinal tract will be attached to the apron.

Shucking Oysters

Shucking is the process of opening a bivalve. Bivalves such as oysters and clams must be shucked to access the edible flesh inside the shell. An oyster knife is used to shuck oysters.

Cleaning Squid

Squid is often shipped cleaned and ready for cooking. If a squid is white and the eyes have been removed, it has already been cleaned. If a squid is dark in color and the eyes are intact, it must be cleaned before being prepared. The tentacles may be left whole or cut into smaller sections. The mantle can be cut into rings, sections, or left whole.

Shucking Clams

Clams are shucked with a clam knife. Clams also may be steamed open to access the flesh.

Procedure for Shucking Oysters

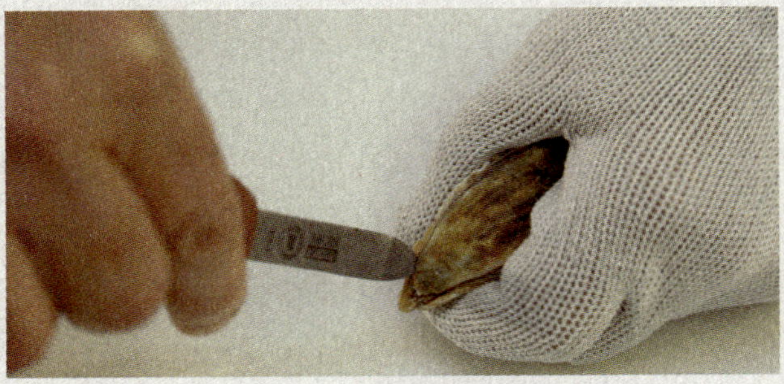

1. While wearing a mesh glove on the guiding hand, firmly hold the oyster in the palm, hinge side up.
2. Insert the tip of an oyster knife near the hinge of the shell.

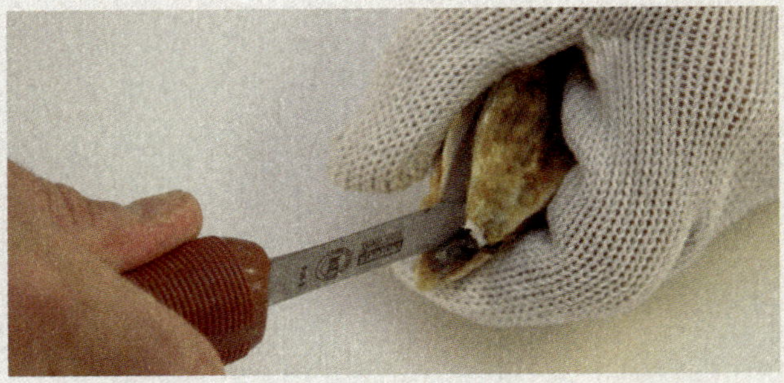

3. Carefully twist the blade to pop the hinge open. *Note:* Care should be taken not to cut the flesh of the oyster.

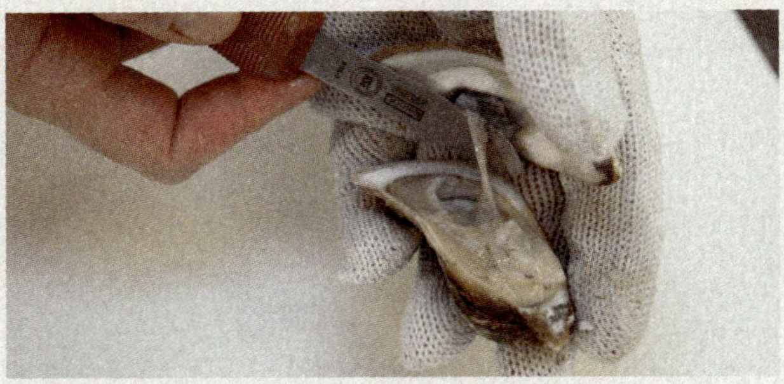

4. Slide the oyster knife under the flesh of the top shell to cut it away from the shell. *Note:* Keep the oyster as level as possible to avoid spilling the juice.

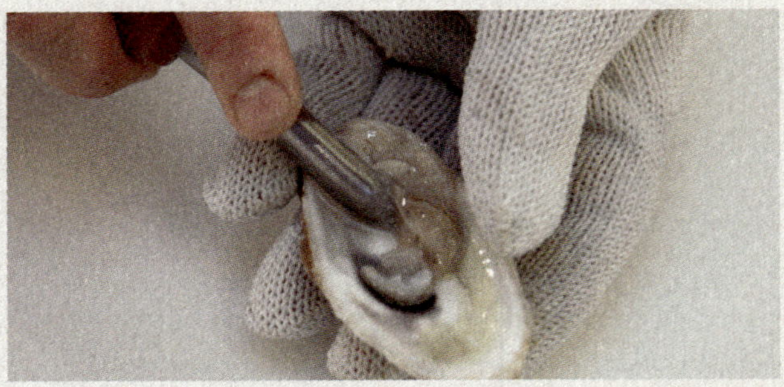

5. Slide the oyster knife under the adductor muscle to separate the flesh from the bottom shell.
6. Remove any shell fragments that may have mixed with the oyster flesh.

Procedure for Cleaning Squid

1. Gently pull the arms and tentacles of the squid away from the body (mantle).
2. Using a boning knife, cut just below each of the eyes to remove the eight arms and two tentacles from the head.
3. Remove the ink sac from the head and reserve.
4. Pull back the tentacles to expose the beak. Carefully pull out the beak and discard.
5. Hold the body (mantle) in guiding hand and carefully pull out the transparent quill with the other hand. Discard the quill.
6. Carefully pull as much skin as possible from the body (mantle) to reveal the lighter-colored flesh below. Discard the skin.
7. Thoroughly rinse all of the squid flesh before use.

Shucking Oysters
Media Clip

Fabricating Squid

Procedure for Shucking Clams

1. While wearing a mesh glove on the guiding hand, firmly hold the clam with the hinge resting against the base of the thumb and the lip facing the fingertips.
2. Insert the blade of a clam knife between the top shell and bottom shell until halfway inserted.
3. Carefully twist the blade to pop the shells open.

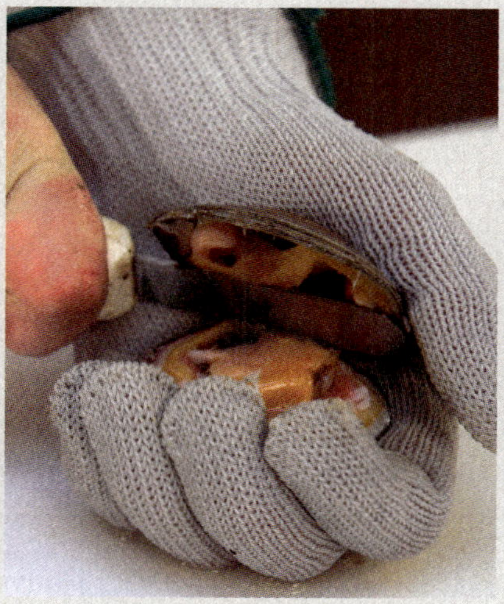

4. Use the tip of the clam knife to separate the flesh from the top shell. Discard the top shell.

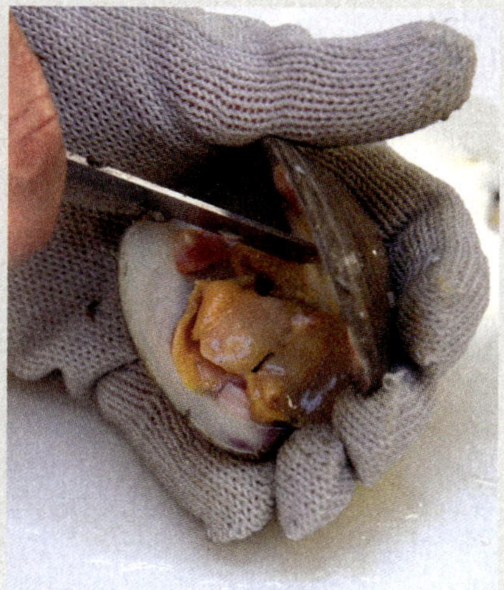

5. Use the knife blade to separate the adductor muscle from the bottom shell.
6. Remove any shell fragments that may have mixed with the clam flesh.

CHECKPOINT 4-9

1. Devein shrimp.
2. Debeard mussels.
3. Explain how to split lobster tails.
4. Split a lobster tail.
5. Explain how to clean soft-shell crabs.
6. Explain how to shuck oysters.
7. Shuck oysters.
8. Explain how to clean squid.
9. Explain how to shuck clams.
10. Shuck clams.

MARKET FORMS OF BEEF

Beef is the flesh of domesticated cattle. The age and gender of the cattle affects the quality and flavor of the meat.

Most beef comes from steers. A *steer* is a male calf that has been castrated prior to reaching sexual maturity. Steers that are 15–24 months old produce the best-quality beef. Most steers are grain-fed, which increases the marbling of the meat. *Grain-fed beef* is meat from cattle that were grain-fed in confined feeding operations for 90 days to 1 year. *Grass-fed beef* is meat from cattle that were raised on grass or hay with little or no special feed. Grass-fed beef contains less saturated fat than grain-fed beef.

Organic beef is produced within the strict guidelines of and audited by the USDA. There are many restrictions to the production and processing of meat from cattle that is certified as organic.

Market Forms of Beef

In order for the term "natural" to appear on a food label, the USDA requires that the product be minimally processed, not contain any artificial ingredients, and not contain any preservatives. The USDA has no specific restriction on management practices during the life of the animal.

Beef is available in a variety of market forms. Knowledge of each of these market forms is necessary for accurate product ordering. Common market forms of beef include partial carcasses, primal cuts, and fabricated cuts.

Partial Carcasses of Beef

Sides are the most common partial carcass of beef. A *side of beef* is a half of a carcass split along the backbone. There are two sides to each carcass. A side can be purchased for less per pound than any primal or fabricated cut. However, in order for purchasing sides of beef to be cost-effective, adequate facilities and equipment for fabrication and storage must be available. In addition, foodservice staff capable of cutting a side of beef is required. Because of the amount of space, time, and labor required to break down a side of beef, sides are rarely purchased by foodservice operations.

A *quarter of beef* is a side of beef that has been divided into two parts between the 12th and 13th ribs. Quarters are sold as forequarters and hindquarters. Quarters are not often purchased by foodservice operations.

Primal and Fabricated Cuts of Beef

A *primal cut* is a large cut from a whole or a partial carcass. A single beef carcass has two of each primal cut (a left side and a right side). Primal cuts of beef are very large. The eight primal cuts of beef are the chuck, rib, short loin, sirloin, round, flank, short plate, and brisket and shank. **See Figure 4-42.**

Primal Cuts of Beef

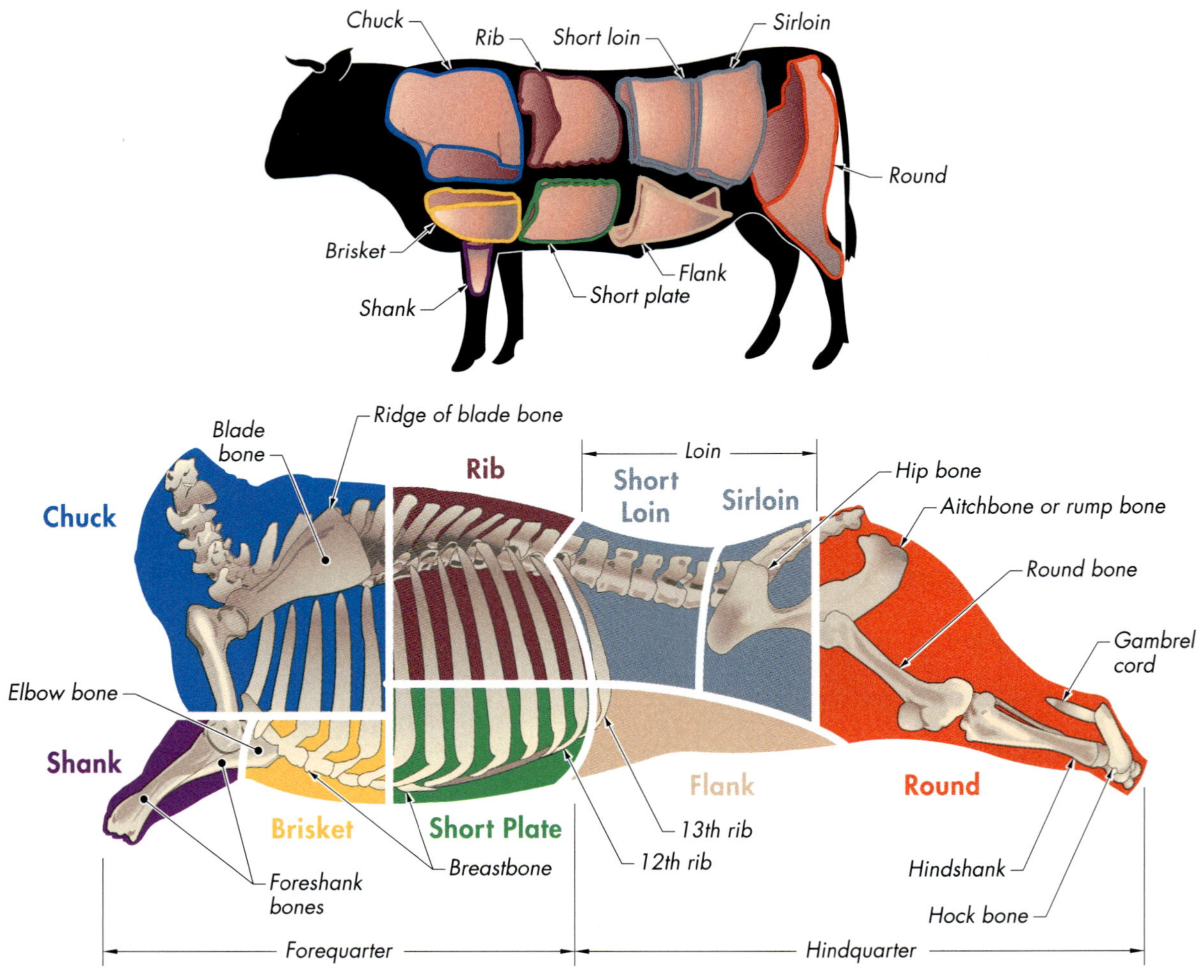

Figure 4-42. The primal cuts of beef include the chuck, rib, short loin, sirloin, round, flank, short plate, and brisket and shank.

Each primal cut of beef is further divided into fabricated cuts. A *fabricated cut* is a ready-to-cook cut that is packaged to certain size and weight specifications. Fabricated beef cuts include short ribs and tenderloins, as well as top sirloin, flat-iron, eye-roll, T-bone, porterhouse, Delmonico, butt, and skirt steaks. Fabricated cuts are a convenient way of providing uniform portions while reducing labor costs. The price per pound for fabricated cuts is higher than the price per pound for primal cuts. Foodservice operations often purchase some primal cuts and some fabricated cuts of beef.

Chucks. A *beef chuck* is a shoulder primal cut of beef that contains the first five rib bones, some of the backbone, and a small amount of the arm and blade bones. The chuck is the largest primal cut and its average weight is approximately 26% of the total carcass weight. The shoulder is one of the most-exercised muscles on the animal, so it is a tough cut of meat with a lot of connective tissue. However, chuck is also quite lean and has an excellent flavor.

Fabricated cuts from the chuck include shoulder clods, clod tenders, chuck rolls, top blade chucks, flat-iron steaks, and strips. **See Figure 4-43.** Flat-iron steaks are the second-most-tender chuck cuts after the tenderloin. Short ribs are produced from the small rib bone ends that are sawed off of the primal rib as the rib roast is removed. Short ribs have a sizable portion of lean meat on them.

Larger pieces of chuck lend themselves to braising and stewing. Smaller pieces of chuck and trimmings produce very flavorful ground meat.

Ribs. A *beef rib* is the primal cut of beef located between the chuck and short loin and contains seven rib bones, from the 6th to the 12th rib. Its average weight is approximately 10% of the total carcass weight. The meat is tender and well marbled. The rib bones (finger bones) make the rib cut a good cut for roasting, because they form a natural rack on which the meat can cook.

A beef rib is often fabricated into a variety of cuts. **See Figure 4-44.** A beef rib contains the prime rib roast. The rib bones can be left on for roasting to produce a moist roast. If the bones are removed, a boneless rib eye roast can be further cut into fabricated bone-in or boneless rib eye steaks. A *rib eye* is a large, eye-shaped muscle within the rib that is a continuation of the sirloin muscle. Meat from eye-shaped muscles, such as the rib eye or tenderloin, is often referred to as eye meat. The 6th to the 12th ribs can be prepared as smoked or barbequed beef ribs. If the rib bones are removed, the meat can be rolled and tied into a rolled-rib roast.

Chuck Cuts of Beef

Canada Beef Inc.

Figure 4-43. A beef chuck is often fabricated into a variety of cuts including shoulder clods, clod tenders, chuck rolls, top blade chucks, flat-iron steaks, and strips.

Canada Beef Inc.

Figure 4-44. A beef rib is often fabricated into a variety of cuts.

Short Loins. A *beef short loin* is a primal cut of beef located just to the rear of the primal rib and includes the 13th rib and a small section of the backbone. Its average weight is approximately 8% of the total carcass weight. The short loin can be cut in cross sections to produce some of the most popular fabricated cuts. **See Figure 4-45.** Cuts from the short loin are commonly grilled, broiled, or roasted.

When the short loin and sirloin are split apart, the smaller portion of tenderloin is part of the short loin. A *beef tenderloin* is an eye-shaped muscle running from the primal rib cut into the primal leg. The tenderloin is located just beneath the strip loin and is the most tender piece of beef. Sometimes the entire tenderloin is removed prior to dividing the short loin and sirloin. The whole tenderloin can be roasted whole or divided into Chateaubriand, filet mignons, and tournedos.

A *beef strip loin* is a short loin without a tenderloin. A strip loin can be cut into boneless strip steaks or roasted whole. Fabricated cuts from the short loin are often aged, as they have ample fat covering and marbling and are very tender. Aging beef loin intensifies the flavor and tenderness of the meat.

The short loin produces many popular fabricated cuts. Starting from the end nearest the primal rib, cross-section cuts produce Delmonico steaks, T-bone steaks, and porterhouse steaks. Delmonico steaks do not include any tenderloin, T-bone steaks include only a small section of tenderloin, and porterhouse steaks include a large section of tenderloin.

Sirloins. A *beef sirloin* is a primal cut of beef situated just behind the short loin and contains some of the backbone and hip bone. Its average weight is approximately 9% of the total carcass weight. With the exception of the butt tenderloin muscle, meat from the sirloin is not quite as tender as meat from the short loin. However, the sirloin can be cut into butt steaks that can be marinated, skewered, and then grilled or broiled. Ball-tip steaks, sirloin flaps, sirloin tri-tips, sirloin tri-tip steaks, top sirloin roasts, top sirloin steaks, top sirloin caps, and top sirloin cap steaks can be fabricated from the sirloin primal cut. **See Figure 4-46.**

Rounds. A *beef round* is a primal cut of beef that includes a large grouping of muscles that represent the hind hip and thigh of the carcass. The average weight of the round is approximately 27% of the total carcass weight. A round contains large bones including the round bone (leg), aitchbone (pelvis), shank, and tailbone.

Short Loin Cuts of Beef

Strip Loin

Tenderloin

Porterhouse Steak

Strip Steaks

Tenderloin Steaks

T-Bone Steak

Canada Beef Inc.

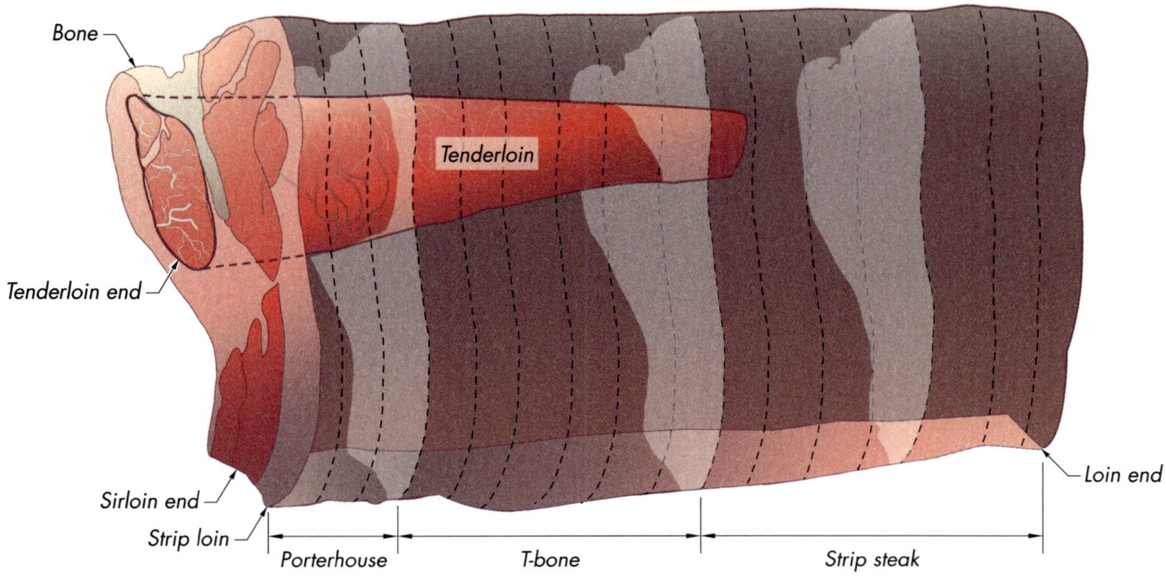

Figure 4-45. The short loin can be cut in cross sections to produce some of the most popular fabricated cuts.

Canada Beef Inc.

Figure 4-46. Ball-tip steaks, sirloin flaps, sirloin tri-tips, sirloin tri-tip steaks, top sirloin roasts, top sirloin steaks, top sirloin caps, and top sirloin cap steaks can be fabricated from the sirloin primal cut.

A round can be slow-roasted whole. However, due to its size, the round is commonly broken down and sold as separate subprimal cuts. Subprimal and fabricated cuts from the round include the top round, bottom round, knuckle, and shank. The bottom round can be cut into the outside round and the eye of round. **See Figure 4-47.**

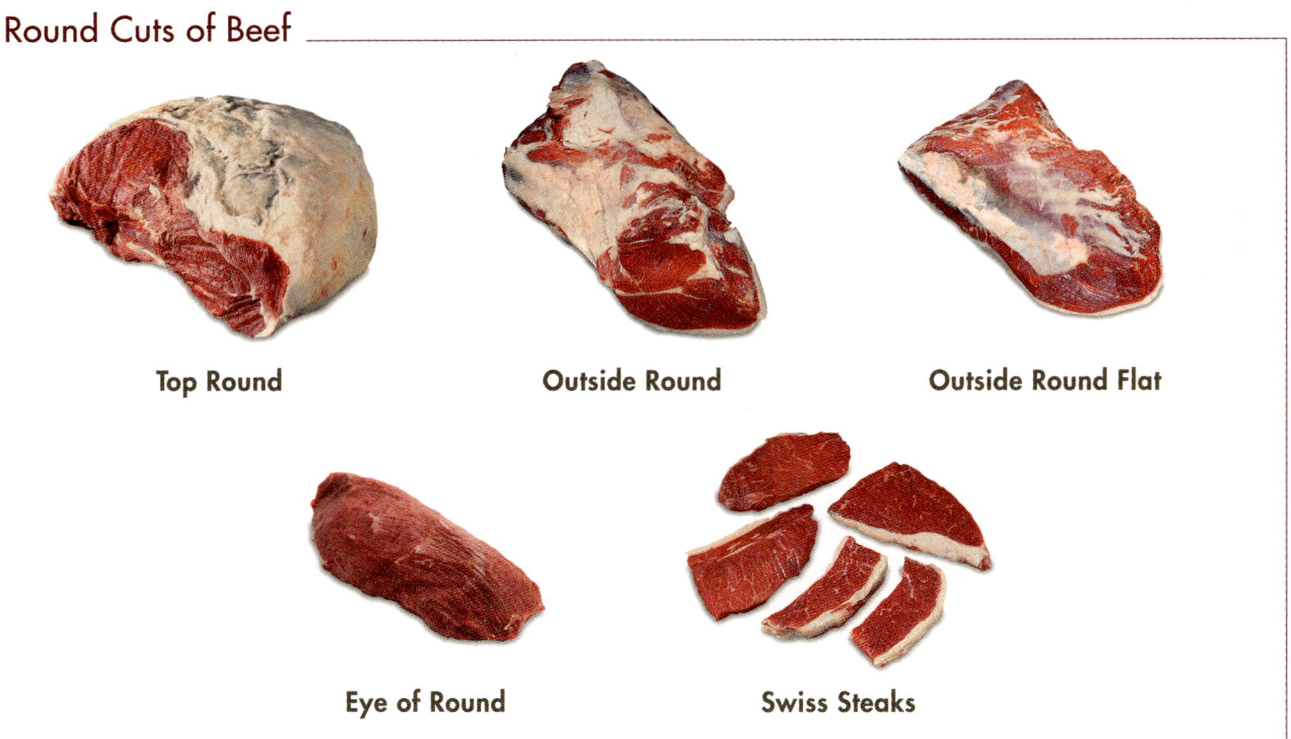

Canada Beef Inc.

Figure 4-47. The round can be broken down into a variety of subprimal and fabricated cuts.

When the round is trimmed for cooking it yields a large amount of usable meat. A *beef rump roast* is a roast cut from the primal round, above the back end of the hip bone. If the bone is left in, it is called a standing rump roast. A boneless rump roast is rolled and tied prior to being sold. The rump roast is a very flavorful cut that is often braised. A *steamship round roast* is the beef round with the shank and rump removed. The top round and knuckle portions can be roasted. The outside round and eye of round are often braised.

Flanks. The *beef flank* is a primal cut of beef that includes the thin, flat section of the hindquarters located beneath the loin. Its average weight is approximately 4% of the total carcass weight. The flank has more fat than lean meat and contains one thin, oval-shaped, boneless flank steak. The flank steak can be scored or cubed before it is cooked. Prior to cooking, the fat covering should be removed and the meat can be marinated to produce a more tender and flavorful piece of meat. If cooked whole, flank steak should be cut across the grain or it will be tough. **See Figure 4-48.**

Flank Cuts of Beef

Canada Beef Inc.

Figure 4-48. A beef flank is fabricated into flank steaks.

Short Plates. A *beef short plate* is a primal cut of beef that includes a thin portion of the beef forequarter located just beneath the rib cut. The average weight of the short plate is approximately 5% of the total carcass weight. The bones attached to this cut are the remaining sections of the rib bones. The small bones from the short plate are called short ribs. Short ribs are not quite as meaty as those from the rib cut. In addition to short ribs, the short plate yields the flavorful skirt steak. **See Figure 4-49.**

Short Plate Cuts of Beef

Canada Beef Inc.

Figure 4-49. A beef short plate is fabricated into skirt steak and short ribs.

Briskets and Shanks. The brisket and shank are two separate muscle groups that make up one primal cut of beef that is located just below the chuck. **See Figure 4-50.** The weight of the brisket and shank is approximately 9% of the total carcass weight. The *beef brisket* is a thin section of beef that contains some of the ribs, the breastbone, and layers of lean muscle, fat, and connective tissue. The ribs and breastbone are always removed prior to cooking the brisket.

Brisket and Shank Cuts of Beef

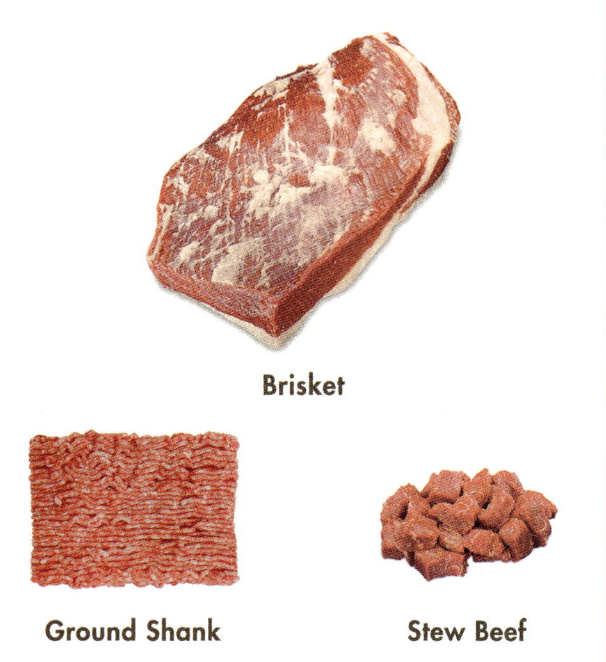

Canada Beef Inc.

Figure 4-50. A brisket and shank are two separate muscle groups that make up one primal cut of beef that is located just below the chuck.

Brisket is a tough cut of beef with excellent flavor that can become tender when cooked properly. It has long muscle fibers that run in several directions, making it difficult to slice. Brisket is often braised or simmered. Brisket can be cured, peppered to make pastrami, braised as sauerbraten, simmered as New England-style brisket, and corned or pickled as corned beef.

The *beef shank* is a bony section of beef that is surrounded by a small amount of very tough but flavorful meat. Shanks are used for making stocks and rich reduction sauces. Shank meat is usually ground to flavor and clarify consommés, because the meat has a high concentration of collagen that converts to gelatin when cooked. Shanks are generally cut perpendicular across the bone and then braised.

Beef Offals

An *offal* is an edible part of an animal that is not part of a primal cut. Commonly used beef offals include liver, tongue, tripe, oxtail, and kidney. **See Figure 4-51.** With the exception of liver, beef offals are prepared using moist-heat cooking methods. Beef liver is covered with a thin membrane that should be removed before slicing. Beef liver is commonly broiled, sautéed, or pan-fried.

Beef Offals

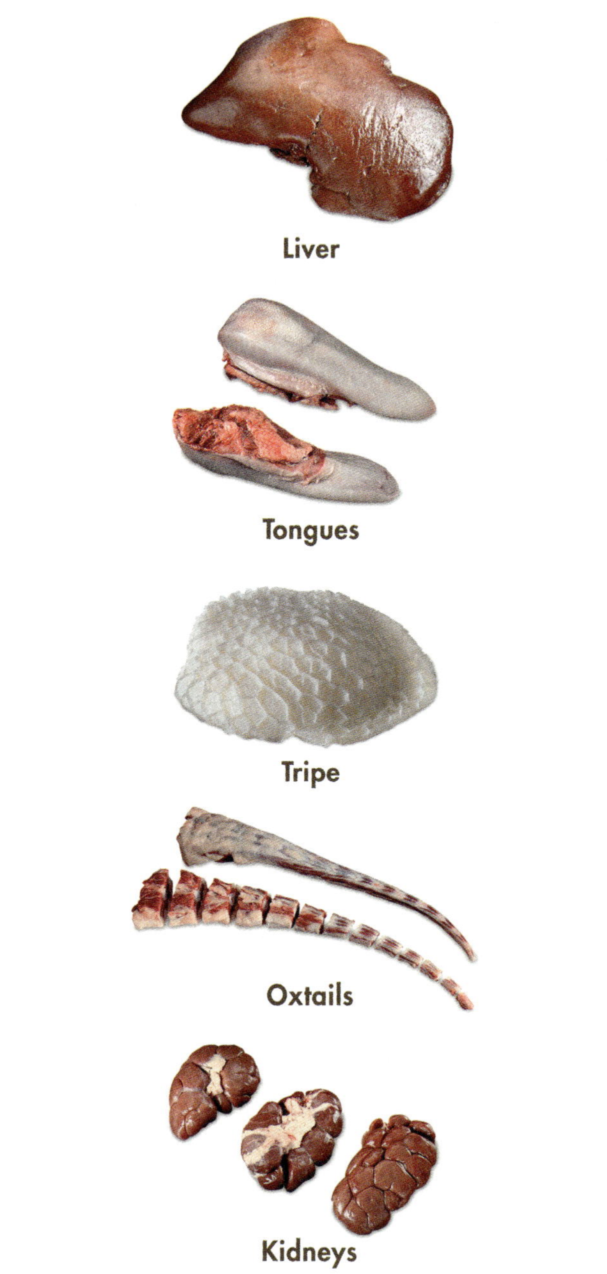

Canada Beef Inc.

Figure 4-51. Beef offals include liver, tongue, tripe, oxtail, and kidney.

Beef tongue is available fresh, smoked, pickled, and corned, but smoked is the most popular form. Tongue is always prepared using moist-heat cooking methods. To determine doneness, the tip of the tongue is felt to test tenderness. When the tip is soft, the tongue is done. Cooked tongue is cooled in cold water and then skinned. Beef tongue can be served cold or can be reheated and served hot. **See Figure 4-52.**

Beef Tongues

U.S. Wellness Meats

Figure 4-52. Beef tongue can be served cold or can be reheated and served hot.

Tripe is the muscular inner lining of a stomach of an animal, such as cattle or sheep. *Honeycomb tripe* is the lining of the second stomach found in cattle. Honeycomb tripe may be purchased fresh, pickled, or canned. It is often fried, creamed, used in soups, or served cold with a vinaigrette dressing.

An *oxtail* is the tail from a cattle carcass. Oxtail contains a considerable amount of bone and a good portion of richly flavored flesh. Oxtail is most often used in stews or braised in rich cooking liquids. The thin end of the oxtail can be used to make oxtail soup.

Beef kidneys have irregularly shaped lobes divided by deep cracks. Before cooking, the kidney is split lengthwise, and the suet (fat) and urinary canals are carefully removed. Kidneys are commonly braised.

CHECKPOINT 4-10

1. Identify the eight primal cuts of beef.
2. Describe the advantages and disadvantages of purchasing fabricated cuts.
3. Identify the cuts fabricated from each primal cut of beef.
4. Explain how to prepare beef offals.

Photo Courtesy of D'Artagnan, Photography by Doug Adams Studio

MARKET FORMS OF VEAL

Veal is the flesh of calves, which are young cattle. Veal is available in a variety of market forms, including whole and partial carcasses, primal cuts, fabricated cuts, and offals. Knowledge of these different market forms is necessary for accurate product ordering. Understanding the skeletal structure of calves can aid in identifying the market forms of veal. **See Figure 4-53.**

Nutrition Note

Veal is a good source of protein and is low in fat. However, veal contains more cholesterol than beef.

Partial Carcasses of Veal

In order for the purchasing of partial carcasses of veal to be cost-effective, skilled labor and storage space are required. Veal is typically not split into sides like beef. Veal is split into head and tail sections known as the foresaddle and hindsaddle. The foresaddle and hindsaddle are split between the 11th and 12th ribs, not down the backbone. A *veal foresaddle* is the front half of a carcass consisting of the primal shoulder, rack, breast, and shank cuts. A *veal hindsaddle* is the rear half of a carcass consisting of the loin and leg.

Primal and Fabricated Cuts of Veal

The left and right primal cuts of veal remain joined together and are sold as a single cut. For example, the leg primal cut has two joined legs. The primal cuts of veal include the shoulder, rack, loin, leg, and foreshank and breast. **See Figure 4-54.**

Veal Facts

Skeletal Structure of Calves

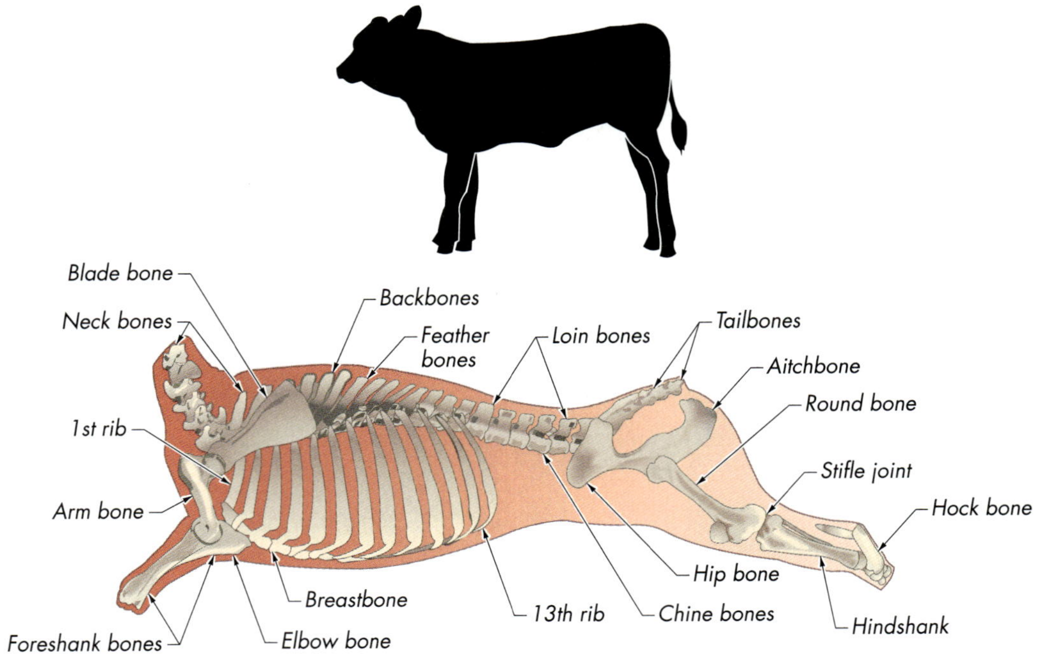

Figure 4-53. Understanding the skeletal structure of calves can aid in identifying the market forms of veal.

Primal Cuts of Veal

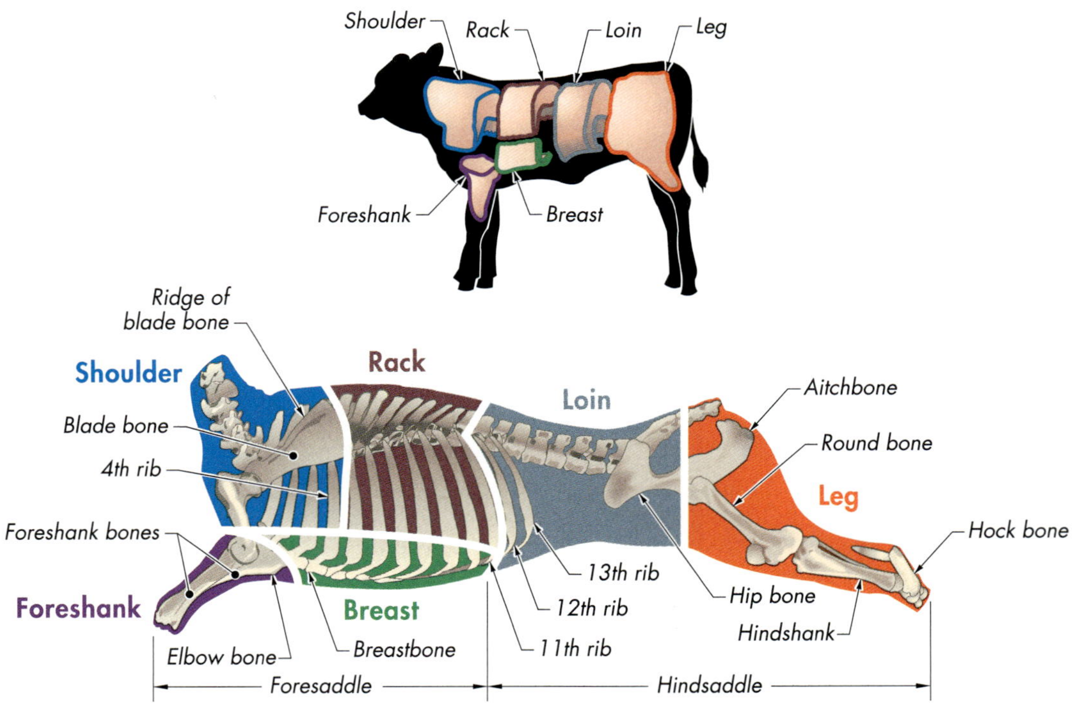

Figure 4-54. The primal cuts of veal include the shoulder, rack, loin, leg, and foreshank and breast.

Each primal cut is further divided into fabricated cuts. Fabricated cuts of veal include frenched chops, baby T-bone steaks, boneless cutlets, and ossobuco-cut shanks. Some foodservice operations may choose to purchase only fabricated cuts of veal. However, the price per pound is higher for fabricated cuts than the price per pound for primal cuts.

Shoulders. A *veal shoulder* is a primal cut that contains the first four rib bones, some of the backbone, and a small amount of the arm and blade bones. The average weight of a veal shoulder is approximately 21% of the total carcass weight. Veal shoulder is tough but flavorful. It can be fabricated into steaks or chops. However, it is most often ground, cut into cubes for stewing, or cooked whole. **See Figure 4-55.**

Racks. The *veal rack* is a primal cut located between the shoulder and loin and contains seven rib bones. Its average weight is approximately 9% of the total carcass weight. The meat is tender and well marbled. A veal rib is different from a beef rib in that veal is not split into two halves along the backbone (2nd chine bones). An unseparated veal rack is called a hotel rack and consists of two very tender veal rib loins. A veal rack can be split into halves and tied into a circle to form a crown rib roast.

Veal racks can be trimmed, frenched, and cut into veal chops. **See Figure 4-56.** Other fabricated cuts from the rack include a small portion of the tenderloin, known as the short tenderloin, and the boneless veal rib eye roast.

Loins. A *veal loin* is a primal cut located between the primal rack and leg and includes the 12th and 13th rib, the loin eye muscle, the center section of the tenderloin, the strip loin, and flank meat. The average weight of the loin is approximately 10% of the total carcass weight. A complete, unsplit primal loin from a veal carcass is commonly referred to as a saddle. Veal loins are often divided into fabricated cuts such as baby T-bones and cutlets. **See Figure 4-57.**

Shoulder Cuts of Veal

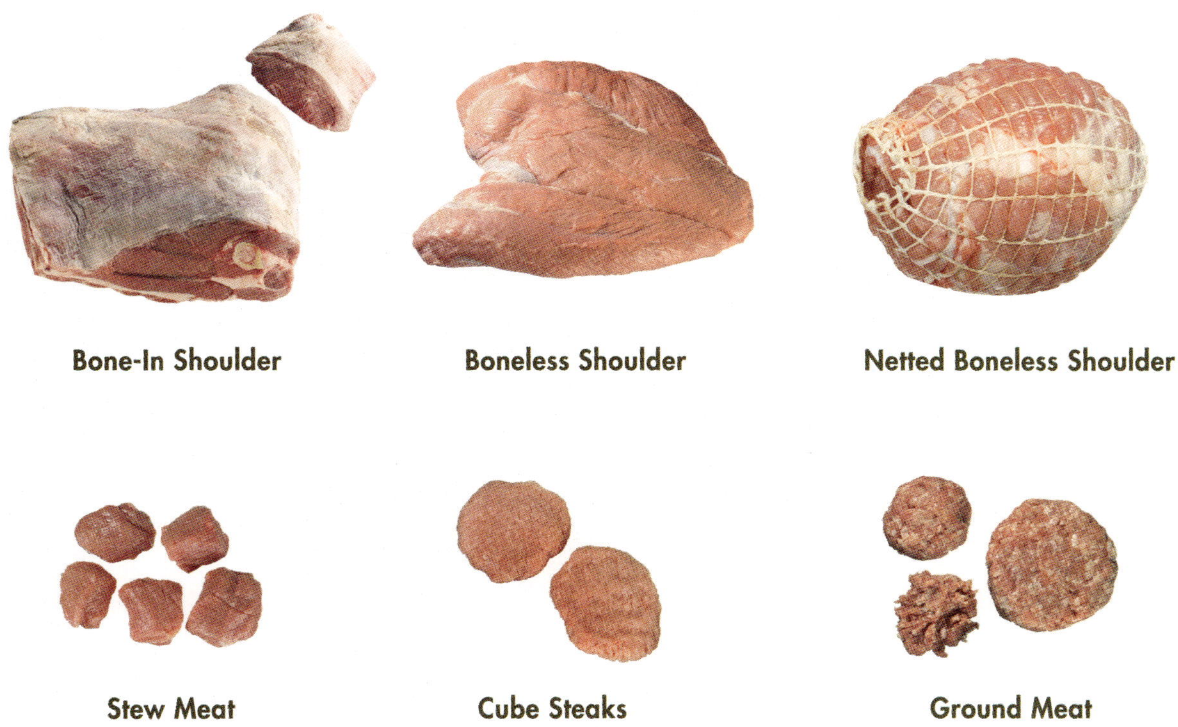

Strauss Free Raised

Figure 4-55. A veal shoulder can be fabricated into steaks or chops. However, it is most often ground, cut into cubes for stewing, or cooked whole.

Rack Cuts of Veal

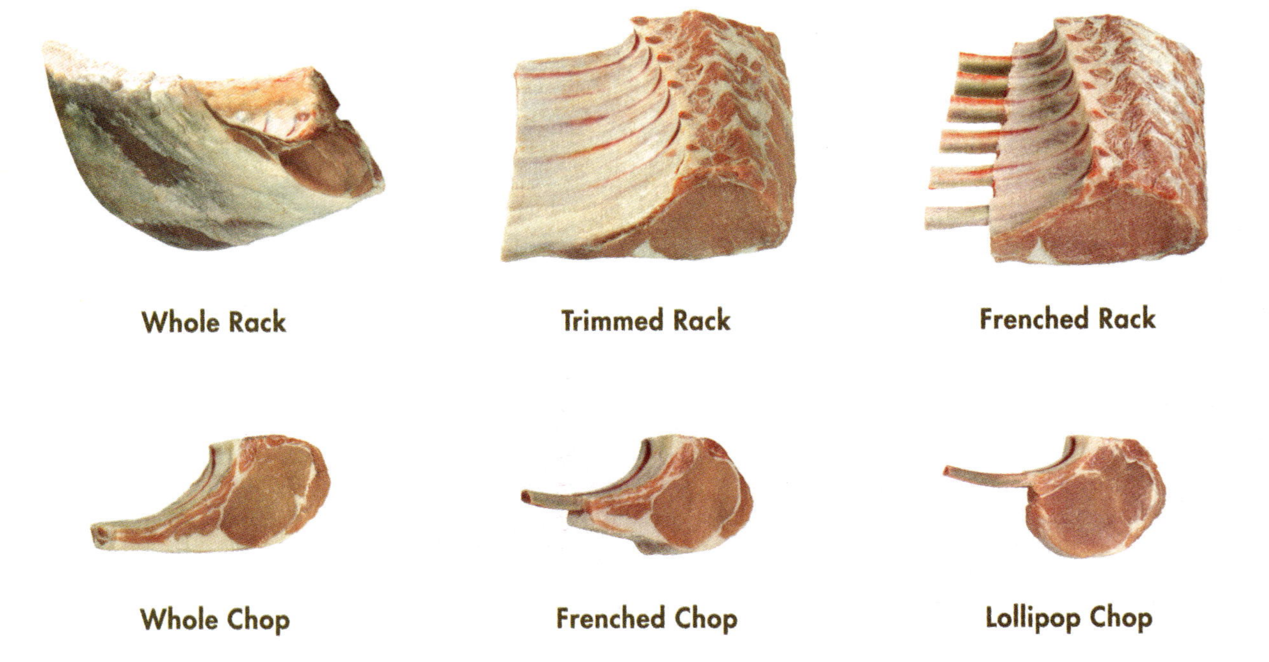

Strauss Free Raised

Figure 4-56. Veal racks can be trimmed, frenched; and cut into veal chops.

Loin Cuts of Veal

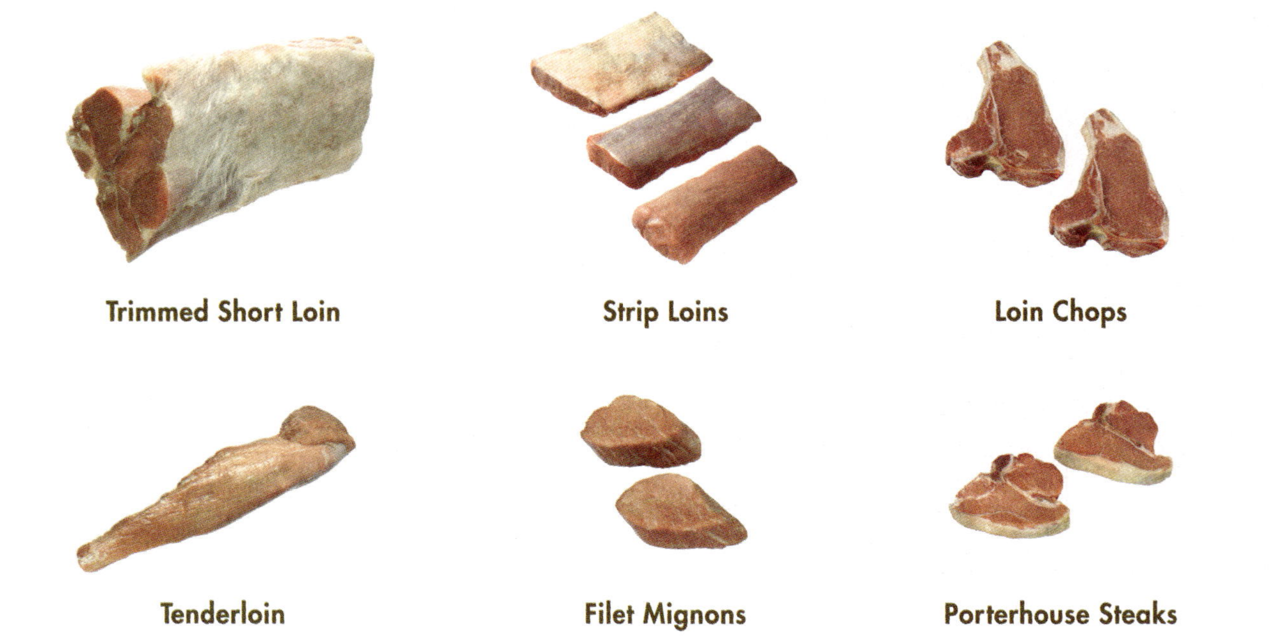

Strauss Free Raised

Figure 4-57. Veal loins are sometimes roasted whole, however they are often divided into fabricated cuts.

A *baby T-bone* is a 6–8 oz steak cut from the loin of veal. It contains loin meat on one side and tenderloin on the other side. Cuts from the loin are usually grilled, broiled, or roasted. A *veal cutlet* is a thin slice of veal. Wiener schnitzels are prepared by pounding veal cutlets until very thin and then breading and frying them. Wiener schnitzels are commonly served with lemon, topped with a variety of vegetables, or topped with woodland cream sauce.

Legs. A *veal leg* is a primal cut from the hind leg that contains the leg, sirloin, last portion of the backbone, pelvis (hip bone and aitchbone), round bone, hindshank, and tailbone. Its average weight is approximately 42% of the total carcass weight. The leg is the most versatile cut of veal because it contains solid, lean, fine-textured meat. Tender meat is located near the sirloin end, and tougher meat is located toward the shank. The entire leg is typically boned and cut into scallops or cutlets rather than being roasted whole. The leg is boned by following the muscle structure of the meat so that pieces of equal tenderness are removed.

Veal legs can be divided into leg, hindshank, ossobuco, inside round, eye round, and scallopini cuts. **See Figure 4-58.** These cuts are commonly sliced against the grain and pounded until thin to tenderize them. Ossobuco-cut shanks are cut perpendicular to the bone. A *scallopini* is a small, ¼ inch thick slice of veal (generally leg meat) that is 2–3 inches in diameter.

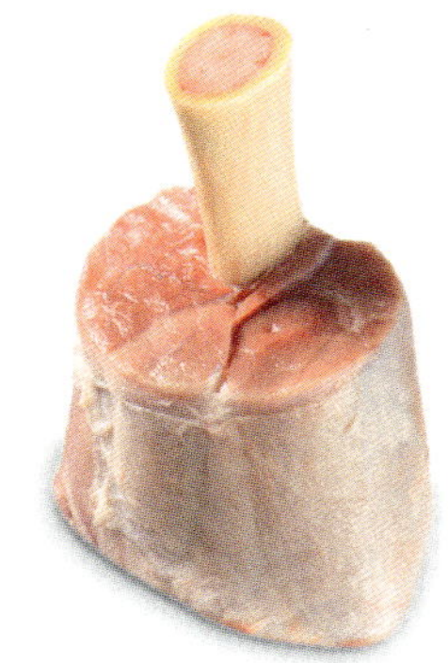

Strauss Free Raised

Foreshanks and Breasts. The foreshank and breast form the primal cut of veal from the lower foresaddle. **See Figure 4-59.** Its average weight is approximately 16% of the total carcass weight. A *veal foreshank* is the upper portion of the front leg of a calf. It can be braised whole or sliced in cross sections across the bone. A *veal breast* is a thin, flat cut of meat located under the shoulder and ribs and contains the breastbone, tips of the rib bones, and cartilage. The breastbone is typically still cartilage because the animal is so young. The thin, flat shape makes the breast easy to stuff, roll, and tie into a tender rolled roast that can be braised to break down the connective tissue.

Leg Cuts of Veal

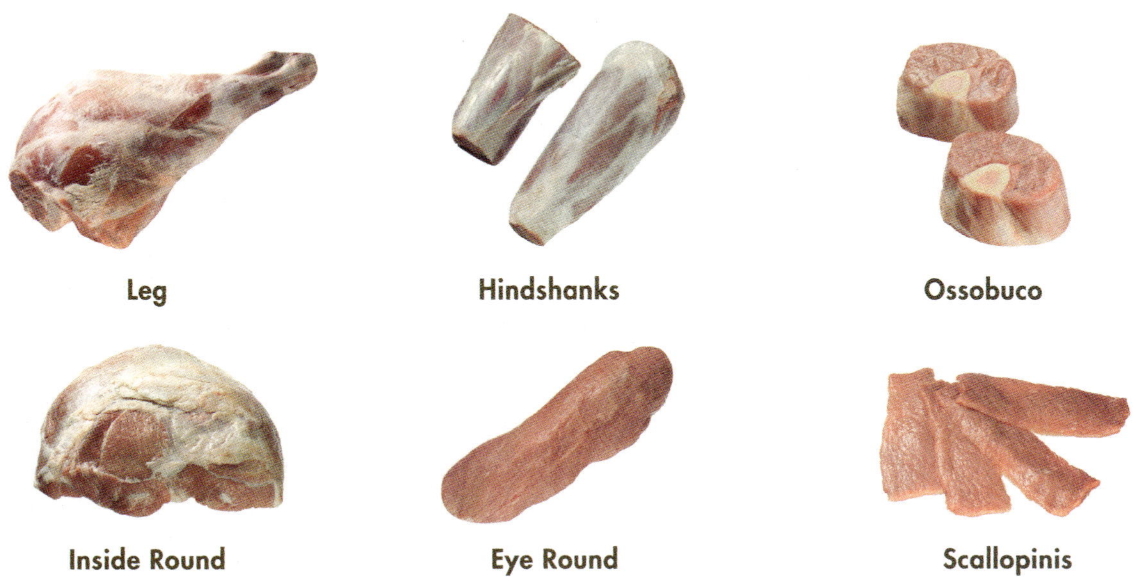

Strauss Free Raised

Figure 4-58. Veal legs can be divided into leg, hindshank, ossobuco, inside round, eye round, and scallopini cuts.

Foreshank and Breast Cuts of Veal

Figure 4-59. The foreshank and breast form the primal cut of veal from the lower foresaddle.

Veal Offals

Veal offals are more tender than beef offals. Common veal offals include sweetbreads, liver, and kidney. *Sweetbreads* are the thymus glands of a calf, located in the neck. High-quality sweetbreads should be plump and somewhat firm with a thin protective membrane. Sweetbreads are often sautéed in brown butter or pan-fried. **See Figure 4-60.**

Veal Offals

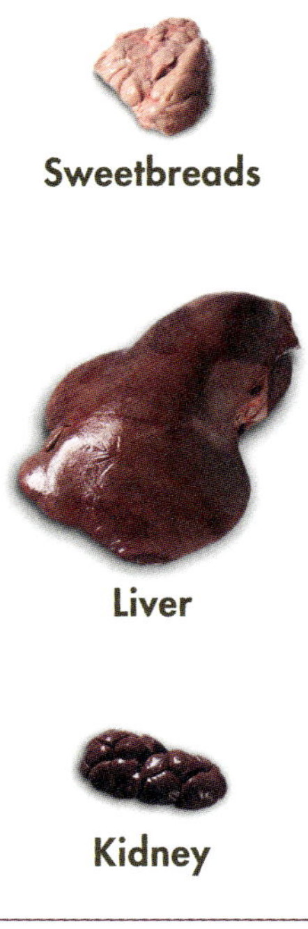

Figure 4-60. Common veal offals include sweetbreads, liver, and kidney.

Veal liver is commonly broiled, pan-fried, or sautéed and should be cooked medium. Veal kidneys can be broiled with excellent results, unlike beef kidneys, which must be cooked with moist heat. Veal kidneys are typically used to prepare entrées such as kidney stew and kidney pie.

CHECKPOINT 4-11

1. Identify the five primal cuts of veal.
2. Identify the cuts fabricated from each primal cut of veal.
3. Explain how to prepare veal offals.

RECEIVING AND STORING BEEF AND VEAL

Beef and veal are potentially hazardous foods that must be checked for color, odor, texture, and temperature upon receipt. **See Figure 4-61.** Beef should be red in color with white fat. Veal should be pink in color with white fat. There should be no odor, and the texture should be firm and not dry or slick. Refrigerated beef and veal should maintain an internal temperature of 41°F or below. If the meat received is vacuum-packed, its temperature can be taken by placing an instant read thermometer between two of the packages. Beef or veal that is in the temperature danger zone should be rejected.

Receiving Beef and Veal

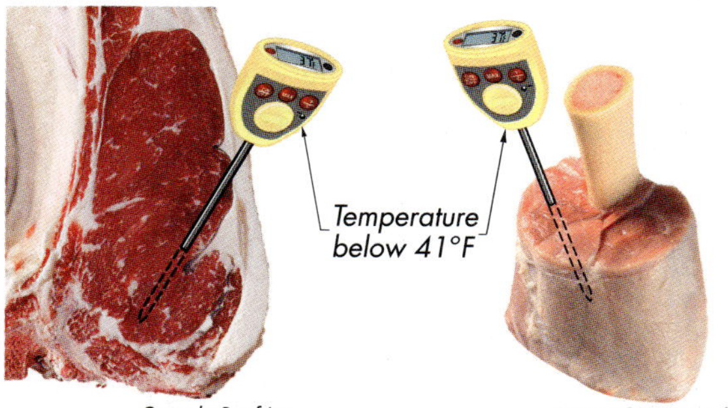

Canada Beef Inc. Strauss Free Raised

Beef **Veal**

Figure 4-61. Beef and veal are potentially hazardous foods that must be checked for color, odor, texture, and temperature upon receipt.

Frozen beef and veal should be kept at temperatures that will allow the meat to remain frozen. When frozen meats need to be thawed, they should be placed in the refrigerator overnight. Larger cuts of meat may take more than one day to thaw under refrigeration.

Vacuum-packed beef and veal should never be opened until needed for service or preparation and can be stored refrigerated for three to four weeks. **See Figure 4-62.** Once the vacuum seal is broken, meat has a shelf life of only two to three days. Cut meats should be rewrapped airtight, refrigerated immediately, and used as soon as possible.

Vacuum-Packed Beef

Edlund Co.

Figure 4-62. Vacuum-packed beef and veal should never be opened until needed for service or preparation and can be stored refrigerated for three to four weeks.

All beef and veal used in foodservice operations must be procured from USDA-inspected plants. At the time of slaughter, the carcasses are stamped to indicate that the animal was slaughtered at a USDA-inspected plant. The USDA inspection stamp will be directly on large cuts or on the packaging of fabricated cuts of beef and veal. **See Figure 4-63.** The number on the stamp identifies the plant where the animal was processed. It does not indicate quality.

USDA Inspection Stamps

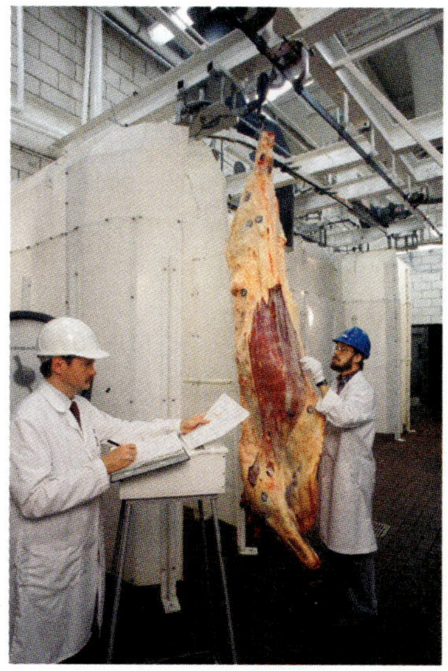

United States Department of Agriculture

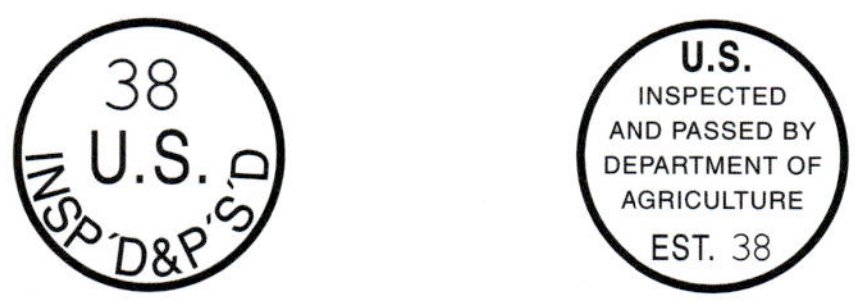

Stamp for Whole Carcass

Stamp for Fabricated or Processed Meats

Figure 4-63. The USDA inspection stamp will be directly on large cuts or on the packaging of fabricated cuts of beef and veal.

Unlike inspection, USDA quality and yield grading is optional for beef and veal producers. Beef can be graded for quality, yield, or both. Veal can only be graded for quality. Quality and yield grading stamps are stamped on carcasses in the same manner as inspection stamps.

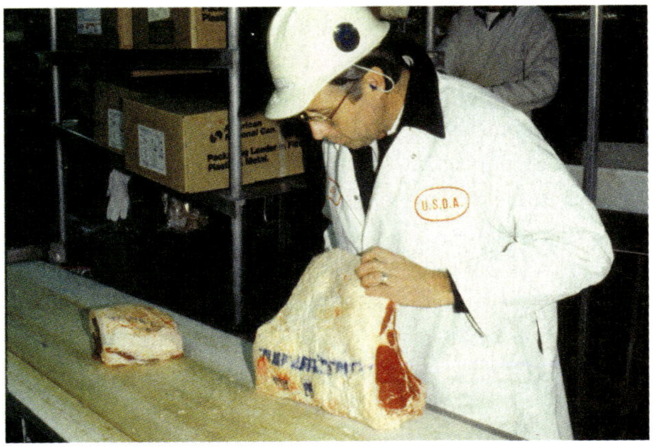

United States Department of Agriculture

Quality Grading

USDA quality grading of beef and veal is voluntary. Beef that has been inspected but not quality graded is referred to as "no roll" beef. Although the grading for beef and veal is based on specific details, the overall terms have relatively the same criteria. However, the quality grade is not a guarantee of quality.

Most veal is USDA Prime or USDA Choice. The grades of beef most commonly used in foodservice operations are USDA Prime, USDA Choice, and USDA Select. **See Figure 4-64.**

USDA Quality Grade Stamps

Figure 4-64. USDA Prime, USDA Choice, and USDA Select beef is commonly used in foodservice operations.

USDA Prime. USDA Prime beef is well marbled and has a thick, firm fat covering. It is the juiciest and most flavorful of all meats, but it is also the most expensive because of the large amount of fat that must be trimmed from it before cooking. Of all the beef marketed in the United States, only a small percentage is graded USDA Prime.

USDA Choice. USDA Choice has slightly less marbling and fat covering than USDA Prime. Most foodservice operations prefer this grade because there is less waste.

USDA Select. USDA Select beef has minimal marbling and a soft fat covering. This grade of beef is often used in quick service operations.

Yield Grading

Yield grades of beef are numbered 1 to 5 and indicate how much usable meat can be obtained from a carcass. A grade of 1 indicates the highest yield of meat, and a grade of 5 indicates the lowest yield of meat. **See Figure 4-65.**

USDA Yield Grades

Grading Beef

Figure 4-65. Most high-quality beef will have a yield grade of 3 or higher.

CHECKPOINT 4-12

1. Identify four traits to check upon receiving beef and veal.
2. Identify the temperature at which refrigerated beef and veal must be kept.
3. Explain why vacuum-sealed packages should only be opened at the time of use.
4. Explain the significance of the USDA inspection stamp.
5. Explain why the quality grade of veal is less important than the quality grade of beef.
6. Describe common USDA quality grades of beef.

FABRICATING BEEF AND VEAL

The size of storage facilities and availability of staff with the fabrication skills often determine whether a foodservice operation purchases primal cuts, fabricated cuts, or a combination. Beef and veal fabrication require trimming, cutting, tenderizing, and tying.

Trimming and Cutting Beef Tenderloin

The tenderloin is often trimmed and then cut into portion-controlled cuts. These cuts include tenderloin tips, Chateaubriand, filet mignons, and tournedos.

Procedure for Trimming and Cutting Beef Tenderloin

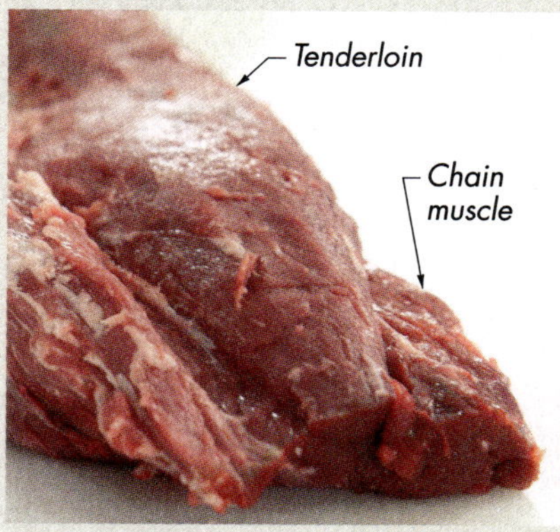

1. With a rigid boning knife, carefully remove the chain muscle from the side of the tenderloin and reserve.

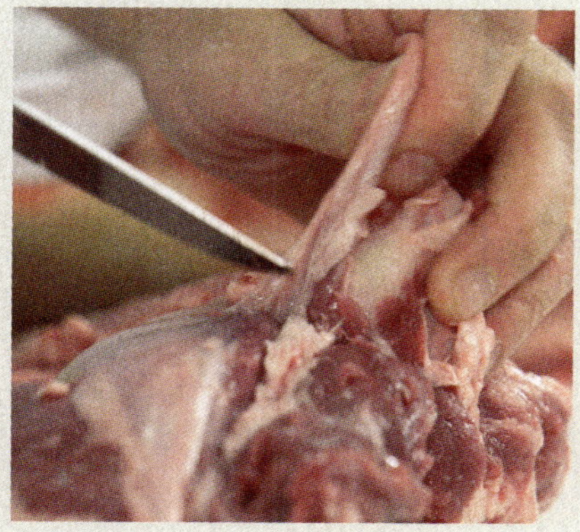

2. Trim and pull the thick fat covering away from the tenderloin.

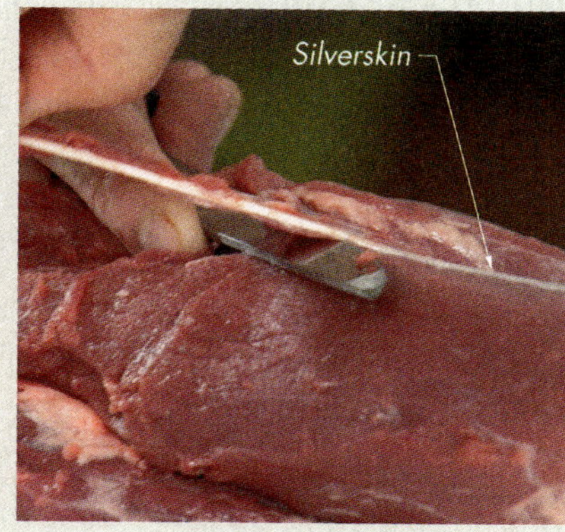

3. Insert the tip of the boning knife just beneath the silverskin at the tail end of the tenderloin. Draw the blade slightly upward along the length of the tenderloin, just beneath the silverskin, toward the head of the tenderloin.

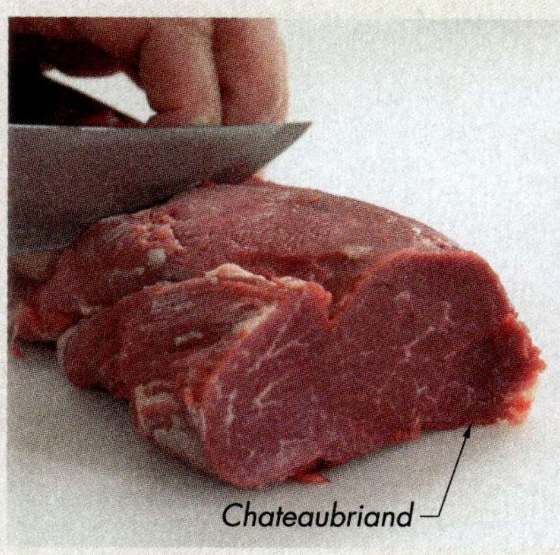

4. Starting at the largest end, cut off the uneven tip of the tenderloin. Cut the tip across the grain into tenderloin tips.
5. Make a cut across the grain just after the large portion ends to remove the Chateaubriand.

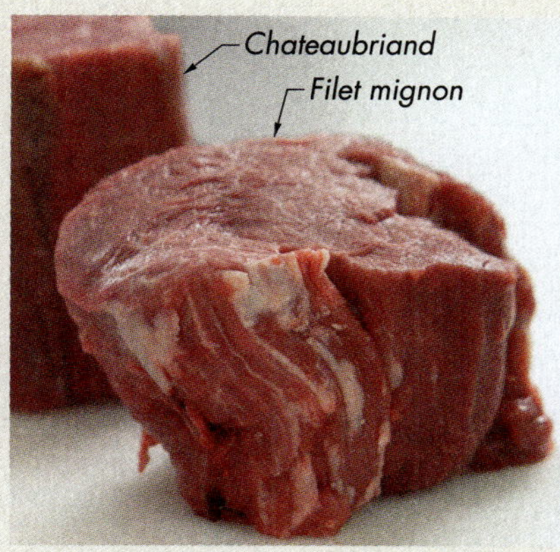

6. Cut the center of the tenderloin across the grain to desired thickness to produce filet mignons.

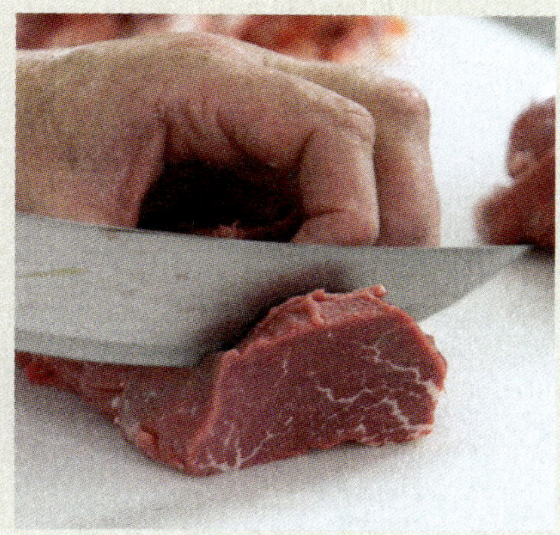

7. Cut the smallest third of the tenderloin across the grain to produce tournedos ½–¾ inch thick and approximately 2½ inches in diameter.

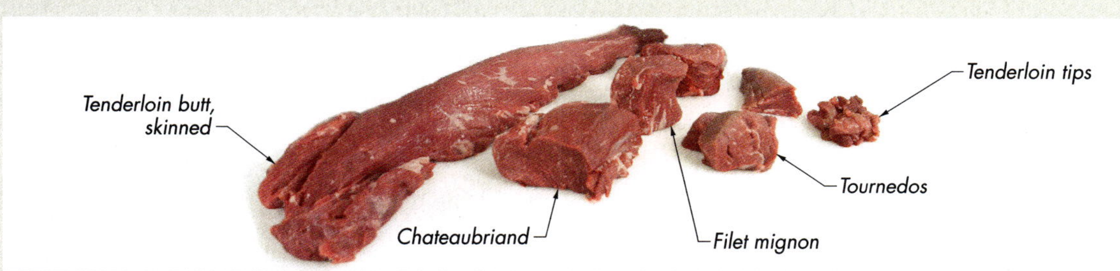

Fabricated beef tenderloin cuts.

Trimming Beef Tenderloin
Media Clip

Cutting Boneless Strip Loin into Steaks

A boneless strip loin is often trimmed and cut into portion-controlled cuts. The sirloin end of the strip loin contains connective tissue that does not break down during cooking. Steaks cut from this end of the loin are referred to as vein steaks. Although the meat is as tender as the rest of the loin, the amount of connective tissue makes vein steaks less desirable than steaks from the rest of the strip loin.

Procedure for Cutting Boneless Strip Loin into Steaks

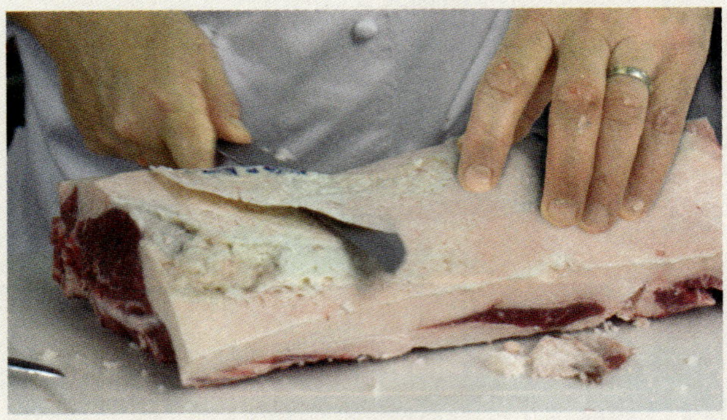

1. Trim the surface fat to approximately ¼ inch thickness.
2. Turn the loin over and trim off any additional fat or connective tissue that could affect the quality of the steaks.

3. Cut the loin across the grain into steaks of desired weight or thickness.

Frenching Veal Chops

Veal chops are a popular menu item and are available boneless or with the bone in. Bone-in veal chops are commonly frenched for service. **See Figure 4-66.**

Cutting Boneless Strip Loin
Media Clip

Frenched Veal Chops

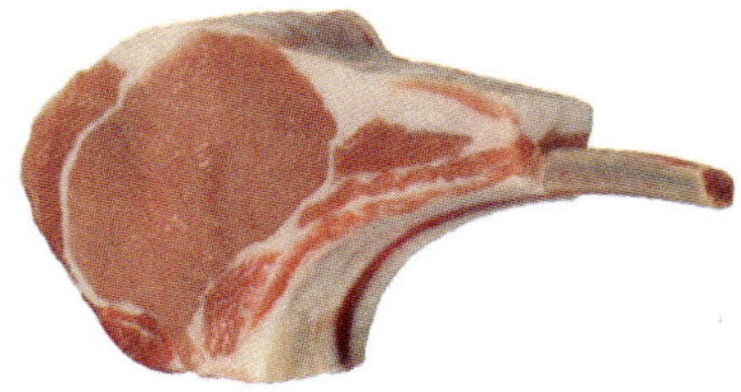

Strauss Free Raised

Figure 4-66. Bone-in veal chops are commonly frenched for service.

Procedure for Frenching Veal Chops

1. Peel back the fat cap where it still is connected to the rack and remove it completely.
2. Trim the entire rack of any excess fat, leaving a thin covering between ¼–⅛ inch thick.
3. Starting on the top side, score the meat by making a straight cut all the way to the bone, approximately 1 inch above the eye meat.
4. Turn the rack over and use the tip of the knife to score the thin membrane on the back of each rib bone from the location of the cut to the tip of the bone.
5. From the scored mark, scrape the meat away from the top of the ribs to expose the bones.
6. Cut between each rib down to the scored mark to remove the meat.

Tenderizing Beef

Beef often benefits from tenderizing. Pounding or cutting meat tenderizes the muscle and connective tissue. **See Figure 4-67.** The following methods are used to tenderize meat:

- pounding meat with a mallet to break up the protein structure and muscle tissue
- using a hand tenderizer with needlelike knives that pierce and gently cut the connective tissues and muscle fibers
- slicing the meat across the grain to produce thin slices with shorter muscle fibers
- grinding the meat to completely break apart strands of connective tissue and muscle tissue

Tenderizing Meat

Pounding with a Mallet

Using a Hand Tenderizer

Slicing Across the Grain

Figure 4-67. Pounding or cutting meat tenderizes the muscle and connective tissue.

Ground beef and veal can be used to make meatballs, sausages, meatloafs, and meat fillings. Meat that is to be ground should be very cold before grinding, as should all parts of the grinder that will contact the meat. Room temperature meat and meat that is only somewhat cool has a tendency to be smashed while grinding, which causes it to lose its texture and appear puréed instead of ground. When grinding meat, it should be passed through progressively smaller dies on the grinder in order to evenly distribute the fat and give the mixture a smooth appearance. **See Figure 4-68.**

Grinding Meats

Figure 4-68. When grinding meat, it should be passed through progressively smaller dies on the grinder in order to evenly distribute the fat and give the mixture a smooth appearance.

Tying Beef and Veal

Tying beef and veal helps maintain a consistent shape and ensures that the meat will cook evenly without falling apart. Some cuts of meat are rolled to form roasts and tied so that they do not unroll during cooking. **See Figure 4-69.** Stuffed veal breast is commonly tied to retain its shape and hold the stuffing.

Tied Roasts

Photo Courtesy of the Beef Checkoff

Figure 4-69. Some cuts of meat are rolled to form roasts and tied so that they do not unroll during cooking.

CHECKPOINT 4-13

1. Describe how to trim and cut a beef tenderloin.
2. Trim and cut a beef tenderloin.
3. Describe how to cut a boneless strip into steaks.
4. Cut a boneless strip into steaks.
5. Describe how to french veal chops.
6. French veal chops.
7. Describe four ways to tenderize beef.
8. Tenderize identical cuts of beef four different ways.
9. Describe how to grind fresh meat.
10. Grind fresh beef.
11. Explain the purpose of tying meat.

MARKET FORMS OF PORK

Many breeds of hogs are used in the production of pork products. Some of the more common breeds are the Duroc, Chester White, Hampshire, Yorkshire, and Berkshire. The Berkshire is most desirable because of its intramuscular fat and juiciness. The Berkshire is sometimes called the American Kurobuta because it has many of the same characteristics as the highly prized Japanese Kagoshima Kurobuta.

Pork is available in a variety of market forms, including whole carcasses, primal cuts, fabricated cuts, and variety meats. Knowledge of these market forms is necessary for accurate product ordering.

Whole Carcasses of Pork

Both whole hog and suckling pig carcasses are available. Purchasing a whole hog carcass allows a chef to be more creative with the menu, rather than being restricted to pre-portioned cuts that can be found in most restaurants. A *suckling pig* is a pig 4–6 weeks old that weighs 20–35 lb dressed. Suckling pig carcasses are sold with the head attached and are priced per pig rather than by the pound.

Primal and Fabricated Cuts of Pork

A primal cut is a large cut from a whole or a partial carcass. A hog carcass has two sets of primal cuts. One set is on the left side and one set is on the right side. Each primal cut is divided into fabricated cuts. The five primal cuts of pork are the picnic shoulder, shoulder butt, loin, leg, and belly. **See Figure 4-70.** Some operations have a butcher on staff and order only primal cuts. Other operations order only fabricated cuts.

A fabricated cut is a ready-to-cook cut that is packaged to certain size and weight specifications. Fabricated cuts of pork include hams, roasts, pork chops, pork cutlets, stew meat, and ground pork. Fabricated cuts are a convenient way of providing uniform portions while reducing labor costs. However, the price per pound for fabricated cuts is higher than the price per pound for primal cuts.

Picnic Shoulders. The *picnic shoulder* is a primal cut of pork that is the lower half of the shoulder of a hog. The average weight of a picnic shoulder is approximately 9% of the total carcass weight. Picnics are fabricated from the picnic shoulder primal cut. **See Figure 4-71.** A *picnic* is a cut of pork fabricated from the upper part of the foreleg that includes a portion of the shoulder. A picnic resembles a ham in shape, but is smaller and contains more bone and less lean meat. Fresh picnic can be used to prepare chop suey, pork patties, or pork sausage. Pulled pork is prepared from smoked picnic meat.

Market Forms of Pork

Primal Cuts of Pork

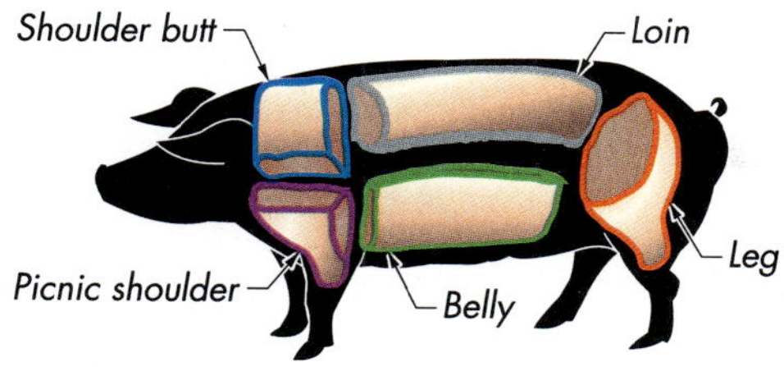

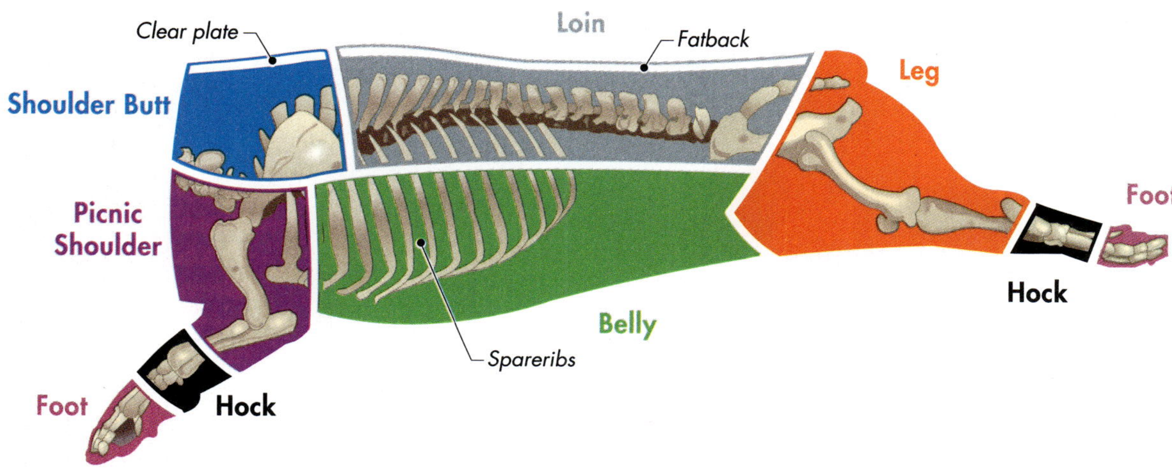

Figure 4-70. The primal cuts of pork include the picnic shoulder, shoulder butt, loin, leg, and belly.

Picnic Shoulder Cut of Pork

National Pork Producers Council

Figure 4-71. Picnic shoulders are fabricated from the picnic shoulder primal cut.

Shoulder Butts. The *pork shoulder butt,* also known as Boston butt, is a square, compact area of the shoulder located just above the front legs of a hog. Its average weight is approximately 8% of the total carcass weight. The shoulder butt contains the blade bone and a large portion of lean meat. It is usually sold fresh with the bone. The meat is moderately tough due to the amount of connective tissue, so it is typically roasted or braised. A shoulder butt can also be fabricated into blade steaks, cottage ham, or ground meat. **See Figure 4-72.** A *cottage ham* is the smoked, boneless meat extracted from the blade section of the shoulder butt. Boneless shoulder butts are often tied with string because they fall apart easily when cooked.

The *clear plate* is a rectangular slab of fat that contains a few strips of lean pork located just above the shoulder butt. Clear plate that has been cured in salt is called salt pork. It is often used as a flavoring ingredient in dishes such as beans and bitter greens. Salt pork is often blanched to extract excess salt before it is used as a flavoring ingredient.

National Pork Producers Council

Figure 4-72. A shoulder butt can also be fabricated into blade steaks, cottage ham, or ground meat.

Loins. The *pork loin* is a primal cut that extends along the backbone, from about the second rib, through the rib and loin area of a hog. Its average weight is approximately 18% of the total carcass weight, and it can be cut into a variety of fabricated cuts. **See Figure 4-73.** The *pork tenderloin* is a fairly long, tapered strip of lean meat taken from the underside of the loin. Tenderloin is the most tender pork cut and can be prepared using any cooking method.

Baby back ribs are the meaty bones on the rib end of the pork loin. These meaty ribs are only 3–6 inches long and are curved where they meet the backbone. A full slab of baby back ribs has 11–13 ribs. A standing rib roast is the whole pork loin muscle with the baby back ribs attached. A boneless loin and pork cutlets are also fabricated cuts from the loin.

Canadian bacon is the trimmed, pressed, and smoked boneless loin of pork. *Fatback* is the layer of fat that runs along the back of the hog. It can be used to flavor dishes such as beans and collard greens or added to sausage or ground pork. Lard is usually rendered from fatback.

Legs. A *pork leg* is a primal cut of pork that is composed of the hind thigh and buttock of a hog. A pork leg is sometimes referred to as a ham because it is the cut that is fabricated from the leg. It contains a high proportion of lean meat and the average weight is approximately 24% of the total carcass weight. The ham is typically cut from the middle of the shank bone to the aitchbone, or hip bone. **See Figure 4-74.**

Unprocessed ham is called fresh ham or green ham. Hams are sold boneless, bone-in, and partially boned. The most popular form of ham is cured in a solution of salt, sodium nitrite, and sugar and then smoked. The skin may be left on a ham or it may be removed.

Ham is often cut into steaks or cutlets and broiled or pan-fried. Ham is also sliced and used to make sandwiches. *Prosciutto* is a type of dry-cured Italian ham. Prosciutto is typically sliced paper thin and used to make hors d'oeuvres or appetizers.

Loin Cuts of Pork

National Pork Producers Council

Figure 4-73. A variety of cuts are fabricated from the loin primal cut.

Leg Cuts of Pork

National Pork Producers Council

Figure 4-74. A primal leg cut is fabricated into several cuts.

Belly. A *belly* is a primal cut of pork that is the lower portion of the hog between the shoulder and the leg. Its average weight is approximately 19% of the total carcass weight. Spareribs, pork belly, and bacon are fabricated cuts from the belly primal cut. **See Figure 4-75.**

Pork spareribs are the long, narrow ribs and breastbone of a hog. They are quite fatty, yet the meat is tender and has an excellent flavor. A full rack contains 14 ribs. Spareribs may be purchased fresh or smoked. Fresh spareribs should be cooked slowly over low heat until the meat is tender and the fat is rendered. Spareribs are typically seasoned prior to cooking with a rub or a marinade. Cooked spareribs can be browned on the grill or under a broiler to caramelize the meat. They are commonly barbequed, broiled, or roasted and served as an appetizer or an entrée.

Pork belly is a fatty slab of meat and skin from the side and belly of a hog. Braised pork belly is a popular dish. Pork belly can also be cured and served as bacon or pancetta. *Bacon* is pork belly that has been cured and usually smoked. There are three basic cuts of bacon: thin, regular, and thick. Thin bacon is sliced into 22–26 strips per lb. Regular bacon is cut into 16–20 strips per lb, and thick bacon is cut into 12–16 strips per lb. Bacon is used to make entrées, appetizers, and sandwiches. It also is used to garnish soups. Bacon and the fat rendered from bacon are often used to season other foods.

Pancetta, also known as Italian bacon, is unsmoked pork belly that has been cured in salt and spices, such as nutmeg and pepper, and then dried for a few months. Pancetta is typically cut into paper-thin slices before being added to a dish. It adds a distinctive pork flavor, especially to pastas, that is not smoky like bacon.

Pork Offals

Unlike most domesticated meat animals, just about every part of a hog is used for consumption. An offal is an edible part of an animal that is not part of a primal cut. Offals are often ground and used to prepare sausage, meatballs, and fillings. Pork that is to be ground should be very cold before grinding, as should all parts of the grinder that come in contact with the pork. Pork that is cool instead of cold has a tendency to lose its texture and can appear puréed instead of ground. Some chefs place pork in the freezer for a short time before grinding it for better results.

In addition to common offals such as the heart, liver, and kidneys, pork variety meats include jowls, hocks, pig feet, and headcheese. A *jowl* is meat from the cheek of a hog. Jowls are cured and smoked in the same manner as bacon. Jowls are also used to flavor items such as baked beans or long beans.

Hocks are cut from the lower part of the front and hind legs of a hog. They have very little meat, but good flavor, with a good amount of fat, bone, and gristle. Hocks are available fresh, cured, or cured and smoked. They are often used to flavor soups or served with bean dishes.

Belly Cuts of Pork

National Pork Producers Council

Figure 4-75. Spareribs, pork belly, and bacon are fabricated cuts from the belly primal cut.

Pig feet, also known as trotters, can be purchased fresh, smoked, or pickled. Fresh pig feet are commonly broiled or served boneless on a cold plate.

Headcheese is the spiced, pressed, and jellied meat from the head of a hog. The tongue, heart, feet, and quality trimmings may also be ground and included in headcheese. It is often served cold like lunchmeat or used as a binding agent in stocks.

Receiving and Storing Pork

Like all meats, pork is a potentially hazardous food. The color, odor, texture, and temperature must be checked when received. The meat should be pink in color and the fat should be white. There should be no odor, and the texture should be firm and not dry or slick. Pork that is in the temperature danger zone should be rejected. Refrigerated pork should maintain an internal temperature of 41°F or below.

Frozen pork should be stored at temperatures below 0°F. Frozen pork that needs to be thawed should be placed in the refrigerator overnight. Larger cuts of pork may take more than one day to thaw under refrigeration. If necessary, pork can be thawed under running water if it is cooked immediately after thawing. Some frozen cuts, such as sausage patties or breaded pork tenderloins, can be thawed as part of the cooking process.

Vacuum-packed pork must not be opened until needed for service or preparation. The airtight plastic of vacuum packaging helps to preserve meat for three to four weeks. **See Figure 4-76.** Once the vacuum seal is broken, the meat has a shelf life of only two or three days. Cut pork that is not vacuum-sealed should be wrapped tightly, refrigerated immediately, and used as soon as possible.

Vacuum-Packaged Pork

Figure 4-76. Airtight plastic of vacuum packaging helps to preserve meat for three to four weeks.

In addition to packaging, pork can also be irradiated to reduce the risk of potentially harmful microorganisms. The irradiation process does not cook pork and does not have an adverse effect on its appearance or taste. The USDA requires all irradiated food be labeled with an irradiation symbol.

Inspection and Grades of Pork

All pork used in foodservice operations must be procured from USDA inspected plants. At the time of slaughter, a hog carcass is stamped with the round USDA inspection stamp, indicating that the hog was slaughtered at an inspected plant. This stamp does not indicate anything about the quality of the meat. **See Figure 4-77.** The purple inspection stamp is used for whole carcasses and all fabricated and processed meats. It is found either on the meat itself or on the case in which it is packed. The number on the stamp identifies the plant where the animal was processed.

USDA Inspection Stamps

Figure 4-77. A round USDA inspection stamp indicates that a hog was slaughtered at a USDA-inspected plant.

Pork is not graded like beef, veal, and lamb. It is produced from young hogs that were bred and fed to produce uniformly tender meat. Quality pork has very little fat covering on the surface. The meat is firm and grayish-pink in color.

CHECKPOINT 4-14

1. Identify common market forms of pork.
2. Explain the advantage of purchasing a whole carcass.
3. Identify the five primal cuts of pork.
4. Identify the cut fabricated from a picnic shoulder.
5. Identify cuts fabricated from a shoulder butt.
6. Identify cuts fabricated from a pork loin.
7. Identify cuts fabricated from a leg of pork.
8. Identify cuts fabricated from a pork belly.
9. Describe four offals that are only fabricated from pork.
10. Identify four traits that should be checked upon receiving pork.
11. Identify the required storage temperature for refrigerated pork and for frozen pork.
12. Explain why vacuum-sealed packages of pork should only be opened at the time of use.
13. Describe the effects of irradiation on pork.
14. Describe the USDA inspection and grading of pork.

FABRICATING PORK

The size of storage facilities and availability of staff with fabrication skills often determine whether a foodservice operation purchases primal cuts, fabricated cuts, or a combination. Some operations purchase whole hog carcasses and fabricate all the cuts in-house, while others only fabricate the loin. Pork fabrication requires the removal, trimming, and tying of meat as well as butterfly cuts.

Removing Pork Tenderloins

Some foodservice operations purchase primal loins and fabricate cuts in-house. If a loin is being fabricated in-house, the tenderloin must first be removed from the loin.

Procedure for Removing the Tenderloin from a Loin

1. Trim the surface fat of the loin to approximately ¼ inch thick on both sides.
2. Slide the knife under the exposed rib and slice through to the backbone to separate the tenderloin muscle from the vertebrae.
3. Carefully cut down the center of the backbone between the muscle and the exposed vertebrae.
4. Continue cutting to the end of the vertebrae until the tenderloin is completely removed from the loin.

Once the tenderloin has been removed from the loin, it needs to be trimmed. Trimming a tenderloin involves removing all the fat and silverskin. **See Figure 4-78.**

Pork Tenderloin

Photo Courtesy of D'Artagnan, Photography by Doug Adams Studio

Figure 4-78. Trimming a tenderloin involves removing all the fat and silverskin.

Procedure for Trimming a Tenderloin

1. With a rigid boning knife, carefully remove the chain muscle from the side of the tenderloin and reserve.
2. Use a rigid boning knife to pull back and trim the thick fat covering from the tenderloin.
3. Position the tenderloin so that the head (wide end) is near the cutting hand and the tail (narrow end) is near the guiding hand. Insert the tip of the boning knife just beneath the silverskin at the tail of the tenderloin.
4. Angle the knife blade upward slightly and draw it along the length of the tenderloin, just beneath the silverskin, toward the head. *Note:* Use the guiding hand to hold the silverskin firmly while cutting.

Tying Boneless Pork Roasts

A boneless pork roast falls apart when cooked. To prevent this, the roast is often rolled and then tied with butcher's twine prior to cooking.

Fabricating Boneless Pork Loin
Media Clip

Procedure for Tying a Boneless Roast

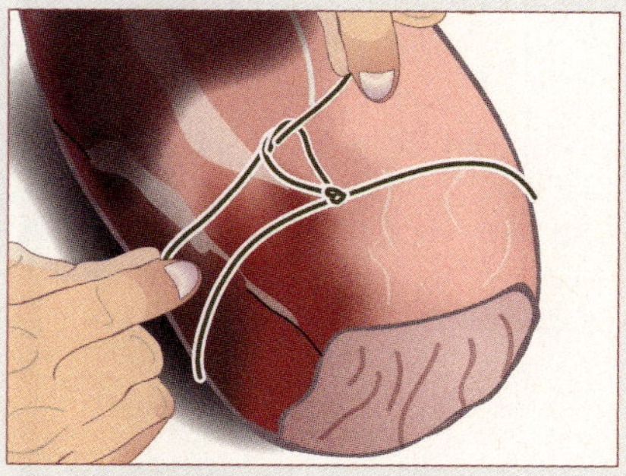

1. Place the outside of the meat facedown and roll the sides tightly toward the center.
2. Wrap butcher's twine around one end of the roast using a slipknot to tighten firmly. Then tie a square knot to secure the tie.

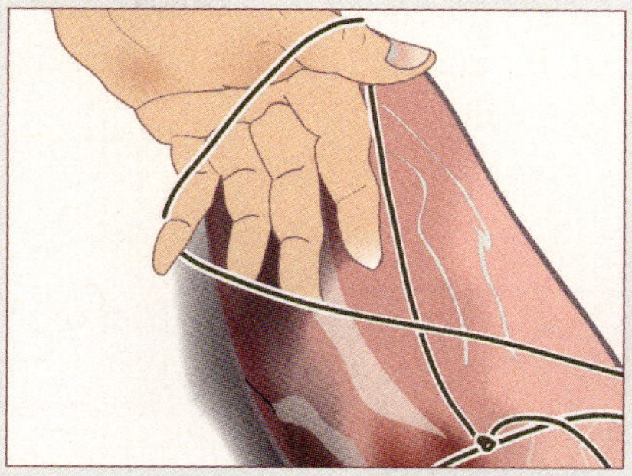

3. Wrap the twine around one hand to form a loop. Then twist the hand over the meat until the palm is facing down.

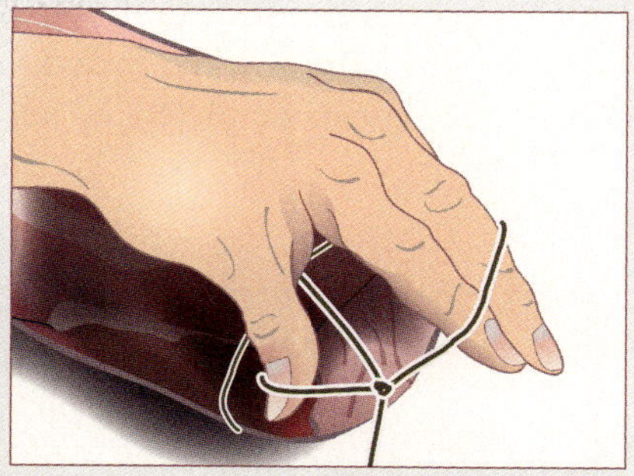

4. Move the palm to the front of the roast and slide the loop over the meat.

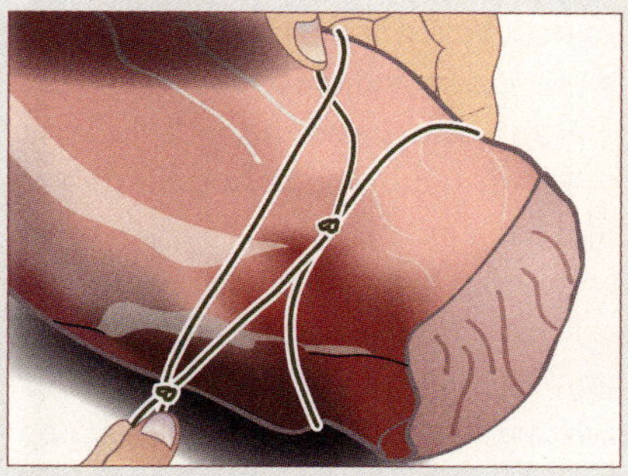

5. Remove the hand from the loop and secure the twine around the meat.

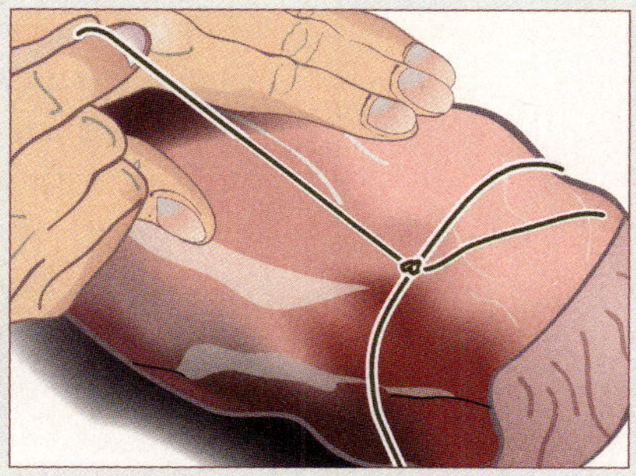

6. Pull tightly to straighten the loop.

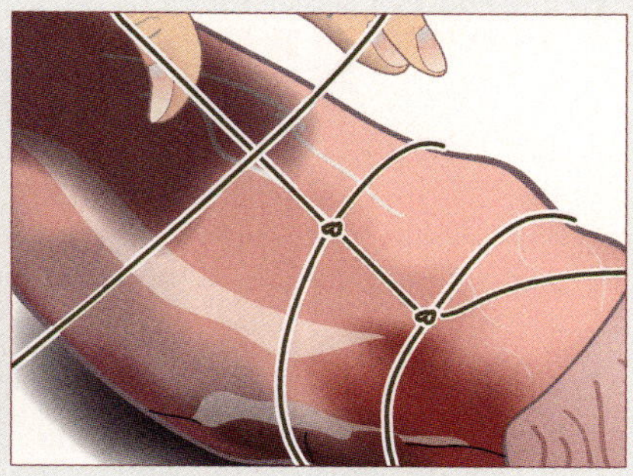

7. Continue making loops in the same fashion 1½ inches apart down the length of the roast. *Note:* Check each loop to make sure it is an even distance from the last loop.

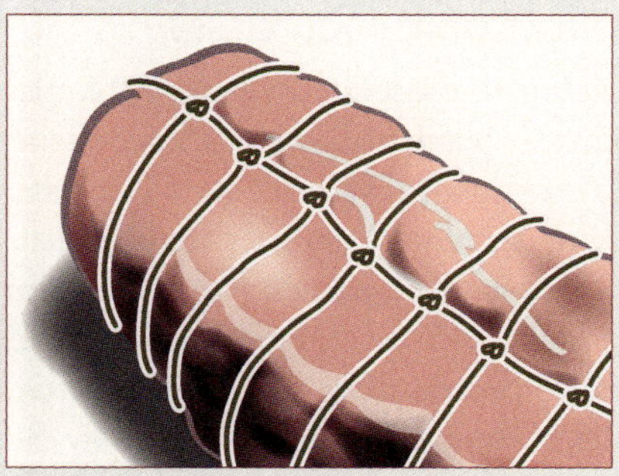

8. Turn the roast over. Then pull the twine over the first loop and tuck it under and around the loop, pulling it toward the front of the roast.

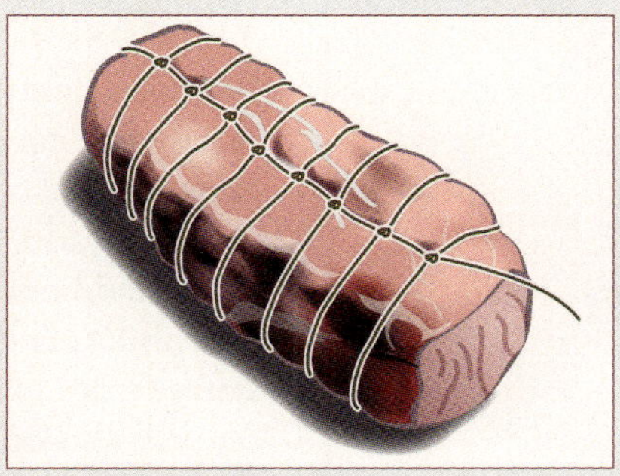

9. Continue pulling over the loops until the underside of the roast is fully tied.

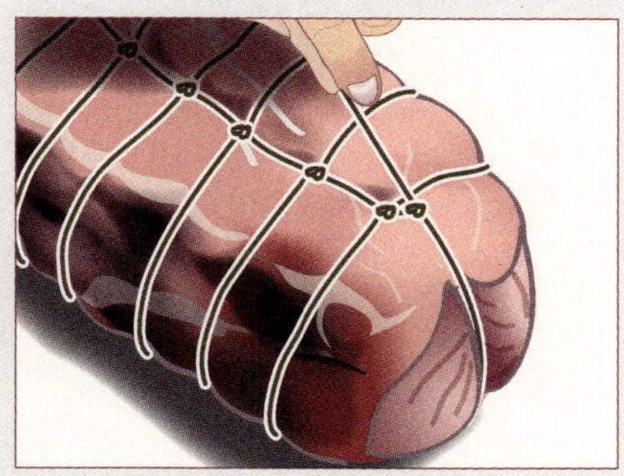

10. Turn the roast over and bring the twine up to tie a square knot to secure the roast. Cut the twine close to the knot.

Tying a Pork Loin
Media Clip

Making Butterfly Cuts

Thicker cuts such as pork chops can be butterfly cut to make a pocket for stuffing or to create a thinner product that will cook more quickly. To make a butterfly cut, the chop is cut almost completely in half horizontally.

Procedure for Making a Butterfly Cut

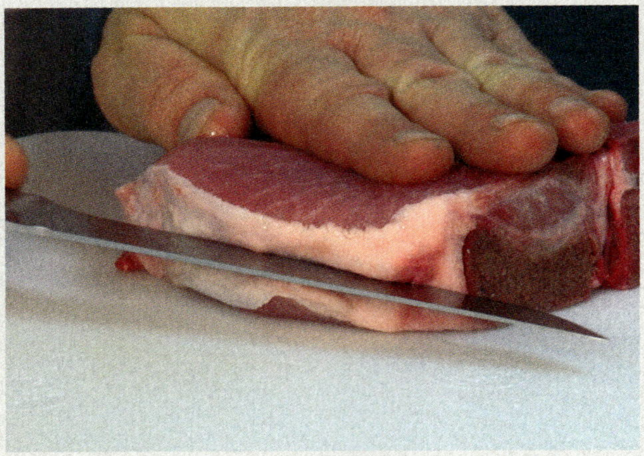

1. Place the pork chop near the edge of a cutting board to prevent the knuckles of the knife hand from hitting the board.
2. Place the guiding hand flat on top of the chop to hold it in place. While holding the knife blade parallel to the cutting board, place the blade edge at the midpoint of the chop.

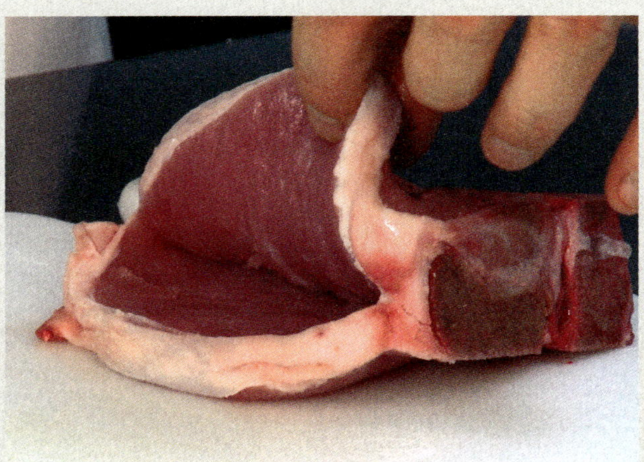

3. Slice the chop almost all the way through, leaving ¼–½ inch connected at the back side.

CHECKPOINT 4-15

1. Describe how to remove the tenderloin from a pork loin.
2. Remove the tenderloin from a pork loin.
3. Describe how to trim a pork tenderloin.
4. Trim a pork tenderloin.
5. Describe how to tie a boneless pork roast.
6. Tie a boneless pork roast.
7. Describe how to butterfly boneless pork chops.
8. Butterfly a boneless pork chop.

MARKET FORMS OF LAMB

Lamb is the meat from slaughtered sheep that are less than a year old. It is important to understand the basic composition and bone structure of a lamb before working with the meat. Lamb meat has smooth grain and is similar in color to beef. It is also very tender and has a mild flavor.

American lamb is weaned to grain, then hay, and finished with a formulated feed that contains wheat, sorghum, and vitamins. American lamb has a light gamey flavor. Australian and New Zealand lambs are smaller in size and have a more pronounced flavor than American lamb. The pronounced flavor has been attributed to their grass diet.

Lamb is available in a variety of market forms, including whole and partial carcasses, primal cuts, fabricated cuts, and offals. Knowledge of these different market forms is necessary for accurate product ordering.

Whole and Partial Lamb Carcasses

Both whole and partial lamb carcasses are available, but partial carcasses are not commonly purchased by foodservice operations. A partial carcass is a primary division of a whole carcass. Partial carcasses of lamb include foresaddles, hindsaddles, backs, and bracelets. Like whole carcasses, partial carcasses of lamb are often not purchased due to the skilled labor and storage space required to process them. Whole and partial carcasses also yield cuts of lamb that may not be used by the foodservice operation and are therefore wasted.

Foresaddle and Hindsaddle. Lamb is typically split into head and tail sections known as the foresaddle and the hindsaddle. The foresaddle and hindsaddle are split between the 12th and 13th ribs, not down the backbone. The *lamb foresaddle* is the front half of the carcass consisting of the primal shoulder, rack, and breast/shank cuts. The *lamb hindsaddle* is the rear half of the carcass consisting of the loin and leg.

Back and Bracelet. Lamb carcasses are also divided into backs and bracelets. A *lamb back* is a lamb rack and loin that are still joined. A back is often purchased by foodservice operations that sell a lot of lamb chops because chops can be cut from the end of the rack to the end of the loin.

A *lamb bracelet* is a hotel rack with the breast still attached. **See Figure 4-79.** A *lamb hotel rack* is a lamb back that remains joined along the backbone. A bracelet is typically fabricated into a hotel rack and a breast. Leaving the breast attached to the rack yields a larger amount of meat and fat.

Primal and Fabricated Cuts of Lamb

Unlike beef, lamb is not split into sides before being divided into primal cuts. The left and right primal cuts remain joined together and are purchased as a single cut. For example, the leg primal cut has two joined legs. The five primal cuts of lamb are the shoulder, rack, loin, leg, and breast/shank. **See Figure 4-80.**

Lamb Bracelet

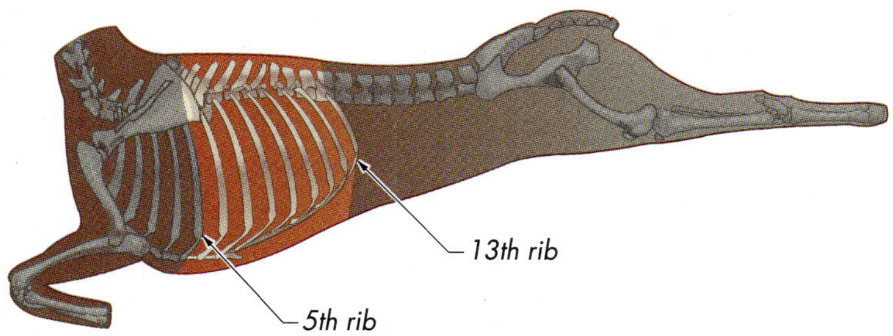

Figure 4-79. A lamb bracelet is a hotel rack with the breast still attached.

Primal Cuts of Lamb

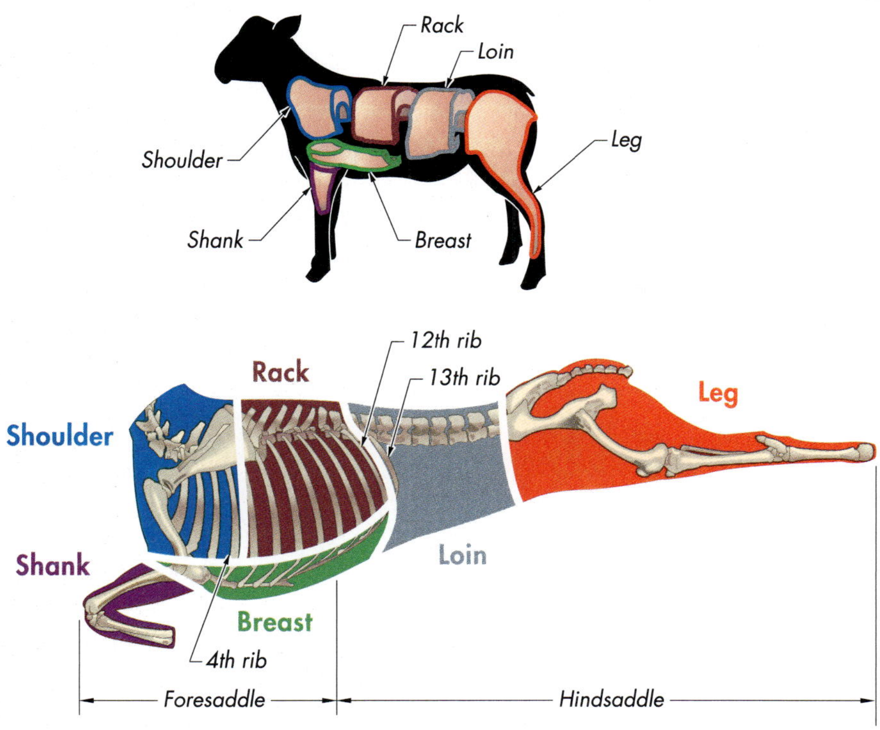

Figure 4-80. The primal cuts of lamb are the shoulder, rack, loin, leg, and breast/shank.

Each primal cut is further divided into fabricated cuts. Some fabricated cuts of lamb include stew meat, ground meat, hotel racks, roasts, chops, and leg of lamb.

Fabricated cuts are a convenient way of providing uniform portions while reducing labor costs. The price per pound for fabricated cuts is higher than the price per pound for primal cuts. Some operations order only primal cuts. Other operations order only fabricated cuts. Most foodservice operations primarily purchase some primal cuts and some fabricated cuts of lamb.

To help ensure that foodservice operations and suppliers communicate efficiently, the USDA publishes the Institutional Meat Purchase Specifications (IMPS) for commonly purchased meats and meat products. Cuts of lamb are numbered by category.

Shoulder. A *lamb shoulder* includes the first four rib bones of each side and the arm and neck bones. Its average weight is about 36% of the total carcass weight. Shoulder meat is quite lean and has excellent flavor. It is seldom cooked whole because of its many small bones and connective tissues. Instead, it is fabricated into a variety of cuts, including roasts, chops, stew meat, and ground meat **See Figure 4-81.** Ground lamb meat can be used in the preparation of sausages, meatballs, and meat fillings.

Rack. A *lamb rack* is eight rib bones located between the shoulder and loin of a lamb. Its average weight is about 8% of the total carcass weight. The meat is tender and well-marbled because it comes from an area of the back where the muscles are not worked much. A rack is often split along the backbone into two racks. A *lamb crown roast* is a hotel rack containing sixteen ribs with the bones frenched, notched, and tied to create a circle to resemble a crown. Paper frills are often used to cover the exposed tip of each rib bone of a crown roast. *Frenching* is a method of removing the meat and fat from the end of a bone and is generally applied to chops. **See Figure 4-82.**

A rack of lamb is sometimes coated with herbs, roasted whole, and then sliced to order. However, in most cases it is cut into single or double chops and grilled or broiled. A *double rib lamb chop* is a rib chop cut to a thickness equal to two standard rib chops. An *English lamb chop* is a 2 inch thick fabricated cut taken along the entire length of the unsplit loin. Chops may also be frenched.

Loin. A *lamb loin* is a primal cut located between the rack and leg that includes the 13th rib, the loin eye muscle, the center section of the tenderloin, the strip loin, and some flank meat. **See Figure 4-83.** An unsplit primal lamb loin is commonly known as a saddle and has an average weight of about 13% of the total carcass weight.

Shoulder Cuts of Lamb

American Lamb Board

Figure 4-81. The shoulder is often fabricated into a variety of cuts, including roasts, chops, stew meat, and ground meat.

Rack Cuts of Lamb

American Lamb Board

Figure 4-82. A rack is fabricated into rib roasts, crown roasts, and rib chops.

Loin Cuts of Lamb

American Lamb Board

Figure 4-83. A primal loin can be fabricated into a loin roast, boneless loin strip, tenderloin, loin chop, and double loin chop.

Fabricated cuts from the loin are best prepared using dry-heat cooking methods such as grilling, broiling, or roasting. Lamb loins are typically cut into boneless or bone-in chops that can be grilled or broiled. The tenderloin can be removed and cut into noisettes or can be roasted whole. A *noisette* is a small, round, boneless medallion of meat.

Leg. Lamb legs are not split into two separate legs. Lamb legs remain joined at the hip. A *leg of lamb* is a primal cut of lamb that contains the last portion of the backbone, hip bone, aitchbone, round bone, hindshank, and tailbone. The aitchbone is the buttock or rump bone and is located at the top of the leg. The leg accounts for approximately 34% of the total carcass weight. Leg of lamb includes part of the sirloin, the top round, bottom round, and knuckle meat. **See Figure 4-84.**

The leg contains lean, fine-textured meat that is more tender near the sirloin end and tougher toward the shank end. Lamb leg is commonly split in two and partially boned, stuffed, and roasted. It can also be split in two and completely boned, rolled, tied, and roasted. Single lamb legs can be purchased boned, rolled, and tied (BRT). Meat from the sirloin end can also be cut into lamb steaks. Meat from the shank end is most commonly used for stew meat or ground for patties. Shank meat is commonly cut in cross-sections, braised, and served in a rich, flavorful sauce.

Breast/Shank. A *lamb breast* is a thin, flat, primal cut of lamb that contains the breastbone, the tips of the rib bones, and cartilage that is located under the shoulder and ribs. The breastbone is actually cartilage because the animal is so young. The breast also includes the shank and weighs about 17% of the total carcass weight. **See Figure 4-85.** A *lamb shank* is a cut of lamb that contains the upper foreshank bones. The small section of seven or more rib tips can also be braised and is often referred to as Denver ribs or lamb riblets. A *lamb riblet* is a rectangular strip of meat cut from the lamb breast that contains part of a rib bone.

Leg Cuts of Lamb

American Lamb Board

Figure 4-84. A primal leg of lamb is fabricated into many cuts.

Breast/Shank Cuts of Lamb

American Lamb Board

Figure 4-85. A primal breast/shank is fabricated into a foreshank, spareribs, and riblets.

The breast is not a popular cut of meat, but the shape and ample connective tissue make the breast a good cut to stuff, roll, tie, and then braise. Braising breaks down the connective tissue and yields a very tender rolled roast.

Lamb Offals

Lamb offals are not commonly used in most foodservice operations. However, lamb tongue, kidneys, liver, heart, and sweetbreads are featured dishes in Mediterranean and Middle Eastern cuisine. Lamb intestines may also be used as casings for small sausages.

Receiving and Storing Lamb

Like all meats, lamb is a potentially hazardous food and the color, odor, texture, and temperature must be checked upon receipt. The meat should be light red in color and the fat should be white. **See Figure 4-86.** There should be no odor, and the texture should be firm and not dry or slick. Refrigerated lamb should maintain an internal temperature of 41°F or below. Lamb that is in the temperature danger zone should be rejected.

Frozen lamb should be stored at temperatures below 0°F. Frozen lamb that needs to be thawed should be placed in the refrigerator overnight. Larger cuts of lamb, such as legs, may take more than one day to thaw under refrigeration.

Vacuum-packed lamb must not be opened until needed for service or preparation. Once the vacuum seal is broken, the meat has a shelf life of only 2–3 days. Cut lamb that is not vacuum-sealed should be wrapped tightly, refrigerated immediately, and used as soon as possible.

Receiving Standards

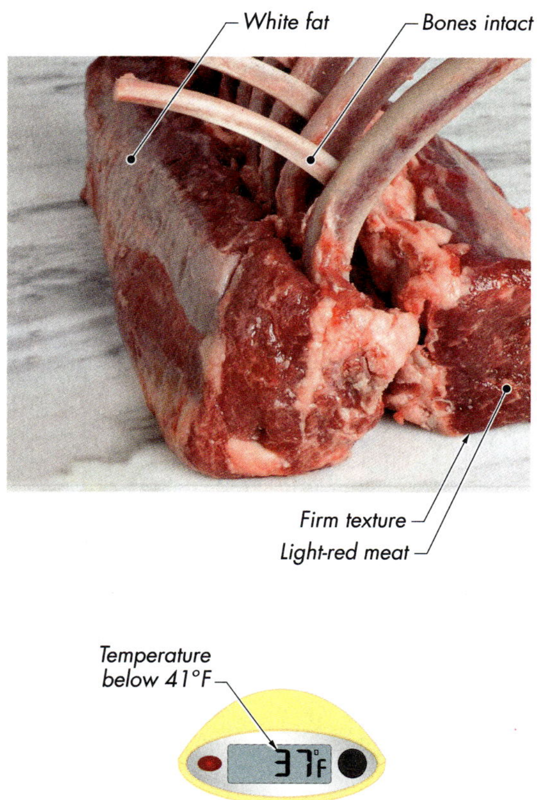

Photo Courtesy of D'Artagnan, Photography by Doug Adams Studio

Figure 4-86. Upon receiving, lamb cuts should be checked for color, odor, texture, and temperature. The bones should be intact.

Inspection of Lamb

All lamb used in foodservice operations must be purchased from USDA-inspected plants. At the time of slaughter, the lamb carcass or the inspection tag is stamped with the round USDA inspection stamp, indicating the lamb was slaughtered at an inspected plant. This stamp does not indicate anything about the quality of the meat. The inspection stamp is used for whole and partial carcasses as well as all fabricated and processed meats. It is found either on the meat itself, on the tag attached to the meat, or on the case in which it is packed. **See Figure 4-87.** The number on the stamp identifies the plant where the animal was processed.

Quality Grades of Lamb

Unlike inspection, USDA quality and yield grading is optional for lamb producers. Quality and yield grading stamps are stamped onto carcasses in the same manner as inspection stamps. Quality grading is based on the overall tenderness, juiciness, and flavor of the meat. However, quality grades do not guarantee these characteristics.

The grades of lamb commonly used in foodservice operations are USDA Prime and USDA Choice. **See Figure 4-88.** USDA Prime lamb is well marbled. USDA Choice lamb has slightly less marbling, but is still the most popular grade of lamb used in foodservice operations.

USDA Quality Grade Stamps

Figure 4-88. USDA Prime and USDA Choice lamb is used in foodservice operations.

Yield Grades of Lamb

Lamb can also be yield graded for the percentage of edible meat to fat and bone. Yield grade shields are numbered 1 to 5 and indicate how much usable meat can be obtained from a lamb carcass. A grade of 1 indicates the highest yield of meat, and a grade of 5 indicates the lowest yield.

USDA Inspection Stamps

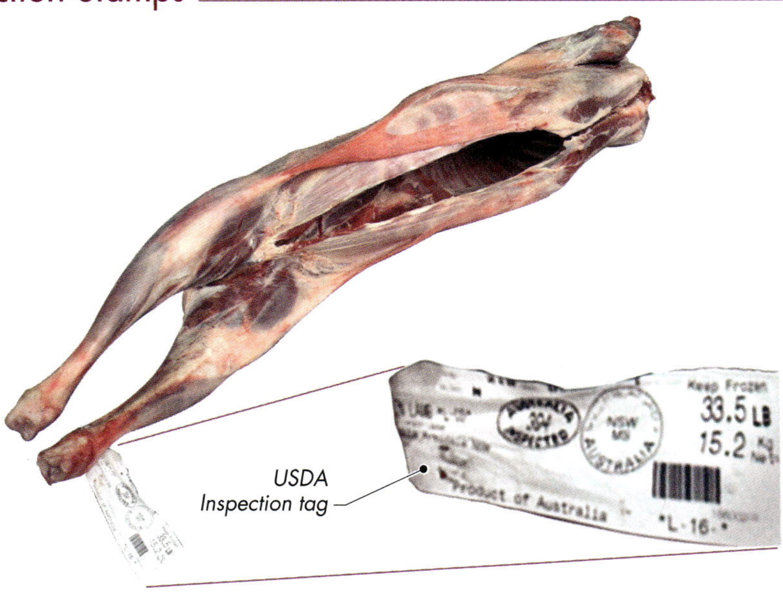

Figure 4-87. A USDA inspection stamp is sometimes attached to a lamb carcass in the form of a tag.

CHECKPOINT 4-16

1. Identify common market forms of lamb.
2. Explain the difference between a foresaddle and a hindsaddle.
3. Explain the difference between a back and a bracelet.
4. Identify the five primal cuts of lamb.
5. Identify the cuts fabricated from a lamb shoulder.
6. Identify the cuts fabricated from a lamb rack.
7. Identify the cuts fabricated from a lamb loin.
8. Identify the cuts fabricated from a leg of lamb.
9. Identify the cuts fabricated from a lamb breast/shank.
10. Identify the lamb offals that are used in some cuisines.
11. Identify four traits that should be checked upon receiving lamb.
12. Identify the required storage temperature for refrigerated lamb and frozen lamb.
13. Describe the USDA inspection and grading of lamb.

FABRICATING LAMB

The size of storage facilities and availability of staff with the fabrication skills often determine whether a foodservice operation purchases primal cuts, fabricated cuts, or a combination. Lamb fabrication involves separating and frenching racks, boning and tying loins and legs, and cutting noisettes.

Fabricating Lamb

Separating Hotel Racks

The first step in fabricating a rack is to separate the hotel rack. This yields two separate racks.

Procedure for Separating a Hotel Rack

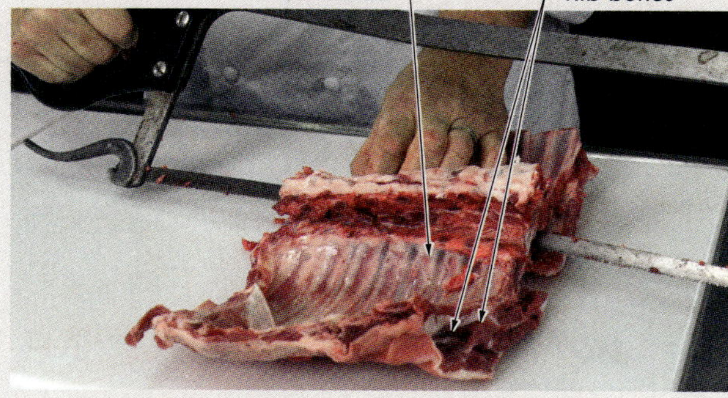

1. Turn the rack upside-down, with the ribs pointing upward.
2. Use a meat saw to cut at a 45° angle between the base of the rib bones and the vertebrae.
3. Cut completely through the bones, using caution not to cut the meat.

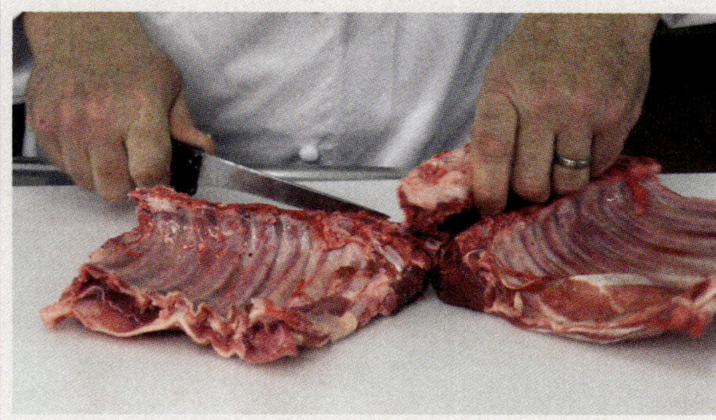

4. Use a knife to finish separating the meat from one side of the vertebrae.
5. Repeat on the other rack.

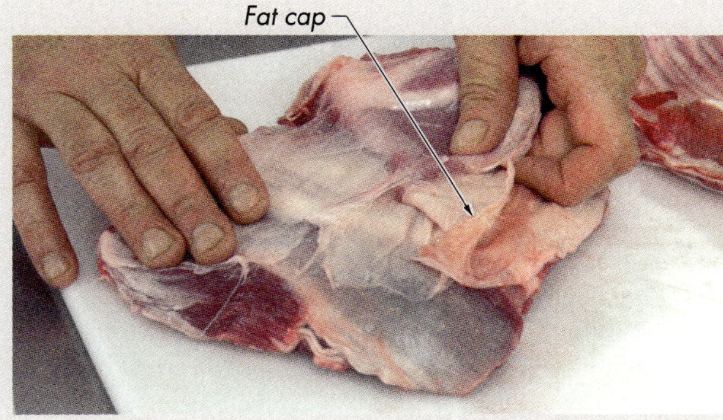

6. Remove the back strap and fat cap. *Note:* The back strap is a tough tendon that runs parallel to the vertebrae.

Frenching Racks

A rack of lamb is often frenched for service. Several techniques are used to french racks of lamb. Regardless of the technique used to french a rack of lamb, the meat is removed from the end of each bone.

Procedure for Frenching Racks of Lamb

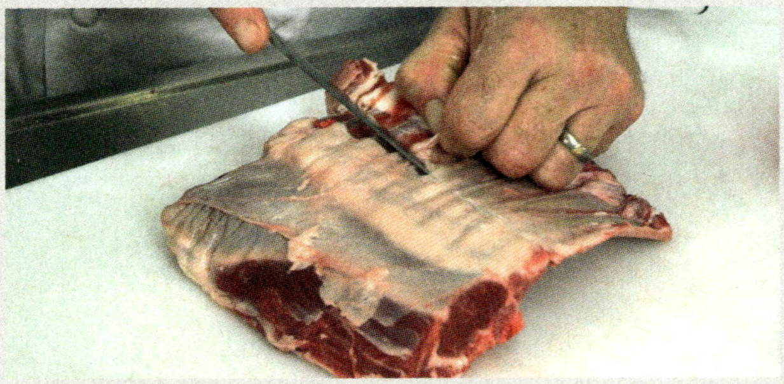

1. Starting on the top side of the rack, score the meat all the way to the bone, 1 inch above the eye meat.
2. Turn the rack over and score the thin membrane on the back of each rib bone from the scored cut to the tip of the bone.

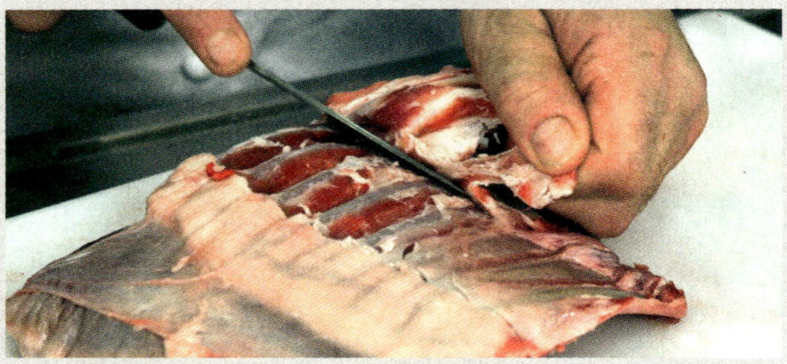

3. From each scored mark, scrape the meat away from the top of the rib to expose the bone.

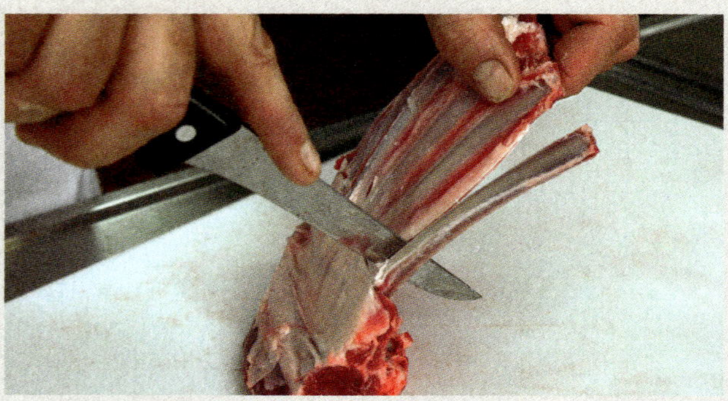

4. Cut between each rib bone down to the scored mark to remove the meat.

Frenching Racks of Lamb
Media Clip

Boning and Tying Lamb Loins

A whole lamb loin is often boned and then tied to form a rolled roast. It is important to remove all connective tissue from the skin side of the loin.

Procedure for Boning a Loin to Form a Rolled Roast

1. Trim the layer of connective tissue from the skin side of the loin.
2. Place the skin side down and remove the fat from around the tenderloin.
3. Cut under the vertebrae to separate the bones from the tenderloin. Be careful not to remove the tenderloin. Repeat this step on the other side.
4. Use a knife to cut under the rib bones to the backbone. This cut will free the rib eye from the bones.
5. Remove the backbone by pulling the meat away from the bones and making small cuts as needed.
6. Check the trimmed piece of loin for bone or connective tissue.
7. With the skin side facing up, closely trim the surface so there is about ¼ inch of fat remaining.
8. Place the outside of the meat face down and roll the sides tightly toward the center.

Rolled loin roasts are commonly tied before roasting. A rolled roast is tied to maintain a consistent shape and ensure even cooking.

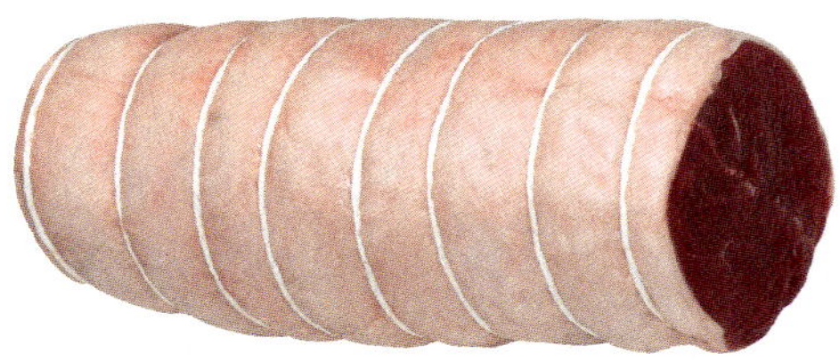

American Lamb Board

Boning and Tying Legs of Lamb

The bone is commonly removed from a leg of lamb before cooking. A boned leg of lamb makes a nice presentation.

A leg of lamb may be tied before roasting. Tying helps the meat maintain a consistent shape and cook evenly during the cooking process.

Procedure for Boning a Leg of Lamb

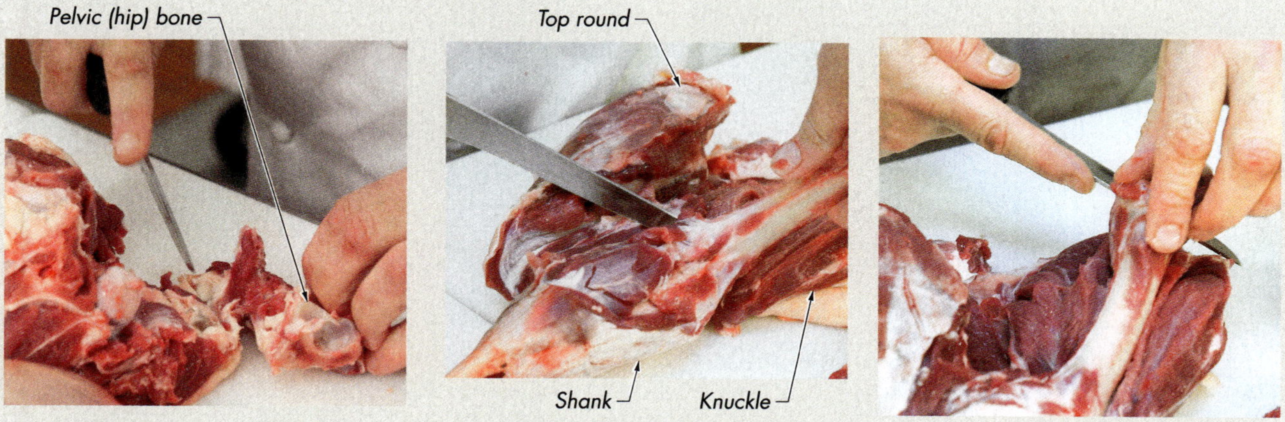

1. Hold the pelvic bone while cutting around the bone to free the pelvic bone from the leg.
2. Carefully remove the top round by cutting lengthwise along the leg bone.
3. Continue trimming around the knee joint to separate the knuckle meat from the bone.
4. Working near the shank end of the leg, continue to cut lengthwise to free the remaining meat from the bone.

Cutting Noisettes

Noisettes are small, round, boneless medallions of meat. A lamb tenderloin may be cut into noisettes. Noisettes may also be cut from a boned loin chop. Once the bone is removed, the loin is rolled and then can be cut into noisettes.

CHECKPOINT 4-17

1. Describe how to separate a hotel rack.
2. Separate a hotel rack.
3. Describe how to french a rack of lamb.
4. French a rack of lamb.
5. Describe how to bone and tie a lamb loin.
6. Bone and tie a lamb loin.
7. Describe how to bone and tie a leg of lamb.
8. Bone and tie a leg of lamb.
9. Describe how to cut noisettes.
10. Cut a lamb tenderloin into noisettes.

Boning a Leg of Lamb
Media Clip

FABRICATING RABBIT

Some menus may require the fabrication of specialty game such as rabbit. Rabbits sold for consumption are often crosses between New Zealand and Belgian varieties, Chinese rabbits, or Scottish hares. Rabbits and other specialty game meats must be purchased from insured vendors and stored with care to ensure food safety. Rabbit meat is often sold frozen. Uncooked rabbit is similar in appearance and texture to uncooked chicken. **See Figure 4-89.**

Uncooked Rabbit

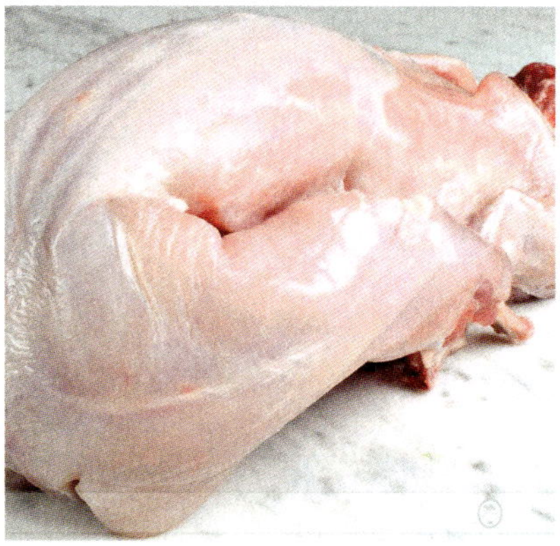

Photo Courtesy of D'Artagnan, Photography by Doug Adams Studio

Figure 4-89. Uncooked rabbit is similar in appearance and texture to uncooked chicken.

Procedure for Fabricating a Rabbit

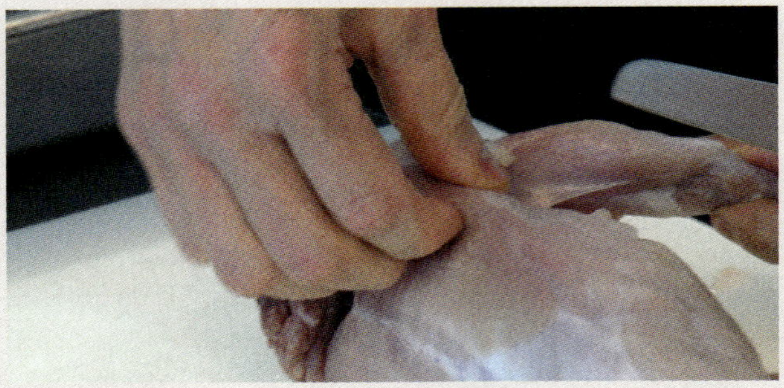

1. Place the rabbit on its side with the forelegs facing the cutting edge of the table. Slice under the joint where one of the forelegs connects to the body. Continue to cut through the joint to remove the foreleg.
2. Repeat the process to remove the other foreleg.
3. Place the rabbit on its back and open the chest cavity.

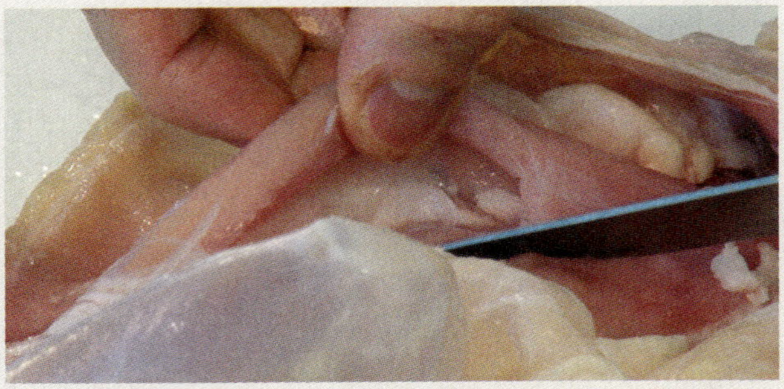

4. To remove the two tenderloins, insert the tip of the knife under one of the tenderloins and make small cuts in order to remove the meat from the rabbit.
5. Repeat the process to remove the second tenderloin.

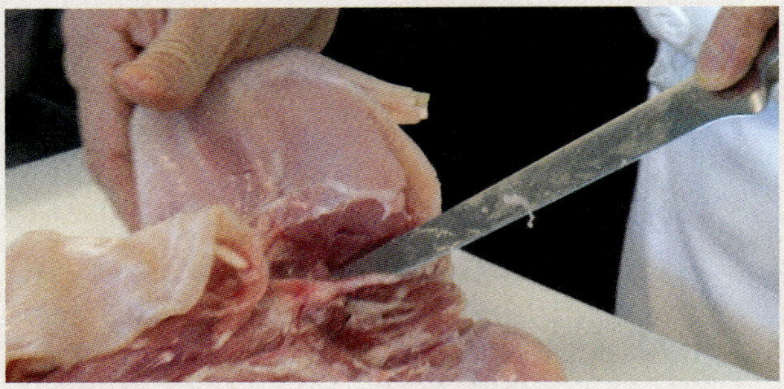

6. To remove the hind legs, firmly hold the back leg and cut at the joint where the leg meets the loin.
7. Peel the meat back to expose the joint and cut through to the backbone. Pull the leg back toward the spine to crack open the joint and cut through the separation.
8. Repeat the process to remove the remaining back leg.

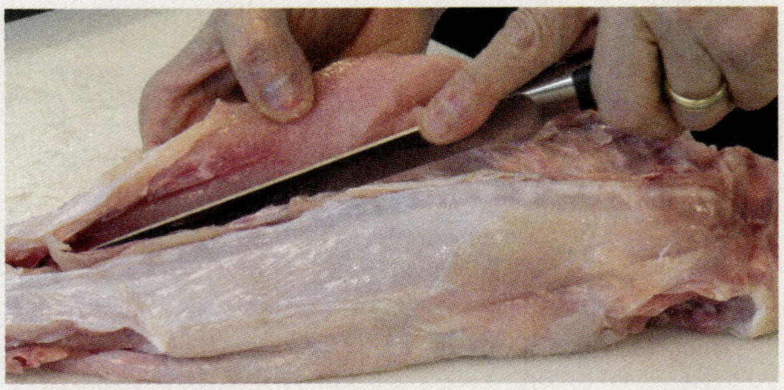

9. To remove the two loins, position the rabbit with the back facing upward. Cut along either side of the spine by making a small incision with the point of the knife and then running the tip along the side of the spine. Repeat this shallow cut, going a little deeper each time.
10. Follow the bone structure of the ribs and continue to cut away the loin muscle from the spine.

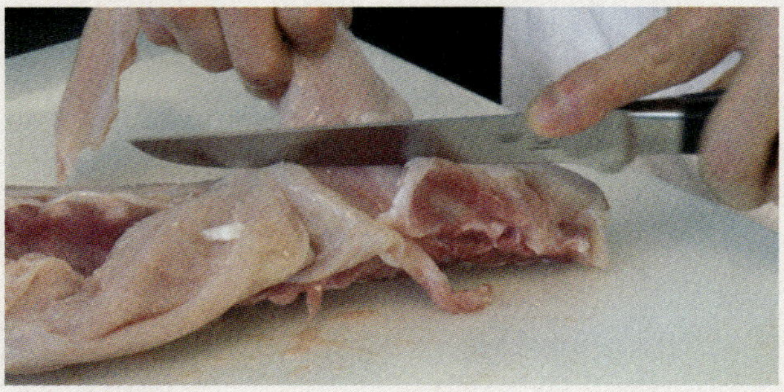

11. To finish separating the loin, cut straight down at the shoulder joint.
12. Repeat the process to remove the loin on the opposite side of the spine.

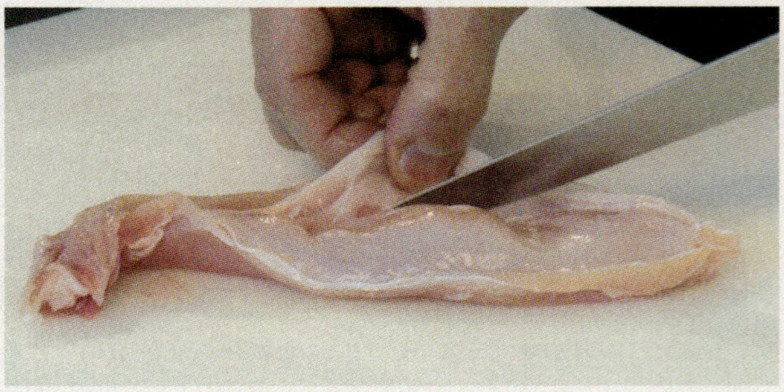

13. Place one loin silverskin-side down. Make a small cut into the flesh at the thicker end of the loin in order to grasp the silverskin. Holding the silverskin, slice along the silverskin to separate it from the loin.
14. Repeat the process to remove the silverskin from the second loin.

SKETCHING EXERCISE

25. Sketch a beef carcass and identify the locations of all of the primal cuts.

Broiler/Char-Grill Station

Introduction

At the broiler/char-grill station, you will be responsible for the final preparation of foods using broiling, grilling, and roasting cooking techniques. Finished plates will be garnished and presented appropriately for service.

As an apprentice, you will also learn to correctly identify the cuts of poultry, seafood, and meats that are most appropriate for various cooking methods. Some of the large pieces of equipment associated with this station are the broiler, grill, and convection oven. You will learn to prepare marinades, wet and dry rubs, and other flavoring agents used when preparing proteins. You will prepare hot hors d'oeuvres such as brochettes, stuffed mushroom caps, oysters Rockefeller, and chicken satay. Preparing hot sandwiches such as Reubens and paninis will provide lessons in the proper use of equipment in the broiler/char-grill station.

Sustainability will also be an important lesson at this station. Descriptive terms such as "free range" will become part of your language. You will learn the pros and cons of buying food locally. Learning to purchase sustainable seafood using the red, yellow, and green seafood watch lists will be an important step toward sustainable practices.

Vulcan-Hart, a division of the ITW Food Equipment Group LLC

The broiler/char-grill station is a focal point of many successful foodservice operations. Proteins are the primary items cooked at this station and account for much of the food cost and the revenue of a foodservice operation. Understanding the various market forms of proteins, how to use flavor enhancers, and how to broil, grill, and roast these items is a requirement for mastery of this station. Many hors d'oeuvres and appetizers are also finished at this station. The proper use of broiling, grilling, and roasting equipment yields flavorful foods for guests to enjoy time after time.

STATION PROTEINS

Poultry, seafood, and meats are prepared at the broiler/char-grill station. Each market form of protein has unique characteristics that can impact the way in which it is prepared and cooked. It is important to know the classifications used for whole and fabricated cuts of poultry, fish, shellfish, beef, veal, pork, and lamb.

Market Forms of Poultry

The six different kinds of poultry recognized by the USDA include chickens, turkeys, ducks, geese, guinea fowls, and pigeons. Each kind of poultry is also subdivided into classes based on age and/or gender. Poultry is available as whole birds as well as fabricated cuts. Market forms of poultry prepared at the broiler/char-grill station include whole poultry, ground poultry, breasts, and legs. **See Figure 5-1.**

Whole Poultry. Purchasing whole poultry allows a chef to be more creative with the menu. It is also less expensive to purchase whole poultry and debone it in-house. When purchasing whole poultry, it is common to find a small bag of giblets (i.e., neck, heart, gizzard, and liver) inside the cavity of the bird. When purchasing poultry without giblets, the acronym WOG is used by suppliers.

Fabricated Poultry. Common fabricated cuts of poultry include breasts, tenders, tenderloins, wings, legs, leg quarters, breast quarters, halves, and ground poultry. Each wing can be cut into wing tips, wing paddles, and drummettes. Each leg can be divided into drumsticks and thighs. Cutlets and sausages are also available.

Poultry and Game Birds

Poultry Cuts Prepared at the Broiler/Char-Grill Station

National Turkey Federation

Figure 5-1. Market forms of poultry prepared at the broiler/char-grill station include whole poultry, ground poultry, breasts, and legs.

Market Forms of Fish

Fish are classified as roundfish, flatfish, or cartilaginous fish, based on their shape and structure. Fish contain protein, carbohydrates, fat, water, and trace amounts of vitamins and minerals. Fish are considered lean or fatty based on the amount of fat they contain in their raw form. The amount of cholesterol in a fish is not related to the amount of fat that a fish contains. **See Figure 5-2.** The fat content of a fish can affect both flavor and the cooking technique required. Lean fish, such as striped bass, contain very little fat and are often prepared using moist-heat cooking techniques such as poaching. Fatty fish contain more fat and are rich in omega-3 fatty acids, vitamin A, and vitamin D. Fatty fish, such as salmon, are best prepared using dry-heat cooking techniques such as grilling and broiling. The fat content of a particular type of fish can vary slightly by season.

Fish are available fresh, frozen, or processed. The more that is done to a fish prior to delivery, the higher the cost per pound. A whole fish is the market form of a fish that is taken from the water and sold as is. Whole fish have the shortest shelf life of any market form because all the internal organs (viscera) are still present. Whole fish cost less per pound, require more preparation, and yield more waste than any other market form. Frozen fish cannot be purchased whole. Market forms of fish prepared at the broiler/char-grill station include drawn fish, pan-dressed fish, fish steaks, and fish fillets. **See Figure 5-3.**

Drawn and Dressed Fish. A drawn fish is a fish that has had only the viscera removed. Drawn fish can be prepared whole. A dressed fish is a fish that has been scaled and has had the viscera, gills, and fins removed. A pan-dressed fish, also known as a headed and gutted fish, is a dressed fish that has had its head removed.

Fish Steaks. A fish steak is a cross section of a dressed fish. Generally, the only bone present in a steak is a small section of the backbone. Steaks from very large fish, such as swordfish, are boneless. Tuna, salmon, swordfish, and sharks are commonly sold as steaks. Fish may also be purchased as loins and wheels. A fish loin is cut lengthwise from either side of the backbone of large roundfish, such as tuna. A fish wheel is a crosswise section cut from behind the head and just short of the tail of large fish. Loins and wheels are commonly broken down into fish steaks.

Approximate Fat and Cholesterol Content of Common Fish

Lean Fish	*Fat Content**	*Cholesterol**	*Fatty Fish*	*Fat Content**	*Cholesterol**
Cod	1 g	31 mg	Farmed catfish	6 g	40 mg
Flounder	1 g	41 mg	Rainbow trout	4 g	50 mg
Halibut	2 g	60 mg	Coho salmon	6 g	43 mg
Striped bass	2 g	68 mg	Shad	10 g	64 mg

* *per 3 oz serving*

Figure 5-2. The amount of cholesterol in a fish is not related to the amount of fat that a fish contains.

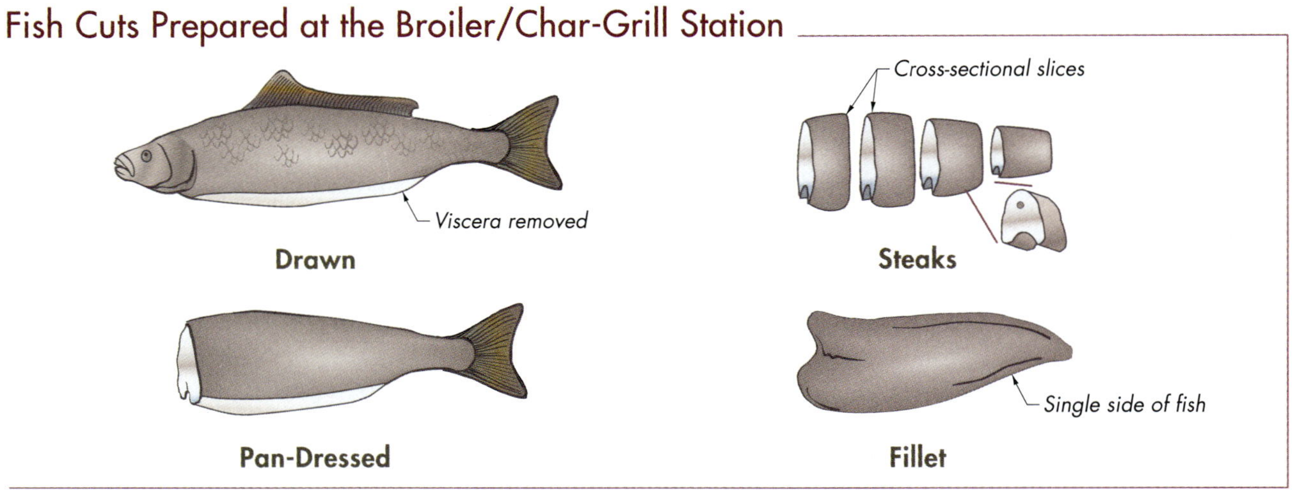

Figure 5-3. Market forms of fish prepared at the broiler/char-grill station include drawn fish, pan-dressed fish, fish steaks, and fish fillets.

Fish Fillets. A fish fillet is the lengthwise piece of flesh cut away from the backbone. Roundfish have two fillets, one on each side. Flatfish have four fillets, two on each side. Fillets can be purchased with or without bones and with or without skin. A butterflied fillet is two single fillets from a dressed fish that are held together by the uncut back or belly of the fish. A fish portion is cut from fillets into smaller pieces.

Market Forms of Shellfish

Shellfish is the classification of aquatic invertebrates that may or may not have a hard, external shell. An external shell functions as a skeleton and is called an exoskeleton. Shellfish are classified as crustaceans or mollusks.

- A crustacean is a shellfish that has a hard, segmented shell that protects soft flesh and does not have an internal bone structure. Crustaceans live in both freshwater and saltwater. Unlike fish, crustaceans can live out of water for a few days if they are kept moist. Crustaceans include shrimp, prawns, lobsters, crayfish, and crabs.
- A mollusk is a shellfish with a soft, nonsegmented body. Some mollusks, such as clams, have a hard, external shell. Other types of mollusks, such as squid, do not have an external shell. Like crustaceans, mollusks do not have an internal skeleton. Mollusks should always be alive prior to cooking, unless they are frozen or canned. If a mollusk dies prior to cooking, it should be discarded. The three classifications of mollusks are univalves, bivalves, and cephalopods. **See Figure 5-4.**

Mollusks

Classification	Description	Examples
Univalves	Single solid shell	• Abalone • Conch • Welk
Bivalves	Two shells that are hinged together	• Clams • Mussels • Oysters • Scallops
Cephalopods	No external shell; some have an internal cuttlebone	• Cuttlefish • Octopus • Squid

Figure 5-4. The three classifications of mollusks are univalves, bivalves, and cephalopods.

Shellfish can be purchased live, shucked, or frozen. Each market form of shellfish has advantages and disadvantages in terms of cost, convenience, and labor. Common forms of shellfish include shrimp, scallops, and squid. **See Figure 5-5.**

Shellfish

Florida Department of Agriculture and Consumer Services

Shrimp

Fortune Fish Company

Scallops

Fortune Fish Company

Squid

Figure 5-5. Common forms of shellfish include shrimp, scallops, and squid.

Sustainable Seafood

Live Shellfish. Lobsters are usually purchased live and come in various sizes. On the coast both hard-shell and soft-shell crabs are sold live. Away from the coast only soft-shell crabs are available live. Live clams, oysters, and scallops can be purchased by the gallon or the pound. They are usually packed by count. A higher count indicates a smaller size. If a container of oysters is marked 350 per gallon and another container is marked 225 per gallon, the 350 count container holds more oysters that are smaller. Live oysters and clams have tightly closed shells. If a shell is open or does not close fully when lightly tapped, the animal is dead and must be discarded. Oyster and clam shells that are unusually heavy probably contain mud and should also be discarded.

Shucked Shellfish. Clams, oysters, and scallops are often sold with the shell removed and are available both fresh and frozen. Shrimp may also be purchased cooked and canned. Cooked shrimp may be purchased by the pound either in the shell or peeled. Crabmeat can be cooked (fresh or frozen) and canned. Crabmeat also may be purchased fresh or frozen in the shell in a variety of cooked forms. Cooked crabmeat should be kept packed in ice and refrigerated until it is used.

Frozen Shellfish. Prepared lobster tails may be purchased frozen. The head and thorax (body) of shrimp and prawns are typically removed, and the tails are frozen while still at sea to ensure maximum freshness. **See Figure 5-6.** Frozen shellfish is commonly packaged by count. For example, shrimp packed and labeled 21/25 count indicates that there is an average of 21–25 shrimp per 1 lb package. Frozen shrimp can be purchased green (uncooked), peeled or unpeeled, cooked and peeled, or peeled, deveined, and breaded. Frozen shrimp are usually sold in 5 lb blocks.

Shrimp Tails

L. Isaacson and Stein Fish Company

Figure 5-6. The head and thorax (body) of shrimp and prawns are typically removed, and the tails are frozen to ensure maximum freshness.

Production Tip

Due to the frozen water in a 5 lb block of shrimp, the total weight of the block is actually 6–7 lb.

Market Forms of Beef and Veal

Beef and veal are available as partial (beef) and whole (veal) carcasses, primal cuts, and fabricated cuts. A single beef carcass has two of each primal cut (a left side and a right side). The eight primal cuts of beef are the chuck, rib, short loin, sirloin, round, flank, short plate, and brisket and shank. Fabricated cuts of beef from each primal cut are prepared at the broiler/char-grill station. **See Figure 5-7.**

The left and right primal cuts of veal remain joined together and are sold as a single cut. For example, the leg primal cut has two joined legs. The primal cuts of veal include the shoulder, rack, loin, leg, and foreshank and breast. Fabricated cuts of veal from each primal cut are prepared at the broiler/char-grill station. **See Figure 5-8.**

Market Forms of Beef

Production Tip

Fabricated cuts are a convenient way of providing uniform portions while reducing labor costs. However, the price per pound for fabricated cuts is higher than the price per pound for primal cuts.

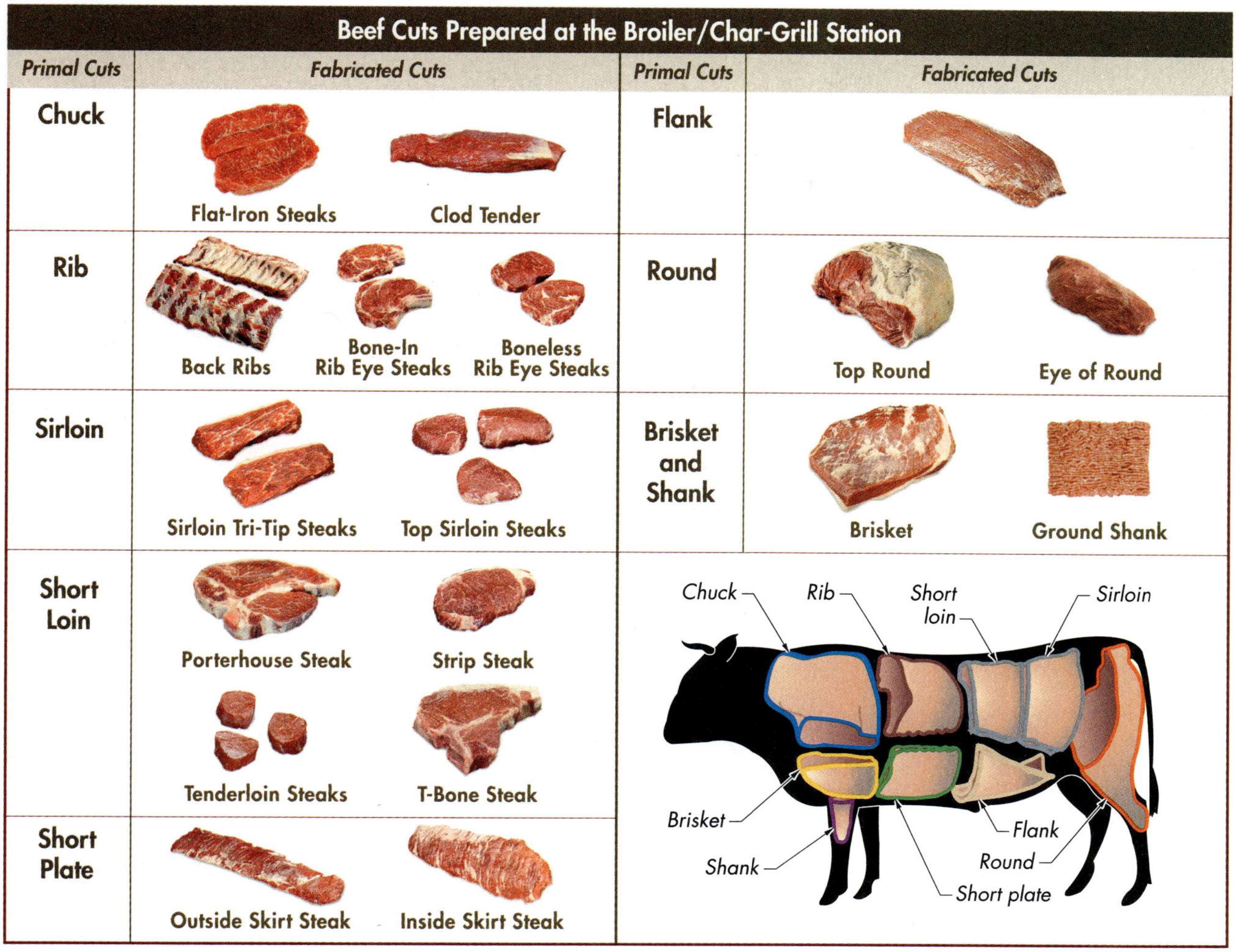

Canada Beef Inc.

Figure 5-7. Fabricated cuts of beef from the chuck, rib, sirloin, short loin, short plate, flank, round, and brisket and shank are prepared at the broiler/char-grill station.

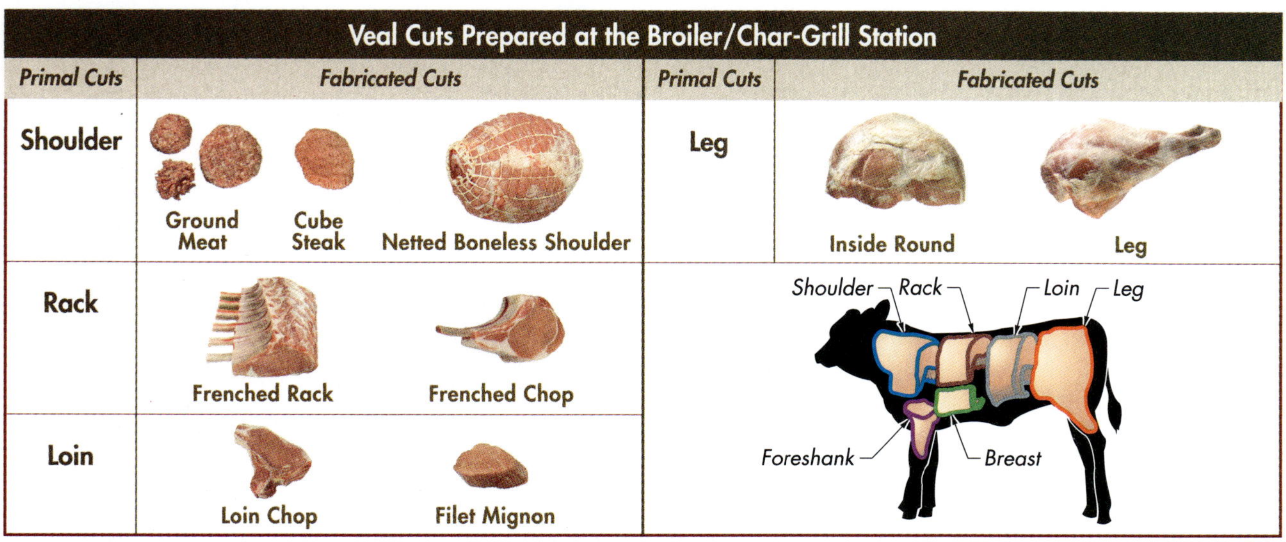

Strauss Free Raised

Figure 5-8. Fabricated cuts of veal from the shoulder, rack, loin, leg, and foreshank and breast are prepared at the broiler/char-grill station.

Market Forms of Pork

Pork is available as whole carcasses, primal cuts, and fabricated cuts. A hog carcass has two sets of primal cuts. One set is on the left side and one set is on the right side. Each primal cut is divided into fabricated cuts. The five primal cuts of pork are the picnic shoulder, shoulder butt, loin, leg, and belly. Fabricated cuts of pork from each primal cut are prepared at the broiler/char-grill station. **See Figure 5-9.**

Market Forms of Lamb

Lamb is available as whole and partial carcasses, primal cuts, and fabricated cuts. The left and right primal cuts remain joined together and are purchased as a single cut. The five primal cuts of lamb are the shoulder, rack, loin, leg, and breast/shank. Fabricated cuts of lamb from each primal cut are prepared at the broiler/char-grill station. **See Figure 5-10.**

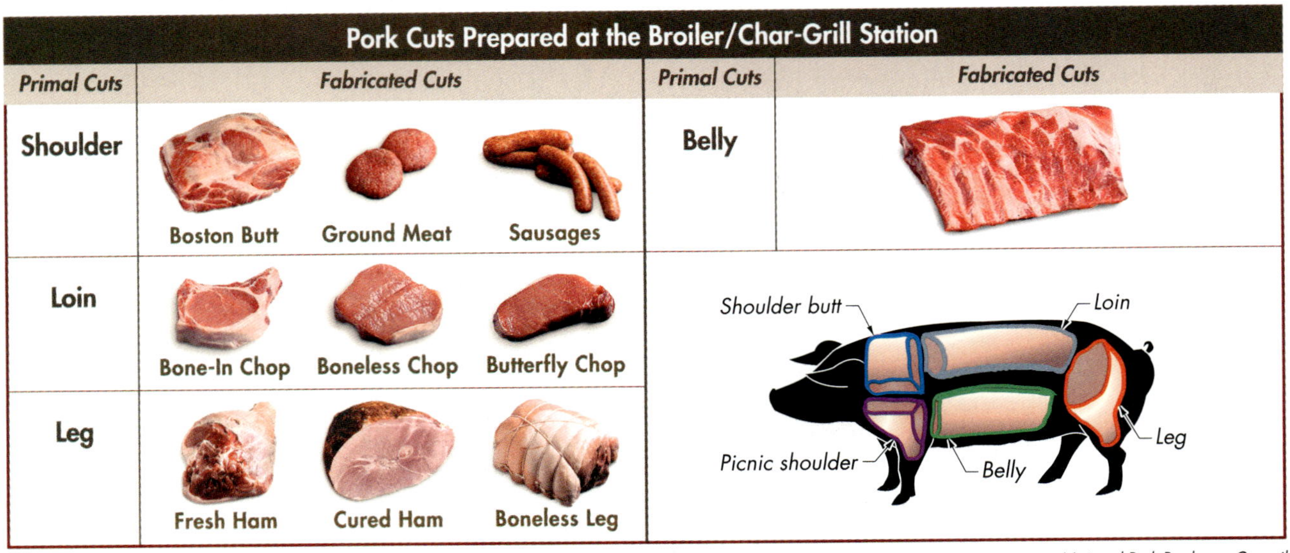

National Pork Producers Council

Figure 5-9. Fabricated cuts of pork from the picnic shoulder, shoulder butt, loin, leg, and belly are prepared at the broiler/char-grill station.

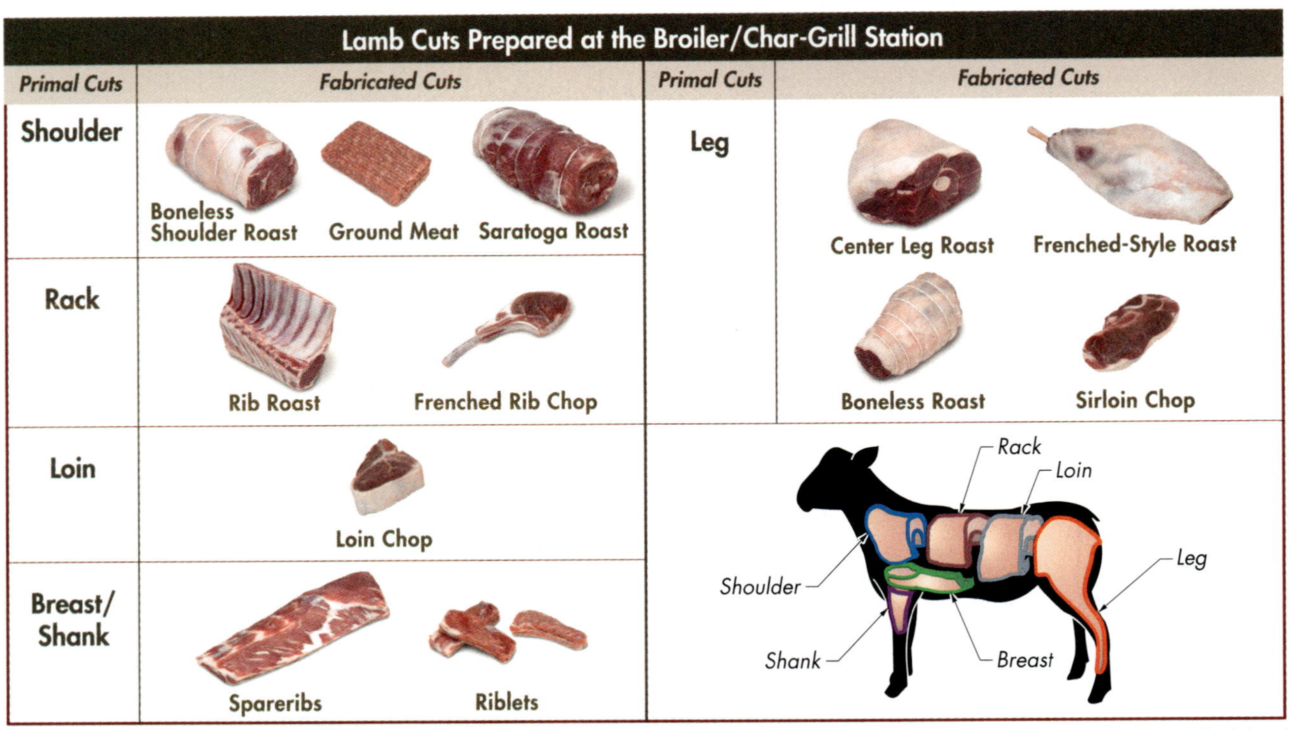

American Lamb Board

Figure 5-10. Fabricated cuts of lamb from the shoulder, rack, loin, breast/shank, and leg are prepared at the broiler/char-grill station.

Production Tip

The cuts of beef, veal, pork, and lamb are listed by category in the USDA Institutional Meat Purchase Specifications.

CHECKPOINT 5-1

1. Explain the advantage of purchasing whole poultry.
2. Identify the cuts of poultry prepared at the broiler/char-grill station.
3. Differentiate between lean and fatty fish.
4. Identify three ways fish is sold.
5. Identify the type of fish that yields the most waste.
6. Describe four market forms of fish prepared at the broiler/char-grill station.
7. Differentiate between crustaceans and mollusks.
8. Describe three classifications of mollusks.
9. Describe the three market forms of shellfish.
10. Identify eight primal cuts and the related fabricated cuts of beef prepared at this station.
11. Identify five primal cuts and the related fabricated cuts of veal prepared at this station.
12. Identify five primal cuts and the related fabricated cuts of pork prepared at this station.
13. Identify five primal cuts and the related fabricated cuts of lamb prepared at this station.

FLAVOR ENHANCERS

The flavors of poultry, seafood, and meats are often enhanced by a variety of methods before or during the cooking process. Common flavor enhancers for poultry, seafood, and meats include aging, larding, barding, curing, brining, and the use of marinades, rubs, stuffings, basting liquids, and herbs and spices.

Aging

Aging is the period of rest that occurs for a length of time after an animal has been slaughtered. Immediately after slaughter, the carcass is very limp. However, within 5–24 hours the carcass goes through a period known as rigor mortis where natural enzymes in the meat cause the tissues to seize and become stiff. This stiffening stage lasts for two to three days under normal refrigeration. If meat is frozen immediately after slaughter, the process is paused and rigor mortis sets in once the meat is thawed. Natural enzymes in the tissue of the meat begin to relax and tenderize the muscles two to three days after slaughter. The meat is then ready to be fabricated, frozen, or cooked.

Veal is typically not aged any longer than the initial two to three days for rigor mortis to leave the meat. Beef, however, benefits from longer aging periods as the enzymes in the meat continue to break down and tenderize the tissues, making it more flavorful. Beef can be wet aged or dry aged for an extended period.

Wet Aging. *Wet aging* is the process of aging meat in vacuum-sealed plastic. Beef carcasses are typically broken down into manageable cuts and placed in vacuum-sealed plastic packaging. Beef is allowed to wet age from one to six weeks in vacuum packaging under normal refrigeration. **See Figure 5-11.** During this period, natural enzymes further break down and tenderize the meat, intensifying its natural flavor.

Wet Aging

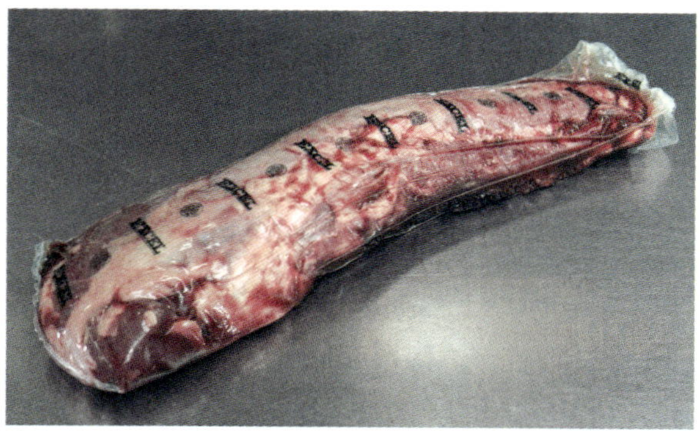

Figure 5-11. Beef is allowed to wet age from one to six weeks in vacuum packaging under normal refrigeration.

Dry Aging. *Dry aging* is the process of aging larger cuts of meat that are hung in a well-controlled environment. During dry aging temperature, humidity, and airflow are monitored around the clock for up to six weeks. Over this time, natural enzymes break down the tissue of the meat, making it more flavorful and tender. Dry aging commonly results in 5–20% weight loss through the evaporation of moisture and the eventual trimming of inedible surface material. Dry-aged meats are typically found only at elite steak houses.

Larding and Barding

Both larding and barding involve the use of natural fats. *Larding* is the process of inserting thin strips of fatback into lean meat with a larding needle. **See Figure 5-12.** The fatback is cut into thin strips and drawn through the beef to increase the juiciness when the meat is roasted. Larded meats are commonly braised to allow ample time for the fat to be rendered and absorbed by the meat.

Larding Needles

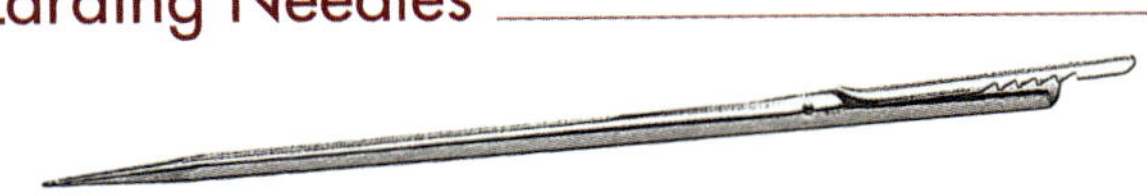

Paderno World Cuisine

Figure 5-12. Larding is the process of inserting thin strips of fatback into lean meat with a larding needle.

Barding is the process of laying a piece of fatback across the surface of a lean cut of meat to add moisture and flavor. **See Figure 5-13.** The fatback is tied to the meat to prevent it from falling off during the roasting process. As the meat cooks, the fat is rendered from the fatback and absorbed into the meat. While the fatback prevents the meat from drying out, it also prevents the meat from browning. Occasionally, the fatback is removed for the last 10 minutes of roasting to allow the item to gain color.

Barding

Figure 5-13. Barding is the process of laying a piece of fatback across the surface of a lean cut of meat to add moisture and flavor.

Curing

Curing is the process of using salt and sodium nitrite alone or with flavorings or sugar to preserve a food item. Summer sausage and beef jerky are common cured-beef products. More than two-thirds of all pork is cured. For example, ham and bacon are cured-pork products. Pork can be cured using a dry, wet, or combination curing method. **See Figure 5-14.** Veal and lamb are not commonly cured.

Dry Curing. *Dry curing* is a curing method that involves the use of salt to dehydrate the protein in food. Meats are dry cured by completely covering the meat with salt or a salt-cure mixture, which includes salt, sugar, seasonings, and pink curing salt. Dry curing is done in stages to maximize the absorption of the curing ingredients. Ham, bacon, and sausages are often dry cured. The length of curing time depends on the size of the meat and its composition. Fat and skin create a barrier for the cure and slow down the process. The curing time takes approximately three days per pound for large cuts and two days per pound for small cuts.

Carlisle FoodService Products

Wet Curing. *Wet curing*, also known as immersion curing, is a curing method in which foods are submerged in a brine. Wet curing is sometimes referred to as brining or sweet pickling. A *brine* is a salt solution that usually consists of 1 cup of salt per 1 gal. of water. A sweet pickle brine is composed of salt, water, and sugar. Sugar is only added when curing is done at refrigerator temperatures because fermentation will start in warmer temperature and spoil the meat. Smaller cuts of pork can be wet-cured in a shorter amount of time than larger cuts.

Brines may also contain other flavorful ingredients such as sugar, molasses, fruits, juices, herbs, and spices. Immersing a raw protein, such as poultry or pork, into a brine solution and refrigerating it for several hours or overnight allows the brine to be absorbed. When the brined protein is cooked, it is especially tender because the brine aids in moisture retention. Brines are also used to preserve foods such as herring and pickles.

Brining

Methods of Curing

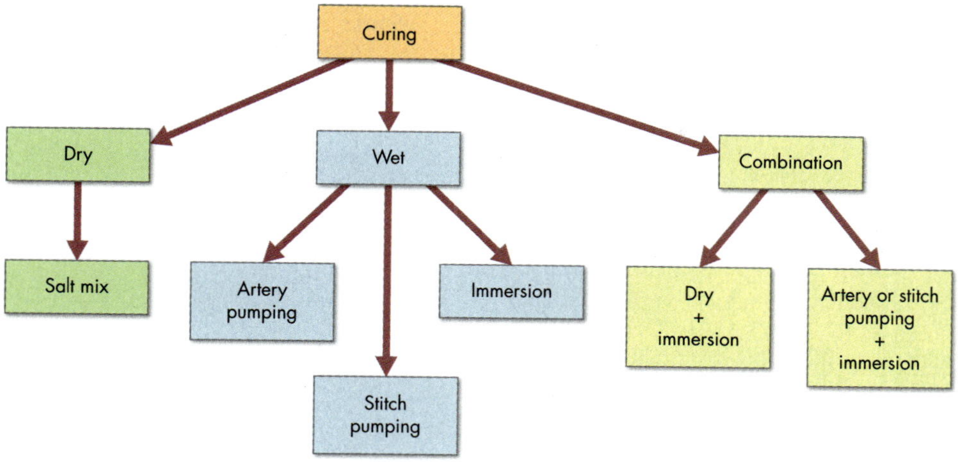

Figure 5-14. Pork can be cured using a dry, wet, or combination method.

Wet curing yields a less salty product with a shorter shelf life than dry curing. Protein absorbs different amounts of sodium in relation to curing time and the amount of surface fat. Fast curing takes 7 to 14 days and requires around a 10% brine solution in relation to the original weight of the meat. Corned beef is fast cured. **See Figure 5-15.** Slow curing takes 30 to 50 days and requires approximately a 5% brine solution in relation to the original weight of the meat.

Combination Curing. *Combination curing* is the process of combining either dry curing or wet curing with artery or stitch pumping to reduce processing time. One option is to rub a dry cure mix into the meat and then use an artery or stitch pump to inject a brine solution. The second option is to use an artery or stitch pump to inject a brine solution into large pieces of meat and then stack them on top of each other inside a container filled with brine. The meat must be weighed down to keep it from surfacing and the pieces need to be turned over each day for the duration of the curing process. The first option is the most efficient.

Corned Beef

Figure 5-15. Corned beef is fast cured.

Marinades

A *marinade* is a flavorful liquid used to soak uncooked foods such as poultry, fish, and meats to impart flavor and sometimes to tenderize. Most marinades contain an acidic liquid base, such as lemon juice, wine, or vinegar. Marinades also contain herbs and spices. The acidic base tenderizes the meat by breaking down the protein structure of the meat. Any herb, spice, or condiment can be added to the base to create the desired flavor. If wine is used, a red wine should be used for beef and a white wine for veal.

Food items are completely submerged in the marinade to provide a consistent flavor. The length of time that meat should be left in the marinade is determined by the size of the piece of meat. **See Figure 5-16.** The intensity of the flavor increases the longer an item sits in a marinade. Leftover marinade is contaminated and must be discarded to prevent cross-contamination. Once a meat has been placed in a marinade, the marinade cannot be used to create a sauce.

Marinades and Rubs

Marinades

Figure 5-16. Unlike rubs, marinades contain a high percentage of liquid ingredients such as vinegars, oils, or fruit juices.

Rubs

A *rub* is a blend of ingredients that is pressed onto the surface of uncooked foods such as poultry, fish, and meat to impart flavor and sometimes to tenderize. An endless variety of ingredients can be combined to create rubs that add flavor without being overbearing. After a rub has been applied for the appropriate amount of time, the food is often broiled, grilled, roasted, or baked. Rubs are classified as dry rubs or wet rubs.

- A *dry rub* is a rub made by grinding herbs and spices into a fine powder and then rubbing the mixture into a protein prior to cooking. **See Figure 5-17.**
- A *wet rub* is a rub made by incorporating wet ingredients, such as Dijon mustard, flavored oils, puréed garlic, or honey, into a dry rub mixture and then rubbing it into a protein prior to cooking.

Dry Rubs

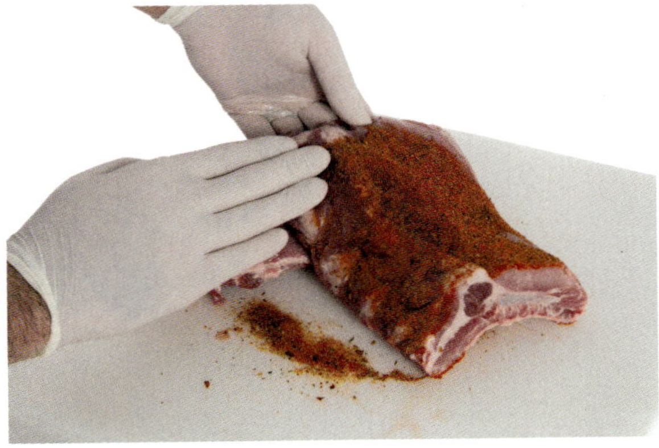

Figure 5-17. A dry rub can be used to add intense flavor to a cut of pork.

Stuffings

Small birds such as Cornish hens and squabs are often stuffed before they are cooked. **See Figure 5-18.** If stuffing is to be served with a large bird, such as a turkey, it is prepared and baked separately.

Stuffed Birds

National Honey Board

Figure 5-18. Small birds, such as Cornish hens and squabs, are often stuffed.

Bastings

Many proteins are basted. *Basting* is the process of using a brush or a ladle to place fat on or pour juices over an item during the cooking process to help retain moisture and enhance flavor. An au jus, a jus lié, or a barbeque sauce is commonly used to baste poultry, fish, shellfish, and meats such as beef, veal, pork, and lamb.

- An *au jus* is the natural, unthickened juices that flow from meat as it cooks. A stock, wine, or other liquid is added to the pan to dissolve the caramelized particles, also known as fond, and then used as a natural sauce for meats.
- A *jus lié,* also known as a fond lié, is a sauce that is made by thickening a brown stock either by adding a cornstarch or arrowroot slurry or simply by a slow reduction. A jus lié may also be created by thickening an au jus.
- A *barbeque sauce* is a type of sauce that is used to baste proteins and that can be served as an accompaniment to a cooked protein. **See Figure 5-19.** The base for a barbeque sauce is often made with tomatoes, onions, mustard, garlic, brown sugar, and vinegar. Regional variations of barbeque sauce may use a mayonnaise or vinegar base and range in flavor from sweet and tangy to smoky and spicy. Barbeque sauces that contain white or brown sugar, honey, or molasses can burn quickly and are diluted with vinegar or water to create a mopping sauce that is applied to the meat as it cooks in order to flavor and moisten the meat without burning it. Barbeque sauce may also be applied at the end of the cooking process.

Barbeque Sauces

Barbeque Sauce

Figure 5-19. A barbeque sauce is a type of sauce that is used to baste proteins and that can be served as an accompaniment to a cooked protein.

Herbs and Spices

Herbs and spices are used to enhance the natural flavor of food. **See Figure 5-20.** If a dish has a long cooking time, spices and resinous herbs should be added early in the process to extract the most flavor. A *resinous herb* is an herb that requires time to release its fragrant aroma. Resinous herbs include rosemary, sage, bay, lemon verbena, thyme, and marjoram. If a dish has a shorter cooking time, herbs such as parsley, chives, tarragon, and chervil are added at the beginning of the cooking process.

Flavor Combinations	
Food	**Flavoring**
Beef	Bay leaf, marjoram, nutmeg, sage, thyme
Corn	Cayenne, chervil, chives, cumin, curry powder, paprika, parsley
Fish	Curry powder, dill, dry mustard, lemon zest and juice, marjoram, paprika
Fruits	Cinnamon, ground cloves, ginger, mace, mint
Lamb	Curry powder, rosemary, mint
Pork	Sage, oregano
Potatoes	Dill, paprika, parsley, sage
Poultry	Ginger, marjoram, oregano, paprika, poultry seasoning, rosemary, sage, tarragon, thyme
Summer squash	Cloves, curry powder, marjoram, nutmeg, rosemary, sage
Tomatoes	Basil, bay leaves, dill, marjoram
Veal	Bay leaves, curry powder, ginger, marjoram, oregano
Winter squash	Cinnamon, ginger, nutmeg

Figure 5-20. Herbs and spices are used to enhance the natural flavor of food.

CHECKPOINT 5-2

1. Contrast wet aging and dry aging.
2. Contrast larding and barding.
3. Identify three commonly cured beef products.
4. Identify three curing methods used on pork cuts.
5. Describe the process of wet (immersion) curing.
6. Explain why combination curing is used on most cuts of pork.
7. Differentiate between marinades and rubs.
8. Explain the use of rubs and marinades in flavor development.
9. Contrast dry rubs and wet rubs.
10. Describe three basting sauces used to baste proteins.
11. Define resinous herbs.
12. Identify which herbs are added based on cooking time.

BROILING

Broiling is a dry-heat cooking technique in which food is cooked directly under or over a heat source. The broiling process reduces the fat content of proteins because some of the fat is rendered due to the high heat used. Poultry, seafood, meats, and some fruits and vegetables can be broiled. Other items, such as bruschetta and French onion soup, are often finished in a broiler. A *broiler* is a large piece of cooking equipment in which the heat source is located above or behind the food instead of below it. The gas, electric, or ceramic heat source cooks the food from above. The hot grates on which the food is placed can be raised or lowered to adjust the intensity of the heat. Broilers can be stand-alone units or combined with an oven. The three basic types of broilers include standard broilers, salamanders, and rotisseries. **See Figure 5-21.**

A *salamander* is a small overhead broiler that is usually attached to an open burner range. The heat source is not nearly as intense as a standard broiler. A salamander is primarily used to brown, glaze, melt, or finish cooking foods. A *rotisserie* is a sideways broiler in which foods are placed on a steel rod or spit that revolves past the heat source to ensure even heating. Instead of the heat being above the food, it is behind it. Rotisseries are most often used to cook poultry and meats.

Broiling Procedures

A broiler station requires easy access to all of the ingredients, tools, and equipment required to prepare menu items for service. Before use, a broiler is scraped clean and then lightly oiled to prevent food from sticking.

Procedure for Preparing a Broiler

1. Preheat the broiler. Use a wire brush to scrape clean the grates of the broiler.

2. Wipe the metal grates with a towel lightly coated in vegetable oil.

Broilers

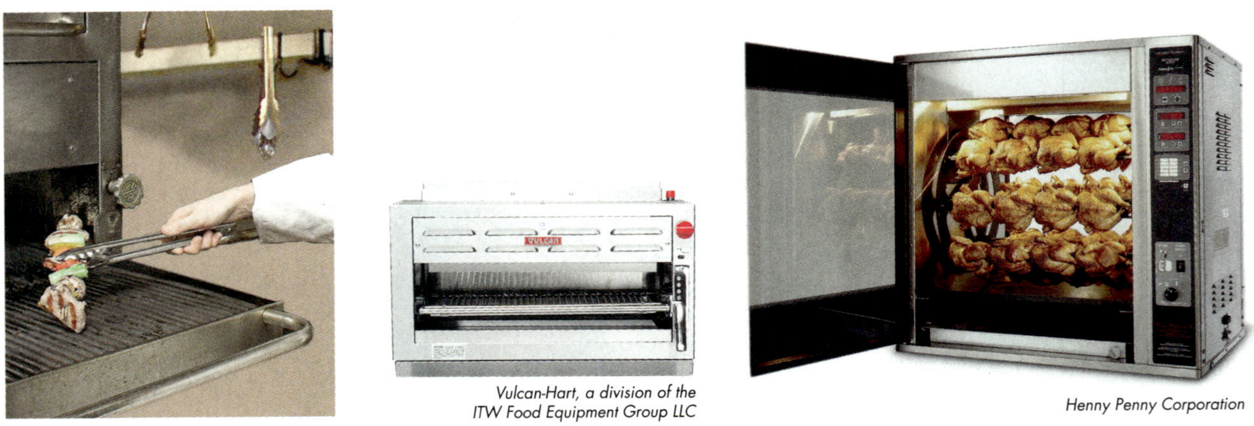

Vulcan-Hart, a division of the ITW Food Equipment Group LLC

Henny Penny Corporation

Broiler **Salamander** **Rotisserie**

Figure 5-21. Three basic types of broilers include the standard broiler, the salamander, and the rotisserie.

Food is placed on the char-broiler at a 45° angle and allowed to cook for 3–4 minutes before being turned 90° to achieve proper grill marks. The item is then turned over after caramelization is achieved. The length of cooking time is based on the degree of doneness required. Sometimes foods are partially cooked in a broiler and then finished in an oven. Finishing foods in the oven provides a less charred surface.

Production Tip

Broiling is a quick cooking technique that works best when the protein items are removed from the refrigerator for a few minutes before cooking.

Broiling Poultry

The heat from a broiler is intense, so care must be taken to not overcook or dry out poultry. Small birds, such as quails, are typically skewered prior to being cooked in a broiler so that they will lie flat on the grates. Cornish hens should be split in half or butterflied before being broiled. Birds larger than Cornish hens should be cut into quarters or eighths before broiling. A whole turkey is too thick to be broiled because the exterior will dry out and burn before the interior can be fully cooked.

Photo Courtesy of Perdue Foodservice, Perdue Farms Incorporated

Procedure for Broiling

1. Brush a small amount of oil on the items to be broiled and season as desired.
2. Using tongs, place the items in the broiler with the presentation side facing down.
3. Allow the items to cook until char lines have developed on the presentation side.
4. If desired, use the tongs to rotate the items 90° to create crosshatch markings on the presentation side.
5. Use tongs to turn the items and cook the other side to the desired degree of doneness.
6. Fish should flake and a thermometer should be used to check the temperature of thicker meats.

Poultry cuts are typically seasoned prior to broiling so that the flavors of the seasoning can penetrate the flesh during the cooking process. It is important to remember that most poultry cuts contain very little fat and are prone to drying out. Broiled poultry should be moist and cooked throughout. **See Figure 5-22.** Duck and squab breasts are an exception. They are cooked medium-rare or medium.

Broiled Poultry

Chicken Breasts

Chicken Fajitas

Photo Courtesy of Perdue Foodservice, Perdue Farms Incorporated

Figure 5-22. Broiled poultry should be moist and cooked throughout.

Nutrition Note

A 4 oz. serving of grilled chicken breast contains 25 g of protein and 4 g of fat. It also provides 7% of the daily-recommended intake of iron and 2% of the daily-recommended intake of calcium.

Broiling Fish

Unlike land animals, fish have delicate connective tissue that is very thin. Fish flesh naturally flakes and comes apart easily. Therefore, thicker fish are broiled. For example, salmon is perfect for broiling because the thick flesh can withstand the intense heat of open flames. **See Figure 5-23.** When cooking fish fillets with the skin left on, the skin should be scored with a sharp knife first so that the fillet does not curl during cooking. Oil should be applied to fish prior to placing it on the broiler to prevent the fish from sticking.

Broiled Fish

Figure 5-23. Salmon is perfect for broiling because the thick flesh can withstand the intense heat of open flames.

Nutrition Note

Broiled fish is an excellent source of protein and omega-3 fatty acids. When oils, sauces, or other garnishes are added to broiled fish, the caloric content increases significantly. For example, a tablespoon of olive oil adds an extra 119 calories to broiled fish.

Broiling Meats

Only tender cuts of meat should be broiled. Well-done beef cooked on a broiler will be somewhat dry, because most of the moisture cooks out as the meat cooks. Removing the fat after cooking can help the meat retain flavor and moisture. Cuts that are broiled are commonly taken from the rib and the loin. **See Figure 5-24.**

Broiling Hors d'Oeuvres and Appetizers

Many savory hors d'oeuvres and appetizers, commonly known as starters, are finished at the broiler/char-grill station. Commonly, starters are assembled, cooked or parcooked, cooled, and then stored as individual portions. When an order is received, the starter is finished in a salamander or an oven. **See Figure 5-25.** In foodservice operations where a large number of servings are needed, starters are prepared ahead of time, placed in hotel pans, and then held in warming cabinets until service.

Broiled Meats

Photo Courtesy of the Beef Checkoff

Beef

National Pork Producers Council

Pork

Lamb

Figure 5-24. Cuts that are broiled are commonly taken from the rib and the loin.

Broiling Pork

Broiled Hors d'Oeuvres and Appetizers

Courtesy of The National Pork Board

Tartlets

Courtesy of Chef Daniel Pliska

Bouchées

National Turkey Federation

Roulades

Figure 5-25. When an order is received, the filled starter will be finished in a salamander or a hot oven.

CHECKPOINT 5-3

1. Differentiate among the three types of broilers.
2. Explain how to prepare a broiler.
3. Describe the broiling procedure.
4. Explain how the size of a protein cut affects the positioning of the food in the broiler.
5. List considerations specific to broiling fish.
6. Broil a fish fillet and evaluate the results.
7. Identify types of meat that are typically broiled.
8. Broil two different sizes and types of meat and evaluate the final products for preparation, color, texture, and flavor.
9. Describe how hors d'oeuvres and appetizers are finished at the broiler/char-grill station.
10. Finish hors d'oeuvres or appetizers in the broiler. Evaluate the results.

GRILLING

Grilling is a dry-heat cooking technique in which food is cooked on open grates above a direct heat source. A *grill* is a cooking unit consisting of a large metal grate placed over a heat source. **See Figure 5-26.** The heat source may be gas or another burning fuel such as charcoal or wood. The heat from a grill is intense, so care must be taken to not overcook or dry out foods. Food is placed on the preheated grate and is turned over halfway through the grilling process to cook the other side. Because of the open cooking surface, foods that are grilled have a smoky flavor, especially when cooked over charcoal. Grilled foods also have char lines where the food came in contact with the hot grill.

Grilling

Vulcan-Hart, a division of the ITW Food Equipment Group LLC

Grill

Photo Courtesy of D'Artagnan, Photography by Doug Adams Studios

Grilling

Figure 5-26. Grilling is a dry-heat cooking technique in which food is cooked on open grates above a direct heat source.

Grilling

Grilling Procedures

Proteins should be removed from the refrigerator a few minutes prior to grilling. If there is any surface moisture on the product, it should be lightly blotted with a clean towel. Moisture retards browning and in some cases can toughen the product by steaming it.

Before use, a grill is scraped clean and then lightly oiled to prevent food from sticking. Crosshatching the presentation side of grilled items enhance the presentation of the dish. To create crosshatch markings when grilling, the food is rotated 90° about halfway through the cooking process.

Procedure for Grilling

1. Brush a small amount of oil on the items to be grilled and season as desired.
2. Using tongs, place the items on the hot grill with the presentation side facing down.

3. Allow the items to cook until char lines have developed on the presentation side.
4. Use the tongs to rotate the items 90° to create crosshatch markings on the presentation side.

5. Turn the items and cook the other side to the desired degree of doneness. *Note:* Crosshatch markings do not need to be added to this side since it faces the plate.

6. With a gloved finger, gently press the item to determine doneness. *Note:* Fish should flake and a thermometer should be used to check the temperature of thicker meats.

Foods may also be grilled on a flattop grill or a Panini grill. A flattop grill resembles a griddle but cooks differently because the heating element is circular. The circular heating element creates an extremely hot cooking surface that spreads heat in a radial fashion across the surface. Flattop-grilling is best for oily foods such as hamburgers and sausages but also works well for steaks, chops, and fish.

A *Panini grill* is an Italian clamshell-style grill made specifically to cook grilled sandwiches. **See Figure 5-27.** This hinged, double-sided grill applies heat to both sides of the food at once, reducing the cooking time required.

Panini Grill

Edward Don & Company

Figure 5-27. A Panini grill is an Italian clamshell-style grill made specifically to cook grilled sandwiches.

Grilling
Media Clip

Many foods are cooked directly on the grill, while others are placed on skewers first. A *brochette* is a food that is speared onto a wooden, metal, or natural skewer and then grilled or broiled. **See Figure 5-28.** Natural skewers can include twigs of rosemary, shoots of lemon grass, and slivers of sugar cane. Classical examples of brochettes include a brochette of fresh fruit, a seafood brochette composed of shrimp and scallops, a chicken or beef satay, and a vegetarian brochette of mushrooms, zucchini, cherry tomatoes, peppers, and onions. Brochettes, also known as shish kebabs or kebabs, are often marinated.

Brochettes

National Honey Board

Bacon-Wrapped Scallops

National Turkey Federation

Turkey Satay

Figure 5-28. A brochette is a food that is speared onto a wooden, metal, or natural skewer and then grilled or broiled.

Production Tip

Soak wooden skewers in water for at least 30 minutes prior to use so they will not burn or catch fire. Spray metal skewers with a nonstick cooking spray prior to use.

Grilling Proteins

Grills are commonly used to cook poultry, seafood, meats, and some fruits and vegetables. **See Figure 5-29.** Most poultry can be grilled in the same manner as it is broiled. Many thick cuts of fish are also grilled. Only rib and loin cuts of beef, veal, pork, or lamb should be grilled.

Grilled Proteins

Florida Department of Agriculture and Consumer Services

Lobster

Entourage

Beef

Courtesy of The National Pork Board

Pork

Figure 5-29. Grills are commonly used to cook poultry, seafood, meats, and some fruits and vegetables.

Grilling Fruits and Vegetables

Some fruits and many vegetables can also be grilled. For example, pears and pineapples are often grilled as accompaniments for entrées. They also make excellent desserts. Vegetables are usually marinated briefly before being grilled. Hard vegetables, such as carrots, also should be blanched before grilling to make them more tender. Grilling fruits and vegetables causes the natural sugars to caramelize and intensify in flavor. **See Figure 5-30.** However, the high heat also results in a loss of vitamin C and folic acid. Despite the vitamin loss, grilling is a healthy cooking technique.

Grilled Fruits and Vegetables

Pear Bureau Northwest

Fruits

Tanimura & Antle®

Vegetables

Figure 5-30. Grilling fruits and vegetables causes the natural sugars to caramelize and intensify in flavor.

Barbequing

Barbequing is a dry-heat cooking technique in which food is slowly cooked over hot coals or smoldering hardwoods. Barbequing may be described as a combination of grilling and roasting, depending on the source of the heat. An indirect heat source from a side smoke box is used to keep the temperature between 180–250°F to impart more smoke flavor and render the fat and collagen more slowly in order to retain more moisture in the finished product. Indirect heat is often used to barbeque pork and beef and may also be used on delicate fish fillets.

Production Tip

Soft woods such as pine and cedar should not be used for barbequing because they impart a pitch to the smoke that contributes to poor flavor.

The internal temperature required of a barbequed protein may be debated, but 180°F is suggested for tough cuts of meat with ample fat content. Barbequing at this temperature should result in a product that pulls apart easily and is tender and juicy. Connective tissues comprised of collagen melt at 180°F. In some regions, a dry rub is added to the meat to impart flavor during the cooking process. Meats may also be basted with a barbeque or mopping sauce to keep them moist during cooking. Some chefs prefer to add barbeque sauce after the protein is cooked.

There are many regional specialty barbeques, commonly referred to as BBQ. For example, Kansas City, North and South Carolina, Memphis, and Texas all claim to serve the best BBQ. The type of wood and sauce used as well as the time the meat is cooked varies and directly affects the flavor of the BBQ.

Kansas City Barbeque. Kansas City BBQ is characterized by a low temperature and a cooking time ranging from 8–16 hours depending on the cut of meat, with oak or hickory most often being used for cooking. Dominant meats include pork shoulder, pork ribs, and brisket, which all benefit from the "low and slow" cooking process, though chicken can be barbequed as well. The meats are usually liberally rubbed with a combination of spices dominated by paprika along with brown sugar, onion powder, and garlic powder. Kansas City BBQ is cooked by indirect heat and often served with a liberal application of a sweet and tangy tomato-based sauce. **See Figure 5-31.** The Kansas City BBQ Society is a strong presence in the promotion of the Kansas City style of BBQ through state, regional, and national competitions.

Kansas City BBQ

Courtesy of The National Pork Board

Figure 5-31. Kansas City BBQ is often served with a liberal application of a sweet and tangy tomato-based sauce.

Carolina Barbeque. Carolina BBQ is characterized by the dominance of pork. In North Carolina, whole hogs may be used, or more simply the pork shoulder. The pork is usually seasoned liberally with salt and pepper, smoked over indirect heat using hardwoods such as apple wood, and then chopped and served alone or as a sandwich with a vinegar-based sauce made with tomatoes or ketchup. Carolina BBQ is often accompanied by vinegar-based coleslaw. In South Carolina, a vinegar-based mustard sauce is served with chopped, pulled, or sliced pork shoulder. In both North and South Carolina, the pork is often mopped with a BBQ sauce diluted with extra vinegar, and extra sauce is usually served on the side. **See Figure 5-32.**

Memphis Barbeque. Memphis BBQ is characterized for its dry-rubbed ribs, though pork shoulder is also popular. **See Figure 5-33.** The meat is cooked slowly over lump charcoal. The rub usually consists of a combination of spices dominated by paprika and chili powder, which adds a spiciness to the rub not seen in most other regions of the country, and a small amount of sugar. A mopping sauce that includes the same spices as the rub plus ketchup and vinegar is often used during the cooking process. The meat may be served with or without a sauce. BBQ spaghetti made from leftover BBQ is a popular dish in this region.

Memphis BBQ

Courtesy of The National Pork Board

Figure 5-33. Memphis BBQ is characterized by its dry-rubbed ribs, though pork shoulder is also popular.

Carolina BBQ

Carolina Pulled Pork

Carolina Pulled Pork with Coleslaw

Courtesy of The National Pork Board

Figure 5-32. In both North and South Carolina, the pork is often mopped with a BBQ sauce diluted with extra vinegar.

Texas Barbeque. Texas BBQ is characterized by beef that is seasoned liberally with salt and pepper and sometimes a little cayenne. Variations within Texas BBQ may use brisket, clod, ribs, or another tough cut of beef. Beef or pork sausages and goat are often barbequed in this region. Pecan, oak, and mesquite woods are used. Sauces are not as popular in Texas as in other regions, but when used, they are often vinegar-based, spicy, and have drippings and occasionally a tomato product added. Because of the lack of a rub and often the lack of sauce, the quality of the meat is seen as most imperative for Texas BBQ. **See Figure 5-34.**

Texas BBQ

Photo Courtesy of the Beef Checkoff

Figure 5-34. Because of the lack of a rub and often the lack of sauce, the quality of the meat is seen as most imperative for Texas BBQ.

Smokers

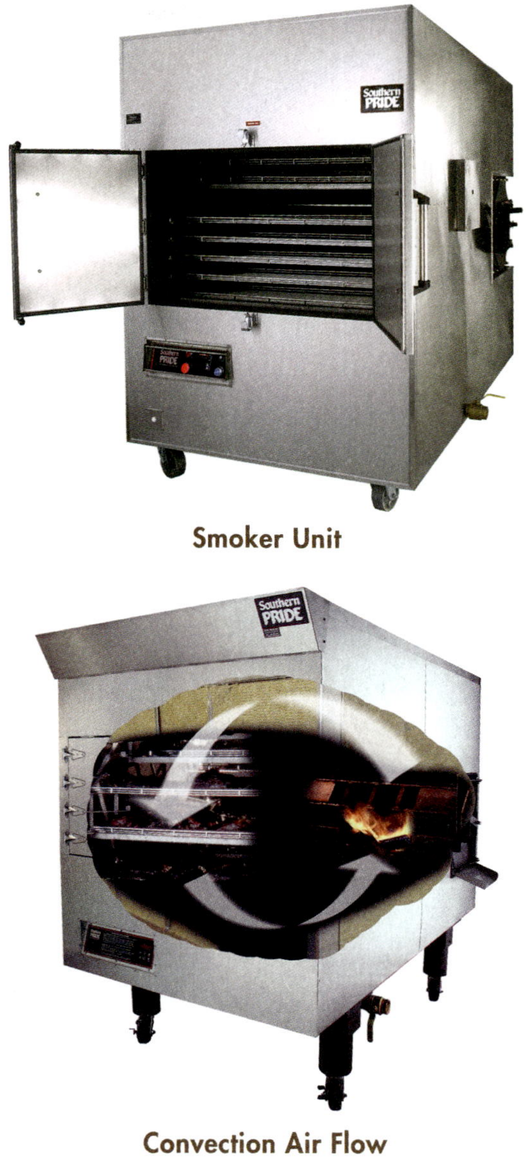

Southern Pride

Figure 5-35. In a professional kitchen, hardwood shavings or chips are used in both gas and electric smokers to produce the desired smoke flavor.

Smoking Proteins

Smoking is a dry-heat cooking technique in which food is slowly cooked over smoldering hardwoods in a vented enclosure. In a professional kitchen, hardwood shavings or chips are used in both gas and electric smokers to produce the desired smoke flavor. **See Figure 5-35.** Mesquite, hickory, maple, apple, and cherry wood are used to impart different smoked flavors. The temperature of a smoker is kept low and proteins are cooked for many hours to yield a flavorful and tender product. Proteins that are smoked contain higher amounts of fat and collagen. Doneness is determined using the fork-tender method.

Nutrition Note

Smoking meats exposes the meat to carcinogens known as polycyclic aromatic hydrocarbons (PAHs). PAHs are chemicals formed during the incomplete burning of meat. PAHs form when the animal fat from the meat hits the heat source and creates smoke. Exposure to high amounts of PAHs can be detrimental to a person's health.

CHECKPOINT 5-4

1. Explain how to prepare a grill.
2. Describe the grilling procedure.
3. Identify proteins that are typically grilled.
4. Grill a piece of poultry and broil an identical piece of poultry. Compare the final products for color, texture, and flavor.
5. Grill one fish fillet and broil another fillet identical in size and variety. Compare the final products for color, texture, and flavor.
6. Grill and broil a cut of beef or veal. Compare the final products for color, texture, and flavor.
7. Grill and broil a cut of pork. Compare the final products for color, texture, and flavor.
8. Grill a cut of lamb and broil an identical cut of lamb. Compare the final products for color, texture, and flavor.
9. Grill a fruit and a vegetable. Evaluate each product for color, texture, and flavor.
10. Contrast Kansas City, Carolina, Memphis, and Texas BBQ.
11. Barbeque the same protein using two different styles of BBQ. Compare the final products for color, texture, and flavor.
12. Describe the smoking process.

ROASTING

Roasting is a dry-heat cooking technique in which food that contains fat or that has fat added to it is cooked uncovered at a high temperature in an oven or on a revolving spit over an open flame. **See Figure 5-36.** Roasting produces a well-browned exterior and a moist interior. Whole poultry, large cuts of meat, and root vegetables are commonly roasted at lower temperatures to prevent shrinkage and lessen moisture loss.

Allowing roasted items to rest after being removed from the oven permits carryover cooking to occur. *Carryover cooking* is the rise in internal temperature of an item after it is removed from a heat source due to residual heat on the surface of the item. The internal temperature will rise between 5°F and 10°F during this period. Carryover cooking helps a cooked item retain moisture that will escape as steam if the item is cut too soon after cooking. Roasted meat should not be cut immediately after being removed from the oven. If it is, the internal moisture will be released as steam and evaporate into the air. Roasted meat that is allowed to rest before it is cut will retain moisture.

Roasting Proteins

Charlie Trotter's

Roasting Process

Photo Courtesy of the Beef Checkoff

Roasted Beef

Figure 5-36. When roasting, food that contains fat or that has fat added to it is cooked uncovered at a high temperature in an oven or on a revolving spit over an open flame.

Roasting Procedures

When roasting a protein, it is important to select a cut that will produce a moist and tender product when the item is cooked to the desired degree of doneness. Roasting reduces the moisture content in larger protein items by shrinking the muscle fibers. By the time the protein has reached an internal temperature of 120°F, the shrinking process is underway. The connective tissues and marbling within the meat actually soften and melt, basting the item in fat and tenderizing the meat as it roasts.

Prior to roasting a protein, it should be trimmed of any excess surface fat and seasoned or marinated. The protein is then placed on a rack in a roasting pan, fat-side up, and the rack is placed in a preheated oven to roast uncovered, allowing the exterior to sear.

Roasting Poultry

Whole chickens, turkeys, and hogs may be skewered and placed on a rotisserie and spit-roasted. *Spit-roasting* is the process of cooking food by skewering it and suspending and rotating it above or next to a heat source. Birds that are 2 lb or less, such as squabs, pheasants, and Cornish hens, are roasted between 375°F and 400°F to produce a crisp, golden-brown skin quickly without drying out the flesh. Very-high-temperature roasting is used for thick-skinned or fatty-skinned birds, such as ducks and geese. Roasting at a temperature between 400°F and 425°F allows the fat to melt out and away from the skin. The fat also insulates the flesh, preventing it from drying out at higher temperatures. As the hot fat seeps out of the skin, it fries the skin slightly, turning the skin crispy.

Procedure for Roasting

1. Trim excess fat from the item and season it. *Note:* Some items, such as turkey, may be brushed with fat before roasting for more even browning.

2. Place the item in a roasting pan on a roasting rack or a bed of mirepoix.

3. Place the uncovered roasting pan in an oven preheated to between 300°F and 350°F, and allow the item to cook.
4. Baste the item as needed.

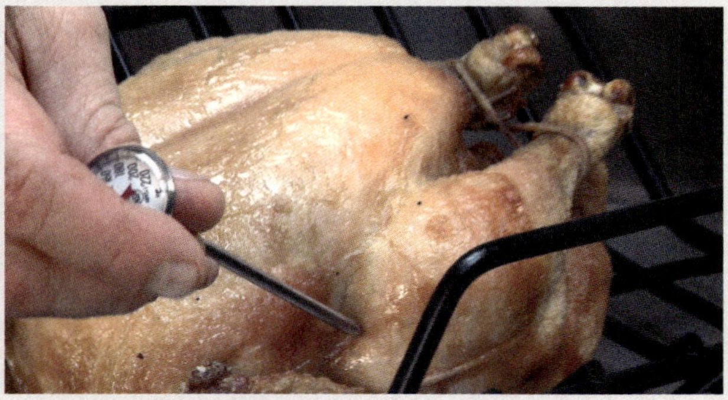

5. Remove the roasted item from the oven approximately 10 minutes before it reaches the desired degree of doneness. Insert a probe thermometer in the thickest portion to check the internal temperature.
6. Allow the item to rest and carryover cooking to complete the cooking process.

Roasting
Media Clip

Production Tip

It is important to prick the skin of fatty-skinned or thick-skinned birds prior to roasting and again during the roasting process to ensure that the skin is not seared in the high heat.

All poultry should be seasoned prior to roasting, but the size of the bird determines where the seasoning should be placed. Whole herbs and mirepoix are added to the internal cavity of a large bird, such as a turkey or capon. These aromatics infuse the flesh with their aromas and flavors as a bird slowly roasts. A smaller bird can be seasoned directly on the skin with salt and pepper. **See Figure 5-37.** Herbs should not be placed on the skin of poultry that is being roasted at a high temperature because the herbs will burn.

Roasted Poultry

National Chicken Council

Figure 5-37. A small bird can be seasoned directly on the skin with salt and pepper.

Poultry is often basted with the fat that escapes from the bird during cooking or with fat that is added to the roasting pan prior to cooking. Fatty birds, such as ducks and geese, are not basted because it makes them unpleasant tasting.

Roasting Meats

Many cuts of beef and veal from the rib, loin, and leg are roasted. Larger roasts require a longer cooking time and should be roasted at lower temperatures, between 275°F and 325°F, to prevent excessive shrinkage. **See Figure 5-38.** Lean cuts of meat will become dry when roasted unless some form of fat is added. Barding or larding can be done to add additional fat.

Larger Roasts

Paderno World Cuisine

Figure 5-38. Larger roasts require a longer cooking time and should be roasted at lower temperatures, between 275°F and 325°F, to prevent excessive shrinkage.

Cuts from the pork loin and leg as well as the lamb rack, loin, and leg are also commonly roasted. **See Figure 5-39.** Smaller roasts should be cooked at higher temperatures, between 400°F and 450°F, to allow them to caramelize on the exterior without overcooking the interior.

Production Tip

It is important to allow for carryover cooking when roasting meats.

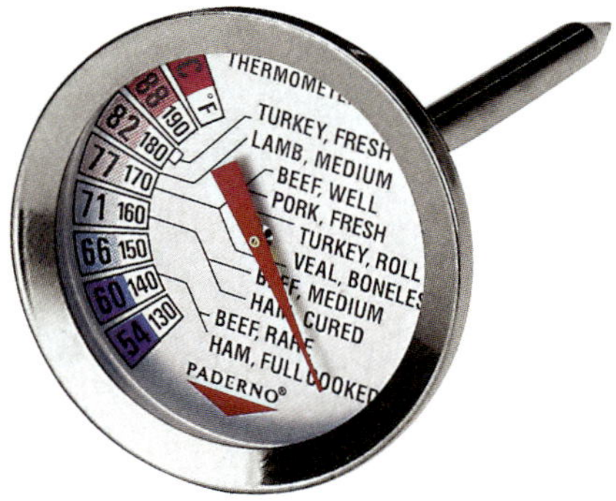

Paderno World Cuisine

Roasting Poultry and Meats

Roasted Pork and Lamb

National Pork Producers Council

Rack of Pork

Rack of Lamb

Figure 5-39. Cuts from a pork loin as well as cuts from a lamb rack are commonly roasted.

Roasted Vegetables

National Onion Association

Figure 5-40. Many root vegetables, such as onions, can be roasted.

CHECKPOINT 5-5

1. Define roasting.
2. Roast one chicken and barbeque another chicken. Compare the final products for color, texture, and flavor.
3. Roast a cut of beef and a similar cut of veal. Compare the final products for color, texture, and flavor.
4. Roast and barbeque a cut of pork. Compare the final products for color, texture, and flavor.
5. Roast a cut of lamb and grill the same cut of lamb. Compare the final products for color, texture, and flavor.
6. Identify the proteins that are typically roasted at high temperatures.
7. Identify ways to add fat to lean cuts of protein.
8. Roast one vegetable and grill a second vegetable of the same type. Compare the final products for color, texture, and flavor.

Roasting Vegetables

Many root vegetables such as onions, carrots, turnips, parsnips, and various types of potatoes are often roasted together or placed alongside a large cut of meat. **See Figure 5-40.** When roasting vegetables, it is important to cut vegetables into uniform sizes to ensure doneness.

American Metalcraft, Inc.

CHECKING FOR DONENESS

It is important to check each broiled, grilled, and roasted protein for doneness. Depending on the protein, the manner in which doneness is checked may vary. Temperature remains the most accurate way to check for doneness, but there are also other ways in which experienced chefs check proteins for doneness.

Doneness of Poultry

Poultry is always cooked well-done, with the exception of duck and squab breasts. Well-done cuts of poultry should still be moist and juicy. The four methods that chefs use to determine the doneness of poultry are temperature, touch, joints, and juices (TTJJ). **See Figure 5-41.**

Checking Poultry for Doneness
Temperature—165°F internal
Touch—Firm and solid
Joints—Soft and tender
Juices—Clear with no signs of blood

Figure 5-41. The doneness of poultry can be determined by temperature, touch, joints, and juices (TTJJ).

Temperature. Poultry should always be cooked to an internal temperature of 165°F. To determine doneness, the temperature is taken by inserting an instant-read thermometer in the innermost part of the thigh, wing, and/or breast. The thermometer should be inserted close to, but not touching, any large bones. The difficulty with using temperature to indicate doneness is that smaller cuts, such as boneless chicken breasts, may be too thin for a thermometer to provide an accurate reading.

Touch. Touch is not as accurate as temperature. However, experienced chefs know that the firmness of the poultry increases in proportion to its doneness. When poultry is thoroughly cooked, the flesh is firm but springs back when touched.

Joints. Testing joints is not as accurate as temperature. The joints of raw or undercooked poultry are firmly connected. In contrast, the joints become soft and tender as the cartilage that holds them together cooks. For example, when the ends of a thoroughly cooked leg and thigh portion are twisted, the bones will separate easily at the knee joint.

Juices. Testing the color of juices from poultry is not as accurate as taking the temperature. However, juices from raw poultry are red, and juices from fully cooked birds are clear. As poultry cooks, the color of the juices changes, which explains why chicken stock is clear. The color of the internal juices, not the external juices, is used to determine doneness.

Doneness of Fish

Determining the doneness of fish takes practice because it involves sight and touch more than temperature. A fish that is done will have a caramelized skin, firm and dense interior edges, and a moist, opaque center. **See Figure 5-42.** An instant-read thermometer that registers 145°F when inserted into the thickest part of a thick-fleshed fish indicates the fish is cooked through. When a fork is used to lightly prod the thickest part of a thin fish, done flesh will be opaque and the juices will be milky white. Undercooked fish is translucent and the juices will be clear and watery. Overcooked fish is dry and falls apart easily.

Determining Doneness of Fish

Daniel NYC / M. Duisterhof

Figure 5-42. A fish that is done will have a caramelized skin, firm and dense interior edges, and a moist, opaque center.

Doneness of Meats

When cooking meat to the desired degree of doneness, the type of meat, the thickness of the meat, the temperature of the meat when it begins to cook, and the intensity of the heat are factors that must be considered. Because of these variables, time is not an accurate way to determine the doneness of meats. The most accurate way to determine the degree of doneness of broiled, grilled, and roasted meats is by measuring the internal temperature of the meat. A probe thermometer is inserted in the thickest part of the meat to determine when the temperature of the meat has reached a safe level. **See Figure 5-43.**

Internal Temperature

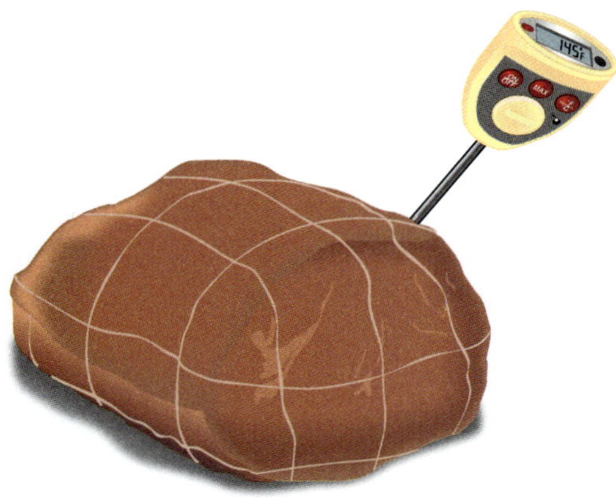

Figure 5-43. A probe thermometer is inserted in the thickest part of the meat to determine when the temperature of the meat has reached a safe level.

Beef, veal, and pork cuts that are done have been cooked to an internal temperature of 145°F for at least 15 seconds and rested for 3 minutes. All types of ground meat must be cooked to an internal temperature of 160°F.

Guests may request that a cut of beef, veal, or lamb be cooked to a specific degree of doneness. It is important to remember that each degree of doneness corresponds to a specific internal temperature. **See Figure 5-44.** For example, a rare steak should be cooked to an internal temperature of 140°F.

Production Tip

Smaller cuts of meat that are grilled or broiled can be tested for doneness with the touch method to check the firmness of cooked meat. For example, a chop that is well done is firm and springs back immediately when gently pressed with the fingertip.

Determining Degree of Doneness

Beef and Veal	
Degree of Doneness	***Internal Temperature***
Very rare	130°F
Rare	140°F
Medium-rare	145°F
Medium	160°F
Medium-well	165°F
Well-done	170°F
Lamb	
Degree of Doneness	***Internal Temperature***
Medium-rare	145°F
Medium	160°F
Medium-well	165°F
Well-done	170°F

Figure 5-44. Guests may request that a cut of beef, veal, or lamb be cooked to a specific degree of doneness.

CHECKPOINT 5-6

1. Explain the four methods used to determine the doneness of poultry.
2. Name the safest method of determining the doneness of poultry.
3. Identify the internal temperature required of cooked poultry (with the exception of duck).
4. Explain how to determine the doneness of fish.
5. Demonstrate how to determine the doneness of fish.
6. Explain the safest way to determine the doneness of meats.
7. Identify the size of meat cuts that can be tested for doneness by the touch method.
8. Broil and grill two similar cuts of poultry, fish, or meat to two different degrees of doneness.
9. Barbeque and roast two similar cuts of poultry, fish, or meat to two different degrees of doneness.

PRESENTING PROTEINS

After a protein has reached its desired degree of doneness it should be removed from the heat and allowed to rest. Carryover cooking will increase the temperature of the meat a few degrees and allow the juices to redistribute, making the meat more tender and easier to carve.

Slicing and Carving Poultry and Meats

Some cuts of beef are roasted whole and then sliced to order. Cuts from the inside round are relatively easy to slice across the grain. Slicing across the grain produces a cut of meat that is more tender than a cut sliced in the same direction as the grain. **See Figure 5-45.** Tender items may be cut thicker than tougher cuts of meat. Tougher cuts of meat should be thinly sliced across the grain.

Carving Poultry

Slicing Across the Grain

Irinox USA

Figure 5-45. Slicing across the grain produces a cut of meat that is more tender than a cut sliced in the direction of the grain.

Production Tip

Thin slices are the best way to minimize waste at an action station. Guests who want several slices or a thicker slice can request it.

Equipment required at a carving station include a sharp knife, a thermometer, a heat lamp, a clean serving tray to catch juices, a sanitation bucket, and a supply of clean towels. Depending on the number of guests, some slices can be carved a few minutes ahead and held under the heat lamp to help expedite service. Additional meat is held in a hot box or chafer. The temperature of the meat must be checked periodically to make sure the meat is at a proper serving temperature.

Some cuts of meat contain bones that must be removed prior to slicing or that require the meat to be sliced off the bone. A bone-in steamship round roast can be difficult to carve because large bones are present in the center of the roasted meat. A steamship round roast is typically prepared for large buffets and special events. A bone-in prime rib is a tender and juicy cut of meat that is often carved before service. **See Figure 5-46.** Slices may be carved with or without the bone.

Carving a Bone-In Prime Rib

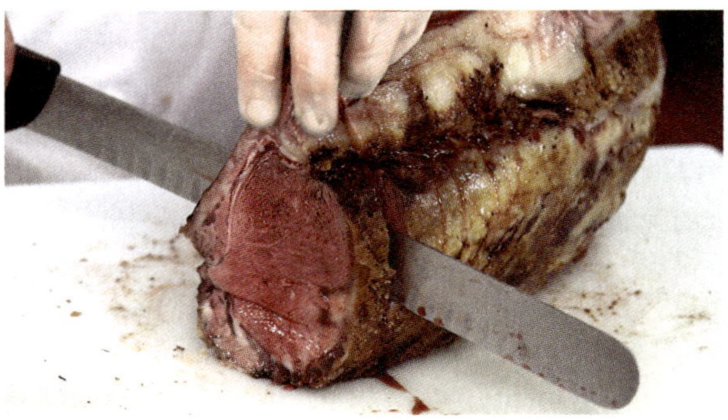

Figure 5-46. A bone-in prime rib is a tender and juicy cut of meat that is often carved before service.

Plating Proteins

Plating refers to placing food on a plate in a visually appealing manner. Plate presentation takes skill and experience. The creativity and skill that went into cooking the protein should also be reflected in the plating. Food should be presented in a natural and appealing manner. The main item in the dish, usually the protein, should be prominent in the presentation. **See Figure 5-47.** A protein can be elevated by placing it on or leaning it up against a starch for a more dramatic presentation.

Plating Proteins

Daniel NYC / N. Clutton

Figure 5-47. The main item in the dish, usually the protein, should be prominent in a presentation.

Production Tip

Spooning sauce under the meat rather than on top allows the surface of the protein to remain crisp while offering a contrasting shape underneath.

Protein items are basically brown or white, so it is always best to brighten the plate with colorful garnishes. A few sprigs of fresh cilantro, julienned peppers, a trio of tobacco onions, a wedge of fruit, toasted black sesame seeds, lemon twists, or broiled tomatoes are examples of garnishes that add color to a protein-based plate. Another popular garnish is a compound butter that is added just prior to service so that the butter is melting as it is received by the guest. A garnish needs to add flavor, texture, and color to the dish. For example, a chicken breast becomes more appealing by placing a few spoonfuls of salsa on top or beside the protein. A garnish should never be an afterthought.

MacFarlane Pheasants, Inc.

CHECKPOINT 5-7

1. Describe the importance of slicing across the grain.
2. Identify the direction of the grain on cuts of poultry, fish, beef, veal, pork, and lamb.
3. Identify the type of slices that minimize waste at an action station.
4. List six items required to properly set up a carving station.
5. Describe how to plate proteins.
6. Identify three contributions that a garnish should provide to a plated dish.

HOT SANDWICHES

Hot sandwiches feature cooked fillings and are often served on warm toasted or grilled bread. Grilled sandwiches, hot open-faced sandwiches, hot closed sandwiches, and hot wrap sandwiches are often prepared at the broiler/char-grill station.

Grilled Sandwiches

A *grilled sandwich* is a hot sandwich that is assembled by adding a precooked filling or cheese to bread and then heated on a griddle, in a sauté pan, or on a grill. **See Figure 5-48.** Grilled sandwiches often include cheese as a filling, as it melts during cooking and holds the bread to the filling. All grilled sandwich fillings, such as bacon, chicken, or beef, must be thoroughly cooked prior to assembly. Roasted root vegetables or fresh fruit can be combined with various cheeses and artisan breads to create tasty grilled sandwiches. Panini are sandwiches that are cooked on both sides using a Panini grill. In Italy, panini are made with a main ingredient and cheese. Outside of Italy, the sandwiches have been adapted to local tastes with various fillings.

Production Tip

Fresh garnishes, such as lettuce, tomatoes, and pickle slices, are placed on hot sandwiches after the hot filling and bread are plated.

Grilled Sandwiches

Photo Courtesy of Perdue Foodservice, Perdue Farms Incorporated

Poultry Fillings

California Strawberry Commission

Meatless Fillings

Figure 5-48. A grilled sandwich is a hot sandwich that is assembled by adding a precooked filling or cheese to bread and then heated on a griddle, in a sauté pan, or on a grill.

Hot Open-Faced Sandwiches

Figure 5-49. A hot open-faced sandwich is often covered with a sauce and is usually eaten with a knife and fork.

Hot Open-Faced Sandwiches

A *hot open-faced sandwich* is a sandwich consisting of one or two slices of fresh, toasted, or grilled bread, topped with one or more hot fillings, and covered with a sauce, gravy, or a melted cheese topping. Often, the hot open-faced sandwich is browned under a broiler just prior to serving. Due to the fact that this variety of sandwich is often covered with a sauce, it is usually eaten with a knife and fork. **See Figure 5-49.** Popular hot open-faced sandwiches include hot turkey, hot roast beef, and various "melts" such as the turkey melt, crab melt, and patty melt.

Hot Closed Sandwiches

A *hot closed sandwich* is a sandwich made by placing one or more hot fillings between two pieces of bread or a split roll or bun. Hot closed sandwiches are often served with cold garnishes, such as lettuce and slices of tomato and onion. Common hot closed sandwiches include the hamburger, veggie burger, fish fillet, grilled chicken, pastrami, chicken parmesan, hot roast beef, meatball marinara, pulled or chopped barbeque, and quesadilla. **See Figure 5-50.**

Hot Sandwiches

Sullivan University

Hot Closed Sandwiches

National Turkey Federation

Turkey Burger

National Turkey Federation

Turkey Pastrami

Courtesy of The National Pork Board

Pork and Mushroom Quesadilla

Figure 5-50. Hamburgers, reubens, and quesadillas are popular types of hot closed sandwiches.

Hot Wrap Sandwiches

A *hot wrap sandwich* is a sandwich made by adding a spread and precooked fillings to a flatbread and then cooking it. Hot wraps can be made using one of two methods. In one method, flatbread is covered with a spread, topped with a cold precooked filling, and rolled up. The resulting wrap sandwich is then baked, fried, or grilled. An alternate method of making a hot wrap is to place flatbread on a heated griddle, cover it with a spread, top it with a hot precooked filling, and then roll it tightly and serve. Hot wrap sandwiches include burritos, tacos, fajitas, and enchiladas. **See Figure 5-51.**

Plating Hot Sandwiches

Plating sandwiches properly can take a basic sandwich to the next level in visual appeal. For example, preparing a wrap using a green spinach-flavored or red tomato-flavored wrap will add visual appeal along with flavor. Sandwiches may be served whole, sliced at an angle, or sliced down the middle. Sandwiches held together with melted cheese or a sauce usually hold together well. Service style should also be considered when plating sandwiches. Deli paper may be used for more casual service. Sometimes a pick may be used to hold the ingredients in place. For messier sandwiches, the use of a knife and fork may be required.

Sandwiches are often plated with traditional accompaniments such as coleslaw, green salad, potato salad, macaroni salad, fruit salad, French fries, potato chips, pickled vegetables like chow-chow, and/or a pickle spear. However, a little creativity can add even more color and appeal to a sandwich plate. For example, a plated sandwich can be enhanced by the addition of a crisp red apple and purple potato salad with fresh dill. This accompaniment adds a contrasting texture to the sandwich bread and fillings. **See Figure 5-52.** It is important that the flavors of each accompaniment complement the flavors of the sandwich and do not overpower the main component of the sandwich.

Hot Wrap Sandwiches

USA Rice Federation

Burrito

Courtesy of The National Pork Board

Hard Tacos

Courtesy of The National Pork Board

Soft Tacos

Figure 5-51. Burritos and tacos are popular hot wrap sandwiches.

Plating Hot Sandwiches

Courtesy of The National Pork Board

Figure 5-52. An accompaniment can add a contrasting texture to a sandwich bread and fillings.

CHECKPOINT 5-8

1. Identify four types of hot sandwiches.
2. Identify four pieces of equipment that can be used to grill sandwiches.
3. Prepare and plate a grilled sandwich. Evaluate color, texture, and flavor.
4. Prepare and plate a hot open-faced sandwich and a hot closed sandwich. Compare the sandwiches for color, texture, and flavor.
5. Prepare and plate a hot wrap. Evaluate the sandwich for color, texture, and flavor.
6. Describe how to plate different types of sandwiches based on service style.
7. Plate two sandwiches with accompaniments that provide contrasting colors, textures, and flavors.

Refer to DVD for **Quick Quiz®** questions

SUSTAINABILITY CORNER

Sustainable seafood is seafood that is fished or farmed using sustainable practices. These practices help maintain production without jeopardizing the ecosystem. However, the sustainable choice for a foodservice operation is not always obvious. It depends on each species, their habitat, and where and when the seafood was caught. In general, species that grow quickly and breed young are considered more sustainable choices than slow-growing species that reproduce late in life.

Also, overfishing has resulted in the decline of virtually all commercial fish populations, destroyed natural habitats, and often leads to bycatch (catching unwanted species). Due to the negative effects of overfishing, aquaculture (the act of farming fish) has become more prevalent. Sustainable aquaculture practices include protecting natural habitats, monitoring water quality to prevent pollution, and using feed produced without antibiotics or hormones.

There are several resources that can help foodservice operations obtain and utilize sustainable seafood. The Marine Stewardship Council (MSC), for example, is a leading certification entity for sustainable wild-caught seafood. The MSC logo helps ensure that the seafood purchased is not a vulnerable species. Another way to incorporate sustainable seafood choices is to become familiar with websites, such as the site for Monterey Bay Aquarium, that include "red," "yellow," and "green" seafood lists. These lists serve as a guide for knowing which species are the most sustainable choices. To promote and enhance sustainability efforts, only green-listed fish should be purchased.

New Zealand Greenshell™ Mussels

In the United States, approximately 65% of all seafood travels through restaurants to the consumer. An additional 24% is sold fresh or frozen through retail outlets. By choosing to purchase and prepare sustainable seafood, restaurants and retail outlets have the opportunity to profoundly impact the health of the oceans. Also, serving sustainable seafood is a practice that can add value for a growing number of environmentally conscious consumers.

Certified Sustainable Seafood

Certification Exam Preparation Questions

1. ___ is the period of rest that occurs for a length of time after an animal is slaughtered.
   A. Aging
   B. Curing
   C. Barding
   D. Larding

2. A ___ is a blend of ingredients that is pressed onto the surface of uncooked foods such as poultry, fish, or meat to impart flavor and sometimes tenderize.
   A. cure
   B. brine
   C. rub
   D. marinade

3. ___ aging is the process of aging meat in vacuum-sealed plastic.
   A. Brine
   B. Cure
   C. Dry
   D. Wet

4. ___ is the most accurate way to determine the degree of doneness of a protein.
   A. Time
   B. Temperature
   C. Touch
   D. Color

5. ___ is a dry-heat cooking technique in which food is slowly cooked over smoldering hardwoods in a vented enclosure.
   A. Barbequing
   B. Broiling
   C. Smoking
   D. Grilling

6. ___ is a dry-heat cooking technique in which food that contains fat or that has fat added to it is cooked uncovered at a high temperature in an oven or on a revolving spit over an open flame.
   A. Barding
   B. Barbequing
   C. Roasting
   D. Smoking

7. ___ is a dry-heat cooking technique in which food is cooked on open grates above a direct heat source.
   A. Barbequing
   B. Broiling
   C. Smoking
   D. Grilling

8. ___ BBQ is characterized by its dry-rubbed ribs, though pork shoulder is also popular.
   A. Kansas City
   B. Carolina
   C. Memphis
   D. Texas

9. Meats that are broiled are usually taken from the ___ and the loin.
   A. round
   B. rib
   C. sirloin
   D. chuck

10. ___ is the process of laying a piece of fatback across the surface of a lean cut of meat to add moisture and flavor.
    A. Basting
    B. Barding
    C. Larding
    D. Marinating

11. A ___ is made by incorporating wet ingredients, such as Dijon mustard, flavored oils, puréed garlic, or honey, into a dry rub mixture and then rubbing it into the meat prior to cooking.
    A. baste
    B. marinade
    C. wet rub
    D. wet cure

12. Slicing across the grain produces a cut of meat that is more ___.
    A. tender
    B. tough
    C. dry
    D. juicy

13. ___ BBQ is cooked by indirect heat and often served with a liberal application of a sweet and tangy tomato-based sauce.
    A. Kansas City
    B. Carolina
    C. Memphis
    D. Texas

14. A taco is a hot ___ sandwich.
    A. grilled
    B. open-faced
    C. closed
    D. wrap

15. ___ is a dry-heat cooking technique in which food is cooked directly under or over a heat source.
    A. Barbequing
    B. Broiling
    C. Smoking
    D. Grilling

16. A ___ is a flavorful liquid used to soak uncooked foods such as poultry, fish, and meats to impart flavor and sometimes tenderize.
    A. baste
    B. brine
    C. marinade
    D. wet cure

17. ___ BBQ is characterized by beef that is seasoned liberally with salt and pepper and sometimes a little cayenne.
    A. Kansas City
    B. Carolina
    C. Memphis
    D. Texas

18. A hot ___ sandwich is a sandwich consisting of one or two slices of bread, topped with one or more hot fillings, and covered with a sauce or melted cheese.
    A. grilled
    B. open-faced
    C. closed
    D. wrap

19. ___ is the process of using salt and sodium nitrate alone or with flavorings or sugar to preserve a food item.
    A. Curing
    B. Basting
    C. Marinating
    D. Brining

20. All types of ground meat must be cooked to an internal temperature of ___.
    A. 145°F
    B. 160°F
    C. 170°F
    D. 175°F

21. Explain the broiling procedure.

____

____

____

____

____

22. Explain the grilling procedure.

____

____

____

____

____

23. Explain the roasting procedure.

____

____

____

____

24. Identify the cooking technique that uses low temperatures and many hours to cook proteins.

____

SKETCHING EXERCISE

25. Sketch a visually appealing plated protein.

Notes

Soup & Sauce Station

Introduction

In the sixth station in the ACFEF apprenticeship program, you will be responsible for the production and presentation of a variety of soups and sauces. This station will teach you more fundamentals than any other station, including many necessary basics that will enable you to become an accomplished culinarian.

Soups and sauces will become the foundation of much of your food preparation. You will start by learning the "classics." You will discover how to thicken stocks with the appropriate white, blond, or brown roux. Your stock-making skills will also lead to creating consommés, bisques, chowders, and specialty soups. As you move on, you will become proficient in the "mother sauces"—béchamel, velouté, espagnole, tomato sauce, and hollandaise. Then, you will create variations from the mother sauces such as a supreme from a velouté and a bordelaise from an espagnole. To complete your sauces studies, you will produce classical sauce garnishes that will enhance the presentation as well as the flavor of your sauces.

Sustainability through water conservation practices is as important as learning the basics of cooking. You will be introduced to the simple steps of how to conserve water by identifying faucet leaks, insulating pipes to save energy, and not using water unnecessarily. All of these steps toward sustainability best practices will round out your sixth station.

A necessary skill is the mastery of preparing quality stocks, soups, and sauces. Flavorful stocks are either reduced or thickened to make countless soups and sauces. Soups are versatile foods that can range from a flavorful broth to a hearty chowder. Soups can set the stage for a wonderful meal and can also be a good way to use high-quality leftovers. The primary purpose of a sauce is to complement and enhance the flavor of a dish.

STOCKS

A *stock* is an unthickened liquid that is flavored by simmering seasonings with vegetables, and often, the bones of meat, poultry, or fish. The ingredients in a fresh stock are simmered to extract flavors, nutrients, and color. However, many foodservice operations use commercially prepared bases to save time and money. Bases are available in highly reduced pastes or dehydrated forms. Quality bases are made from the meat, bones, and juices of the product they represent. Other bases rely on high-sodium ingredients for flavor. Bases may be used to add flavor to a fresh stock. Likewise, fresh vegetables and bones may be added to a stock made from a base to make the stock taste fresher.

Stock Composition

Stocks are composed of water, a flavoring component, mirepoix, and aromatics. The composition of a stock can vary, but is generally 10 parts water, 5 parts flavoring component, and 1 part mirepoix, plus aromatics. **See Figure 6-1.** A stock is often reduced to further concentrate its flavor, especially when used to make sauces. Salt is never added to stock in the early stages because the resulting flavor is too intense. If desired, salt is added at the final stage of preparation.

Stock Composition

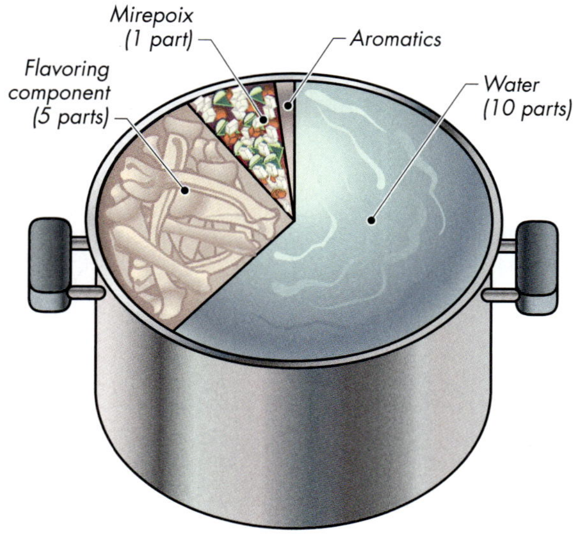

Figure 6-1. The initial composition of a stock is approximately 10 parts water, 5 parts flavoring component, and 1 part mirepoix, plus aromatics.

Water. Cold water is used to start a stock. As the water heats, it slowly draws out impurities and blood from the bones which then coagulate and float to the surface where they can be removed by skimming. If hot water is used, the impurities will coagulate too quickly and stick to the bones, causing the stock to become cloudy as it simmers.

Flavoring Ingredients. Each flavoring ingredient adds to the flavor and nutritional makeup of a stock. Bones or trimmings are the primary flavoring ingredients used to make meat, poultry, or fish stocks. Animal bones and trimmings contain collagen, a fibrous protein that is converted to gelatin when heated. In order to more effectively expose the collagen, bones are cut into smaller pieces or split in half lengthwise. The resulting gelatin that forms when the collagen cooks out of the bones and trimmings provides the stock with nutrients and body. A stock that is rich in gelatin will thicken naturally as it is reduced.

Bones can also be roasted or blanched before being used in a stock. **See Figure 6-2.** The blanching process removes any loose blood proteins and impurities from the bones that could cloud the stock. After the bones have been blanched, the blanching water is discarded and the bones are covered with fresh cold water to start a stock.

Mirepoix. *Mirepoix* is a mixture of 50% onions, 25% celery, and 25% carrots, roughly cut into the appropriate size for the stock being produced. **See Figure 6-3.** Because the mirepoix is strained from the finished stock, a perfect dice is not required. A white mirepoix is used to make a white stock. A *white mirepoix* is a mirepoix made of onions, celery, and leeks or parsnips instead of carrots.

Roasted Beef Bones

Sullivan University

Figure 6-2. When roasting bones, they must be placed in a single layer on an oiled roasting pan.

Mirepoix

Figure 6-3. Mirepoix is a mixture of 50% onions, 25% celery, and 25% carrots cut into the appropriate size for the stock being produced.

Procedure for Blanching Bones

1. Place the bones in a pot of cold water.
2. Bring water to a boil over high heat.
3. Reduce heat to a simmer and cook for a few minutes.
4. Remove the bones and plunge them into cold water.
5. Discard the blanching water.

Vegetables must be thoroughly cleaned and trimmed prior to being used in a stock. Most chefs remove the peelings from carrots because the peels will break down and discolor the stock. Celery leaves are not used in stock because the leaves are often bitter. Onion peels are sometimes used in brown stocks, but never in white stocks because the peels may darken the stock. Stock should always be relatively clear, not cloudy. Care should be taken to remove any impurities that could cloud the stock, such as overcooked vegetables.

A stock made from large bones, such as beef stock, requires large cuts of mirepoix, usually 1–2 inch square pieces. Fish and chicken stocks that cook for a shorter time and contain smaller, thinner bones require smaller cuts of mirepoix, typically ½ inch square pieces. The smaller the cut, the quicker the mirepoix cooks. If a small-cut mirepoix is added to a beef stock, the vegetables will overcook before the stock is done. Overcooked mirepoix can make the stock cloudy.

Sometimes a matignon is used in place of mirepoix. A *matignon* is a uniformly cut mixture of onions, carrots, and celery and may also contain smoked bacon or ham. A white matignon includes parsnips instead of carrots. Matignon is often referred to as "edible mirepoix" because it is cut uniformly and left in the completed dish.

Aromatics. An *aromatic* is an ingredient such as an herb, spice, or vegetable added to a food to enhance its natural flavors and aromas. There are two standard methods for adding aromatics to a stock. **See Figure 6-4.** A *bouquet garni* is a bundle of herbs and vegetables tied together with twine that is used to flavor stocks and soups. It usually consists of a sprig of thyme, several parsley stems, a dried bay leaf, and leek leaves or a celery stalk cut in half. The bundle is tied to the handle of the stockpot so that it can be easily retrieved.

A *sachet d'épices*, or sachet, is a mixture of spices and herbs placed in a piece of cheesecloth and tied with butcher's twine. A sachet often contains several parsley stems, a dried bay leaf, a sprig of thyme, a teaspoon of cracked peppercorns, and a clove of garlic. Sometimes spices such as cardamom, ginger, or cinnamon are added. The sachet is tied to the handle of the stockpot for easy retrieval. Aromatics infuse their flavor and aroma, adding depth to the flavor of the stock.

Aromatics

Bouquet Garni

Bouquet Garni

Sachet d'Épices

Figure 6-4. A bouquet garni and a sachet d'épices are aromatics that enhance the flavor and aroma of a stock.

Guidelines for Preparing Stocks

Although there are many different types of stocks used in the professional kitchen, there are basic guidelines that can be applied to any stock. For example, stock is always prepared in a pot that is taller than it is wide to allow for slower evaporation. Following these general guidelines yields consistent stocks that are easy to produce.

1. **Begin with cold water.** Cold water added to bones at the beginning stages of stock preparation helps remove any impurities and blood proteins that could later cloud a stock. As the stock is heated, these impurities coagulate or thicken and float to the surface. The coagulated impurities are commonly referred to as "scum," which can be skimmed from the surface and discarded.
2. **Cover bones completely.** Bones should be completely submerged in water during preparation. A bone that is above the surface of the water does not add any flavor or nutrients to the stock. Stocks made with larger and thicker bones, such as a beef stock, require more cooking time than stocks made with small, thin bones. A longer cooking time is needed to extract the nutrients of dense bones than is needed for small bones. Beef stocks usually simmer for about 8 hours. A chicken stock made with many small, less dense bones usually cooks for about 4 hours. A fish stock made from thin, fragile fish bones usually requires a cooking time of approximately 45 minutes. Vegetable stocks only require about 30 minutes of cooking time. **See Figure 6-5.**

Cooking Times for Stocks

Type	*Cooking Time**
Beef	8 hours
Chicken	4 hours
Fish	45 minutes
Vegetable	30 minutes

* *approximate*

Figure 6-5. Required cooking times for stocks vary depending upon the size of the bones used.

3. **Simmer stock.** Simmering stock at a temperature between 180°F and 185°F helps draw flavor and nutrients from the bones. Boiling a stock (even if only for a short time) breaks up coagulated impurities as they are released from bones and causes the impurities to break up and cloud the stock.

4. **Do not stir while simmering.** The solid components of a stock should not be disturbed. Stirring the stock during cooking causes impurities in and around the solid ingredients to be broken up throughout the stock and results in cloudiness.
5. **Remove impurities from the top.** The solid material that rises to the surface of a stock should be skimmed off continuously during the cooking process. Care should be taken not to disturb any material remaining on the surface while draining the stock. Many stockpots have a spigot drain that makes it easy to drain all liquid out through the bottom of the stockpot without needing to tip the pot. This gentle draining of unstrained stock helps ensure that impurities remain in the pot and are not disturbed, which would cause them to break up and cloud the stock. If a stockpot with a drain spigot is not available, a large ladle is used to gently remove the stock.
6. **Strain stock thoroughly.** A chinois or a cheesecloth-lined china cap is used to strain stock. The cheesecloth helps remove small particles that can cloud the stock.
7. **Cool stock rapidly.** Stock is first cooled to 70°F and then to below 41°F before being stored. The most common methods for cooling stocks are in an ice bath and with a cooling paddle. When using an ice bath, the hot stockpot is placed on top of an inverted perforated hotel pan inside a clean sink filled with ice cubes and cold water. Then, the stock is stirred to speed the cooling process.

   Another method for cooling a stock is to use a cooling paddle. A *cooling paddle* is a heavy-gauge, hollow plastic paddle with a screw-on cap at the top of the handle that is filled with water and frozen prior to use. **See Figure 6-6.** The cooling paddle is inserted into the hot stock and used to stir the stock until it is cooled.

Proper Cooling of Stocks

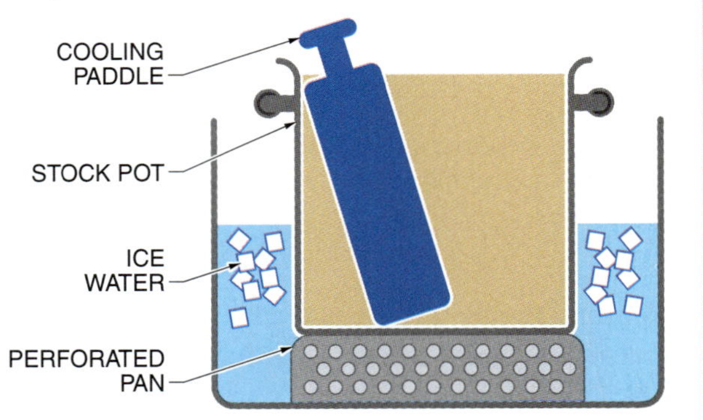

Figure 6-6. Strained stock should be cooled using proper cooling techniques.

8. **Follow proper storage procedures.** When a stock has cooled, any remaining fat will have risen to the surface as a solid and can be easily removed. Many chefs use this solidified fat in place of butter or oil when sautéing meats to add additional flavor to various dishes. Once the fat is removed, the stock is covered, labeled, dated, and stored in a refrigeration or freezer unit.

The way a stock is prepared determines the outcome of the stock. Some stocks are made by simply simmering uncooked ingredients together. Other stocks include ingredients that are roasted prior to use, which produces a rich, brown color and roasted flavor. The taste of a stock should be identifiable, but should not overpower the taste of the dish that it is used in, such as a soup or a sauce.

Brown Stocks

A *brown stock* is a stock produced by simmering roasted meat, poultry, or game bones with mirepoix and an optional tomato product. When roasting the bones, they must be placed in a single layer in an oiled roasting pan. The bones should be roasted in a 400°F oven for 45–60 minutes and stirred occasionally to produce even browning. The roasted bones are then transferred to a stockpot, covered with cold water, and brought to a simmer. The mirepoix is caramelized in the same roasting pan and then added to the pot. The roasted bones and vegetables add the rich, brown color, flavor, and high-gelatin content that are characteristic of a well-made brown stock.

Preparing Brown Stock
Media Clip

Procedure for Preparing Brown Stocks

1. Roast bones in a roasting pan until evenly brown. Transfer the roasted bones to a stockpot and cover with cold water. Reserve the rendered fat in the roasting pan.
2. Begin heating the contents of the stockpot. Then, sauté the mirepoix in the reserved rendered fat until it is well caramelized. Stir the mirepoix continuously to avoid burning.
3. Pour off excess fat and reserve for later use.

4. If desired, add a small amount of tomato sauce or paste to the mirepoix and cook until the tomato product caramelizes.
5. Deglaze the roasting pan.

6. Once the water in the stockpot has reached a simmer, skim the impurities from the surface and then add the contents of the roasting pan to the stockpot.
7. Return the contents of the stockpot to a simmer and continue cooking and skimming impurities from the surface until done. Do not let the stock boil.

8. Strain the stock with a chinois or cheesecloth-lined china cap.
9. Quickly cool the strained stock in an ice bath or with a cooling paddle and refrigerate or freeze.
10. Label and date the stock and refrigerate or freeze until needed.

White Stocks

A *white stock* is a light-colored stock produced by gently simmering poultry, veal, or fish bones in water with vegetables and herbs. A white mirepoix is used to avoid a possible color change in the stock. Chefs often disagree about whether or not bones should be blanched prior to being used in a white stock. Some chefs think that blanching removes many impurities from the bones. Other chefs think that blanching bones wastes flavor and nutrients. It is common practice to simply rinse the bones in clean, cold water to remove any loose impurities that are present on the surface of the bones before using them to prepare a white stock.

Production Tip

Using a white stock instead of water when cooking grains will add flavor without drastically changing their colors.

Procedure for Preparing White Stocks

1. Rinse the bones in cold water and place in a stockpot. Completely cover the bones with cold water.
2. Bring water to a simmer. Continually skim off impurities as they rise to the surface.
3. Add a white mirepoix and a sachet to the pot. Maintain a simmer and continually skim off impurities.
4. Strain the stock with a chinois or cheesecloth-lined china cap.
5. Quickly cool the strained stock in an ice bath or with a cooling paddle.
6. Label the stock with the correct product name and date before storing in the refrigerator or freezer.

Fish Stocks

A *fish stock* is a basic stock prepared by adding fish bones or shellfish shells, vegetables, a sachet, and cold water to a stockpot and bringing it to a simmer over medium heat. The resulting stock is clear and mild in flavor. The cooking time for fish stocks is much shorter than white or brown stocks because fish bones and shellfish shells are thinner and more delicate than the bones used for other stocks. Bones used for a fish stock should be those from a lean fish, such as sole or snapper, as opposed to an oily fish, such as salmon. **See Figure 6-7.** Fish heads with the eyes and gills removed may also be used.

Fish bones and shellfish shells must be completely rinsed prior to use in a fish stock in order to make a clear stock. Vegetables used in a mirepoix for a fish stock are cut very small due to the quick cooking time. Often mushroom stems or trimmings are added to fish stocks to increase the flavor and aromatic characteristics.

Procedure for Preparing Fish Stocks

1. Rinse fish bones or shellfish shells in cold water to remove any surface debris.
2. Place the bones or shells, a small diced mirepoix (including mushroom pieces if desired), a sachet, and cold water in a stockpot over medium heat and bring to a simmer.
3. Simmer for 30–45 minutes, using caution to never let the stock reach a boil. Continually skim the surface to remove impurities.
4. Strain the stock carefully with a chinois or cheesecloth-lined china cap.
5. Quickly cool the strained stock in an ice bath or with a cooling paddle.
6. Label and date the stock before storing in the refrigerator or freezer.

Fish Stocks

Figure 6-7. Bones used for a fish stock should be from a lean fish.

Fumets

A *fumet* is a concentrated stock made from fish bones or shellfish shells and vegetables. A fumet differs from a fish stock in that the mirepoix, bones, and shells are sweated in a stockpot over low-to-medium heat prior to use. Care is taken to not discolor the vegetables, bones, or shells in the sweating process in order to produce a flavorful yet relatively colorless stock. Fish stock or water and a sachet are added to the sweated ingredients, and wine and/or lemon juice is added for flavor.

Procedure for Preparing Fumets

1. Rinse fish bones and shellfish shells in cold water to remove any surface debris.
2. Melt a small amount of butter in a stockpot over low-to-medium heat. Add onions, parsley stems, thyme, mushroom pieces, garlic, trimmings, bones, and shells and slowly sweat them, being careful not to allow them to gain color.
3. When the onions are translucent, deglaze the pot with a small amount of lemon juice and white wine.
4. Add fish stock or water and a few lemon slices and bring to a simmer.
5. Simmer for about 30 minutes. Continually skim the surface to remove impurities.
6. Quickly cool the stock in an ice bath or with a cooling paddle.
7. Label and date the stock and refrigerate or freeze until needed.

Essences

An *essence* is a concentrated fish stock that includes a large amount of aromatic ingredients such as celery, morels, a bouquet garni, a sachet, and fennel root. The flavor of an essence is more concentrated than a fumet and extremely aromatic.

Vegetable Stocks

A *vegetable stock* is a clear, light-colored stock produced by gently simmering vegetables with a white mirepoix and a sachet. A vegetable stock can be used in place of a meat-based stock in many recipes. This substitution option is helpful when a chef chooses to offer a more nutritious sauce or soup. Almost any combination of vegetables may be used, but strongly flavored vegetables such as asparagus and spinach should be avoided when making an all-purpose vegetable stock. **See Figure 6-8.** Starchy vegetables such as potatoes should also be avoided unless a cloudy stock is not a concern.

Vegetable Stocks

All-Clad Metalcrafters

Figure 6-8. A vegetable stock can be used in place of a meat-based stock in many recipes.

Glaces

A *glace* is a highly reduced stock that results in an intense flavor. For example, one gallon of a stock can be slowly reduced to one-eighth or one-tenth of its original volume to yield just one or two cups of glace.

Procedure for Preparing Glaces

1. Simmer the stock slowly to allow a slow reduction. Continually check the bottom of the saucepan to ensure that the reduction does not begin to scorch or burn.
2. Continually skim any surface impurities.
3. As the stock reduces, strain and transfer it into progressively smaller pans. Straining the stock each time it is transferred helps ensure the final glace is as free from impurities as possible.
4. When the reduction is finished, strain the glace carefully with a chinois or cheesecloth-lined china cap.
5. Quickly cool the strained glace in an ice bath.
6. Label and date the glace before storing in the refrigerator or freezer.

Once completely cooled, glace has a rubbery texture due to the concentrated amount of gelatin that results from the reduction process. The glace can then be cut into cubes, which are easy to store. **See Figure 6-9.** It is important to avoid handling glace with bare hands as this will cause the glace to spoil quickly. Improper storage and handling can greatly shorten the amount of time that glace can be stored. When properly prepared, stored, and handled, a glace will last for several months. Classic glaces are added to soups or sauces to intensify their flavor.

Glaces

Figure 6-9. Cold glace has a rubbery texture that allows it to be easily cut into cubes and stored for future use.

Remouillages

Remouillage means "rewetting" in French. A *remouillage* is a stock made from using bones that have already been used once to make a stock. Used bones are placed in water with vegetables and aromatics and simmered to extract any additional flavor they may contain. The resulting stock is less flavorful than the original stock made from the same bones. A remouillage is often used in place of water when starting a new stock as it has more flavor than plain water.

Bouillons

Bouillon is the liquid that is strained off after cooking vegetables, poultry, meat, or seafood in water. It is not a stock, but bouillon can form the base for sauces and soups.

Likewise, court bouillon is also not a stock. *Court bouillon* is a highly flavored, aromatic vegetable broth made from simmering vegetables with herbs and a small amount of an acidic liquid (usually vinegar or wine). Court bouillon is used to poach fish, seafood, and vegetables.

CHECKPOINT 6-1

1. Identify the components of a stock.
2. List eight guidelines for preparing stocks.
3. Describe how to prepare brown stock.
4. List the main ingredients of a white stock.
5. Prepare a brown stock and a white stock. Compare the appearance and flavor of each.
6. Describe how to prepare fish stock.
7. Prepare a fish stock. Evaluate the appearance and flavor of the stock.
8. Explain how a fumet differs from an essence.
9. Describe how to prepare vegetable stock.
10. Prepare a vegetable stock. Evaluate the appearance and flavor of the stock.
11. Describe two different techniques for storing a glace.
12. Explain the primary use of a remouillage.
13. Prepare a glace and a remouillage. Compare the appearance and flavor of the two stocks.
14. Describe a bouillon and a court bouillon.
15. Contrast the appearance and flavor of a bouillon and a court bouillon.

CONVENIENCE PRODUCTS

Many chefs use bases instead of preparing house-made stocks, soups, or sauces. These products are available as either ready-to-use (RTU) products or as a base. Convenience products are becoming more common but have not eliminated the need for house-made preparations. Stock, soup, and sauce bases are used to save labor costs.

Convenience products are available in several different forms. The most common forms include pastes and liquid, dehydrated, and frozen products. **See Figure 6-10.** One form is not necessarily superior to another. The best guide when evaluating the quality of bases is to taste the product to ensure it meets the flavor desired for the dish being prepared. The first indicator of how natural the flavor of a base may be is the amount of sodium per serving. This information is available on the side of the container, box, or bag. The amount of sodium may be as low as 85 mg and sometimes more than 1000 mg. It is important to remember that salt may be added to a dish but it cannot be removed from a finished product.

Stock Convenience Products

A wide variety of bases are available to quickly prepare any type of stock. The most readily available form of stock bases is the paste form. There are three basic reasons to use a stock base. First, when preportioned meats and seafood are used, trimmings or bones are not available for making stocks. Instead, bases can provide the added flavor to create a stock. They also can be used if there is an insufficient amount of house-made stock. Lastly, a stock base can also be added to enhance flavor in place of a reduction.

Soup Convenience Products

Convenience soups are available in cans, frozen vacuum-packed containers, or dry mixes that need only the addition of liquid and heat. Fresh ingredients are often added to a soup base prior to service. For example, many produce houses carry precut vegetables that can be added to a soup base to make a fresher product. Soup bases can be customized by adding fresh herbs and spices to enhance the flavor profile of a soup. Also, adding fresh meats, poultry, or seafood can yield a fresher flavor.

Sauce Convenience Products

Canned tomato sauces are the most widely accepted form of sauce base used in the professional kitchen. Sometimes sauce bases are in the form of a powder. The advantage of powdered bases is long shelf life and minimal space required for storage. Some powdered sauce bases have a shelf life of two years. The modified starches that are used in bases allow sauces to be held longer and at a higher temperature without breaking. Sauce bases may be used as a finished product or may be improved by the addition of fresh ingredients.

Convenience Products

Figure 6-10. The most common forms of convenience products include pastes and liquid, dehydrated, and frozen products.

CHECKPOINT 6-2

1. Name the four most common forms of convenience products used to prepare stocks, soups, and sauces.
2. Name the two ways of determining how natural a base will taste.
3. Name the three basic reasons to use a stock base.
4. Identify the three advantages of using a sauce base.
5. Prepare a stock, soup, or sauce using a base. Prepare the same type of house-made stock, soup, or sauce. Compare the appearance and flavor of the two stocks, soups, or sauces.

SOUP BASICS

A *soup* is a food that is made with a stock and pieces of meat, poultry, seafood, and/or vegetables. Many soups are produced by using a high-quality stock as a base and adding other ingredients to make the dish unique. Soups can be made using fresh and seasonal ingredients or high-quality leftovers such as the meat removed from baked chicken. **See Figure 6-11.**

Chicken Soups

National Chicken Council

Figure 6-11. Many soups are produced by using a high-quality stock as a base and adding other ingredients to make the dish unique.

Nutrition Note

To reduce the fat added to some soups, yogurt may be substituted for heavy cream or sour cream.

Each soup ingredient must be properly prepared and added in the appropriate sequence to attain a balance of flavors. All of the ingredients need to complement one another. The stock or base should be flavorful, but not strong enough to mask the flavors of the other ingredients.

A garnish may be cooked into a soup or added before service. Common soup garnishes include vegetables, meats, pastas, grains, croutons, cheese, herbs, sour cream, and unsweetened whipped cream. **See Figure 6-12.** A soup garnish supplies flavor and enhances the presentation.

Garnishes

All-Clad Metalcrafters

Figure 6-12. A garnish may be cooked into the soup or added before service.

National Turkey Federation

Different varieties of soups are characterized by the overall finished appearance and consistency of the soup and the preparation and cooking methods used to make the soup. Most soups can be classified as clear, thick, or specialty. Specialty soups include bisques, chowders, and gumbos, as well as unique regional varieties. Although many specialty soups are thick in consistency, they are produced using different methods than those used to make traditional thick soups.

Soup is one of the most versatile categories of food and allows chefs to be creative while demonstrating classical preparation techniques. Countless ingredient combinations are used to create soup varieties in every culture around the world. **See Figure 16-13.**

Soup Varieties Around the World

Soup	Origin
Asopao de Gandules	Puerto Rico
Avgolemono	Greece
Borscht	Ukraine
Bouillabaisse	France
Bredie	South Africa
Caldo Verde	Portugal
Callaloo	Trinidad and Tobago
Erwtensoep	Netherlands
Fanesca	Ecuador
Fruktsuppe	Norway
Fufu and Egusi	Ghana
Gazpacho	Spain
Ginataan	Philippines
Goulash	Hungary
Harira	North Africa
Menudo	Mexico
Minestrone	Italy
Miso	Japan
Pho	Vietnam
Pozole	Colombia
Sambar	India
Shchi	Russia
Shark Fin	China
Tarator	Bulgaria
Tom Yam	Thailand
Trahana	Turkey
Vatapá	Brazil
Vichyssoise	France
Zurek	Poland

Figure 6-13. A wide variety of ingredients are used to create soups in every culture around the world.

Soup Recipes

CHECKPOINT 6-3

1. Identify the most common garnishes used with soups.
2. Name the three general classifications of soups.
3. Locate recipes for at least two of the soups listed on the "Soup Varieties Around the World" chart.

CLEAR SOUPS

A *clear soup* is a stock-based soup with a thin, watery consistency. Clear soups include broths and consommés. Broths, which are produced from well-made stocks, can be made from meat, poultry, seafood, or vegetables. When meat is used, a broth will have a much more concentrated flavor and color than a stock made from bones. Consommés are made from high-quality broths that have been further clarified to remove all impurities and surface fat.

Broths

A *broth* is a flavorful liquid made by simmering stock along with meat, poultry, seafood, or vegetables and seasonings. A stock is used as the liquid ingredient in a broth, as opposed to water. Meat or vegetables may be added to the stock and simmered to create a more intense flavor, color, and overall body. A broth can be served as a finished product or made into a broth-based soup or consommé.

Tougher cuts of meat or poultry are typically used to produce a full-flavored, rich broth. In addition, not allowing the broth to boil vigorously, continually skimming impurities from the surface, and carefully straining the cooked broth produces a broth of excellent quality.

The mirepoix used to make a broth may be cut larger or smaller depending on the time required to cook the main flavoring ingredient. For example, if the broth contains a cut of meat that requires a lengthy cooking time, a large-cut mirepoix can be added halfway through the cooking process. Beef broth, for example, typically simmers for 2 or more hours. However, if the main flavoring ingredient cooks more quickly, a smaller mirepoix can be added at the beginning of the cooking process. Chicken broth, for example, is typically done within 30–40 minutes. The size of the mirepoix should be such that it is cooked thoroughly in the allotted time, but not overcooked.

If a broth is needed for service immediately, the completed broth is brought to a boil. Seasonings are adjusted to taste. Fat droplets on the surface of the broth can be removed with a clean paper towel. The broth is then garnished and served.

Procedure for Preparing Broths

1. Cut the main flavoring ingredient to desired size or truss if needed.

2. If a dark color and roasted flavor are desired, carefully brown the meat and the mirepoix. If no color is desired, simply sweat the mirepoix.

3. Completely cover the main ingredient and mirepoix with cold stock. Add a sachet d'épices.

4. Bring to a simmer and continually skim off impurities as they rise to the surface.
5. Reduce heat to a low simmer and cook until the main ingredient is tender and the desired flavor has developed. Occasionally taste the broth to monitor progress.

6. Strain the broth with a chinois or cheesecloth-lined china cap.

7. Quickly cool the strained broth using an ice bath or a cooling paddle.
8. When the broth is cold, remove excess fat and grease.

Consommés

A *consommé* is a very rich and flavorful broth that has been further clarified to remove any impurities or particles that could cloud the finished product. The flavor of the main ingredient should be pronounced, and the color of the consommé should be very rich. Consommés made from poultry should have a deep golden or amber color, while consommés made from red meat should have a dark, roasted meat color. It is very important that a consommé be made with the highest-quality ingredients and be properly prepared in a nonreactive pot, such as stainless steel, that is taller than it is wide. A finished consommé is only as good as the broth and ingredients used.

Consommés were once used as a way to display a chef's skill to people of nobility or wealth. A perfectly clear consommé showed that the chef had refined soup-making skills. Classical consommés can be made with any meat, poultry, or seafood and are often named after their main flavoring ingredient. There are countless appropriate garnishes for a consommé. It is important to precisely prepare the garnish because it can be seen through a clear consommé. **See Figure 6-14.**

Garnished Consommé

Figure 6-14. It is important to precisely prepare the garnish because it can be seen through a clear consommé.

Although the clarification process adds a small bit of flavor, the bulk of a consommé's flavor is derived from the quality of the broth. To *clarify* is to remove impurities, sediment, cloudiness, and particles to leave a clear, pure liquid. For this to occur, very cold stock or broth is combined with a mixture known as a clearmeat. A *clearmeat* is a cold, lean ground meat, fish, or poultry that is combined with an acid (such as wine, lemon juice, or a tomato product), ground mirepoix, egg whites, and an oignon brûlé. **See Figure 6-15.** An *oignon brûlé* is half an onion that is charred on the cut side. The term oignon brûlé is French for "burnt onion." An oignon brûlé is used to give an intense roasted flavor and deeper color to a darker-colored consommé.

Clearmeats

Figure 6-15. A very cold stock or broth is combined with a mixture known as a clearmeat.

Clarifying Consommés. To clarify a consommé, the clearmeat is mixed well and then whisked into the cold broth and brought to a low simmer. As the temperature increases, the mixture is occasionally stirred until the clearmeat begins to coagulate. Once the clearmeat begins to form a solid mass, the mixture cannot be disturbed or stirred as the clearmeat could break apart and cloud the consommé. As the clearmeat gently simmers, it rises to the surface and, in doing so, strains the impurities from the consommé. When the clearmeat rises, it is known as a raft. A *raft* is the clearmeat that has risen to the surface. In addition to clarifying the consommé, the raft enhances the flavor, color, and richness of the final consommé. **See Figure 6-16.**

Rafts

Figure 6-16. When the clearmeat rises, it is known as a raft.

Clarifying Consommés
Media Clip

Consommés

Procedure for Clarifying Consommés

1. Prepare the clearmeat by combining freshly ground lean meat, mirepoix, and egg whites with an acid (such as a tomato product or lemon juice) and an oignon brûlé.
2. Blend the cold clearmeat well with a small amount of cold broth to loosen the mixture. Both the clearmeat and broth should be between 32°F and 35°F. A small amount of crushed ice can be added to bring down the temperature, if needed.
3. In a heavy-bottomed stockpot (preferably with a drain spigot), combine the clarification mixture and cold broth. Stir well.
4. While occasionally stirring, bring the mixture to a simmer over medium heat. The clearmeat will begin to form a solid mass, or raft. Do not allow to boil.
5. Once a solid mass forms, stop stirring and gently poke a hole in the top of the newly formed raft. This hole allows the progress of the consommé to be monitored as it cooks.
6. Allow the consommé to gently simmer for 1–2 hours until a desired flavor and color are achieved. Occasionally baste the raft with consommé to prevent the raft from drying or breaking apart.
7. Carefully strain the consommé through a chinois or china cap lined with multiple layers of moistened cheesecloth. Degrease completely.

Once the raft forms, a hole is gently poked into the top of it. This hole allows steam to escape without disrupting the raft. It also allows the progress of the consommé to be monitored as it cooks. After gently simmering for 1–2 hours, the consommé is removed through the hole in the raft using a ladle or from the bottom of a stockpot through the drain spigot. The consommé is then strained through multiple layers of dampened cheesecloth to remove any small impurities that may not have been trapped by the raft. The consommé should be cooled immediately and stored properly. Once completely cooled, any remaining fat will solidify on the surface and can be easily removed. The resulting liquid should be a clear and intensely flavored consommé.

Rescuing Cloudy Consommés. On rare occasions, factors such as accidentally boiling the consommé instead of simmering, stirring the consommé after the raft has formed, and starting the consommé with clearmeat and broth that are not cold enough can cause the final consommé to be cloudy. When this happens, it is necessary to conduct a secondary clarification.

To perform a secondary clarification on a gallon of consommé, a mixture consisting of 4 lightly beaten egg whites, ¼ cup of finely ground mirepoix, and either 1 tbsp tomato purée or 2 tsp lemon juice is added to the consommé. The mixture is added to the consommé and the clarification procedure is repeated to remove any remaining impurities.

CHECKPOINT 6-4

1. Name two types of clear soups.
2. Describe the procedure for preparing broths.
3. Prepare a broth. Evaluate the appearance and flavor of the broth.
4. Describe the function of an oignon brûlé.
5. Describe the procedure for clarifying consommés.
6. Clarify a consommé. Evaluate the clarity and flavor of the consommé.
7. Identify factors that contribute to an unsuccessfully clarified consommé.

THICK SOUPS

A *thick soup* is a soup having a thick texture and consistency. **See Figure 6-17.** The two basic types of thick soups are cream soups and purée soups. The main difference between a cream soup and a purée soup is that most cream soups are thickened by an added starch, such as the flour in a roux. Purée soups, on the other hand, are thickened by puréeing all of the ingredients. Additionally, purée soups are generally thicker and coarser than cream soups and are seldom strained. Due to the slightly lower starch content of vegetable purée soups, rice or potatoes are sometimes added to increase the texture and add body to the soup.

Thick Soups

Daniel NYC / T. Schauer

Figure 6-17. A thick soup is a soup having a thick texture and consistency.

Cream Soups

Cream soups are traditionally made using one of two preparation methods. The first method uses velouté, and the second uses a roux. These two preparation methods work in much the same way, and the choice of which to use is a matter of preference. Regardless of the method chosen, the main ingredients need to be simmered until tender in order to be puréed smooth.

Velouté Preparation Method. The velouté preparation method consists of sweating a white matignon and the flavoring ingredient, such as broccoli for a cream of broccoli soup, in fat.

Next, a velouté (a white stock thickened with a roux) is added and the mixture is simmered until the flavoring ingredient is tender. Then the mixture is puréed. Finally, the soup is seasoned to taste and finished with hot cream.

Procedure for Preparing Cream Soups (Velouté Method)

1. In a heavy-bottomed saucepot, sweat a white matignon and the flavoring ingredient in fat until slightly tender.

2. Add prepared velouté sauce and simmer until the flavoring ingredient is completely tender, stirring often.

3. Carefully purée the soup until smooth.

4. Strain through a chinois or cheesecloth-lined china cap.

5. Return strained soup to a simmer and thin with stock if needed.
6. Season to taste and finish with hot cream.

7. Serve immediately.

Preparing Cream Soups (Velouté Method)
Media Clip

Healthy Cream Soups

Roux Preparation Method. The roux preparation method consists of sweating a white matignon and the flavoring ingredient in fat. Next, flour is added to the mixture to make a roux. Then hot stock is added to the roux, and the mixture is allowed to simmer and thicken. Finally, the soup is puréed until smooth, seasoned to taste, and finished with hot cream.

Procedure for Preparing Cream Soups (Roux Method)

1. In a heavy-bottomed saucepot, sweat a white matignon and the main flavoring ingredient in enough fat to also make the roux.
2. When the aromatics and main ingredient are slightly tender, add enough flour to make a roux.
3. Cook the roux just long enough to achieve a pale golden color and then add warm white stock while whisking vigorously. Bring to a slight simmer and cook until the taste of flour is gone and the flavoring ingredient is completely tender.
4. Carefully purée the soup until smooth.
5. Strain through a chinois or cheesecloth-lined china cap.
6. Return strained soup to a simmer and thin with stock if needed.
7. Finish with hot cream and season to taste.
8. Serve immediately.

Purée Soups

Purée soups are made by cooking vegetables such as potatoes, squash, turnips, or carrots or dried legumes such as lentils, beans, or split peas in a broth until tender. Preparing vegetables in a relatively consistent size ensures uniform cooking. The cooked ingredients are puréed. **See Figure 6-18.** Often a portion of the main ingredient is reserved and added into the purée to add texture to the soup. Purée soups can have a completely smooth texture for a sophisticated look or a coarse texture for a more rustic look. Purée soups can contain many different ingredients, from roasted vegetables and chilies to bacon, salt pork, and smoked ham.

Production Tip

Reserve some of the main ingredient that is to be puréed and use as a garnish.

Preparing Cream Soups (Roux Method) Media Clip

Purée Soups

Figure 6-18. The cooked ingredients are puréed.

Procedure for Preparing Purée Soups

1. Per the recipe, either sweat or lightly brown the mirepoix in fat. Add the cooking liquid and bring to a simmer.
2. Add the main ingredients and sachet d'épices. Simmer until the main ingredients are tender enough to purée. Do not allow the soup to scorch or burn on the bottom.
3. Remove the sachet d'épices and reserve about 20% of the liquid.
4. Purée the remainder of the ingredients until the desired texture is achieved. Use the reserved liquid to thin the soup as needed.
5. Return the soup to a simmer and season to taste.
6. Garnish and serve immediately.

Purée soups tend to further thicken if they are made in advance and stored. When it is time to reheat, a small amount of hot water, stock, broth, milk, or cream may be added to thin the soup to the desired consistency. If a purée soup is too thin, a small amount of roux, beurre manié, or cornstarch slurry may be whisked in and the soup brought to a simmer.

Melissa's Produce

Preparing Purée Soups Media Clip

CHECKPOINT 6-5

1. Name two types of thick soups.
2. Describe the difference between the velouté method and the roux method of preparing cream soups.
3. Prepare two cream soups using the same main ingredient. Use the velouté method to prepare one cream soup and the roux method to prepare the other cream soup. Compare the appearance, flavor, and texture of the two cream soups.

SPECIALTY SOUPS

Many soups do not fall into the traditional categories of clear or thick soups. These other soups are considered specialty soups and include bisques, chowders, gumbos, and cold soups.

Nutrition Note

Soups contain carbohydrates, protein, fat, vitamins, and minerals in varying proportions. The amount of fat and sodium in house-made soups can be adjusted by using fat-free unflavored yogurt, fat-free sour cream, or half-and-half and a wide variety of herbs and spices.

Bisques

A *bisque* is a form of cream soup that is typically made from shellfish. The name bisque traditionally applies to soup made from shellfish and thickened with rice. Today bisques are commonly thickened with a roux or cream instead of rice to achieve a smoother and richer consistency. Many puréed soups are referred to as bisques. Bisques are typically prepared using either a roux preparation method or a cream preparation method.

Roux Preparation Method. The roux preparation method consists of caramelizing an aromatic mirepoix in a small amount of fat. Then flavoring ingredients are added and cooked until they caramelize. Next, tomato paste is added to the pan and lightly sautéed. The pan is deglazed with white wine, brandy, or cognac. A sachet d'épices and a white stock are added before simmering until a strong flavor and color are achieved. Roux is then whisked in to thicken the bisque. Finally, the bisque simmers for 15–20 minutes or until the taste of flour is gone.

Procedure for Preparing Bisques (Roux Preparation Method)

1. In a heavy-bottomed saucepot, caramelize the mirepoix in a small amount of fat.
2. Add flavoring ingredients and cook until caramelized.
3. Add a small amount of tomato paste to the pan and sauté lightly.
4. Deglaze the pan with a small amount of white wine, brandy, or cognac.
5. Add sachet d'épices and a white stock. Simmer for 20–25 minutes or until a strong flavor and color is achieved.
6. Add roux to thicken while whisking vigorously.
7. Simmer gently for 15–20 minutes or until the taste of flour is gone.
8. Remove the sachet and purée the bisque.
9. Strain the bisque through a chinois or cheesecloth-lined china cap.
10. Return strained soup to saucepot and bring to a simmer. If needed, thin the bisque with hot stock and bring to a simmer again.
11. Finish with hot cream and season to taste.
12. Garnish and serve immediately.

Cream Preparation Method. The cream preparation method involves sweating a mirepoix in a small amount of fat. Then flavoring ingredients are added and cooked until they caramelize. Next, tomato paste is added to the pan and lightly sautéed. The pan is deglazed with white wine, brandy, or cognac. A sachet d'épices and small amount of white stock are added before allowing the bisque to simmer until reduced by half. Heavy cream is added before simmering until the soup is reduced to a nappe consistency. The sachet d'épices is removed and the soup is puréed slightly. The puréed soup is strained and then brought to a simmer.

Procedure for Preparing Bisques (Cream Preparation Method)

1. In a heavy-bottomed saucepot, caramelize the mirepoix in a small amount of fat.

2. Add flavoring ingredients and cook until caramelized.

3. Add a small amount of tomato paste to the pan and sauté lightly.

4. Deglaze the pan with a small amount of white wine, brandy, or cognac.

5. Add sachet d'épices and a small amount of white stock. Simmer 15–20 minutes or until reduced by half.

6. Add heavy cream. Simmer gently until reduced to a nappe consistency.

7. Remove the sachet and purée the bisque.

8. Strain the bisque through a chinois or cheesecloth-lined china cap.

9. Return strained soup to saucepot and bring to a simmer. If needed, thin the bisque with hot stock and bring to a simmer again. Season to taste.

10. Garnish and serve immediately.

Preparing Bisques (Cream Method)
Media Clip

Chowders

A *chowder* is a hearty soup that contains visibly large chunks of the main ingredients. Although most cream-based chowders are thick, a few chowders are broth-based and therefore thinner. The primary difference between cream soups and chowders is that chowders are not puréed. The majority of cream-based chowders are thickened with a roux.

Florida Department of Agriculture and Consumer Services, Bureau of Seafood and Aquaculture Marketing

Procedure for Preparing Chowders

1. Render diced bacon or salt pork until it begins to caramelize.

2. Add mirepoix and sweat until onions are translucent.

3. Add flour to make a roux.

4. Add stock or flavoring liquid.
5. Add the main flavoring ingredients and a sachet d'épices.

6. Simmer until main ingredients are fully cooked.
7. Remove the sachet.

8. Finish with hot cream.
9. Serve immediately.

Preparing Chowders
Media Clip

Gumbos

A *gumbo* is a rich Creole soup made of broth, onions, celery, green peppers, okra, tomatoes, and rice. **See Figure 6-19.** Creole cuisine refers to the flavorful, French-influenced foods originating from Louisiana. Gumbos are similar to vegetable soups and chowders in appearance. Typically, they also include seafood, chicken, or sausage. The stock is very flavorful and complements the main ingredient. For example, seafood gumbos are made with a seafood stock, while chicken gumbos are made with a chicken stock.

A dark roux, a filé powder, and/or okra are the ingredients most often used to thicken a gumbo. A dark roux imparts a deep rich flavor. *Filé powder* is a leaf herb that is made from the ground leaves of the sassafras tree. Okra contains a juice that thickens when it is cooked.

Gumbos

Figure 6-19. A gumbo is a rich Creole soup made of broth, onions, celery, green peppers, okra, tomatoes, and rice.

Gumbo

Cold Soups

There are two basic types of cold soups, those that require cooking the main ingredients and those that use raw ingredients that have been puréed. Cold soups include traditional cold cream soups such as vichyssoise, a cold potato and leek soup. Cold soups can also be made from a purée of fruit or vegetables and yogurt. For example, a refreshing cantaloupe soup can be made by puréeing fresh cantaloupe with mint and yogurt. Many cold soups are made using fruit or vegetable juice.

CHECKPOINT 6-6

1. Identify four types of specialty soups.
2. Describe the difference between the roux method and the cream method used to prepare bisques.
3. Prepare two bisques using the same main ingredient. Use the roux method to prepare one bisque and the cream method to prepare the other bisque. Compare the appearance, flavor, and texture of the two bisques.
4. Describe how to prepare a chowder.
5. Prepare a chowder. Evaluate the appearance, flavor, and texture of the chowder.
6. Name three ingredients often used to thicken gumbos.
7. Prepare a gumbo. Evaluate the appearance, flavor, and texture of the gumbo.
8. Identify common types of cold soups.
9. Prepare a cold soup. Evaluate the appearance, flavor, and texture of the cold soup.

SAUCE BASICS

Many chefs consider a sauce to be the unifying component of a dish. A *sauce* is an accompaniment that is served with a food to complete or enhance the flavor and/or moistness of a dish. Sauces also add texture to a dish and can be the ingredient that pulls a dish together. **See Figure 6-20.** Sauces provide a complementary coating, rather than disguising the food. A sauce should not overpower the flavor of the food with which it is served. The ability to pair a less expensive entrée with a well-executed sauce enables chefs to be more creative while maintaining cost control. Adding an excellent sauce to a good dish creates something extraordinary.

Several key ingredients are used to make a sauce, including liquids, flavoring components, and thickening agents. Liquids, which may be stocks, broths, dairy products, or juices, form the base that is used in making sauces. When making sauce, a decision needs to be made about what liquid to use to achieve the flavor desired.

Sauces

National Turkey Federation

Figure 6-20. Sauces can be the ingredient that pulls a dish together.

Proper sauce preparation yields a smooth sauce with a nappe consistency. *Nappe* is the consistency of a liquid that thinly coats the back of a spoon and ensures that a sauce will cling lightly to another food. **See Figure 6-21.** To achieve a nappe, sauces are either thickened by reduction or by the addition of a thickening agent. Simmering can reduce a sauce by up to three-fourths of its original volume. This reduction results in concentrated flavor and lessens or eliminates the need for a thickening agent.

Nappe

Figure 6-21. Nappe is the consistency of a liquid that is thick enough to coat the back of a spoon.

Flavoring Components

Adding flavoring components to a sauce requires knowledge of how they will affect the flavor of the sauce. The flavoring components must never overpower the other ingredients used to make a sauce, except in preparations that require a dominant flavor, such as curries. Spices, herbs, vegetables, and wine are common flavoring components.

Spices such as curry, paprika, and dry mustard should be worked into a roux or dissolved in a liquid for a more uniform distribution and a smoother sauce. If whole herbs or spices are used, they should be removed from the sauce when the desired flavor is obtained. If left in, the herbs or spices will continue to disperse flavor and potentially overpower the sauce. Fresh or dried herbs may be used in a sauce. Fresh herbs are normally tied into a bouquet garni or a sachet and are removed from the finished sauce.

Barilla America, Inc.

Nutrition Note

Tomatoes are an excellent source of vitamins, minerals, and the antioxidant lycopene. Cooked tomato products and tomato juice contain high amounts of lycopene. Tomatoes are also very low in sodium, fat, and cholesterol.

Using vegetables such as onions or mushrooms in a sauce provides a desirable flavor. However, it is important to prevent the onions from overpowering the flavor of the sauce. A mild onion flavor can be obtained by using leeks or chives. Mushrooms are added to sauces for their unique flavor. The earthy flavor of mushrooms combines well with a wide range of flavors.

Adding wine to a sauce at the end of the cooking process produces the best results. Some sauce recipes call for wine to be added earlier in order to reduce and therefore concentrate the flavor. A recipe should always be carefully followed when adding wine to a sauce.

Thickening Agents

A *thickening agent* is a substance that adds body to a hot liquid. For example, a thickening agent is added to chicken stock to create chicken gravy. Common thickening agents include flour, cornstarch, arrowroot, roux, beurre manié, and a liaison. The first three of these thickeners are starches, which undergo gelatinization when added to a liquid. *Gelatinization* is the process of a heated starch absorbing moisture and swelling, which thickens the liquid. **See Figure 6-22.** When using a thickening agent, the final sauce should be strained prior to serving. It is always possible that some of the thickening agent did not dissolve completely. Straining removes any unappetizing lumps from the finished sauce.

Gelatinization

Figure 6-22. The starch granules absorb water and swell, thickening the liquid.

Starch Thickeners

Flour. Flour is the primary thickening agent that is often combined with fat before being added to a liquid. Pastry and cake flour are optimal choices when making products that are used to thicken sauces and soups.

Cornstarch. *Cornstarch* is the white, powdery, pure starch derived from corn. Unlike roux, which is only used to thicken hot foods, cornstarch can be used to thicken hot or cold foods. Cornstarch has double the thickening power of flour, so less is needed to thicken a liquid. However, cornstarch-thickened sauces have a tendency to break down if held over heat for long periods of time. It is not recommended to reheat a sauce thickened with cornstarch.

Cornstarch must be made into a slurry before it can be used to thicken a liquid. A *slurry* is a mixture of equal parts of cool liquid and a starch that is used to thicken other liquids. Making a slurry allows the starch granules to absorb liquid most efficiently. If cornstarch alone is sprinkled into a hot liquid, the starch granules will clump together and form lumps. These lumps cannot be whisked or completely dissolved and have to be removed by straining.

The slurry is then added to either a hot or cold liquid with a whisk and heated thoroughly. Once the mixture reaches a boil, the slurry thickens to its full potential. After reaching a boil, it takes about 5 minutes of simmering to completely cook out the taste of the cornstarch. A cornstarch slurry is often used to thicken sauces and gravies because a slurry does not require fat and is therefore less expensive.

Arrowroot. *Arrowroot* is a thickening agent that is the edible starch from the rootstock of the arrowroot plant. Like cornstarch, it is mixed with a cool liquid before being heated or added to hot liquids. However, a sauce thickened with arrowroot is clearer than a sauce made with cornstarch. Arrowroot is flavorless, but more expensive than cornstarch. Arrowroot, potato flour, and tapioca can be substituted as gluten-free thickeners.

Roux. A *roux* is a thickening agent made by cooking a mixture of equal amounts, by weight, of flour and fat and is used to thicken sauces and soups. Pastry flour and cake flour are the most effective types of flour to use in a roux as they both contain more starch than bread flour. All-purpose flour, which is a combination of pastry flour and bread flour, may also be used.

Production Tip

Many chefs make rouxs as part of their mise en place to save time and ensure a consistent flavor profile.

Cooking flour in fat coats the starch granules in fat. This coating of fat helps prevent the starch from lumping when added to a hot liquid. If flour is added by itself to a liquid such as a hot stock, clumps of flour will stick together as the starch goes though the gelatinization process. Once flour begins to clump, these improperly gelatinized flour pieces will never be broken down, no matter how much the mixture is whisked. When a roux is made properly, by cooking the flour and fat together first, the fat-coated starch granules dissolve smoothly in the hot stock.

There are three types of roux made in the professional kitchen: white, blonde, and brown. **See Figure 6-23.**

- White roux is made by cooking equal amounts, by weight, of fat and flour until a slightly foamy appearance is achieved. It is then immediately removed from the heat and allowed to rest for 5–10 minutes before it is added to a liquid. A white roux has a faint aroma of freshly baked bread. White roux is used in white sauces when no color is desired.

Procedure for Preparing Roux

1. Using a heavy-bottomed saucepan to prevent scorching, heat fat until hot.
2. Add an equal amount, by weight, of sifted pastry or cake flour to the hot fat.

3. Stir to form a pasty consistency while cooking over medium heat.

4. Continue stirring until desired color of roux is achieved.

Types of Roux

Figure 6-23. The three types of roux made in the professional kitchen are white, blonde, and brown.

Preparing Roux
Media Clip

- Blonde roux is made by cooking the same mixture of fat and flour until the roux exhibits a slightly golden or tan color. A blonde roux also develops an aroma similar to freshly popped popcorn. Blonde roux is used in sauces that have a slight golden or tan color such as those made with chicken or white veal stock. Blonde roux not only adds some of its color to a sauce but also adds a slightly toasted flavor from starches in the flour beginning to caramelize.
- Brown roux is made by cooking the same mixture of fat and flour for an even longer period of time until the roux changes to a deep brown color and develops a nutty aroma. Due to its intense flavor and color, a brown roux is used with sauces that are made with a rich veal or beef stock. It is important to use caution when making a brown roux to avoid overcooking it. A roux that is too dark has an unpleasant, bitter taste and will leave black flecks in a prepared dish.

Adding a roux to a liquid can be a challenging task. Adding the roux correctly results in a smooth and silky sauce that has a nappe consistency. Using the proper proportions of flour and fat is the first step. **See Figure 6-24.** When adding roux to a liquid, the roux and liquid must be at different temperatures. If they are both either extremely hot or extremely cold, the roux may develop lumps and may have a floury taste.

Proportions of Roux Ingredients to Liquid

Sauce Consistency	*= (Flour**	*+ Fat*	*= Roux)*	*+ Liquid*
Light	6 oz	6 oz	12 oz	1 gal.
Medium	8 oz	8 oz	16 oz	1 gal.
Heavy	10 oz	10 oz	20 oz	1 gal.

Light refers to the consistency of most fine and delicate sauces.

Medium refers to the consistency of a cream soup or standard smooth gravy.

Heavy refers to the consistency of a binding sauce, such as the sauce used in a turkey pot pie.

* *The above proportions ratios are for use with pastry flour or cake flour. If using all-purpose or bread flour, more flour may be needed to achieve the same results.*

Figure 6-24. A sauce can be made light, medium, or heavy depending on the amount of roux added in proportion to the amount of liquid.

There are two accepted methods for adding roux, and the appropriate method depends upon the temperature of the roux. With a hot roux, use a cold or room-temperature stock. The roux should be in a pot large enough to incorporate the cold stock. The cold stock is carefully added to the hot roux while whisking vigorously. If using a roux that is at room temperature or colder, a hot stock should be used. Carefully add room-temperature roux to the hot stock while whisking vigorously. Many restaurants make large batches of roux and store it for future use, rather than making individual batches frequently.

Beurre Manié. Beurre is the French word for butter. A *beurre manié* is a thickening agent made by kneading equal amounts, by weight, of pastry or cake flour and softened butter and can be whisked into a sauce just before service. Unlike roux, a beurre manié is not cooked. Instead, the flour and butter are kneaded together until a smooth texture is achieved. The mixture is then separated into small balls and whisked into a sauce to finish it. The butter component adds additional sheen and richness to the sauce. The process of kneading butter and flour together coats each starch granule with fat, allowing the starch to better absorb moisture without lumping. A beurre manié is often used to thicken a small amount of sauce, as when making a pan sauce, while a roux is more often used to thicken larger quantities of sauce.

Production Tip

When added to a dish, beurre manié should be simmered for 3–5 minutes to reach maximum thickening with no apparent flour taste.

Liaison. A *liaison* is a thickening agent that is a mixture of egg yolks and heavy cream used to thicken sauces. Because a liaison does not contain a starch ingredient, it does not use the process of gelatinization. A liaison is typically used for cream- or milk-based dishes such as custards, puddings, or rich cream sauces.

A liaison is made using a 1:3 ratio, composed of 1 part egg yolk and 3 parts heavy cream. The yolk and cream are whisked together after the liquid to be thickened has been heated to a simmer and removed from the heat. A small amount of the heated liquid is slowly added to the yolk mixture causing coagulation. *Coagulation* is the process of a protein changing from a liquid to a semisolid or a solid state when heat or friction is applied. Egg yolks begin to coagulate at temperatures above 145°F, but the addition of cream raises the coagulation point of the liaison to approximately 180°F. The difference in temperature between the liaison and a hot liquid can cause the yolks to coagulate. Therefore, only a small amount of the hot liquid to be thickened is whisked into the yolk mixture.

Procedure for Adding a Liaison

1. Heat liquid to be thickened to a simmer. Do not overheat the liquid.
2. Remove the liquid from the heat.

3. Whisk together 1 part egg yolk and 3 parts heavy cream.

4. Add a small amount of the heated liquid to the liaison while whisking rapidly. Continue adding the hot liquid while whisking until the liaison is fairly warm.

5. Add the warmed liaison into the remainder of the heated liquid while whisking vigorously.

6. Carefully heat the thickened sauce to a temperature between 140°F and 185°F.

Note: A thickened sauce should have a nappe consistency.

After the liaison has been whisked well, it is added to the hot liquid while whisking vigorously. If the liaison-thickened sauce needs to be heated, the temperature should be kept between 140°F and 185°F. The sauce must be heated to a minimum of 140°F to cook the egg yolks to a safe temperature for consumption. The liaison should not be heated above 185°F as the yolks will coagulate to the point of appearing curdled. If the yolks coagulate, the liaison must not be used or the sauce will also appear curdled.

Production Tip

Pasteurized eggs are used in many kitchens. The pasteurization process takes place before the egg is packaged, either in the shell or as a liquid egg product. The use of pasteurized eggs and egg products reduces the potential of a foodborne illness being contracted from eating foods that contain uncooked or partially cooked eggs. Pasteurized eggs are a good option for preparing egg dishes that are not fully cooked.

Adding a Liaison
Media Clip

CHECKPOINT 6-7

1. Contrast common thickening agents used to prepare sauces.
2. Explain the purpose of a slurry.
3. Describe how to prepare a roux.
4. Contrast the different uses for white, blonde, and brown roux.
5. Prepare a white, blonde, and brown roux. Evaluate each roux for appearance, flavor, and texture.
6. Contrast a beurre manié and a roux.
7. Prepare a slurry using cornstarch and another slurry using arrowroot. Compare the slurries.
8. Describe how to add a liaison to a liquid.
9. Demonstrate how to add a liaison to a liquid.

CLASSICAL SAUCES

Most classical sauces are composed of a flavorful liquid, a thickening agent, and additional flavorings and seasonings. Classical sauces are often referred to as mother sauces and small sauces. A *mother sauce,* also known as a leading or grand sauce, is one of five sauces from which the small classical sauces described by Escoffier are produced. The five mother sauces are béchamel, velouté, espagnole, tomato, and hollandaise. With the exception of hollandaise, these sauces are rarely used by themselves.

Sauce Pairings

Béchamel

Béchamel, also known as a cream sauce or a white sauce, is a mother sauce that is made by thickening milk with a white roux and seasonings. Traditionally, a béchamel was made by adding hot, heavy cream to a thick, reduced white veal stock. Today, the simpler version of béchamel prepared with scalded milk and a white roux is used in many vegetable and gratin dishes.

A well-made béchamel should be silky smooth and have a nappe consistency. It has a very mild flavor with a subtle taste of onion. The onion taste comes from the onion piquet that is added to the milk and then removed after the milk is scalded. An *onion piquet* is half of an onion studded with cloves and a bay leaf. **See Figure 6-25.** Then, a white roux is whisked in. The sauce is finished with a little nutmeg and seasoned to taste. A béchamel can be made lighter or heavier by adding slightly less or slightly more roux. Béchamel is strained prior to use in a dish.

Onion Piquet

Figure 6-25. An onion piquet is half of an onion studded with cloves and a bay leaf.

With the addition of different seasonings, ingredients, and garnishes, a simple but well-made béchamel can be transformed into countless small sauces. For example, a Mornay sauce is made by adding grated cheese to a béchamel, while a soubise sauce is made by adding puréed cooked onions to a béchamel.

Velouté

Velouté is a mother sauce made from a flavorful white stock (either veal, chicken, or fish stock) and a blonde roux. Traditionally, only veal stock was used. Today, chicken stock is typically used instead of veal stock because of its versatility and low cost. Fish stock is not often used because the flavor may overpower a dish. No matter what type of stock is used, a velouté should have a very smooth and silky texture similar to a béchamel. A velouté is typically made into a middle, or intermediate, sauce before being used for a small sauce. There are two middle sauces, suprême and allemande.

- A *suprême sauce* is a chicken velouté with the addition of cream. A suprême sauce can be wonderful by itself. It also is used as a binding sauce in many casseroles.
- An *allemande sauce* is made by adding fresh lemon juice and a yolk-and-cream liaison to velouté. The liaison makes allemande a rich, wonderful sauce.

Suprême and allemande sauces can be used to make a large number of small sauces. Because both middle sauces are derived from velouté, almost any small sauce made from one type of middle sauce can also be made with the other type. For example, an aurora sauce is made from a velouté with a tomato purée, a bercy sauce is made from a fish velouté, and a poulette sauce is made from an allemande sauce flavored with mushroom essence.

Espagnole

Espagnole is a mother sauce made from a full-bodied brown stock, brown roux, tomato purée, and a hearty caramelized mirepoix. An espagnole is typically made into a middle sauce such as a demi-glace or a jus lié.

- A *demi-glace* is a sauce made by adding equal parts of espagnole and brown stock together and reducing the mixture by half. A demi-glace is the most widely used middle sauce. This richly flavored reduction is often finished with a small amount of Madeira or sherry to further boost its flavor. A properly made demi-glace should be a nappe consistency when hot and somewhat rubbery when cold. Espagnole is converted into a demi-glace before being made into one of many small sauces. **See Figure 6-26.**
- A *jus lié,* also known as a fond lié, is a sauce that is made by thickening a brown stock either by adding a cornstarch or arrowroot slurry or simply by a slow reduction.

Some chefs prefer a jus lié to a demi-glace because it does not have the taste of roux and is easier to make. A disadvantage of a jus lié is that it will only be as rich and flavorful as the brown stock from which it was produced because it is thickened with only a cornstarch or arrowroot slurry. In comparison, a demi-glace has a richer taste because of the butter in the roux. When completed, a jus lié is not quite as thick as a demi-glace.

Tomato Sauce

A *tomato sauce* is a mother sauce made by sautéing mirepoix and tomatoes, adding white stock, and thickening with a roux. A well-made tomato sauce is thick, flavorful, and tastes like ripe tomatoes. It is coarser than the other classical sauces, but should have similarly rich characteristics that allow it to cling to the foods it is served with. Typically, plum (Roma) tomatoes are used to make tomato sauce, as they have less water, fewer seeds, and more meat than other varieties. Tomatoes are most often peeled and seeded before use in the sauce.

Some chefs do not use a roux to thicken a tomato sauce. Instead, tomato sauces are either thickened with a slurry, or are puréed or ground in a food mill, resulting in a hearty sauce. A frequent problem with modern sauces, however, is that water can separate from the puréed pulp and form an unattractive pool beneath pasta on a prepared plate. This can be avoided by using a thickening agent such as a roux or a slurry, which helps by causing the water that separates from the tomato pulp to bind with the starch. A properly thickened sauce makes an attractive presentation on a pasta dish, such as spaghetti. **See Figure 6-27.**

Espagnole Sauces

Sauce	Description
Bordelaise	Demi-glace flavored with red wine, herbs, and shallots
Châteaubriand	Demi-glace flavored with white wine, shallots, tarragon, cayenne, and lemon juice
Madeira	Demi-glace flavored with shallots, Madeira wine, and butter
Périgueux	Demi-glace flavored with Madeira wine and truffles
Robert	Demi-glace flavored with onions, white wine, vinegar, and Dijon mustard

Entourage

Figure 6-26. Espagnole is converted into a demi-glace before being made into one of many small sauces.

Spaghetti with Tomato Sauce

Barilla America, Inc.

Figure 6-27. A properly thickened sauce makes an attractive presentation on a pasta dish.

Many chefs rely heavily on the flavor of seasonal, ripe tomatoes cooked for a relatively short period of time to produce a rich and fresh tomato sauce. Others use canned tomato products sautéed with mirepoix, roasted pork or veal bones, and rendered bacon and cook the sauce for a long time over low heat. Some tomato sauces are chunky, while others are puréed smooth.

Many chefs also add a small amount of gastrique to a tomato sauce to mask the acidic qualities of the sauce. *Gastrique* is a sugar syrup made by caramelizing a small amount of granulated sugar in a saucepan and deglazing the pan with a small amount of vinegar. The gastrique does not reduce the acid in a tomato sauce, but merely softens the harshness of the acidic qualities. To reduce acid in a tomato sauce, baking soda can be added while the sauce simmers. Tomato sauce immediately foams when baking soda is added, because of chemical reactions taking place as the soda neutralizes some of the acid. When the foam disappears, the acid in the sauce has been reduced.

Tomato sauce can be made into many small sauces, the majority of which are still called tomato sauce. Tomato sauce variations have an ethnic description attached to identify the sauce, such as Creole, Milanaise, Portuguese, or Spanish.

Hollandaise

Hollandaise is a mother sauce made by thickening melted butter with egg yolks. **See Figure 6-28.** A classical hollandaise is started with a flavorful reduction of dry white wine, white wine vinegar, minced shallots, and freshly cracked peppercorns, reduced over medium heat. Hollandaise relies on an emulsification rather than coagulation and is therefore different from other sauces thickened with yolks. *Emulsification* is the process of temporarily binding two liquids that do not combine easily, such as oil and vinegar.

Hollandaise

Figure 6-28. Hollandaise is a mother sauce made by thickening melted butter with egg yolks.

For example, emulsification occurs when warm, clarified butter is rapidly whisked into heat-tempered egg yolks. **See Figure 6-29.** Egg yolks contain a natural emulsifier called lecithin, which coats each fat droplet, suspending the fat in the mixture. When hollandaise is made properly, each yolk should completely emulsify 4 oz of clarified butter. The resulting sauce should be light and silky, a very pale lemon color, and should have a well-rounded flavor that is more complex than that of just butter.

Emulsification

Figure 6-29. Hollandaise relies on emulsification rather than coagulation.

Procedure for Preparing Classical Hollandaise Sauce

1. Add dry white wine and white wine vinegar, minced shallots, and freshly cracked peppercorns to a saucepan. Reduce over medium heat until almost all of the liquid has evaporated.
2. Strain and slightly cool the reduction.
3. Transfer the reduction to a stainless steel bowl. Add egg yolks and whisk vigorously until light and foamy.
4. Add a small amount of warm water (1 oz for every 6 yolks) and whisk to incorporate.
5. Place the mixture over a gently simmering water bath and whisk rapidly until the mixture reaches approximately 145°F and has at least doubled in volume. *Note:* The bowl may be momentarily removed from the water bath to allow a few seconds of thorough whisking without overcooking the yolks.
6. Remove the bowl from the water bath and place on a moistened towel. Begin gradually whisking in warm clarified butter very slowly in a fine, steady stream. As the clarified butter is incorporated, the mixture will begin to thicken. If the sauce begins to get too thick, add a small amount (1–2 tsp) of warm water to thin the sauce slightly. Continue whisking in the clarified butter until about 4 oz of clarified butter per yolk has been added.
7. Add a small amount of lemon juice to finish the sauce.
8. Serve immediately.

When making a hollandaise for the first time, the sauce may break and appear thin, curdled, or separated. This most often happens either because the yolks are heated at too high of a temperature, resulting in overcooked yolks, or because butter is added too quickly and does not emulsify properly. Sometimes if a hollandaise is kept too cold it becomes thick but does not break until it is served over a warm menu item, such as grilled asparagus.

To repair a broken hollandaise, the first step is to determine whether it is broken because of overcooked yolks or from holding a completed sauce at too low a temperature. If the hollandaise is broken because of overcooked yolks, add 1 yolk to a clean mixing bowl and whisk with 1 tsp of warm water until thick. Slowly add the broken hollandaise to the yolk while whisking rapidly to incorporate. Continue adding broken sauce to the mixture until all of the broken hollandaise has been incorporated.

Production Tip

Hollandaise should be held for no more than 2 hours at 145°F. If the sauce is held above 150°F, the emulsification breaks down because egg yolks curdle and solidify at higher temperatures. If the sauce temperature falls below 100°F, the butter will solidify and the sauce will break.

If the hollandaise breaks from being held at too low of a temperature, reheat the sauce over a double boiler while whisking continuously until the sauce is soft and warm. In a separate stainless steel bowl, add 2 tsp of warm water and a few tablespoons of the broken sauce. Whisk rapidly to emulsify and then slowly whisk in the remainder of the broken hollandaise sauce until all of the sauce has been added.

A béarnaise sauce is very similar to a hollandaise, except for the tarragon flavor. Béarnaise sauce can also be used to make small sauces such as a choron sauce, which is made by blending a béarnaise with a tomato purée. Hollandaise sauce can also be used to make small sauces, such as a mousseline sauce, by blending in unsweetened heavy cream or a Maltaise sauce by blending in Maltese orange juice and grated orange rind.

CHECKPOINT 6-8

1. Describe the five classical mother sauces.
2. Name the most widely used middle sauce.
3. Describe the procedure for preparing a hollandaise sauce.
4. Prepare a hollandaise sauce. Evaluate the sauce for appearance, flavor, and texture of the sauce.

BUTTER SAUCES

Not all sauces are derived from the classical mother sauces. There are also many sauces made from butter. Butter sauces may be as simple as browned butter or as complex as an emulsion. Three common types of butter sauces are compound butters, beurre blanc, and broken butters.

Compound Butters

A *compound butter* is a flavorful butter made by mixing cold, softened butter with flavoring ingredients such as fresh herbs, garlic, vegetable purées, dried fruits, preserves, or wine reductions. Typically, compound butters are prepared, rolled into a cylinder shape on a sheet of plastic wrap or parchment paper, and refrigerated until needed. When needed, a ¼–½ inch thick disk is cut and then placed on top of a freshly cooked item. The dish is served immediately so that the butter finishes melting in front of the customer.

Procedure for Preparing Compound Butters

1. In a food processor, combine and process the flavoring ingredients.
2. Add whole butter, salt, and pepper, and purée until all ingredients are well incorporated.
3. Remove butter and place on a sheet of parchment or plastic wrap. Roll the mixture into a cylinder, approximately 1–1½ inch thick.
4. Place in refrigerator until needed, or freeze for later use.

Compound butters are often prepared in advance and held in the refrigerator for a few days or in the freezer for longer periods. Compound butters may be served alone or over grilled meats, fish, or vegetable dishes. They can also be added to other sauces for additional flavor. Although the variations of compound butters are virtually endless, there are several common preparations. **See Figure 6-30.**

Beurre Blanc

A *beurre blanc,* also known as white butter sauce, is a butter-based emulsified sauce made by whisking cold, softened butter into a wine, white-wine vinegar, shallot, and peppercorn reduction. The butter becomes the sauce, while the shallots and wine are the primary flavoring ingredients. Beurre blanc sauce is lighter and thinner than hollandaise but slightly thicker than heavy cream. It is often served with fish or seafood.

In the traditional preparation, after a fish is poached, a small amount of the poaching liquid is reduced with shallots, white wine, and a small amount of white-wine vinegar. The reduction is removed from the stove and small pieces of whole butter are whisked into the reduction. Trace amounts of lecithin and other emulsifiers naturally found in butter help to create an emulsification. The key to a successful beurre blanc sauce is constantly whisking while adding the butter to the reduction.

The simplicity of a beurre blanc makes it a popular sauce for many different dishes. A beurre blanc sauce can be easily transformed into similar sauces by substituting or adding different flavoring ingredients.

Compound Butters	
Butter	***Preparation***
Almond	Finely grind 4 oz of almonds until very smooth; add 8 oz of whole butter and mix well
Anchovy	Soak 1 oz of anchovies in cold water for 15 minutes to remove excess oil and salt; pat anchovies with paper towels to dry; place anchovies in food processor with 8 oz of whole butter; purée until smooth
Bercy	Reduce 1 cup of white wine and 2 minced shallots until only about 2 tbsp of liquid remain; cool mixture; add 8 oz of whole butter and 2 tbsp of diced parsley; purée until smooth
Colbert	Place 1 tbsp of chopped fresh tarragon, 3 tbsp of glace de viande, 3 tbsp of chopped fresh parsley, and 1 tsp of lemon juice in a food processor and purée until smooth
Herb	Add 2–4 tbsp of desired chopped fresh herb to 8 oz of whole butter and purée until smooth
Lobster, shrimp, or crayfish	In a meat grinder, grind 6 oz of lobster, shrimp, or crayfish shells with 8 oz of whole butter; place in small saucepan over low heat until butter melts; keep on very low heat for 30 minutes and strain; chill the melted butter until solid; season to taste and purée until smooth
Maître d'hôtel	Place 3 tbsp of freshly chopped parsley, 1 tsp of fresh lemon juice, and 8 oz of whole butter in a food processor and purée until smooth (*Note:* This is the most commonly used compound butter.)
Marchand de Vin	Combine 1 cup of dry red wine, 1 tbsp of glace de viande, 1 minced shallot, 1 clove garlic, and 6 cracked black peppercorns; reduce until about 2 tbsp of liquid remain and chill; blend in a food processor with 8 oz of whole butter

Figure 6-30. Compound butters can be made with flavoring ingredients such as fresh herbs, garlic, vegetable purées, dried fruits, preserves, or wine reductions.

Procedure for Preparing a Beurre Blanc

1. In a small saucepan, combine dry white wine, white wine vinegar, shallots, and a few whole peppercorns.
2. Over medium heat, reduce the mixture until only a few tablespoons of liquid remain. If too much of the reduction remains, the resulting sauce will be too thin.
3. Add tablespoon-size pieces of cold butter over a very low heat while whisking constantly. *Note:* Cold butter is whisked into the mixture to keep the sauce between 100°F and 120°F. At temperatures above 120°F, the emulsification can break down. At temperatures below 100°F, the sauce becomes too thick.
4. Strain the sauce with a chinois or cheesecloth-lined china cap.
5. If desired, finish with a small amount of warm heavy cream.

Broken Butters

A *broken butter* is a butter sauce made by heating butter until the fat, milk solids, and water separate or "break." Broken butters are finished with a small amount of an acid, such as lemon juice or wine vinegar. Broken butter sauces are easy to prepare and are intended to be made à la minute, reusing the pan that an item was cooked in. By reusing the pan, flavors from the dish are incorporated into the sauce. A classic dish served with a broken butter is Dover sole à la Meunière.

Meunière Sauce

Almost any flavoring ingredient can be added to a broken butter sauce to change the flavor profile. For example, chopped herbs, julienned leeks, crushed garlic, or sliced truffles can each be added to give a broken butter sauce an exciting flavor. Two common broken butters are beurre noisette and beurre noir.

- Beurre noisette is commonly referred to as "nut-brown butter" because of its color and nutty aroma. To prepare a beurre noisette, whole butter is cooked over medium heat until the white milk solids begin to turn brown. As the solids change color, they add a toasted nut flavor to the butter. The butter is then allowed to cool slightly and fresh lemon juice is added. The juice from one lemon is sufficient for one cup of butter.
- Beurre noir is commonly referred to as "black butter." In a beurre noir, the lemon juice is replaced with wine vinegar and the butter is cooked a bit longer than a beurre noisette, allowing the milk solids to turn a rich, dark brown. Different kinds of vinegar may be used, however, sherry vinegar, champagne vinegar, and balsamic vinegar provide the best overall flavors.

Fortune Fish Company

Preparing a Beurre Blanc
Media Clip

CHECKPOINT 6-9

1. Contrast the three common types of butter sauces.
2. Describe how to prepare a compound butter.
3. Prepare a compound butter, a beurre blanc, and a broken butter. Compare the appearance, flavor, and texture of each butter sauce.
4. List the three main components of a beurre blanc sauce.
5. Explain what is meant by "broken" butter.

CONTEMPORARY SAUCES

Although classical sauces and butter sauces remain popular, many contemporary sauces have gained acceptance on menus around the globe. Some contemporary sauces include gravies, coulis, and nages.

Gravies

A *gravy* is a sauce that has the same flavor as the food it accompanies when served. Gravies are typically prepared from the drippings and juices of roasted meats. **See Figure 6-31.** Meat drippings and juices evaporate during the roasting period. Supplementing the drippings with a brown sauce can increase the flavor and volume. The brown sauce is prepared by browning and boiling bones of the type of animal that is being roasted. Brown sauce also enhances the gravies of meats such as pork and veal, which have delicate flavors.

Brown Gravies

Figure 6-31. A gravy is a sauce that has the same flavor as the food it accompanies.

Salt and pepper are the primary seasonings added to gravies because they enhance natural flavors. A hint of cloves or rosemary in gravy will enhance the flavor of pork, while marjoram enhances lamb. The best thickener to use when preparing gravy is roux.

Coulis

A *coulis* is a sauce typically made from either raw or cooked puréed fruits or vegetables. A coulis can be served either warm or cold over grilled or sautéed items, as well as desserts. The texture of a coulis can range from fairly smooth to slightly coarse. A fruit coulis is usually made from frozen fruit that is puréed with simple syrup and a hint of lemon or vanilla.

Coulis

Nages

A *nage* is an aromatic court bouillon that is often used as a finishing sauce. A nage is prepared by reserving and straining the liquid in which the main item was cooked. Vegetables and aromatics are added to the broth, and the broth is reduced by up to half of the original volume. The resulting liquid is an aromatic broth that is often served to accompany fish and shellfish. Some chefs choose to clarify the broth, while others may whisk in whole butter or a bit of heavy cream to add richness. Only a small amount of a nage is needed to add moisture to a dish. **See Figure 6-32.**

Nages

Figure 6-32. As a finishing sauce, a nage adds an aromatic quality to a dish.

CHECKPOINT 6-10

1. Name the type of sauce used to supplement the flavor and volume when preparing a gravy.
2. Prepare two different types of gravies. Compare the appearance and flavor of the two gravies.
3. Describe three ways that a coulis can be served.
4. Prepare a coulis. Evaluate the appearance, flavor, and texture of the sauce.
5. Describe how a nage is prepared.
6. Prepare a nage. Evaluate the appearance, flavor, and texture of the sauce.

Daniel NYC/N. Clutton

Refer to DVD for **Quick Quiz®** questions

Pantry & Cold Foods Station

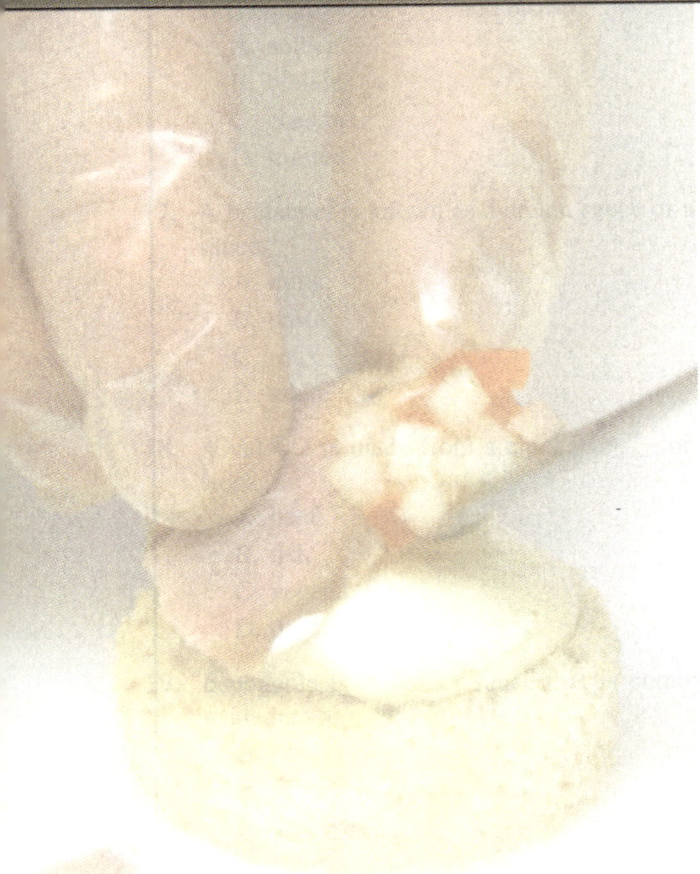

Introduction

In the seventh station of the ACFEF apprenticeship program, you will learn that there is more to cold food production than preparing salad greens. Pantry and cold foods items include, but are not limited to, herbs, spices, seasonings, vinegars, oils, salad greens, nuts, and cheeses. Techniques for keeping foods and serving vessels cold will become second nature. You will also draw upon your knowledge of fruits and vegetables as you create salads using a variety of flavorings and seasonings as well as ingredients such as fruits, starches, poultry, seafood, meats, legumes, eggs, cheeses, and nuts. The production of salad dressings will test your skill at making emulsions and balancing flavors to complement a variety of salads.

You will gain a better understanding of cheese by learning to identify different types of cheeses and how they ripen. You will also learn how to configure sandwich components in order to create a signature sandwich. Then you will prepare a variety of sandwiches, condiment sauces such as aioli, and cold soups.

You will be introduced to charcuterie and prepare items such as pâtés, terrines, and sausages. You will also prepare showpiece cold platters using aspic and chaud froid. With practice, you will be able to decorate buffets with fruit-and-vegetable sculptures and ice sculptures.

Sustainability practices will focus on purchasing, storage, and waste reduction. You will learn when it is best to purchase produce locally and whether or not a local source for nonfood items has value.

Kolpak

Carlisle FoodService Products

Garde manger is a French term used to describe both the chef and the professional kitchen station responsible for preparing cold pantry items such as salads, salad dressings, cold soups, cheeses, sandwiches, condiment sauces and dips, hors d'oeuvres, and appetizers. The cold kitchen is also responsible for buffet presentations such as cold platters, charcuterie items, and display pieces. The mise en place for many additional dishes may also be prepped by the cold kitchen staff prior to being placed in full production at a hot station.

Browne Foodservice

FLAVORINGS AND SEASONINGS

A wide variety of ingredients are used to make salads, salad dressings, cold soups, sandwiches, condiment sauces and dips, hors d'oeuvres, appetizers, and charcuterie items. The complex flavor of each dish produced in the pantry and cold foods station is developed by layering and combining ingredients in different amounts at different times during the preparation process.

Flavors can be enhanced, highlighted, or balanced by adding ingredients at various stages of a recipe. A critical part of layering flavors is selecting ingredients that are complementary, meaning they contrast with one another. For example, salty complements sweet. These contrasts highlight the flavors present in each ingredient. In addition, flavorings and seasonings can be used to bring out the unique qualities of individual ingredients.

A *flavoring* is an item that alters or enhances the natural flavor of food. Herbs, spices, rubs, marinades, condiments, sauces, and nuts are considered flavorings. Herbs and spices are used to enhance, not disguise, the natural flavor of food. Exceptions to this rule include curry and chili because the defining characteristics of these dishes rely on the use of specific herbs and spices. Many herbs and spices should be added several hours before serving uncooked dishes, such as salad dressings, marinades, and fruit beverages, to allow time for flavors to develop.

A *seasoning* is an item that intensifies or improves the flavor of food. Salts, peppercorns, citrus zest, vinegars, and oils are called seasonings. Seasonings can be added closer to the end of the cooking process or at service because their effect is fairly immediate. Moderation is key when using seasonings. More can always be added but removal is impossible.

Combining flavors offers a better aroma, taste, and texture experience than can be achieved by eating the same foods separately. **See Figure 7-1.** Each ingredient should have a purpose in developing the flavor of the dish or it should be omitted. Chefs and cooks need to take every opportunity to bring out the unique qualities of each ingredient by using appropriate flavorings and seasonings.

Food	Flavoring	Food	Flavoring
Beef	Bay leaf, marjoram, nutmeg, sage, thyme	**Potatoes**	Dill, paprika, parsley, sage
Carrots	Cinnamon, cloves, marjoram, nutmeg, rosemary, sage	**Poultry**	Ginger, marjoram, oregano, paprika, poultry seasoning, rosemary, sage, tarragon, thyme
Corn	Cayenne, chervil, chives, cumin, curry powder, paprika, parsley	**Salads**	Basil, celery seed, chervil, chives, cilantro, dill, oregano, rosemary, sage, tarragon, thyme
Eggs	Basil, cilantro, cumin, savory, tarragon, turmeric	**Soups**	Bay leaves, cayenne, chervil, chili powder, cilantro, cumin, curry, dill, marjoram, nutmeg, oregano, rosemary, sage, savory, thyme
Fish	Curry powder, dill, dry mustard, lemon zest and juice, marjoram, paprika		
Fruits	Cinnamon, ground cloves, ginger, mace, mint	**Summer squash**	Cloves, curry powder, marjoram, nutmeg, rosemary, sage
Green beans	Dill, curry powder, lemon juice, marjoram, oregano, tarragon, thyme	**Tomatoes**	Basil, bay leaves, dill, marjoram
Lamb	Curry powder, rosemary, mint	**Veal**	Bay leaves, curry powder, ginger, marjoram, oregano
Peas	Ginger, marjoram, parsley, sage	**Winter squash**	Cinnamon, ginger, nutmeg
Pork	Sage, oregano		

Figure 7-1. Combining flavors offers a better aroma, taste, and texture experience than can be achieved by eating the same foods separately.

Herbs

An *herb* is a flavoring derived from the leaves or stem of a very aromatic plant. The majority of herbs are grown in temperate climates. Although dried herbs are available throughout the year, the flavors and aromas of herbs change slightly when dried. For this reason, chefs prefer to use fresh herbs. When used and stored effectively, herbs can be incorporated into dishes in ways that gratify the senses with pleasing flavors and aromas.

It is important to add fresh herbs at the appropriate time to ensure maximum flavor and aroma. **See Figure 7-2.** Fresh herbs should be tested for quality when delivered. They should have a bright, rich color and fresh appearance. A small amount of an herb can be rubbed in the palm of the hand and smelled. The scent should be fresh and fairly strong. Freshly cut herbs should be wrapped loosely in damp paper or cloth and placed in a plastic bag to prevent them from drying out and wilting. Properly wrapped fresh herbs should be refrigerated between 35°F and 45°F.

When substituting dried herbs for fresh herbs, a smaller amount of dried herbs should be used than the amount of fresh herbs called for in the recipe. Good color, robust flavor, and strong aroma are important points to consider when buying dried herbs. Dried herbs should be stored in airtight containers and out of direct light to help preserve flavors and aromas. Dried herbs tend to lose flavor and aroma the longer they are stored. The two general categories of herbs are leaf herbs and stem herbs.

Herbs

Fresh Herbs

Messermeister

Figure 7-2. It is important to add fresh herbs at the appropriate time to ensure maximum flavor and aroma.

Leaf Herbs. A *leaf herb* is a type of herb derived from the leaf portion of a plant. The size, color, and shape of the leaf can vary. Some leaves are delicate and fragile, some have curled or frayed edges, and some are smooth and elongated. In addition to a unique appearance, the leaf of each herb produces an equally unique flavor. There are many leaf herbs. Commonly used leaf herbs include basil, bay leaves, chervil, cilantro, curry leaves, dill, filé powder, lavender, marjoram, mint, oregano, parsley, rosemary, sage, savory, tarragon, tea, and thyme. **See Figure 7-3.**

Leaf Herbs	
Basil  *Shenandoah Growers*	*Basil* is a leaf herb with a pointy green leaf and is a member of the mint family. Sweet basil is the most common type of basil and is traditionally used in tomato sauces and pesto. It is often used in vinaigrettes, in infused oils, and to add flavor to pasta, vegetable, seafood, chicken, and egg dishes. Opal basil leaves are crinkled, vibrant plum in color, and slightly firmer and a little milder in flavor than sweet basil. Other varieties of basil have hints of lemon, garlic, cinnamon, clove, and even chocolate flavor.
Bay Leaves  *Shenandoah Growers*	A *bay leaf* is a thick, aromatic leaf herb that comes from the evergreen bay laurel tree. Bay leaves are used for flavoring soups, roasts, stews, gravies, and meats and for pickling. Bay leaves are always removed and discarded from a dish before serving.
Chervil  *HerbThyme Farms*	*Chervil* is a leaf herb with dark-green, curly leaves with a flavor similar to parsley but more delicate and with a hint of licorice. Chervil is used in salad, soup, egg, and cheese dishes and as a garnish.
Cilantro  *Shenandoah Growers*	*Cilantro*, also known as Chinese parsley, is a leaf herb that comes from the stem and leaves of the coriander plant. Cilantro has a distinct flavor that is slightly lemony. It is usually added just before serving because heating cilantro nearly destroys its flavor. Cilantro is often used to flavor salsas, soups, salads, and sandwiches.
Curry Leaves  *The Spice House*	A *curry leaf* is a leaf herb with a small, shiny green leaf from the curry tree. Curry leaves have a pungent aroma and slightly lemony flavor. Curry leaves are often used in Indian and Southeast Asian dishes.
Dill  *Shenandoah Growers*	*Dill*, also known as dill weed, is a leaf herb with feathery, blue-green leaves and is a member of the parsley family. Dill leaves have a slight licorice flavor. Dill is one of the primary herbs used in pickling foods. It may also be added to salads, vegetables, meats, and sauces.

Figure 7-3. (continued on next page)

Leaf Herbs	
Filé Powder 	*Filé powder* is a leaf herb that is made from the ground leaves of the sassafras tree. Its flavor resembles that of root beer and is known for its use as a thickener and flavor enhancer in gumbos. Prolonged cooking can cause filé powder to become stringy and tough so it should be added after a dish is removed from the heat.
Lavender 	*Lavender* is a leaf herb with pale-green leaves and purple flowering tops and is a member of the mint family. Both the leaves and flowers of lavender can be used. Lavender is most commonly known for its aromatic qualities and its mild, lemony flavor. It is used in classical French cuisine and often added to salads.
Marjoram  *Shenandoah Growers*	*Marjoram* is a leaf herb with short, oval, pale-green leaves and is a member of the mint family. Marjoram has a flavor that is similar to a cross between fresh oregano and thyme. Marjoram is used to flavor soups, stews, sausages, cheese dishes, and lamb dishes.
Mint 	*Mint* is a general term used to describe a family of similar leaf herbs, such as peppermint and spearmint. The bright-green leaves found on the purple-tinted stems of peppermint have a pungent mint flavor that is somewhat peppery. Spearmint has gray-green to pure-green leaves and is mild in aroma and flavor compared to peppermint. Mint is widely known for its cool, refreshing flavor and is used in sweet and savory dishes as well as beverages.
Oregano  *Shenandoah Growers*	*Oregano*, also known as wild marjoram, is a leaf herb with small, dark-green, slightly curled leaves and is a member of the mint family. Oregano has a pungent, peppery flavor that is slightly stronger than marjoram. It is often used in Italian and Greek dishes.
Parsley  *Tanimura & Antle®*	*Parsley* is a leaf herb with curly or flat dark-green leaves and is used as both a flavoring and a garnish. Italian parsley is called flat-leaf parsley. Parsley has a tangy flavor that is most prominent in the stems. Parsley can be used in a classic bouquet garni, soups, salads, stews, sauces, and vegetable dishes. Parsley is often used as a garnish.

Figure 7-3. (continued on next page)

Leaf Herbs	
Rosemary  Shenandoah Growers	*Rosemary* is a leaf herb with needlelike leaves and is a member of the evergreen family. Rosemary is one of the two strongest herbs, the other being sage. Fresh rosemary is considered more desirable than dried. The dried form is hard and brittle. Rosemary smells of fresh pine with a hint of mint and is used with grilled or roasted meats. Because its needles are secured to a branch, rosemary can easily be removed from a dish.
Sage  Shenandoah Growers	*Sage* is a leaf herb with narrow, velvety, green-gray leaves and is a member of the mint family. It can be purchased as whole leaves, crushed, or ground. Sage is often used to flavor stews, sausages, and bean or tomato preparations.
Savory  Shenandoah Growers	*Savory* is a leaf herb with smooth, slightly narrow leaves and is a member of the mint family. Summer savory is considered superior because it is harvested when the leaves are at peak quality, producing a delicate, spicy-sweet flavor. Savory is often used to flavor bean, egg, fish, and meat dishes as well as some baked products.
Tarragon  Shenandoah Growers	*Tarragon* is a leaf herb with smooth, slightly elongated leaves and is best known as the flavoring in béarnaise sauce. Like chervil, tarragon has a flavor that suggests a touch of licorice. Tarragon blends well with seafood, poultry, and tomato dishes and is often added to salads and salad dressings.
Tea 	*Tea* is a leaf herb with jagged leathery leaves and comes from an evergreen bush in the magnolia family. Black teas are the strongest in aroma and flavor, and green teas are the mildest. Other herbs, spices, and some fruits are often added to tea to create diverse blends. In addition to beverages, tea dust can be used to create sweet and savory dishes. Tea dust can be baked into desserts and used to flavor soups, sauces, and meats.
Thyme 	*Thyme* is a leaf herb with very small gray-green leaves and is a member of the mint family. Thyme is a key ingredient in a classic bouquet garni. Some varieties of thyme have a slight taste of nutmeg, lemon, mint, and sage. Thyme is often added to stews, chowders, and poultry and vegetable dishes.

Figure 7-3. A leaf herb is a type of herb derived from the leaf portion of a plant.

Seed Spices	
Coriander 	*Coriander* is a seed spice made from a light-brown, ridged seed of the coriander plant. It has a flavor similar to a blend of lemon, sage, and caraway. Coriander seeds are commonly used in pickling. Coriander powder is used in many baked products, hot dogs, sausages, and curry dishes.
Cumin 	*Cumin* is a seed spice made from a crescent-shaped amber, white, or black seed of the cumin plant and is a member of the parsley family. Cumin has a slightly bitter and warm flavor and is an essential ingredient in curry powder and chili powder. Cumin is used to flavor chili, soup, tamales, and rice and cheese dishes.
Fennel Seeds 	*Fennel seed* is a seed spice made from an oval, light-brown, and green seed of the fennel plant. Fennel seeds resemble anise in flavor and aroma. They are commonly used in Scandinavian dishes. Fennel seeds also enhance the flavor of roast duck, chicken, and pork dishes and are often used in sausage preparation.
Fenugreek  *The Spice House*	*Fenugreek* is a seed spice made from the pebble-shaped seed of the fenugreek plant in the pea family. Its flavor is similar to burnt sugar and leaves a bitter aftertaste. The orange color of fenugreek seeps into dishes in which the spice is used. Fenugreek is commonly used in Indian curries, chutneys, and vegetable preparations.
Mace 	*Mace* is a seed spice made from the lacy red-orange covering of the nutmeg kernel. Mace turns dark yellow when it is dried. Its flavor resembles that of nutmeg, although it is not as pungent. Mace is often used in pound cake, sweet dough, chocolate dishes, oyster stew, spinach, and pickling.
Mustard Seeds  *The Spice House*	A *mustard seed* is an extremely tiny seed from the mustard plant and is used as a spice. Mustard seeds have virtually no aroma but release a hot, pungent flavor when ground and mixed with hot water. Yellow, brown, and black mustard seeds are available whole and ground. Yellow mustard seeds release the mildest flavor, while black seeds are the most intense. Mustard seeds enhance the flavor of pickles, cabbage, beets, sauerkraut, sauces, salad dressings, ham, and cheese.
Nutmeg 	*Nutmeg* is a seed spice made from an oval, gray-brown seed found in the yellow, nectarine-shaped fruit of a large tropical evergreen. When the nutmeg fruit is ripe, the outer hull splits open, exposing the sister spice mace, which partially covers the nutmeg kernel. The sweet, warm, and spicy flavor of nutmeg can be used to enhance cream soups, potato dishes, sauces, and desserts.
Poppy Seeds  *The Spice House*	A *poppy seed* is a very small, blue-gray seed of the poppy plant. Poppy seeds have a mild, nutty flavor and are best known for garnishing breads. They may also be used in butter sauces for fish, vegetables, and noodles.

Figure 7-6. (continued on next page)

Seed Spices	
Sesame Seeds 	A *sesame seed* is a small, flat, white or black seed found inside the pod on a sesame plant and is used as a spice. Often baked on rolls, bread, and buns, sesame seeds supply a nutty flavor. The seeds are also toasted and stirred into butter and served over fish, noodles, and vegetables. Sesame seeds are ground to form tahini, or toasted and pressed to make sesame oil.
Star Anise 	*Star anise*, also known as Chinese anise, is a star-shaped seed fruit from the Chinese magnolia tree. Star anise is not related to the anise spice. Although similar in flavor to anise, the flavor of star anise is more intense and slightly bitter. Star anise is one of the main components in a Chinese five-spice blend.

Figure 7-6. A seed spice is a type of spice derived from the seed portion of a plant. Seed spices are available whole or ground.

Root Spices. A *root spice* is a type of spice derived from the root portion of a plant that grows underground. If they are used fresh, roots often have to be peeled and then sliced, grated, or shredded. Roots can also be dried and ground into powders. Root spices include ginger, horseradish, turmeric, and wasabi. **See Figure** 7-7.

Root Spices	
Ginger 	*Ginger* is a root spice made from the large bumpy root of a tropical plant. When the plant is about a year old, the roots are dug up and sold fresh, dried, powdered, crystallized, candied, or pickled. Ginger has a warm, pungent, spicy flavor that adds zest to desserts, fruits, and meat preparations.
Horseradish 	*Horseradish* is a root spice made from the large, brown-skinned root of a shrub related to the radish. Horseradish has a pungent, hot, and spicy flavor. When horseradish root is cut or grated, enzymes from the damaged root cells break down and produce mustard oil. If not used immediately or placed in vinegar, the root darkens and becomes bitter when exposed to air and heat. Grated horseradish is often served with grilled or roasted meats and seafood dishes or incorporated into cream sauces and compound butters.
Turmeric 	*Turmeric* is a root spice made from the root of a lily-like plant in the ginger family. When the roots are ground, a bright-yellow powder is produced that is sometimes referred to as Indian saffron. Turmeric has a taste similar to mustard and is an ingredient in curry powder and some prepared mustards.
Wasabi 	*Wasabi* is a root spice made from the light-green root of an Asian plant. Wasabi has a hot and tangy flavor similar to horseradish. It is commonly sold in a dried powder form or as a paste. When wasabi powder is mixed with water, a fiery, green-colored paste is produced.

Figure 7-7. A root spice is a type of spice derived from the root portion of a plant that grows underground.

Flower Spices. A *flower spice* is a type of spice derived from the flower of a plant. Capers and cloves are taken from the bud of unopened flowers. Saffron is obtained from the stigma of a flower. **See Figure 7-8.**

Berry Spices. A *berry spice* is a type of spice made from the berry portion of a plant. Berries used for spices are usually the unripe part of a plant that later develops into a fruit. Berry spices are commonly dried and then crushed or ground. Examples of berry spices include allspice, cardamom, cayenne pepper, crushed red pepper, juniper berries, paprika, and Szechuan pepper. **See Figure 7-9.**

Flower Spices	
Capers 	A *caper* is a flower spice from the unopened flower bud of a shrub and is only used after being pickled in strongly salted white vinegar. Capers should be stored in their original liquid and will spoil if water or additional vinegar is added. Capers are commonly used in fish and poultry dishes and are one of the key ingredients in tartar sauce.
Cloves 	A *clove* is a flower spice made from the dried, unopened bud of a tropical evergreen tree. Cloves are picked when the buds turn red. Dried cloves are dark brown. Cloves possess the most pungent flavor of all the spices and are sometimes referred to as the nail-shaped spice because of their shape. Cloves are used in pickling and in the preparation of roast pork, corned beef, baked ham, soups, applesauce, and baked products.
Saffron 	*Saffron* is a flower spice made from the dried, bright-red stigmas of the purple crocus flower. It takes over 75,000 flowers to produce 1 lb of saffron. Saffron strands are steeped in warm or hot water before use to release their color, aroma, and flavor. Saffron imparts a perfumelike aroma and deep-yellow color that is desired in certain rice dishes and baked products.

Figure 7-8. A flower spice is a type of spice derived from the flower of a plant.

Berry Spices	
Allspice 	*Allspice*, also known as Jamaican pepper, is a berry spice made from the dried, unripe fruit of a small pimiento tree. Because its complex flavor suggests a combination of cinnamon, nutmeg, and cloves, this pea-shaped spice is called allspice. Allspice is used in both whole and ground form in the preparation of pies, cakes, puddings, stews, soups, preserved fruit, curries, relishes, and gravies.
Cardamom 	*Cardamom* is a berry spice made from the dried, immature fruit of a tropical bush in the ginger family. A yellow pod about the size of a cranberry holds the dark, aromatic cardamom seeds. Cardamom has a lemony flavor and a pleasant aroma. It is available in whole or ground form. Cardamom is commonly used in curries, pastries, and pickled items.
Cayenne Pepper 	*Cayenne pepper*, also known as red pepper, is a berry spice made from dried, ground berries of certain varieties of hot peppers. Cayenne pepper is used in soups and meat, fish, and egg dishes.

Figure 7-9. (continued on next page)

Berry Spices	
Crushed Red Pepper 	*Crushed red pepper*, also known as crushed chiles or chili flakes, is a berry spice made from a blend of dried, crushed berries from hot chili peppers. Crushed red pepper is often used to flavor soups and pizza.
Juniper Berries 	A *juniper berry* is a berry spice made from a small, purple berry of an evergreen bush. It has an aromatic flavor similar to rosemary but can be slightly tart. Juniper berries are most commonly crushed and used in the preparation of wild game dishes and are also the key ingredient used to make gin alcohol.
Paprika 	*Paprika* is a berry spice made from a dried, ground sweet red pepper berry that is often used as a colorful garnish. Spanish paprika is mild in flavor and has a bright-red color. Hungarian paprika is darker in color and has a more pungent flavor. Paprika is used to prepare goulash, chicken paprika, Newburg sauce, and French dressing.
Szechuan Pepper 	*Szechuan pepper* is a berry spice made from the dried berries of an ash tree. Szechuan pepper is a fiery-flavored powder used in many Chinese dishes. It is also a component of Chinese five-spice powder.

Figure 7-9. Berry spices are commonly dried and then crushed or ground.

Spice and Herb Blends

A variety of spices and herbs can be combined to create complementary blends that add appealing flavors to food. Some blends may be used in unintended and creative ways. For example, a blend typically used for pickling can be used to enhance the flavor of a pot roast. Commonly used spice and herb blends include Cajun seasoning, chili powder, Chinese five-spice powder, curry powder, fines herbes, herbes de Provence, jerk seasoning, pickling spice, and poultry seasoning. **See Figure 7-10.**

Spice and Herb Blends	
Name of Blend	***Common Ingredients***
Cajun seasoning	Zesty spice blend; ground chiles, fennel seeds, garlic, oregano, paprika, red and black pepper, salt, and thyme
Chili powder	Ground chiles, cloves, coriander, cumin, garlic, and oregano
Chinese five-spice powder	Equal proportions of ground cinnamon, cloves, fennel seeds, star anise, and Szechuan pepper
Curry powder	Mild to spicy blend; can consist of more than 20 spices including cardamom, cinnamon, cloves, coriander, cumin, fenugreek, ginger, mace, nutmeg, red and black pepper, and turmeric
Fines herbes	Chopped fresh chervil, chives, parsley, and tarragon; may include marjoram, savory, or thyme
Herbes de Provence	Dried basil, fennel seed, lavender, marjoram, rosemary, sage, savory, and thyme
Jerk seasoning	Spicy blend; ground allspice, chiles, cinnamon, cloves, garlic, and ginger
Pickling spice	Whole and coarsely ground allspice, bay leaves, cinnamon, cloves, dill, fennel seeds, ginger, mace, mustard, nutmeg, peppercorns, and red pepper
Poultry seasoning	Black pepper, marjoram, nutmeg, rosemary, sage, and thyme; may include celery salt, mustard powder, or oregano

Figure 7-10. A variety of spices and herbs can be combined to create complementary blends that add appealing flavors to food.

Salts

Salt is the most widely used seasoning in the professional kitchen. *Salt* is a crystalline solid composed mainly of sodium chloride and is used as a seasoning and a preservative. It has a more intense taste flavor on cold foods than it does on hot foods. Salt also has the unique ability to make sweet items taste more full-bodied, sour items taste more pronounced, and bland items taste much more flavorful. Common varieties of salt include kosher salt, pickling salt, and sea salt. **See Figure 7-11.**

Specialty salts are mined from oceans and saltwater beds that evaporated thousands of years ago. These intensely flavored salts carry the characteristics of the soil where they are mined as well as the color. Specialty salts include artisan-mined salts such as Himalayan pink, Murray River flake, and kala namak as well as various smoked salts. **See Figure 7-12.** Smoked salts are produced by placing a specialty salt over a fire to gently impart a subtle smoky flavor of an ingredient such as bamboo, banana leaves, coconut shells, or wood.

Common Salts	
Kosher Salt 	*Kosher salt* is a salt used to cure, season, and prepare kosher foods. For example, kosher salt can be rubbed into meat to tenderize it and to draw water out of the meat. Kosher salt crystals are larger than other salts and do not contain iodine or anticaking additives. Kosher salt is lighter in weight than table salt and sticks to food very well.
Pickling Salt 	*Pickling salt*, also known as canning salt, is a pure form of salt that contains no residual dust, iodine, or other additives. The purity of pickling salt keeps the salt from clouding the brine or settling at the bottom of the jar.
Sea Salt 	A *sea salt* is salt produced through the evaporation of seawater. Large, shallow reservoirs (pans) are filled with seawater. As the pans of seawater evaporate, a thick layer of crystallized sea salt is formed. The resulting sea salt is then raked from the salt beds. Sea salt contains calcium, potassium, and magnesium. High concentrations of minerals give sea salt a more intense flavor. Sea salt is available in fine and coarse crystals and is used to cook and preserve food.

Figure 7-11. Salt is a crystalline solid composed mainly of sodium chloride and is used as a seasoning and a preservative.

Sample Specialty Salts

Name	*Characteristics*	*Attributes*
Himalayan pink	Pink-colored salt mined from deep within the Himalayan Mountains and harvested from ancient sea salt deposits; contains over 84 minerals and trace elements	Mild, pleasant flavor
Murray River flake	Peach-colored or apricot-colored salt harvested from the Murray River in the foothills of the Australian Alps; produced from natural underground brine pools	Crunchy texture; a great finishing salt
Kala namak (also known as Indian black)	Black-colored salt mined from rock salt in an area where the soil is high in minerals such as sulfur	Strong sulfur odor; flavor similar to cooked egg yolks

Figure 7-12. Specialty salts include mined salts such as Himalayan pink, Murray River flake, kala namak, and various smoked salts.

Peppercorns

Pepper is the second most widely used seasoning in the professional kitchen. A *peppercorn* is the dried berry of a climbing vine known as the Piper nigrum and is used whole, ground, or crushed. The Piper nigrum never grows farther than 20° from the equator. Although all peppercorns come from the same plant, their variations in color are a direct result of when they are harvested and how they are processed. Black, green, and pink peppercorns are commonly used in the professional kitchen. **See Figure 7-13.**

White pepper is made from black peppercorns. Ripe berries are soaked to soften the outer orange-red husk. The husk is removed when it begins to wrinkle, also removing the spicy characteristic of black peppercorns. The remaining portion of the berries is white in color. White pepper is sold whole as white peppercorns and ground as white pepper.

Citrus Zests

Citrus zests can be used to add flavor and color to salads and salad dressings. *Zest* is the colored, outermost layer of the peel of a citrus fruit, which contains a high concentration of oil. When removing the zest from a citrus fruit, it is important to remove only the colored portion of the peel because the white pith beneath the peel is bitter tasting. The zests of lemons and limes are often used in both sweet and savory applications. **See Figure 7-14.** Only a small amount of zest is needed in any dish.

Citrus Zests

Browne Foodservice

Figure 7-14. The zests of lemons and limes are often used in both sweet and savory applications.

Peppercorns	
Black Peppercorns  Pepper-Passion Inc.	Black peppercorns are the most common variety of peppercorn used in cooking. The berries are harvested when they are just starting to ripen. At this stage they are a yellowish color but still immature. The berries are blanched in boiling water and placed on screens in the sun to ferment and fully dry. The intense heat of the sun causes the berries to completely dry out, shrivel, and turn black. Black peppercorns have a very strong, spicy-hot, and aromatic quality. Peppercorns grown in different parts of the world will express different characteristics of flavor, aroma, and pungency. Black peppercorns are sold as whole peppercorns, cracked pepper, or ground pepper. Black peppercorn varieties with the best flavor include Tellicherry, Malabar, and Lampong.
Green Peppercorns  Pepper-Passion Inc.	Green peppercorns are immature berries that are picked well before they have ripened. They are either pickled in a salted brine or vinegar-based solution, freeze-dried, or air-dried to retain their green color, fresh flavor, and aromatic qualities. Green peppercorns are the mildest of the peppercorn varieties. Pickled green peppercorns are soft and have a taste and texture similar to that of capers. The pickled version is most commonly used in sauces and in condiments. The freeze-dried and air-dried versions are often cracked or ground as a seasoning.
Pink Peppercorns  Pepper-Passion Inc.	Pink peppercorns are not actual peppercorns. They are the dried berries from the baies rose plant. They are best cracked instead of ground. Their flavor is pungent and slightly sweet, and they are often used with pork and seafood dishes.

Figure 7-13. A peppercorn is the dried berry of a climbing vine known as the Piper nigrum and is used whole, ground, or crushed.

Fruit Juices

Fruit juices are often used to add moisture, flavor, and acid to salad dressings. Raspberry, cranberry, strawberry, and blackberry are a few of the many berry juices used in salad dressings and gelatin salads. Citrus juices have a high acid content and are typically used to enhance the flavor of other ingredients.

Wines and Liquors

Wines and liquors are used to flavor salad dressings. A *wine* is the fermented juice of grapes or other fruit. Fortified wines such as sherry, port, Madeira, and Marsala are grape-based wines to which additional alcohol has been added. A *liquor* is a distilled liquid made from potatoes, corn, or other grains that has a high alcohol content.

Vinegars

Vinegar is a sour, acidic liquid made from fermented alcohol. Many vinegars are made from wine, but other alcohols can also be used to produce vinegar. High-quality alcohols yield high-quality vinegars. Vinegars can range from strong and pungent to sweet and mild. Vinegar is an essential ingredient in pickled items, mustards, and vinaigrettes. Common vinegars used in the professional kitchen include balsamic, beer, Champagne, cider, distilled, flavored, fruit, kombucha, malt, rice, sherry, and wine vinegar. **See Figure 7-15.**

Vinegars	
Balsamic Vinegar	*Balsamic vinegar* is a vinegar made by aging trebbiano wine vinegar in a succession of casks made from a variety of woods for many years. The vinegar picks up characteristics, such as color and aroma, from the type of wood in which it is stored. After aging, the vinegar darkens in color and becomes sweeter. Although balsamic vinegar has a higher acid content than red wine vinegar, its sweet characteristics mellow its acidic qualities. Balsamic vinegar adds an appealing flavor when drizzled over grilled vegetables or salads.
Beer Vinegar	*Beer vinegar* is a light-golden-colored vinegar made from beer with a strong malt taste. Beer vinegar carries the flavor of the beer or ale that it is made from.
Champagne Vinegar	*Champagne vinegar* is a vinegar made from Champagne grapes. It is a mildly flavored vinegar that pairs well with salads composed of delicate greens.
Cider Vinegar	*Cider vinegar* is honey-colored vinegar made by fermenting unpasteurized apple juice or cider until the sugars are converted into alcohol. It has a mild acidity and a subtle apple taste. Cider vinegar is often used in salad dressings.
Distilled Vinegars	*Distilled vinegar*, also known as white vinegar, is vinegar made by fermenting diluted, distilled grain alcohol. It is most often used in pickling due to its high acidic quality.
Flavored Vinegars	A *flavored vinegar* is any vinegar in which other items such as herbs, spices, garlic, fruits, vegetables, or flowers are added. Raspberry, pear, blueberry, and mango are examples of fruit-flavored vinegars that add a pleasant sweetness and unique flavor to salads. Tarragon, rosemary, and peppercorns are commonly used to produce savory vinegars.
Fruit Vinegar	*Fruit vinegar* is a vinegar made from fruit juice that is first turned into fruit wine and then fermented by acetobacters (bacteria), which turns the wine into vinegar. Real fruit vinegar is fragile and heating it will destroy the true fruit nature of this vinegar. Fruit vinegar is often sprinkled on salads or appetizers such as tuna or beef carpaccio, baked or steamed chicken or seafood, and dessert items such as fruit and ice cream.

Figure 7-15. (continued on next page)

Vinegars	
Kombucha Vinegar	*Kombucha vinegar* is vinegar made with tea, acetobacters, yeast, and berries or other fruit. Kombucha vinegar is used in salad dressings and dipping sauces.
Malt Vinegar	*Malt vinegar* is vinegar made from malted barley. It has a robust lemony flavor and a lower acidity than other vinegars. Malt vinegar is often served with fish and chips.
Rice Vinegar	*Rice vinegar* is vinegar made from rice wine. Rice vinegar is slightly sweet and mildly acidic. It is most often used in Asian dishes and dipping sauces.
Sherry Vinegar	*Sherry vinegar* is vinegar made from sherry wine. The strong flavor of sherry comes through in sherry vinegar.
Wine Vinegar	*Wine vinegar* is vinegar made from red or white wine. Red wine vinegars are aged longer than white wine vinegars and are slightly more pungent.

Figure 7-15. Vinegars can range from strong and pungent to sweet and mild.

Nutrition Note

Most types of vinegar are fat-free and low in sodium and calories. Vinegars are devoid of vitamins but do contain potassium, magnesium, sugar, and carbohydrates. Balsamic vinegar is much higher in sodium and sugar than other vinegars.

Oils

An *oil* is a fat that remains in a liquid state at room temperature. Many oils used in the professional kitchen are produced from vegetables, seeds, nuts, or fruit. These oils impart flavors that can be subtle and mild or rich and robust. Because many oils are plant-based products, they do not contain cholesterol and are often high in unsaturated fats. Oils have a limited shelf life and should be stored in a cool, dry place out of direct sunlight in order to prolong freshness.

Different oils are suited for different tasks. For example, some oils have a high smoke point, making them suitable for frying foods. Other oils are more flavorful, yet have low smoke points. For example, olive oil is best used to season foods or for sautéing. Canola, corn, and peanut oil are often used for frying. Oils commonly used in the cold kitchen include avocado oil, canola oil, corn oil, grapeseed oil, hazelnut oil, infused oils, olive oil, rice bran oil, sesame oil, soybean oil, and walnut oil. **See Figure 7-16.**

Oils

Barilla America, Inc.

Oils	
Avocado Oil	*Avocado oil* is a type of oil produced by pressing avocados. This smooth, nutty-flavored oil is more than 50% monounsaturated, making it a healthy choice for salads as well as for sautéing.
Canola Oil	*Canola oil*, also called rapeseed oil, is a type of oil produced from rapeseeds. It contains omega-3 fatty acids and is very low in saturated fat, making it a healthy choice for dressings, cooking, or baking. Canola oil has a neutral flavor that is perfect for making mayonnaise.
Corn Oil	*Corn oil* is a type of oil produced from the germ of corn kernels. Corn oil has a light golden color and is often used in pepper- or garlic-flavored salad dressings and dips.
Grapeseed Oil	*Grapeseed oil* is a type of oil produced from grape seeds. Grapeseed oil has a delicate flavor and is used in dips and salsas. It may also be mixed with garlic and basil and drizzled on toasted bread.
Hazelnut Oil	*Hazelnut oil* is a type of oil produced from hazelnuts. Due to its distinct flavor and aroma, hazelnut oil is primarily used in salad dressings and garnishing applications.
Infused Oils	An *infused oil* is any variety of oil with added herbs, spices, or additional ingredients that increase the flavor. Infused oils are produced by heating neutral-flavored oil until it is warm and then infusing it with ingredients such as garlic, lemongrass, truffles, or various spices. Infused oils are commonly used as dipping sauces.
Olive Oils	*Olive oil* is a type of oil produced from olives. Olive oil can differ slightly in taste, color, and consistency. *Extra-virgin olive oil* is a type of olive oil produced from the first pressing of the olives without the use of heat or chemicals and has an acid content of less than 1%. It is considered the highest quality of olive oil and its rich flavor makes it desirable for salad dressings and garnishing applications. *Virgin olive oil* is a type of olive oil produced from the first pressing of the olives without the use of heat or chemicals and has an acid content of as much as 3%. *Pure olive oil* is a type of olive oil produced using heat, and often chemicals, to extract additional oils from the olive pulp after the first pressing. *Olive-pomace oil* is a type of olive oil produced using heat, and often chemicals, to extract additional oils from the olive pulp and olive pits after the first pressing.
Rice Bran Oil	*Rice bran oil* is a type of oil produced from the bran (outer layer) of rice kernels. This healthy oil is a good source of vitamins, minerals, amino acids, essential fatty acids, and antioxidants that help fight disease and promote good health. Rice bran oil has a light, buttery taste and a light viscosity that emulsifies well in salad dressings. The 490°F smoke point also makes this oil a good choice for savory applications.
Sesame Oil	*Sesame oil* is a type of oil produced from sesame seeds. Light sesame oil has a mild, nutty flavor and is often used in salad dressings and spreads. Dark sesame oil is stronger in flavor because the sesame seeds are toasted before the oil is produced. Dark sesame oil is often used to accent Asian dishes.

Figure 7-16. (continued on next page)

Oils	
Soybean Oil	*Soybean oil* is a type of oil produced from soy beans. Soybean oil contains natural antioxidants and is low in saturated fat. Soybean oil is used in salad dressings, mayonnaise, and sandwich spreads.
Walnut Oil	*Walnut oil* is a type of oil produced from walnuts. Due to the distinct aroma and flavor of walnut oil, it is often used in salad dressings and in garnishing applications.

Figure 7-16. Many oils used in the professional kitchen are produced from vegetables, seeds, nuts, or fruit.

CHECKPOINT 7-1

1. Explain how flavors are developed in food.
2. Identify 18 common leaf herbs.
3. Identify three common stem herbs.
4. Use herbs to flavor a food. Evaluate the food for contrast, balance, and flavor.
5. Identify two common bark spices.
6. Identify 14 common seed spices.
7. Identify four common root spices.
8. Identify three common flower spices.
9. Identify seven common berry spices.
10. Use spices to flavor a food. Evaluate the food for contrast, balance, and flavor.
11. Identify nine common spice and herb blends.
12. Describe three common salts used in the professional kitchen.
13. Explain why peppercorns vary in color.
14. Explain how citrus zests, fruit juices, wines, and liquors are used in salads.
15. Identify 12 commonly used vinegars.
16. Explain how to select the appropriate oil to use for a given preparation.
17. Identify 11 commonly used oils.
18. Use herbs, spices, salts, peppercorns, citrus zests, fruit juices, wines, liquors, vinegars, and oils to flavor a variety of salads. Evaluate the salads for contrast, balance, and flavor.

Tanimura & Antle®

SALADS

Salads are popular menu items that offer limitless opportunities when combining ingredients and flavors. Salads may contain raw or partially cooked ingredients and can be served cold or hot. Salads are often served with a dressing, which may be sweet or savory.

Although many salads contain lettuce, salads do not require lettuce as an ingredient. Salads can be presented as an appetizer, a main course, a separate course, a side salad, or a dessert, depending on the ingredients used. **See Figure 7-17.**

Salad Presentations	
Appetizer Salads  *Daniel NYC/P. Medilek*	An *appetizer salad* is a salad that is served as a starter to a meal. Appetizer salads are the most basic type of salad and often feature varieties of lettuce, tomato wedges, cucumber slices, croutons, and a dressing.
Main Course Salads  *Courtesy of The National Pork Board*	A *main course salad* is a large salad containing an assortment of vegetables and a protein such as poultry, seafood, meat, eggs, or legumes that is served as an entrée. A Caesar salad with grilled chicken is a main course salad.
Separate Course Salads  *Daniel NYC/N. Clutton*	A *separate course salad* is a salad served as a course of its own, typically following the main course. The purpose of a separate course salad is to refresh and cleanse the palate prior to dessert. These salads are usually light and are often served with a vinaigrette dressing.
Side Salads 	A *side salad* is a small salad served to accompany a main course. If a main course is hearty or heavy, a side salad is typically light and refreshing. If a main course is light, a heavier side salad such as a pasta, grain, or potato salad is served. A small amount of salad may be served alongside an entrée.
Dessert Salads  *Frieda's Specialty Produce*	A *dessert salad* is a sweet salad usually consisting of nuts, fruits, or sweeter vegetables such as carrots. Dessert salads are often served with yogurt, whipped cream, or a citrus-flavored dressing.

Figure 7-17. Salads can be presented as an appetizer, a main course, a separate course, a side salad, or a dessert.

A variety of ingredients can be used to create endless combinations of salads that delight the palate. It is important to remember that a salad should complement other courses. The most common ingredients used in salads are salad greens, vegetables, starches, fruits, and proteins.

Nutrition Note

Salads that offer a variety of flavors are an appealing dish. However, using an overabundance of cheese, meat, nuts, croutons, and/or salad dressing can turn a nutritious salad into a fat-laden dish.

Salad Greens

A *salad green* is an edible leaf used in raw salads or as a garnish. Salad greens can be varying shades of green as well as red, yellow, white, and rusty brown. **See Figure 7-18.** Salad greens include crisphead, butterhead, romaine, and looseleaf lettuces and spinach, chard, arugula, cabbage, chicory, dandelion greens, mesclun greens, and garnishing greens. Salad greens are low-calorie, low-fat ingredients that are high in vitamin A, vitamin C, and iron. However, when items such as meat, cheese, croutons, and salad dressing are added to salad greens, a salad can quickly become a high-calorie, high-fat dish.

Salad Greens

Figure 7-18. A salad green is an edible leaf used in raw salads or as a garnish.

Trimming Salad Greens. Solidly packed head lettuces, such as iceberg, must have the core removed and be cut to an appropriate size before washing the leaves. If the leaves are not cut to an appropriate size, it is nearly impossible for water to reach the center of the tightly packed heads to effectively remove dirt, pests, and chemicals.

Edible Flowers

Procedure for Removing the Core from Head Lettuce

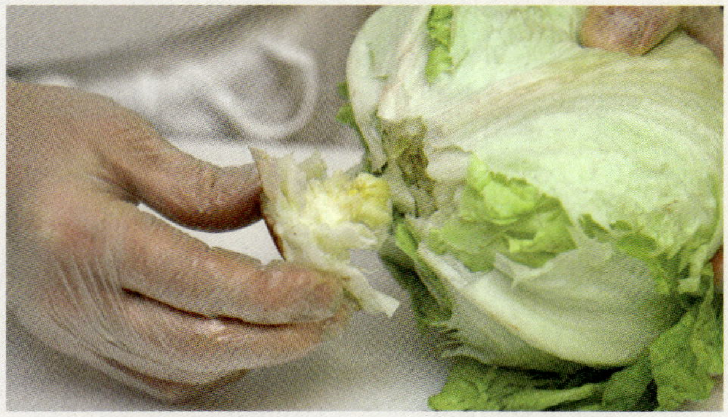

1. Hold the head of lettuce in the palm of one hand with the core facing down.
2. Carefully hit the core against a clean work surface to free the core from the rest of the head.
3. Grab the core and pull it to remove it from the head.
4. Cut the lettuce and then wash, dry, and store it.

Romaine lettuce needs to be cut or trimmed prior to being washed to make sure that all dirt, pests, and chemicals have been washed away. The flavorful ribs are not removed. Loose greens, such as chicory and spinach, often collect dirt in the center rib. The ribs of loose greens must be removed prior to preparation.

Procedure for Removing Ribs from Loose Greens

1. Fold each leaf in half so that the two sides of the top surface meet, exposing the rib.
2. With the opposite hand, pinch the rib and carefully pull it apart from the leaf. Discard the rib.
3. Wash, dry, and store leaves.

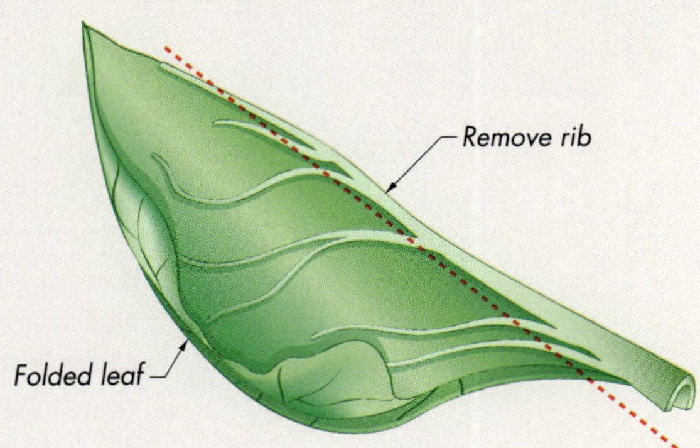

Washing Salad Greens. It is essential to properly wash salad greens before use. Even if salad greens look clean, insects, dust, dirt, pesticides, and fertilizers may be hidden between the leaves.

To wash salad greens, they should first be cut or torn to the desired size and then completely submerged in a sink of clean, cold water. The greens are gently stirred to rinse away dirt and then removed from the dirty water. The sink is rinsed completely and refilled with clean, cold water. The rinsing process is repeated until the water in the sink remains clean.

If the leaves are to be kept whole, each leaf should be washed. This task must be done gently because tender greens can bruise easily. It may also be necessary to cut elongated heads in half lengthwise in order to remove all of the dirt, grit, and sand.

All salad greens must be dried completely after washing. Wet greens become limp in a short amount of time. Also, oil-based vinaigrettes or mayonnaise-based dressings do not stick to wet leaves. Using a salad spinner is the best way to dry lettuce.

Storing Salad Greens. All salad greens should be stored between 35°F and 38°F, which is colder than the storage temperature of other produce. If any part of a green is frozen, it should be discarded. Delicate greens, such as butterhead lettuces, deteriorate faster than firmer varieties such as romaine.

Salad greens must be stored away from tomatoes and fruits that emit ethylene gas. Storing salad greens near these items will cause the greens to wilt and deteriorate rapidly. Washed salad greens should be stored in a perforated stainless steel pan with a second solid pan as an under-liner to hold the water drippings. The salad greens should be covered with a damp paper towel or plastic wrap to retain crispness.

Vegetables

A *vegetable* is an edible root, bulb, tuber, stem, leaf, flower, or seed of a nonwoody plant. Vegetables, fruit-vegetables, sea vegetables, and mushrooms make nice additions to salad greens. **See Figure 7-19.** Vegetables add color and flavor to salad greens and are also low in fat and calories. For example, bell peppers (sweet peppers) can add color and crispness to salads. Before adding bell peppers to a salad the core and seeds must be removed. To accomplish this, the rolling method is often used. Once the seeds have been removed, bell peppers are typically julienned or diced before being added to a salad.

Procedure for Washing Salad Greens

1. Cut or tear greens to the appropriate size.
2. Submerge greens completely in a sink of cold water and gently stir greens to rinse.
3. Remove greens from the dirty water and drain the sink. Rinse the sink and refill it with cold water.
4. Repeat the process until there is no dirt on the bottom of the sink.
5. Remove greens from the water and spin dry.
6. Store greens in a stainless steel pan with a perforated pan insert and cover with damp paper towels or plastic wrap until needed.

Vegetable Ingredients	
Type	*Ingredients*
Roots	Carrots, parsnips, salsify, radishes, turnips, rutabagas, beets, celeriac, jicamas, and lotus roots
Bulbs	Garlic, shallots, onions, scallions, ramps, and leeks
Tubers	Potatoes, sweet potatoes, yams, ocas, sunchokes, and water chestnuts
Stems	Asparagus, celery, fennel, rhubarb, kohlrabi, and hearts of palm
Leaves	Cabbages, bok choy, Brussels sprouts, lettuces, chicory, watercress, spinach, sorrel, chard, kale, collards, mustard greens, turnip greens, beet greens, dandelion greens, nopales, tatsoi, and fiddlehead ferns
Flowers	Squash blossoms, broccoli, cauliflower, and artichokes
Seeds	Legumes (beans, peas, and lentils) and sprouts
Fruit-vegetables	Tomatoes, cucumbers, eggplants, bell peppers, chiles, okra, sweet corn, squashes, tomatillos, and pumpkins
Sea vegetables	Nori, kombu, arame, wakame, and dulse
Mushrooms*	Button, portobello, enokitake, wood ear, shiitake, oyster, chanterelle, morel, and porcini

** Mushrooms are fungi, not vegetables, but can be added to salads in the same manner as vegetables.*

Figure 7-19. Vegetables, fruit-vegetables, sea vegetables, and mushrooms make nice additions to salad greens.

Procedure for Coring Peppers

1. Cut off the top of the pepper just below the stem. Remove the stem and reserve the top piece.

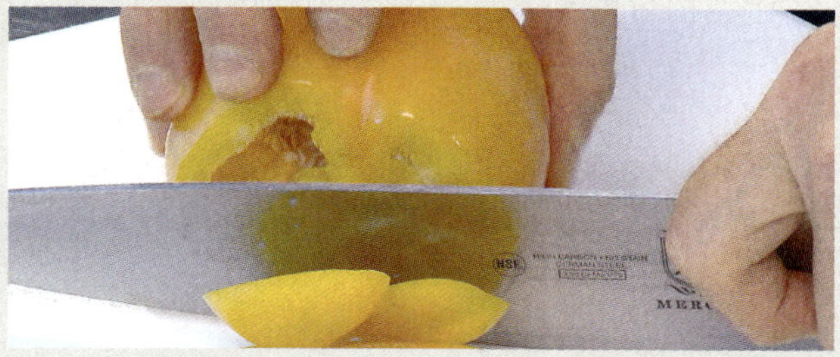

2. Cut off the bottom of the pepper, so that the pepper sits flat on the cutting board. Reserve the bottom piece.

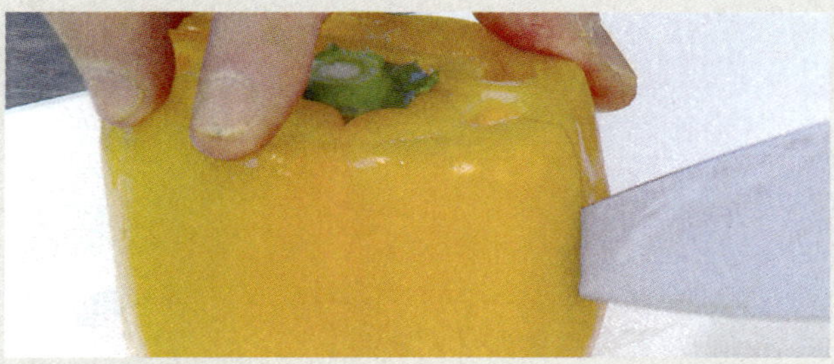

3. Cut a vertical slice to create an opening in the exterior of the pepper.

4. Turn the pepper on its side and insert the top of the knife blade between the outer skin and the internal ribs.

5. Slowly move the blade along the inside of the skin while rolling the pepper in one continuous motion.

6. Discard the center portion of the pepper. Julienne or dice the cored pepper.

Coring Peppers
Media Clip

Vegetables used in salads can be raw or cooked. Hard vegetables, such as cauliflower, should be blanched prior to use in a salad. An entire salad can be made from a single vegetable, such as beet salad made from blanched or roasted beets. Soft vegetables, such as tomatoes, can be added raw. A cucumber and onion salad is a raw vegetable salad.

The petals of an artichoke, called leaves, are the edible part of the bud. The tips of the larger leaves can be very sharp with a spiny thorn that must be removed during preparation. The dense area where the leaves attach to the stem is known as the heart. As the flower matures, a fuzzy, sometimes thorny, center called the choke develops just above the center of the heart. The choke must be removed before the artichoke is served.

Fruits

When used in a salad, fruits can be fresh, dried, or roasted. Fruits add texture, flavor, and aroma to a salad. **See Figure 7-20.** Fruits can impart acidic qualities, sweetness, chewiness, or crispness. For example, a microgreen salad could be topped with dried cherries, fresh pears, and a balsamic vinaigrette.

Procedure for Preparing Artichokes

1. With a French knife, remove the entire top half of the artichoke, exposing the lighter-colored choke in the center. Discard the top half.

2. Cut off the top third (the thorn) of each of the larger exposed leaves. Discard the removed tips.

3. Grasp the bottom of the artichoke, and use both thumbs to spread open the center and expose the choke.

4. With a spoon, thoroughly scrape the choke from the solid heart. Discard the choke.

5. Cut off any extra stem.

6. A small amount of lemon juice can be squeezed onto the exposed heart to deter browning.

Preparing Artichokes
Media Clip

Fruit Ingredients

National Honey Board

Figure 7-20. Fruits add texture, flavor, and aroma to a salad.

Nutrition Note

Fruit salads typically contain no more than five different fruits and are served as a side dish or a dessert. A fresh fruit salad provides essential vitamins and minerals along with a significant amount of carbohydrates in the form of natural sugars. Fruit salads do not contain protein or fat.

The flesh of a citrus fruit is naturally divided into wedge-shaped segments that are separated by a thin membrane. These segments are often separated or cut to form "supremes" when preparing citrus fruits for use in a dish or as a garnish. A supreme is the flesh from a segment of a citrus fruit that has been cut away from the membrane.

Procedure for Cutting Citrus Supremes

1. Using a chef's knife, carefully slice off the top and bottom of the fruit. Place the fruit on the cutting board so the bottom is lying flat.
2. Place the tip of the knife at the top edge of the fruit and cut from top to bottom using a slight sawing motion to remove a strip of peel and pith.

3. Following the contour of the fruit, use a slight sawing motion to cut away a small slice of peel and pith. Rotate the fruit clockwise and removed another small slice. Repeat until the entire peel and pith have been removed.

4. When the peel has been removed, check to make sure that none of the pith remains. Carefully cut off any remaining pith from the fruit.

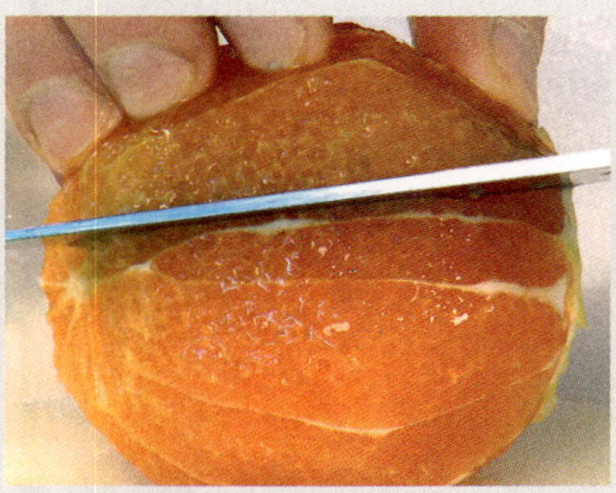

5. Hold the fruit against the cutting board and cut along the membrane in a "V" to remove a supreme. Hold back the membrane as each cut is made.

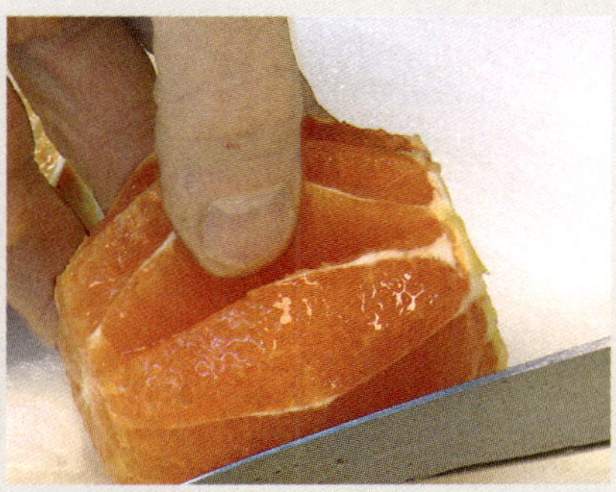

6. Continue to cut each segment until all the supremes have been separated from the fruit. *Note:* The remainder of the fruit can be juiced to flavor a dressing or a sauce.

The supremes are ready for use.

Cutting Citrus Supremes
Media Clip

Starches

Starches such as grains, pastas, and breads can add variety, texture, and nutritional value to many salads. Wheat berries, barley, and quinoa are used in salads. Grains used in a salad should be cooked slightly less than al dente so they retain their shape and are firm. Overcooked grains become mushy as they absorb dressing.

Pasta salads may be traditional, such as rotini pasta tossed with a vinaigrette, or contemporary, such as couscous salad with pomegranate, mint, and curry. Pasta used in salads should be slightly undercooked so that it does not become soggy or break apart when tossed. **See Figure 7-21.**

Starch Ingredients

Tanimura & Antle®

Italian Chop Salad

Courtesy of The National Pork Board

Rotini Pork Salad

Figure 7-21. Pasta used in salads should be slightly undercooked so that it does not become soggy or break apart when tossed.

Breads, such as toasted croutons, are often used as a salad garnish. Breads can also be an ingredient in a tossed salad. In panzanella, also known as bread salad, crusty bread is toasted or grilled and then torn into bite-sized pieces and tossed with tomatoes, basil, other ingredients, and a red wine vinaigrette.

Poultry, Seafood, and Meats

Common proteins used in salads include cooked poultry, seafood, meats, legumes, nuts, and cheese. Any protein can be used as the main ingredient of a salad or to make a salad more hearty and flavorful. **See Figure 7-22.**

Protein Ingredients

Courtesy of The National Pork Board

Figure 7-22. Any protein can be used as the main ingredient of a salad or to make a salad more hearty and flavorful.

Legumes

Legumes are used in salads because they are rich in protein and fiber and contain little or no fat. A *legume* is the edible seed of a nonwoody plant and grows in multiples within a pod. Beans, peas, pulses, and lentils are legumes. **See Figure 7-23.**

Legume Salads

Figure 7-23. Legumes are used in salads because they are rich in protein and fiber and contain little or no fat.

Eggs

Cooked eggs may be sliced, diced, quartered, or served whole in salads. Although the common term for cooking eggs in the shell is "boiled eggs," eggs in the shell should be simmered, never boiled. Boiling tends to toughen the texture of the egg and can create a green sulfur ring around the outside of the yolk. Simmering eggs in the shell at a temperature of 195°F is recommended.

Procedure for Preparing Eggs in the Shell

1. Place the room temperature eggs in a pot and cover them with cold water.
2. Bring the water to a gentle boil over medium heat.
3. Reduce the heat to a simmer and begin timing. Simmer eggs to desired doneness (i.e., soft, medium, or hard).

Eggs cooked in the shell for 3 minutes are considered soft-cooked. Medium-cooked eggs in the shell are cooked for 4–6 minutes. Hard-cooked eggs in the shell are cooked for 9–10 minutes and have a bright, solid white and a pale, crumbly yolk. **See Figure 7-24.** Hard-cooked eggs are plunged into cold water to stop the cooking process. If hard-cooked eggs are to be held for use in a cold preparation, such as egg salad, they should be cooled in ice-cold water for about 5 minutes, peeled and placed in a bain-marie, covered with cold water, and then refrigerated for up to a day. This process will prevent the eggs from drying out. Cooked eggs in the shell can be refrigerated for up to a week.

Hard-Cooked Eggs

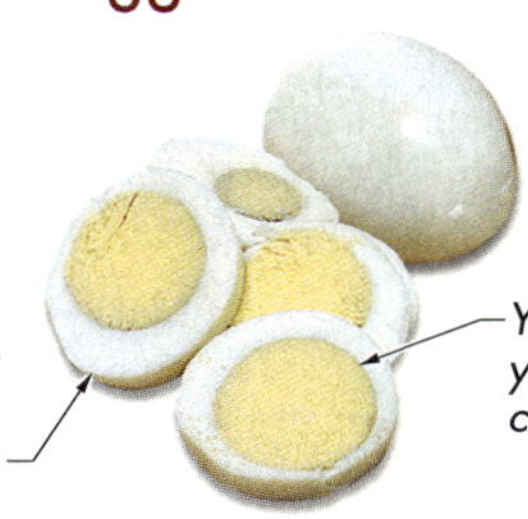

American Metalcraft, Inc.

Figure 7-24. Hard-cooked eggs in the shell have a bright, solid white and a pale, crumbly yolk.

Production Tip

Eggs cooked in the shell are easier to peel from the large end to the narrow end. Keeping the eggs submerged in cold water while peeling them will help loosen the shells.

Cheeses

Cheeses may be sliced, cubed, crumbled, or grated for use in salads. Cheeses add flavor and textures to salads and serve as a protein component as well as a garnish. **See Figure 7-25.** For example, a Caprese salad is typically served with fresh mozzarella, sliced heirloom tomatoes, fresh basil, and a balsamic vinegar dressing. Parmesan is usually grated in a Caesar salad and is also part of the dressing. Cheeses can be prepared ahead of service and should be wrapped and kept refrigerated.

Salads Garnished with Cheese

U.S. Apple® Association

Figure 7-25. Cheeses add flavor and texture to salads and serve as a protein component as well as a garnish.

Nuts

Nuts enhance salads by adding a crunchy texture and protein and provide a sweet, salty, or spicy flavor. For example, maple-glazed walnuts and blue cheese add protein to a salad. A ***nut*** is a hard-shelled, dry fruit or seed that contains an inner kernel. Nuts are often added to foods to impart texture or a nutty flavor. For example, salads are often tossed with toasted nuts for added flavor and texture. Nuts are also used in many baked products and a number of poultry, seafood, and meat dishes.

Nuts commonly used in the professional kitchen include almonds, Brazil nuts, cashews, chestnuts, hazelnuts, macadamia nuts, peanuts, pecans, pine nuts, pistachios, soy nuts, and walnuts. Some nuts, such as peanuts and soy nuts, are not nuts. They are actually legumes that are used like nuts. **See Figure 7-26.**

Nuts	
Almonds 	An *almond* is a teardrop-shaped fruit that grows on small almond trees. Almonds are marketed whole, chopped, slivered, sliced, ground, or as almond paste.
Brazil Nuts 	A *Brazil nut* is a 1–2 inch long, white, richly flavored seed of a very large fruit that grows on a Brazilian nut tree.
Cashews 	A *cashew* is a nut from a kidney-shaped kernel of the fruit of the cashew tree and has a buttery flavor. There is a toxin present in the shell, which is why cashews are never sold in shells.
Chestnuts  *Trails End Chestnuts*	A *chestnut* is a nut from a rounded-off, triangular-shaped kernel found inside the burrlike fruit of the chestnut tree. Chestnuts are higher in starch and lower in fat than any other nut. When chestnuts are roasted or boiled, they become sweet. Chestnuts are available dried, canned, or as a sweetened paste.
Hazelnuts 	A *hazelnut*, also known as a filbert, is a grape-sized nut found in the fuzzy outer husks of the hazel tree. Shelled hazelnuts have bitter, brown, papery skins that cover their surface. The skins are removed prior to service. Skinned hazelnuts have a sweet flavor.
Macadamia Nuts 	A *macadamia nut* is a tan, marble-sized nut with a hard shell and is the fruit of the macadamia tree. Macadamia nuts have a very high fat content and a sweet, creamy taste.
Peanuts 	A *peanut* is a legume that is contained in a thin, netted, tan-colored pod that grows underground. Unshelled peanuts are often sold roasted or salted. Peanut butter is made from ground peanuts.
Pecans 	A *pecan* is a nut with a light-brown kernel from the pecan tree. Pecans are 70% fat, making them the fattiest of all nuts and quick to deteriorate. Pecans are often candied or seasoned with spices.
Pine Nuts 	A *pine nut* is an ivory-colored, torpedo-shaped nut from the pine cone of various types of pine trees.

Figure 7-26. (continued on next page)

Nuts	
Pistachios 	A *pistachio* is a pale-green, bean-shaped nut from the pistachio tree.
Soy Nuts 	A *soy nut* is a legume from the soybean pod that has been soaked in water, drained, and then roasted or baked. Although not technically a nut, soy nuts have a crunchy, nutlike texture.
Walnuts 	A *walnut* is a nut from the fruit of the walnut tree. Two popular varieties of walnuts are English walnuts and black walnuts. The English walnut has a rich sweet flavor. The black walnut has a stronger, somewhat bitter flavor.

Figure 7-26. Nuts enhance salads by adding a crunchy texture and protein.

CHECKPOINT 7-2

1. Define salad greens.
2. Explain how to remove the core from head lettuce.
3. Remove the core from a head of lettuce.
4. Remove the rib from loose greens.
5. Describe how to properly store salad greens.
6. Wash and store salad greens.
7. Identify eight types of salad ingredients other than salad greens.
8. Identify the cooking times for preparing soft-cooked, medium-cooked, and hard-cooked eggs in the shell.
9. Prepare and plate eggs in the shell. Evaluate the appearance, flavor, and texture of the eggs.
10. Identify two ways that cheeses can be used in salads.
11. Identify 12 common nuts used to flavor food.
12. Use nuts to add flavor to a dish. Evaluate the dish for contrast, balance, and flavor.

SALAD TYPES

The term "leafy green salad" can refer to any salad made with a base of greens. Leafy green salads can be as basic as iceberg lettuce, a wedge of tomato, and grated carrots topped with a choice of salad dressing. Leafy green salads can also be as complex as mixed baby field greens with diced pears, Gorgonzola cheese, glazed pecans, and balsamic vinaigrette. All salads fall into one of six types based on how the salad is to be plated or served. The six types of salads are tossed, composed, bound, vegetable, fruit, and gelatin.

Tossed Salads

A *tossed salad* is a mixture of leafy greens, such as lettuce, spinach, chicory, or fresh herbs, and other ingredients, such as fruits, vegetables, nuts, cheese, meats, and croutons, served with a dressing. There are endless combinations of ingredients that can be used to construct tossed salads, but care should be taken to choose complementary flavors and textures. The dressing should not overpower or conflict with the ingredients. To prepare a tossed salad, four simple steps can be followed.

Carlisle FoodService Products

Procedure for Preparing Tossed Salads

1. Select a variety of leafy greens so the salad will be colorful and have a variety of textures and flavors.
2. Cut, wash, dry, and store the greens until ready for use.

3. Prepare and store any additional ingredients and garnishes that will be added to the salad.

4. Carefully combine the greens, additional ingredients, garnishes, and dressing in a mixing bowl and toss gently to distribute all components evenly.

Production Tip

Adding salad dressing along the perimeter of the bowl prior to tossing a salad will help minimize overdressing.

Composed Salads

A *composed salad* is a salad that consists of a base, body, garnish, and dressing attractively arranged on a plate. **See Figure 7-27.**

The Base. The base of a composed salad serves as the foundation on which a salad is built. A base typically is a bed of salad greens, fruits, or vegetables. The base is not only an edible component of the salad, but a colorful liner between the rest of the salad and the plate. For example, a base may be a cup-shaped leaf in which the rest of the ingredients can be placed.

The Body. The body of a composed salad consists of the main ingredients. Tossed salad greens, a fruit medley, vegetables, cooked grains, cooked pastas, or precooked proteins such as chicken salad can be used as the body of a composed salad.

The Garnish. The garnish for a composed salad should add color, flavor, and texture to a dish. It may be as simple as adding a tomato wedge or a fanned strawberry or as substantial as a sliced hard-cooked egg.

The Dressing. A dressing should bring all the flavors, textures, and components of a salad together. It should complement the ingredients and not mask the other flavors in the salad. Dressing can be applied to a composed salad by tossing the salad in the dressing, by ladling it on top of the salad, or by being sprayed on the salad.

Courtesy of The National Pork Board

Bound Salads

A *bound salad* is a salad made by combining a main ingredient, often a protein, with a binding agent such as mayonnaise or yogurt and other flavoring ingredients. **See Figure 7-28.** Popular bound salads include egg salad, chicken salad, tuna salad, potato salad, and kidney bean salad. Many ingredients can be used for the main item in a bound salad including hard-cooked eggs, cooked poultry, seafood, meats, fruits, vegetables, legumes, potatoes, grains, or pastas. Minced chives, onions, and celery are common flavoring ingredients, and mayonnaise is often used as a binding agent. Other binding agents include vinaigrettes, yogurt, and creamy dressings.

Composed Salads

Charlie Trotter's

Figure 7-27. A composed salad is a salad that consists of a base, body, garnish, and dressing attractively arranged on a plate.

Bound Salads

American Egg Board

Figure 7-28. A bound salad is a salad made by combining a main ingredient, often a protein, with a binding agent such as mayonnaise or yogurt and other flavoring ingredients.

Some bound salads are served as a side dish, while others are primarily served as a filling. For example, potato salad is often served as side dish. In contrast, chicken salad is often served as a filling but is not typically served as a side dish.

Vegetable Salads

A *vegetable salad* is a salad that is primarily made of vegetables. **See Figure 7-29.** Raw or precooked vegetables can be used. Raw vegetables must be washed and trimmed to the desired size. Examples of vegetable salads include coleslaw, cucumber and onion, carrot and raisin, green bean salad, and roasted beet and goat cheese salad.

When making vegetable salads, knowing how the vegetables react to acidity is important. For example, asparagus and broccoli can turn yellow and unappealing if left in an acidic solution, such as a vinaigrette, too long. For this reason, when making a mushroom, zucchini, red pepper, and asparagus salad tossed with a vinaigrette, it is best to leave the asparagus out of the salad until immediately prior to service.

Vegetable Salads

Daniel NYC / B. Milne

Beet and Radicchio

Daniel NYC / M. Price

Artichoke and Frisée

Figure 7-29. A vegetable salad is a salad that is primarily made of vegetables.

Fruit Salads

A *fruit salad* is a salad that is primarily made of fruits. Fruit salads can be prepared from fresh, frozen, or canned fruits. **See Figure 7-30.** Fresh fruits are superior to frozen and canned fruits in taste and texture. However, professional kitchens often rely on the convenience of frozen or canned products in order to meet production demand and reduce labor costs.

Fruit Salads

Figure 7-30. Fruit salads can be prepared from fresh, frozen, or canned fruits.

Some fruit salads can be mixed or tossed, but most require the fruit to be arranged in an attractive manner, like a composed salad. Fruit salads are fragile, can discolor rapidly, and become soft when cut fruit is exposed to air too long. They should be prepared as close to serving time as possible and served chilled.

Some fresh fruits, such as apples, avocados, bananas, and pears, discolor when cut and exposed to air. To prevent rapid discoloration, these fruits should be dipped or lightly tossed in liquids that contain citric acid, such as lemon or lime juice. Using a nonreactive stainless steel knife also helps prevent discoloration.

In place of a dressing, fruit salads can be tossed with a small amount of fruit juice or fruit purée, a chiffonade of fresh mint, or a splash of a sweet liqueur. If a dressing is used with a fruit salad, it is usually a lighter dressing made with yogurt, honey, or whipped cream mixed into a fruit purée.

Gelatin Salads

A *gelatin salad* is a salad made from flavored gelatin. Gelatin salads can be presented in many different forms, colors, and flavors, and are easy to prepare. **See Figure 7-31.** In any gelatin preparation, the correct ratio of gelatin powder to liquid must be used. Fruit-flavored gelatin powder packaged for commercial use typically contains

1 lb 8 oz of gelatin powder. A total of 1 gal. of water or fruit juice should be added. Half of the water or juice should be boiling liquid, which is used to dissolve the gelatin. The other half should be cold liquid, which is used to cool and set the mixture.

Gelatin Salads

Figure 7-31. Gelatin salads are often presented in different colors and flavors.

Many gelatin recipes call for whipped cream or beaten pasteurized egg whites to be folded in to give the gelatin a light and spongy texture. The gelatin mixture must be chilled until slightly thickened before adding whipped cream, beaten egg whites, mayonnaise, sour cream, grated cheese, or fruits. This ensures even distribution of the added ingredients.

When layering two or more gelatin mixtures in the same pan or mold, each layer poured into the pan or mold must be chilled until slightly firm before adding the next layer. If one layer is set too firm, the layer placed on top may slip off or separate from the completed mold when the entire salad is unmolded. Before adding gelatin to a layer that has already set, be sure the gelatin being added is completely cool. If the gelatin being added is too warm, it can melt the set layer and the two mixtures will run together.

When preparing a gelatin mold for display on a buffet, the selected mold is filled with water, and then the water is poured into a liquid measure to determine the volume of the mold. If the amount of water held by the mold is equal to or less than the recipe yield, the mold will hold the amount of gelatin stated on the recipe.

Plating Salads

When plating any type of salad, the plates or bowls must be large enough so that the salad does not look crowded. Placing greens on a plate in a twisting motion will ensure that they sit high on the plate. Salad plates or bowls should also be served ice cold. The salad should contain a mixture of textures and colors. **See Figure 7-32.** For example, greens such as arugula, frisée, and fennel fronds add flavor and contrast to a salad. Likewise, items such as candied nuts, crunchy pancetta, and crisp toast points offer additional color, flavor, and crunch. A fresh chiffonade of herbs can also enhance a salad.

Plating Salads

United States Potato Board

Appetizer Salad

Courtesy of The National Pork Board

Entrée Salad

Figure 7-32. A plated salad should contain a mixture of textures and colors.

CHECKPOINT 7-3

1. Name six types of salads.
2. Prepare a tossed salad. Evaluate the quality of the prepared dish.
3. Identify the four parts of composed salads.
4. Prepare a composed salad. Evaluate the quality of the prepared dish.
5. Identify five common bound salads.
6. Prepare a bound salad. Evaluate the quality of the prepared dish.
7. Explain why it is important to know how vegetables will react to acidity when making vegetable salads.
8. Prepare a vegetable salad. Evaluate the quality of the prepared dish.
9. Explain how to prevent fruits from discoloring while making fruit salads.
10. Prepare a fruit salad. Evaluate the quality of the prepared dish.
11. Describe the ratio of gelatin powder to liquid used to make gelatin salads.
12. Explain why it is important to chill gelatin layers until slightly firm before combining layers.
13. Prepare a gelatin salad. Evaluate the quality of the prepared dish.
14. Plate a salad that showcases texture, color, and crunch.

National Honey Board

SALAD DRESSINGS

A *salad dressing* is a sauce for a salad. A low-quality dressing can ruin a well-prepared salad. This makes it essential to prepare salad dressings using the finest ingredients in the correct proportions. Most dressings are prepared by making an emulsion that forms a vinaigrette or mayonnaise base. An *emulsion* is a combination of two unlike liquids that have been forced to bond with each other. **See Figure 7-33.** The result is a creamy, smooth product with a uniform appearance. Emulsions can be either temporary or permanent.

Emulsions

Figure 7-33. An emulsion is a combination of two unlike liquids that have been forced to bond with each other.

Rapidly whisking oil and vinegar together creates a temporary emulsion. As the oil is broken into tiny droplets, the droplets are surrounded with a coating of vinegar. If the mixture is allowed to rest, it will separate into a pool of oil floating on a pool of vinegar. A temporary emulsion that separates can be emulsified again by shaking or stirring it thoroughly.

Mayonnaise is an example of a permanent emulsion. A permanent emulsion is formed when an emulsifier, such as egg yolk, is added to stabilize the combined liquids.

Vinaigrettes

A *vinaigrette* is a temporary emulsion of oil and vinegar that may also include additional flavorings and seasonings. The acid in a vinaigrette can liven up the overall flavor of a dish. In addition to salads, vinaigrettes are often served over broiled or grilled vegetables, poultry, and seafood.

A basic vinaigrette is commonly referred to as "basic French" in the foodservice industry and should not be confused with orange-red French or Russian dressing. In a basic French vinaigrette, the ratio of oil to vinegar is 3:1, or 3 parts oil to 1 part vinegar. When using intensely flavored vinegar, more than 3 parts oil is recommended to balance the flavors. If a mildly flavored vinegar is used, less than 3 parts oil is needed. When preparing citrus vinaigrettes where half or more of the vinegar is replaced with a citrus juice, less than 3 parts oil is needed.

An emulsified vinaigrette is more stable than a basic French vinaigrette. It takes more time to prepare an emulsified vinaigrette and the ingredients must be at room temperature before starting the preparation.

Production Tip

Canola oil is best paired with a red, balsamic, cider, flavored, or infused vinegar when creating a mustard vinaigrette.

Flavors, seasonings, and other ingredients can be added to basic French vinaigrette or emulsified vinaigrette to produce flavored dressings. Minced garlic, minced shallots, Dijon mustard, and sugar are common additions to vinaigrettes. Garlic and shallots are used to add a savory flavor, mustard adds a touch of zest, and sugar mellows the sharpness of vinegar. Ingredients such as mustard and paprika help stabilize the emulsion. Fresh herbs, spices, or acidic fruit juices can be added to further intensify flavor.

Melissa's Produce

Creamy Dressings

Mayonnaise, yogurt, or sour cream is used as the primary ingredient in creamy dressings. Mayonnaise is the foundation for creamy dressing such as ranch, thousand island, blue cheese, and green goddess. Mayonnaise is a permanent emulsion formed by dripping oil into a small amount of vinegar and egg yolks while whisking rapidly and continuously. The oil naturally repels the water in the vinegar, but the egg yolks act as an emulsifier, preventing the separation of the two opposing liquids.

Procedure for Preparing a Basic French Vinaigrette

1. Select a vinegar (or another acid) and an oil that pair well and that complement the item they are intended to dress.
2. Place seasonings, flavoring ingredients, and vinegar in a mixing bowl and whisk to incorporate.
3. Slowly add the oil in a fine stream while whisking continuously to form a temporary emulsion.
4. Let the vinaigrette rest 1–2 hours to marry the flavors.
5. Whisk to incorporate oil and vinegar prior to use.

Creaming Garlic
Media Clip

Procedure for Preparing Emulsified Vinaigrettes

1. Allow all ingredients to sit at room temperature for 1 hour prior to preparation.
2. Place pasteurized yolks or pasteurized whole eggs in a mixing bowl and whisk until foamy and light.
3. Add all dry ingredients and any flavoring ingredients to the eggs and whisk to incorporate.
4. Combine any liquid ingredients, except the oil. Add one-fourth of the liquid mixture and whisk well.
5. While whisking rapidly, begin adding the oil very slowly in a fine stream until a smooth emulsion forms.
6. Continue to add the oil very slowly in a fine stream until the mixture begins to thicken.
7. Add a little more of the liquid mixture, then add a little more oil while whisking rapidly to keep the oil and vinegar from separating. Repeat until all ingredients are incorporated.

Procedure for Preparing Mayonnaise

1. Allow all ingredients to sit at room temperature for 1 hour prior to preparation.
2. In a nonreactive mixing bowl, whisk the seasonings and half of the vinegar into the yolks until the mixture is lightly colored and foamy.
3. Add oil very slowly (one drop at a time), until the emulsion begins to form and the mixture thickens slightly.
4. Once one-fourth of the oil has been added, add the oil in a fine stream while whisking rapidly to incorporate.
5. As the mixture thickens and becomes harder to whisk, add a small amount of vinegar to thin it slightly.
6. Continue alternating the oil and vinegar while whisking constantly until all the oil and vinegar have been added. Refrigerate until needed.

Yogurt and sour cream may be used instead of mayonnaise to make creamy dressings. Yogurt-based salad dressings are typically made from fruit juice, honey, mint, and yogurt in place of mayonnaise. Yogurt-based dressings are often served with fresh fruit. Romanoff dressing is made with sour cream, whipping cream, brown sugar, and vanilla. Romanoff dressing is typically served with strawberries or peach slices.

Ranch. Ranch dressing is typically made from mayonnaise, buttermilk, sour cream, green onions, and a variety of herbs and spices. Ranch dressing is also used as a dip, spread, marinade, and sauce.

Thousand Island. Thousand Island dressing is made from mayonnaise, ketchup, onions, pickle relish, and vinegar. Thousand Island dressing is also used as a dip and a sandwich spread.

Caesar. Caesar dressing is comprised of Parmesan cheese, lemon juice, olive oil, egg, Worcestershire sauce, and black pepper. Some recipes also add anchovies. Romaine lettuce is tossed with Caesar dressing and served with toasted garlic croutons and grated Parmesan cheese to make a Caesar salad.

Honey Mustard. Honey mustard salad dressing contains lemon juice, lemon zest, honey, Dijon or a coarse-grain mustard, and olive or canola oil or mayonnaise. Honey mustard dressing may also contain a variety of herbs or spices.

Green Goddess. A classic green goddess dressing is made with mayonnaise, sour cream, tarragon, chervil, chives, anchovies, lemon juice, and pepper. This creamy dressing is named for its green color and smooth appeal to the palate.

Procedure for Preparing Green Goddess Dressing

1. Place the premeasured chopped garlic, minced anchovies, chervil, tarragon, chopped chives, and parsley in a food processor. Blend the ingredients until the parsley is finely chopped.

2. Add the premeasured Worcestershire sauce, lemon juice, and sour cream to the food processor. Blend the ingredients.

3. Add the premeasured mayonnaise to the mixture and blend until thoroughly incorporated.

4. Season the salad dressing to taste with sea salt and pepper.
5. Pour the finished dressing into a container and refrigerate until needed.

Preparing Green Goddess Dressing
Media Clip

Fruit Dressings

Fruit or fruit juice is the foundation for many fruit dressings, such as lemon dill, orange vinaigrette, ginger lime, raspberry vinaigrette, and avocado ranch. Fruits offer a natural flavor and great aroma to salad dressings. **See Figure 7-34.** Vinaigrette fruit dressings can offer a nutritional alternative to heavier creamy dressings.

Fruit-Based Dressings

National Turkey Federation

Figure 7-34. Fruits offer a natural flavor and great aroma to salad dressings.

Convenience Products

While most fine dining operations offer house-made signature dressings and dips, other foodservice operations use convenience products to save time and labor. Convenience dressings can be customized by adding a flavoring ingredient.

Plating Salad Dressings

Salad dressings can be tossed with the greens, drizzled over the top of the salad, or plated to the side of the salad. Salad dressings are often served in a small dish or pitcher. **See Figure 7-35.**

Dressing Greens

Plated Salad Dressings

Vita-Mix® Corporation

Figure 7-35. Salad dressings are often served in a small dish or pitcher.

CHECKPOINT 7-4

1. Differentiate between a temporary emulsion and a permanent emulsion.
2. Explain how to prepare a basic French vinaigrette.
3. Prepare a basic French vinaigrette. Evaluate the quality of the vinaigrette.
4. Explain how to prepare an emulsified vinaigrette.
5. Prepare an emulsified vinaigrette. Evaluate the quality of the vinaigrette.
6. Explain how to prepare a mayonnaise.
7. Prepare a mayonnaise. Evaluate the quality of the mayonnaise.
8. Differentiate between mayonnaise-based, yogurt-based, and sour-cream-based cream dressings.
9. Plate a salad that has been dressed and plate a separate salad with dressing on the side.

COLD SOUPS

Cold soups are easy to prepare and can be savory or sweet. Borscht, which is often served as a hot savory soup, can also be served as a cold soup. Classic cold soups such as gazpacho or cucumber are made by blending vegetables with water, stock, vinegar, or juice. Cold fruit soups are sweet in flavor and often enhanced with a flavorful spirit such as kirshwasser, plum wine, or fruit brandy. All cold soups are either chilled in a refrigerator or over an ice bath and then served in a chilled bowl or glass.

Savory Cold Soups

Savory cold soups can be served as an appetizer to wake up the palate or as an entrée. They can be either clear or thick and are usually made with a stock base. Some thick soups are puréed and others are served thick and chunky.

Vichyssoise. *Vichyssoise* is a puréed cold soup made with milk, potatoes, leeks, onions, and chicken stock and finished with heavy cream. **See Figure 7-36.** Vichyssoise is garnished with snipped chives or slivers of toasted leeks. Dairy-based vichyssoise is also used as a base for other cold soups.

Vichyssoise

Vita-Mix® Corporation

Figure 7-36. Vichyssoise is a puréed cold soup made with milk, potatoes, leeks, onions, and chicken stock and finished with heavy cream.

Gazpacho. *Gazpacho* is a cold soup made from stale cubes of bread, vinegar, water, olive oil, and diced tomatoes, cucumbers, and onions. **See Figure 7-37.** White gazpacho is made from cubed bread, almonds, garlic, olive oil, and water and is garnished with grapes.

Cold Soups

Gazpacho

Figure 7-37. Gazpacho is a cold soup made from stale cubes of bread, vinegar, water, olive oil, and diced tomatoes, cucumbers, and onions.

Sweet Cold Soups

A sweet cold soup is served as a soup course or in some cases as a dessert. Sweet cold soups can be made from puréed or diced vegetables or fruits. Sweet fruit soups are favored in Northern and Eastern Europe as well as in Russia.

Borscht. Cold borscht is made from red beets and garnished with cucumbers and chopped hard-cooked egg yolks and whites. The garnishes are often sweetened with brown or granulated sugar and are topped with kefir, sour cream, or yogurt. **See Figure 7-38.** Kefir is a creamy drink made from fermented milk. Borscht that is served cold is commonly called holodnik.

Cold Borscht

Figure 7-38. Cold borscht is made from red beets.

Fruit Soups. Fruit soups can be made from berries, melons, pears, citrus fruits, and tropical fruits. **See Figure 7-39.** Berry soups are usually puréed and topped with whipped cream, sour cream, or yogurt. Melon and tropical fruit soups often include coconut milk or yogurt, honey, and spices such as curry powder, or cooling herbs such as peppermint or lemongrass.

Fruit Soups

Pear Bureau Northwest

Figure 7-39. Pears and squash can be puréed together to make a flavorful cold soup.

Plating Cold Soups

Cold soups are served in chilled bowls or small glass coupes to maintain their cold temperature. Cold soups can be garnished with minced herbs, dollops of whipped cream, fresh berries, mini melon balls, fruit caviar, or toasted coconut. Small quenelles of mascarpone or goat cheese may be floated on top of the cold soup. A sauce of the same consistency as the soup can be drizzled on top and then swirled to make an attractive pattern for presentation.

CHECKPOINT 7-5

1. Name two types of cold soups.
2. Prepare a savory cold soup. Evaluate the quality of the prepared dish.
3. Prepare a cold fruit soup. Evaluate the quality of the prepared dish.
4. Explain why it is important to serve cold soups in chilled bowls or coupes.
5. Plate a cold soup using a small quenelle of cheese. Evaluate the plated dish.

CHEESES

Most cheeses used in the professional kitchen are made from the milk of cows, sheep, or goats. Cheese is most often made from milk that has been coagulated or curdled using rennet. *Rennet* is a substance that contains acid-producing enzymes or an acid. After milk curdles, it is separated into the cheese curds and whey. The *curd* is the thick, casein-rich part of coagulated milk. The *whey* is the watery part of milk. The flavor and texture of each cheese differs depending on what type of animal produced the milk, the diet of the animal, the percentage of butterfat in the cheese, and how long the cheese was aged. Cheeses are generally classified as fresh, soft, semisoft, blue-veined, hard, or grating cheese. Cheese products are processed foods made of natural cheeses that also include emulsifiers.

Fresh Cheeses

A *fresh cheese* is a cheese that is not aged or allowed to ripen. Fresh cheeses spoil easily. Baker's cheese, chèvre, cottage cheese, cream cheese, feta, mascarpone, mozzarella, Neufchâtel, and ricotta are fresh cheeses. **See Figure 7-40.** Fresh cheeses should be used soon after they are purchased.

Fresh Cheeses	
Baker's Cheese 	*Baker's cheese* is a fresh cheese made from skim milk that is like cottage cheese but softer and finer grained. Baker's cheese can be kept moist by placing it in a container, smoothing the surface until even, and covering it with a thin layer of fine sugar. Baker's cheese is used in making cheesecakes, pies, and pastries.
Chèvre 	*Chèvre* is a fresh cheese made from goat milk that varies in texture from soft to firm. Chèvre has a pure white color and a soft, spreadable texture that is slightly dry. It has a mild, slightly peppery flavor and is often blended with herbs or spices. It can be used in cooking and is also used as a spread. Montrachet is among the most popular varieties.

Figure 7-40. (continued on next page)

Cheeses

Fresh Cheeses	
Cottage Cheese  Wisconsin Milk Marketing Board, Inc.	*Cottage cheese* is a pebble-shaped fresh cheese with a mildly sour taste. It is marketed in small, large, flake, and home-style curds. Cottage cheese is often called Dutch cheese or smearcase. Cottage cheese is highly perishable and must always be refrigerated.
Cream Cheese  Wisconsin Milk Marketing Board, Inc.	*Cream cheese* is a soft, fresh cheese with a rich, mild flavor and smooth consistency that is made from cream or a mixture of cream and milk. Gum arabic is often used as a stabilizer to extend the product life of cream cheese. Cream cheese is used in spreads, salad dressings, sandwiches, and cheesecakes.
Feta 	*Feta* is a fresh cheese made from sheep milk or goat milk. Feta is slightly cured and has a salty taste. When aged properly, feta has a creamy texture and a soft to semisoft consistency. The smell is similar to cider vinegar and the taste has a faint trace of olives. Feta can be used in salads, sandwiches, and cooked dishes such as casseroles and omelets.
Mascarpone  Wisconsin Milk Marketing Board, Inc.	*Mascarpone* is a fresh cheese that has a smooth texture, is white or pale yellow in color, and has a buttery, somewhat sweet flavor. It can be used in savory dishes but is most often seen in desserts, such as tiramisu. Mascarpone works well as a spread or can be served on its own with fruit, a liqueur, or a dusting of cocoa.
Mozzarella  Wisconsin Milk Marketing Board, Inc.	*Mozzarella* is a very tender fresh cheese with a soft, elastic-like curd. Mozzarella is primarily made from cow milk. In the process of making mozzarella, the whey is drained from the curd and reserved for making ricotta cheese. Mozzarella is commonly used in pizza and lasagna. *Mozzarella di bufala* is a fresh cheese made from the milk of water buffalo or a combination of cow and water buffalo milk. It is prized for its rich, slightly sour flavor.

Figure 7-40. (continued on next page)

Fresh Cheeses	
Neufchâtel 	*Neufchâtel* is a soft, fresh cheese made from whole or skim milk or a mixture of milk and cream. Neufchâtel is similar to cream cheese, but has more moisture and less fat. Neufchâtel has a mild, slightly salty flavor that becomes more pungent as the cheese ripens. Neufchâtel is used in spreads, salad dressings, and desserts.
Ricotta  Wisconsin Milk Marketing Board, Inc.	*Ricotta* is a creamy fresh cheese that looks similar to cottage cheese but has a bland, yet slightly sweet flavor. Ricotta made in Europe is made from the whey of other cheeses, while ricotta made in North America is made using a mixture of whey and whole or skim milk. Ricotta blends well with the flavors and textures of other foods such as lasagna and manicotti.

Figure 7-40. Fresh cheeses are not aged or ripened.

Soft Cheeses

A *soft cheese,* also known as a rind-ripened cheese, is a cheese that has been sprayed with a harmless live mold to produce a thin skin or rind. A *rind* is an exterior layer of a food. The mold ripens the soft cheese by reacting with the rind. The result is a soft, suede-like outer coating and a soft interior. A soft cheese will become extremely soft and somewhat runny once it is fully ripened. Brie and Camembert are soft cheeses. **See Figure 7-41.**

Soft Cheeses	
Brie  Wisconsin Milk Marketing Board, Inc.	*Brie* is a creamy, white soft cheese with a strong odor and a sharp taste. Brie is usually produced in small disks and is best known as a buffet or dessert cheese. It is also used in hors d'oeuvres, salads, sandwiches, and hot entrées.
Camembert  Wisconsin Milk Marketing Board, Inc.	*Camembert* is a soft cheese made from cow milk and has a yellow color and a waxy, creamy consistency. Camembert is similar to Brie, but there are differences in flavor and aroma due to variations in manufacturing and ripening. The thin rind has the appearance of felt. Camembert is most often served as a dessert accompanied by crackers and fruit. Camembert should be held at room temperature prior to serving.

Figure 7-41. Soft cheeses have a thin skin, or rind.

Dry-Rind Semisoft Cheeses

A *dry-rind cheese* is a semisoft cheese that is allowed to ripen through exposure to air. The air dries out the exterior, producing a dry, almost woody rind. Although the rind becomes hard and dry, the interior of the cheese remains tender and smooth. Common examples of dry-rind semisoft cheeses are bel paese, Havarti, and Monterey Jack. **See Figure 7-42.**

Washed-Rind Semisoft Cheeses

A *washed-rind cheese* is a semisoft cheese with an exterior rind that is washed with a brine, wine, olive oil, nut oil, or fruit juice. Washing the rind generates the growth of harmless bacteria that penetrate the cheese to ripen the interior. Common examples of washed-rind semisoft cheeses are brick, Limburger, Muenster, and Port Salut. **See Figure 7-43.**

Dry-Rind Semisoft Cheeses	
Bel Paese 	*Bel paese* is a lightly colored, semisoft dry-rind cheese with a buttery flavor that melts easily. It is allowed to mature for nearly eight weeks. Bel paese can be substituted for mozzarella.
Havarti 	*Havarti* is a Danish, semisoft dry-rind cheese made from cow milk and has a buttery, somewhat sharp flavor. During the aging process, havarti develops very small holes, similar to those of Swiss cheese, but smaller.
Monterey Jack 	*Monterey Jack* is a semisoft dry-rind cheese that has a smooth texture, a creamy white color, and a mild taste. Monterey Jack that is aged for a long period of time becomes harder in texture and zestier in flavor.

Figure 7-42. Dry-rind semisoft cheeses have a dry, almost woody rind.

Washed-Rind Semisoft Cheeses	
Brick  Wisconsin Milk Marketing Board, Inc.	*Brick cheese* is a washed-rind semisoft cheese made from cow milk. It has a mild, sweet flavor and a texture that is firm, yet elastic, with many small holes. Brick cheese is sold in a brick shape. It slices and melts well and is often used for cheese platters and sandwiches.
Limburger  Wisconsin Milk Marketing Board, Inc.	*Limburger* is a washed-rind semisoft cheese with a strong aroma and flavor. Limburger cheese is made from whole or skim cow milk. It has a creamy texture that is developed through ripening in a damp atmosphere for a period of two months. Limburger cheese is often served with crackers and onions as a dessert or buffet cheese.

Figure 7-43. (continued on next page)

Washed-Rind Semisoft Cheeses	
Muenster  *Wisconsin Milk Marketing Board, Inc.*	*Muenster* is a washed-rind semisoft cheese with a flavor between that of brick cheese and Limburger. European Muenster ranges from mild to sharp in taste and may have a strong aroma due to aging. A version of this cheese is produced in North America and is comparatively mild. Muenster is often used as a buffet or sandwich cheese.
Port Salut 	*Port Salut* is a washed-rind semisoft cheese that has a soft, smooth, orange-colored rind and a glossy, ivory-colored interior. Its flavor can range from mellow to robust, depending on the age of the cheese. The aroma is often compared to a mild Limburger, and the flavor is compared to Gouda. Port Salut is used both as an appetizer and a dessert cheese.

Figure 7-43. Washed-rind semisoft cheeses have an exterior rind that has been washed with a brine, wine, olive oil, nut oil, or fruit juice.

Waxed-Rind Semisoft Cheeses

A *waxed-rind cheese* is a semisoft cheese produced by dipping a wheel of freshly made cheese into a liquid wax and allowing the wax to harden. The cheese ripens while encased in the wax. Although the outer coating is hard, the cheese inside the wax stays soft. Common waxed-rind semisoft cheeses include Edam, fontina, and Gouda. **See Figure 7-44.**

Waxed-Rind Semisoft Cheeses	
Edam  *Wisconsin Milk Marketing Board, Inc.*	*Edam* is a waxed-rind semisoft cheese made from cow milk and has a firm, crumbly texture. It is usually shaped like a ball with a slightly flattened top and bottom and has a distinctive red wax coating. Edam cheese made in the Netherlands is rubbed with oil but is not colored. However, the cheese that is exported is coated in red wax before it is shipped. Edam that is made in the United States is coated in red wax. Edam is used most often as a dessert cheese.
Fontina  *Wisconsin Milk Marketing Board, Inc.*	*Fontina* is a waxed-rind semisoft cheese made from cow milk. Young fontina cheese is somewhat soft and becomes harder as it ages. It has a nutty, herbal flavor and is perfect for use in fondues and on platters and buffets.
Gouda  *Wisconsin Milk Marketing Board, Inc.*	*Gouda* is a waxed-rind semisoft cheese that is similar to Edam but contains more fat. Gouda is usually shaped like a flattened ball or formed into a loaf. It is generally served as a dessert or buffet cheese.

Figure 7-44. Waxed-rind semisoft cheeses are produced by dipping a wheel of freshly made cheese into a liquid wax and allowing the wax to harden.

Blue-Veined Cheeses

A *blue-veined cheese* is a cheese produced by inserting harmless live mold spores into the center of ripening cheese with a needle. The blue vein that runs through these cheeses indicates where the needle was inserted and the mold spores were released. The mold decreases the time required for the cheese to ripen and is safe to eat. It also adds a distinctive flavor to the cheese. After the mold spores are injected, the exterior of the cheese wheel is salted to help keep the surface dry and prevent the mold from overtaking the exterior. Common blue-veined cheeses include blue cheese, Gorgonzola, Roquefort, and Stilton. **See Figure 7-45.**

Hard Cheeses

A *hard cheese* is a firm, somewhat pliable and supple cheese with a slightly dry texture and buttery flavor. Because hard cheeses have a firmer overall texture than softer varieties, they slice well, making them well suited for sandwiches. All hard cheeses grate well. Cheddar, Cheshire, Gruyère, Manchego, provolone, and Swiss are hard cheeses. **See Figure 7-46.**

Blue-Veined Cheeses	
Blue Cheese  Wisconsin Milk Marketing Board, Inc.	*Blue cheese*, also known as bleu cheese, is a blue-veined cheese made from cow milk and is characterized by the presence of a blue-green mold. Blue cheese is generally produced in wheels weighing approximately 7 lb. Blue cheese is used in blue cheese dressing, on sandwiches, and on cheese platters.
Gorgonzola  Wisconsin Milk Marketing Board, Inc.	*Gorgonzola* is a blue-veined cheese that is mottled with blue-green veins. Gorgonzola is generally cured for a period of 6–12 months. Gorgonzola is used in salads and salad dressings and as a dessert and buffet cheese.
Roquefort 	*Roquefort* is a blue-veined cheese made from sheep milk and is characterized by a sharp, tangy flavor. The blue-green veins are created by spreading a powdered bread mold over the curd before ripening. The cheese is aged for a period of two to five months, depending on the sharpness desired. Roquefort is used primarily as a dessert cheese but is also used in salads.
Stilton 	*Stilton* is a blue-veined cheese made from cow milk with a flavor that is milder than Roquefort or Gorgonzola. Stilton cheese has a crumbly texture and veins of blue-green mold running through the curd and wrinkled rind. Stilton is primarily used as a dessert and buffet cheese.

Figure 7-45. Blue-veined cheeses are produced by inserting harmless live mold spores into the center of ripening cheese with a needle.

Hard Cheeses	
Cheddar  *Wisconsin Milk Marketing Board, Inc.*	*Cheddar* is a hard, aged cheese that is yellow or white and ranges in taste from mild to sharp. The sharpness of Cheddar cheese varies based on the length of the aging period. Longer aging produces a sharper taste. Cheddar is used in the preparation of fondue, soufflés, spreads, and sandwiches.
Cheshire 	*Cheshire* is a hard cheese and is the oldest of the English-named cheeses. Cheshire is often sold as a Cheshire-Cheddar blend. Cheshire derives its unique, mildly salty taste from the soil in the Cheshire area of England. Cheshire is similar to Cheddar but has a more crumbly and less dense texture. Cheshire cheese is used in the same manner as Cheddar and can be used for fondue.
Gruyère  *Wisconsin Milk Marketing Board, Inc.*	*Gruyère* is a hard cheese that is similar in texture to Emmentaler but has a sharper flavor. Gruyère is an excellent cheese for fondue and veal cordon bleu.
Manchego 	*Manchego* is a hard cheese with a tangy, slightly salty flavor and is produced from sheep milk. It is aged for at least three months and has a rich golden color. Manchego is the most well-known cheese from Spain. The criss-crossed textured rind is the result of the cheese being aged in grass molds.
Provolone 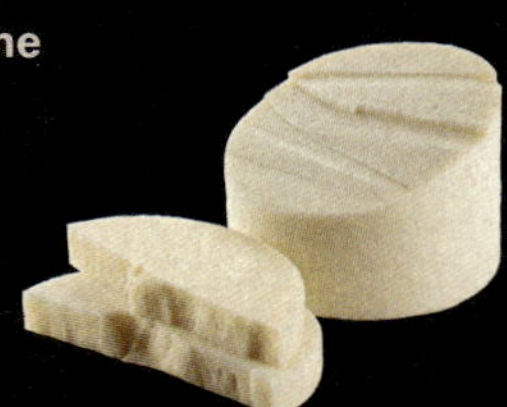 *Wisconsin Milk Marketing Board, Inc.*	*Provolone* is a hard cheese with an elastic texture and a mild to sharp taste depending on the age. It is light in color and can be cut without crumbling. A distinguishing characteristic of provolone is that it is often formed into the shape of a pear, sausage, or cone and corded for hanging. Provolone is used in the preparation of many Italian dishes, especially pizza.
Swiss Cheese  *Wisconsin Milk Marketing Board, Inc.*	*Swiss cheese* is the name used to describe varieties of hard cheeses with large holes, also known as eyes. Swiss cheese has an elastic body and a mild, sweet flavor. The holes in Swiss cheese are developed by gas-producing bacteria that release carbon dioxide during the ripening period, forming bubbles that result in the holes. Emmentaler is a Swiss-made cheese that has larger holes and a more pronounced flavor. Swiss cheese is a traditional cheese for fondue and is used in many preparations, from sandwiches to stuffed veal chops.

Figure 7-46. Hard cheeses slice and grate well.

Grating Cheeses

A *grating cheese* is a hard, crumbly, dry cheese grated or shaved onto food prior to service. The crumbly texture makes it difficult to slice. Grating cheeses are usually produced as large wheels, many weighing close to 100 lb. The most common grating cheeses are Asiago, Parmesan, and Romano. **See Figure 7-47.**

Cheese Products

A *cheese product* is a processed food made of natural cheeses that may include additional ingredients such as emulsifiers. Cheese products, such as cold-pack cheese and processed cheese, often contain other ingredients in order to increase shelf life and to produce a product that spreads and melts easily. **See Figure 7-48.**

Grating Cheeses	
Asiago  Wisconsin Milk Marketing Board, Inc.	*Asiago* is a grating cheese with a nutty, toastlike flavor. High-quality Asiago often contains a crunchy material that results from an amino acid in milk that crystallizes during the aging process. Italian Asiago is produced from the milk of cows that graze in fields where herbs and flowers grow, which gives it a distinct flavor.
Parmesan  Wisconsin Milk Marketing Board, Inc.	*Parmesan* is a grating cheese that is extremely hard, has a granular texture, and can keep indefinitely when produced properly. Parmesan from Italy is considered superior. It is available in grated or whole form and is considered a seasoning cheese. Parmesan is used to season items such as soups and pasta dishes.
Romano  Wisconsin Milk Marketing Board, Inc.	*Romano* is a grating cheese that is similar to Parmesan but softer in texture and made from cow or goat milk. Romano has a granular texture, sharp flavor, and brittle, black rind. It can be aged from 5–12 months. A longer aging period sharpens the flavor. Romano is used in various ways, including as a seasoning in pasta dishes and as an au gratin topping.

Figure 7-47. Grating cheeses have a crumbly texture that makes slicing difficult.

Cheese Products	
Cold-Pack Cheeses  Wisconsin Milk Marketing Board, Inc.	A *cold-pack cheese* is a creamy cheese product made by blending natural cheeses without the addition of heat. Cold-pack cheeses usually consist of two or more varieties of mild and sharp natural cheeses that are ground together. Cold-pack cheeses range from white to orange in color.
Processed Cheeses  Wisconsin Milk Marketing Board, Inc.	A *processed cheese* is a blend of fresh and aged natural cheeses that are heated and melted together. The result is a cheese product that can be packaged in just about any shape or size and is uniform in flavor and texture.

Figure 7-48. Cheese products are processed food made of natural cheeses and emulsifiers.

Storing Cheese

Cheese is a dairy product that contains a considerable amount of butterfat, absorbs odors easily, spoils relatively quickly, and can dry out if left exposed to air. In order to serve cheese at its best, it is important to understand proper storage and serving requirements. Cheese is perishable and should be kept tightly wrapped in plastic wrap or plastic bags and stored in the refrigerator. Wrapping cheese helps it retain moisture and keeps out odors from other foods. When refrigerated properly, cheese should retain freshness as follows:

- Fresh cheeses keep for seven to ten days.
- Soft cheeses keep for about two weeks.
- Semisoft cheeses keep for two to three weeks.
- Hard cheeses keep for about one month.
- Grating cheeses keep for several months.

Plating Cheese

With the exception of fresh and soft cheeses, cheeses are served at room temperature to maximize aroma and flavor. Many types of serving plates and platters are used to plate cheese. Mirrors and platters are often used for large scale cheese displays. Small plates are used for amenities. Cheese for a buffet may be served on a special cheese board.

When plating cheeses, it is important to offer a balance of flavors, textures, colors, aromas, and types of cheese. At least three different types of cheese should be plated together, and five different types are commonly used for a platter or large mirror. **See Figure 7-49.** The cheese should be different styles of cheese such as creamy Camembert, a well-aged Gorgonzola, and a hard crumbly Manchego or Parmigiano-Reggiano. Cheeses should be served with accompaniments such as toasted nuts, fresh or dried fruits, jams or spreads, and some type of bread or cracker.

Plating Cheese

Figure 7-49. When plating cheeses, the cheeses should be of different types.

An ideal portion for an appetizer or dessert cheese plate is ¾ to 1 oz of each cheese type per person. Up to 4 oz of cheese per guest is optimum for a cheese board or platter. If cheese is to be served after the main course, it should consist of bite-sized pieces of soothing flavors to finish the meal.

CHECKPOINT 7-6

1. Explain the purpose of rennet in making cheese.
2. Differentiate between curds and whey.
3. Identify four factors that determine the flavor and texture of a cheese.
4. Describe nine types of fresh cheese.
5. Describe two types of soft cheese.
6. Identify three ways semisoft cheeses are ripened.
7. Describe three types of dry-rind cheese.
8. Describe four types of washed-rind cheese.
9. Describe three types of waxed-rind cheese.
10. Describe four types of blue-veined cheese.
11. Describe six types of hard cheese.
12. Describe three types of grating cheese.
13. Describe how to store cheese for maximum freshness.
14. Identify five characteristics to balance when plating cheeses.

SANDWICH COMPONENTS

Sandwiches have become the symbol of a quick and tasty meal on the go. They can be made from many combinations of ingredients and can be served hot or cold. Sandwiches can be simple or elaborate. Sandwiches are one of the most versatile categories of food and allow chefs to use their imagination. Special ingredient combinations are used to create signature sandwiches around the world. **See Figure 7-50.**

Sandwich Varieties Around the World		
Sandwich	***Origin***	***Key Ingredients***
Arepa	Venezuela	A corn cake filled with chicken, avocado, cheese, shredded beef, eggs, or other ingredients; an arepa is often toasted on a griddle
Banh Mi	Vietnam	A Vietnamese-style baguette filled with pork meatballs, shredded barbeque pork, cold cut pork, grilled chicken, sardines, scrambled eggs, or julienned vegetables; garnishes may include liver pâté, shredded pickled carrots, shredded daikon, cilantro, cucumber slices, and/or jalapeños
Cubano	Cuba	Cuban bread filled with ham and roast pork marinated in mojo sauce (garlic, salt, pepper, oregano, and citrus juice), dill pickles, and Swiss cheese
Falafel	Middle East	A pita filled with a fried patty or ball of spiced chickpeas with lettuce, tomatoes, cucumbers, and onions, along with a tahini or yogurt-based sauce
Gyro	Greece	A pita filled with roasted lamb or beef cut from a vertical spit; garnishes include lettuce, tomatoes, onions, cucumbers, and a yogurt-based tzatziki sauce
Muffuletta	New Orleans	A Sicilian round bread filled with a marinated olive salad, mortadella or smoked ham, salami, and provolone cheese; the olive salad contains olives, assorted pickled vegetables, celery, pimentos, garlic, and cocktail onions
Pan Bagnat	France	A French roll filled with cooked tuna, red onions, hard-cooked eggs, olives, capers, tomatoes, anchovies, and red wine vinegar
Po' Boy	Louisiana	A baguette typically filled with fried shrimp, oysters, or catfish; ham, turkey, or roast beef are considered alternate fillings; may also include lettuce, tomatoes, and remoulade or tartar sauce
Roti	India	An Indian flatbread filled with curried meats, potatoes, and vegetables
Shawarma	Middle East	A pita filled with lamb or chicken flavored with vinegar and spices such as cardamom, cinnamon, and nutmeg; garnishes include onions, tomatoes, lettuce, pickled turnips, and cucumbers; a shawarma is often topped with hummus, tahini, or hot sauce
Torta	Mexico	A crusty or a soft roll filled with marinated steak, fried pork, or marinated pork; garnishes include avocados, lettuce, jalapeños, tomatoes, and cheese; a torta is often grilled

Figure 7-50. Signature sandwiches are created by combining ingredients that reflect flavor combinations from particular regions or cultures.

Most sandwiches consist of four main components: a base, a spread, one or more fillings, and one or more garnishes. Although some sandwiches may have all four of these components, it is not mandatory for a sandwich to include every one of them. Understanding each of the main sandwich components helps a chef to choose interesting combinations that complement each other and deliver satisfying flavors and an enticing presentation. Due to the vast variety of bases, spreads, fillings, and garnishes available, the potential number of sandwich combinations is endless.

Sandwich Bases

A *sandwich base* is the edible packaging that holds the contents of a sandwich. Bread is the most common and easy-to-handle sandwich base. Other bases include rice paper wrappers, nori, and various types of leaves such as lettuce, grape leaves, and banana leaves. The base determines the shape of the sandwich and adds flavor, color, and nutritional value. **See Figure 7-51.**

Courtesy of The National Pork Board

Figure 7-51. Sandwich bases come in many forms, such as baguettes, pitas, tortillas, and lettuce leaves.

Breads commonly used for sandwiches include various rolls such as Kaiser, French, and Italian, as well as tortillas, pitas, croissants, bagels, baguettes, sourdough bread, rye bread, focaccia bread, ciabatta bread, whole-grain breads, dried cherry or walnut breads, fruit breads, and savory breads such as onion rye or rosemary garlic. **See Figure 7-52.** Whole-grain and multigrain breads provide more nutrients than refined breads. Bread also quickly satisfies the appetite.

Sandwich Breads

Figure 7-52. Breads are available in a wide range of textures, flavors, and colors.

Some bread varieties have a very soft texture while others, such as sourdough, are more firm and chewy. Baguettes and other artisan breads such as ciabatta and pane campagnolo have a hard crusty surface and a chewy interior. When choosing a sandwich bread, consider the following points:

- The flavor of a sandwich bread should complement the other sandwich components, not overpower them.
- The thickness of the bread should not make the fillings appear skimpy.
- The bread should not tear when a spread is applied nor be too tough to bite or chew.

Sandwich breads should only be ordered as needed to offer the freshest bread possible. Other than crusty varieties, breads should be wrapped airtight to prevent them from becoming stale. Crusty varieties are not wrapped airtight as this causes the crust to become soft and lose its crispness. It is important to remember that crusty breads stale quickly and should be used within a day of purchase.

Sandwich Spreads

A *sandwich spread* is a slightly moist, flavorful substance that seals the pores of the bread and creates a thin moisture barrier. **See Figure 7-53.** However, because spreads are slightly moist, they also can make breads soggy if added too far in advance of service. Common types of sandwich spreads include mayonnaise, butters, salad dressings, purées, and variety spreads. A spread should complement the flavor of the main sandwich ingredient.

Spreads as Moisture Barriers

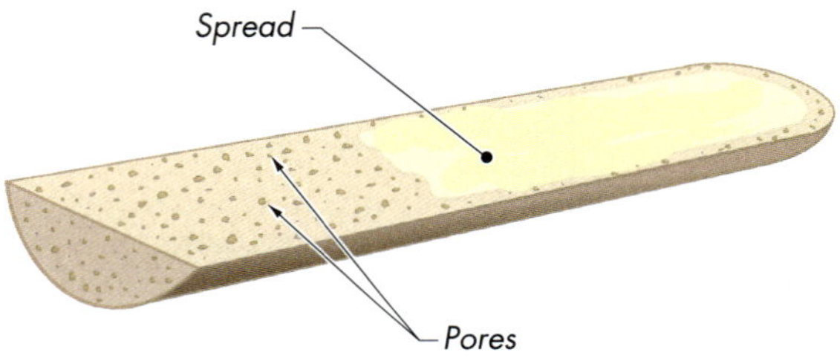

Figure 7-53. A sandwich spread seals the pores of the bread, creating a thin moisture barrier.

Mayonnaise. Mayonnaise is the most commonly used spread. Mayonnaise is an uncooked emulsion made by combining oil, egg yolks, and lemon juice or vinegar. An emulsion is a combination of two unlike liquids that have been forced to bond with each other. Rapid mixing separates tiny droplets of each ingredient, allowing them to combine. As an oil-based spread, mayonnaise acts as a moisture blocker, helping to prevent bread from becoming soggy. However, butter is a better moisture blocker than mayonnaise.

Mayonnaise is often used as a binding agent to hold sandwich fillings together, such as with egg salad, chicken salad, tuna salad, and potato salad. **See Figure 7-54.** Mayonnaise also can be flavored in the same manner as a compound butter by adding garlic, mustard, hot chili paste, curry, lemon, dill, capers, wasabi, or horseradish. Because mayonnaise is much less dense than butter, it does not have to be whipped to incorporate flavoring ingredients.

Spreads as Binding Agents

Alpha Baking Co., Inc.

Figure 7-54. Mayonnaise is used as a binding agent for sandwich fillings, such as with egg salad.

Butters. Butter is one of the most common sandwich spreads. It adds richness and flavor to a sandwich and also acts as a barrier to prevent moisture from saturating the bread. Butters used on sandwiches should be somewhat soft or whipped before use to make the butter spread more evenly without tearing the bread. Chefs will often whip a small amount of water into butter to lighten its texture and density before using it as a sandwich spread. This technique also increases the volume of the butter as air is incorporated during the whipping process.

Salad Dressings. Salad dressing refers to sandwich spreads similar in flavor, texture, and appearance to mayonnaise. The primary difference between salad dressing and mayonnaise is that salad dressing does not contain egg yolks and it is typically sweeter. Like mayonnaise, salad dressing can also be flavored with other ingredients.

Purées. Purées of fruits, vegetables, and legumes can also be used as flavorful sandwich spreads. Purées offer lower fat, cholesterol, and calorie alternatives to other types of spreads. **See Figure 7-55.** For example, a cilantro hummus purée made with garbanzo beans, cilantro, lemon, tahini paste, and olive oil enhances the flavor of a grilled southwest chicken sandwich. A sundried tomato, roasted garlic, and pine nut purée complements a grilled Portobello mushroom sandwich. Likewise, a spicy jalapeño and basil pesto purée spread pairs well with a grilled cumin-rubbed steak sandwich. A roasted pineapple honey mustard purée goes nicely with a grilled mahi mahi sandwich. Other types of purées include guacamole, refried beans, baba ghanoush (eggplant purée), and spreadable cooked pâtés.

Variety Spreads. A *variety spread* is any other food mixture of a spreadable consistency that can be added to a sandwich to complement or increase flavor and moisture. Spreadable cheeses, an olive tapenade, a spicy green olive jardinière, and jams are examples of variety spreads that enhance sandwiches.

Nutritional Comparison of Spreads (1 oz each)

Nutrition Facts	Butter	Mayonnaise	Cream Cheese	Fresh Avocado	Hummus
Calories	200	110	100	50	50
Total Fat	25 g	10 g	10 g	5 g	5 g
Saturated Fat	14 g	1 g	6 g	1 g	0 g
Cholesterol	60 mg	10 mg	30 mg	0 mg	0 mg

Figure 7-55. Purées of fruits, such as avocado, and legumes, such as chickpeas (hummus), offer lower fat, cholesterol, and calorie spreads compared to butter, mayonnaise, or cream cheese.

Sandwich Fillings

A *sandwich filling* is the main ingredient in a sandwich and is stacked, layered, or folded on top of the base to form a sandwich. The filling is often the reason people order a particular sandwich. Many sandwiches are named for their filling, such as Thai turkey wraps and Italian sausage and peppers. **See Figure 7-56.** Types of sandwich fillings include poultry, seafood, meats, cheeses, bound salads, vegetables, or a combination of these items.

Some sandwiches may be composed of several fillings. For example, the Reuben contains corned beef, sauerkraut, and Swiss cheese. When multiple fillings are used, it is important to select flavors that complement one another.

Since the filling is the main attraction of a sandwich, it is essential that the filling be prepared and served properly. The filling served should never be at room temperature. The temperature of the filling should be appropriately hot or cold, depending on the type of sandwich. For example, when serving a hot fish sandwich, the fish should be hot, moist, flaky, and not overcooked or undercooked. It is also important to place the driest fillings next to the bread to increase the time it takes for the bread to become soggy from the moisture-rich fillings.

Sandwiches

Filling-Named Sandwiches

Photo Courtesy of Perdue Foodservice, Perdue Farms Incorporated

Thai Turkey Wrap

Courtesy of The National Pork Board

Italian Sausage and Peppers

Figure 7-56. Many sandwiches are named for their filling, such as Thai turkey wraps and Italian sausage and peppers.

Poultry. Poultry used on sandwiches can be served either precooked and chilled or cooked to order. Precooked poultry includes sliced or diced breast meat, grilled strips, smoked breast meat, and deli-style meat. A chicken breast is grilled, broiled, or fried before being added to a sandwich. **See Figure 7-57.** Chicken breasts also may be marinated, seasoned, or barbequed to add flavor and moisture during the cooking process.

Turkey-based meat substitution products that are lower in fat and in cholesterol than other meats are also popular sandwich fillings. Examples of turkey-based meat substitution products include turkey sausage, turkey bacon, turkey bologna, turkey pastrami, turkey ham, ground turkey burgers, and turkey hot dogs.

Poultry Fillings

Photo Courtesy of Perdue Foodservice, Perdue Farms Incorporated

Figure 7-57. A chicken breast can be grilled, broiled, or fried before being used as a sandwich filling.

Fish and Shellfish. Fish and shellfish are also common protein fillings that are used to make sandwiches. Flake tuna and salmon, crab meat, surimi (imitation crab meat), and cooked shrimp are commonly used to make bound salad fillings. **See Figure 7-58.** These items are either fresh or canned and have been fully cooked before being added to a bound salad.

Fish fillets are also common sandwich fillings. The fillets can be grilled, broiled, sautéed, or fried. Other seafood products may be cured or smoked, such as smoked salmon or lox, and simply sliced and served.

Seafood Fillings

Alpha Baking Co., Inc.

Figure 7-58. Seafood fillings include crab meat and surimi.

Meats. Many meat varieties used in sandwiches are precooked and chilled prior to slicing and serving. This is because attempting to cook them to order in a quick amount of time would not be possible as they come from larger cuts that take hours to prepare properly. Some of them are referred to as deli meats or processed meats. Common examples of fully cooked and cold sandwich style meats include ham, bacon, roast beef, pastrami, salami, and corned beef. Some precooked meats can also be reheated and served hot such as roast beef, hot dogs, pulled or barbecued pork, sausages, corned beef, and bologna.

A common mistake is slicing sandwich meats thick instead of thin. A few thick slices of meat stacked on a sandwich will have far less volume and therefore look like less than an equal amount of thinly sliced meat. **See Figure 7-59.** Also, thicker cut meats tend to be chewier, especially when cold. Thinly sliced sandwich meats make the sandwich portion appear larger, are more tender, and are generally easier to eat.

Some sandwich meats are cooked to order. These meat fillings are individually portioned and have a short cooking time. Examples of à la carte sandwich meats include hamburgers, sausages, steaks, and fish fillets.

Cheeses. Cheese is a wonderful sandwich filling that adds color, flavor, and nutrients. Cheese complements almost any other type of filling. Sliced cheese dries out easily so it is important to keep it covered in an airtight container or in plastic wrap. Gloves should always be worn when handling cheese to prevent mold growth.

Cream cheese can be used as a filling or a spread. The neutral flavor of cream cheese lends itself to the addition of a variety of flavoring ingredients. Common cream cheese variations include veggie cream cheese (includes puréed or grated vegetables such as carrots, zucchini, and onions), fruit cream cheese (includes puréed fruit such as strawberries or pineapple), and caper cream cheese, which often accompanies lox and bagels.

Vegetables and Fruits. Vegetables and fruits can also be used as the main filling for a sandwich. Grilled vegetable or fruit sandwiches provide a meatless option and are full of flavor, color, and nutrients. **See Figure 7-60.** Vegetables and fruits also add texture, flavor, nutrients, and moisture to meat-filled sandwiches.

Meatless Fillings

California Strawberry Commission

Figure 7-60. Grilled fruit sandwiches, such as a strawberry panini, are a meatless sandwich option full of flavor, color, and nutrients.

Meat Fillings

Photo Courtesy of Perdue Foodservice, Perdue Farms Incorporated

Figure 7-59. Thinly sliced meat gives a sandwich more volume than thick slices of meat.

Vegetables and fruits can also serve as a garnish for a sandwich. Sliced tomatoes, lettuce, and pickles are commonly served on the side of a hot sandwich so they do not get limp from the heat. These same vegetables add color and contrast to cold sandwich fillings. Diced vegetables and fruits such as celery and apples are often folded into bound salads to add crunch and flavor. Fresh vegetables and fruits add a fresh, crisp taste to any sandwich.

Eggs. Eggs are a versatile filling for many types of sandwiches and can be served at any mealtime. **See Figure 7-61.** Hard-cooked eggs are cooled, peeled, chopped, and then tossed with mayonnaise, seasonings, and diced vegetables to make a bound egg salad that is perfect on toasted or fresh sliced bread. Hard-cooked eggs can also be sliced and used as a sandwich garnish.

Egg Fillings

Photo Courtesy of Perdue Foodservice, Perdue Farms Incorporated

Figure 7-61. Eggs are a versatile filling for many types of sandwiches, such as a pastrami egg muffin.

Bound Salads. A *bound salad* is a salad made by combining a main ingredient, often a protein, with a binding agent such as mayonnaise or yogurt and other flavoring ingredients. Popular bound salads include egg, chicken, ham, and tuna salad. **See Figure 7-62.** A vegetable bound salad can be used to create a meatless sandwich.

Nutrition Note

Using yogurt instead of mayonnaise as a bound salad dressing offers a healthy alternative. Blending plain low-fat yogurt with Dijon mustard and cilantro or chives yields a flavorful and creamy dressing without a lot of saturated fat.

Bound Salad Fillings

Figure 7-62. Popular bound salads include egg, chicken, ham, and tuna salad.

Sandwich Garnishes

Garnishes add color, texture, and nutrition to sandwiches and should be carefully chosen. A *sandwich garnish* is a complementary food item that is served on or with a sandwich. **See Figure 7-63.** Examples of garnishes include lettuces, tomato and onion slices, pickle slices or spears, olives, raw vegetables (crudités), grilled peppers, crumbled or shredded cheese, sliced fresh fruit, pasta salad, and coleslaw. The number of garnishes that can be used alone or in various combinations to complement sandwiches is endless. It is important to choose garnishes that do not overpower or overshadow the main filling.

Sandwich Garnishes

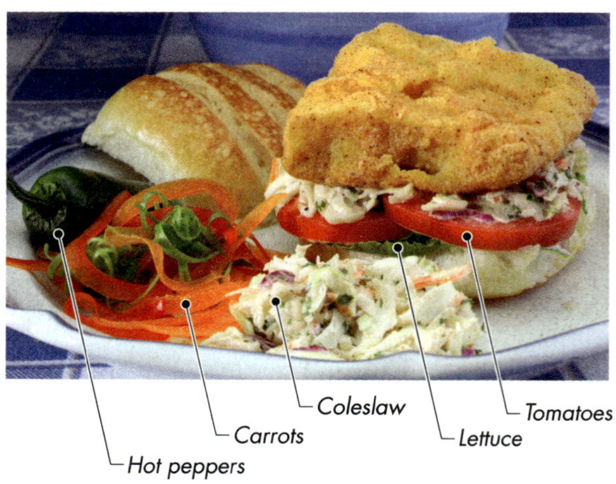

Florida Department of Agriculture and Consumer Services

Figure 7-63. Garnishes complement sandwiches by adding color, texture, and nutrition.

CHECKPOINT 7-7

1. Identify the four main sandwich components.
2. Identify common types of sandwich bases.
3. Identify five common types of sandwich spreads.
4. Prepare a puréed sandwich spread. Evaluate the texture and flavor of the spread.
5. Identify common types of sandwich fillings.
6. Prepare a cold meat-filled sandwich and a cold vegetarian sandwich. Compare the results.
7. Prepare a bound salad sandwich filling. Evaluate the texture and flavor of the filling.
8. Identify common sandwich garnishes.
9. Prepare a variety of sandwich garnishes.
10. Create a signature sandwich.

COLD SANDWICHES

Cold sandwiches are so called because their fillings are served below room temperature. Often the fillings are cooked ahead of time and then refrigerated, such as with roast beef, turkey, or ham luncheon meats. Cold sandwiches can be grouped into four distinct types: cold open-faced sandwiches, cold closed sandwiches, cold wrap sandwiches, and tea sandwiches. Although each type has some or all of the four basic sandwich components, they are quite different in presentation.

Cold Open-Faced Sandwiches

A *cold open-faced sandwich* is a variety of cold sandwich that consists of a single slice of bread that is often toasted or grilled and then coated with a spread and topped with thin slices of poultry, seafood, meat, partially cooked or raw vegetables, or a thin layer of a bound salad and a garnish. Cold open-faced sandwiches have a more upscale appearance than closed sandwiches. A lox and cream cheese sandwich is a popular cold open-faced sandwich. **See Figure 7-64.**

Cold Open-Faced Sandwiches

Alpha Baking Co., Inc.

Figure 7-64. A lox and cream cheese sandwich is a popular cold open-faced sandwich.

Cold Closed Sandwiches

Cold closed sandwiches are the quickest to prepare and the most commonly served. A *cold closed sandwich* is a variety of cold sandwich that consists of two pieces of bread, or the top and bottom of a bun or roll, coated with a spread and loaded with one or more fillings and garnishes. Examples of cold closed sandwiches include hero, submarine, and liverwurst sandwiches. **See Figure 7-65.**

Cold Closed Sandwiches

Photo Courtesy of Perdue Foodservice, Perdue Farms Incorporated

Figure 7-65. A cold closed sandwich, such as an Italian submarine, consists of two pieces of bread, or the top and bottom of a bun or roll, coated with a spread and loaded with one or more fillings and garnishes.

Multidecker sandwiches, such as the turkey club, are a special type of cold closed sandwich. A *multidecker sandwich* is a sandwich that consists of three pieces of bread, a spread, and at least two layers of garnishes and fillings. **See Figure 7-66.**

Multidecker Sandwiches

Figure 7-66. Multidecker sandwiches consist of three pieces of bread, a spread, and at least two layers of garnishes and fillings.

Cold Wrap Sandwiches

A *cold wrap sandwich* is a variety of cold sandwich in which a flat bread or tortilla is coated with a spread, topped with one or more fillings and garnishes, and rolled tightly. It is typically wrapped in parchment or waxed paper and cut in half on the bias to reveal the filling. A cold wrap sandwich is one of the most convenient sandwiches to eat, because the person eating it can hold on to the paper rather than the sandwich.

Procedure for Preparing Cold Wraps

1. Lightly coat ⅔ of a piece of flatbread (such as a tortilla) with a spread, staying ¾ inch from the outside edge.
2. Arrange desired fillings and garnishes on top of the spread.
3. Roll the flatbread around the fillings about ⅓ of the way up the flatbread.
4. Fold in both sides of the flatbread around the rolled portion to seal the ends of the wrap.
5. Continue to roll the flatbread tightly around the fillings.
6. Roll the sandwich in deli wrap or parchment paper in the same manner that the flatbread was rolled.
7. Cut the wrap in half on the bias and serve.

An *aram sandwich*, also known as a levant, is another type of a cold wrap. An aram sandwich is created by spreading cream cheese on softened cracker bread called lahvosh and then layering thin slices of meats, cheeses, lettuce, and pickles before rolling it up, wrapping it tightly in plastic wrap, and refrigerating it for several hours. This cold wrap sandwich is cut into 1 inch slices before being served.

Preparing Multidecker Sandwiches
Media Clip

Tea Sandwiches

A *tea sandwich* is a petite and delicate cold sandwich with a trimmed crust and a soft filling. The tea sandwich originated in England as a snack served with afternoon tea. The breads used for tea sandwiches are light and soft in texture. The crusts are trimmed off to make them easy to eat without creating a lot of crumbs. The spreads used in tea sandwiches are usually flavorful, and the fillings are usually delicate in flavor, such as with cucumber or watercress fillings. Tea sandwiches are often cut into rectangles, squares, triangles, or circles or can be rolled and sliced into pinwheels. **See Figure 7-67.**

Tea Sandwich Cutting Patterns

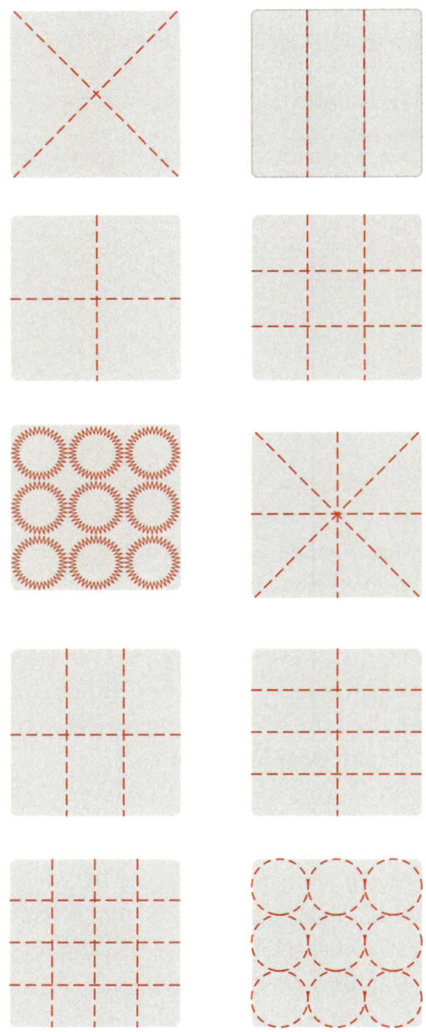

Figure 7-67. Tea sandwiches are often cut into rectangles, squares, triangles, or circles or can be rolled and sliced into pinwheels.

Sandwich Preparation

Preparing sandwiches is a skill that relies on organization, consistency, and speed. Whether preparing sandwiches in large quantities, such as for a banquet, or individually to order, the sandwich station must be well organized to reduce the amount of movement needed and maximize efficiency. If needed, each ingredient in the sandwich station must be washed, dried, sliced, cut, mixed, and portioned in advance.

The equipment needs of a sandwich station are determined by the menu. For example, if deli meats and cheeses are bought sliced, there is no need for an electric slicer at the sandwich station. If there are hot items, such as a hot roast beef sandwich with au jus, the station needs to have a steam table to keep the items hot. Equipment found at a sandwich station includes storage equipment, hand tools, portioning equipment, and cooking equipment. **See Figure 7-68.**

- Storage equipment at a sandwich station typically includes refrigerators, steam tables, and holding cabinets.
- Hand tools such as knives, cutting boards, spatulas, and spreaders are basic necessities for a sandwich station.
- Portioning equipment such as digital or mechanical scales and portion control scoops are essential for making consistent sandwiches.
- Cooking equipment at a sandwich station may include a griddle, grill, broiler, oven, toaster, and a fryer.

Range of Motion. Because much of sandwich making is assembly work, it is important that the individual making the sandwiches use both hands for efficient production. A sandwich maker can determine a comfortable range of motion by extending both arms directly out in front and then sweeping each arm in an arc to each side. **See Figure 7-69.** All of the ingredients in a sandwich station should be within this range of motion to maximize efficiency and speed.

Range of Motion

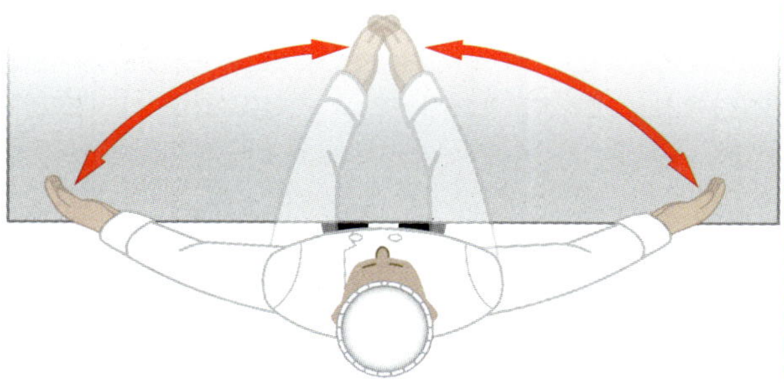

Figure 7-69. A comfortable range of motion can be determined by extending both arms directly out in front and then sweeping each arm in an arc to each side.

Sandwich Station Equipment

Barker Company

Figure 7-68. A sandwich station includes storage equipment, hand tools, portioning equipment, and cooking equipment.

Right-handed sandwich makers should place the bread supply to their left, and left-handed sandwich makers should place the bread supply to their right. Spreads and filling ingredients should be placed directly in front of the sandwich maker. An organized workspace enables efficient production and reduces the time that ingredients are exposed to the temperature danger zone.

Quantities of sandwiches can be made using an assembly procedure. This procedure includes grilling or toasting the bread (if desired) and coating one side of each slice with a spread, placing garnishes and hot or cold fillings on top of the spread, adding the top piece of bread, and then securing the sandwiches before cutting them for service.

Hobart

Procedure for Preparing Sandwiches in Quantity

1. Arrange bread or toast slices on a sheet pan lined with parchment paper.
2. Use a spatula to evenly coat each slice with a spread.

3. Arrange garnishes on top of the bread or toast in the two center rows.

4. Place portioned fillings on top of the garnishes, making sure the bread is well covered.
5. Place the remaining slices of bread on the top of slices holding the fillings and garnishes.

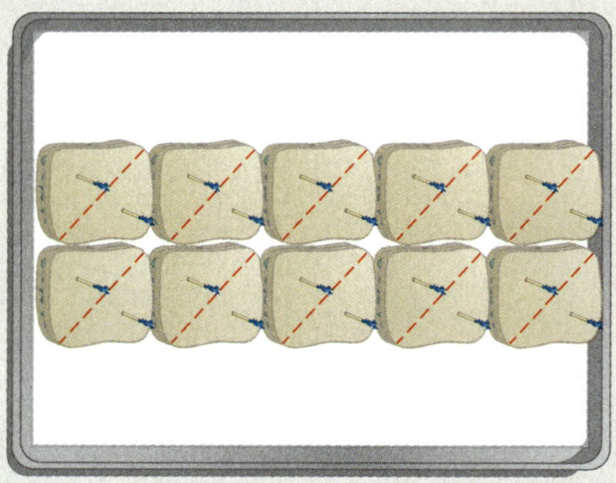

6. Secure each sandwich with frill picks and then cut each sandwich in half.

Preparing Sandwiches in Quantity
Media Clip

Portion Control. All items for sandwich making should be portioned prior to service. All meats should be sliced and portioned by weight. Individual portions can either be placed in small plastic bags or in small piles separated by individual squares of waxed paper and then stacked or layered in a storage container. Items that are served by the slice, such as cheese, should be sliced to an exact thickness to ensure a consistent portion.

Practicing proper portion control helps maintain accurate food costs, as the menu price of an item is based on that item's portion size. If portions are not consistent from one sandwich to the next, the actual cost of making each sandwich will also not be consistent. **See Figure 7-70.**

Portion Control

Edlund Co.

Figure 7-70. If portions are not consistent from one sandwich to the next, the actual cost of making each sandwich will also not be consistent.

Proper portion control also ensures that customers receive the same value for their money. For example, if two customers at the same table order corned beef sandwiches and one customer receives more meat than the other customer, the customer who receives less meat will not be happy with the sandwich.

Sandwich Stabilizers

An assembled sandwich should have a neat appearance and should be easy to eat. While different types of bread, buns, and flatbreads make handling sandwiches easier, a sandwich often needs to be secured to prevent the fillings from falling out while the sandwich is being held. Cheeses, wrappers, and frill picks serve as stabilizers that secure sandwich components together. **See Figure 7-71.**

Sandwich Stabilizers

Alpha Baking Co., Inc.

Cheeses

Photo Courtesy of Perdue Foodservice, Perdue Farms Incorporated

Wrappers

Photo Courtesy of Perdue Foodservice, Perdue Farms Incorporated

Frill Picks

Figure 7-71. Cheeses, wrappers, and frill picks serve as stabilizers that secure sandwich components together.

Cheeses. In addition to being a sandwich filling, cheese also serves as a stabilizer that makes a sandwich easier to hold and eat. The cheeses used on many hot closed sandwiches bind the bread to the fillings due to the heat of the main filling.

Wrappers. One of the easiest ways to secure a sandwich is with a wrapper. Sandwiches on a bun can be served in a paper sleeve to make them easier and less messy to handle. Likewise, sandwich wraps can be rolled in a piece of waxed paper or parchment paper. Having a paper wrapper around a sandwich not only keeps the hands cleaner but is a fast, safe, and convenient way to hold the entire sandwich together.

Frill Picks. Cold closed sandwiches are commonly held together with frill picks, which are slightly longer toothpicks that have a frilly plastic decoration at one end and a sharp point on the other end. The pointed end is inserted into the top of a sandwich to secure the top and bottom breads to the contents inside. The frill picks are quite visible to prevent a choking accident. A frill pick may also be used to secure the breads to the fillings of some hot closed sandwiches.

Condiment Sauces and Dips

Sandwiches are often served with a condiment sauce or a dip that enhances or complements the flavor of the sandwich. **See Figure 7-72.** Condiment sauces and dips can be sweet or savory, smooth or chunky, and served cold or hot.

Condiment Sauces

Alpha Baking

Figure 7-72. Sandwiches are often served with condiment sauces or dips.

Cold dips often have sour cream, yogurt, mayonnaise, or cream cheese as the base ingredient. Other cold dips, such as guacamole and hummus, have a purée as the main ingredient. Almost any flavoring ingredient, such as crabmeat or cooked spinach, can be added to the base. Cold dips can be served in a small dish, bread bowl, or hollowed vegetable.

Hot dips often have a Mornay, tomato, or béchamel sauce as the base ingredient. A variety of items can be added to the base for flavoring. For example, jalapeños add a punch of flavor to a cheddar-based Mornay sauce. Goat cheese is an appetizing addition to a fresh tomato sauce. Hot dips need to be held at the appropriate temperature and then plated immediately.

Condiment sauces and dips, such as ketchup and ranch dressing, have stable flavor profiles. Other condiment sauces and dips contain layers of flavor, as in the case of pestos and salsas. Ketchups, cocktail sauce, barbeque sauces, mayonnaises, mustards, aiolis, tartar sauces, yogurts, pestos, salsas, pepper sauces, and Asian sauces are used as condiment sauces and dips.

Ketchups. *Ketchup* is a thick, tomato-based product that usually includes vinegar, sugar, salt, and spices. Because vinegar and sugar are two of the main ingredients in ketchup, it has a slightly sweet-and-sour acidic flavor. Once opened, ketchup will retain its freshness longer if it is refrigerated, although it is not necessary. Ketchup loses color and quality as it deteriorates with age.

Cocktail Sauce. *Cocktail sauce* is a thick tomato based product that usually contains vinegar, sugar, salt, spices, grated horseradish, hot sauce, Worcestershire sauce, and lemon juice. It is used as a condiment with oysters, shrimp, or crabmeat in a cocktail glass and is also served with fried seafood of all types.

Barbeque Sauces. A *barbeque sauce* is a type of sauce that is used to baste proteins and that can be served as an accompaniment to a cooked protein. The flavors of barbeque sauces range from sweet and tangy to smoky and spicy. Barbeque sauces can feature a mayonnaise, tomato, mustard, or vinegar base. Barbeque sauces are served hot or room temperature when used as a condiment sauce or dip.

Mayonnaises. *Mayonnaise* is a thick, uncooked emulsion formed by combining oil with egg yolks and lemon juice or vinegar. Although often exceptional in quality, house-made mayonnaise only lasts up to four days, while commercially prepared mayonnaise has a shelf life of approximately six months. Mayonnaise is often used as a sandwich spread and as a base for salad dressings, bound salads, and sauces.

Mustards. *Mustard* is a pungent powder or paste made from the seeds of the mustard plant. In addition to mustard seeds, ingredients commonly added to mustards include vinegar or wine, salt, and various spices. Mustard comes in many different varieties. Some mustards are made from ground seeds, resulting in a very smooth product. Other mustards use cracked or whole seeds for added texture and a more rustic appearance. Depending on the type of seed used to prepare mustard, the flavor can range from mild and tangy to hot and spicy. Common mustards used as condiments include Dijon mustard, English mustard, and prepared yellow mustard. **See Figure 7-73.**

Mustards	
Dijon Mustard	*Dijon mustard* is a light-tan mustard that has a strong, tangy flavor and is made from brown or black mustard seeds, vinegar, white wine, sugar, and salt. Because of its ingredient blend, Dijon mustard has a less acidic taste than prepared yellow mustard. In addition to use as a condiment, Dijon mustard is often used in salad dressings, sauces, wet rubs, and marinades.
English Mustard	*English mustard* is a type of yellow mustard that has a hot, spicy flavor and is made from ground yellow and brown or black mustard seeds, wheat flour, and turmeric. Because English mustards are extremely hot and spicy, only a small amount of this condiment is needed.
Prepared Yellow Mustard	A *prepared yellow mustard* is a type of bright-yellow mustard that is mild in flavor and made from ground yellow mustard seeds, vinegar, and turmeric. Because it has a high acid content, prepared yellow mustard does not spoil. However, it can dry out and darken on the surface, and the flavor deteriorates with age. Prepared yellow mustard is often served with sandwiches and is also used in wet rubs.

Figure 7-73. Depending on the type of seed used to prepare mustard, the flavor can range from mild and tangy to hot and spicy.

Aiolis. *Aioli* is a thick emulsion made by slowly blending olive oil with raw garlic and egg yolks or cooked potatoes. Aioli tastes like a mayonnaise that has been flavored with garlic. Aioli is often used as a dip or a sauce for steamed or grilled vegetables, poultry, fish, and shellfish.

Tartar Sauces. *Tartar sauce* is a mayonnaise-based sauce that contains finely chopped relish or pickles, capers, grated onions or minced chives, and fresh parsley. It may also contain, minced fresh dill, tarragon, and lemon juice. It is used as a garnishing sauce for fried fish and seafood.

Yogurts. *Yogurt* is a semisolid food made from milk that has fermented due to the addition of bacteria. Some yogurts are sweetened and/or flavored. Plain yogurt is often mixed with herbs to make a variety of condiment sauces and dips. Yogurt-based sauces and dips are frequently served with Middle Eastern, Greek, and Indian dishes.

Pestos. A *pesto* is a sauce made from fresh ingredients that have been crushed with a mortar and pestle or finely chopped in a food processor before being mixed with olive oil. Cilantro, basil, mint, pine nuts, and parmesan cheese are common pesto ingredients.

Salsas. A *salsa* is a sauce made by mixing or puréeing diced vegetables or fruits, herbs, and spices together. Salsas offer low fat content and high nutrient values. **See Figure 7-74.** Both salsas and relishes have textures that range from coarse to puréed. Mangoes are often used to make salsas. The peel and stone are removed when preparing mangoes for use in a dish. The flesh of a mango clings to a large flat stone in the center of the fruit.

Salsas

Tomato Salsa

Salsa Verde

Vita-Mix® Corporation

Figure 7-74. Salsas offer nutritious options that are low in fat.

Procedure for Preparing Mangoes

1. Using a chef's knife, carefully slice off the top and bottom of the mango. Place the mango on the cutting board so the bottom is lying flat. *Note:* This should slightly expose the stone inside the mango.

2. Using the tip of the chef's knife, start at the top edge of the mango and cut down in a slight sawing motion to remove a strip of the peel. *Note:* Try not to remove too much of the flesh.

3. Following the contour of the mango, use a slight sawing motion to cut away another small slice of peel. Rotate the mango clockwise and repeat the cutting motion until the entire peel has been removed.

4. Once the peel been removed, position the knife about ¼ inch on one side of the stone. *Note:* The stone is a flat oval about ½ inch thick and about 1–1½ inches wide.

5. Slice down from the top to the bottom of the mango to remove the flesh on one side of the stone. Repeat this same cut along the other side of the stone.

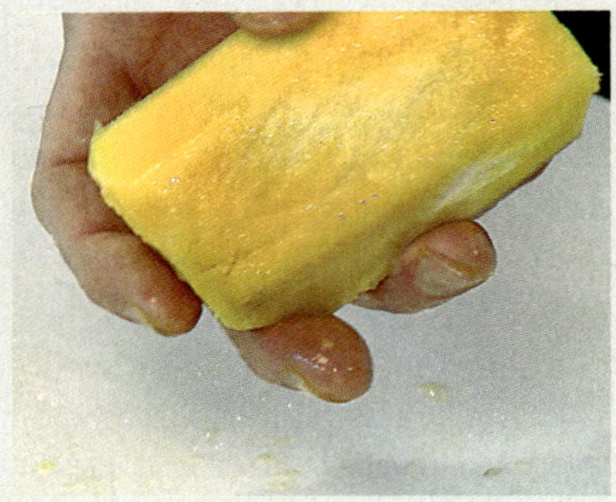

6. Reserve the mango halves and place the piece containing the stone on the cutting board.

7. Cut the flesh away from the stone by slicing at angles. Discard the stone.

8. Slice or dice the flesh of the mango.

Use the diced mango in a recipe of choice.

Preparing Mangoes
Media Clip

Pepper Sauces. A *pepper sauce* is a type of hot and spicy sauce made from various types of chili peppers and typically includes vinegar. Pepper sauces impart heat and flavor. The burn that is felt when pepper sauces are consumed is caused by capsaicin. *Capsaicin* is a potent compound that gives chiles their hot flavor. Capsaicin is found in the seeds and the veins or membranes of a chili pepper. Scoville units are used to measure the heat level of chili peppers. **See Figure 7-75.**

Asian Sauces. Many Asian sauces used in rice, noodle, and vegetable dishes can also be used as dipping sauces. Common Asian sauces used in the professional kitchen include Asian chili sauce, fish sauce, hoisin sauce, and soy sauce. **See Figure 7-76.**

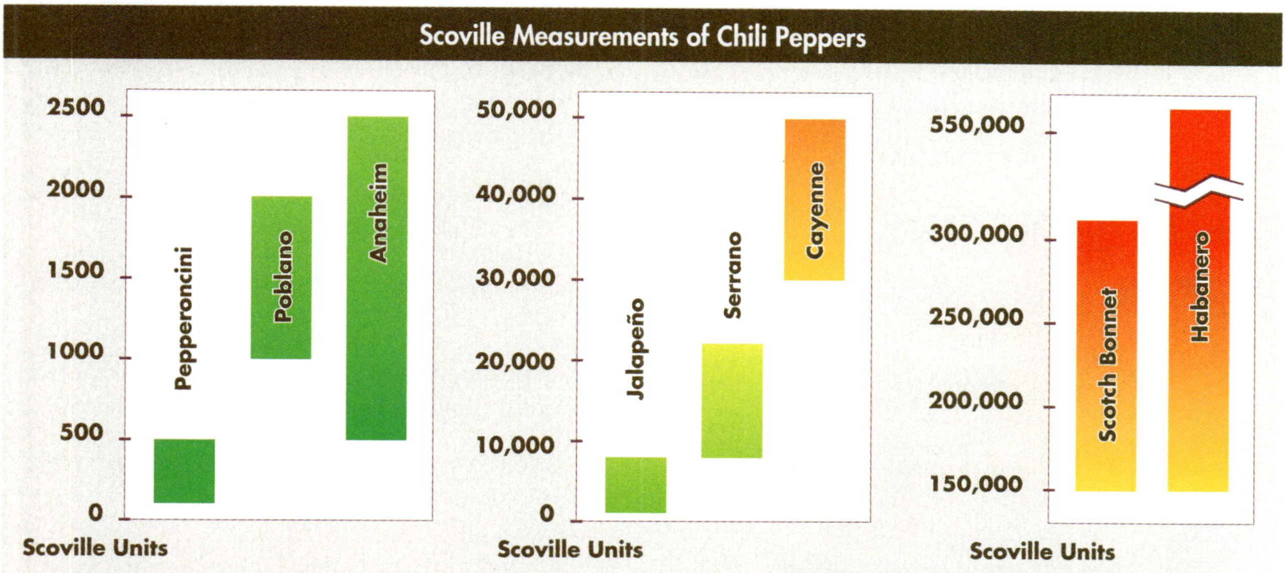

Figure 7-75. Different chiles contain varying amounts of capsaicin, and Scoville units are used to measure the heat level of hot peppers.

Asian Sauces	
Asian Chili Sauce	*Asian chili sauce* is a red-colored Asian sauce made from puréed red chiles and garlic. Asian chili sauce has a spicy, savory flavor and is marketed as Sriracha, sambal, or sambal oelek.
Fish Sauce	*Fish sauce* is a type of Asian sauce made from a liquid drained from fermented, salted fish. Chiles or sugar can be added to the sauce. Fish sauce has a pungent, salty flavor that adds depth of flavor to dishes. Fish sauce can be mixed with sesame seed oil and served as a dip.
Hoisin Sauce	*Hoisin sauce* is a type of Asian sauce made from fermented soybean paste, garlic, vinegar, chiles, and sugar. Hoisin sauce is a rich amber color and has a sweet and spicy flavor. Hoisin sauce is served as a dipping sauce for meats, dumplings, and fried items.
Soy Sauce	*Soy sauce* is a type of Asian sauce made from mashed soybeans, wheat, salt, and water. Tamari soy sauce is made with little or no wheat and is darker, thicker, and milder in taste than regular soy sauce. Soy sauce is a common dipping sauce for Asian dishes.

Figure 7-76. Many Asian sauces used in rice, noodle, and vegetable dishes can also be used as dipping sauces.

Plating Sandwiches

Consistent sandwich presentation is important. Cold sandwiches are typically cut into halves or quarters to expose the interior of the sandwich and to make it easier to eat.

Most sandwiches are served with a side dish that complements that particular sandwich. Side dishes such as coleslaw, pasta salad, potato salad, French fries, sweet potato fries, potato or vegetable chips, pickled ginger, cut fruit, or a side salad add visual appeal, flavor, and nutrition to plated sandwiches. **See Figure 7-77.** Any garnish used should complement the sandwich.

Plated Sandwiches

Photo Courtesy of McCain Foods USA

Figure 7-77. Common side dishes such as sweet potato fries add visual appeal and flavor to a plated sandwich.

CHECKPOINT 7-8

1. Identify four types of cold sandwiches.
2. Prepare and plate a cold open-faced sandwich and a cold closed sandwich. Evaluate the appearance and flavor of the two sandwiches.
3. Prepare and plate a multidecker sandwich. Evaluate the appearance and flavor of the sandwich.
4. Prepare and plate a cold wrap. Evaluate the appearance and flavor of the sandwich.
5. Prepare and plate tea sandwiches using two different cutting patterns. Evaluate the appearance of the two cutting patterns.
6. Explain the importance of range of motion at a sandwich station.
7. Describe how to prepare large quantities of sandwiches.
8. Prepare and plate a quantity of sandwiches using a bound salad filling. Evaluate the appearance and flavor of the sandwiches.
9. Describe common ways to portion meats and cheeses at a sandwich station.
10. Identify 12 types of condiment sauces and dips.
11. Prepare a condiment sauce or dip for a plated dish. Evaluate the resulting flavor combination.
12. Identify common side dishes served with plated sandwiches.

Courtesy of The National Pork Board

HORS D'OEUVRES AND APPETIZERS

The terms hors d'oeuvre and appetizer are often used interchangeably. However, there is a difference. An *hors d'oeuvre* is an elegant, bite-size portion of food that is creatively presented and served apart from a meal. **See Figure 7-78.** An *appetizer* is food that is larger than a single bite and is typically served as the first course of a meal. Most appetizers are meant to be shared. Because hors d'oeuvres and appetizers commonly precede a meal, their flavor should complement the courses to come. They should also be appropriately seasoned so that no further seasoning is necessary.

Hors d'Oeuvres

Hors d'Oeuvres

Photo Courtesy of Perdue Foodservice, Perdue Farms Incorporated

Figure 7-78. An hors d'oeuvre is an elegant bite-size portion of food that is creatively presented and served apart from a meal.

A guest should never need a knife to eat an hors d'oeuvre. However, some hors d'oeuvres may require a cocktail fork or toothpick because they are coated in a sauce, making them difficult to eat with the fingers. Hors d'oeuvres can be made from a countless number of fillings, spreads, and garnishes and may be served hot or cold. Hors d'oeuvres may also be served buffet style. Buffet presentations usually combine hot and cold varieties of hors d'oeuvres. If a food was prepared hot, it should be served hot or held in a manner that would enable it to be eaten hot. The same is true for cold items. Serving items at the appropriate temperature signifies a commitment to providing guests with high-quality food and enhances the dining experience.

Most appetizers are served with a dipping sauce and may require the use of one or more utensils. **See Figure 7-79.** Appetizers are often simple foods but can also be elegant. Common appetizers include fried mozzarella sticks, battered onion rings, nachos, and Buffalo wings. More elegant appetizers include beef Carpaccio, spring rolls, or bruschetta.

Canapés and starters are terms used to describe various types of hors d'oeuvres and appetizers. Regardless of the term used, small bites are popular menu items.

Canapés

A *canapé* is an hors d'oeuvre that looks like a miniature open-faced sandwich. **See Figure 7-80.** Canapés are often served at receptions, teas, and other events where a lighter fare is desired over a full meal. Canapés are composed of a base, a spread, and an edible garnish.

Canapés

Indian Harvest Specialtifoods, Inc./Rob Yuretich

Figure 7-80. A canapé is an hors d'oeuvre that looks like a miniature open-faced sandwich.

Appetizers

Photo Courtesy of Perdue Foodservice, Perdue Farms Incorporated

Shared

Daniel NYC / E. Laignel

Individual

Figure 7-79. Most appetizers are served with a dipping sauce and may require the use of one or more utensils.

Procedure for Preparing Canapés Using Toasted Bread

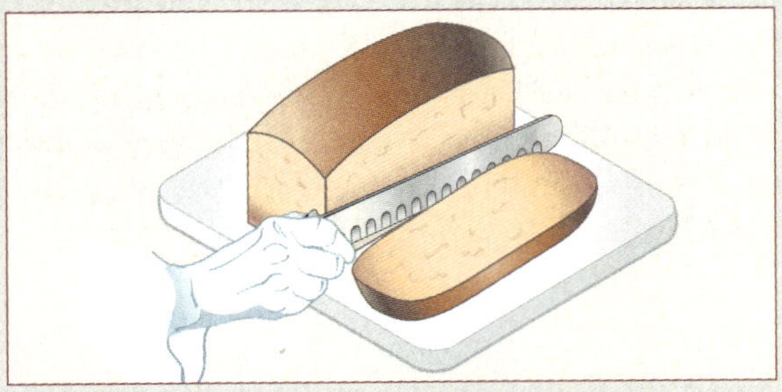

1. Remove the crusts and reserve for making croutons or bread crumbs.
2. Slice the bread lengthwise into evenly sliced pieces.

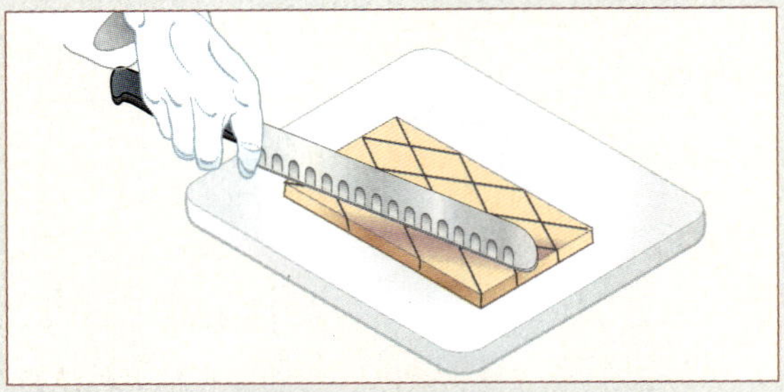

3. Cut the bread into desired shapes.
4. Brush both sides of the cutouts with melted butter and place in a 450°F oven for 8–10 minutes.
5. When evenly golden, remove slices from oven and cool completely.

6. Cover the toasted bread with a thin layer of spread.

7. Garnish as desired.

Procedure for Preparing Canapés Using Untoasted Bread

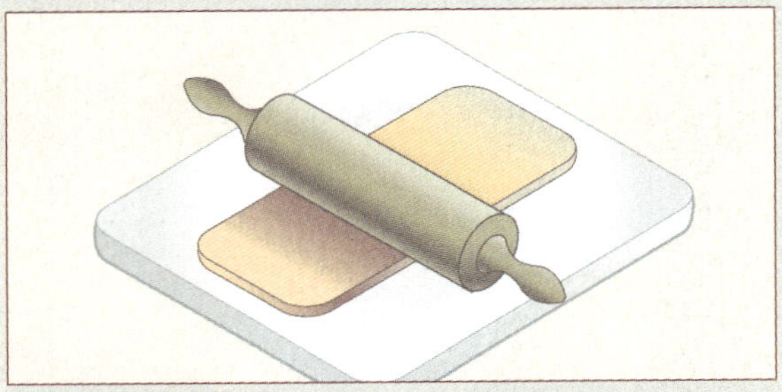

1. Remove the crusts and reserve for making croutons or bread crumbs.
2. Slice the bread lengthwise into evenly sliced pieces.
3. Using a rolling pin, firmly roll over the bread slices to flatten them.

4. Cover the flattened bread with a thin layer of spread.

5. Cut the bread into desired shapes.

6. Garnish as desired.

Canapé Bases. Bread is the most common base for canapés. Canapé bases also can be made from slices of fresh, firm vegetables such as cucumbers, squash, or carrots. The vegetables are thinly sliced into rounds and topped with a spread. However, toasted bread can handle more moisture and holds its shape better than fresh bread. If freshly sliced bread is used, it must be firm enough to hold the topping without becoming soggy. When making canapés using bread as a base, a decision has to be made regarding the use of toasted or fresh bread. Slightly different procedures are used for making canapés with toasted and untoasted bread.

Canapé Spreads. A canapé spread can be smooth and spreadable or thick with texture. Spreads start with a main ingredient, such as butter, cream cheese, or mayonnaise, or a puréed food, such as meats, seafood, vegetables, or legumes. Flavoring ingredients, such as lemon juice, herbs, Dijon mustard, or roasted garlic, are then blended with the main ingredient to create a spread. Smooth spreads include compound butters, pâtés, and hummus. A pastry bag is often used to apply spreads to a base. To prevent the pastry tip from clogging and to keep a smooth texture, solid ingredients should be puréed.

Textured spreads include bound salads made from vegetables, legumes, poultry, seafood, or meats. The salad ingredients are finely chopped or slightly puréed and mixed with a binding agent such as mayonnaise, vinaigrette, or yogurt. Examples of bound salads include chicken salad, tuna salad, and grilled vegetable salad.

Canapé Garnishes. A canapé garnish should be delicate and provide an elegant finished appearance. Examples of garnishes include the leaf of a delicate herb, a thin slice of smoked salmon, or a dab of caviar or fried capers.

Production Tip

Canapés can be made to reflect almost any dish. For example, a sliver of rare beef can be draped over a crostini and topped with horseradish cream, or a fudge brownie can be topped with raspberry preserves and crème fraîche.

Cold Starters

Some well-known cold starters include cocktails, crudités, and relishes. These starters are served in a range of foodservice operations from casual restaurants to banquets and buffets.

Cocktails. A *cocktail* is a chilled appetizer served in a stemmed glass. Poached and chilled shrimp placed on the edge of a glass surrounding cocktail sauce is a well-known cocktail. **See Figure 7-81.**

Cocktails

Harbor Seafood, Inc.

Figure 7-81. Poached and chilled shrimp placed on the edge of a glass is a well-known cocktail.

A cocktail can be as simple as cubed fruit or as complex as seafood tossed with avocado, jalapeño, cilantro, lime juice, and chile sauce. Cocktails should always be served as fresh and as cold as possible.

Crudités. Crudité is a French term meaning "raw things." A *crudité* is a group of raw vegetables arranged on a platter and served with a dipping sauce. Any vegetable that is acceptable to eat raw can be cut into bite-size pieces and arranged beautifully on a serving platter as crudités. A crudité platter may consist of broccoli, carrots, celery, bell peppers, snow peas, and mushrooms. **See Figure 7-82.** Cherry tomatoes, cauliflower, radishes, zucchini, yellow squash, cucumbers, and asparagus are also popular crudités.

Relishes. A *relish* is an assortment of uncooked vegetables that are served raw, marinated, or pickled. **See Figure 7-83.** A variety of vegetables are suitable for marinating or pickling. Examples of these vegetables include cucumbers, olives, beets, peppers, artichoke hearts, baby corn, and mushrooms. Marinated or pickled vegetables can range in flavor from sour and tangy to sweet and spicy. Relishes should be served cold with a small amount of the marinade or pickling liquid.

Raw Meat and Seafood Starters

Raw meats and seafood are often served as starters. Only fresh meats and seafood that have been handled with extreme care can be served raw. Raw meats and seafood are popular dishes that include items such as carpaccio, tartar, caviar, and sushi. Some foodservice operations have raw bars that feature a variety of raw seafood presentations.

Crudités

Carlisle FoodService Products

Figure 7-82. A crudité platter may consist of broccoli, carrots, celery, bell peppers, snow peas, and mushrooms.

Relishes

Carlisle FoodService Products

Figure 7-83. A relish is an assortment of raw, marinated, or pickled vegetables.

Carpaccio

Courtesy of Chef Gui Alinat

Figure 7-84. Carpaccio is thin slices of meats or seafood that are served raw.

Carpaccio. *Carpaccio* is thin slices of meats or seafood that are served raw. **See Figure 7-84.** Typically, the meat or seafood is partially frozen in order to make slicing it easier. The thinly sliced meat is then attractively arranged on a serving plate, seasoned with salt and pepper, and often drizzled with extra virgin olive oil or lemon or lime juice.

Tartares. *Tartare* is freshly ground or chopped raw meat or seafood. Tartare is usually seasoned and served with toast points or another cracker-like accompaniment. Steak tartare and tuna tartare are often served as starters.

Caviars. *Caviar* is the harvested roe of sturgeon. Although roe, or eggs, are harvested from other varieties of fish, such as salmon and whitefish, only sturgeon roe can be labeled as caviar. Other fish roe must be labeled by the type of fish it comes from, such as salmon caviar or whitefish caviar. The highest quality caviar is harvested from beluga, osetra, and sevruga. **See Figure 7-85.** Caviar is often served on toast points, blinis, or chilled, cooked fingerling potatoes with a dollop of sour cream. Standard accompaniments for caviar include finely minced hard-cooked eggs and red onions.

Caviar

Tru, Chicago

Figure 7-85. The highest quality caviar is harvested from beluga, osetra, and sevruga.

Production Tip

Caviar should only be handled with a mother of pearl, china, or wooden spoon. Metal spoons cause a chemical reaction which gives the caviar a metallic flavor.

Sushi. *Sushi* is a vinegar-seasoned rice dish garnished with raw fish, cooked seafood, eggs, or vegetables. **See Figure 7-86.** The term sushi is derived from "zushi," the Japanese term for cold, vinegar-seasoned, cooked rice. Sushi rice is made by steaming or simmering a sticky, short-grain rice until it is slightly firm and chewy. The cooked rice is gently mixed with rice wine vinegar and sugar syrup. Once the sushi rice cools, it is formed by hand into various shapes. Common forms of sushi include nigirizushi, makizushi, inarizushi, and chirashizushi.

- *Nigirizushi* is a type of sushi made with small, hand-formed mounds of vinegar-seasoned cooked rice that is garnished with a topping.
- *Makizushi*, also known as a maki roll, is a type of sushi made with vinegar-seasoned cooked rice layered with other ingredients and rolled in a dried seaweed paper called nori. Maki rolls form a cylinder shape that is then sliced into bite-size portions.
- *Inarizushi* is a type of sushi made with vinegar-seasoned rice and toppings stuffed into a small purse of fried tofu.
- *Chirashizushi,* also known as scattered sushi, is a type of sushi made by scattering toppings over or mixing toppings into vinegar-seasoned rice.

Sushi

Nigirizushi

Maki Rolls

Figure 7-86. Sushi is a vinegar-seasoned rice dish garnished with raw fish, cooked seafood, eggs, or vegetables.

Sushi

Raw Bars. A *raw bar* is a presentation of a variety of raw and steamed seafood presented and served on a bed of ice. **See Figure 7-87.** Common items found on raw bars include shrimp, crab legs, clams, mussels, and oysters. Shellfish shooters are also popular. Shellfish shooters consist of raw or cooked seafood presented in a shot glass or on the half shell with a flavorful sauce or liquid. Common shellfish shooters include oysters in horseradish cocktail sauce, mussels on the half shell with a creamy vinaigrette, and shrimp in a roasted pepper and spicy tomato juice cocktail.

Raw Bars

Carlisle FoodService Products

Figure 7-87. A raw bar is a presentation of a variety of raw and steamed seafood served on a bed of ice.

Production Tip

Using ice-cold metal serving bowls and utensils is essential to maintaining safe serving temperatures on a raw bar. The temperature of the raw items needs to be checked hourly. Plenty of chilled serving plates need to be provided so guests will not reuse plates when returning to the raw bar.

Plating Hors d'Oeuvres and Appetizers

Hors d'oeuvres and appetizers should be plated in odd numbers and kept approximately ½ inch apart so that each piece can be seen as well as picked up without disturbing the other pieces. **See Figure 7-88.** Ceramic squares, rounds, or rhombus-shaped small plates or ceramic arches, plastic triangles, or rectangular plates with built-in sauce holders are used to plate hors d'oeuvres and appetizers. The plates used and the food served on the plates should be easy for guests to handle. Small serving forks, picks, or pins are used to hold items together as needed. Some of these serving items may be offered separately to guests.

CHECKPOINT 7-9

1. Differentiate between hors d'oeuvres and appetizers.
2. Prepare canapés using both toasted and untoasted bread.
3. Describe three types of cold starters.
4. Prepare a cocktail, a crudité tray, and a relish tray. Evaluate the quality of each prepared dish.
5. Describe five types of raw meat and seafood starters.
6. Plate a tray of hors d'oeuvres. Evaluate the tray for spacing, presentation, and flavor.
7. Plate an appetizer. Evaluate the dish for presentation and flavor.

Plated Hors d'Oeuvres and Appetizers

Pieces

Spoons

Pops

Push Pumps

Skewers

Courtesy of Chef Eric LeVine

Figure 7-88. Plated hors d'oeuvres and appetizers should be easy for guests to handle.

CHARCUTERIE

Charcuterie is the art of making sausages and other preserved items such as pâtés, terrines, galantines, and ballotines. Charcuterie began in the 15th century as a way of preserving pork products prior to refrigeration. Items were smoked, cured, or processed with a high amount of fat to preserve and protect them from being exposed to the air and spoiling. Today, charcuterie refers to cured, smoked, or preserved items made from a variety of meats, poultry, game, seafood, and vegetables. **See Figure 7-89.**

Charcuterie

Daniel NYC / E. Laignel

Figure 7-89. Charcuterie refers to cured, smoked, or preserved items made from a variety of meats, poultry, game, seafood, and vegetables.

Charcuterie

Common forms of preserving charcuterie items include curing and smoking. Cured charcuterie items are often served as hors d'oeuvres or appetizers. Prosciutto, gravlax, and kippered herring are examples of cured meats. Smoked charcuterie items can be prepared by either hot or cold smoking. While poultry, seafood, and meats are commonly smoked, other foods such as cheeses and vegetables can also be smoked. The length of time an item is exposed to smoke determines the intensity of the smoke flavor.

Forcemeat Ingredients

Forcemeat is the main ingredient of sausages, pâtés, terrines, galantines, and ballotines. A *forcemeat* is a mixture of raw or cooked meat, poultry, seafood, or vegetables, fat, seasonings, and sometimes other ingredients. Additional ingredients may include cubed bread, heavy cream, eggs, and edible garnishes that are strategically placed inside the forcemeat for both flavor and appearance. Depending on the intended use, a forcemeat can be ground smooth like a purée, coarsely, or somewhere in between. When uncooked, properly prepared forcemeat should have a shiny and slick appearance. When cooked, it should appear moist and have a smooth texture.

A *grinder* is a tool that consists of an auger, which pushes the forcemeat forward, and a cutting blade that cuts the forcemeat as it is pushed through perforated stainless steel dies. **See Figure 7-90.** Dies come in several sizes, including small, medium, and large. Forcemeat is usually ground in a progressive style, using a larger die for the first grind and then replacing it with a smaller die to make the next grind and so forth. When grinding forcemeats, it is important to refrigerate all parts of the grinder after it has been thoroughly cleaned and sanitized.

Grinders

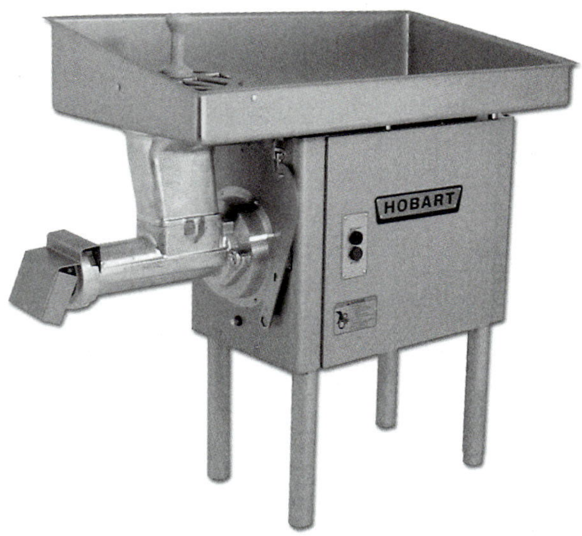

Hobart

Figure 7-90. A grinder cuts the forcemeat as it is pushed through perforated stainless steel dies.

Meats, Poultry, Game, or Seafood. The main protein used in most charcuterie preparations is meat, poultry, game, or seafood. Almost all forcemeats incorporate some amount of pork, because pork adds a smooth texture to forcemeats that would otherwise be dry or rubbery. Pork is usually added to forcemeats without a pork base in a ratio of 1 part pork to 2 parts other protein. Veal, game meats, or vegetables may also be used.

Livers. Liver is often added to forcemeats as a binding agent and as a flavoring. The most common types of liver used in forcemeats are from poultry. Poultry livers are small, tender, and milder in flavor as compared to pork, lamb, or beef liver. It is important to process livers appropriately before grinding.

Procedure for Preparing Liver for Use in a Forcemeat

1. Trim liver of all connective tissue and cut liver into 1 inch pieces.
2. Wash liver thoroughly under cold running water and drain well.
3. Place liver in a bowl, cover completely with milk, and soak for at least 24 hours.
4. Drain the liver from the milk. Wash the liver well under cold running water and drain again. *Note:* At this stage the liver can be added to pâtés or terrines where larger pieces of liver may be desirable.
5. Purée the liver in a blender.
6. Strain the liver purée through a chinois or cheesecloth-lined china cap.
7. Add the liver purée to a forcemeat.

Fats. Fat adds moisture and flavor to a forcemeat. In classical forcemeat recipes, the percentage of fat to lean meat is typically 50% meat and 50% fat. However, the protein may be as high as 65%. Higher percentages of protein result in a dry product.

A hard fat, such as pork fatback, is the most desirable fat to use in a forcemeat. The dense structure of fatback prevents it from melting as easily as a softer fat, allowing it to easily bind with the meat. When forcemeats are not processed properly, the fat and any additional moisture will separate from the meat when it is cooked and yield a dry and grainy forcemeat. A forcemeat with a perfect emulsification will be smooth, moist, and have a pleasant texture.

Binding Agents. Forcemeats that include liquids, such as wine, brandy, or heavy cream, may need an additional binding agent other than the ground protein from which they are made. The addition of a liquid may overwhelm the protein's binding and emulsification properties, causing the fat and liquids to leak out of the forcemeat during the cooking process. Common binding agents include whole eggs, egg whites, dried bread, grains, and flour. Binding agents help bind protein to liquids, resulting in a better quality forcemeat. **See Figure 7-91.**

Binding Agents

Figure 7-91. Binding agents, such as eggs, help bind protein to liquids, resulting in a better quality forcemeat.

Seasonings and Flavorings. Forcemeats must be seasoned prior to cooking or baking. Almost any herb or spice may be incorporated in a forcemeat, depending on the desired flavor profile. However, salt is the most important seasoning to add to a forcemeat. Salt will intensify the flavor of the other ingredients in the forcemeat and facilitate the emulsification of the protein and the fat. Salt also acts as an agitator during the grinding process.

Curing salt, also known as TCM, is a combination of table salt and sodium nitrite. **See Figure 7-92.** Curing salts are used to reduce the spoilage caused by harmful bacteria and enhance the flavor of the meat being cured. Forcemeats that do not include curing salt can have a grayish appearance, whereas those produced with curing salt look fresher.

Marinades add flavor and moisture to a forcemeat. Meat absorbs the aromas, flavors, and moisture from a marinade. Oils, alcohol, such as wine or brandy, herbs, spices, citrus zest, and aromatic vegetables are common marinade ingredients.

Curing Salts

Figure 7-92. Curing salt, also known as TCM, is a combination of table salt and sodium nitrite.

Garnishes. Garnishes, such as batonnet of blanched carrots, bell pepper julienne, pistachios, hard-cooked eggs, whole eggs, peeled eggs, slices of lean cooked tenderloin, cooked seafood, diced ham, or truffles, can be placed in the middle of a forcemeat in a uniform pattern prior to cooking. When a garnished forcemeat is sliced, the colorful garnishes make a beautiful presentation. **See Figure 7-93.** Garnishes must be cooked prior to being added to a forcemeat because shrinkage occurs during the cooking process and can create unattractive air gaps in the finished product.

Forcemeat Garnishes

Daniel NYC / E. Kheraj

Figure 7-93. When a garnished forcemeat is sliced, the colorful garnishes make a beautiful presentation.

Forcemeat Varieties

Varieties of forcemeats include straight, country-style, mousseline-style, emulsified, and gratin. Straight, country-style, and mousseline forcemeats are commonly used for the production of pâtés, terrines, and galantines. Emulsified and gratin style forcemeats are not commonly used in foodservice operations.

Straight Forcemeats. A *straight forcemeat* is a forcemeat that consists of seasoned ground meat emulsified with ground fat. Some straight forcemeats may incorporate liver or eggs. Pâtés and terrines are typically prepared using a straight forcemeat.

Country-Style Forcemeats. A *country-style forcemeat,* also known as pâté de campagne, is a coarsely ground forcemeat with a strong flavor. **See Figure 7-94.** Country-style forcemeats are typically marinated with a substantial amount of garlic, onion, bay leaves, juniper berries, freshly ground pepper, and wine or brandy. They also often include pork or chicken liver. Most varieties of sausage are country-style forcemeats.

Country-Style Forcemeats

Photo Courtesy of D'Artagnan, Photography by Doug Adams Studio

Figure 7-94. A country-style forcemeat, also known as pâté de campagne, is a coarsely ground forcemeat with a strong flavor.

Mousseline-Style Forcemeats. A *mousseline-style forcemeat* is a forcemeat made from less-fibrous proteins such as poultry, seafood, veal, or pork. The protein is commonly ground first and then puréed. It is finished by folding in heavy cream and eggs or egg whites. A mousseline-style forcemeat is commonly used to make seafood sausages, hot or cold timbales, and terrines. When properly prepared, mousseline-style forcemeats have a light and tender consistency and a more finished appearance than the other styles of forcemeats. Adding too much cream results in too fragile a forcemeat, and too many eggs or egg whites results in a rubbery forcemeat.

Emulsified Forcemeats. An *emulsified forcemeat,* also known as the 5-4-3 style, is a forcemeat made from a ratio of five parts protein, to four parts fat, to three parts crushed ice. Hot dogs, knockwurst, and mortadella are examples of emulsified forcemeats.

Gratin Forcemeats. A *gratin forcemeat* is a forcemeat made from meat that is partially seared prior to grinding and a starchy binding agent. Dried white bread or a cooked grain that has been soaked in milk, known as a panada, is typically ground with the meat to help emulsification occur.

Forcemeat Preparation

When preparing forcemeats all of the ingredients and equipment must be ice cold. Having everything cold will help the meat grind easier and prevent it from becoming tough during processing. Even the part of the grinder used to process the forcemeat needs to be refrigerated prior to grinding to ensure it does not warm the forcemeat as it is ground. Because charcuterie products have a high ratio of fat added while grinding, keeping items cold helps prevent the fat from getting soft or melting and results in a better emulsification.

Procedure for Preparing Forcemeats

1. Completely chill all ingredients and equipment at or below 41°F.
2. Trim meat of all connective tissue and cut into approximately 2 inch cubes.
3. Mix meat and fat in a marinade and refrigerate until needed. *Note:* Meat can be partially frozen as the cold will help the emulsification process.
4. Cut fatback into 2 inch cubes and place the cubes in the freezer until needed.
5. If the specific recipe calls for liver, trim the liver of all veins and connective tissue. Grind the liver to a fine texture and force it through a fine-meshed strainer or tamis. Place the ground liver in the freezer until needed.
6. Place a medium die in the cold grinder and grind all of the meat, including the liver (if applicable).
7. Change to a small die and grind the meat again.
8. Grind the partially frozen fatback and then mix well with the ground meat. *Note:* When using a food processor, add a small amount of crushed ice to offset the heat generated by the friction of the blades.
9. Add other ingredients such as eggs, panada, and seasonings and mix well to incorporate.
10. Check forcemeat for consistency and overall flavor by making and tasting a quenelle.
11. Refrigerate forcemeat until needed.

CHECKPOINT 7-10

1. Explain the purpose of curing and smoking foods.
2. Identify the main ingredient of sausages, pâtés, terrines, galantines, and ballotines.
3. Explain why forcemeat is ground using a progressive style of grinding.
4. Explain the role of fat as a forcemeat ingredient.
5. Explain the role of a binding agent as a forcemeat ingredient.
6. Identify the two main ingredients in a curing salt.
7. Explain why garnishes have to be cooked prior to being added to a forcemeat.
8. Name three styles of forcemeats used in foodservice operations.
9. Describe how forcemeat is prepared.
10. Prepare a forcemeat and then poach a quenelle to test the texture and flavor of the forcemeat. Evaluate the quality of the forcemeat.

CHARCUTERIE APPLICATIONS

A forcemeat is used as the primary component in making sausages, pâtés, terrines, galantines, and ballotines. Forcemeats may be molded into various shapes such as cylinders, ovals, and rectangles. The forcemeat used must be an emulsion that holds together when it is sliced. Forcemeats should always have a rich taste and rest well on the palate.

Sausages

A *sausage* is a forcemeat that is typically stuffed into a casing or shaped into a patty. The common ratio of fat to lean meat in most sausages is 1 to 3 or about 33% fat. Fat percentages lower than 25% will result in a dry and mealy sausage. Adding a starch, such as rice or oats, and crushed ice to a lean sausage forcemeat helps lower the fat percentage without drying out the sausage because the starch absorbs the moisture.

Sausage seasonings vary depending on the variety of sausage being prepared and the flavor profile desired. Curing salts can be either added to the diced meat during the marinating process or can be added during the grinding procedure. Some sausages are simply composed of ground meat, fat, and seasonings, while others may include garnishes such as a chiffonade of fresh oregano, roasted red peppers, feta cheese crumbles, or sliced black truffles. Garnishes are always added after the grinding process to keep the identity of the garnish intact.

Sausages can be produced from pork, veal, seafood, poultry, beef, game meats, or vegetables. The three basic categories of sausages are fresh, cured, and smoked.

Fresh Sausages. *Fresh sausage* is a type of sausage that is freshly made and has not been cured, smoked, dried, or further processed in any way. According to the USDA, a fresh sausage contains no nitrites, which are added to other categories of sausages to help preserve them and aid in the processing. Fresh sausages are made solely of seasoned ground forcemeat. Game meats such as rabbit and venison are commonly used to make fresh sausages.

Cured Sausages. *Cured sausage* is a type of raw sausage that contains sodium nitrite mixed into the forcemeat. Salt and sodium nitrite help to greatly extend the shelf life of cured sausages as they work to prevent the development of foodborne pathogens. Cured sausages can be sold in many forms from raw and soft, such as with Italian sausage, to hard, dry, and firm, such as with salami. **See Figure 7-95.**

Cured Sausages

Photo Courtesy of D'Artagnan, Photography by Doug Adams Studio

Figure 7-95. Cured sausages can be sold in many forms from raw and soft, such as with Italian sausage, to hard, dry, and firm, such as with salami.

Smoked Sausages. *Smoked sausage* is a type of cured sausage that is smoked during the curing process. Smoked sausages can be either hot smoked or cold smoked. Hot-smoked sausages are placed in a heated smoker, which cooks the sausages as they are being smoked. Cold-smoked sausages are placed in a smoker set below 100°F so the sausages are not cooked during the smoking process. Cold-smoked sausages are then cured or cooked in another manner before being served to guests.

Sausage Casings

Sausage casings are used for almost every type of sausage produced. A *natural casing* is a casing produced from the intestines of sheep, hogs, or cattle. Natural casings are cleaned thoroughly and then salted heavily before being stuffed. They are sold in salt-packed bundles called hanks. Sheep casings are the most tender natural casings and are no more than 1 inch in diameter. Hog casings are a little tougher than sheep casings and about 1½ inches in diameter. Cattle casings range from between 1¾–4 inches in diameter and are used for specialty style meats.

Procedure for Cleaning Natural Casings

1. Carefully remove desired amount of casing from a salt pack to avoid creating knots that cannot be easily undone. *Note:* It is always a good idea to prepare more casing than might be needed to allow for holes or unusable sections.
2. Place a large bowl or bain-marie under the faucet in a clean, sanitized sink.
3. Place the untangled casing in the bowl with one end near the faucet opening. Turn on the faucet and flush the casing to rinse out the salt and impurities.
4. Search for any large holes in the casing. *Note:* Small holes the size of a pin are natural. If large holes are discovered, cut the casing and tie it off at the hole to make that the end of the link.

To store unused rinsed casings, fill a container about 1 inch from the top with cold water and add the casings one at a time while holding on to one end of each casing. Drape one end of each casing over the top edge of the container and then secure the lid. This allows one end of each casing to be found without searching through the whole tub and helps prevent the casings from becoming tangled.

Collagen Casings. A *collagen casing* is a casing produced from collagen. It is uniform in size and more durable than natural casings. The consistent size of collagen casings makes them ideal for portion control. Collagen casings are produced from collagen that has been removed from the hides of cattle. Collagen casings come in different strengths to accommodate fresh sausages, which use more delicate casings, and sausages that are hung for smoking, which require durable casings. **See Figure 7-96.** Collagen casings are often dipped in water before use to make them more pliable.

Smoked Sausages

Figure 7-96. Sausages that are hung for smoking require durable casings.

Synthetic Casings. A *synthetic fibrous casing* is a casing produced from a plastic-like synthetic material. Synthetic fibrous casings are not edible and must be peeled off prior to eating. Since they are a plastic composite, they do not need to be refrigerated. However, they need to be submerged in water for a few minutes before use to make them slightly more pliable.

Stuffing Sausages

Foodservice operations can prepare an assortment of sausages using a meat grinder and a sausage stuffer. A *sausage stuffer* is a manual or electric piece of equipment that uses a piston to pump forcemeat through a nozzle into casings. Casings are fed onto the nozzle, and the forcemeat is pumped from the stuffer reservoir to the casing to create a sausage link.

Procedure for Stuffing Sausages

1. Completely chill the sausage stuffer to prevent the forcemeat from heating up during processing.
2. Add the forcemeat to the sausage stuffer.
3. Inspect the cleaned casings before use. Then, slide the sausage casing over the stuffing nozzle.

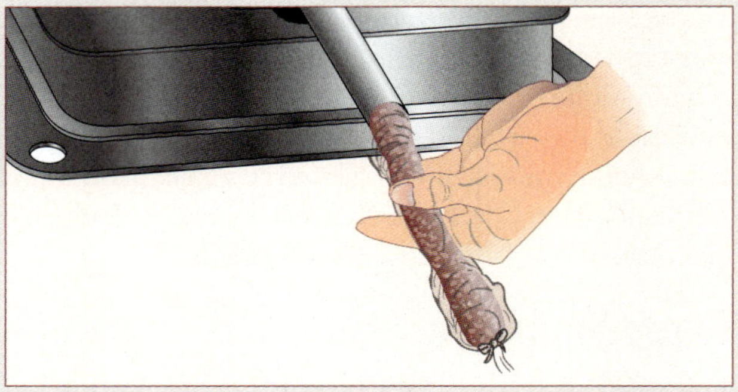

4. Tie a knot in the end of the casing hanging off the nozzle and then poke a small hole in it with a toothpick to allow air to escape while the casing is being filled.

5. Hold the end of the casing on the stuffer nozzle to guide the casing as it fills with forcemeat.
6. Fill all of the casing with forcemeat.

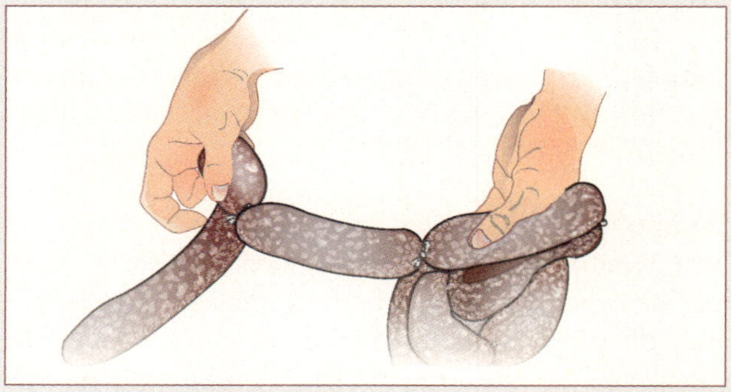

7. Twist each long link into individually-sized sausage links.

Pâtés

A *pâté* is a forcemeat that is typically layered with garnishes and baked in a mold. Pâté is often encased in a pastry crust and baked in a hinged mold. Pâtés prepared with a crust are commonly referred to as pâté en croûte, which translates to pâté in a crust. **See Figure 7-97.** The most common shapes of pâté molds are oval and rectangular.

Pâtés en Croûte

Figure 7-97. Pâtés prepared with a crust are commonly referred to as pâté en croûte, which translates to pâté in a crust.

Terrines

A *terrine* is a forcemeat that is baked in a mold without a crust. A terrine can be made using any style of forcemeat. Some terrines are a combination of products, such as game birds or vegetables. **See Figure 7-98.** Rectangular and oval molds have always been the most popular shapes for terrines. Terrines are baked in earthenware, glass, or enamel-coated cast-iron molds. Classically, terrine molds were lined with thin sheets of fatback and then the forcemeat was placed inside and encased in fatback on all sides. Today many chefs omit the fatback or replace it with a much thinner piece of caul fat.

Terrines

Photo Courtesy of D'Artagnan, Photography by Doug Adams Studio

Pheasant Terrine

Vegetable Terrine

Figure 7-98. Some terrines are a combination of products, such as game birds or vegetables.

Galantines and Ballotines

Galantines and ballotines are often confused because they are made in a similar manner. A *galantine* is forcemeat that is wrapped in the skin of the animal the meat was taken from, shaped into a cylinder, trussed, and poached in a flavorful stock. For example, a duck galantine would be a duck forcemeat wrapped in duck skin. Galantines may be roasted instead of poached. A cooked galantine is cooled, sliced, and glazed with gelatin.

A traditional galantine is made from boned-out poultry stuffed with poultry forcemeat. It is chilled, glazed with aspic, and garnished with items included in the filling, such as olives or truffles. Galantines are always served cold, whereas ballotines may be served hot or cold. **See Figure 7-99.** A *ballotine* is a boned poultry leg that has been stuffed with forcemeat. The stuffed leg is then trussed to maintain its shape. A ballotine is braised or roasted and is normally served hot but can be served cold.

Galantines and Ballotines

Galantine

National Turkey Federation

Ballotine

Figure 7-99. Galantines are always served cold, whereas ballotines may be served hot or cold.

Glazes

Many charcuterie items are presented on platters for elaborate buffets. Pâtés, terrines, roasted meats, hors d'oeuvres, and other items can be sliced and arranged on platters. Since the exposed surface of the terrine, pâté, or other item is exposed to air as it sits on a platter, there is a possibility that it may dry out, darken in color, and lose quality. Glazing items with aspic or chaud froid will help to preserve them, keep them from losing quality, and give them a more beautiful finished appearance. **See Figure 7-100.**

Aspic. *Aspic,* also known as aspic gelée, is a savory jelly made from a consommé or clarified meat, fish, or vegetable stock that produces a clear finish when coating foods. It can be made from the natural gelatin extracted from bones, a packaged gelatin, or a combination of both. The flavor of the aspic should complement the flavor of the dish without overwhelming the dish itself.

Aspic is used as a protective coating and as a binding agent. As a protective coating, aspic can be used to preserve the color of foods as well as prevent them from drying out due to exposure to air. Food with an aspic glaze has an artificial appearance because it has such a shiny, clear finish. It is important to always lift aspic-coated foods with an offset spatula because aspic retains fingerprints.

Glazes

Aspic

Chaud Froid

2008 ACF Culinary Team USA

Figure 7-100. Glazing items with aspic or chaud froid will help to preserve them, keep them from losing quality, and give them a more beautiful finished appearance.

Procedure for Preparing Aspic

1. Prepare a rich white or brown stock from poultry, meat, or fish bones and mirepoix.
2. Test the stock for its gelatin properties by ladling a small amount onto a cold plate and then refrigerating the plate until the stock is completely chilled.
3. Remove the plate from the refrigerator and tilt it from side to side to see if the aspic has set completely. If the stock is too watery, dilute instant gelatin in a small amount of cold water and add it to the stock. Clarify the stock in the same way as when preparing a consommé. Then, strain the stock to remove all traces of fat from the surface. Retest the stock to ensure that it will set properly.

Chaud Froid. *Chaud froid* is a French term meaning "hot-cold." Chaud froid is similar to clear aspic except that it is bright white in color. Like aspic, chaud froid protects foods from drying out when exposed to air. An additional benefit to coating foods with chaud froid is that it serves as a white canvas for colorful decorations or garnishes.

Preparing a chaud froid can be as simple as adding gelatin to heavy cream and a white stock. It also can be prepared using a roux or an egg yolk liaison. Although chaud froid can be eaten, it is typically removed from food prior to eating. Chaud froid may also be used as a gelatin-based binding in terrines. Any charcuterie item can be glazed with aspic or chaud froid.

Procedure for Glazing

1. Lay items on a glazing screen presentation-side up. Tightly cover the items and refrigerate them until needed.
2. Heat the aspic or chaud froid until it has completely melted.
3. Place the hot glaze in a nonreactive bowl over an ice bath and stir constantly while allowing it to cool until it becomes slightly thicker in consistency but not gelatinous. *Note:* If the glaze cools too much and becomes gelatinous it will not yield a smooth coating. If the glaze is too thin or warm, the glaze will not be thick enough to coat the items.
4. Remove items from the refrigerator. Using a spoon or ladle, completely cover the item with aspic or chaud froid.
5. Place glazed items in the refrigerator to allow the aspic or chaud froid to set.
6. Repeat steps 4 and 5 at least three times to completely glaze each item.

Plating Charcuterie

Charcuterie platters are a colorful display of items such as thinly sliced cured meats, pâtés, terrines, and duck confit. Depending upon the theme, three main items are commonly presented in an elegant manner. Charcuterie is traditionally served with piquant or acidic accompaniments to counterbalance the richness of the forcemeats. Pickles such as small dill cornichons or sweet gherkins are often chosen as an accompaniment. Sauces such as onion marmalade, cranberry relish, fruit conserves, and savory mustards offer contrasting flavors and have a palate-cleansing effect. Flatbread or crackers may also be served as garnishments. **See Figure 7-101.**

Charcuterie Platters

Daniel NYC / B. Milne

For Service

2008 ACF Culinary Team USA

For Competition

Figure 7-101. Charcuterie items are commonly presented on platters.

CHECKPOINT 7-11

1. Identify the common ratio of fat to lean meat in most sausages.
2. Identify the three categories of sausages.
3. Explain how sausages are stuffed into clean casings and made into links.
4. Stuff and then smoke sausages. Evaluate the quality of the finished items.
5. Differentiate between a pâté and a terrine.
6. Differentiate between a galantine and a ballotine.
7. Prepare a pâté and a terrine, galantine, or ballotine. Evaluate the quality of the finished items
8. Explain the purpose of glazes.
9. Differentiate between aspic and chaud froid.
10. Prepare and glaze charcuterie items with aspic or chaud froid and then plate them for presentation. Evaluate the quality of the glazed items and the way in which they are presented.

BUFFET PRESENTATIONS

Buffets are an efficient way to serve a large number of people in a short amount of time. A successful buffet is attractive, accessible, and offers foods at appropriate temperatures. The flow of food at a buffet guides guests in a sequential manner as they build their meals. **See Figure 7-102.** Plates are located at the beginnitng of a buffet and need to be kept stocked throughout service. Salads, sides, and starches are placed toward the beginning of the buffet and the proteins are placed near the end along with sauces and condiments. Some buffets may include manned action stations where a chef serves guests a special dish.

Buffet Layouts

Buffets offer a set menu for a predetermined amount of time. The layout of a buffet is based on the requirements and theme of the event, the number of expected guests, the type of food to be served, and the venue. Theme-based centerpieces and decorations can add to the visual appeal of a buffet. Buffet table tops are often covered with linens and the sides are skirted. Most buffet tables are of equal height. However, taller tables may be used to showcase centerpieces or special dishes. Risers are frequently used to give height to displayed foods such as cheese boards. Plates are placed at the beginning of a buffet and always kept well stocked.

Buffet Presentations

Courtesy of Lauren Frisch

Figure 7-102. The flow of food at a buffet guides guests in a sequential manner as they build their meals.

Buffets

When setting up buffets, guests must be allowed enough room at the buffet table to dress salads and assemble items such as sandwiches. When planning a buffet, allowances should be made for the following:

- how guests will move from one food item to the next and from one table to another
- how traffic will flow from the buffet to the seating area
- how beverage traffic will be separated from food traffic
- how the kitchen can be accessed from each buffet table
- how the ongoing replenishment of food can be made unobtrusive to guests

All buffet layouts should be planned and organized for optimal efficiency and movement for both guests and service staff. Placing accompaniments on a separate table is an efficient way to keep guests moving along a buffet. Rectangular, round, and serpentine table combinations can be configured to offer a wide variety of wall-based, round, or two-sided buffets. **See Figure 7-103.**

Sample Buffet Layouts

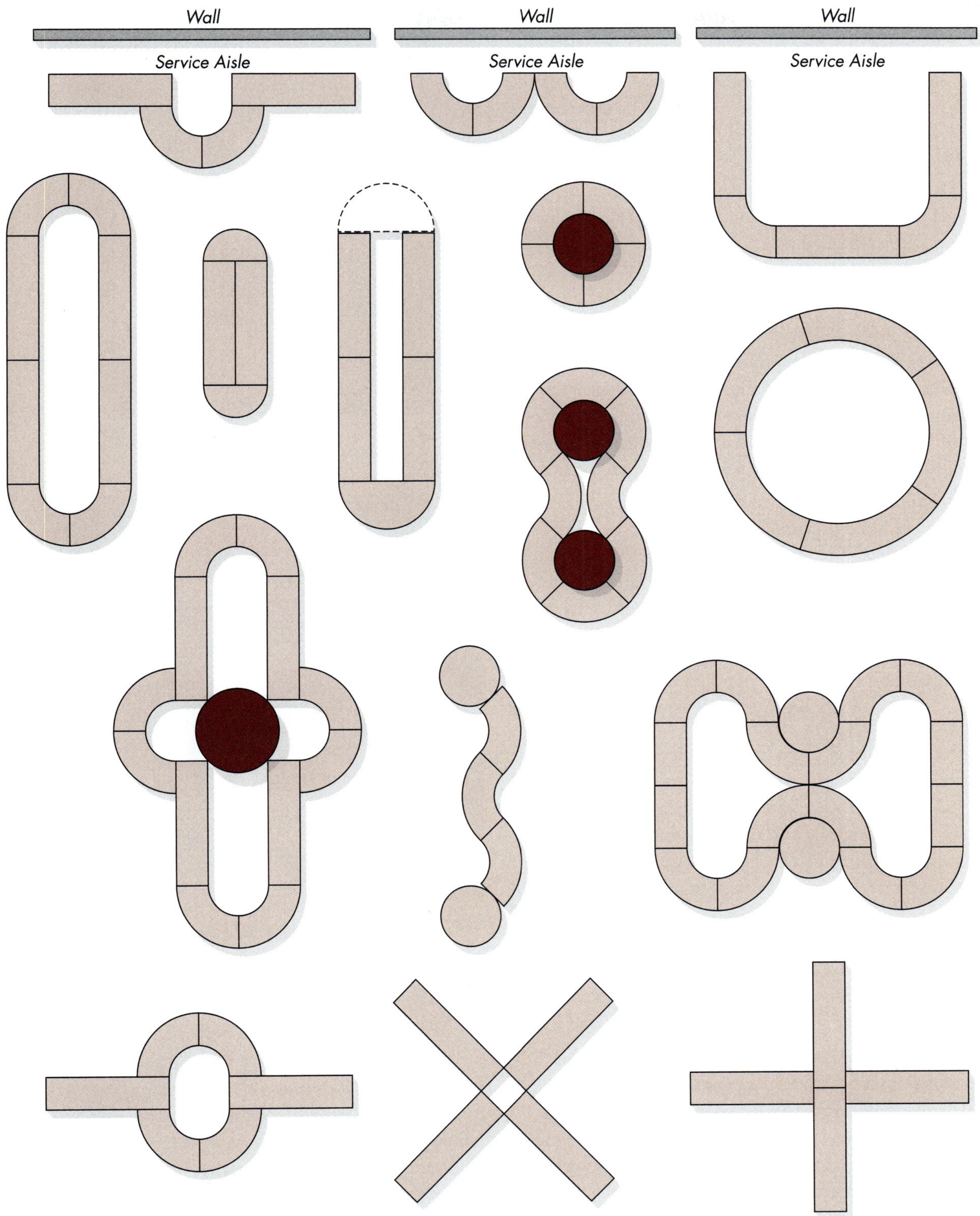

Figure 7-103. Rectangular, round, and serpentine table combinations can be configured to offer a wide variety of wall-based, round, or two-sided buffets.

Wall-Based Buffets. To structure a wall-based buffet, tables are placed near a wall so that guests only have access to one side of the buffet. In a short wall-based buffet arrangement, guests start at one end and finish at the other. Longer wall-based buffet arrangements may offer identical foods at the starting point on each end of the buffet. This arrangement also works when a carving station is located at the center of the buffet where guests can be served a special dish by a chef.

Round Buffets. Round buffets are commonly used when there are many different buffet stations set up in the same room. Special buffet stations or zones may be set up for items such as pasta or sushi. Round buffets are often used for desserts, cheese boards, and beverage service.

Two-Sided Buffets. Two-sided buffets, also known as double-sided buffets, can be presented as rectangles, serpentines, half rounds, or a combination thereof. Identical food offerings are located on each side of the buffet, allowing guests to be seated more quickly. An action station, such as a carving station, may also be part of a two-sided buffet.

Combination Buffets. Wall-based, round, and two-sided buffets may be used together in very large venues. When combining buffets, it is important to remember that guests should not be overwhelmed by the myriad of configurations to the point that they meander around the room without direction. The traffic pattern should be easy for guests to discern.

Presenting Cold Platters

Cold platters are used to present food to a large number of guests. Very large plates, pieces of marble or granite, woven trays, and silver platters are used to display cold foods. Oval, rectangular, and teardrop-shaped platters in clear, white, and shiny black Lexan® plastic as well as glass platters and racetrack mirrors are also used to present foods on buffets. Tiny appetizers, petits fours, individual desserts, and large displays of tropical fruits are often served on mirrors or reflective trays. In many hotel and banquet facilities, cold platters are presented as elegant pieces of edible art. **See Figure 7-104.**

Preparing Whole Meats. When planning for the buffet, a variety of meats for both cold platters and hot buffet service may be required. Beef eye of round, sirloin, strip loin, rib eye, and boneless pork loin work well for sliced sandwich bars and cold platters. A bone-in turkey breast that has been roasted and then had the bone removed is also popular. Boneless ham is often sliced for a cold platter presentation. Commercially prepared meats such as roast turkey, honey ham, pastrami, corned beef, and Italian-style roast beef can be sliced thin and layered on a mirror or cold platter.

Edible Art

ManYee DeSandies/Pics4Learning, Inc.

Carved Centerpiece

Photo Courtesy of Lauren Frisch

Buffet Display

Figure 7-104. Cold platters can be presented as elegant pieces of edible art.

When preparing for the hot buffet carving station, whole steamship rounds of beef, standing beef rib roast, roasted whole turkeys or turkey breasts, simmered or baked whole corned beef, and roasted whole hams are often used. Slow-roasting meats and poultry in a combi oven or a cook-and-hold oven are recommended for maximum retention of natural juices. Seasonings may be applied to roasts the day before service to enhance the flavor of the meat.

Creating Garnishes

Carving Meats. There are two ways of serving meats on a buffet. Cold roast meats, cold cuts, turkey, and other cooked and chilled meats are traditionally served presliced and presented on a platter. Hot meats such as steamship rounds, prime ribs, legs of lamb, whole hams, and roasted turkeys are typically carved at an action station. **See Figure 7-105.**

Carving Meats

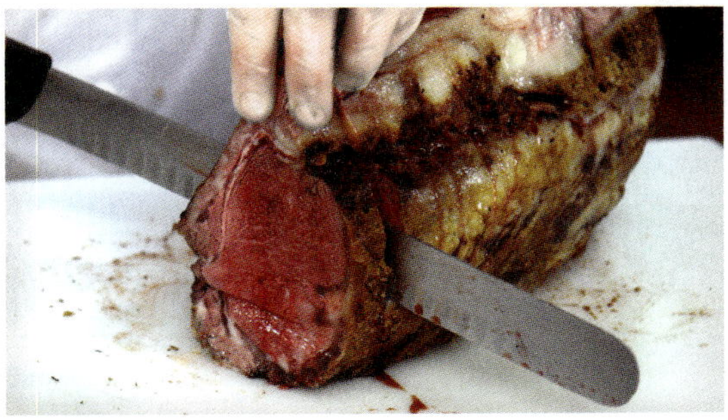

Figure 7-105. Hot meats such as steamship rounds, prime ribs, legs of lamb, whole hams, and roasted turkeys are typically carved at an action station.

A carving station requires a cutting board with a pan underneath to catch any drippings, a carving knife, a fork and steel, tongs, a pan for the trimmings, and a heat lamp to keep the meat at the proper temperature. The au jus and sauces such as horseradish are held in a bain marie along with a ladle or serving spoon.

Preparing Seafood. When preparing seafood, the freshest fish and shellfish should be used. Fresh fish may be poached in a flavorful court bouillon and placed in the refrigerator to cool in the poaching liquid for added flavor. The fish is then placed on wire racks in the refrigerator to cool. It can be coated with aspic or left as is, depending on the venue and the menu. Firm fish, such as mahi mahi, can be grill-marked on one side and then finished in the oven to control the cooking time. The cooked fish is then chilled and garnished for the cold platter.

Shellfish such as shrimp, crab, and lobster should be cooked in a flavorful shellfish broth until tender. Overcooked shellfish is tough. Shrimp should be deveined and cooked with or without the shells, depending on the menu. Live crab and lobster should be used for a buffet unless a frozen prepared product better meets the menu requirements. Frozen shellfish should be defrosted in the refrigerator overnight or under cold running water and then poached or steamed until tender. The shellfish is then chilled and served with a complementary sauce and/or a colorful garnish. **See Figure 7-106.**

Seafood Presentations

Florida Department of Agriculture and Consumer Services, Bureau of Seafood and Aquaculture Marketing

Oysters

New Zealand Greenshell™ Mussels

Mussels

Figure 7-106. Shellfish are chilled and served with a complementary sauce and/or a colorful garnish.

Meat and Seafood Platters. Protein is the most expensive menu item on a buffet and great care must be given to portion size and advance planning. Meats and seafood used on platters must be fresh and tender, sliced into appropriate portions, and accompanied with complementary sauces. Seafood platters can be challenging as the flesh of seafood is more tender than poultry or meat and breaks apart easily. Whole poached salmon or another firm fleshed fish such as halibut may be placed on a large platter and used as an edible centerpiece.

Mollusks such as oysters and clams are traditionally served on the half shell over crushed ice or surrounding an ice carving. Crustaceans such as crawfish, crabs, and shrimp may be served on a bed of ice or in bowls. It is important to either provide the necessary tools for extracting the flesh or to make sure guests can eat the seafood without having to crack or remove shells.

Meat and Cheese Platters. Meat and cheese platters are often the central focus of a buffet. Meat and cheese can be plated separately or together. **See Figure 7-107.** A large block of cheese or a vegetable or fruit carving can be used as a focal point with cheese cubes or slices arranged in a geometric pattern around it. For a luncheon buffet, 2 oz of sliced meat and 2 oz of sliced cheese per person is adequate. For a dinner buffet, up to 4 oz of meat and 3 oz of cheese per guest is ideal.

Meat and Cheese Platters

Photo Courtesy of Perdue Foodservice, Perdue Farms Incorporated

Figure 7-107. Meat and cheese can be plated together.

Slices of rare beef, ham, turkey, pork loin, or leg of lamb, or cured meats such as salami or prosciutto can be sliced and placed on a platter surrounding a bowl of olives and gherkins or other appropriate condiments. A large piece of meat such as a roasted turkey breast can serve as the centerpiece of a platter. **See Figure 7-108.**

Meat Platters

Photo Courtesy of Perdue Foodservice, Perdue Farms Incorporated

Figure 7-108. A large piece of meat such as a roasted turkey breast can serve as the centerpiece of a platter.

Presenting Salads

Salads are always a colorful addition to a buffet. When presenting buffet salads it is important to remember that the salads have to remain appealing throughout the service. Salads should be made with the freshest ingredients and contain at least three different colors to catch the eye. Salads can either be pregarnished or can have garnishes such as croutons, crispy bacon bits, nuts, dried fruit, and seeds placed next to them at the buffet table. Salad bowls should be easy to refill or replace. Serving tongs can be placed next to each salad bowl. Salad dressings should be chilled and served in separate bowls, pitchers, or crocks, with appropriate serving utensils placed beside them.

Carlisle FoodService Products

Presenting Breads

Breads for the buffet table should complement the other items on the menu. Breads should be served in baskets or in woven trays lined with cloth. Breads such as baguettes or sour dough should be presliced into ½ inch thick rounds. Whole wheat, rye, and pumpernickel rolls are offered on many buffets. Crackers such as lahvosh or pita wedges are useful for dips or spreads. **See Figure 7-109.** Breads for use on cheese platters should be cut into bite-sized pieces.

Presenting Breads

Figure 7-109. Crackers such as lahvosh or pita wedges are useful for dips and spreads.

Carved Display Pieces

Carved display pieces can elevate the elegance of a buffet and showcase a particular theme. All carved items should be made from an edible food source. Carvings can be made from large blocks of cheese, such as a barrel carved from a 50 lb block of cheddar, butter, or breads. Many carved display pieces are made from fruits and vegetables, ice, or tallow.

Fruit and Vegetable Sculptures. The exteriors of fruits and vegetables can be carved into geometric and floral designs as well as cartoon figures or human faces. **See Figure 7-110.** Whole fruits such as watermelon and honeydew or vegetables such as pumpkins and winter squash can be carved into attractive centerpieces as well as serving bowls and baskets. Designs are stenciled or traced on the skin and carved in relief. The flesh can be left as is or carefully dyed with food-based colorants.

Ice Sculptures. An ice sculpture is a stunning art form that can be the focal point of a buffet. A 300 lb block of ice can be carved into almost any shape. **See Figure 7-111.** Multiple blocks can be used to make room sized sculptures, such as the Eiffel Tower, for a special event or banquet hall. Blocks of ice can also be made into "luges" for bar service. To make an ice sculpture, the ice block is scored and then carved with chain saws, ice routers, chisels, picks, and an assortment of ice grinding tools.

Fruit and Vegetable Sculptures

Carving Cantaloupe

Cantaloupe and Turnip

Watermelon

Mango and Beet

ManYee DeSandies/Pics4Learning, Inc.

Figure 7-110. The exterior of fruits and vegetables can be carved into geometric and floral designs as well as cartoon figures and human faces.

Ice Sculptures

Figure 7-111. A 300 lb block of ice can be carved into almost any imaginable shape.

Production Tip

Safety glasses, ice aprons, ice chaps, waterproof boots, gloves, and hearing protection must be worn when carving ice sculptures.

Tallow Sculptures. A tallow sculpture is formed from tallow, which is a mixture of purified beef fat, beeswax, and paraffin that have been melted together. Tallow sculptures are often used for buffet tables in large ballrooms, as centerpieces in hotel restaurant entrances, and for culinary competitions. **See Figure 7-112.** To create a tallow sculpture, the design of the sculpture is drawn and then an armature is built of wire and attached to a stable base material. The armature is then wrapped with cheese cloth. The tallow is applied in layers to build the structure outward, and then the defining details are carved into the tallow.

Tallow is sold premade in 25 lb tubs and 50 lb blocks. Tallow comes in many colors and can be dyed using fat-based food dyes or poured into molds or forms. Tallow that is kept clean and dry can be reused.

Tallow Sculptures

2008 ACF Culinary Team USA

Figure 7-112. Tallow sculptures are often used for buffet tables in large ballrooms, as centerpieces in hotel restaurant entrances, and for culinary competitions.

CHECKPOINT 7-12

1. Explain the purpose of buffets.
2. Name at least five items to keep in mind when planning buffet layouts.
3. Describe three common types of buffet table layouts.
4. Identify five ways cold platters are presented on buffets.
5. Prepare and present a cold platter.
6. Describe how to present salads on a buffet.
7. Describe how to present breads on a buffet.
8. Plan a buffet layout and detail the placement of menu items on the layout.
9. Describe three common types of carved display pieces.
10. Carve a display piece for a buffet.

Refer to DVD for **Quick Quiz®** questions

SUSTAINABILITY CORNER

A certain amount of food waste is inevitable within the foodservice industry. However, in some foodservice operations, as much as 70% of the operation's refuse is made up of food waste. Although recycling and composting can help, the majority of food waste can be reduced through proper purchasing and storage practices.

Purchasing includes ordering adequate amounts of items and building positive relationships with suppliers. Communicating with suppliers about where products are harvested and the length of time products spend in transit can provide valuable information regarding freshness, usability, and storage options. Suppliers willing to consolidate items to limit packaging and those who use recyclable packaging materials can further enhance sustainability efforts.

Once products arrive, they have to be stored properly. To promote freshness, walk-ins should be organized to facilitate easy access to foods. This includes eliminating stacked crates of product and using a first-in, first-out (FIFO) inventory rotation. FIFO is the process of dating new items as they are placed in inventory and placing them behind or below older items to ensure that the older items are used first. Freshness can also be prolonged by storing produce in airtight containers away from condenser units. Products stored in the freezer should be tightly wrapped, dated, and used in a timely manner to minimize waste due to freezer burn. Limiting food waste saves money and reinforces sustainable practices.

Another way to save money and support sustainable practices is to grow a facility garden. Fruits, vegetables, and herbs can be grown on rooftop gardens, window boxes, container gardens, and raised beds. A garden that yields fresh produce can cut costs while offering premium ingredients harvested at the peak of flavor.

HerbThyme Farms

Planting a facility garden as well as communicating with suppliers in order to attain local products such as cheese, bread, beef, eggs, or produce can have a positive impact on the economy, the environment, and the dining experience of the customer. Buying local supports the growth of small farms and businesses, creates jobs, and preserves the natural landscape. Items purchased from local producers can also add value to the dining experience by offering customers foods with the freshest taste and most vibrant colors. Implementing a philosophy where foodservice operators and suppliers work together for the betterment of a community can not only enhance the local economy, but also provide the opportunity to help sustain and strengthen the Earth's valuable resources.

Seasonal Food Guide

Certification Exam Preparation Questions

1. A ___ salad is a salad that consists of a base, body, garnish, and dressing attractively arranged on a plate.
   A. bound
   B. tossed
   C. composed
   D. gelatin

2. Brick, Limburger, Muenster, and Port Salut are ___ semisoft cheeses.
   A. dry-rind
   B. wet-rind
   C. washed-rind
   D. waxed-rind

3. A(n) ___ is an item that alters or enhances the natural flavor of food.
   A. herb
   B. spice
   C. seasoning
   D. flavoring

4. ___ should be served with accompaniments such as toasted nuts, fresh or dried fruits, and jams or spreads.
   A. Salads
   B. Cheeses
   C. Sandwiches
   D. Appetizers

5. Depending on the ingredients used, ___ can be served as appetizers, entrées, sides, or desserts.
   A. salads
   B. cheeses
   C. charcuterie
   D. sandwiches

6. A salad green is an edible ___ used in raw salads or as a garnish.
   A. stem
   B. leaf
   C. flower
   D. root

7. ___ is the second most widely used seasoning in the professional kitchen.
   A. Vinegar
   B. Wine
   C. Lemon
   D. Pepper

8. A clove is a ___ spice made from the dried, unopened bud of a tropical evergreen tree.
   A. leaf
   B. stem
   C. flower
   D. root

9. ___ cold soups can be served as an appetizer or as an entrée.
   A. Salty
   B. Bitter
   C. Sweet
   D. Savory

10. A ___ is a temporary emulsion of oil and vinegar that may also include additional flavorings and seasonings.
    A. salad dressing
    B. vinaigrette
    C. mayonnaise
    D. none of the above

11. A ___ cheese is a cheese produced by inserting harmless live mold spores into the center of ripening cheese with a needle.
    A. soft
    B. hard
    C. blue-veined
    D. waxed-rind

12. When making vegetable salads, knowing how the vegetables react to ___ is important.
    A. acidity
    B. heat
    C. cold
    D. none of the above

13. Many ___ used in the professional kitchen are produced from vegetables, seeds, nuts, or fruit.
A. zests
B. vinegars
C. wines
D. oils

14. All of the ingredients in a sandwich station should be ___ the range of motion to maximize efficiency and speed.
A. outside
B. within
C. above
D. below

15. ___ is the main ingredient of sausages, pâtés, terrines, galantines, and ballotines.
A. Aspic
B. Chaud froid
C. Charcuterie
D. Forcemeat

16. A(n) ___ is a flavoring derived from the bark, seeds, roots, flowers, berries, or beans of aromatic plants.
A. herb
B. spice
C. seasoning
D. flavoring

17. Most ___ consist of a base, a spread, one or more fillings, and one or more garnishes.
A. salads
B. soups
C. sandwiches
D. appetizers

18. A(n) ___ is an elegant, bite-sized portion of food that is creatively presented and served apart from a meal.
A. hors d'oeuvre
B. appetizer
C. crudité
D. tartare

19. Types of sandwich ___ include poultry, seafood, meats, cheeses, bound salads, vegetables, or a combination of these items.
A. bases
B. spreads
C. fillings
D. garnishes

20. A(n) ___ sandwich is a petite and delicate cold sandwich with a trimmed crust and a soft filling.
A. open-faced
B. closed
C. wrap
D. tea

21. Explain how sausages are stuffed.

___
___
___
___
___

22. Identify items that are often used to carve display pieces.

___
___
___
___
___

23. Describe five varieties of forcemeats.

___
___
___
___
___

24. Identify five allowances to make when planning a buffet.

___
___
___
___
___

SKETCHING EXERCISE

25. Sketch a cold foods buffet and designate the placement of the serving plates, cold platters, salads, salad dressings, cheeses, breads, charcuterie, and centerpieces.

Notes

Cooking Techniques Station

Introduction

In the eighth station in the ACFEF apprenticeship program, you will be responsible for the production and service of hot foods such as entreés. From braising veal shanks to pan-frying fish, you will learn cooking techniques that you will use for the rest of your culinary career.

Cooking techniques such as sautéing, pan-frying, stir-frying, deep-frying, roasting, baking, poaching, simmering, boiling, steaming, stewing, and braising enable you to prepare a wide variety of poultry, seafood, meat, vegetable, and fruit dishes. Your skills will be enhanced as you learn how to clarify butter, bread items, stuff meats, and batter foods for frying. Your sautéing skills will be put to good use when creating entreés such as veal scaloppini, beef stroganoff, and poached fish. Rounding out your repertoire of cooking techniques is a lesson in deep-frying foods, including identifying the smoke points of commonly used oils.

Sustainability is practiced through recycling disposable goods including used oils. You will learn to identify products that can be recycled, as well as how to find recycling resources in your community. Paper towels are used in excess in the food industry, so you will research alternative choices. These and other sustainable best practices will give you a blueprint for making more environmentally conscious decisions.

Cooking involves a thorough knowledge of cooking techniques and how to develop flavors in foods. The cooking technique chosen depends on the type of food being cooked and the flavors the chef is developing in the dish. Successfully cooking foods will yield dishes that have appealing textures and the correct degree of doneness. After properly cooking the food and developing the flavors, attractive platings will enhance the guest experience. Forethought given to designing a beautiful presentation will set the stage for the food that is to be enjoyed. Cooking techniques are the foundation for creating memorable dishes that guests will enjoy.

Conduction

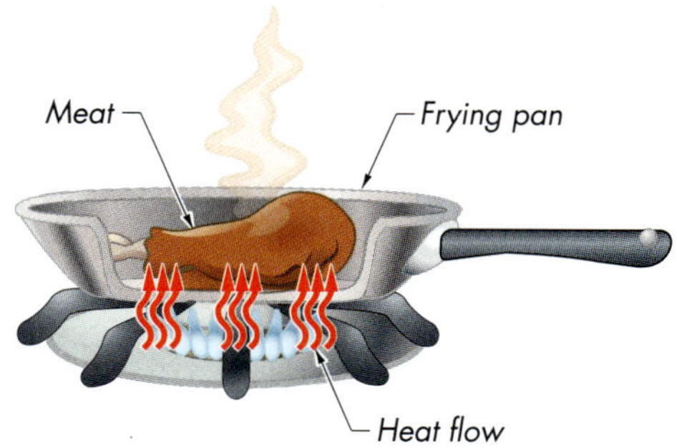

Figure 8-1. Conduction is a type of heat transfer in which heat passes from one object to another through direct contact.

Heat Transfer

THE COOKING PROCESS

Cooking is the process of heating foods in order to make them taste better, make them easier to digest, and kill harmful microorganisms that may be present in the food. *Heat* is energy that is transferred between two objects or substances of different temperatures. Heat typically flows from a warmer material to a cooler material. When heat is transferred, the particles inside an object increase their movement as the temperature increases. The three different methods of heat transfer used with food are conduction, convection, and radiation. Each type of heat transfer cooks food in a different manner and produces different results.

Conduction

Conduction is a type of heat transfer in which heat passes from one object to another through direct contact. **See Figure 8-1.** Heat flows from a warm object to a cool object. Direct contact enables the heat to be transferred from the heat source to the pan and then to the food in the pan. Metal is a good conductor of heat, while other materials, such as glass, wood, and plastic, are not.

Convection

Convection is a type of heat transfer that occurs due to the circular movement of a fluid or a gas. **See Figure 8-2.** Heating a pot of water is an example of convection. The pot itself becomes heated by conduction. As the water heats it expands and rises up through the surrounding cooler water, causing the cooler water to sink to the bottom of the pot where it will become heated. This circulating path of the warmer water rising and the cooler water sinking is called a convection current.

A convection oven circulates hot air around food items, cooking them more quickly and evenly than a conventional oven. It also uses a reduced cooking temperature compared to a conventional oven. The difference in cooking temperature is needed because circulating air transfers heat more quickly than still air of the same temperature.

An impingement oven also operates on convection heat transfer. Impingement ovens are often used to toast breads, heat sandwiches, and cook pizzas. Impingement ovens circulate hot air above and below food as it moves across a conveyer.

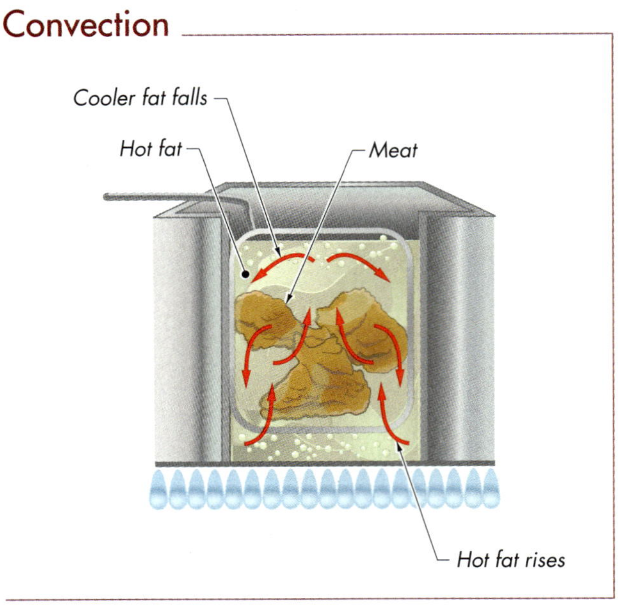

Figure 8-2. Convection is a type of heat transfer that occurs due to the circular movement of a liquid or gas.

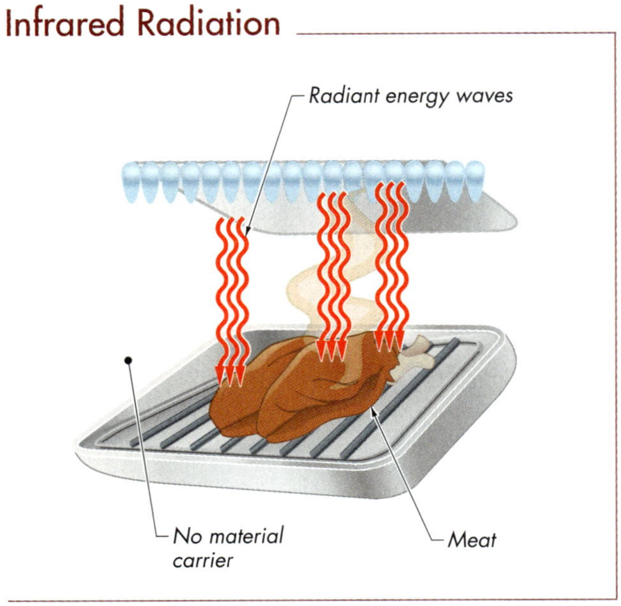

Figure 8-3. The food does not cook from direct contact with the cooking surface but cooks from the heat given off by the heating element.

Radiation

Radiation is a type of heat transfer that uses electromagnetic waves to transfer energy. When electromagnetic waves strike an object, the energy carried by the waves is transferred to the object. This is the reason that the sun feels warm. Some materials are better at absorbing radiation than others. Aluminum foil is shiny and therefore reflects rather than absorbs radiation. The three types of radiation heat transfer used to heat foods in the professional kitchen are infrared, microwave, and induction.

Infrared. Infrared radiation can be seen in a toaster or a broiler when the heating element glows red. When food is placed in infrared cooking equipment, such as a broiler, the food is cooked as it absorbs the electromagnetic waves. For example, a steak placed in a broiler cooks on top first, even though the bottom of the steak is in direct contact with the broiler rack. The food does not cook from direct contact with the cooking surface but cooks from the heat given off by the heating element. **See Figure 8-3.**

Production Tip

When reheating food in a microwave, the internal temperature of the food must reach 165°F for 15 seconds. The food should rest for two minutes in order for the temperature to distribute evenly.

Microwave. Microwave radiation uses electromagnetic waves to heat the water, fat, and sugar molecules in food. Microwaved food cooks quickly but does not brown. However, it is important to remember that as the food heats up it also conducts heat to the plate through direct contact. Microwaves are generally used to quickly heat items in the professional kitchen. **See Figure 8-4.**

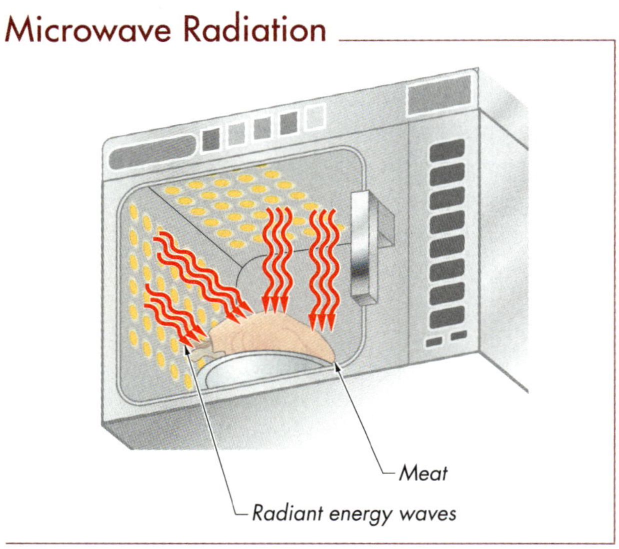

Figure 8-4. Microwaves are generally used to quickly heat items in the professional kitchen.

Induction. Induction radiation uses electromagnetic current to heat magnetic cookware. The cookware heats the food and the cooktop remains cool. Copper coils create an electromagnetic field beneath the smooth ceramic surface of an induction burner. The magnetic pulses cause the molecules in the magnetic cookware to move quickly and generate heat. Induction cooking is rapid, and when the pan is removed from the cooktop, cooking slows quickly.

The cookware used on an induction cooktop must be flat-bottomed and made of a material that will conduct electricity, such as cast iron, stainless steel, copper, or aluminum. Induction radiation gives a chef maximum control over a heated pan. **See Figure 8-5.**

Induction Radiation

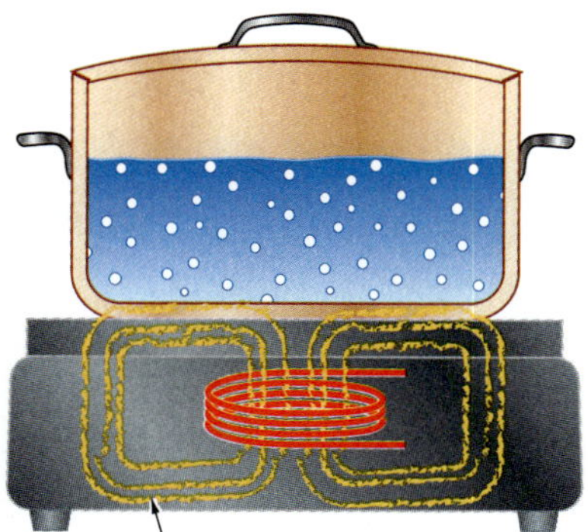

Figure 8-5. Induction radiation gives a chef maximum control over the heated pan.

Changes in Color and Texture

Before learning specific cooking techniques, it is important to understand how heat changes the color and texture of food. In general, the longer a food cooks, the more color and texture changes. The Maillard reaction, caramelization, reduction, gelatinization, and coagulation are a few ways to change the color and texture of food.

Maillard Reaction. The *Maillard reaction* is a reaction that occurs when the proteins and sugars in a food are exposed to heat and merge together to form a brown exterior surface. **See Figure 8-6.** For example, the Maillard reaction occurs when meat is seared. *Searing* is the process of using high heat to quickly brown the surface of a food. Searing is done with little or no fat in a very hot sauté pan on a hot grill or griddle. Some foods are seared before cooking, and others may be seared after they are cooked to add color. Searing may also be the entire cooking process. Fish, such as tuna, and some meats are often served seared. Grilling, frying, and toasting foods also causes the Maillard reaction.

Maillard Reaction

Raw Meat

Seared Meat

Figure 8-6. The Maillard reaction is a reaction that occurs when the proteins and sugars are exposed to heat and merge together to form a brown exterior surface.

Maillard

Photo Courtesy of the Beef Checkoff

Caramelization. Many recipes call for caramelized onions. *Caramelization* is a reaction that occurs when sugars are exposed to high heat and produce browning and a change in flavor. **See Figure 8-7.**

Caramelization

Figure 8-7. Caramelization is a reaction that occurs when sugars are exposed to high heat and produce browning and a change in flavor.

The first step toward caramelization is sweating. *Sweating* is the process of slowly cooking food to soften its texture. Many recipes require the sweating of vegetables such as onions and garlic. In the second step, the sugars begin to darken as the temperature of the food rises, which changes the aroma and color. In the final step, the flavor further develops and completes the caramelization. The top of a crème brûlée and the crust of a freshly baked loaf of bread are examples of caramelization.

Eloma Combi Ovens

Reduction

Figure 8-8. Reduction is the process of gently simmering a liquid until it reduces in volume and results in a thicker liquid with a more concentrated flavor.

Reduction. *Reduction* is the process of gently simmering a liquid until it reduces in volume and results in a thicker liquid with a more concentrated flavor. **See Figure 8-8.** Sauces, stocks, soups and pan drippings are often reduced. Pan drippings left over after cooking meat can be simmered to create a densely flavored sauce. Wine, water, cream, or balsamic vinegar can be added to pan drippings to create a reduction.

Reductions

Gelatinization. *Gelatinization* is the process of a heated starch absorbing moisture and swelling, which thickens the liquid. As starches (complex carbohydrates) are heated, they begin to absorb liquid and swell. This is why muffin batter is wet in its raw form but expands and dries out as it bakes. The starch molecules in the flour absorb moisture from the wet batter and create steam during the baking process. Starches also absorb moisture when a sauce is made. Adding starch to a stock thickens the stock, and a sauce is formed as the starch absorbs some of the water from the stock.

Coagulation. When proteins are exposed to heat, they begin to firm and coagulate. *Coagulation* is the process of a protein changing from a liquid to a semisolid or solid state when heat or friction is applied. Coagulation takes foods from a high-moisture state to a low-moisture state, such as when a raw egg begins to coagulate as it is poached. **See Figure 8-9.** Egg whites also coagulate as they are beaten to form a meringue.

Coagulation

Figure 8-9. Coagulation takes foods from a high-moisture state to a low-moisture state, such as when a raw egg begins to coagulate as it is poached.

Production Tip

Eggs that are to be poached are first parcooked and shocked in an ice bath. Then the eggs are poached to finish the cooking process.

Changes in Nutrient Values

Foods are made of various amounts of proteins, carbohydrates, lipids, vitamins, minerals, and water. The nutrient values of food can be altered by the way it is prepared, cooked, and stored. For example, water-soluble vitamins and minerals are easily destroyed by excess heat. However, nutrient-rich blanching water can be used to make soups. The length of cooking time also determines the level of nutrient loss or gain. Cooking bright-green vegetables, such as broccoli, causes them to change color as the chlorophyll is destroyed by heat. **See Figure 8-10.** Likewise, the crisp texture of vegetables decreases as the heat exposure increases.

While heat can destroy vitamins, cooking can also increase the nutrient value of some foods, such as tomatoes. Cooked tomatoes contain more lycopene than uncooked tomatoes. Likewise, heated cinnamon has more antioxidant power than raw cinnamon. It is important to be aware of the impact of heat on nutrient values.

Changes in Nutrient Value

Steamed

Overcooked

Figure 8-10. Cooking bright-green vegetables, such as broccoli, causes them to change color as the chlorophyll is destroyed by heat.

Nutrient Retention

CHECKPOINT 8-1

1. Describe the three techniques of heat transfer used to cook food.
2. Identify three types of radiation heat transfer used to cook food.
3. Describe five reactions that change the color or texture of food.
4. Identify the two nutrients most often destroyed by heat.

DRY-HEAT COOKING TECHNIQUES

The three general categories of cooking techniques are dry-heat cooking, moist-heat cooking, and combination cooking. *Dry-heat cooking* is any cooking technique that uses hot air, hot metal, a flame, or hot fat to conduct heat and brown food. Sautéing, frying, roasting, baking, grilling, and broiling are dry-heat cooking techniques. Tender meats, ground meats, and some fruits and vegetables are often cooked using dry-heat cooking techniques.

Sautéing

Sautéing is a dry-heat cooking technique in which food is cooked quickly in a sauté pan over direct heat using a small amount of fat. Sautéing is done over high heat, and caution must be used to not allow the small amount of fat in the pan to burn. Butter is often used to sauté foods, but butter can easily brown or burn. Butter is clarified to raise the temperature that butter browns. *Clarified butter* is butter fat that has been separated from the milk solids and water found in whole butter.

Procedure for Clarifying Butter

1. Melt the butter in a sauce pan over low heat. Do not stir the butter as it melts.
2. When the butter has completely melted, skim the milk solids that collect on top of the butter.
3. Raise the temperature slightly so the water will evaporate.
4. Transfer the clarified butter to another container. *Note:* Avoid transferring any milk solids that may still be on the bottom or on the sides of the pan.

Harbor Seafood, Inc.

Some foods are dredged before they are sautéed. *Dredging* is the process of lightly dusting an item in seasoned flour or fine bread crumbs.

Procedure for Dredging

1. Coat items in flour or finely ground crumbs.

2. Shake off excess coating.
3. Immediately cook coated items.

Fresh produce is often sautéed. Items such as vegetables are often flipped using a wrist-flicking motion to ensure even cooking without burning the item. Many proteins are also sautéed. Tongs are typically used to turn large protein items in a sauté pan.

Production Tip

When sautéing, the pan should be hot and the small amount of fat that is added should be heated. Starting with a cold pan will cause the food to stick and absorb more fat.

Procedure for Sautéing Small Items

1. Place sauté pan over high heat and add a small amount of fat to coat the bottom of the pan.

2. When the fat is hot, add the items to the pan, using caution to not overfill the pan.

3. Once the items are lightly browned on one side, use a wrist-flicking motion to toss and flip the items. Flip until the items are cooked on all sides.

Procedure for Sautéing Large Items

1. Prepare and season items. If desired, dredge items and shake off excess flour.

2. Place sauté pan on burner over high heat and add a small amount of fat to coat the bottom of the pan.

3. When the fat is hot, place the items flat in the pan, using caution to not overfill the pan. *Note:* If the pan is too full, the items will steam or simmer rather than sauté.

4. Once the edges change color and begin to brown, turn the items.

5. Cook the items until golden brown on both sides and cooked through.

Sautéing Small Items
Media Clip

Sautéing Large Items
Media Clip

Stir-Frying

Stir-frying is the process of quickly cooking items in a heated wok with a very small amount of fat while constantly stirring the items. Foods that are stir-fried are crisp, yet tender when cooked. **See Figure 8-11.** The time required to stir-fry an item depends on its thickness. For example, denser vegetables, such as broccoli and carrots, require more cooking time than less dense vegetables, such as bok choy and snow peas. Once the fat is hot, dry herbs and spices are added to the hot wok. Then the proteins are seared before the vegetables and any liquid ingredients are added.

Stir-Frying

Photo Courtesy of the Beef Checkoff

Figure 8-11. Foods that are stir-fried are crisp, yet tender when cooked.

Pan-Frying

Pan-frying is a dry-heat cooking technique in which food is cooked in a pan of hot fat. Eggplant parmigiana and country-fried steak are typically pan-fried. Foods that are pan-fried are typically dredged or breaded and then placed in a pan of hot fat.

Foods are typically breaded before they are pan-fried or deep-fried. *Breading* is a three-step procedure used to coat and seal an item before it is fried. The item is first dredged in flour, then dipped into a mixture of beaten eggs and a liquid such as milk, and finally dipped into a bread crumb mixture. The bread crumbs adhere to the flour and egg mixture, sealing the food item. When breading items, one hand is used exclusively to coat items with dry ingredients and the other hand is used exclusively to coat items in the wet ingredients. By following this guideline, the hands do not become breaded while moving food through the breading procedure.

Procedure for Breading

1. Use the designated dry hand to dredge item in seasoned flour and then shake off the excess.

2. Use the dry hand to gently place the dredged item in an egg wash (a mixture of beaten eggs and milk or water).
3. Use the designated wet hand to coat both sides of the dredged item in the egg wash.
4. Use the wet hand to remove the item from the egg wash and lay it carefully in the bread crumbs.

5. Use the dry hand to coat the surface and edges of the item with bread crumbs.
6. Use the dry hand to gently shake off excess crumbs before setting it aside.

Procedure for Pan-Frying

1. Place enough oil in a pan to come halfway up the thickness of the item to be pan-fried. Heat the oil to 350°F.

2. Dredge or bread the item.

3. Gently place the item in the hot oil.

4. Pan-fry the item until cooked halfway up its side.

5. Turn the item over and cook the other side. *Note:* If the item is too thick, it may need to be finished in the oven.

6. Remove the item from the hot oil and drain on a screen.

Pan-frying allows items to lie flat in the pan, preventing them from curling. Fish fillets, boneless chicken breasts, and pork chops are often pan-fried.

When pan-frying, it is common for the fat to be preheated and then added to a hot pan. Preheating will expedite the cooking process and produce a less greasy product. The oil should not be hot enough for the outside of the food to burn before the inside is cooked. A common cooking temperature for pan-frying is between 325°F and 375°F.

Pan-Frying
Media Clip

Deep-Frying

Deep-frying is a dry-heat cooking technique in which food is completely submerged in very hot fat. The fat used for deep-frying must have a high smoke point. Fat heated to the right temperature will produce a crisp exterior and a juicy interior. The average temperature for deep-frying is 375°F, but recipes differ depending on the type of food.

When fats are exposed to high heat they begin to break down and smoke. A *smoke point* is the temperature at which an oil begins to smoke and give off an odor. Smoke points vary depending on the type, age, and clarity of the oil. The higher the smoke point, the better suited the oil is for frying. **See Figure 8-12.** The smoke point for butter is around 350°F, whereas vegetable oils begin to smoke around 440°F. The smoke point for olive oil is about 375°F, making it less desirable for frying.

Smoke Points of Oils and Fats	
Butter	350°F
Olive oil	375°F
Soybean oil, refined	440°F
Canola oil	450°F
Peanut oil, refined	460°F
Rice bran oil	490°F

Figure 8-12. The higher the smoke point, the better suited the oil is for frying.

Battering. Most deep-fried foods are breaded or battered. *Battering* is the process of dipping an item in a wet mixture of flour, liquid, and fat for frying. Fish and seafood are typically battered before deep-frying. *Tempura* is a very light batter that is used on vegetables, poultry, seafood, and meats served in Asian cuisine.

Procedure for Battering

1. Prepare a batter. *Note:* If the items to be battered are damp, lightly dredge the items in flour before placing them in the batter.
2. Dip the items in the batter and slowly add the battered items to the fryer.
3. Fry battered items until crispy and cooked through.
4. Remove items and drain to remove excess fat.

Deep-fryers should be heated to the required temperature prior to adding food. If the fat is not at the correct temperature it may cause the foods to stick together or become very greasy. Overfilling the deep-fryer with food can also produce these same results. The temperature of the fat drops as food is added. Adding too much food can cause the temperature of the fat to drop too low. A temperature drop will trigger the fryer to begin heating and if the recovery time is too long, the finished product will not be acceptable. *Recovery time* is the time it takes for the fat to return to the set temperature.

Photo Courtesy of McCain Foods USA

Foods that are deep-fried are added to a preheated fryer by either the swimming method or the basket method. In the swimming method an item is slowly lowered into the hot fat without the use of a fryer basket. The swimming method is used for battered items because battered foods stick together as they fry and can also stick to the basket. By slowly adding each battered item to the fryer, the item first sinks to the bottom of the fryer and then floats back up to the top. As the side that is facing down turns brown, the item is flipped over to brown the other side.

In the basket method, items are added to a fryer basket that is sitting on top of a pan, not over the fryer. Frozen items are often added using the basket method because wet crumbs can fall into the fryer and shorten the usable life of the fat. Once the basket is filled, it is submerged in the hot fat. All of the items in the basket should be submerged when the basket is lowered. When the items are fully cooked, they are removed from the fryer and placed in a drain pan to allow excess fat to drain off. It is important to strain the used fat of all food particles after frying. This process will extend the life of the fat.

Thinly cut potatoes, such as potato chips and shoestring fries, can be deep-fried in a single step, usually in hot oil between 350°F and 375°F. French fries need to be blanched and cooled before being deep-fried. Blanching potatoes in 275°F oil removes some of the moisture, causing them to cook more evenly when they are deep-fried. Potatoes that are cooked in 375°F oil without being blanched first instantly sear, preventing their interior moisture from escaping. The trapped internal moisture causes the potatoes to become limp and soggy soon after frying.

Healthy Frying

Procedure for Deep-Frying

1. Prepare the items to be deep-fried and bread or batter them.
2. Use either the swimming method or the basket method to add the items to the preheated fryer.
3. Fry the items until they are golden brown. *Note:* Some items float on top of the fat when done.
4. Remove the items from the fryer and place them on paper towels or a drain pan to allow excess fat to drain away.

Procedure for Deep-Frying Potatoes

1. Wash, peel (if desired), and cut potatoes to the desired size and place them in cold water to avoid browning. When ready for blanching, drain the precut potatoes and carefully pat dry.
2. Blanch the potatoes in 275°F oil until slightly undercooked.
3. Remove the potatoes from the hot oil and drain well.
4. Place the parcooked potatoes on sheet pans and refrigerate until cool. *Note:* Potatoes can be held in the refrigerator for a few hours or can be frozen for approximately one month.
5. When the potatoes are ready to be fried, place the parcooked potatoes in 375°F oil and fry until golden-brown and crispy. *Note:* Do not overfill the frying basket.

Deep-Frying Potatoes
Media Clip

Convenience Foods. Many items are available prefabricated for frying. These foods are commonly referred to as convenience foods. The most common is the preblanched frozen French fry. Other common convenience foods include breaded chicken, vegetables, and cheeses. These deep-fryer convenience foods are typically purchased frozen, and many times these foods are cooked from the frozen state. The advantages of using prefabricated foods include labor savings and a minimal amount of loss due to fewer spoiled or expired products.

Roasting

In the past, roasting meant cooking food over an open flame, typically on a spit over a fire. *Roasting* is a dry-heat cooking technique in which food that contains fat or that has fat added to it is cooked uncovered at a high temperature in an oven or on a revolving spit over an open flame. Leaving items uncovered while roasting enables caramelization to occur. Roasting produces a well-browned exterior and a moist interior. Whole poultry, large cuts of meat, and root vegetables are commonly roasted at lower temperatures to prevent shrinkage and lessen moisture loss.

Whole chickens, turkeys, and hogs may be skewered and placed on a rotisserie and roasted. *Spit-roasting* is the process of cooking food by skewering it and suspending and rotating it above or next to a heat source. Roasting may require periodic basting during the cooking process. *Basting* is the process of using a brush or a ladle to place fat on or pour juices over an item during the cooking process to help retain moisture and enhance flavor.

Procedure for Roasting

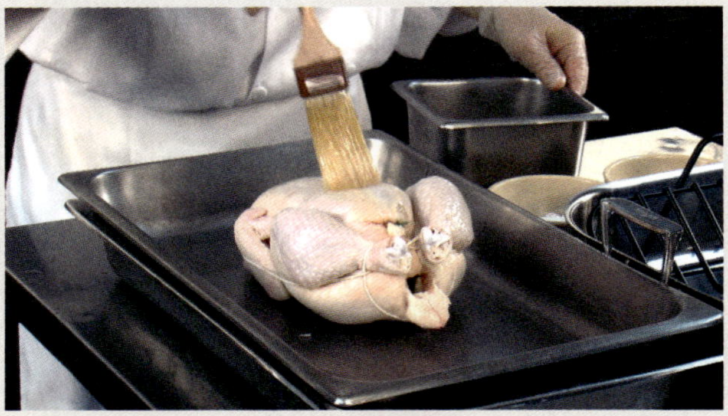

1. Trim excess fat from the item and season it. *Note:* Some items, such as turkey, may be brushed with fat before roasting for more even browning.

2. Place the item in a roasting pan on a roasting rack or a bed of mirepoix.

3. Place the uncovered roasting pan in an oven preheated to between 300°F and 350°F, and allow the item to cook.
4. Baste the item as needed.

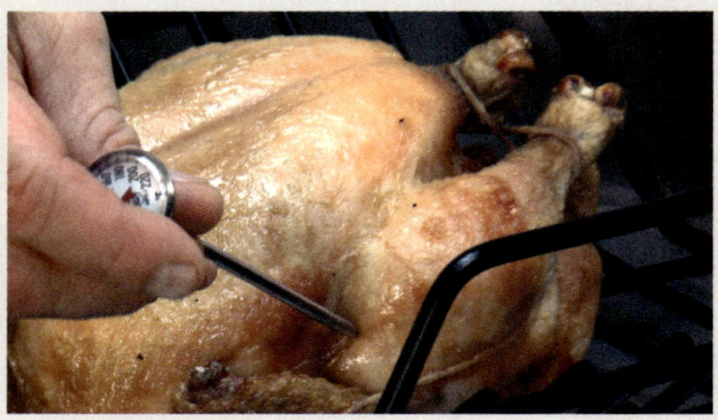

5. Remove the roasted item from the oven approximately 10 minutes before it reaches the desired degree of doneness. Insert a probe thermometer in the thickest portion to check the internal temperature.
6. Allow the item to rest and carryover cooking to complete the cooking process.

Roasting
Media Clip

Allowing roasted items to rest after being removed from the oven permits carryover cooking to occur. *Carryover cooking* is the rise in internal temperature of an item after it is removed from a heat source due to residual heat on the surface of the item. The internal temperature will rise between 5°F and 10°F during this period. Carryover cooking helps a cooked item retain moisture that will escape as steam if the item is cut too soon after cooking. Roasted meat should not be cut immediately after being removed from the oven. If it is, the internal moisture will be released as steam and evaporate into the air. Roasted meat that is allowed to rest before it is cut will retain moisture.

Baking

Baking is the dry-heat cooking technique in which food is cooked uncovered in an oven. Leaving items uncovered while they bake enables caramelization to occur. Like roasting, baking produces a well-browned exterior and a moist interior. Baking time, like roasting time, varies depending upon the type and size of the item, the temperature of the oven, and the ingredients used. Baking is the primary cooking technique used to cook yeast breads, quick breads, cookies, cakes, pies, and pastries. Casseroles, such as lasagna, moussaka, and cassoulet, are baked. **See Figure 8-13.**

Lasagna

Barilla America, Inc.

Figure 8-13. Casseroles, such as lasagna, moussaka, and cassoulet, are baked.

Procedure for Baking

1. Preheat an oven to the desired temperature.
2. Place items in parchment-lined or oiled pans per recipe directions.
3. Place pans in the oven and bake for the allotted time and until fully cooked.

Before placing items in the oven to bake, the oven must be preheated to the correct temperature. When baked items are removed from the oven, the items need to rest before being cut. Carryover cooking helps baked items retain their moisture while the baking process is completed. **See Figure 8-14.** If a loaf of bread is cut as soon as it comes out of the oven, it will immediately begin to dry out and lose much of its freshness.

Carryover Cooking

Paderno World Cuisine

Figure 8-14. Carryover cooking helps baked items retain their moisture while the baking process is completed.

Grilling

Grilling is a dry-heat cooking technique in which food is cooked on open grates above a direct heat source. A grill is scraped clean and then seasoned with fat before use to prevent food from sticking.

Poultry, seafood, meats, and some fruits and vegetables are often grilled. Many foods are cooked directly on the grill, while others are placed on skewers first. Because of the open cooking surface, foods that are grilled have a smoky flavor. Grilled foods have char lines where the food came in contact with the hot grill. Crosshatching the presentation side of grilled food enhances the presentation of the dish. To create a crosshatch marking when grilling, the food is rotated 90° about halfway through the cooking process. **See Figure 8-15.**

Grill Marks

Photo Courtesy of the Beef Checkoff

Figure 8-15. To create crosshatch marking when grilling, the food is rotated 90° about halfway through the cooking process.

Grilling and griddling are similar cooking techniques. *Griddling* is a dry-heat cooking technique in which food is cooked on a solid metal cooking surface called a griddle. The heat comes from below the cooking surface, but items do not come in contact with a flame. Because of the flat surface, a griddle generates less smoke than a grill. A small amount of oil is placed on a hot griddle to prevent foods from sticking. Pancakes, eggs, breakfast meats, and hot sandwich meats are often cooked on a griddle. A *French grill* is a griddle with raised ridges that creates grill marks where the food touches the ridges.

Broiling

Broiling is a dry-heat cooking technique in which food is cooked directly under or over a heat source. A broiler is typically scraped clean and then seasoned with fat before use to prevent food from sticking.

Poultry, seafood, meats, and some fruits and vegetables are broiled. Other items, such as bruschetta and French onion soup, are often finished in the broiler. Broiled food is exposed to an overhead flame or an overhead electric heating element when it is broiled. The hot grates on which the food is placed create distinctive markings on broiled items. The grates can be raised or lowered to adjust the intensity of the heat.

CHECKPOINT 8-2

1. Identify eight dry-heat cooking techniques.
2. Describe how to sauté food.
3. Sauté a protein and evaluate the final product for color, texture, and flavor.
4. Sauté a vegetable and evaluate the final product for color, texture, and flavor.
5. Define stir-frying.
6. Stir-fry a dish and evaluate the final product for color, texture, and flavor.
7. Describe the standard breading procedure.
8. Bread and pan-fry a protein and evaluate the final product for color, texture, and flavor.
9. Contrast pan-frying and deep-frying.
10. Explain why it is important to know the smoke point of oils used for frying.
11. Batter and deep-fry a potato and evaluate the final product for color, texture, and flavor.
12. Contrast roasting and baking.
13. Roast a food and evaluate the final product for color, texture, and flavor.
14. Bake a food and evaluate the final product for color, texture, and flavor.

MOIST-HEAT COOKING TECHNIQUES

Moist-heat cooking is any cooking technique that uses liquid or steam as the cooking medium. This involves cooking foods either by submerging them in hot liquid or by exposing them to steam. Because of this, often the natural flavor and aroma of the food is highlighted. It is important to understand the different types of moist-heat cooking to know which technique is best for a given preparation. Moist-heat cooking techniques include poaching, simmering, boiling, blanching, and steaming. **See Figure 8-16.**

Moist-Heat Cooking Techniques			
Techniques	*Temperature of Cooking Liquid*	*Appearance of Cooking Liquid*	*Common Uses*
Poaching	160°F to 180°F	Very tiny bubbles may have formed on bottom of pot, but they do not break the surface; liquid shows very little movement and emits slight steam	Delicate fish fillets, eggs out of the shell, soft-fleshed fruits, and poultry
Simmering	185°F to 205°F	Small bubbles have formed and come to the surface	Stews, sauces, soups, meats, and poultry
Boiling	212°F (at sea level)	Rapid motion in water as large bubbles break the surface	Potatoes, grains, and pasta
Blanching	212°F (at sea level)	Rapid motion in water as large bubbles break the surface	Vegetables, fruits, and removal of impurities from bones for stocks
Steaming	Greater than 212°F	Moist hot air emitted rapidly from boiling liquid	Vegetables, eggs in the shell, shellfish, some fish, and poultry

Figure 8-16. Moist-heat cooking techniques include poaching, simmering, boiling, blanching, and steaming.

Sullivan University

Poaching

Poaching is a moist-heat cooking technique in which food is cooked in a liquid that is held between 160°F and 180°F. Poaching imparts some of the flavor of the poaching liquid into the item being cooked. The type and amount of liquid used depends on the food being poached. For example, eggs are poached in lightly salted water and vinegar, and fish are commonly poached in court bouillon. A mirepoix is often added to a poaching liquid. Poaching liquids are often reduced and incorporated into a sauce that is served with the finished dish. The sauce is enriched with the nutrients of the food that were lost during poaching.

The two methods of poaching are submersion and shallow poaching. Submersion poaching requires the food to be completely covered by the poaching liquid. Eggs are poached using submersion poaching.

In shallow poaching the poaching liquid does not cover the food being cooked. Fish are typically shallow poached.

Procedure for Submersion Poaching

1. Bring the appropriate amount of water, salt, and vinegar to between 160°F and 180°F. *Note:* Bubbles should not break the surface of the water.
2. Crack the eggs one at a time into a bowl and then gently lower each one into the poaching water.

3. Cook the eggs to desired doneness.

4. Remove the poached eggs with a perforated spoon and rest them on a clean towel for a few seconds to remove excess water.

5. The finished egg should have a firm white and a soft yolk that is somewhat liquid in the center.

Procedure for Shallow Poaching

1. Bring the appropriate amount of poaching liquid to 180°F.

2. Gently lower the cleaned and prepared fish into the poaching liquid.
3. Cook the fish until almost cooked through.

4. Remove from the heat and allow the fish to rest in the liquid for a few minutes to finish cooking.

Shallow Poaching
Media Clip

Simmering

Simmering is a moist-heat cooking technique in which food is gently cooked in a liquid that is between 185°F and 205°F. Simmering can be identified by tiny bubbles that reach the surface, but do not break into a full boil. It is important to maintain a constant and even temperature when simmering foods. If a protein-rich food, such as a corned beef brisket, is boiled instead of simmered, the brisket will shrink and become tough. In contrast, simmering a brisket produces a tender product.

Procedure for Simmering

1. Add enough liquid to a pan to completely cover the item.
2. Bring the liquid to between 185°F and 205°F.
3. Add mirepoix, if desired.
4. Carefully place the cleaned and prepared item in the hot liquid and maintain a steady temperature until done. *Note:* Doneness is based on tenderness or cooking time.

Boiling

Boiling is a moist-heat cooking technique in which food is cooked by heating a liquid to its boiling point. Different liquids boil at different temperatures. Depending on the liquid, the boiling point may be more or less than 212°F (the boiling point of water at sea level). As the temperature of a liquid rises, it reaches a poaching temperature, then a simmer, and finally a boil. When a liquid reaches full boil, large bubbles that have formed at the bottom of the pan rise rapidly and then break on the surface of the liquid. Potatoes, grains, and pasta are often boiled in the professional kitchen.

Steaming

Steaming is a moist-heat cooking technique in which food is placed in a container that prevents steam from escaping. **See Figure 8-17.** Food is typically placed in perforated pans in a pressure steamer. The movement of the steam cooks the food gently and evenly. Steaming retains the shape, texture, flavor, and many of the vitamins and minerals. Many vegetables, such as broccoli, green beans, carrots, and asparagus, are often steamed. Pot stickers may also be prepared with the steaming method.

Steaming

Vulcan-Hart, a division of the ITW Food Equipment Group LLC

Figure 8-17. Steaming is a moist-heat technique in which food is placed in a container that prevents steam from escaping.

Steaming Techniques

Aromatics may be added to flavor the food that is being steamed. An aromatic is an ingredient added to a food to enhance its natural flavors and aromas. Aromatics, such as wine, herbs, and spices, release flavors and odors into the steamer that are absorbed by the food as it cooks.

CHECKPOINT 8-3

1. Identify five moist-heat cooking techniques.
2. Contrast submersion poaching and shallow poaching.
3. Poach an egg and evaluate the final product for color, texture, and flavor.
4. Poach a piece of fish and evaluate the final product for color, texture, and flavor.
5. Describe how to simmer food.
6. Simmer a grain and evaluate the final product for preparation, color, texture, and flavor.
7. Describe how to know when a liquid has reached a full boil.
8. Boil a potato and evaluate the final product for color, texture, and flavor.
9. Describe how to steam food.
10. Steam a vegetable and evaluate the final product for color, texture, and flavor.

COMBINATION COOKING TECHNIQUES

Combination cooking includes any cooking technique that uses both moist and dry heat. Combination cooking methods are often used on tougher cuts of meat to make them more tender and flavorful. The two most common combination cooking techniques are braising and stewing. In both methods the protein is seared and then simmered. The main differences between these two techniques are the sizes of the proteins cooked and the amount of liquid used to cook them.

Cooking Techniques

Braising

Braising is a combination cooking technique in which food is browned in fat and then cooked, tightly covered, in a small amount of liquid for a long period of time. The slow cooking process tenderizes the food by breaking down fibers. Braising is most often used for larger, roast-sized pieces of meat. First, the meat is seared on all sides in a small amount of fat. Aromatic vegetables are often added during the searing process. Then, a flavorful stock or liquid is added about halfway up the side of the seared piece of meat. The seared vegetables intensify the flavor of the braising liquid as it is brought to a simmer. The pot is covered with a lid and left to cook slowly until tender.

Nutrition Note

Braising and stewing result in a lower loss of vitamins than boiling or steaming because the cooking liquid is retained.

Stewing

Stewing is a combination cooking technique in which bite-sized pieces of food are browned in fat, barely covered with a liquid, and simmered for a long period of time in a tightly covered pot. Stewing requires more liquid than braising, because the food is completely covered in liquid. The pieces of food need to be very similar in size so they cook evenly.

When stewing, the pot is kept at a simmer on the stove or covered and placed in an oven at a low temperature. Diced vegetables are usually added when the meat is about three-fourths done. Sometimes the vegetables are cooked separately and added at the end. The cooking liquid, usually a stock, develops even more flavor from the caramelized meat and the slow cooking process. The meat and vegetables are removed from the stewing liquid, and a roux is added to thicken the stewing liquid to the desired consistency. The meat and vegetables are then added back to the stewing liquid and the dish is plated for service.

Foods can also be stewed using a technique called fricassee. A *fricassee* is a combination cooking technique in which food is sautéed in butter, barely covered with a liquid, and simmered for a long period of time in a tightly covered pot. Fricassees are commonly white in color but can also be dark, such as with coq au vin. Chicken, veal, and rabbit are commonly used to make fricassees.

Procedure for Braising

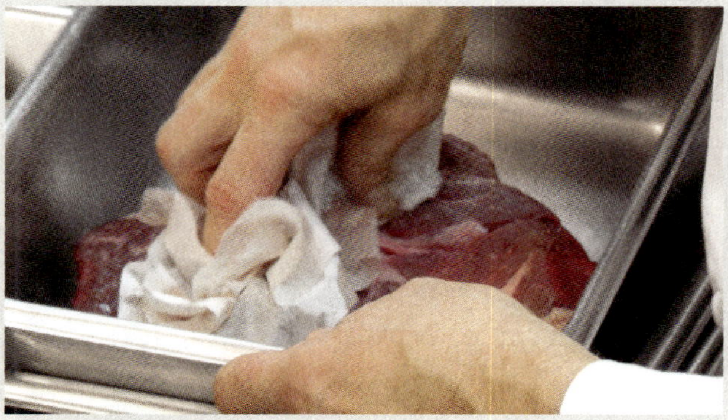

1. Clean, trim, and dry the meat with paper towels.

2. Coat the bottom of a preheated braising pan with a small amount of fat.

3. Add the meat and sear all sides. Remove the meat from the pan. *Note:* If desired, the meat can be dredged in flour prior to searing.

4. Add mirepoix to the pan and sauté until caramelized. Then, add a small amount of tomato product and cook until caramelized.

5. Deglaze the pan with a small amount of stock and add the meat back to the pan.

6. Bring the ingredients to a simmer and cover the pan. Cook until the meat is fork tender.

7. Remove the meat and strain the liquid to remove the mirepoix.

8. Thicken the liquid with brown roux until it reaches a nappe consistency.

9. Slice the meat and serve with the thickened braising liquid.

Braising
Media Clip

Procedure for Stewing

1. Add a small amount of fat to the bottom of a hot pot. Then, add a single layer of trimmed and cubed meat (1½ – 2 inch cubes).

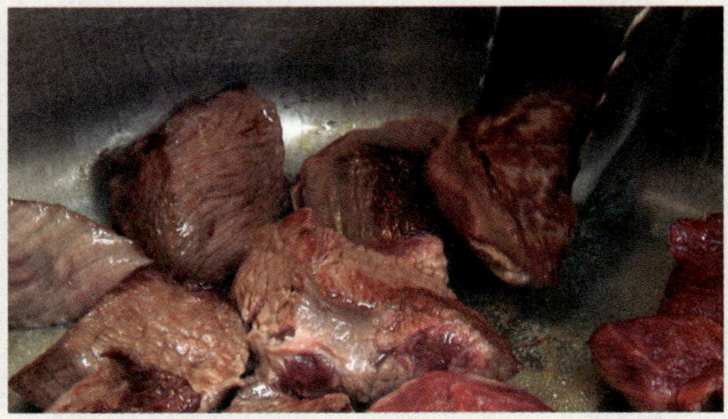

2. Sear the meat on all sides. *Note:* If desired, the meat can be dredged in flour prior to searing.

3. Add onions or mirepoix and cook until caramelized. Then add a tomato product and stir until it has caramelized.

4. Add stock until it just covers the meat. If desired, add a sachet d'épices.
5. Cover the pot and simmer until the meat is fork tender. Simmer for 30 minutes longer.
6. Add diced potatoes and cover the pot. Simmer for 10 minutes. The potatoes should be fork tender.

7. Remove the meat and vegetables from the liquid.
8. Add a brown roux to thicken the liquid to a nappe consistency.

9. Add the meat and vegetables back to the thickened liquid. Adjust seasonings and serve.

Stewing
Media Clip

CHECKPOINT 8-4

1. Define combination cooking.
2. Contrast braising and stewing.
3. Braise a meat and evaluate the final product for color, texture, and flavor.
4. Stew a meat and evaluate the final product for color, texture, and flavor.
5. Fricassee a meat and evaluate the final product for color, texture, and flavor.

PLATED PRESENTATIONS

Food presentation is a combination of shapes, colors, and textures. A beautifully plated meal is first experienced with the eyes. The plated food needs to be visually appealing. The presentation immediately prompts a judgment about how the meal is going to taste.

Using the appropriate plate is very important. For example, white plates are often used to highlight the contrasting colors and textures of food. **See Figure 8-18.** Temperature is also important. Cold foods should be presented on chilled plates and hot foods on warmed plates.

White Plates

United States Department of Agriculture

Figure 8-18. White plates highlight the contrasting colors and textures of food.

Portion Sizes

Guests have become accustomed to seeing plates overfilled with food and often perceive larger portions as a better value. Presenting food on smaller plates is one way to alter this perception. **See Figure 8-19.** For example, a 12 inch plate can be used instead of a 14 inch plate, and an 8 inch plate can be used in place of a 10 inch plate.

Smaller Plates

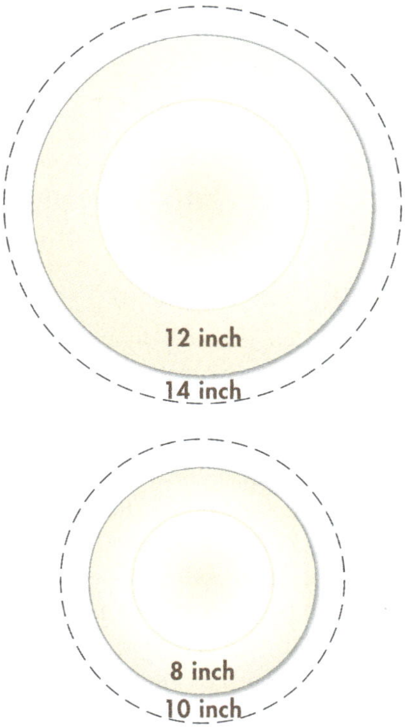

Figure 8-19. Presenting food on smaller plates is one way to alter the perception that larger portions are a better value.

Menus that offer smaller portions, such as one piece of grilled chicken instead of two, can promote healthy choices. For example, instead of offering a 12 oz steak, a 6 oz steak with savory roasted potatoes and a large portion of steamed vegetables can be offered. The USDA created the ChooseMyPlate website to help people understand how to meet the Dietary Guidelines and maintain their health and weight. **See Figure 8-20.** The ChooseMyPlate program emphasizes the need to eat a variety of vegetables and fruits, whole grains, lean proteins, and low-fat dairy products.

MyPlate Icon

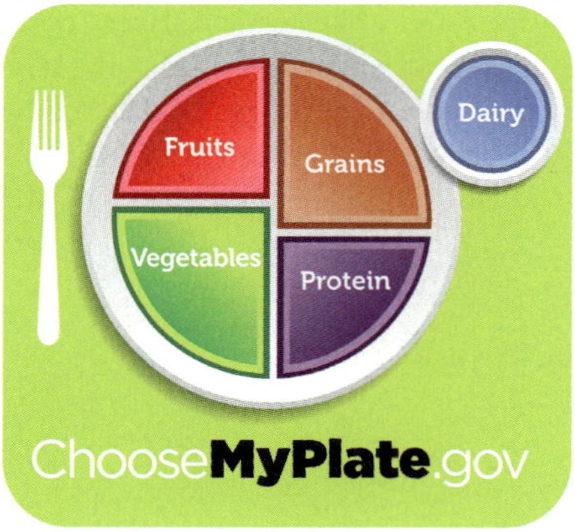

United States Department of Agriculture

Figure 8-20. The MyPlate icon is the symbol of the USDA ChooseMyPlate program.

Display and Garnishment

Careful consideration should be given to making every plate look appealing to the guest. Appealing plates display food in a visually appealing manner and use carefully selected garnishes. A *garnish* is an edible item that complements the food presented on the plate while adding visual interest through the use of color, shape, and texture. Garnishes can be raw, as in the case of watercress, or cooked, as with a caramelized lemon. **See Figure 8-21.** The contrasting colors, textures, and heights of the foods should be enhanced by the garnishes.

The way in which foods are cut also impacts the visual presentation. Cutting foods on the bias and using piped potatoes and molds to present grains adds dimension to the finished plate. For example, a well-seared piece of roasted meat displayed alongside a portion of vibrant steamed vegetables and piped mashed potatoes can be garnished with a spoonful of chutney and some chives.

Production Tip

Once the layout and garnish has been determined for a specific plate, a photo should be taken and then displayed in the kitchen to maintain plating consistency.

Garnishes

All-Clad Metalcrafters

Figure 8-21. Garnishes can be raw, as in the case of watercress, or cooked, as with a caramelized lemon.

Plated Presentations

CHECKPOINT 8-5

1. Explain how plating food on smaller plates alters perception.
2. Plate an entrée on two different sized plates. Compare the appearance of the two plates.
3. Explain how garnishes are utilized in the plating of food.

Refer to DVD for **Quick Quiz®** questions

SUSTAINABILITY CORNER

Many materials found in foodservice operations, such as grease, glass, aluminum, plastic, and paper, can be recycled into usable products. These materials are often referred to as post-consumer waste. Post-consumer waste is the garbage that individuals discard. By recycling post-consumer waste products and using recycled products, foodservice operations can reduce costs, attract environmentally conscious customers, and save natural resources.

Food scraps and trimmings are examples of post-consumer waste. These organic materials can often be composted to produce nutrient-rich soil that promotes plant growth. Other food-related items suitable for recycling include fats, oils, and grease (FOG). If disposed of improperly, FOG can pollute the municipal wastewater stream or clog water pipes through buildup. Investing in grease traps and disposal services helps the environment and may be less expensive than repairing clogged drains. Some disposal collectors will provide bins and haul FOG waste away free of charge or even pay foodservice operations for their FOG. The FOG collected from foodservice operations can then be used to make cosmetics, soaps, animal feed, paints, and biodiesel fuel.

In addition to food scraps, trimmings, and FOG, there are numerous items found in foodservice operations that are suitable for recycling. Salad dressing and cooking-oil containers, milk and juice jugs, canned goods, disposable tableware, carry-out containers, and egg cartons are examples of items made out of recyclable materials. Often these items are received in cardboard boxes that are also recyclable. In addition, many products can be purchased in packaging made from post-consumer recycled (PCR) materials.

Frymaster

Certain cities and states require batteries to be recycled. A variety of items found in foodservice operations, such as thermometers, table pagers, bar code scanners, emergency/security lights, smoke detectors, two-way radios, remote controls, and company vehicles, use batteries. Recycling batteries and replacing them with rechargeable batteries when possible helps reduce landfill waste, keeps hazardous materials from entering the environment, and preserves natural resources.

Recycling not only has positive effects for the planet, it also has the potential to decrease costs and increase revenue. To find local sources that recycle, contact the Environmental Protection Agency (EPA), search the Internet for recycling centers, research websites such as Earth911®, and visit local foodservice operators.

Recycling

Certification Exam Preparation Questions

1. The three different methods of ___ used when cooking food are conduction, convection, and radiation.
   A. caramelization
   B. induction
   C. moisture transfer
   D. heat transfer

2. An impinger oven operates using ___ heat transfer.
   A. induction
   B. radiation
   C. convection
   D. conduction

3. The Maillard reaction is the reaction that occurs when the ___ and sugars in a food are exposed to heat and merge together to form a brown exterior surface.
   A. lipids
   B. proteins
   C. carbohydrates
   D. starches

4. Caramelization is a reaction that occurs when sugars are exposed to high heat and produce ___ and a change in flavor.
   A. browning
   B. sweating
   C. coagulation
   D. gelatinization

5. The three steps to caramelizing are ___, browning, and caramelizing.
   A. cooking
   B. heating
   C. sweating
   D. reducing

6. Reduction is the process of gently simmering a liquid until its volume decreases, resulting in a ___ liquid with a more concentrated flavor.
   A. clearer
   B. clarified
   C. lighter
   D. thicker

7. ___ cooking is any technique that uses hot air, hot metal, a flame, or hot fat to conduct heat and brown food.
   A. Conduction
   B. Convection
   C. Dry-heat
   D. Moist-heat

8. ___ is a dry-heat cooking technique in which food is cooked quickly in a sauté pan over direct heat using a small amount of fat.
   A. Sautéing
   B. Stir-frying
   C. Pan-frying
   D. Deep-frying

9. ___ butter is butter fat that has been separated from the milk solids and water found in whole butter.
   A. Clarified
   B. Broken
   C. Compound
   D. none of the above

10. Using one hand for dry ingredients and the other hand for wet ingredients, the hands do not become ___ while moving food through the breading procedure.
    A. battered
    B. breaded
    C. dredged
    D. none of the above

11. Pan-frying is a ___ cooking technique in which food is cooked in a pan of hot fat.
    A. moist-heat
    B. dry-heat
    C. combination
    D. none of the above

12. The average temperature of fat for deep-frying is ___.
    A. 275°F
    B. 325°F
    C. 375°F
    D. 400°F

13. Foods being deep-fried are added to hot fat using the basket method or the ___ method.
A. floating
B. swimming
C. breading
D. battering

14. ___ is the dry-heat cooking technique in which food is cooked uncovered in an oven.
A. Baking
B. Roasting
C. Broiling
D. Grilling

15. ___ is a dry-heat cooking technique in which food is cooked on open grates above a direct heat source.
A. Broiling
B. Grilling
C. Griddling
D. Roasting

16. Poaching is a moist-heat cooking technique in which food is cooked in a liquid that is held between 160°F and ___.
A. 170°F
B. 175°F
C. 180°F
D. 185°F

17. The time required to stir-fry an item depends on its ___.
A. color
B. aroma
C. thickness
D. texture

18. ___ retains the shape, texture, flavor, and many of the vitamins and minerals of vegetables.
A. Baking
B. Boiling
C. Simmering
D. Steaming

19. Combination cooking methods are often used on ___ cuts of meat to make them more tender and flavorful.
A. expensive
B. less expensive
C. tougher
D. fatty

20. The main difference between braising and stewing is the size of the proteins cooked and the amount of ___ needed to cook them.
A. fat
B. liquid
C. time
D. heat

21. Describe the procedure for sautéing large items.

____
____
____
____

22. Describe the breading procedure.

____
____
____
____

23. Describe the procedure for shallow poaching.

____
____
____
____
____
____
____

24. Describe the procedure for braising.

____
____
____
____
____
____
____

SKETCHING EXERCISE

25. Sketch an appealing and appropriately portioned plate presentation.

Notes

Baking & Pastry Station

Introduction

In the ninth station of the ACFEF apprenticeship program, you will cover all aspects of baking, from equipment and tools, to preparation, to the production and presentation of baked goods. You will become skilled in the basics of preparing classical and contemporary baked products and pastries by preparing breads, cakes, cookies, pies, tarts, pastries, and specialty desserts.

Equipment is also a focus in the baking and pastry station because many of the techniques require the use of special machines and tools. Measuring, mixing, assembling, and decorating as well as docking, proofing, and scoring are just a few of the techniques you will learn. The basics of using pastillage, marzipan, chocolate, and blown and spun sugar will become part of your everyday vocabulary.

The food industry uses more energy than any other industry. Sustainable practices can begin with simple changes, such as replacing traditional incandescent light bulbs with compact fluorescent light bulbs (CFLs) to reduce power bills and minimize energy use. You will also learn how to lower energy use through the proper maintenance of equipment and when it may be best to replace equipment with more energy-efficient models. In addition, you will learn how to calculate the cost savings of using energy-efficient equipment over time.

Photo Courtesy of Bobby Watson

Irinox USA

Yeast breads, quick breads, cakes, icings, cookies, pies, pastries, frozen desserts, and specialty desserts are produced in the baking and pastry station. This is also the station where chocolate and sugar showpieces are made. Baked products are either made completely in-house or made using a combination of convenience products and scratch ingredients and then portioned and plated for service. It is important to understand how to use bakeshop tools and equipment and how bakeshop ingredients interact to produce formula-based products. Some apprentices find this station so appealing that they choose to pursue a pastry chef certification path.

Sullivan University

BAKESHOP TOOLS AND EQUIPMENT

A variety of ovens, refrigeration units, and work tables are used in the baking and pastry station. Work tables provide a surface with the perfect height for the vast number of preparation tasks performed in the bakeshop. Stainless steel tables, marble- or quartz-topped, and wood-topped work tables are all used in the bakeshop. In addition, special tools, bakeware, and equipment are used to prepare yeast breads, quick breads, cakes, cookies, pies, pastries, and specialty desserts.

Bakeshop Tools

Countless tools are used in the bakeshop. However, some common tools are kept close at hand due to their frequent use. Commonly used bakeshop tools include bench brushes, dough cutters, dough dockers, markers, palette knives, pastry bags and tips, pastry brushes, pastry wheels, rolling pins, silicone mats, and tamis. **See Figure 9-1.**

Commonly Used Bakeshop Tools	
Bench Brushes  American Metalcraft, Inc.	A *bench brush* is a brush with long bristles set in vulcanized rubber attached to a handle. Bench brushes are used to brush excess flour from the baker's bench.
Dough Cutters  Paderno World Cuisine	A *dough cutter*, also known as a bench knife, is a flat, stainless steel blade attached to a sturdy handle. Dough cutters are used to cut dough into portions and to scrape dough off the surface of the baker's bench.

Figure 9-1. (continued on next page)

Commonly Used Bakeshop Tools	
Dough Dockers  *American Metalcraft, Inc.*	A *dough docker* is a rolling tool with small spikes that is used to perforate dough so that it will bake evenly without blistering in the oven heat.
Markers  *American Metalcraft, Inc.*	A *marker* is a round tool that has wire guides that leave marks indicating where to cut pies, round cakes, or pizzas into equal portions. Markers are available in various diameters and portion sizes.
Palette Knives  *Dexter-Russell, Inc.*	A *palette knife*, also known as a cake spatula, is a flat or offset narrow knife with a rounded, 3½–12 inch blade that varies in flexibility. Both offset and flat palette knives are used to ice cakes.
Pastry Bags and Tips  *Messermeister*	A *pastry bag* is a cone-shaped paper, canvas, or plastic bag that is fitted with a pastry tip. Pastry bags are used to pipe icings or soft foods such as whipped potatoes. A *pastry tip* is a cone-shaped tip that is fitted into the narrow end of a pastry bag.
Pastry Brushes 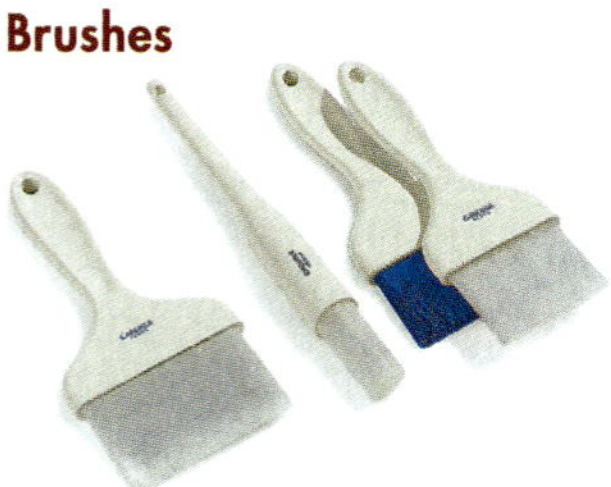 *Carlisle FoodService Products*	A *pastry brush* is a small, narrow brush that is used to apply liquids, such as egg wash or butter, onto baked products. Pastry brushes are available in natural, nylon, or silicone bristles. Nylon brushes cannot be used with hot items. Silicone bristles can withstand temperatures up to 650°F.
Pastry Wheels 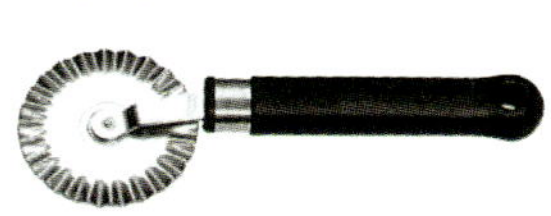 *Messermeister*	A *pastry wheel* is a dough-cutting tool with a rotating disk attached to a handle. Pastry wheels are used to cut dough into desired shapes.

Figure 9-1. (continued on next page)

Commonly Used Bakeshop Tools	
Rolling Pins  *Paderno World Cuisine*	A *rolling pin* is a slim cylinder that is used to flatten pastry dough, bread crumbs, or other foods. Rolling pins are available in wood, marble, ceramic, and metal and may have a handle on each end. French rolling pins have tapered ends instead of handles.
Silicone Mats 	A *silicone mat* is a woven, nonstick mat that may be used in the refrigerator, freezer, or oven and can withstand temperatures between –40°F and 580°F.
Tamis  *Paderno World Cuisine*	A *tamis*, also known as a drum sieve, is a flat, round sieve with an aluminum or wood frame and a mesh screen bottom. A tamis is used to sift lumps from dry ingredients. It may also be used to purée soft foods by using a spatula or bowl scraper to force the food through the wire mesh bottom.

Figure 9-1. A variety of tools are used to prepare items in the bakeshop.

Bakeware

Professional bakeware is typically made from metal or silicone. Bakeware is available in a variety of shapes and sizes, including rectangular, round, and square pans. The size and shape of the pans used depends on the items being prepared. Bakeware commonly used in the bakeshop includes cake pans, loaf pans, muffin pans, pie pans, sheet pans, springform pans, and tart pans. **See Figure 9-2.**

Commonly Used Bakeware	
Cake Pans  *Lincoln Foodservice*	A *cake pan* is a round, square, or specially shaped pan with short or tall sides and is used to bake cakes. Several cake pans may be required to make multiple layers of a cake.
Loaf Pans  *Carlisle FoodService Products*	A *loaf pan* is a deep rectangular pan that is used to bake loaves of bread.

Figure 9-2. (continued on next page)

Commonly Used Bakeware	
Muffin Pans 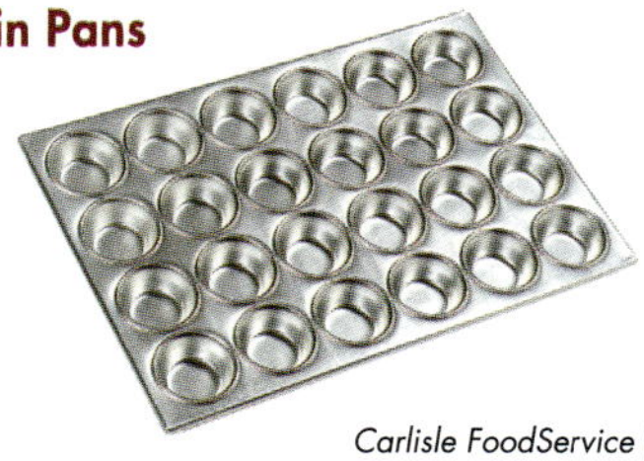 *Carlisle FoodService Products*	A *muffin pan* is a rectangular pan with cuplike wells and is used to bake teacakes, muffins, or cupcakes. The diameter of the cuplike wells varies from miniature to jumbo in size.
Pie Pans  *Lincoln Foodservice*	A *pie pan* is a round, shallow pan with sloped sides and is used for baking pies.
Sheet Pans  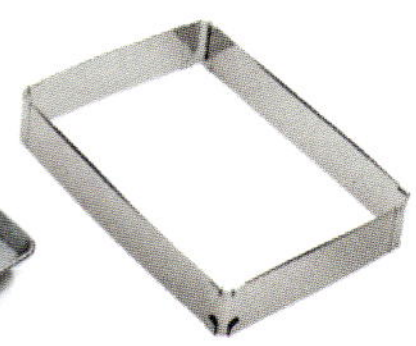 *Carlisle FoodService Products* *Paderno World Cuisine*	A *sheet pan* is a flat aluminum pan with very low sides. Sheet pans come in full pan (18 × 26 × ¾ inches) and half pan (18 × 13 × ¾ inches) sizes. Sheet pans are used for cooking meats and for baking rolls, scones, cookies, and sheet cakes. A *sheet pan extender* is an adjustable frame that extends the height of a sheet pan to accommodate items such as sheet cakes.
Springform Pans  *Paderno World Cuisine*	A *springform pan* is a round pan with a metal clamp on the side that allows the bottom of the pan to be separated from the sides. Springform pans are typically used to bake cheesecakes.
Tart Pans 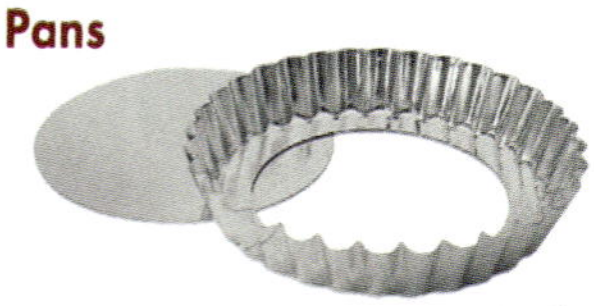 *Matfer Bourgeat USA*	A *tart pan* is a round, shallow baking pan with sloped sides that are smooth or fluted and may have a removable bottom.

Figure 9-2. Professional bakeware is typically made from stainless steel, aluminum, or silicone.

Bakeshop Equipment

Several different types of refrigeration and freezer units, ovens, and specialized equipment are used in the bakeshop. Walk-in, roll-in, and reach-in units as well as blast chillers are used in the bakeshop.

Walk-in Units. A *walk-in unit* is a room-sized insulated storage unit used to store bulk quantities of cold or frozen food. Walk-in units are often outfitted with adjustable shelving. Speed racks can be wheeled directly into walk-in units.

Roll-in Units. A *roll-in unit* is a refrigeration unit that allows speed racks to be rolled in and out of the unit through a door opening that is just above floor height. Roll-in units have a small ramp at the door opening to allow speed racks to be easily rolled in and out.

Reach-in Units. A *reach-in unit* is a temperature-controlled cabinet for storing cold or frozen food items. Reach-in units can have one, two, or three doors, each with shelves the size of a standard sheet pan. A *lowboy* is a reach-in refrigerated unit located beneath a work surface. This allows the foodservice worker to prepare foods on the work surface and have adequate refrigerated storage beneath.

Blast Chillers. A *blast chiller* is a specialized cooling unit that rapidly reduces the temperature of foods, rendering them safe for immediate storage. Items also can be quickly frozen in a blast chiller, resulting in a high-quality frozen product.

Irinox USA

Convection ovens, combi ovens, deck ovens, revolving tray ovens, and rotating rack ovens are commonly used in the bakeshop.

Convection Ovens. A *convection oven* is a gas or electric oven with an interior fan that circulates dry, hot air throughout the cabinet. This fan creates intense and even heating as it keeps the air moving while items are cooking. Cooking temperatures are 50°F lower in a convection oven than in a conventional oven. The airflow leads to more efficient cooking, reduced shrinkage, and cooks more uniformly than a conventional oven.

Combi Ovens. A *combi oven,* also known as a combination oven, is an oven that has both convection and steaming capabilities. Combi ovens can steam and circulate hot air at the same time. They also decrease cooking time, save energy, and enable even roasting and baking.

Deck Ovens. A *deck oven* is a drawer-like oven that is commonly stacked one on top of another, providing multiple-temperature baking shelves. Items such as breads are placed directly on the floor of a deck oven to bake.

Revolving Tray Ovens. A *revolving tray oven* is a large oven with shelves that rotate like a ferris wheel to allow a large number of items to be baked at one time.

Rotating Rack Ovens. A *rotating rack oven,* also known as a reel oven, is a large oven that rotates 10–80 pans of food as it cooks. Some rotating rack ovens are available with a steam kettle, hood, and trough used for cooking bagels.

In addition to refrigeration and freezers units and ovens, there are pieces of specialty equipment that are invaluable in the bakeshop. Commonly used pieces of bakeshop equipment include dough sheeters, mixers, proofing cabinets, scales, speed racks, and storage bins. **See Figure 9-3.**

Dough Sheeters. A *dough sheeter* is an electric appliance used to sheet, or roll out, doughs to the desired thickness.

Mixers. A *mixer* is a versatile appliance with U-shaped arms that secure a stainless steel mixing bowl under a rotating head that accommodates various attachments. Both bench and floor models are used in the bakeshop. Bench mixers range in size from 4½–20 qt capacities. Floor mixers range in size from 30–140 qt capacities. A paddle attachment is used for mixing and creaming. A whip attachment is used for whipping volume into products. A dough hook attachment is used for kneading dough.

Proofing Cabinets. A *proofing cabinet,* also known as a proofer, is a holding cabinet that contains both temperature and humidity controls. A proofing cabinet is used for proofing dough. *Proofing* is the process of letting yeast dough rise in a warm (85°F) and moist (80% humidity) environment until the dough doubles in size.

Scales. A baker's scale or a digital scale can be used to measure the weight of ingredients in the bakeshop. A *baker's scale,* also known as a balance scale, is a scale with two platforms that use a counterbalance system to measure weight. The food to be weighed is placed on one platform, and a weight is placed on the other platform. Weight is added to, or removed from, the second platform until the scale balances. A baker's scale is used to measure most baking ingredients. Digital scales are used when weighing very small amounts, such as ⅛ oz, of an ingredient.

Speed Racks. A *speed rack,* also known as a tallboy, is a tall cart on wheels with rails that hold entire sheet pans of food. These storage units can be moved to various areas of the kitchen as needed.

Storage Bins. Large polyurethane storage bins on wheels are used to store bulk items such as flour, sugar, and grains. Bins can easily be moved to where they are needed. Each bin should be labeled with the contents and the date the contents were placed in the bin.

Figure 9-3. Frequently used pieces of bakeshop equipment include scales, dough sheeters, mixers, proofing cabinets, speed racks, and storage bins.

CHECKPOINT 9-1

1. Describe common tools used in the bakeshop.
2. Describe seven common types of bakeware.
3. Describe four types of refrigeration and freezer units.
4. Describe five types of ovens used in the bakeshop.
5. Explain the purpose of a dough sheeter.
6. Explain the purpose of the paddle, whip, and dough hook mixer attachments.
7. Differentiate between a proofing cabinet and a speed rack.
8. Use a baker's (balance) scale to measure 1 lb 3 oz of flour or sugar.
9. Use a digital scale to measure ⅛ oz of baking powder or salt.
10. Explain the purpose of storage bins.

BAKESHOP INGREDIENTS

Baking is based upon chemical reactions among ingredients. Once a product is placed in the oven, there is no opportunity to make any changes. Because of this, bakeshop recipes are called formulas because they require precise measurements of specific ingredients in order to ensure successful end products. Formulas help to ensure that measurements are performed consistently and accurately. For example, substituting bread flour for cake flour will result in a product with an entirely different look, texture, and flavor.

Bakeshop ingredients are measured by weight or by volume. Although weight is more accurate than volume, it is important to understand both. Weight is measured in grams (g), pounds (lb), and ounces (oz). Volume is measured in liters (l), pints (pt), cups (c), tablespoons (tbsp), and teaspoons (tsp).

The weight of liquid is measured in ounces (oz) and the volume of liquid is measured in fluid ounces (fl oz). However, 1 oz does not equal 1 fl oz. Because different ingredients have different densities, two items of the same volume can have different weights. For example, 1 cup of honey weighs more than 1 cup of water.

Different types of flours, sugars and sweeteners, fats, eggs, and milk are used to produce baked products. Thickeners and leavening agents alter the texture and weight of baked products. The flavoring ingredient used often determines the flavor of the baked product.

Flours

Flour is a fine powder that is created by grinding grains or nuts. Flour supplies structure and nutritional value to baked products. It also acts as an absorbing agent. Flour is characterized as being hard or soft, depending on the protein content of the grain from which it is ground. *Gluten* is a type of protein found in grains such as wheat, rye, barley, and some varieties of oats. Gluten is the protein in flour that, when combined with water, gives a baked product its structure. **See Figure 9-4.** Common flours used in the bakeshop include bread flour, cake flour, pastry flour, rye flour, and whole-wheat flour.

Gluten Formation

Figure 9-4. Gluten is the protein in flour that, when combined with water, gives a baked product its structure.

Bread Flour. Bread flour, also known as hard wheat flour, has low amounts of starch and high amounts of protein that produce a large amount of gluten when water is added. Bread flour is typically used to make bread doughs.

Flours

Cake Flour. Cake flour, also known as soft wheat flour, contains low amounts of protein and high amounts of starch. Cake flour does not form gluten as readily as bread flour, which results in a more delicate, tender crumb. Cake flour is typically used to make cakes, cookies, pies, and pastries. Most cake flour is bleached during processing, which results in flour that is more acidic. The acidity of cake flour causes starches to gelatinize and, in turn, reduces baking time.

Pastry Flour. Pastry flour contains the ideal amount of gluten for pie doughs. If pastry flour is not available, a blend of 60% cake flour and 40% bread flour can be used. Pastry flour has a tendency to pack and form lumps, so it should always be sifted before use. Some bakers and pastry chefs chill pastry flour in order to keep pie dough below 70°F during the mixing period.

Rye Flour. Rye flour is a dark-colored flour milled from rye seeds. Rye flour lacks protein, so it is almost always combined with wheat flour when making bread. Variations of rye flour are made from different parts of rye seeds and include white, cream, dark, and pumpernickel flour. Pumpernickel flour is a meal form of rye made by grinding the whole grain.

Whole-Wheat Flour. Whole-wheat flour is unbleached flour that still contains the bran, endosperm, and germ of the wheat kernel. It has a nuttier flavor than other flours and is used to make breads. It has more nutritional value than other types of wheat flours.

Flours are also made from corn, soybeans, rice, barley, oats, millet, potatoes, buckwheat, spelt, quinoa, and nuts. Almond and hazelnut flours are commonly used by bakeshops. Nut flours lack gluten-forming properties, so they must be combined with wheat flour when making breads.

Sugars and Natural Sweeteners

Sugars and natural sweeteners add color, texture, and flavor to baked products. Sugar stimulates the growth of yeast, supplies moisture, and helps prolong freshness. Sugar also adds flavor, color, and tenderness to baked products through caramelization. Sugar and natural sweeteners come from a variety of plant sources and are available in many different forms. Agave nectar, brown sugar, coarse sugar, confectioners' sugar, granulated sugar, honey, molasses, and turbinado sugar are some of the sugars and natural sweeteners used in the bakeshop. **See Figure 9-5.**

Baking with Honey

Sugars and Natural Sweeteners	
Agave Nectar 	*Agave nectar*, sometimes called agave syrup, is the nectar from an agave plant. The blue agave contains a high percentage of fructose in its nectar and produces the premium form of agave nectar. Lighter and darker varieties of agave nectar are made from the same plant. The flavor of agave nectar is comparable, but not identical, to honey.
Brown Sugar 	Brown sugar is available in two forms: light and dark. *Light brown sugar* is a moist sugar product that contains approximately 3.5% molasses. *Dark brown sugar* is a moist sugar product that contains approximately 7% molasses. Brown sugar should be kept in an airtight container or it will dry out and turn rock hard.
Coarse Sugar 	*Coarse sugar*, also known as decorators' sugar, is a white sugar composed of coarse-grain crystals that are about four times larger than granulated sugar.

Figure 9-5. (continued on next page)

Sugars and Natural Sweeteners	
Confectioners' Sugar 	*Confectioners' sugar*, also known as powdered sugar, is granulated sugar that has been ground into a very fine powder. A small amount of cornstarch is added to confectioners' sugar to prevent lumping. Confectioners' sugar is often used to make icings and meringues.
Granulated Sugar 	*Granulated sugar* is a white sugar composed of small, uniformly sized crystals.
Honey 	*Honey* is a sweet, thick fluid made by honeybees from flower nectar and is approximately 1½ times as sweet as granulated sugar. The flavor of honey depends on the flowers from which the nectar was derived. Clover, buckwheat, and lavender each yield a different flavor of honey.
Molasses 	*Molasses* is the dark syrup that is left after sugar cane has been processed. Cane syrup is processed three to four times, yielding a more concentrated grade of molasses each time. Blackstrap molasses is the strongest and most processed form of molasses. Molasses is often added to bread, cake, and cookie recipes because of its distinct flavor.
Turbinado Sugar 	*Turbinado sugar* is a blond sugar that is purified and cleaned by steam before being dried into coarse-grain crystals.

Figure 9-5. Agave nectar, brown sugar, coarse sugar, confectioners' sugar, granulated sugar, honey, molasses, and turbinado sugar are commonly used in the bakeshop.

Baker's sugar is used by professional bakers and pastry chefs to produce products with a very smooth texture and delicate crumb. The ultra-fine grain of baker's sugar mixes, blends, and melts more evenly than granulated sugar. Superfine sugar, also known as castor sugar, dissolves almost instantly and is often used to make meringues.

Fats

Fat is an essential ingredient in the bakeshop. Fats add richness and tenderness and improve the grain and texture of baked products. Butter, oil, and shortening are common types of fat used to make tender and moist baked products. **See Figure 9-6.**

Fats	
Butters 	*Butter* is a fat made from cream that has a butterfat content of at least 80%. European-style butter contains 82% to 86% butterfat and is considered premium butter. Unsalted butter does not contain salt and is preferred for baking.
Oils 	*Oil* is a fat that remains in a liquid state at room temperature. Most oils are taken from plant sources and help maintain moisture in baked products. Mildly flavored oils, such as vegetable and canola oil, are preferred for baking. Stronger flavored oils, such as olive oil, can overpower the flavor of baked products.
Shortenings 	A *shortening* is a flavorless fat made by chemically changing a vegetable oil into a solid form. *Hydrogenation* is a process that changes the molecular structure of oil into a solid. When hydrogenation occurs, the melting point of fat is raised, making it more pliable. Hydrogenated shortening does not contain any water but is 10% air by weight. Both hydrogenated and emulsified shortenings contain trans fats. *Hi-ratio shortening* is an emulsified shortening that can absorb more sugar and moisture than hydrogenated shortenings. Hi-ratio shortening gives cakes that contain more sugar than flour a finer texture and keeps icings stable.

Figure 9-6. Butter, oil, and shortening are types of fat used to make tender and moist baked products.

Eggs

Both fresh and frozen eggs are used in bakeshops. **See Figure 9-7.** In addition to adding color and flavor, the protein in eggs gives structure to baked products. Eggs also function as a thickener in custards and add moisture, leavening, and nutritive value to cake batters. Beaten and whipped eggs trap pockets of air inside doughs and batters that expand when heated, helping baked products rise.

Egg yolks are natural emulsifiers, meaning that they hold or bind ingredients together that normally would not combine, such as water and oil. Egg yolks also add a rich flavor and golden color to baked products.

Eggs

Fresh

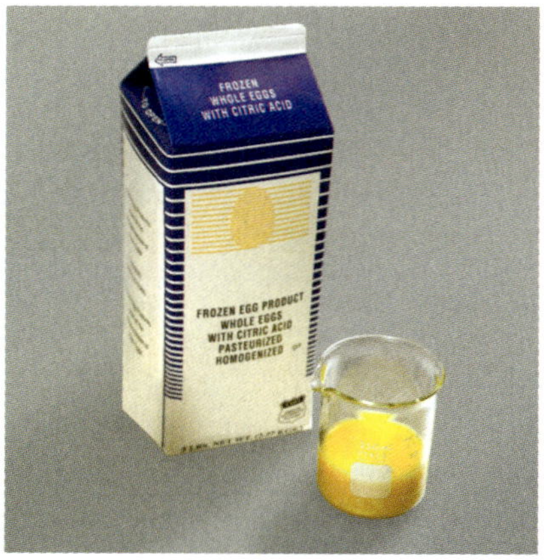

American Egg Board

Frozen

Figure 9-7. Both fresh and frozen eggs are used in the bakeshop.

Eggs

Milks

Milk and other dairy products have many different effects on baked products. Milk improves the texture and flavor, supplies moisture, and adds nutrients such as calcium, vitamin D, and protein. The proteins in milk assist in producing a finer crumb in yeast breads. When breads or pastry items containing milk are baked, the proteins and sugars in the milk absorb moisture, producing a softer crust. These same proteins and lactose (milk sugar) break down during baking and begin to caramelize, which adds a rich color to the crusts of baked products.

Milk used in the preparation of breads and sweet doughs may be liquid or dried. **See Figure 9-8.** Dry milk must be reconstituted before or during the mixing period. Milk-like products, such as soy milk, rice milk, and almond milk, can be used in place of milk. Water also can be substituted for milk in most bread recipes.

Thickening Agents

Thickening agents provide structure and stability to baked products. Each thickening agent has properties that make it more favorable for use in some items than others. Common thickening agents used in the bakeshop include arrowroot, cornstarch, flours, gelatin, tapioca, and modified starches. **See Figure 9-9.** A *modified starch* is a combination of starch and one or more vegetable gums. Modified starches gelatinize quickly when cooked and offset the action of fruit acids. Modified starches such as ClearJel® produce a finished product with a high sheen that maintains its color, has a smooth consistency, and does not cloud when refrigerated.

Milks

Liquid Dried

Figure 9-8. Milk used in the preparation of breads and sweet doughs may be liquid or dried.

Sullivan University

Common Thickening Agents	
Arrowroot 	*Arrowroot* is a thickening agent that is the edible starch from the rootstock of the arrowroot plant. It is often used to make pie fillings and custards. Like cornstarch, arrowroot is made into a slurry before being added to a hot mixture.
Cornstarch 	*Cornstarch* is the white, powdery, pure starch derived from corn. Cornstarch is a popular thickening agent because it will not cloud a clear sauce and is preferred for use in pie fillings. Cornstarch should be mixed with a small amount of cold water to make a slurry before it is added to a hot mixture. Cornstarch should be thoroughly cooked or it will leave an unpleasant aftertaste.

Figure 9-9. (continued on next page)

Common Thickening Agents	
Flours 	A low-protein flour works best for use as a thickening agent. However, it requires twice as much flour as cornstarch or arrowroot to thicken a product. Any product thickened with flour should be cooked thoroughly or the flavor of raw flour will remain.
Gelatin Sheets 	*Gelatin* is a flavorless thickening agent made from animal protein. It is available in powdered and sheet forms. Powdered gelatin comes in bulk form. Gelatin sheets are sold in a standard size that is equivalent to 1 tsp of powdered gelatin. Gelatin must be dissolved or soaked in cold water and then gently heated before being added to a product.
Tapioca 	*Tapioca* is a thickening agent derived from the cassava plant. It is available in round granules or as ground flour. Tapioca leaves fillings clear and glossy and has twice the thickening power of flour. When using granular tapioca, it is important to have enough liquid in a filling to ensure that the granules will dissolve.

Figure 9-9. Thickening agents used in the bakeshop include arrowroot, cornstarch, flours, gelatin, and tapioca.

Leavening Agents

A *leavening agent* is any ingredient that causes a baked product to rise by the action of air, steam, chemicals, or yeast. The most basic way to leaven baked products is to incorporate air or steam into a dough or batter. Because air is lighter than the surrounding dough or batter, the air forces the product to rise as it tries to escape. Steam is created when liquids in the dough or batter evaporate and force a bread to expand. Air pockets also can be created by whipping, creaming, or folding ingredients.

Both natural and chemical leavening agents are used in the bakeshop. In addition to air and steam, ammonium bicarbonate, baking powder, baking soda, cream of tartar, and yeast are used to leaven breads, cakes, cookies, and pastry doughs. **See Figure 9-10.**

Common Leavening Agents	
Ammonium Bicarbonate 	*Ammonium bicarbonate* is a chemical leavening agent that reacts as soon as it is heated without the help of an acid or base. Ammonium bicarbonate is used mainly in cookie and cracker recipes. It is important to bake any product containing ammonium bicarbonate thoroughly so that all of the gases are released. Otherwise, an ammonia smell will remain in the finished product.
Baking Powder 	*Baking powder* is a chemical leavening agent that is a combination of baking soda and cream of tartar or sodium aluminum sulfate. The alkalinity of baking soda reacts with the acidity of the other ingredients to produce carbon dioxide. Double-acting baking powder reacts when it is mixed with liquids, releasing some carbon dioxide, and then reacts again when it is heated. Double-acting baking powder allows a batter to be prepared in advance without the risk that it will not rise when baked at a later time.

Figure 9-10. (continued on next page)

Common Leavening Agents	
Baking Soda 	*Baking soda*, also known as sodium bicarbonate, is an alkaline chemical leavening agent that reacts to an acidic dough or batter by releasing carbon dioxide without the addition of heat. Baking soda is often used in conjunction with acidic liquids such as buttermilk, lemon juice, or molasses. Using too much baking soda can produce a yellow color, cause brown spots, and leave a bitter aftertaste. Baking soda should not be kept longer than one year, because it absorbs moisture from the air, which reduces its strength over time.
Cream of Tartar 	*Cream of tartar* is a chemical leavening agent derived from tartaric acid, a by-product of wine production. When used in recipes, it is often combined with baking soda, which is an alkaline. When the two are combined, they neutralize each other and release carbon dioxide. Cream of tartar is often added to angel food cake batter because the acid helps to stabilize the egg whites.
Yeast 	*Yeast* is a microscopic, living, single-celled fungus that releases carbon dioxide and alcohol through a process called fermentation when provided with food (sugar) in a warm, moist environment. Yeast increases volume, improves flavor, and adds texture to dough.

Figure 9-10. In addition to air and steam, ammonium bicarbonate, baking powder, baking soda, cream of tartar, and yeast are used to leaven breads, cakes, cookies, and pastry doughs.

Yeast is very sensitive to temperature and, depending on the type of yeast, is activated between 70°F and 130°F. **See Figure 9-11.** If the temperature is too cold, the yeast remains dormant. If the temperature is too hot, the yeast dies. Common varieties of yeast used in the bakeshop include active dry yeast, compressed yeast, instant yeast, and yeast starters.

Yeast Activation

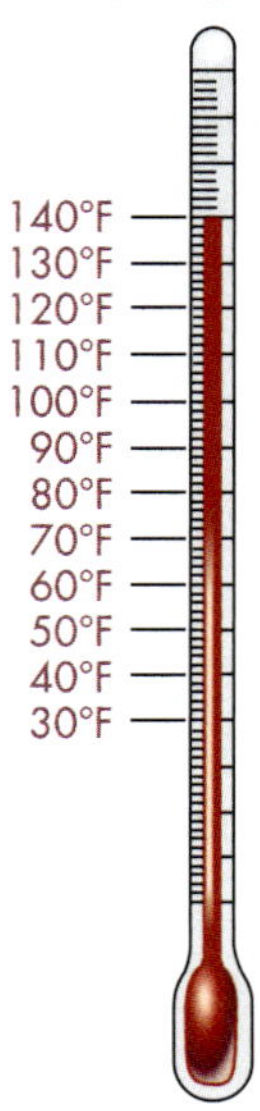

Activation Temperatures	
Death	140°F
Fast action	90°F to 100°F
Normal action	80°F to 85°F
Slow action	60°F to 75°F
Storage stage	30°F to 40°F

Figure 9-11. Yeast is very sensitive to temperature and, depending on the type of yeast, is activated between 70°F and 130°F.

- *Active dry yeast* is a form of yeast that has been dehydrated and looks like small granules. It is sold in ¼ oz packages or in 1 lb vacuum-sealed packages. Because active dry yeast is lighter than compressed yeast, only half as much dry yeast (by weight) is needed. It produces the least amount of carbon dioxide. It is best activated in warm water between 105°F and 115°F. It has a shelf life of one year.
- *Compressed yeast,* also known as fresh yeast, is yeast that has approximately 70% moisture and is available in 1 lb cakes or blocks. Compressed yeast produces the most carbon dioxide of any yeast. Once opened, compressed yeast should be wrapped tightly in plastic wrap, refrigerated, and used within two weeks. It also can be frozen for up to four months. Its yeasty aroma will turn ammonia-like when old or spoiled. The best method for using compressed yeast is to dissolve it in double its weight of lukewarm water (between 90°F and 100°F) prior to adding it to a bread recipe.
- *Instant yeast* is a form of yeast that does not need to be hydrated and may be directly added to a warm flour mixture. It is often labeled as rapid-rise yeast and is available in both dried and vacuum-packaged form. Instant yeast is a more aggressive yeast than compressed or active dry yeast. Because it is so active, only one-fourth the amount of instant yeast is needed compared to compressed yeast and one-half the amount compared to active dry yeast.
- A *yeast starter* is a mixture of flour, yeast, sugar, and water. This mixture is set in a warm place until the yeast ferments, causing the mixture to foam. A portion of a yeast starter can be used as a leavener and as a base for some breads. For example, a sourdough starter is used to make sourdough bread. The active yeast spores in the starter give sourdough bread its distinct tangy flavor. A yeast starter can be kept going for years if held in the right environment and by adding equal parts of flour and water each time a portion of the starter is used.

Flavorings

Flavorings are generally divided into two categories: natural and artificial. Natural flavorings yield a more pleasing flavor than imitation flavorings. Chocolates, extracts, flavoring emulsions, salts, seeds, and vanilla beans are a few of the key flavorings used in the bakeshop.

Chocolates. *Chocolate* is a food made from roasted and ground cacao beans that is often used to flavor other foods and beverages. The cacao beans are ground into a paste called chocolate liquor. This paste is then combined with various ingredients to create the desired flavor, from bitter to sweet. Chocolates commonly used in the bakeshop include cocoa powder, dark chocolate, milk chocolate, unsweetened chocolate, and white chocolate. **See Figure 9-12.**

Chocolates	
Cocoa Powder 	*Cocoa powder* is a reddish-brown, unsweetened powder extracted from ground cacao beans. *Dutch-processed cocoa powder* is a dark, unsweetened cocoa powder processed with an alkali to neutralize the natural acidity. Dutch-processed cocoa powder adds a smooth flavor to baked products.
Dark Chocolate 	*Dark chocolate* is chocolate that contains at least 35% chocolate liquor in addition to sugar, vanilla, and cocoa butter. Bittersweet and semisweet dark chocolate are used in the bakeshop.
Milk Chocolate 	*Milk chocolate* is a chocolate product that contains at least 12% milk solids and 10% chocolate liquor.

Figure 9-12. (continued on next page)

Chocolates	
Unsweetened Chocolate 	*Unsweetened chocolate* is pure chocolate liquor that contains 50% to 58% cocoa butter. Unsweetened chocolate is melted and used in many cake and icing recipes.
White Chocolate 	*White chocolate* is a confectionary product made from cocoa butter, sugar, milk solids, and flavorings. White chocolate does not contain any cocoa solids.

Figure 9-12. Cocoa powder, dark chocolate, milk chocolate, unsweetened chocolate, and white chocolate are a few of the types of chocolate used in the bakeshop.

Production Tip

Chocolate used for dipped or molded chocolates needs to be tempered. Tempered chocolate has a shiny finish and snaps when broken.

Extracts. An *extract* is a flavorful oil that has been mixed with alcohol. Vanilla, lemon, almond, coffee, mint, and rum are a few of the extracts used in the bakeshop. Only a small amount of an extract is used since its flavor is concentrated.

Extracts

Flavoring Emulsions. A *flavoring emulsion* is a flavored oil that has been mixed with water and an emulsifier. Emulsions are stronger than extracts and should be used sparingly. Lemon and orange emulsions are commonly used in the bakeshop.

Salts. *Salt* is a crystalline solid composed mainly of sodium chloride and is used as a seasoning and a preservative. Salt is used to balance the sweetness in baked products while adding additional flavor. Salt should be dissolved in liquid ingredients to ensure even distribution. When used in doughs, salt improves the flavor and controls yeast growth. When used to make piecrusts, salt shortens the gluten strands in the flour, which helps to create tender crusts.

Seeds. A *seed* is the fruit or unripened ovule of a nonwoody plant. Sesame seeds, poppy seeds, caraway seeds, and flax seeds are often used to add flavor to yeast breads, quick breads, and cookies. For example, sesame seeds are often used to garnish rolls and bagels. **See Figure 9-13.** Caraway seeds are commonly used to flavor rye bread. Ground flax seeds may be baked into yeast doughs and muffin and cookie batters.

Seed Use

Figure 9-13. Sesame seeds are often used to garnish rolls and bagels.

Vanilla Beans. A *vanilla bean* is the dark-brown pod of a tropical orchid. **See Figure 9-14.** The bean is split open and the inside is often scraped directly into a mixture. The split bean is often steeped in milk prior to use in a cream application. Vanilla is one of the most commonly used flavorings in the bakeshop and is difficult to substitute.

Vanilla Beans

Figure 9-14. A vanilla bean is the dark-brown pod of a tropical orchid.

Baking Terminology

Terms such as blend, cream, cut-in, fold, sift, stir, and whip are used to describe different methods of combining ingredients in the bakeshop. Understanding baking terms is essential to the proper preparation of yeast breads, quick breads, cakes, cookies, pies, pastries, and specialty desserts. **See Figure 9-15.**

Bakeshop Terminology	
Term	***Description***
Blend	To *blend* is to mix two or more ingredients together until they are evenly distributed.
Cream	To *cream* is to combine fat and sugar in order to add air.
Cut-in	To *cut-in* is to incorporate a solid fat into a dry ingredient.
Fold	To *fold* is to gently incorporate light ingredients into heavier ones using a smooth and gentle fold-over motion.
Sift	To *sift* is to pass dry ingredients through a sieve in order to remove lumps and incorporate air.
Stir	To *stir* is to gently mix two or more ingredients together until they are evenly combined.
Whip	To *whip* is to agitate ingredients in order to incorporate air.

Figure 9-15. Understanding baking terms is essential to the proper preparation of yeast breads, quick breads, cakes, cookies, pies, pastries, and specialty desserts.

CHECKPOINT 9-2

1. Identify how ingredients are measured in the bakeshop.
2. Explain how two ingredients can have the same volume but different weights.
3. Identify the protein in flour that gives a baked product structure.
4. Contrast five types of flour commonly used in the bakeshop.
5. Explain how sugars and natural sweeteners affect baked products.
6. Identify eight types of sugars and natural sweeteners used in the bakeshop.
7. Explain how fats affect baked products.
8. Explain the process of hydrogenation.
9. Differentiate between hydrogenated and hi-ratio shortenings.
10. Explain how eggs affect baked products.
11. Explain how milk and other dairy products affect baked products.
12. Explain how thickening agents affect baked products.
13. Identify five common thickening agents.
14. Explain the purpose of a leavening agent.
15. Identify five common leavening agents.
16. Identify four varieties of yeast used to leaven baked products.
17. Explain how flavorings affect baked products.
18. Identify five types of chocolate used in the bakeshop.
19. Differentiate between an extract and a flavoring emulsion.
20. Explain how vanilla bean is added to a mixture.
21. Define seven common terms used to describe methods of combining ingredients.

BAKER'S PERCENTAGES

A *baker's percentage* is the weight of an ingredient expressed as a percentage of the weight of the main ingredient in a formula. **See Figure 9-16.** If flour is the main ingredient in a formula, its baker's percentage is set to 100% and all other ingredients in the formula are expressed as a percentage based on the weight of flour. For example, if the formula also uses sugar and the baker's percentage of the sugar in the formula is 50%, no matter how much flour is used, the weight of the sugar will always be equal to 50% of the weight of flour. It does not mean that 50% of the total formula is made up of sugar. In fact, the sum of the baker's percentages in a formula will always be greater than 100%.

Bakeshop Formulas		
Ingredient	**Ingredient Weight**	**Baker's Percentage**
Bread flour	7 lb 8 oz	100%
Salt	3 oz	3%
Sugar	3½ oz	3%
Shortening	3 oz	3%
Egg whites	3 oz	3%
Water	4 lb 8 oz	60%
Yeast	4½ oz	4%

Figure 9-16. A baker's percentage is the weight of an ingredient expressed as a percentage of the weight of the main ingredient in a formula.

If a formula uses two types of flour, the two types of flour combined are considered the main ingredient and will have a total baker's percentage of 100%. For example, a formula for rye bread may contain white flour with a baker's percentage of 75% and rye flour with a baker's percentage of 25%. The two types of flour will have a total baker's percentage of 100% (75% + 25% = 100%).

Two important rules must be followed when using baker's percentages. Ingredients must be weighed (not measured by volume) and all of the measurements must have the same unit of measure. The first step when calculating the weight of each ingredient in a formula is to determine the weight of the main ingredient. The next step is to convert all of the ingredient amounts in the recipe to the same weight unit of measure. Then, the weight of each ingredient is divided by the weight of the main ingredient. The result, in decimal form, is then converted to a percentage. A baker's percentage is calculated by using the following equation:

$$BP = (WI \div WM) \times 100$$

where

BP = baker's percentage

WI = weight of ingredient

WM = weight of main ingredient

Once the weight of the main ingredient is determined, the remaining ingredient weights are calculated. This is done by multiplying the baker's percentage for each of the remaining ingredients in the formula by the weight of the main ingredient. **See Figure 9-17.**

The weights of the remaining ingredients are calculated using the following equation:

$$WI = WM \times (BP \div 100)$$

where

WI = weight of ingredient

WM = weight of main ingredient

BP = baker's percentage

For example, in a hard-roll dough formula, bread flour is the main ingredient. If 16.5 lb of bread flour is used to make the dough, and the baker's percentage for salt in this formula is 3%, the weight of salt required is calculated as follows:

$$WI = WM \times (BP \div 100)$$

$$WI = 16.5 \text{ lb} \times (3\% \div 100)$$

$$WI = 16.5 \text{ lb} \times 0.03$$

$$WI = \mathbf{0.5 \text{ lb}}$$

Using a Baker's Percentage		
Ingredients	**Baker's Percentage**	**Ingredient Weight***
Bread flour (main ingredient)	100%	16.5 × (100% ÷ 100) = 16.5 lb × 1.00 = **16.5 lb**
Salt	3%	16.5 × (3% ÷ 100) = 16.5 lb × 0.03 = **0.5 lb**
Sugar	5%	16.5 × (5% ÷ 100) = 16.5 lb × 0.05 = **0.8 lb**
Shortening	6%	16.5 × (6% ÷ 100) = 16.5 lb × 0.06 = **1.0 lb**
Egg whites	3%	16.5 × (3% ÷ 100) = 16.5 lb × 0.03 = **0.5 lb**
Water	60%	16.5 × (60% ÷ 100) = 16.5 lb × 0.60 = **9.9 lb**
Yeast	5%	16.5 × (5% ÷ 100) = 16.5 lb × 0.05 = **0.8 lb**
Total	**182%**	**30 lb**

* *Results rounded to the tenths place*

Figure 9-17. When using a formula, ingredient weights are calculated by multiplying the baker's percentage for each ingredient by the weight of the main ingredient.

Formulas and Percentages

Flour is not always the main ingredient in a formula. In fact, some formulas will not contain any flour. For example, dried apricots are the main ingredient in a recipe for an apricot filling. If the baker's percentage for water is 20% and the filling is to be made using 10 lb of dried apricots, the weight of water required is calculated as follows:

WI = *WM* × *(BP* ÷ *100)*
WI = 10 lb × (20% ÷ 100)
WI = 10 lb × 0.20
WI = **2.0 lb**

Sometimes, the weight of the main ingredient is stated directly. In other cases, the weight of the main ingredient may need to be calculated based on the desired yield of the formula. When a specific total yield is desired, the weight of the main ingredient can be calculated using the following equation:

WM = *DY* ÷ *(TBP* ÷ *100)*
where
WM = weight of main ingredient
DY = desired yield
TBP = total baker's percentage

For example, if the total of the baker's percentages in a hard-roll dough recipe is 182%, and the desired yield is 30 lb of dough, the weight of the main ingredient (bread flour) is calculated as follows:

WM = *DY* ÷ *(TBP* ÷ *100)*
WM = 30 lb ÷ (182% ÷ 100)
WM = 30 lb ÷ 1.82
WM = **16.5 lb**

A recipe can be converted into a formula by converting all of the ingredient quantities to a common weight unit of measure and then dividing each individual ingredient weight by the weight of the main ingredient. **See Figure 9-18.**

For example, a white cake recipe may call for 5 lb of cake flour, 3.5 lb of shortening, 5 oz of baking powder, and a number of other ingredients. To create a formula from this recipe, all of the original ingredient amounts need to be converted to a common unit of measure. First, the amount of the main ingredient, cake flour, is converted from 5 lb to 80 oz (5 lb × 16 = 80 oz). To calculate the baker's percentage of shortening, the weight of shortening is first converted to ounces (3.5 lb × 16 = 56 oz). Then the baker's percentage is calculated as follows:

BP = *(WI* ÷ *WM)* × *100*
BP = 56 oz ÷ 80 oz
BP = 0.70
BP = **70%**

The same process is repeated for every ingredient in the recipe. *Note:* Conversion calculations are not necessary for the ingredient measurements that are already provided in ounces.

Converting Recipes into Formulas

Ingredient	*Ingredient Weight*	*Ingredient Weight Converted to Ounces*	*Baker's Percentage*
Cake flour (main ingredient)	**5 lb**	5 lb × 16 oz/lb = **80 oz**	80 oz ÷ 80 oz = 1.0 × 100 = **100%**
Shortening	3.5 lb	3.5 lb × 16 oz/lb = **56 oz**	56 oz ÷ 80 oz = 0.7 × 100 = **70%**
Sugar	6.25 lb	6.25 lb × 16 oz/lb = **100 oz**	100 oz ÷ 80 oz = 1.25 × 100 = **125%**
Salt	3 oz	**3 oz**	3 oz ÷ 80 oz = 0.0375 × 100 = **3.75%**
Baking powder	5 oz	**5 oz**	5 oz ÷ 80 oz = 0.0625 × 100 = **6.25%**
Water	1.75 lb	1.75 lb × 16 oz/lb = **28 oz**	28 oz ÷ 80 oz = 0.35 × 100 = **35%**
Nonfat dry milk	5 oz	**5 oz**	5 oz ÷ 80 oz = 0.0625 × 100 = **6.25%**
Eggs	1.25 lb	1.25 lb × 16 oz/lb = **20 oz**	20 oz ÷ 80 oz = 0.25 × 100 = **25%**
Egg whites	2 lb	2 lb × 16 oz/lb = **32 oz**	32 oz ÷ 80 oz = 0.4 × 100 = **40%**
Vanilla extract	2 oz	**2 oz**	2 oz ÷ 80 oz = 0.025 × 100 = **2.5%**

Figure 9-18. A recipe can be converted into a formula by converting the ingredient quantities to a common weight unit of measure and then dividing that weight by the weight of the main ingredient.

CHECKPOINT 9-3

1. Define baker's percentage.
2. Explain how ingredients in a formula must be weighed.
3. Calculate the baker's percentage for each ingredient of a formula.
4. Explain how to convert a recipe into a formula.
5. Use baker's percentages to convert a recipe into a formula.

YEAST BREAD PREPARATION

Yeast bread is a versatile staple that can be served at any meal. Most yeast bread doughs contain flour, liquid, sugar, salt, and yeast. Many also contain fat and eggs. Flour gives strength to dough and acts as an absorbing agent. Liquid, usually milk or water, supplies moisture and helps form gluten. Fat supplies tenderness and improves shelf life. Eggs supply structure to dough and add color. Sugar acts as a stimulant to the yeast. Salt brings out the flavor in the dough.

There are three types of yeast doughs. The three types of yeast doughs include lean doughs, rich doughs, and rolled-in doughs. **See Figure 9-19.**

- A *lean dough* is a yeast dough that is low in fat and sugar. Hard rolls, baguettes, and rye bread are made from lean doughs.
- A *rich dough* is a yeast dough that incorporates a lot of fat, sugar, and eggs into a heavy, soft structure. The finished product may be faintly yellow in color due to the large number of eggs that are used. Cinnamon rolls and doughnuts are made from rich doughs.
- A *rolled-in dough* is a yeast dough with a flaky texture that results from the incorporation of fat through a rolling and folding procedure. By alternating the layers, a very light and flaky texture is achieved in the finished product. Rolled-in doughs may be sweet, as with Danishes, or may not be sweet, as with croissants.

Yeast dough production involves scaling ingredients, mixing, kneading, fermenting, punching, scaling dough, rounding, panning, proofing, baking, cooling, and storing. These steps generally apply to all yeast doughs with slight variations depending on the product.

Yeast Doughs

Types of Yeast Doughs

National Honey Board

Lean | Rich | Rolled-In

Figure 9-19. The three types of yeast doughs include lean doughs, rich doughs, and rolled-in doughs.

Scaling Ingredients

All ingredients must be scaled on either a baker's scale or a digital scale. *Scaling* is the process of weighing an ingredient using a scale. A scale must be properly balanced before use and again after adding measured ingredients.

Procedure for Scaling Ingredients

1. Set the scale scoop on the left side of the scale.
2. Balance the scale by placing weights on the right side until the scale is balanced.
3. Adjust the scale to the desired weight of the ingredient being measured by adding the appropriate amount of weight to the right side of the scale and adjusting the ounce weight on the bar as needed.

4. Add the ingredient to the scoop until the scale is balanced.

Production Tip

Honey can replace up to one-half of the granulated sugar in many formulas. If honey is used instead of sugar, the liquid called for in the formula must be reduced by ¼ cup for each 1 cup of honey used. The oven temperature must also be reduced by 25°F to prevent over-browning.

Mixing Doughs

Mixing is very important for gluten development and uniform distribution of yeast throughout the dough. A bench or floor mixer is used to mix yeast doughs and cake batters. Commonly used mixer attachments include the paddle, hook, and whip. The three methods commonly used to mix yeast doughs are the straight-dough method, the modified straight-dough method, and the sponge method.

- The *straight-dough method* is a method of mixing yeast dough that combines yeast with warm water and then adds it to the rest of the ingredients, mixing them together all at once.
- The *modified straight-dough method* is a method of mixing yeast dough in which the ingredients are added in sequential steps.
- The *sponge method* is a method of mixing yeast dough in two steps. In the first step, 50% of the flour is added to all of the liquid and yeast to form a batter that looks like a sponge. As the sponge forms and increases in size, the flour, yeast, and liquid begin to ferment and air bubbles develop throughout the batter. In the second step, the rest of the flour and all of the fat, salt, and sugar are added. Yeast breads made using the sponge method usually have an intense flavor and a light texture.

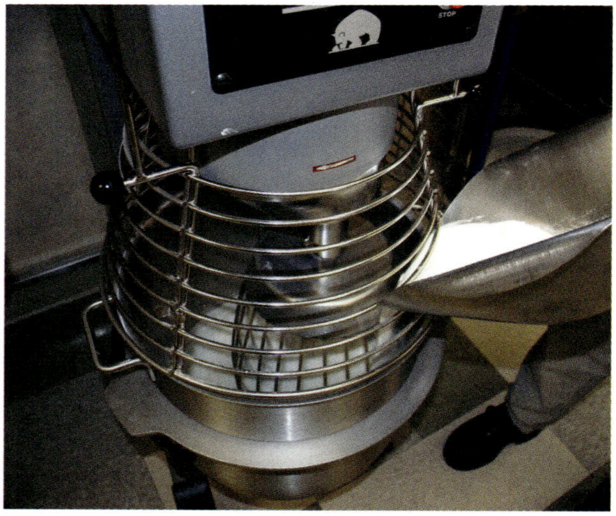

Kneading Doughs

Kneading is the process of pushing and folding dough until it is smooth and elastic. The kneading process aids gluten development and the uniform distribution of yeast throughout the dough. Lean doughs require less kneading than sweet doughs because they contain less fat and sugar.

Procedure for Kneading Yeast Dough

1. Place the dough on a lightly floured surface.
2. Using the heel of the hand, push the dough down and away from the body.
3. Fold the dough in half, bringing the farthest edge up and on top of the nearest edge.
4. Repeat steps 2 and 3 until the dough is smooth.

Fermenting Doughs

Fermentation is the process by which yeast converts sugar into carbon dioxide and alcohol. During fermentation, yeast dough should be covered and allowed to rest until it doubles in size or until it no longer springs back when pressed. **See Figure 9-20.**

Punching Doughs

After the dough has doubled, it must be gently deflated to relax the gluten, dispel some of the carbon dioxide, and redistribute the yeast. *Punching* is the process of deflating risen yeast dough to allow carbon dioxide to escape. **See Figure 9-21.** Punching also helps redistribute yeast.

Fermentation

Figure 9-20. During fermentation, yeast dough should be covered and allowed to rest until it doubles in size or until it no longer springs back when pressed.

Punching

Figure 9-21. Punching is the process of deflating risen yeast dough to allow carbon dioxide to escape.

Kneading Yeast Dough
Media Clip

Scaling Doughs

Scaling dough is the process of weighing pieces of risen dough to ensure products will be consistent in size. **See Figure 9-22.** A dough cutter is used to cut dough into pieces of a desired size before they are weighed. It is important to use a dough cutter instead of tearing the dough, which destroys some of the gluten that has developed. Dough pieces can be added or taken away until the desired weight is achieved for each portion of dough.

Scaling Doughs

Edlund Co.

Figure 9-22. Scaling dough is the process of weighing pieces of risen dough to ensure products will be consistent in size.

Rounding Doughs

Rounding is the process of shaping scaled dough into smooth balls. Rounding enables the dough to proof evenly and have a smooth outer surface. To round dough, the dough is shaped into a ball using the fingers to form the shape, and then the dough is rotated on a flat surface while applying pressure. **See Figure 9-23.**

Rounding Doughs

Figure 9-23. To round dough, the dough is shaped into a ball using the fingers to form the shape, and then the dough is rotated on a flat surface while applying pressure.

When shaping rolled-out dough into a loaf, it needs to be rolled tightly. The end of the roll is referred to as a seam and should be pinched tightly to prevent it from unrolling during proofing or baking. The seam is always placed facing the bottom of the pan.

Shaping Doughs

A variety of rolls can be made by using different shaping techniques. **See Figure 9-24.** Artisan shaped rolls offer a beautiful presentation in baskets, on buffets, and à la carte. Standard shapes can be made using both hard and soft yeast doughs. In addition to the ball-shaped rolls, both hard and soft doughs can be shaped into Kaiser rolls and crescent rolls. Soft doughs can also be shaped into knots, cloverleaf rolls, and parker house rolls. Soft sweetened doughs are often made into donuts.

Shaped Breads

Eloma Combi Ovens

Figure 9-24. A variety of rolls can be made by using different shaping techniques.

Kaiser Rolls. Kaiser rolls can be made two ways. One method of making Kaiser rolls is to form the dough into a ball shape and then let the balls rest and rise. Each ball is indented with a special tool that makes a swirl shape on the finished bun. The second method of making Kaiser rolls is done by flattening the balls into a disk shape and then folding and pressing the thumb of one hand into the middle of the ball while folding and turning over the edges into the middle to form five cushions on the edges of the ball. In both methods, the finished rolls are turned top down onto a layer of poppy seeds or corn meal, allowed to rise, and then turned back over and baked.

Crescent Rolls. Crescent rolls are made from a large disk of dough that is cut into wedges and then rolled up with the same technique used to make croissants. Crescent rolls are then egg washed, proofed, and baked.

Knot-Shaped Rolls. A variety of knot-shaped rolls such as overhand knots, braids, and figure eights are frequently formed, proofed, egg washed, and then baked.

Cloverleaf Rolls. Cloverleaf-shaped rolls are made from a rope-shaped length of dough that is cut into small lengths and then shaped into small balls. Three balls are placed in each greased muffin tin, proofed, egg washed, and then baked to make cloverleaf rolls.

Parker House Rolls. Parker house rolls are made by flattening a ball of dough and then using a rolling pin to roll it into an elongated oval. The oval shape is then folded back onto itself, proofed, egg washed, and then baked.

Donuts. Donuts can be made with yeast dough or cake batter. Sweetened yeast dough is rolled out and cut into circles with a cookie cutter. The donuts are allowed to rise and are sometimes filled with pastry cream or jellies. After they have risen, the donuts are fried in hot oil on both sides. Donuts can be left plain, powdered, glazed, or iced. A classic example of a raised donut is the German specialty known as the Berliner, which is filled with jelly and elaborately decorated with fondant, marzipan, and chocolate.

Panning Doughs

Panning is the process of placing pieces of dough into pans. **See Figure 9-25.** Some yeast breads are baked in loaf pans. Rolls are typically baked on sheet pans or, in the case of cloverleaf rolls, in muffin pans. It is important to leave enough space in the pan for the yeast bread to proof, or rise. Yeast breads typically double in size during the proofing stage.

Panning Doughs

Figure 9-25. Panning is the process of placing rounded pieces of dough into pans.

Production Tip

A single knot can be made by rolling a rounded piece of dough into a 9 inch strand and then forming the strand into the number "9." The next step is to thread the long end of the "9" under and through the loop to form a knot.

Proofing Doughs

Proofing is the process of letting yeast dough rise in a warm (85°F) and moist (80% humidity) environment until the dough doubles in size. **See Figure 9-26.** A proofing cabinet helps reduce the time required for proofing and ensures a consistent product because the temperature and humidity are controlled.

Proofing is an extension of the fermentation process, but it is a different process. Fermentation is done at a lower temperature than proofing. During proofing, the yeast bread doubles in size again and should spring back when pressed. It is important not to overproof or underproof yeast doughs as this will result in poor texture.

Proofing Doughs

Bridgford Foods Corporation

Figure 9-26. Proofing is the process of letting yeast dough rise in a warm (85°F) and moist (80% humidity) environment until the dough doubles in size.

Scoring and Docking Doughs

After proofing, some yeast breads require scoring or docking. *Scoring* is the process of making shallow, angled cuts across the top of unbaked yeast bread to allow carbon dioxide to escape during baking. A special knife called a lame, also known as a baker's knife, is used to score yeast doughs. **See Figure 9-27.**

Docking is the process of making small holes in yeast dough before it is baked to allow steam to escape and to promote even baking. A dough docker or the tines of a fork can be used to dock yeast doughs.

Scoring Doughs

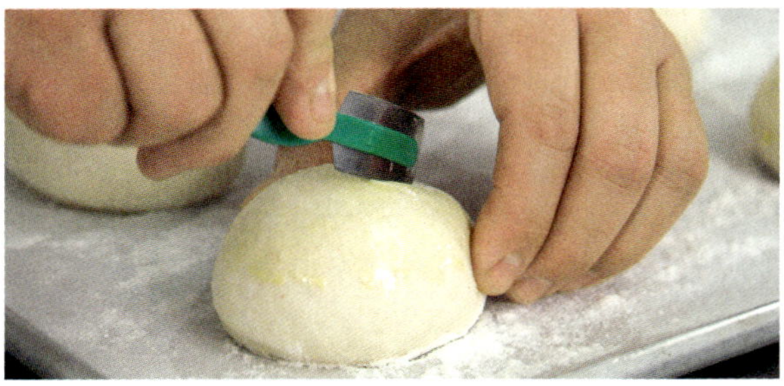

Figure 9-27. A lame is used to score yeast doughs.

Washing Doughs

Washes are often used to create color, make the outer surface shiny or dull, or make other ingredients, such as seeds, stick to the dough. A *wash* is a liquid that is brushed onto the surface of dough prior to baking. **See Figure 9-28.** The most common type of wash is an egg wash, which consists of whole eggs whisked with a small amount of water or milk.

Washing Doughs

Figure 9-28. A wash is a liquid that is brushed on the surface of dough prior to baking.

Baking Yeast Breads

Yeast breads expand quickly when first placed in a hot oven. *Oven spring* is the rapid expansion of yeast dough in the oven, resulting from the expansion of gases within the dough. **See Figure 9-29.** When the temperature of the dough reaches 140°F, the yeast dies and there is no further expansion.

Baking Yeast Breads

Paderno World Cuisine

Figure 9-29. Oven spring is the rapid expansion of yeast dough in the oven, resulting from the expansion of gases within the dough.

Bakeries typically have a steam injector connected to their ovens. *Steam injection* is the process of adding water directly in the hot cavity of an oven so that steam is created. As the water sprays on the interior surface of the oven, the water vaporizes and turns into steam. Steam gives yeast breads, such as French bread, a light, crispy crust while the inner crumb remains moist and chewy.

Blodgett Oven Company

Cooling and Storing Yeast Breads

After baking, yeast breads must be removed from the pans to cool. Typically, breads are placed on cooling racks. If breads are left in the pans to cool, the hot bread sweats and becomes soggy.

Cooled yeast breads should be stored in airtight containers or plastic bags. Hot bread should never be bagged until almost room temperature. If bread is bagged before it has cooled, steam will be trapped in the bag and the bread will become soggy.

Hard-crusted yeast breads, such as baguettes and French bread, should not be wrapped because the crust will become soft. Baguettes are often displayed in baskets or open paper bags. **See Figure 9-30.**

Baguettes

Paderno World Cuisine

Figure 9-30. Baguettes are often displayed in baskets or open paper bags.

Production Tip

Yeast breads should be stored at room temperature or frozen. Refrigeration causes breads to stale.

CHECKPOINT 9-4

1. Contrast lean dough, rich dough, and rolled-in yeast dough.
2. Prepare a lean yeast dough. Evaluate the appearance, texture, and color.
3. Prepare a rich yeast dough. Evaluate the appearance, texture, and color.
4. Prepare a rolled-in yeast dough. Evaluate the appearance, texture, and color.
5. Explain how to scale ingredients.
6. Describe three methods of mixing yeast dough.
7. Explain how to knead yeast dough.
8. Define fermentation.
9. Explain why yeast dough needs to be punched.
10. Explain how to scale yeast dough.
11. Explain why yeast dough may need to be rounded.
12. Describe six ways dough can be shaped.
13. Explain how to pan yeast dough.
14. Describe the purpose of proofing yeast dough.
15. Explain how to score yeast dough.
16. Explain how to dock yeast dough.
17. Explain how to wash yeast dough.
18. Explain the cause of oven spring and the temperature at which it stops.
19. Describe steam injection.
20. Explain why some yeast breads should not be wrapped for storage.

QUICK BREAD PREPARATION

A *quick bread* is a baked product that is made with a quick-acting leavening agent such as baking powder, baking soda, or steam. Common quick breads include biscuits, muffins, quick bread loaves, and corn bread.

Biscuits

A *biscuit* is a quick bread made by mixing solid fat, baking powder or baking soda, salt, and milk with flour. **See Figure 9-31.** Some biscuit recipes also call for sugar. Butter may be added to improve the tenderness and richness of biscuits. Flavorings may be added to create specialty products, such as buttermilk biscuits, orange biscuits, or cheese biscuits.

Biscuits

Figure 9-31. Biscuits are made by mixing solid fat, baking powder or baking soda, salt, and milk with flour.

Muffins

A *muffin* is a quick bread that is made with eggs, flour, and either oil, butter, or margarine. Muffins are shaped like cupcakes and have a pebbled top and a cake-like texture. **See Figure 9-32.** Muffins can be served with any meal.

Muffins

Figure 9-32. Muffins are shaped like cupcakes and have a pebbled top and cake-like texture.

Quick Bread Loaves

A *quick bread loaf* is a loaf of quick bread made from a batter similar to muffin batter that has been flavored with fruits, vegetables, nuts, and/or spices. **See Figure 9-33.** The shape of the loaf is determined by the type of loaf pan that is used. Banana nut bread, blueberry pecan, pumpkin bread, zucchini bread, and lemon tea bread are common quick bread loaves. Some quick bread loaves are topped with icing. Quick bread loaves can be served warm or cold.

Quick Bread Loaves

Figure 9-33. Quick bread loaves are made from a batter that has been flavored with fruits, vegetables, nuts, and/or spices.

Corn Bread

Corn bread is a quick bread made from a batter containing cornmeal, eggs, milk, and oil. **See Figure 9-34.** Cornmeal is dried, ground corn kernels. Available types include yellow corn or white corn and coarse grain or fine grain. Corn flour is finely ground cornmeal. Corn flour can be used in combination with wheat flour to make cakes and crêpes.

Corn Bread

Figure 9-34. Corn bread is quick bread made from a batter containing cornmeal, eggs, milk, and oil.

When making corn breads, the liquid must be added slowly because cornmeal does not absorb liquid quickly. If the liquid is added too rapidly, lumps will form. It is also important to avoid overmixing. A corn bread mixture should not be beaten. The batter should only be stirred until the ingredients are mixed. Corn bread batter can be poured into molds prior to baking to make corn sticks. Corn bread is sometimes prepared in a cast iron skillet.

Quick Breads

Mixing Quick Breads

Quick breads are easier to prepare than yeast breads because the quick-acting leavening agents allow a quick bread to be mixed and baked without waiting for the dough to rise. Quick breads are made using the biscuit method, the muffin method, or the creaming method.

- The *biscuit mixing method* is a mixing method used to produce a light and flaky baked product. Quick bread formulas that include a cold, solid fat, such as butter, are typically mixed using the biscuit mixing method.

Procedure for Preparing Biscuits

1. Scale all ingredients.
2. Sift all dry ingredients together.

3. Cut the fat into the dry ingredients using a pastry blender or a mixer with a paddle attachment.

4. Combine liquid ingredients in a separate bowl.
5. Add liquid ingredients to the dry ingredients and mix until the mixture holds together.

6. Place the dough on a flat, floured surface and knead lightly for 30–45 seconds. *Note:* Overkneading will toughen biscuits.

7. Roll the dough out on a floured surface.
8. Using a biscuit cutter, cut biscuits from the rolled out dough as close together as possible.

9. Place biscuits on a greased or parchment-lined sheet pan.
10. Brush the tops of the biscuits with melted butter or an egg wash for added color.

- The *muffin mixing method* is a mixing method that uses liquid fats, such as vegetable oil or melted butter, to produce a rich and tender baked product. To prevent the gluten from becoming tough, the dry and liquid ingredients should be placed in a mixer with a paddle attachment and mixed at a slow speed. The batter should be mixed until it is moistened, leaving it with a slightly rough appearance. Muffin batter should not be smooth. Overmixing the ingredients will create unsightly air tunnels inside the muffins.
- The *creaming mixing method* is a mixing method used to produce a baked product with a fine crumb. Room temperature fat and sugar are combined in a mixer using the paddle attachment until light and fluffy. Eggs are then added one at a time and thoroughly incorporated into the creamed mixture. Liquid ingredients and dry ingredients are alternately added in thirds while continually mixing. The creaming method creates a finer crumb than the muffin mixing method.

Vita-Mix® Corporation

Procedure for Preparing Muffins

1. Scale all ingredients.
2. Sift dry ingredients together.
3. Combine liquid ingredients, including any melted butter or oil.
4. Add the liquid ingredients to the dry ingredients and mix just until moistened. *Note:* The batter should appear lumpy. Do not overmix.
5. Grease muffin pans or use paper liners.
6. Use a portion control scoop to add batter to pans for a uniform size.

Procedure for Preparing Quick Bread Loaves

1. Scale all ingredients.
2. Combine fat and sugar in a mixing bowl and mix using the paddle attachment on medium speed until light and fluffy.

3. Add approximately one beaten egg at a time, mixing well after each addition.

4. Sift dry ingredients together.

5. Combine liquid ingredients in a separate bowl.

6. Alternately add dry and liquid ingredients to the creamed mixture. Do not overmix.

7. Portion batter into greased or lined pans and bake.

Baking Quick Breads

Quick breads can be baked in a variety of loaf pans to create muffins, loaves, or sticks of various sizes. When baking quick breads, pans should be filled only three-quarters full. If the pans are overfilled, the batter will spill over the top as it rises during the baking process. To check for doneness, a wire tester or a toothpick is inserted into the center of the quick bread. **See Figure 9-35.** If the toothpick comes out clean, the quick bread is done. Cracks may form on the surface of baked quick breads due to the quick leavening action that occurs during the baking process.

Testing Quick Breads for Doneness

Figure 9-35. A wire tester or a toothpick is inserted into a quick bread to test for doneness.

Cooling and Storing Quick Breads

If quick breads are not served immediately, they should be refrigerated after cooling. To extend the freshness of quick breads, they should be wrapped tightly as soon as they cool. This allows them to be sliced more easily without crumbling. Quick breads may also be frozen with excellent results.

CHECKPOINT 9-5

1. Identify four common types of quick breads.
2. Describe three methods of mixing quick breads.
3. Prepare biscuits. Evaluate the appearance, texture, color, and flavor.
4. Prepare muffins. Evaluate the appearance, texture, color, and flavor.
5. Prepare quick bread loaves. Evaluate the appearance, texture, color, and flavor.
6. List guidelines for baking quick breads and checking them for doneness.
7. Explain how to cool and store quick breads.

CAKE PREPARATION

Before mixing cake batter, all of the ingredients should be at room temperature. Each ingredient should be weighed separately on a baker's scale or a digital scale for maximum accuracy. It is important to mix cakes using the appropriate method. **See Figure 9-36.** Common methods of mixing cake batters include the two-stage method, the creaming method, the chiffon method, and the sponge method.

Preparing Cakes

Figure 9-36. Cake batters need to be mixed using the appropriate mixing method.

Two-Stage Mixing Method

The two-stage mixing method is the easiest method of mixing cake batter. After being carefully weighed, the dry ingredients are sifted together and blended with the fat and part of the liquid. In a separate bowl, the rest of the liquid and eggs are combined. The two mixtures are then mixed together until evenly distributed. At intervals throughout the mixing process, the sides of the bowl are scraped down so all ingredients are blended and a smooth batter is obtained.

Cake Preparation

Procedure for Mixing Cake Batter Using the Two-Stage Method

1. Weigh all ingredients carefully.
2. Place all dry ingredients, fat, and part of the milk in the mixing bowl. Blend on slow speed.
3. In a separate bowl, blend eggs and the remaining milk.
4. Add the egg mixture to the batter in thirds, blending well after each addition to ensure a smooth, uniform batter.

Creaming Mixing Method

The creaming mixing method produces a lighter product with a much finer crumb. Cream room temperature fat and sugar in a mixer with a paddle attachment until the batter is light and fluffy. The eggs are then added one at a time and incorporated into the creamed fat and sugar. Next, liquid ingredients and dry ingredients are alternately added in small amounts and creamed. At intervals throughout the mixing process, the sides of the bowl are scraped down so all ingredients are blended and a smooth batter is obtained.

Procedure for Mixing Cake Batter Using the Creaming Method

1. Scale all ingredients.
2. Combine the fat and sugar in a mixing bowl and mix with a paddle attachment on medium speed until light and fluffy.
3. Add eggs, one at a time, mixing well after each addition.
4. Sift dry ingredients together.
5. Combine liquid ingredients in a separate bowl.
6. Alternately add dry and liquid ingredients to the creamed mixture. Do not overmix.
7. Portion batter into greased or lined pans and bake.

Chiffon Mixing Method

Figure 9-37. The chiffon mixing method involves folding whipped egg whites into a batter made from flour, egg yolks, and fat.

Chiffon Mixing Method

The chiffon mixing method involves folding whipped egg whites into a batter made from flour, egg yolks, and fat. **See Figure 9-37.** When folding in whipped products, such as egg whites, meringues, or whipped cream, it is important to add the ingredients in thirds. Each third is added using a circular folding motion, which gently incorporates air to produce a very light cake batter. It is important to make sure that each third of a whipped ingredient is fully incorporated before the next third is added and to take care not to deflate the air that has been incorporated.

Mixing Cake Batter Using the Creaming Method
Media Clip

Sponge Mixing Method

The sponge mixing method produces a light, fluffy batter. Although there are many variations of the sponge mixing method, the most common method is referred to as a génoise. The eggs and sugar are warmed and whipped to create volume and incorporate air before any other ingredients are added. This airy whipped yolk and sugar mixture is referred to as a foam. After the foam is created, the flour and melted butter are carefully folded in order to avoid deflating the egg foam.

Procedure for Mixing Cake Batter Using the Sponge Method

1. Warm the eggs (whole eggs, egg whites, or egg yolks as specified) and sugar to between 100°F and 105°F over a hot water bath and whisk continually.
2. Remove the mixture from the water bath and place in the mixer.

3. Using the whip attachment, whip on medium to high speed until the mixture peaks in volume and develops a thick foam. *Note:* The mixture should be thick enough to form a ribbon.
4. Turn the mixer to low and slowly add any liquid and flavorings.

5. Gently fold in the sifted dry ingredients with a spatula. Be careful not to deflate the egg foam. *Note:* If fat is to be added, it should be done after the dry ingredients have been folded in.

Baking Cakes

Most cake batters can be used to produce cakes in a variety of shapes and sizes. The amount of cake batter required for a cake varies with the type and size of the cake. **See Figure 9-38.** Many cake pans are prepared by covering the bottom of the pan with parchment paper and greasing the sides lightly.

Baking Cakes		
Cake Shape	*Scaling Weight*	*Pan Size*
Layer	13–14 oz	8 inch diameter
Bar	5–6 oz	2¾ inches wide × 10 inches long
Ring	10–14 oz	6½ inch diameter
Loaf	11–24 oz	3¼ inches wide × 7⅛ inches long
Oval Loaf	8 oz	6¼ inches long
Sheet	6–7 lb	17¾ inches wide × 25 inches long

Figure 9-38. The amount of cake batter required for a cake varies with the type and size of the cake.

Ovens must be preheated to the required temperature. If the oven temperature is set too low, the cake will rise and then fall, producing a dense, heavy texture. If the temperature is set too high, the outside of the cake will bake too rapidly and form a crust that will burst when heat reaches the center of the cake and causes the cake to expand. Most cake recipes are developed to produce good results when baked at or near sea level. If a product is baked in a high-altitude area, adjustments in ingredient amounts are often required.

Whenever possible, cakes should be placed in the center of the oven where the heat is evenly distributed. Leaving approximately 2–3 inches around each of the pans and between the pans and oven walls allows the heat to circulate freely around each pan. Cakes bake in four stages:

- Stage 1—The cake is placed in the oven and starts to rise. At this stage, the lowest oven temperature called for in the baking instructions should be used to prevent overly quick browning and to keep a crust from forming.

- Stage 2—The cake continues to rise and the top surface begins to brown. The oven door should not be opened at this stage. The heat can be increased at this time if the recipe calls for it.
- Stage 3—The rising stops but the surface of the cake continues to brown. The heat can be reduced if the cake is browning too quickly.
- Stage 4—The cake starts to shrink and separates slightly from the sides of the pan. At this stage, the oven door can be opened and the cake can be tested for doneness.

Testing Cakes for Doneness

Dense cakes are tested for doneness by inserting a wire tester or a toothpick in the center of the cake. The cake is done if the tester is dry when removed, with no batter adhering to it. The touch method is used for lighter cakes, such as sponge cakes. The top surface of the cake is pressed with a finger, and if the surface feels firm and the finger impression does not remain the cake is done.

Cooling and Storing Cakes

When a cake is removed from the oven, it should be placed on a wire rack or shelf so that air circulates around the pan. **See Figure 9-39.** The cake should be allowed to cool for a minimum of 5 minutes before it is removed from the pan. The pan can then be flipped over and the cake removed from the pan. If wax or parchment paper was used on the bottom of the pan, it is also removed at this time. The cake is then placed back on the wire rack or shelf and allowed to cool thoroughly.

Cooling Cakes

Figure 9-39. When a cake is removed from the oven, it should be placed on a wire rack or shelf so that air can circulate around the pan and cool the cake.

CHECKPOINT 9-6

1. Describe the two-stage method of mixing cake batters.
2. Use the two-stage method to mix a cake batter. Evaluate the appearance and texture.
3. Describe the creaming method of mixing cake batters.
4. Use the creaming method to mix a cake batter. Evaluate the appearance and texture.
5. Describe the chiffon method of mixing cake batters.
6. Use the chiffon method to mix a cake batter. Evaluate the appearance and texture.
7. Describe the sponge method of mixing cake batters.
8. Use the sponge method to mix a cake batter. Evaluate the appearance and texture.
9. Describe the four stages of baking cakes.
10. Describe how to test cakes for doneness.

ICING PREPARATION

Like cake recipes, variations of icing recipes are used to obtain different flavors and textures. *Icing,* also known as frosting, is a sugar-based coating often spread on the outside of and between the layers of a baked product. Icings have three main functions when applied to a baked product such as a cake. Icings form a protective coating to seal in moisture and flavor, improve the flavor, and add eye appeal. Common types of icing include whipped cream, glaze, foam, flat, royal, buttercream, fondant, fudge, and ganache.

It is important to obtain the proper consistency before applying any type of icing. The consistency needed depends on the intended use of the icing. In most cases, the consistency of icings can be controlled by adding more or less confectioners' sugar than the recipe specifies.

Whipped Cream Icings

Whipped cream icings are made by beating heavy cream until stiff peaks form. It is critical to chill the bowl, the beaters, and the cream before making whipped cream. Whipped cream icings may be sweetened by adding confectioners' sugar. Whipped cream icings are often piped onto fruit-based cakes and pies. **See Figure 9-40.**

Whipped Cream Icings

Paderno World Cuisine

Figure 9-40. Whipped cream icings are often piped onto fruit-based cakes and pies.

Glaze Icings

A glaze icing is a thin, transparent coating of icing that is poured over a baked product. Many glaze icings are made by thinning out jams or fruit purées. Apricot glaze is the most common fruit glaze used in the professional bakeshop. It has a mild flavor and a nice sheen. Chocolate glazes are made by thinning the chocolate with cream or butter, and in some recipes coffee. Glazes extend shelf life by sealing in moisture.

Foam Icings

Foam icing, also known as boiled icing, is prepared by boiling a glucose and water mixture to approximately 240°F, and then adding the resulting syrup to an egg white meringue. If a thin syrup is added, the result is a thin icing. If a heavy syrup is added to the meringue, a heavy icing is produced. Foam icing is applied in generous amounts and often worked into peaks. **See Figure 9-41.**

Foam (Boiled) Icings

Carlisle FoodService Products

Figure 9-41. Foam icing is applied in generous amounts and often worked into peaks.

Flat Icings

Flat icings are applied to sweet rolls, doughnuts, Danish pastries, and other baked products. A flat icing is made by combining water, confectioners' sugar, corn syrup, and flavoring and then heating the mixture to approximately 100°F. Flat icing should always be heated in a double boiler because overheating causes it to lose its gloss when it cools. Flat icing should be covered with a thin coating of water, plastic wrap, or wax paper before it is stored to prevent it from drying out. Previously stored flat icing must be heated to approximately 100°F over a water bath before use.

Royal Icings

A royal icing is similar to a flat icing, but the addition of egg whites produces a thicker icing that hardens to a brittle texture. Royal icing hardens when exposed to air, so it must be kept covered with a damp towel when not in use. It is often used for making decorations for cakes and sugar sculptures.

Buttercream Icings

Buttercream icing is a type of icing made by creaming together shortening or butter, confectioners' sugar, and vanilla. **See Figure 9-42.** Some types of buttercream icings incorporate eggs, giving them a richer flavor. Buttercreams are light because air cells are trapped in the icing when mixed using the creaming method. They should be stored in a cool place and covered with plastic wrap or wax paper to avoid crusting. For best results, they should not be stored in the refrigerator, because refrigeration causes them to harden.

Icing Preparation

Buttercream Icings

Paderno World Cuisine

Figure 9-42. Buttercream icing is made by creaming together shortening or butter, confectioners' sugar, and vanilla.

Fondant Icings

Fondant is a rich, white, cooked icing that hardens when exposed to the air. It is prepared by cooking a sugar, glucose, and water mixture to a temperature of 240°F, letting it cool to 150°F, and then mixing it until it is creamy and smooth. Rolled fondants have a consistency similar to dough and can be rolled out and used as a thick covering for cakes and to make cake decorations. **See Figure 9-43.**

Fondant can be colored and flavored as desired and used as a base for other icings. Fondant must be kept covered with a very thin coat of water, plastic wrap, or wax paper and stored in a cool place. It can be refrigerated, but it may lose some of its gloss when reheated.

When needed for use, fondant is heated to about 100°F in a double boiler while stirring constantly. This thins the icing so that it flows freely over the item to be covered. To successfully cover an item with fondant, the consistency of the icing must be correct. If the fondant is too heavy after it is heated, it can be thinned down by using a glaze consisting of 1 part glucose to 2 parts water, or a regular simple syrup may be used. Fondant is the most time-consuming icing to prepare. However, only water and a small amount of glucose need to be added to a powdered product called Drifond® to produce fondant.

Fudge Icings

Fudge icing is a rich, heavy-bodied icing. It is usually prepared by adding a hot liquid or syrup to the other ingredients required while whipping to obtain smoothness. Fudge icing should be used warm. If it cools, it should be reheated in a double boiler before use. Fudge icing is generally used to ice layer cakes, sheet cakes, and cupcakes. **See Figure 9-44.** Fudge icing should be covered with plastic wrap and stored in the refrigerator. Fudge icing can be reheated over a water bath before use.

Fudge Icings

Entourage

Figure 9-44. Fudge icing is generally used to ice layer cakes, sheet cakes, and cupcakes.

Using Fondant Icings

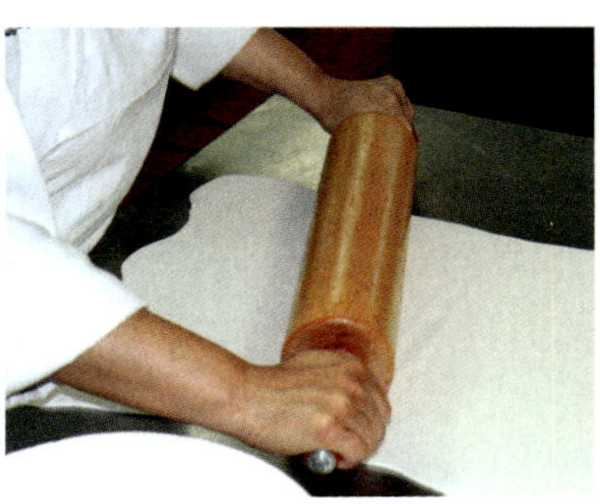

Rolling Fondant

Draping Fondant

Fondant-Iced Cake

Figure 9-43. Rolled fondants have a consistency similar to dough and can be rolled out and used as a thick covering for cakes and to make cake decorations.

Ganache Icings

Ganache icing is a rich chocolate icing made by heating chocolate and whipping cream until the chocolate melts. The ratio (in weight) of chocolate to cream can be 1:1 for a light ganache or 2:1 or more for a firm ganache. In its liquid state, a light ganache can be used to enrobe cakes and pastries. Firm ganache can be chilled and then whipped to make a creamy filling for bon bons and cakes. In its solid state, ganache is the easiest way to make chocolate truffles. **See Figure 9-45.** Ganache can be made with dark, milk, or white chocolate and may also include butter, glucose, liquor, or a fruit purée.

In its simplest form, ganache is made with only chocolate and heavy cream. Ganache is made by first bringing the cream to a boil and then removing the cream from the heat, adding chopped chocolate, and stirring until it is incorporated. Another way to prepare ganache is by pouring hot cream over chopped chocolate and blending it until smooth.

Filling Pastry Bags

A *pastry bag* is a cone-shaped bag made of parchment paper, silicone, canvas, or plastic. Paper cones are simple to make and easy to handle. Regardless of the materials, a separate cone should be used for each color of icing.

To fill a pastry bag with icing, the bag is held in one hand with a light grip and the top half of the bag is folded over the hand. A spatula is then used to insert the icing into the center of the bag. **See Figure 9-46.**

Filling Pastry Bags

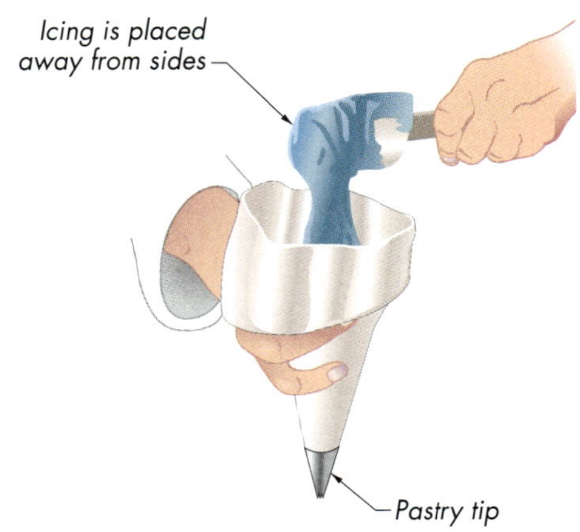

Figure 9-46. A spatula is used to insert icing into the center of a pastry bag.

Ganaches

Irinox USA

Icing | **Truffles**

Figure 9-45. In its liquid state, a light ganache can be used to coat cakes and pastries. In its solid state, ganache is the easiest way to make chocolate truffles.

Procedure for Forming a Paper Pastry Bag

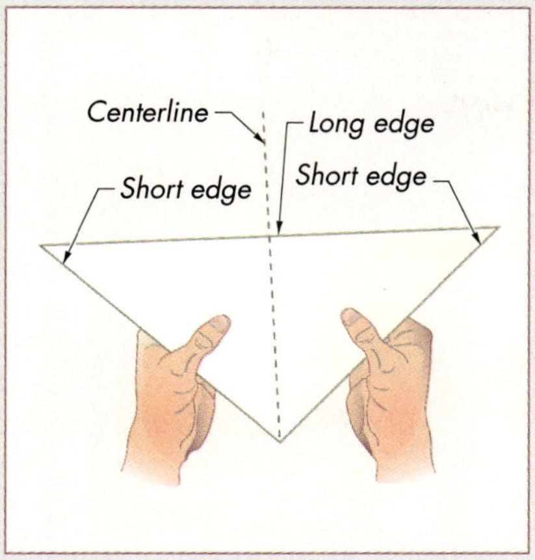

1. Cut a square of parchment paper into a large triangle.

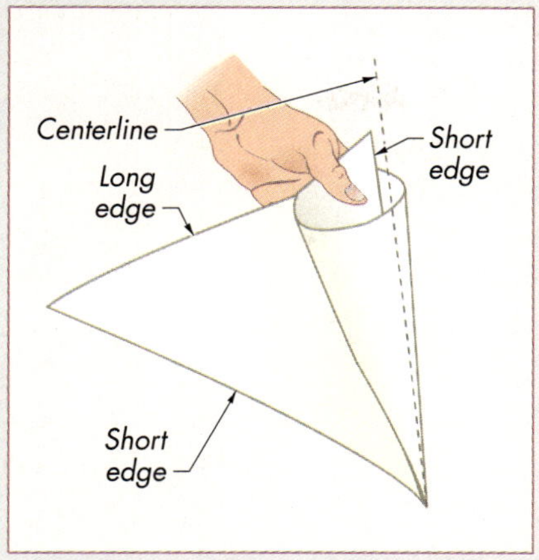

2. With the long edge on top, start to roll the paper by turning one short edge towards the centerline of the triangle.

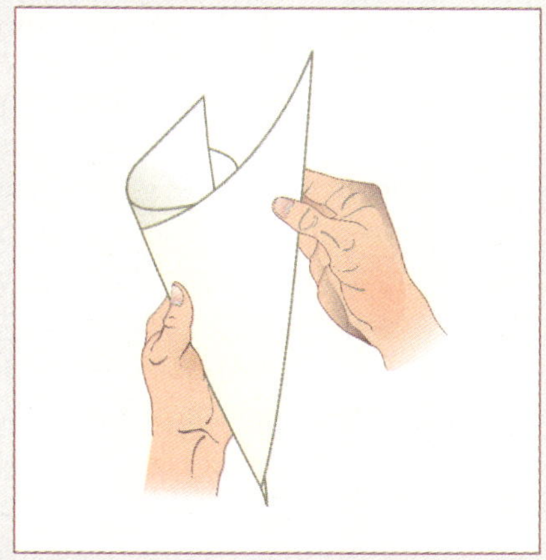

3. Continue rolling the paper into a cone shape across to the far short edge.

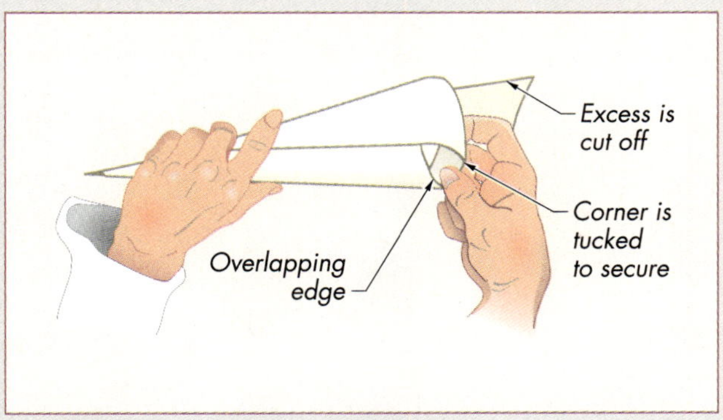

4. Tuck the top corner of the far short edge into an overlapping edge of the cone to secure. Cut off excess paper at the top of the cone.

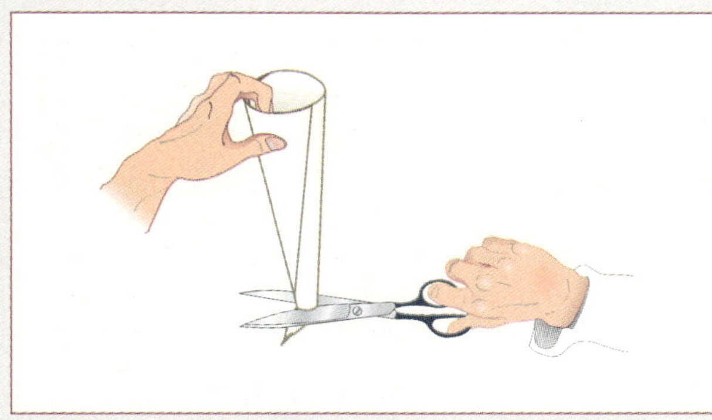

5. Cut off the bottom tip of the cone so a pastry tip can be inserted. *Note:* If too much paper is cut away, the hole will be too large to hold the pastry tip.

Each time the spatula is withdrawn, the icing is gently squeezed down into the bag. Once the pastry bag is halfway filled, it should be tightly twisted just above the filling. The twisted portion of the bag becomes the handle while piping the icing. As the bag empties, it can be twisted further to push the icing toward the tip.

Production Tip

Air pockets within the icing or along the sides of the pastry bag can ruin a decorating job. The icing should be constantly squeezed toward the pastry tip to push out any air pockets.

Piping Icing

In order to create decorative patterns and shapes with icing, a metal or plastic pastry tip is inserted into the pastry bag before it is filled. A vast array of pastry tips can be used to make most any design. **See Figure 9-47.**

When decorating, the top of the bag is held with one hand while the other hand is used to lightly grip the lower half. The hand at the top of the bag applies the pressure that causes the icing to flow. The hand on the lower half of the pastry bag is used to guide. The pastry bag must be held at the proper angle and the proper amount of constant pressure applied to obtain a smooth, even flow. **See Figure 9-48.**

Production Tip

To make piping dots, hold a pastry bag fitted with a writing tip at a 90° angle and squeeze while keeping the tip down until a dot forms. Then lift the tip straight up.

Using Pastry Bags

Figure 9-48. The pastry bag must be held at the proper angle and the proper amount of pressure applied to obtain a smooth, even flow.

Common Pastry Tips

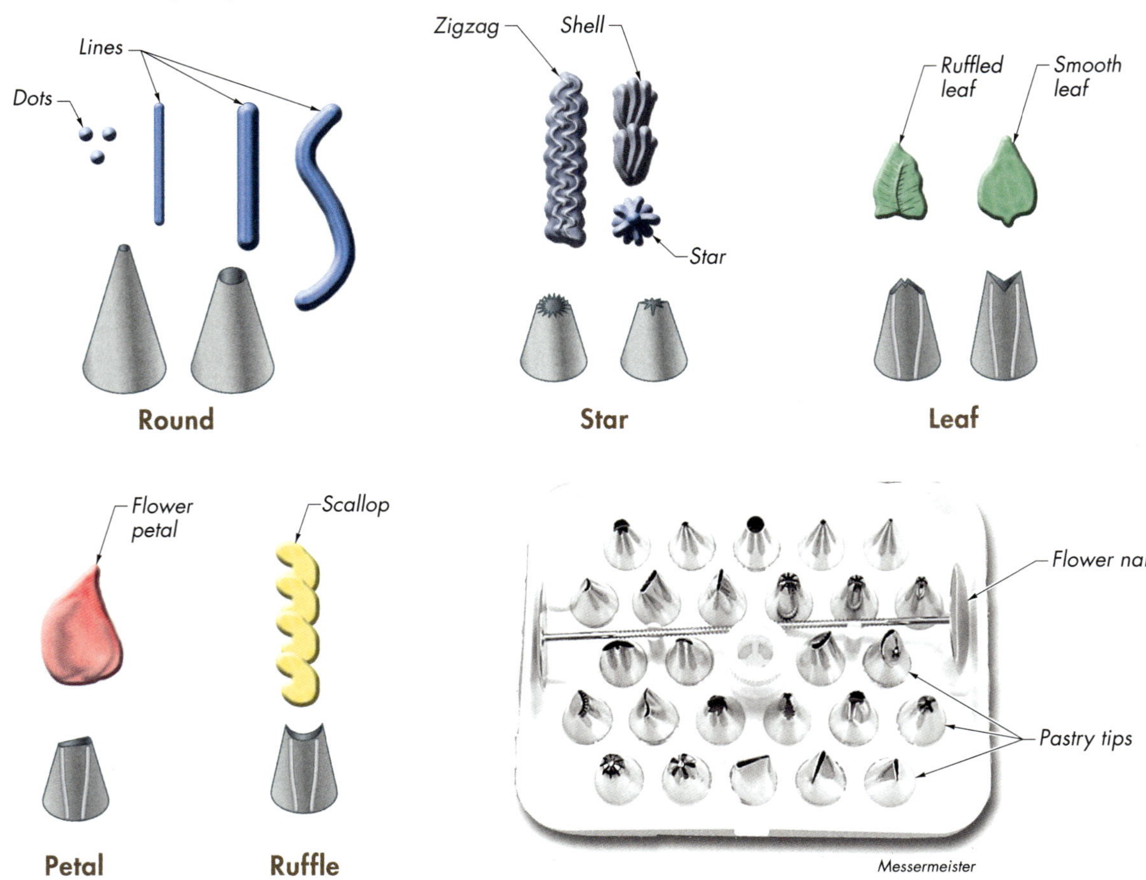

Figure 9-47. A vast array of pastry tips can be used to make most any design.

Icing Techniques

While there are many ways to ice cakes, some procedures are considered standard. The following guidelines address the basics of icing cakes:

- Layer cakes are built in two or more layers. The top of each layer is iced before being topped with the next layer. The top and sides of a layered cake are masked with a thin layer of icing in order to create a crumb barrier before a finish coat of icing is applied. After a layer cake has been masked, it can be enrobed with fondant, marzipan, or ganache. Layer cakes may also be iced with buttercream icing using a turntable and a palette knife. **See Figure 9-49.** A cake comb is sometimes used to leave grooves in the icing on the sides or top of the cake before any decorations are applied.

Icing Layer Cakes

Figure 9-49. Layer cakes may be iced with buttercream icing using a turntable and a palette knife.

- The sides of a cake may be partially or entirely covered with garnishes such as crushed nuts or coconut. Decorative chocolate plaques may be added to the sides of a cake.
- The tops of cakes are commonly finished in one of two styles. One style is to decorate the cake as a whole using a central garnish, such as flowers made of a sugar gum paste or piped flowers made of buttercream icing. A second style of decorating is to mark the cake into 12–16 portions and then to decorate each portion individually using a combination of piping and garnishes, such as nuts, shaved chocolate, or candied fruit.

CHECKPOINT 9-7

1. Name the three functions of icings.
2. Describe nine different types of icings.
3. Prepare two different types of icing. Evaluate the texture, color, and flavor of each.
4. Fill a pastry bag with icing and pipe script lettering on a cake. Evaluate the results.
5. Ice and then decorate a layer cake. Evaluate the results.

COOKIE PREPARATION

Cookies are a popular dessert that can be served alone, as part of another dessert, or with fruit. The ingredients used to make cookies are similar to those used in cakes. However, cookie dough generally contains more fat and less moisture than cake batter. Cookies are usually baked at higher temperatures and for less time than cakes. Cookies are classified by texture as either soft or crisp. **See Figure 9-50.** A *soft cookie* is a cookie prepared from dough that contains a lot of moisture. A *crisp cookie* is a cookie prepared from dough that contains a high percentage of sugar.

Cookies

Carlisle FoodService Products

Figure 9-50. Cookies are classified by texture as either soft or crisp.

Cookies can be made using the creaming mixing method or the sponge mixing method. The creaming method makes a dough and the sponge method creates a batter. In the creaming method of making cookies, butter is mixed together with sugar to incorporate air and make a creamy mixture. Then the eggs and any liquids used are added. Finally, the flour and any leavening agent used are added. The cookie dough is then formed into cookies and baked.

The sponge method of making cookies can be executed three different ways. In the first option, sugar and eggs or yolks are whipped until a light foam is achieved. The foam is the correct consistency when it can be whipped into a ribbon. Then, the flour and flavoring are folded in. The resulting batter is piped into cookies and baked. A second option involves mixing a small portion of an egg foam with flour and softened butter. Then the remainder of the egg foam and the flour are folded in. The resulting batter is piped into molds and baked.

In the third option, the egg yolks and whites are separated. The yolks are whipped together with a small amount of sugar. The whites are whipped with the remainder of the sugar to form a meringue. The meringue is folded into the yolk mixture and then the flour is folded in. This method is used to make lady fingers. Lady fingers can be served as a cookie, used to make tiramisu, or as a fence surrounding a mold filled with Bavarian cream to make a classic Charlotte Russe.

Types of Cookies

Cookies are often named for their preparation method. Bar, drop, molded, pressed, refrigerator, and rolled cookies are six types of cookies named for the way they are prepared. The method used is often determined by the consistency of the cookie dough or batter. **See Figure 9-51.**

Cookie Preparation	
Bar Method 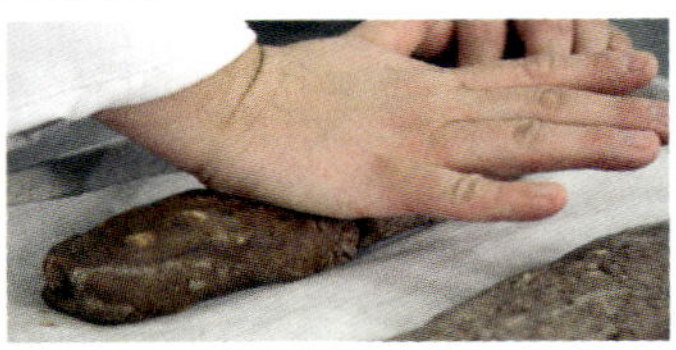	Bar cookies are made from a fairly stiff, dry cookie dough that is scaled into 1 lb pieces, refrigerated until thoroughly chilled, and rolled into strips the length of a sheet pan. Three strips are placed an equal distance apart on a parchment-lined sheet pan, flattened by hand, and then brushed with an egg wash. The bar cookies are baked and then cut after they have cooled.
Drop Method 	Drop cookies are formed into consistently sized balls. The cookies are prepared from a moist, soft batter. The batter should be at room temperature and then dropped in uniform balls onto parchment-lined sheet pans and baked.
Molded Method 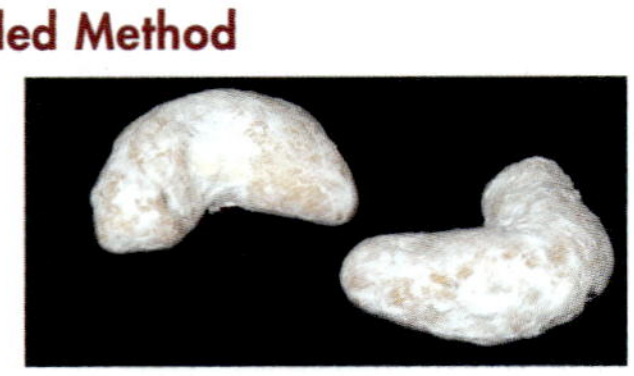	Molded cookies are made by molding dough into shapes, such as small balls or crescents, by hand. The molded cookies are placed on parchment-lined pans and baked.
Pressed Method 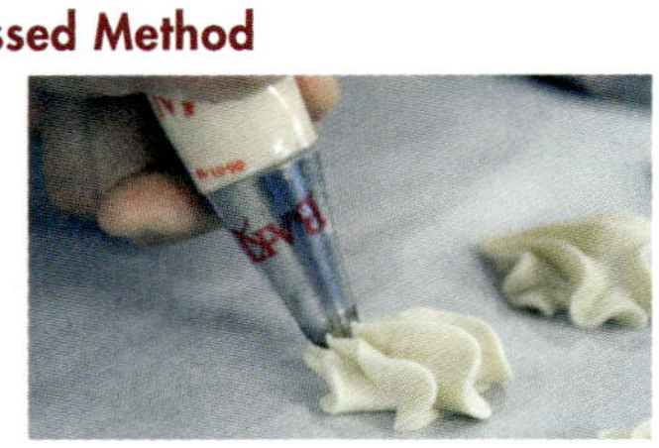	Pressed cookies are made by placing soft, moist dough in a cookie press or a pastry bag to create shapes such as leaves, snowflakes, or stars. The cookies are placed on parchment-lined sheet pans and then baked.

Figure 9-51. (continued on next page)

Preparing Drop Cookies
Media Clip

Cookie Preparation	
Refrigerator Method 	Refrigerator cookies are made by scaling a fairly stiff, dry dough into 1–1¼ lb rolls approximately 16 inches in length. The rolls are then wrapped in parchment paper and refrigerated overnight. The next day the dough is cut into ¼ inch thick slices, placed on parchment-lined sheet pans, and then baked. Thinner slices can be cut to produce crispier cookies.
Rolled Method 	Rolled cookies are made from a fairly stiff, dry cookie dough that is refrigerated until chilled thoroughly and then rolled out on a floured surface to ⅛ inch thickness. The dough is then cut into desired shapes with a cookie cutter, placed on parchment-lined sheet pans, and baked.

Figure 9-51. Bar, drop, molded, pressed, refrigerator, and rolled cookies are named for the way they are prepared.

Cooling and Storing Cookies

Soft cookies should be placed in an airtight tin container. Crisp cookies should be stored in a tin container with a loose-fitting top and placed in a dry location. Crisp cookies can be warmed in a 225°F oven for 5 minutes just prior to serving.

CHECKPOINT 9-8

1. Identify six methods of cookie preparation.
2. Prepare six different types of cookies. Compare the appearance, texture, color, and flavor of each type of cookie.

PIE AND TART PREPARATION

Pies and tarts are popular desserts. The two basic types of pies are single-crust pies and double-crust pies. **See Figure 9-52.** A single-crust pie consists of one crust on the bottom and a filling. Tarts also have a single-crust. A double-crust pie consists of two crusts, one on the bottom and one on the top, with a filling in between. The tenderness of the crust often determines how a pie is received. The filling may be of excellent quality, but a tough crust can ruin the entire preparation. Different types of pie dough vary in preparation techniques but have similar ingredients.

Pie Types

Single Crust

Double Crust

Figure 9-52. The two basic types of pies are single-crust pies and double-crust pies.

Most pie doughs are made primarily of flour and fat. However, the types of dough differ in how the flour and fat are mixed together and the amount of liquid added to the dough. Pie dough can be made by hand, in small batches using a pastry blender, in a food processer, or in a mixing machine. Pie dough must be mixed properly to achieve the best results. The correct amount and mixing of ingredients produces a good piecrust. The salt, sugar, and cold liquid should be blended together until the salt and sugar are thoroughly dissolved. The liquid mixture is then poured into the flour and fat mixture and mixed only until the flour absorbs the liquid.

Many pie dough recipes are written as ratios. A *ratio* is a mathematical way to represent the relationship between two or more numbers or quantities. In the professional kitchen, a ratio is expressed in terms of parts. For example, a ratio of 3 parts flour to 1 part shortening is often expressed as 3:1. A ratio describes how the ingredients are used in relation to one another. The simplest ratio used in the professional kitchen is a ratio of 1:1, also known as equal parts.

When a ratio must be used in terms of weight only or volume only, the ratio will be followed by the words "by weight" or "by volume." For example, a pie dough recipe has a ratio of 3 parts flour to 2 parts butter to 1 part water (3:2:1) by weight.

When using a ratio, ingredient quantities can be calculated by converting the ratio into fractions and multiplying the fractions by the desired yield. **See Figure 9-53.** The first step is to convert the ratio to fractions by dividing each number in the ratio by the total number of parts (3 + 2 + 1 = 6). The fraction for flour would be 3/6 (which reduces to ½), the fraction for butter would be 2/6 (which reduces to ⅓), and the fraction for water would be ⅙. By multiplying each of the fractions by the total desired amount of dough (12 lb), it can be calculated that the recipe would require 6 lb of flour (½ × 12 lb = 6 lb), 4 lb of butter (⅓ × 12 lb = 4 lb), and 2 lb of water (⅙ × 12 lb = 2 lb).

When making pie dough, most mistakes occur while the dough is being mixed. First, the flour must be sifted before the fat is cut-in, or incorporated into the flour. Another mistake is adding extra flour or water to pie dough, causing overmixing and resulting in a tough dough.

Types of Piecrusts

Piecrusts are generally classified as mealy or flaky. Other types of piecrusts include crumb and specialty. The type of filling to be added often determines which type of dough is the best choice.

Mealy Piecrusts. A *mealy piecrust* is a low-moisture piecrust prepared by rubbing fat into flour until the mixture resembles fine cornmeal. The flour is then unable to absorb a large amount of liquid or moisture from the filling because the flour granules are coated in fat. Mealy piecrust is used mostly for bottom-crust pies, such as fruit or custard pies, because it has high resistance to moisture.

Flaky Piecrusts. A *flaky piecrust* is a piecrust prepared by cutting fat into flour until pea-size pieces of dough form. When flaky piecrust is made with butter it is known as pâte brisée. The flour in flaky piecrust is not completely blended with the fat like it is in a mealy piecrust. There are two types of flaky crust: short-flake and long-flake.

- A *short-flake crust* is a piecrust that absorbs slightly more liquid than mealy piecrust. When making short-flake crusts, the flour and fat are cut-in only until no flour spots are evident and the fat is in pea-size particles. The flour of a short-flake crust is not coated with fat to the same degree as mealy piecrust, so the flour is able to absorb slightly more liquid. A short-flake crust is the most common type of flaky piecrust and is commonly used to make lattice tops and the bottom crust of a pie filled with a cooked filling.

Using Pie Dough Ratios

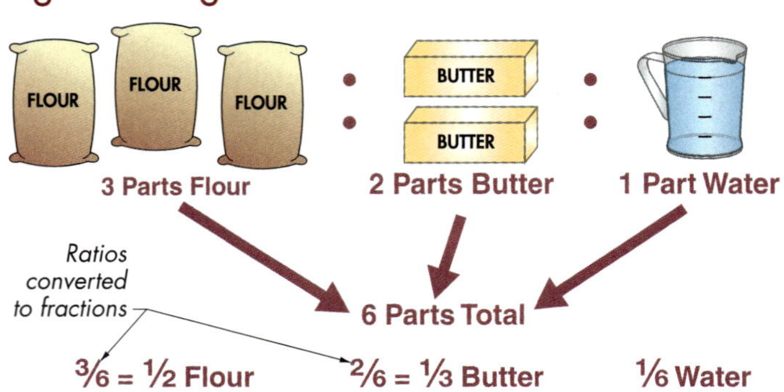

Desired recipe yield

12 lb of Pie Dough	
Ingredients	**Quantity Required**
Flour	½ × 12 lb = 12 lb/2 = 6 lb
Butter	⅓ × 12 lb = 12 lb/3 = 4 lb
Water	⅙ × 12 lb = 12 lb/6 = 2 lb

Figure 9-53. When using a ratio, ingredient quantities can be calculated by converting the ratio into fractions and multiplying the fractions by the desired yield.

- A *long-flake crust* is a piecrust that absorbs a large amount of liquid because the flour and fat are rubbed together very lightly, leaving the fat in chunks about the size of a walnut. Long-flake piecrust is commonly used for top crusts as well as prebaked bottom crusts. Tarts often have a long-flake crust.

The basic difference between mealy and flaky piecrusts is how the fat is combined with the flour. **See Figure 9-54.** Mealy crusts and short-flake crusts are handled in the same manner after mixing. Once the dough is mixed, it is wrapped with plastic wrap and refrigerated until it is firm enough to be rolled out (about 45–60 minutes). Long-flake crust must be refrigerated for a longer period, usually several hours or overnight. If long-flake dough is not refrigerated long enough, it will be too soft and difficult to roll out. When the pie dough is firm, it is removed from the refrigerator and divided into 8 oz units. Each 8 oz unit provides enough dough for one bottom or one top crust for an 8 inch or 9 inch pie.

Piecrusts		
Type	*Characteristics*	*Mixing*
Mealy	• High resistance to absorbing moisture • Used for bottom crust of pies	Cut-in fat and flour until fat is completely covered
Short-flake	• Most common type of flaky piecrust	Cut-in fat and flour until fat is pea size
Long-flake	• Absorbs the most moisture of any piecrust	Cut-in fat and flour until fat is walnut size

Figure 9-54. Piecrusts are classified as mealy or flaky.

To keep pie dough from drying out, only one unit of dough should be rolled out at a time. **See Figure 9-55.** After the dough is rolled and the bottom or top crust is formed, any remaining scraps can be pressed together and reused. Any unneeded 8 oz units should be returned to the refrigerator until ready to be rolled out.

Crumb Crusts. A *crumb crust* is a piecrust made from crumbled cookies or crackers that are held together with melted butter. Crumb crusts can be made from various ingredients including graham crackers, vanilla wafers, or ginger snaps. Ground nuts may be added along with sugar. Crumb crusts are often baked for a short time to stabilize the crust. Crumb crusts are most often used for cheesecakes, unbaked pies such as cream or chiffon pies, and some tarts. **See Figure 9-56.**

Crumb Crusts

Wisconsin Milk Marketing Board, Inc.

Figure 9-56. A crumb crust is made from crumbled cookies or crackers that are held together with melted butter.

Specialty Crusts. Specialty piecrust may be prepared by adding spices, ground nuts, or other items to the pie dough. The added ingredient usually replaces up to 20% of the flour, except in the case of spices.

Rolling Piecrusts

Figure 9-55. To keep pie dough from drying out, only one unit of dough should be rolled out at a time.

Pie Preparation

Blind Baking Crusts

Sometimes pie dough is baked for a short period before it is filled. Blind-baked crusts are often used for pies with unbaked fillings, such as chiffon or cream. *Blind baking* is a process in which a piecrust is baked for 10–15 minutes before a filling is added.

Procedure for Blind Baking Pie Crusts

1. Roll the dough out to the desired size and thickness and place it in the bottom of a pie pan.

2. Use a dough docker or a fork to prick small holes in the pie dough to allow steam to escape and prevent the dough from rising in the oven.
3. Cover the bottom, sides, and edges of the dough with parchment paper.
4. Place pie weights or dried beans on top of the paper to prevent the crust from rising as it bakes.
5. Bake the crust at 350°F for 10–15 minutes or until light golden brown.

Pie Fillings

A pie filling must be thickened to the proper consistency and flavored appropriately. A thickener is typically added to a filling while it is being cooked. Approximately 3–5 oz of starch is required for each quart of liquid. More thickener is required for high-acidity fruits and juices. A pregelatinized starch can thicken a liquid without additional cooking because it does not require heat to absorb liquid and gelatinize.

The proper amount of filling required for each pie is determined by weight. The prepared piecrust is placed on a baker's scale and the scale is balanced. The baker's scale is then set for the required amount of filling. Filling is then added to the piecrust until the scale balances a second time. Common types of pie fillings include fruit, cream, soft, and chiffon fillings.

Fresh Fruit Pie Fillings. Fruit fillings are the most popular type of pie filling. **See Figure 9-57.** Fruit used as pie filling can be fresh, frozen, canned, or dried. Dried fruits, such as raisins, currants, or figs, need to be softened by cooking before being baked in a piecrust. Each form of fruit requires different preparation for use in a pie filling.

Fruit Pie Fillings

Wisconsin Milk Marketing Board, Inc.

Figure 9-57. Fruit fillings are the most popular type of pie filling.

Procedure for Preparing Cooked-Fruit Pie Fillings

1. Place fruit, sugar, and a small amount of juice in saucepan with the desired spices. Bring to a boil.

2. Dissolve starch in cold water and pour slowly into the boiling fruit and juice mixture while stirring constantly.
3. Bring the mixture back to a boil and cook until clear.

4. Add salt and color (if desired) and stir until thoroughly blended.
5. Cool slightly and pour the filling into unbaked piecrusts.

The amount of water and sugar to use in a fresh fruit filling is based on the amount and type of fresh fruit used. Typically, 65% to 70% water in comparison to the weight of the fruit is sufficient. For example, if 10 lb of fruit is used, 6½–7 lb of water is needed. Most fresh fruit fillings (except those made with fresh berries) are prepared using the cooked-fruit method, which only requires a few simple steps.

Fruit flan is another name for a fresh fruit tart that is quite striking in appearance. **See Figure 9-58.** A fruit flan or tart is made by blind baking a sweet crust (called a pâte sucrée) in a parchment-lined fluted tart pan or a straight-sided flan ring. The next step is to use a piping bag to fill the crust with pastry cream. After the tart is filled, the surface is covered with a variety of fruits such as strawberries, kiwis, and raspberries. Sometimes the fruit is poached first, as in the case of pears. After the fruit is decoratively laid out to cover the surface of the pastry cream, it is then brushed with a hot glaze (often made by thinning apricot jam with water). When the glaze sets it gives the flan or tart a shiny finish and keeps the fruit from drying out. The tart may be crumbed on the edge with crushed nuts before it is sliced and served.

Preparing Cooked-Fruit Pie Fillings
Media Clip

Fruit Flan

Sullivan University

Figure 9-58. Fruit flan is another name for a fresh fruit tart that is quite striking in appearance.

Frozen Fruit Pie Fillings. Frozen fruit is commonly used in commercial pie fillings. **See Figure 9-59.** Frozen fruit has the same flavor characteristics and appearance as fresh fruit, except that it is available year-round. Either raw or parboiled fruit is mixed with natural juices and sugar and then packed in plastic buckets or vacuum-sealed in plastic and then frozen. Frozen fruit is commonly sold in 30 lb buckets.

Frozen Fruit

Figure 9-59. Frozen fruit is commonly used in commercial pie fillings.

Frozen fruit must be completely defrosted before it is used in a pie filling. The best method for defrosting frozen fruit is to place the unopened container in the refrigerator for one day. An opened container of fruit can be set in a hot water bath and stirred constantly to completely defrost fruit for immediate use. After defrosting, the fruit is strained to remove the juice. The strained juice is reserved, thickened, and added back to the fruit. Thickening the juices and adding them back to the fruit prevents the fruit from bleeding.

Canned Fruit Pie Fillings. Canned fruit is also used for commercial pie filling because it is available year-round. However, canned fruit fillings are inferior to fresh and frozen fruit fillings because canned fruit is overcooked during the canning process. Canned fruit can be purchased packed in water, syrup, or in a solid pack. A *solid pack* is a canned product with little or no water added. A solid pack is preferred for pie fillings because it contains more fruit and less sugar. More sugar can be added after the juice is thickened, which produces better results.

Dried Fruit Pie Fillings. Dried fruit is occasionally used for pie fillings. Dried fruit must be rehydrated in water or juice to restore natural moisture that was removed in the drying process. In some cases, the liquid and fruit may be brought to a boil to plump the fruit. The dried fruit is then soaked as it cools. After soaking, the liquid is drained from the fruit, thickened, flavored, and poured back over the fruit.

Cream Pie Fillings. Smooth, flavorful, cream pie fillings only require a few simple steps to prepare. After the cream filling is prepared, it is placed in a prebaked piecrust and topped with meringue or a cream topping. Popular cream fillings are chocolate, coconut, butterscotch, and banana.

Procedure for Preparing Cream Pie Fillings

1. Place milk in the top of a double boiler and heat.
2. In a separate container, beat eggs and add sugar, salt, and starch or flour.
3. Add cold milk while stirring constantly until a thin paste forms.
4. Add a small amount of warmed milk to the paste while whisking in the eggs.
5. Add the egg mixture to the remaining milk, whisking constantly until the mixture thickens and becomes smooth.
6. Cook until starch or flour is completely incorporated. Remove from heat.
7. Stir in flavoring and butter.
8. Pour into prebaked piecrusts and let cool.

Soft Fillings. A *soft filling* is an uncooked pie filling that is baked in an unbaked piecrust. Soft-filling pies include pumpkin pie, custard pie, and pecan pie. **See Figure 9-60.** Soft fillings are the most difficult pie fillings to make. The challenge lies in baking the filling and crust to the proper temperature without underbaking or overbaking either part. Common mistakes can result in a soggy crust or a filling that separates or is watery.

Soft Pie Fillings

Wisconsin Milk Marketing Board, Inc.

Figure 9-60. Soft-filling pies include pumpkin pie, custard pie, and pecan pie.

Generally, soft-filling pies are baked at 400°F for the first 10–15 minutes of baking, and then the temperature is reduced to 350°F or 325°F for the remaining baking time. The pie is removed from the oven as soon as the filling sets, which helps prevent the crust from becoming soggy. To check the filling for doneness, a knife is inserted 1 inch from the center. If the knife comes out clean, the filling is done. Another method of checking for doneness is to gently shake the pie. The center of the soft filling may move slightly, but carryover cooking will continue to cook the filling after the pie is removed from the oven.

Chiffon Fillings. A *chiffon filling* is a light, fluffy pie filling prepared by folding a meringue into a puréed fruit or a cream pie filling. In most cases, a small amount of plain gelatin is added to the fruit- or cream-based filling to help the chiffon filling set as it cools.

Procedure for Preparing Chiffon Pie Fillings

1. Prepare either a fruit or cream filling.
2. Soak plain gelatin for 5 minutes in cold water. Add gelatin to the hot fruit or cream filling, stirring until the gelatin is thoroughly dissolved.
3. Place the filling in a fairly shallow pan and let cool. Refrigerate until the filling begins to set. Stir the filling occasionally while cooling so it cools evenly. *Note:* Do not allow the filling to set completely because it will be difficult to fold in the egg whites uniformly.
4. Prepare a meringue by whipping egg whites and sugar together until the mixture forms stiff peaks.
5. Fold the meringue into the jellied fruit or cream mixture gently, preserving as many of the air cells as possible. *Note:* This step should be done quickly so that the gelatin does not set before the folding is finished.
6. Pour the chiffon filling into a baked piecrust and refrigerate until set.

Meringues

Meringues are most often used to top pies. A *meringue* is a mixture of egg whites and sugar. Meringues can be soft or hard, depending on the ratio of sugar to egg whites. The greater the amount of sugar in proportion to egg whites, the harder (more stable) the resulting meringue. A soft meringue typically has a 1:1 ratio, 1 part sugar to 1 part egg whites. A hard meringue has a 2:1 ratio, 2 parts sugar to 1 part egg whites. Soft meringues are often used as a dessert topping, such as with lemon meringue pie.

If a meringue-topped pie is held prior to eating, a hard meringue with a ratio of 1½:1 (1½ parts sugar to 1 part egg white) would work best because it will not weep. Hard meringues can also be used to make crisp disks that hold soft desserts. French meringues, Swiss meringues, and Italian meringues are used in a variety of desserts. **See Figure 9-61.**

Meringues

Meringue	Preparation	Stability
French	Egg whites are whipped with an acid (lemon juice or cream of tartar) and sugar is gradually added	Least stable
Swiss	Egg whites and sugar are warmed over a hot water bath and then beaten to desired stiffness	Moderately stable
Italian	Sugar and water are heated to 240°F, cooled to 220°F, poured into egg whites, and whipped to desired stiffness	Most stable

Figure 9-61. French meringues, Swiss meringues, and Italian meringues are used in a variety of desserts.

French Meringue. French meringue, also known as common meringue, is made by whipping egg whites with a bit of lemon juice or cream of tartar. Then sugar is gradually added until the desired stiffness is achieved.

Swiss Meringue. Swiss meringue is a more difficult meringue to make than French meringue. Swiss meringue is made by gently warming egg whites and sugar over a hot water bath and then beating them until the desired stiffness is achieved. After the whites and sugar have been heated in a double boiler or bain-marie to between 110°F and 165°F (depending on the use), the sugar dissolves thoroughly, making this meringue more stable and less prone to weeping. However, if the egg white mixture gets too hot, the egg whites will not whip properly.

Italian Meringue. Italian meringue is most commonly used in making buttercream frosting. Italian meringue is made by heating a mixture of sugar, water, and cream of tartar to 242°F. This mixture is then cooled to 220°F and poured in a slow, steady stream into egg whites while beating continuously. The cooking of the egg whites with the very hot sugar mixture makes this meringue the most stable type of meringue. However, cooking the whites also results in a taffy-like texture that is not as light as the texture of a French or Swiss meringue.

Nutrition Note

Adding cream of tartar to beaten egg whites helps firm the resulting meringue. A teaspoon of cream of tartar has about 7 calories, 1 mg of sodium, and 495 mg of potassium.

CHECKPOINT 9-9

1. Name four types of piecrusts.
2. Differentiate between mealy and flaky piecrusts.
3. Prepare a mealy pie dough and a flaky pie dough. Compare the appearance and texture of each dough.
4. Differentiate between crumb crusts and specialty crusts.
5. Prepare a crumb crust and a specialty crust. Evaluate the appearance, texture, and flavor of each crust.
6. Explain the purpose of blind baking.
7. Blind bake a piecrust. Evaluate the results.
8. Contrast fruit, cream, soft, and chiffon pie fillings.
9. Prepare a fruit, cream, soft, and chiffon pie filling. Evaluate each filling based on appearance, texture, color, and flavor.
10. Explain how the ratio of sugar to egg whites affects meringues.
11. Describe three common types of meringue.
12. Prepare three different meringues. Evaluate each meringue based on appearance, stiffness, and flavor.

PASTRY PREPARATION

Pastries are commonly thought of as something sweet made with a light, flaky bread or crust. However, pastries can also be savory dishes depending on the ingredients used. Many different sweets, hors d'oeuvres, and savory dishes can be prepared using pastries. The word pastry comes from the word "paste," which means a mixture of flour, liquid, and fat, and typically refers to products made from pastes. Most pastries are made using laminated doughs.

Laminated Doughs

Croissants, puff pastries, and Danishes are made from laminated doughs. A *laminated dough* is a dough that is layered with butter and used to make pastries, such as croissants, turnovers, and strudels. To make laminated dough, the dough is wrapped around butter like a package and then rolled out and folded over. This process is called making a turn and is repeated over and over until the butter and the dough are one cohesive piece. The more turns a laminated dough is given, the more flaky the finished pastry will be. The thin layers of butter in laminated dough cause the dough to puff and rise during baking and give pastries a layered and crispy, but light, appearance. **See Figure 9-62.** The butter also causes the pastry to brown.

Laminated Doughs

Figure 9-62. The thin layers of butter in laminated dough cause the dough to puff and rise during baking and give pastries a layered and crispy, but light, appearance.

Puff Pastry. *Puff pastry dough,* also known as pâte feuilletée, is a laminated unleavened dough that rises as a result of the steam created when moisture in the layers of dough is released during baking. Puff pastry requires special attention during all stages of preparation to yield a good finished product. **See Figure 9-63.** The following tips will help yield a beautiful puff pastry:

- Chill the dough before rolling it out and let the dough rest prior to baking it.
- Cut the dough with a sharp knife or a cutter that does not crimp the edges, which will hinder the release of steam in the oven.
- Start the baking process at 375°F to 400°F for a good initial rise. Once the pastry has fully expanded, reduce the oven temperature to 300°F and bake the pastry until fully crisp. Properly baked puff pastry makes a crunching sound when eaten.

Puff Pastries

Irinox USA

Figure 9-63. Puff pastry requires special attention during all stages of preparation to yield a good finished product.

Puff pastry is used for both sweet and savory applications. In all of the following applications, puff pastry made with butter yields the best-tasting product.

- individual pastries such as turnovers, pinwheels, bouchées, and cream horns
- pastry used to layer Napoleons and to make both rolled and enrobed strudels
- pastry wrappings for meat or fish, such as beef Wellington (referred to as en croûte)
- pastry covering a cup filled with soup and baked in the oven (referred to as en dôme)

Phyllo. *Phyllo* is paper-thin sheets of dough that are brushed with butter and used to make pastries. Frozen phyllo dough is often used in the professional kitchen. Once thawed, it must be kept moist under a damp cloth. One sheet of phyllo at a time is brushed with butter because the dough dries out very quickly. Buttered phyllo sheets are layered to produce pastries with many thin, flaky, buttery layers. **See Figure 9-64.**

Phyllo Pastries

Wisconsin Milk Marketing Board, Inc.

Figure 9-64. Buttered phyllo sheets are layered to produce pastries with many thin, flaky, buttery layers.

Danishes. Danishes are made from laminated dough. A yeast dough is prepared and allowed to rest. After the dough has risen sufficiently, it is punched down and then rolled out into a long rectangle. Slightly softened butter is spread over two-thirds of the rectangle. The dough is then folded to form a three-layer block. This is done by folding the third of the dough that is not covered with butter onto itself and then folding the other side of the dough over the top. The block is then allowed to rest in the refrigerator until it is chilled.

The next step is to give the layered block a 90° turn and then roll it back out into a rectangle three times its original size. The layered-block folding technique is then repeated and the dough is chilled again. The entire process is repeated again before the dough can be used to make Danishes. Danishes are shaped into snecks, snails, and braids. Snecks are the easiest shape of Danish to make. **See Figure 9-65.**

- To make snecks, laminated dough is rolled into a rectangle and then a filling such as almond paste is spread on the dough. The rectangle is rolled up into a log and the log is cut into thick circles and placed on sheet pan. A small amount of fruit filling is placed in the center of each circle and then allowed to proof. Then the pastry is egg washed and baked. After cooling, the snecks are glazed or iced.

Danishes

Carlisle FoodService Products

Figure 9-65. Snecks are the most well-known and easiest shape of Danish to make.

- Snails are made by first rolling out laminated dough and covering it with a spread. The dough is then folded over and cut into thin strips about 4 inches in length. Each strip is rolled into a rope and twisted. The twisted ropes are then formed into tight spirals. The spirals are filled and then baked. Snails are finished in the same manner as snecks.
- Braids are made by rolling out laminated dough and covering it with a spread. The dough is then cut into strips 12–15 inches in length. The strips are braided and then proofed. The braids are egg washed and sometimes sprinkled with chopped nuts or course ground sugar and baked. After cooling, braids are glazed and cut into servings.

Pâte à Choux. *Pâte à choux,* also known as éclair paste, is a pastry made by beating eggs into a paste of boiled water, butter, and flour. Pâte à choux is used to make cream puffs, éclairs, and profiteroles. A *cream puff* is a round pastry filled with sweetened whipped cream. An *éclair* is an oblong pastry filled with pastry cream and topped with a chocolate or fondant glaze. **See Figure 9-66.** A *profiterole* is a round pastry filled with ice cream or savory fillings.

Éclairs

Figure 9-66. Éclairs are oblong pastries filled with pastry cream and topped with a chocolate or fondant glaze.

Procedure for Preparing Pâte à Choux

1. Combine water, butter, and salt and bring to a boil. *Note:* The liquid needs to reach a rapid boil in order for the fat to disperse properly.
2. Add the flour and stir until the paste pulls away from the sides of the pan and forms a ball.
3. Allow the paste to rest and cool for a few minutes.

4. Add the eggs one at a time, vigorously beating after each addition. Stop adding eggs when the dough begins to fall away from the mixing paddle. *Note:* The eggs may not all be used, depending upon the moisture content of the flour and the size of the eggs.

5. Fill a pastry bag with the pâte à choux and pipe it into desired shapes on parchment-lined pans.
6. Bake the paste until it is completely dry.

7. Let the pâte à choux cool. Fill with desired filling and serve.

Napoleons. A classical Napoleon is a three-layered pastry filled with pastry cream, covered with fondant icing, and decorated with chocolate. However, there are countless versions of Napoleons for both sweet and savory applications. **See Figure 9-67.** The classical version is made by prebaking a sheet of puff pastry that has been weighed down with a few sheet pans to keep the dough from expanding. After the puff pastry has baked until it is crisp, it is cooled and cut into thirds lengthwise. Each strip should be about 3–4 inches wide.

A thick layer of vanilla pastry cream or diplomat cream is piped onto the first layer and topped with a second layer of puff pastry. The second layer is covered with pastry cream and a third layer is used as the top. The top is covered with a thin layer of warm fondant icing. Before the icing has set, tempered chocolate lines are piped along the strip about ⅜–½ inch apart. A paring knife is quickly drawn across the chocolate at an angle and then in the opposite direction to pull the chocolate lines through the icing to give the top layer a traditional Napoleon appearance.

Napoleons

Charlie Trotter's

Figure 9-67. Many versions of Napoleons are used in both sweet and savory applications.

Custards and Creams

Custards and creams come in many varieties. Custards and creams include baked custards, soufflés, crèmes anglaises, pastry creams, sabayons, mousses, Bavarian creams, and chiffons.

Baked custards, also known as baked puddings, are most often baked in a water bath at a temperature between 275°F and 300°F until the custard is set, but not completely cooked. Custard continues to cook after it has been removed from the oven. A water bath is used to prevent custard from drying out during cooking. Baked custard is generally a solid that retains the shape of its container and is firm enough to slice. Another method used to prepare baked custard is to cook it in a steamer. Baked custards include crèmes brûlées, crèmes caramels, cheesecakes, bread puddings, and soufflés.

Crèmes Brûlées. *Crème brûlée* is a baked custard that has the texture of a stirred custard beneath a hardened sugar surface. **See Figure 9-68.** A sweet custard mixture is poured into an ovenproof, individually sized casserole dish and baked in a water bath. After it has set, it is cooled completely. For service, crème brûlée is topped with granulated sugar, and the sugar is caramelized with a torch to produce a thin, glasslike candy covering. Crème brûlée can be made savory with the addition of savory ingredients.

Crèmes Brûlées

National Honey Board

Figure 9-68. Crème brûlée is a baked custard that has the texture of a stirred custard beneath a hardened sugar surface.

Crèmes Caramels. *Crème caramel,* also known as flan, is a baked custard served upside down and topped with hot, caramelized sugar. To make crème caramel, a small amount of hot caramelized sugar is poured into the bottom of a serving dish. An uncooked custard is added, and then the dish is baked in a water bath and removed when set. To serve, crème caramel is unmolded and served upside down so that the caramel sauce runs down over the baked custard. **See Figure 9-69.**

Crèmes Caramels

Figure 9-69. Crème caramel is unmolded and served upside down so that the caramel sauce runs down over the baked custard.

Cheesecakes. *Cheesecake* is a variety of baked custard that typically has a graham cracker or cookie crust. Cheesecake is made in a springform pan. There are two basic varieties of cheesecake. Italian cheesecake is light, fluffy, and creamy. New York cheesecake is more dense and rich than Italian cheesecake. **See Figure 9-70.**

Cheesecakes

Planet Hollywood International, Inc.

Italian

Wisconsin Milk Marketing Board, Inc.

New York

Figure 9-70. The two basic varieties of cheesecake are Italian cheesecake and New York cheesecake.

Bread Puddings. *Bread pudding* is a baked custard made by pouring a custard mixture over chunks of bread and baking it in an oven. **See Figure 9-71.** Sweet bread pudding can be flavored with vanilla, cinnamon, nutmeg, or other spices and baked. Savory bread pudding is made without sugar and flavored with herbs such as chives, sage, or tarragon.

Bread Puddings

Figure 9-71. A bread pudding is made by pouring a custard mixture over chunks of bread and baking it in an oven.

Soufflés. A *soufflé* is an airy and fragile custard made from a thick egg yolk sauce that has been lightened by stiffly beaten egg whites and then baked. Soufflés can be sweet or savory and served hot or cold. A sweet soufflé is prepared by using a cooked custard base (pastry cream) combined with flavors such as chocolate, vanilla, or fruit. This base is then lightened with egg whites that have been whipped with sugar and carefully folded in.

Production Tip

Egg whites whip better when not chilled, and having them at the same (room) temperature as the custard base allows the base and the whites to mix more easily.

The soufflé is carefully poured into a ramekin that has been buttered and coated with granulated sugar all the way to the rim. A properly baked soufflé rises high above the rim of the ramekin and has a golden-brown surface. **See Figure 9-72.** If any part of the interior or rim of the ramekin is not buttered and sugared, the soufflé will stick and not rise properly.

Soufflés

Daniel NYC / T. Schauer

Figure 9-72. A properly baked soufflé rises high above the rim of the ramekin and has a golden brown surface.

As a soufflé bakes, the air expansion in the egg foam causes the custard to rise as it is heated. However, soufflés are not stable like cakes and easily collapse once removed from the oven. For this reason, soufflés are made to order and served immediately.

Procedure for Preparing Soufflés

1. Prepare ramekins in advance by thoroughly buttering the interior and rim of each dish and then thoroughly coating the buttered surface in granulated sugar. *Note:* If any part of the ramekin is not completely buttered and sugared, it will prevent the soufflé from rising.
2. Mix the desired flavoring base into the prepared pastry cream.
3. Whip the egg whites. When soft peaks have formed, begin slowly adding sugar while continually whipping until all the sugar has been added and stiff peaks have formed.
4. Carefully fold one-third of the whipped egg whites into the soufflé base and mix well to incorporate and lighten the base. Continue to gently fold in the remaining whipped egg whites in thirds to prevent a loss of volume.
5. Pour the soufflé mixture into the prepared ramekins within ¼ inch from the top and bake.
6. Remove soufflés when they have risen well above the surface of the ramekin and turned golden brown. Serve immediately.

Crèmes Anglaises. *Crème anglaise* is a pourable vanilla custard sauce. Half of the sugar is whisked into already whisked egg yolks. The other half of the sugar is whisked into scalded milk or cream. The scalded milk and sugar mixture is then whisked rapidly into the sugar and yolk mixture in three intervals in order to temper the yolks and prevent curdling. The sauce is then put back on the stove and heated until thick and smooth. Crème anglaise is fully cooked when it reaches 175°F and should be a nappe consistency, or able to smoothly coat the back of a spoon. Crème anglaise is well known for its use as a sauce and as a base for desserts such as ice cream and Bavarian cream. Crème anglaise also freezes well.

Production Tip

Crème anglaise should be made in a nonreactive saucepan or stainless steel double boiler to prevent the eggs from curdling.

Procedure for Preparing Crème Anglaise

1. Place milk (or cream or half and half, as specified), half of the granulated sugar, and salt in a nonreactive saucepan and bring just to a scald to completely dissolve the sugar. *Note:* If vanilla bean is used, it can be added to the milk in this step. If an extract is used, it can be added in step 6.
2. In a stainless steel mixing bowl, whisk egg yolks and the remainder of granulated sugar until light and airy.
3. Temper the yolks by adding about one-third of the hot milk mixture to the yolk and sugar mixture while whisking rapidly.
4. When milk and yolks are mixed well, add the remainder of the hot milk mixture to the yolks while whisking.
5. Pour the complete mixture back into the saucepan and continue to cook while whisking until the mixture reaches 175°F or coats the back of a spoon.
6. Add vanilla extract (or another type of extract for additional flavoring).
7. Strain the sauce into a bain-marie to remove any lumps and immediately stir over an ice bath until cooled.
8. Refrigerate crème anglaise until needed.

Wisconsin Milk Marketing Board, Inc.

Pastry Creams. A *pastry cream,* also known as crème pâtissière, is a thick cornstarch- or flour-based egg custard made with egg yolks, sugar, and milk. A properly prepared pastry cream will have the consistency and appearance of a smooth pudding. Unlike crème anglaise, pastry cream can be brought to a boil and simmered because the starch protects the egg yolks from curdling. Boiling allows the starch to fully gelatinize and removes any raw starch flavor from the finished product. Pastry cream is commonly used to fill a tart crust and then top it with sliced fresh fruit. Pastry cream is also used to fill napoleons, cream puffs, éclairs, and many other sweet pastries. Pastry cream is often flavored with chocolate, mocha, or nut pastes such as hazelnut, praline, or pistachio.

Procedure for Preparing Pastry Cream

1. Place three-fourths of the required milk, half of the required sugar, and the salt in a stainless steel saucepan and bring to a scald in order to completely dissolve the sugar.
2. Place yolks and half of the sugar in the mixing machine. Using the whip attachment, whip on high speed until the mixture thickens and turns pale yellow in color.
3. Whisk flour or cornstarch into yolk mixture and mix well.
4. Add remaining one-fourth of the required milk and mix well until smooth.
5. Pour one-third of the hot milk mixture into the yolk mixture while whisking continuously to temper the yolks.
6. Add the tempered yolk mixture back to the saucepan of milk while stirring rapidly. Return to the heat while stirring constantly.
7. Bring mixture to a boil while whisking vigorously. Continue to whisk for 1 minute. *Note:* As the mixture reaches the boiling point, the continuous whisking will break apart lumps and the pastry cream will appear smooth.
8. Remove the cooked pastry cream from the heat and place in a stainless steel bain-marie or mixing bowl.
9. Add softened butter to the hot cream and mix on high speed until the mixture reaches room temperature.
10. Cover the entire surface of the pastry cream with plastic wrap and refrigerate until needed.

Sabayons. A *sabayon,* also known as a zabaglione, is a light and foamy custard prepared by whisking sugar and egg yolks over a double boiler into a silky foam. **See Figure 9-73.** It is made in a similar manner to stirred custards. Champagne or a high-quality sweet wine, such as Marsala, is commonly added to increase the flavor of the sauce. Heating the eggs while whisking cooks them and makes the resulting foam stable enough to retain a shape. Sabayon is commonly served as a sauce and can be served warm or cold. A classic use for sabayon is over a dish of fresh berries, because the rich, sweet sauce is balanced by the acidic quality of the berries.

Sabayons

All-Clad Metalcrafters

Figure 9-73. A sabayon is a light and foamy custard.

Procedure for Preparing Sabayon

1. Place sugar, yolks, salt, and champagne or Marsala wine in a mixer and whisk until lighter in color and thick.
2. Remove bowl from mixer and place over a pot of simmering water. Whisk continuously until the mixture is thick, is lighter in color, forms a ribbon, and reaches a temperature of 175°F.
3. Serve while warm, or return the bowl to the mixer and whip until completely cooled.
4. When cool, remove the sabayon and place it in a nonreactive bowl covered with plastic wrap pressed over the surface to prevent a skin from forming.

Mousses. A *mousse* is a rich and airy foam that has been lightened with whipped cream, whipped pasteurized egg whites, or both. **See Figure 9-74.** A mousse contains only a small amount of gelatin, if any at all. A mousse can be served as a dessert and can also be used as a cake filling.

Mousses

Cape Cod Cranberry Growers' Association

Cranberry Mousse

Daniel NYC / P. Medilek

Frozen Chocolate Mousse

Figure 9-74. A mousse is a rich and airy foam that has been lightened with whipped cream, whipped pasteurized egg whites, or both.

Bavarian Creams. *Bavarian cream* is a light dessert made with a flavored custard sauce that has been stabilized with gelatin while it is still warm. After it has almost completely cooled, whipped cream is folded in. Bavarian creams can be flavored with chocolate or fruit purées. The mixture is then placed in a mold to set. Once a Bavarian cream has set up completely, the mold is dipped into warm water for a few seconds and then inverted on to a plate. The mold can then be removed and the dessert served.

Chiffons. A *chiffon* is an airy custard sauce most often used as a pie filling. A chiffon is made by folding whipped egg whites into a cooled custard sauce. A chiffon can be placed in a mold, but is most often used as a filling, as in the case of lemon chiffon pie.

Fancy Pastries

Fancy pastries are best described as composed individual pastries that are garnished with rich fillings and can be eaten in two or three bites. **See Figure 9-75.** An assortment of fancy pastries is usually served. The different colors, decorations, and shapes of the pastries make a beautiful display. Examples of fancy pastries include the following:

- small butter cream roulades individually enrobed in ganache, striped with chocolate, and decorated with buttercream
- multilayered cake slices, such as opera torte made with an almond sponge cake, coffee butter cream, and ganache icing
- round, nut, meringue-layered pastries called Japonaise, and the Swiss specialty, zuger kirschtorte, which is layered with buttercream encasing a genoise flavored with cherry brandy
- individual Bavarian-cream-filled tortes encased in decorative stenciled cake wraps known as jaconde sponge, which are often decorated with small chocolate plaques or filigree
- small boat-shaped tarts (barquettes) filled with whipped gianduja ganache, enrobed with chocolate, and decorated with chocolate-dipped hazelnuts
- small, marzipan-enrobed petits fours decorated with marzipan rosebuds
- small, filled, dome-shaped mousse desserts with a lightly colored fruit glaze

Fancy Pastries

Edward Don & Company

Figure 9-75. Fancy pastries are garnished pastries with rich fillings that can be eaten in two or three bites.

CHECKPOINT 9-10

1. Explain how the number of turns used to make laminated dough affects the flakiness of a pastry.
2. Explain the purpose of phyllo dough.
3. Describe three shapes used to make Danishes.
4. Prepare a type of Danish. Evaluate the appearance, texture, color, and flavor.
5. Prepare a puff pastry and pâte à choux. Evaluate each based on appearance, texture, color, and flavor.
6. Prepare a classical Napoleon. Evaluate the appearance, texture, color, and flavor.
7. Describe five types of baked custards.
8. Prepare two baked custards. Evaluate each custard for appearance, texture, color, and flavor.
9. Prepare a soufflé. Evaluate the soufflé for appearance, texture, color, and flavor.
10. Differentiate between crème anglaise and pastry cream.
11. Prepare a pastry cream. Evaluate the appearance, texture, color, and flavor.
12. Contrast a sabayon and a mousse.
13. Prepare a mousse. Evaluate the appearance, texture, color, and flavor.
14. Differentiate between a Bavarian cream and a chiffon.
15. Prepare a tray of fancy pastries. Evaluate the appearance, texture, color, and flavor of each pastry.

FROZEN DESSERTS

Frozen desserts are specialty desserts that may be served alongside or on top of other desserts or eaten as a separate dish. Frozen desserts include ice creams, fried ice cream, bombes, baked Alaska, gelati, sorbets, sherbets, and granités.

Ice Creams

Ice cream can be the main attraction for a dessert course or can be an accompaniment, as with pie à la mode. **See Figure 9-76.** *Ice cream* is a frozen dessert made from cream, butterfat, sugar, and sometimes eggs. There are two general types of ice cream bases. The first incorporates eggs to make a crème anglaise base that produces a rich, creamy dessert. The second is a simple mixture of milk, cream, sugar, and flavorings that is heated to dissolve the sugar and then chilled prior to churning. An ice cream base must be refrigerated for at least a few hours to allow excess moisture to be absorbed and to bind to the sugar.

Pie à la Mode

Planet Hollywood International, Inc.

Figure 9-76. Ice cream can be the main attraction for a dessert course or can be an accompaniment, as with pie à la mode.

The proper balance of ingredients forms the overall texture and flavor that makes ice cream so popular. For example, eggs add richness to ice cream but also act as an emulsifier. Excess moisture is bound by the lecithin in egg yolks, which helps to prevent ice crystals from forming on the surface and throughout the ice cream. Both the cream and eggs allow air to be whipped into the product during the churning process, making the end product smoother and creamier. In addition to adding sweetness, sugar also helps to lower the freezing point of an ice cream base, which keeps it from freezing rock-hard. However, too much sugar can prevent ice cream from freezing, and too little sugar results in a product that is difficult to scoop.

Ice cream increases in volume due to overrun. *Overrun* is an increase in the volume of a frozen product as a result of the incorporation of air during churning and freezing. An overrun of 100% means that an ice cream doubled in size during churning and freezing and contains 50% air by volume. The FDA allows a maximum of 100% overrun, although the best ice creams have less. Too much overrun makes ice cream feel frothy and thin in the mouth, rather than creamy and smooth.

Fried Ice Cream

Fried ice cream is a dessert that is made by dipping rock solid frozen ice cream balls into a light batter and then deep-frying them in very hot oil. The batter used is somewhat like a tempura batter that is sweetened and sometimes flavored with cinnamon. Fried ice cream is served dusted with confectioners' sugar.

Bombes

A *bombe*, also known as a glacée, is an ice cream dessert with at least two varieties of ice cream layered inside a spherical mold that has been frozen and then unmolded and decorated for service. **See Figure 9-77.** A frozen, unmolded bombe can be dipped in tempered chocolate to create a surface shell.

Bombes

Irinox USA

Figure 9-77. A bombe consists of at least two varieties of ice cream layered inside a spherical mold that has been frozen and then unmolded and decorated for service.

Bombes are traditionally made with different layers comprised of a number of frozen dessert creams, such as ice cream, parfait glacé, sorbet, and frozen mousse. To make a bombe, the mold is filled with a layer of soft ice cream and then frozen while the second filling is being made. The second filling is then piped into the hollow center of the frozen ice cream, a thin layer of sponge cake or a baked disk of sweet short dough is placed on top of the mold, and the sphere is refrozen.

To finish the bombe, the mold is heated slightly by dipping it into warm water, inverted onto a plate or cardboard circle, and then removed. The bombe is then decorated with sweetened whipped cream and garnished with chocolate decorations, candied fruit, or sugar decorations. The decorated bombe is placed back into the freezer for a short time to firm up and then the frozen dessert is presented to guests, sliced into portions, and served.

Frozen Desserts

Baked Alaska

Baked Alaska is a composed frozen dessert made with a sponge cake soaked in a liqueur-flavored simple syrup that is topped with ice cream and covered with an Italian meringue. Baked Alaska can be browned in a hot oven, browned with a propane torch, or flambéed at the table. **See Figure 9-78.**

Baked Alaska

Daniel NYC / B. Milne

Figure 9-78. Baked Alaska is browned in a hot oven or flambéed at service.

Gelati

A *gelato* is an Italian ice cream that has a creamier, denser texture than standard ice creams. **See Figure 9-79.** Gelato is churned for a shorter period of time, incorporating less air and resulting in a denser product.

Gelati

National Honey Board

Figure 9-79. A gelato has a creamier, denser texture than a standard ice cream.

Sorbets

A *sorbet* is a frozen dessert made by combining puréed fresh fruit and simple syrup. **See Figure 9-80.** It is churned in an ice cream maker in the same manner as ice cream. A sorbet is naturally lower in calories and fat than ice cream because of the absence of heavy cream and egg yolks. Sorbet is traditionally served before the main course at formal dinners to cleanse the palate. Pear, watermelon, blackberry, and raspberry are popular flavors of sorbets.

Sorbets

Vita-Mix® Corporation

Figure 9-80. A sorbet is made by combining puréed fresh fruit and simple syrup.

Sherbets

A *sherbet* is a frozen dessert made by combining puréed fresh fruit, simple syrup, and milk. **See Figure 9-81.** A sherbet is similar to a sorbet and originated from an ancient Persian, fruit-flavored, frozen dessert called sharbat. Along with the change in name came a slight change to the recipe. Milk is now added during the churning process, making sherbets creamier than sorbets. Sherbets are lower in fat than ice cream but, because of the presence of milk, still have a creamy texture compared to the icy texture of sorbet.

Sherbets

National Cancer Institute
Renee Comet (photographer)

Figure 9-81. A sherbet is made by combining puréed fresh fruit, simple syrup, and milk.

Granités

A *granité,* also known as a granita, is a frozen dessert made by frequently stirring a mixture of water, sugar, and flavorings, such as fruit juice or wine, as it is freezing. A granité has a grainy texture and ice crystals that are larger than those of a sorbet. **See Figure 9-82.** A granité is often used to cleanse the palate between courses. It is also a refreshing dessert.

Granités

National Honey Board

Figure 9-82. A granité has a grainy texture and ice crystals that are larger than those of a sorbet.

CHECKPOINT 9-11

1. Contrast eight types of frozen desserts.
2. Prepare an ice cream base. Evaluate the base for appearance, color, texture, and flavor.
3. Differentiate between a gelato, a sorbet, and a sherbet.
4. Prepare two different types of frozen desserts. Evaluate each dessert based on appearance, texture, color, and flavor.

SPECIAL DESSERTS AND SHOWPIECES

Desserts come in many shapes, sizes, and flavors. Dessert crêpes and fruit-based desserts offer a multitude of combinations that can be tailored to any menu. Likewise, pastillage, marzipan, nougats, and chocolates as well as cast, spun, pulled, and blown sugar pieces can be formed into spectacular showpieces.

Dessert Crêpes

Dessert crêpes are often served both hot and cold. As a hot dessert, they are most well-known for being used in crêpes Suzette. In this dessert, the crêpes are folded in half and then in half again to form a triangle. A sweet orange-flavored butter sauce is then prepared, and the folded crêpes are added to the sauce and flambéed with strong alcohol such as 151 rum.

Crêpes are served as a hot dessert, folded and served with other sauces, or stuffed with fruit or other fillings, such as a sweetened lemon ricotta cheese. **See Figure 9-83.** Crêpes are also served hot when filled with a soufflé mixture, baked in the oven, and served immediately before they deflate. Crêpes are served cold when filled with chocolate mousse or other cold dessert fillings.

Fruit-Based Desserts

Due to their many varieties and exotic flavors, fruits have always been a favorite in desserts and pastries. **See Figure 9-84.** Fruit desserts can be simple, such as a bowl of fresh berries or poached fruit served cold with ice cream. Fruit desserts can also be served hot and are sometimes prepared or flambéed tableside.

Fresh berries, compotes, coupes, and flaming dishes are popular fruit-based desserts.

- Fresh berries are often tossed with sugar or served with a sabayon. For example, strawberries Romanoff is tossed with sugar and an orange liqueur such as Grand Marnier or Curacao.
- Fruit compotes are served cold and usually consist of poached fruit in a simple syrup, sometimes flavored with exotic spices and fortified liqueurs.
- Cold fruit served with ice cream is called a coupe. Peach Melba is served with raspberry sauce and ice cream, and pear belle Hélène is served with chocolate sauce and vanilla ice cream.
- Hot fruit desserts such as bananas Foster are often prepared for display tableside. Bananas Foster is served with a sauce made of brown sugar, orange juice, banana liqueur, rum, and butter. Hot fruit desserts such as baked apples stuffed with brown sugar and nuts are often wrapped in puff pastry.

Dessert Crêpes

Edward Don & Company

Crêpe Iron

Paderno World Cuisine

Crêpe Pan

Frieda's Specialty Produce

Plated Crêpes

Figure 9-83. Crêpes can be served as a hot dessert, folded and served with other sauces, or stuffed with fruit or other fillings such as soft cheeses.

Fruit-Based Desserts

Wisconsin Milk Marketing Board, Inc.

Berries and Ice Cream

Daniel NYC / T. Schauer

Strawberry Tart

Carlisle FoodService Products

Layered Strawberry Creme Cake

Figure 9-84. Fruit has always been a favorite in desserts and pastries.

Pastillage and Gum Paste

Pastillage, also known as sugar paste, is a malleable white paste that dries to a matte finish and is used to make decorative forms. Pastillage is made by mixing confectioners' sugar, water, gelatin, a small amount of acid such as lemon juice, and sometimes cornstarch. The paste is kneaded, rolled into sheets, cut, and then dried. Once the pastillage dries, it sets up hard enough for any rough spots to be sanded. Pastillage can also be shaped and colored or painted to make decorative forms for use on showpieces.

Working with pastillage takes skill and practice due to its fragile nature. Every ripple, smear, or flaw shows on dried pastillage. Dried pastillage can be carefully glued together with royal icing. Due to its matte finish, pastillage is often used as a contrast to shiny pulled-sugar decorations.

Like pastillage, gum paste is also used to make decorative forms for use on fancy cakes and showpieces. *Gum paste* is a very soft and extremely malleable white paste that dries to a porcelain-like finish and is used to make delicate decorative forms. Gum paste contains gum tragacanth instead of gelatin. Gum tragacanth absorbs 15–20 times its weight in water, making gum paste softer and malleable for a longer period than pastillage, allowing more time to shape it into different forms. Gum paste is often used to make paper-thin flower petals. **See Figure 9-85.**

Gum Paste Flowers

Figure 9-85. Gum paste is often used to make paper-thin flower petals.

Marzipan

Marzipan is a candy made from almond paste and sugar or corn syrup. Marzipan can be colored and is often used as a decoration for cakes and pastries. Marzipan is sometimes mixed with a liqueur, such as Curacao, and dipped into chocolate. Marzipan is also used to enrobe cakes and pastries. Marzipan that is shaped into flowers or figurines can be used in showpieces.

Nougats

A *nougat* is a candy made from caramelized sugar and often contains sliced nuts or fruit. **See Figure 9-86.** When and how the caramelized sugar and nuts are mixed and rolled out varies by country. Nougats used for decoration and to build showpieces in France are called nougatines. The French blend this same mixture to form a paste called praline. In Germany, the caramelized sugar and nut mixture is called croquant and is often coarsely crushed and used in pastries and chocolate candies. When this caramelized sugar and nut mix is made into a paste and mixed with chocolate, it is called chocolate nougat. The Italians and the French call this same mixture gianduja and use it as a center for chocolate candies.

Nougats

Figure 9-86. Nougats are made from caramelized sugar and often contain sliced nuts or fruit.

Chocolate

Chocolate is made into a wide variety of candies, pastries, desserts, and showpieces. **See Figure 9-87.** In the hands of a skilled chocolatier or pastry chef, chocolate can be made into very elaborate showpieces as well as decorations for cakes, pastries, and specialty desserts. To be successful in using chocolate for these purposes, a proficiency in tempering (melting) chocolate to the correct temperature is required. Tempering is necessary due to the complex fats that chocolate contains. If chocolate is not tempered correctly, it will not set properly and instead appear grey and streaked, which is referred to as bloomed. Once chocolate is tempered, several different techniques can be used to present it.

Chocolate Desserts

Sullivan University

Chocolate Candies

Irinox USA

Chocolate Torte

Daniel NYC / E. Laignel

Chocolate Coffee Tart

Daniel NYC / T. Schauer

Chocolate Clafoutis

Figure 9-87. Chocolate is made into a wide variety of candies, pastries, desserts, and showpieces.

Molded Chocolates. Chocolate can be poured into special plastic molds of many shapes and sizes. Some common shapes include half spheres, half eggs, and nut shells. After the molds are filled, they are placed in the refrigerator until the chocolate has set, and then the chocolate is removed from the molds. The type of plastic used in the molds yields a very shiny finish to the chocolate. Molded chocolates can be used to create chocolate sculptures.

Piped Chocolates. Chocolate can be piped into fine-line decorations called filigrees. The chocolate is piped onto parchment paper or translucent acetate sheets. After the filigrees set, they are carefully removed with a palate knife and placed on a cake, pastry, or specialty dessert.

Chocolate Plaques. Thin chocolate wafers are often made with white, milk, and dark chocolate or colored cocoa butter for use as decorations. The tempered chocolate is poured onto an acetate sheet and then spread into a thin layer with an offset palate knife. When the chocolate has almost set, it is cut into shapes and then turned over onto a parchment-lined sheet pan and refrigerated until firmly set. The acetate sheet is then removed and the pieces are placed on the cakes, pastries, specialty desserts, or used on showpieces. Special premade transfer sheets of plaques can be embossed with special patterns using colored cocoa butter.

Chocolate Ribbons. Tempered chocolate is often made into ribbons. To make chocolate ribbons, acetate tape is cut into 8–12 inch strips and one end of each strip is attached to a sheet pan. The tape strips are then coated with tempered chocolate. When the chocolate has almost set, the loose end of the tape is formed into a loop and then taped onto the sheet pan and refrigerated. Once the loops completely set, the acetate is removed, yielding chocolate ribbons.

Chocolate Candies

The production of chocolate candies is a specialized field practiced by a chocolatier. **See Figure 9-88.** Specialized tools and equipment are used to efficiently make delicious candies. Chocolate melters, enrobing machines, guitars (steel wire cutters), and dipping forks are all used to make chocolate candies. Chocolate candies can be hand-formed, molded, or enrobed.

Truffles are often hand-formed by shaping ganache cream into balls and then dipping them into tempered chocolate. Special plastic molds are lined with tempered chocolate, piped with ganache or another liquid filling, and then sealed with more chocolate to create chocolate bon bons. Chocolate caramels and giandujas are enrobed candies. Chocolate caramels are made by pouring liquid caramel onto a marble slab between steel square caramel rods. After the liquid has set, it is cut into individual pieces with a wet knife, and then a chocolate fork is used to dip each piece into tempered chocolate.

Chocolate Candies

Irinox USA

Figure 9-88. The production of chocolate candies is a specialized field that is practiced by a chocolatier.

Chocolate marzipan candies are also enrobed, but they are made with a different technique than chocolate caramels and giandujas. Flavored marzipan is rolled out, cut into shapes with a knife or circle cutter, and then each piece is hand-dipped in tempered chocolate. Marzipan candies are often garnished with a toasted nut.

Sugar Sculptures

Sugar is the most complex medium used by pastry chefs. The color, shine, and glasslike appearance of spun, cast, pulled, and blown sugar pieces are created from a cooked sugar-and-water solution. **See Figure 9-89.** The formula for a sugar-and-water solution can vary slightly, but it usually contains an acid such as cream of tartar or vinegar along with an inverted sugar like glucose. The acid and the glucose help keep the cooked sugar solutions from crystallizing. A product that many pastry chefs use instead of granulated sugar is isomalt, which does not crystallize because of its chemical makeup.

Working with Sugar

Spinning Sugar

Sugar Sculpture

Figure 9-89. Spun, cast, pulled, and blown sugar pieces are created from a cooked sugar and water solution.

Production Tip

Extra precautions are called for when working with hot sugar. Sugar syrup reaches temperatures in excess of 300°F and sticks to the skin easily.

Care must be taken in cooking the sugar solution to specific temperatures as well as when manipulating the sugar after it is cooked. Working with sugar can cause serious burns. The correct amount of color and combination of colors contribute to the appearance of finished sugar pieces. Spun and cast sugar are easier to produce than pulled or blown sugar.

Spun Sugar. To spin sugar, a sugar solution is cooked to the required temperature and then the pot is dipped into cold water to stop the sugar from continuing to cook. A rakelike tool is dipped into the sugar solution and then quickly waved over two dowel rods attached to the back of a 4 inch hotel pan to catch the sugar wisps. This is done repeatedly until the sugar is too cold to use.

Cast Sugar. To cast sugar, a sugar solution is cooked to the required temperature and then color is added. The sugar is then poured into premade forms or frames. After the sugar has cooled and hardened, the frames are removed and a marbleized sugar base has been formed.

Pulled Sugar. To pull sugar, the cooked solution, which is often colored, is poured onto a marble slab or a silicone mat. The sugar is then turned back onto itself until it is cool enough to pick up. Rubber gloves should be worn because the sugar is still very hot at this stage. The sugar is pulled and folded back onto itself a number of times. This process improves the luster and sheen of the finished sugar pieces. Once the sugar is pulled, it is formed into a design. Since pulled sugar cools quickly at this point, the sugar is held under a heat lamp while the design is finished. This keeps the sugar malleable while being worked. Pulled sugar is often used to make flowers and multicolored ribbons to decorate showpieces.

Blown Sugar. Blown sugar is started in the same way as pulled sugar. Once the sugar is pulled and folded onto itself to achieve the proper luster, a small piece is removed, formed into a ball, and placed on a hollow tip attached to a handheld air pump. The air pump is used to blow air into the ball, causing it to expand. While the ball is expanding, it is formed into the desired shape. This technique is typically used to make balls and bird-shaped pieces. In the bird pieces, the wings are adhered to the body using pulled sugar. Items such as eyes are painted on. Blown sugar may be sprayed different colors.

Regardless of the medium used, sugar showpieces need to be free of imperfections. Chocolate should be tempered correctly and sugar must be shiny and thin with good luster. Skill and practice lead to success when working with chocolate and sugar.

Plating Desserts

Desserts are plated for individual presentations such as a dessert course as well as for a buffet. The three plating styles often used for desserts are based on the center of the plate, duos or trios, and vertical building. Each style of plating is best served on white plates without borders that detract from the beauty of the dessert presentation.

Center of the Plate. In the center-of-the-plate method, the dessert is plated in the center of the plate and any sauce or garnish flows from or around the center to create the focal point of the presentation. **See Figure 9-90.** Round plates or trays are normally used for this style of plating.

Center-Plated Desserts

Daniel NYC / S. Bokalders

Chocolate Soufflé

Wisconsin Milk Marketing Board, Inc.

Blue Cheese Cheesecake

Figure 9-90. Desserts are often placed in the center of the plate and have garnishes surrounding the main item.

Duos and Trios. The concept of plating duos and trios is typically used when there is more than one dessert served on the plate. **See Figure 9-91.** The portions are smaller and look better when presented on rectangular plates in a clean line. All of the desserts need to be relatively the same size but garnished differently. Individual desserts can also be offered on smaller plates or in small glasses or ramekins and then placed in duos or trios on a larger plate.

Plated Dessert Duos and Trios

Daniel NYC / M. Duisterhof

Mousse Duos

Daniel NYC / T. Schauer

Fig Trios

Figure 9-91. The concept of plating duos and trios is typically used when there is more than one dessert served on a plate.

Vertical Building. Vertical building refers to the stacking of components and garnishes that make up the total plate presentation. **See Figure 9-92.** This style of plating adds height to the appearance of a plated dessert.

Additionally, when the dessert is cut into bites by the guest, all of the components of the dessert are eaten together in each bite. For successful vertical building, heavier components must be used as the base, followed by layers of lighter components, and topped off with delicate garnishes.

Vertically Plated Dessert

Daniel NYC / B. Milne

Figure 9-92. Vertical building refers to the stacking of components and garnishes that make up the total plate presentation.

CHECKPOINT 9-12

1. Prepare and plate dessert crêpes. Evaluate the appearance, texture, color, and flavor.
2. Describe four types of fruit-based desserts.
3. Explain how pastillage is made.
4. Differentiate between marzipan and nougats.
5. Describe four ways to shape tempered chocolate.
6. Name three ways to make chocolate candies.
7. Temper chocolate and form a candy.
8. Contrast chocolate caramels and chocolate marzipan candies.
9. Describe four sugar techniques used to build showpieces.
10. Describe three styles used to plate desserts.
11. Plate a cold dessert. Evaluate the plate for appearance, texture, color, and flavor.
12. Plate a hot dessert. Evaluate the plate for appearance, texture, color, and flavor.

Refer to DVD for **Quick Quiz®** questions

SUSTAINABILITY CORNER

Purchasing, using, and maintaining equipment is essential in the foodservice industry. Purchasing equipment bearing the ENERGY STAR® logo is considered a sustainable best practice that can result in significant savings over the long term while contributing to the preservation of natural resources.

Equipment with the ENERGY STAR® logo ensures a product has met a specific set of criteria and has been found to be energy efficient. For example, commercial ice machines with an ENERGY STAR® rating are, on average, 15% more energy efficient and 10% more water efficient than standard models. To determine efficiency, an equipment life cycle calculator, found at websites such as fishnick.com, can be used to estimate energy, water, and cost savings for various types of equipment used in the foodservice industry.

The type of equipment used in professional kitchens can also have an impact on energy efficiency. For example, ovens are more efficient than rotisseries, and griddles are more efficient than broilers. Examining which types of equipment are most efficient may result in a need to change cooking methods. However, this can be an opportunity to broaden culinary skills while saving money. Using equipment to save money and energy can further be achieved by securely closing doors, cooking with a full load, and cutting idle running time by turning equipment down or off during slow times. Also, investing in combination ovens that allow the user to boil, steam, poach, bake, grill, and fry in one unit may be an effective energy-efficient option.

To ensure that equipment is running at peak performance, proper maintenance is essential. All equipment should be inspected and cleaned according to an established schedule in order to promote longevity and efficiency. This includes tasks such as calibrating thermometers regularly, verifying equipment doors close tightly, and checking timers to ensure equipment is turned on or off at appropriate times.

Equipment Life Cycle Calculators

Baking & Pastry Station

Certification Exam Preparation Questions

1. A ___ is a specialized cooling unit that rapidly reduces the temperature of foods.
   A. reach-in unit
   B. walk-in unit
   C. blast chiller
   D. lowboy

2. A ___ is a mathematical way to represent the relationship between two or more numbers or quantities.
   A. recipe
   B. formula
   C. ratio
   D. scale

3. ___ stimulates the growth of yeast, supplies moisture, and helps prolong freshness.
   A. Flour
   B. Sugar
   C. Milk
   D. Flavoring

4. ___ is a rich, white, cooked icing that hardens when exposed to the air.
   A. Buttercream
   B. Whipped cream
   C. Fondant
   D. Ganache

5. A dough ___ is an electric appliance used to sheet, or roll out, doughs to the desired thickness.
   A. sheeter
   B. mixer
   C. marker
   D. tamis

6. A ___ meringue has a 2:1 ratio, 2 parts sugar to 1 part egg whites.
   A. soft
   B. hard
   C. Swiss
   D. French

7. ___ fillings are the most popular type of pie fillings.
   A. Cream
   B. Fruit
   C. Chiffon
   D. none of the above

8. Because different ingredients have different ___, two items of the same volume can have different weights.
   A. formulas
   B. densities
   C. thicknesses
   D. colors

9. A ___ dough is layered with butter and used to make pastries, such as croissants, turnovers, and strudels.
   A. straight
   B. laminated
   C. modified
   D. rolled-in

10. If chocolate is not ___ correctly, it will not set properly and instead appear grey and streaked, which is referred to as bloomed.
    A. tempered
    B. pulled
    C. molded
    D. piped

11. A ___ is any ingredient that causes a baked product to rise by the action of air, steam, chemicals, or yeast.
    A. thickening agent
    B. leavening agent
    C. flavoring
    D. sweetener

12. ___ is the protein in flour that, when combined with water, gives a baked product its structure.
    A. Arrowroot
    B. Gelatin
    C. Tapioca
    D. Gluten

13. ___ is a process that changes the molecular structure of oil into a solid.

A. Hydrogenation
B. Emulsification
C. Leavening
D. Thickening

14. A ___ crust is commonly used to make lattice tops and the bottom crust of a pie filled with a cooked filling.

A. long-flake
B. short-flake
C. crumb
D. mealy

15. ___, also known as sugar paste, is a malleable white paste that dries to a matte finish.

A. Sabayon
B. Fondant
C. Marzipan
D. Pastillage

16. ___ is a composed frozen dessert made with a sponge cake soaked in a liqueur-flavored simple syrup that is topped with ice cream and covered with an Italian meringue.

A. Baked Alaska
B. Crème anglaise
C. Pâte à choux
D. Puff pastry

17. A ___ is the weight of an ingredient expressed as a percentage of the weight of the main ingredient in a formula.

A. confectioners' ratio
B. confectioners' percentage
C. baker's ratio
D. baker's percentage

18. The ___ mixing method involves folding whipped egg whites into a batter made from flour, egg yolks, and fat.

A. chiffon
B. sponge
C. creaming
D. two-stage

19. When baking quick breads, pans should be filled only ___ full.

A. one-quarter
B. three-quarters
C. one-third
D. one-half

20. Yeast is very sensitive to temperature and, depending on the type of yeast, is activated between 70°F and ___.

A. 120°F
B. 130°F
C. 140°F
D. 150°F

21. Explain how to convert a recipe into a formula.

____

____

____

22. List the twelve steps of yeast dough production.

____

____

____

____

____

____

23. Describe the process for making a bombe.

____

____

____

____

24. Describe four methods used to mix cakes.

____

____

____

____

SKETCHING EXERCISE

25. Sketch a visually appealing plated dessert using a center-of-the-plate, vertical building, and duos or trios plating style.

Notes

Supervisory Development Station

Introduction

The tenth station in the ACFEF apprenticeship program involves the development of supervisory skills as well as menu planning and cost control skills. As an apprentice, you will be responsible for the planning and completion of a banquet function and training a new employee to work the line. You will attend an in-house managers' meeting with the chef and supervise food production. In addition, you will learn how to manage employee issues such as workers' compensation.

With an understanding of menu planning and cost control, you will be able to demonstrate a working knowledge of menu selection by suggesting menu items appropriate for your establishment and clientele. You will design a three-course menu that includes recipes and a complete listing of food costs and profit margins by item.

Sustainability through leadership involves providing staff members with a better understanding of the importance of sustainable practices. Sustainability initiatives include encouraging energy awareness during staff training, turning off lights, and implementing a power-up/power-down schedule. As a supervisor, you will lead by example to make your workplace as sustainable as possible.

NSF International

A foodservice operation is only as good as the people who operate it. Developing supervisory skills signifies a commitment to the industry and to an evolving career in food service. Supervising others is an important responsibility. Supervision requires a high level of interpersonal and technical skills. These include good people skills and a solid understanding of kitchen tasks. The ability to guide others as they perform assigned tasks in the professional kitchen is the first step toward a management role.

SUPERVISORY SKILLS

A *supervisor* is a person who oversees the work done by others and who has the responsibility of directing, training, and evaluating staff. A supervisor typically supervises a team of people, whereas a manager is in charge of several teams of people. A supervisor is responsible for the day-to-day performance of a small group. It may be a team, or a shift. The supervisor has experience in what the group does but is not necessarily better at it than those being supervised. A supervisor's job is to guide the team or shift toward its goals, see that all staff members are productive, and solve problems as they arise.

A supervisor usually does not have the power to hire, fire, or promote staff. A supervisor does, however, often have the authority to change the work roles of the staff, such as deciding which person will be assigned to a station. To be effective, a supervisor must have the interpersonal skills and the technical skills required to both supervise and perform the tasks of staff members. Foodservice supervisors need to be effective communicators, team leaders, time managers, and delegators.

Effective Communication

Foodservice operations are totally dependent upon communication and interactions among staff and guests. Effective communication skills are used to send and receive message instructions, share information, and provide feedback. **See Figure 10-1.** In order for communication to be successful, each message must be clearly understood. Effective communication requires listening, speaking, observing, and writing.

Figure 10-1. Effective communication skills are used to give and understand instructions, share information, and provide feedback.

Listening. An effective listener shows respect by allowing not only guests, but fellow staff to speak without being interrupted. Making eye contact with the speaker shows the person that the message being conveyed is important. Effective listeners also summarize what they hear in order to reinforce the message and allow for any required clarification.

Speaking. Supervisors should always speak in a clear and pleasant tone that is loud enough to be heard, yet not loud enough to disturb others. To develop speaking skills, individuals can practice by speaking slowly and properly enunciating each word so that they are easily understood. A supervisor's speaking tone should always be professional, respectful, and sound welcoming.

Observing. Observation is the ability to read the nonverbal cues of coworkers by taking note of body language. For example, an expeditor may signal that an order is ready by nodding. Likewise, a line cook may look flustered when orders are stacking up at the griddle station. Correctly interpreting and acting on nonverbal cues leads to positive working relationships among staff.

Writing. Written instructions must accurately convey information. Purchase orders, memos, résumés, and job applications are forms of written communication used in foodservice operations. It is also important to write legibly so that the message is not misunderstood.

Teamwork

Effective foodservice supervisors are team leaders who respect the talents of each team member. Teamwork among staff creates an environment that fosters mutual respect and collaboration. **See Figure 10-2.** Working in a cooperative manner creates an atmosphere in which staff show respect for one another. By treating each staff member equally, supervisors demonstrate how to work together toward common goals.

Teamwork

Figure 10-2. Teamwork among staff creates an environment that fosters mutual respect and collaboration.

Supervisors and fellow staff count on one another to be present, prepared, and on task in order to run a successful foodservice operation. Taking the initiative to perform tasks without being asked is a hallmark of teamwork. Recognizing a need and responding with appropriate actions, such as organizing a work station before being asked, demonstrates initiative. Effective supervisors recognize staff who demonstrate teamwork skills.

Time Management

Time management is the ability to use a range of skills, tools, and techniques to effectively accomplish tasks within a given amount of time. Time management involves the effective use of one's own time as well as planning for the efficient use of time for others. For example, time management skills are used to schedule the correct number of staff members for a shift.

Prioritizing Tasks. One way to ensure that tasks are performed in a timely manner is to prioritize the tasks. The tasks are arranged based on what needs to be done and when. Tasks that need to be accomplished in the short term or immediately are placed at the top of the list. Tasks that are not an immediate priority are placed lower on the list.

Scheduling Staff. Preparing effective schedules helps to ensure that the appropriate amount of staff is scheduled for each shift. **See Figure 10-3.** Scheduling fewer staff members than needed for a shift could result in poor service to guests. Scheduling too many staff members could result in the staff not being motivated because there is not enough work to keep everyone busy. Having too many staff members scheduled also increases the labor expenses, which cut into the profit of the operation.

Staffing requirements can change from one day to the next. In order to develop effective staff schedules in spite of fluctuating demand, historical data is often used to predict staffing needs. Historical data refers to information recorded for the same period of the prior year or years, such as the number of guests served and the revenue generated during a given period. This data helps a supervisor estimate how much product is needed and how many staff members are required to handle the expected number of guest orders.

A supervisor also determines the type and number of staff members needed for each shift. Salaried staff members are employees who are paid salaries or shift pay. Their hours are guaranteed for each pay period. Hourly staff members are scheduled on an as-needed basis, depending on how busy the operation is expected to be. The number of hours that hourly staff members are scheduled to work often varies by pay period.

Staff Schedules

Employee Schedule Week of ________								
Staff	**Sunday 10A - 10P**	**Monday 4P - 9P**	**Tuesday 4P - 9P**	**Wednesday 4P - 9P**	**Thursday 4P - 11P**	**Friday 4P - 11P**	**Saturday 10A - 11P**	**Total Hours**
General Manager		2 P - 9 P	2 P - 9 P	8 A - 3 P	8A - 4 P	8 A - 5 P		40
Assistant Manager	12 A - 5 P			2 P - 9 P	2 P - 11 P	2 P - 11 P	12 A - 8 P	40
Executive Chef	12 P - 8 P			12 P - 8 P	12 P - 8 P	12 P - 8 P	12 P - 8 P	40
Banquet Manager	8 A - 5 P		9 A - 5 P		4 P - 11 P	4 P - 11 P	8 A - 5 P	40
Catering/Events Manager	12 P - 8 P	9 A - 5 P		9 A - 5 P		2 P - 11 P	2 P - 11 P	40
Sous Chef	8 A - 5 P	2 P - 9 P			2 P - 11 P		8 A - 5 P	34
Sous Chef	2 P - 10 P			2 P - 9 P	6 P - 11 P	2 P - 11 P		29
Sous Chef			2 P - 9 P	VACATION	VACATION	6 P - 11 P	6 P - 11 P	17
Garde Manger	12 P - 8 P			12 P - 8 P	12 P - 8 P	12 P - 8 P	12 P - 8 P	40
Line Cook/Expediter	8 A - 5 P	2 P - 9 P			2 P - 11 P		8 A - 5 P	34
Line Cook/Expediter		2 P - 9 P	2 P - 9 P	2 P - 9 P		2 P - 8 P	2 P - 9 P	28
Line Cook/Expediter	2 P - 10 P			2 P - 9 P	2 P - 11 P	2 P - 11 P		33
Tournant/Expediter	3 P - 10 P		2 P - 9 P			6 P - 11 P	6 P - 11 P	22
Prep Cook	8 A - 5 P	2 P - 9 P			2 P - 11 P		8 A - 5 P	34
Prep Cook	2 P - 10 P			2 P - 9 P	2 P - 11 P	2 P - 11 P		33
Prep Cook	VACATION	VACATION	2 P - 9 P			6 P - 11 P	2 P - 11 P	19
Pastry Chef	12 P - 8 P			12 P - 8 P	12 P - 8 P	12 P - 8 P	12 P - 8 P	40
Baker		12 P - 8 P	12 P - 8 P		12 P - 8 P	12 P - 8 P	8 A - 5 P	37
Baker	8 A - 5 P	12 P - 8 P	12 P - 8 P	12 P - 8 P		12 P - 8 P		37
Dishwasher	10 A - 6 P	4 P - 9 P					10 A - 6 P	21
Dishwasher			4 P - 9 P		4 P - 11 P		3 P - 11 P	20
Dishwasher	2 P - 10 P			4 P - 9 P		4 P - 11 P		20

Figure 10-3. Foodservice supervisors need to prepare schedules that effectively staff the operation.

Delegation

To be successful, supervisors must know how to delegate. *Delegating* is the act of assigning a task to another person while remaining responsible for the results. Each day orders need to be placed, deliveries arrive and need to be checked in, and menus need to be developed. It is impossible to perform all of these tasks effectively without the assistance of qualified individuals. Delegating is even a greater necessity during busy periods and frees the supervisor to make fast decisions and perform critical tasks.

To delegate effectively, supervisors must clearly communicate the expected results to the individuals assigned to the task. **See Figure 10-4.** The supervisor should also be confident that the individuals can perform the tasks satisfactorily. Monitoring the progress of these individuals reduces the possibility that tasks will not be finished according to the schedule.

Delegating also allows individuals to demonstrate new capabilities in preparation for increased responsibilities within the foodservice operation. Identifying tasks that can be delegated, and selecting the individuals capable of performing the delegated tasks, is a skill acquired through supervisory experience. Acknowledging success in completion of a task that was delegated encourages the growth of skills that make the individual even more valuable to the foodservice operation.

Delegation

Charlie Trotter's

Figure 10-4. To delegate effectively, supervisors must clearly communicate the expected results to the individuals assigned to the tasks.

CHECKPOINT 10-1

1. Identify four types of communication required to effectively interact with and supervise others.
2. Explain the effects that teamwork can have on a foodservice operation.
3. Define time management.
4. Explain the importance of delegating tasks.

SUPERVISORY RESPONSIBILITIES

Supervising staff is one of the primary functions of a foodservice supervisor. Key elements that contribute to the effective supervision of staff include professionalism, mentoring staff, on-the-job training, workplace policies, staff evaluation, and disciplinary actions. When these elements are present and practiced, both the supervisor and the staff have a clear understanding of expectations.

Professionalism

Supervisors are not only expected to be competent in their field, but also to set an example by acting in a professional manner. Supervisors exhibit professionalism by observing good personal hygiene practices, wearing work-appropriate attire, maintaining a positive attitude, and behaving in an ethical manner.

Personal Hygiene. Personal hygiene practices such as bathing daily and following proper handwashing techniques are essential for all foodservice staff. It is also important to refrain from working when seriously ill to prevent jeopardizing the health of others. Foodservice staff must always practice good personal hygiene to help eliminate the spread of pathogens. **See Figure 10-5.** Maintaining personal health minimizes the frequency of illness and requires adequate rest, nutrition, and exercise. The following are universally accepted personal hygiene practices for foodservice staff:

- Shower or bathe daily.
- Keep hair washed, combed, and covered as required.
- Keep facial hair clean, trimmed, and covered as required.
- Wear the proper clean clothing for the job.
- Do not wear nail polish or artificial fingernails.
- Except for a plain wedding band, do not wear jewelry.
- Keep hands and fingernails clean.
- Handle food only as required and always with clean or gloved hands.
- Eat in designated areas.
- Do not allow dirty utensils or equipment to touch food.
- Never work around food with open cuts or sores. Bandage cuts and sores, and wear gloves over injured hands.
- Do not cough, spit, or sneeze near food or in food preparation areas. Always cover your mouth with your arm when coughing, and sneeze into a handkerchief. Wash hands immediately after using a handkerchief or coughing.
- Stay home when sick or if experiencing vomiting, diarrhea, jaundice, or sore throat and fever.

Personal Hygiene

Sullivan University

Figure 10-5. Foodservice staff must always practice good personal hygiene to help eliminate the spread of pathogens.

Attire. Wearing appropriate work attire represents a commitment to professionalism by the individual and the foodservice operation. Work attire should always be clean and neatly pressed as well as fit properly. **See Figure 10-6.** Staff that looks professional makes a positive first impression that reflects the quality of service guests can expect during their dining experience.

Professional Attire

Vulcan-Hart, a division of the ITW Food Equipment Group LLC

Figure 10-6. Staff attire should always be clean and neatly pressed as well as fit properly.

Attitude. Having a positive attitude in the workplace makes a staff member stand out. Instead of seeing obstacles, individuals with a positive attitude see challenges with solutions. A positive attitude leads to increased productivity, influences the attitude of others, and results in the best products and services.

Ethics. An ethical person demonstrates honesty, trustworthiness, character, and responsibility. Exhibiting ethics in the workplace helps build a foundation of respect that can promote success and advancement. Supervisors make difficult decisions on a daily basis, yet the best decision may not always be clear. Being ethical involves making decisions based on honesty and fairness in order to determine the best outcome in any questionable situation. Some actions should not be taken even though they may not violate operational policies. For example, vendors may offer a supervisor gifts to gain favor or special treatment over the competition. Although it may be tempting to accept gifts from vendors, this action is considered unethical.

Mentoring Staff

Supervisors are often mentors. A *mentor* is an experienced person who coaches and guides a less experienced person. **See Figure 10-7.** The responsibilities of a mentor go beyond those of a trainer. Mentors are not only knowledgeable and offer advice based on their professional experience, but also assist in the professional growth of the person mentored. The responsibilities of a mentor include the following:

- assessing strengths and weaknesses
- offering constructive feedback
- providing support and direction
- developing long-term career objectives
- developing strategies for achieving goals

Supervisory Skills

Mentoring

Figure 10-7. A mentor is an experienced person who coaches and guides a less experienced person.

On-the-Job Training

On-the-job training (OJT) is a training method in which staff members are trained to perform tasks while already on the job. During OJT, new staff members are given a task to be completed, shown how to perform the task, and are then expected to complete the task independently. **See Figure 10-8.** New foodservice staff may be asked to perform tasks such as purchasing, receiving, storing, preparing, and serving food.

Supervisory Forms

On-the-Job Training (OJT)

Cres Cor

Figure 10-8. During OJT, new staff members are shown tasks and how to perform them.

A successful supervisor must have the technical skills to perform the primary tasks of the staff members being supervised. Having the technical skills to perform the tasks that staff members are required to perform enables a supervisor to effectively train staff through demonstration and to earn credibility. Supervisors may need to train new staff how to develop profitable menus and to purchase, receive, store, prepare, and serve food.

Planning Menus for Profit. A supervisor may be required to develop a menu. A *menu* is a document that markets a foodservice operation to guests. It lists products that satisfy guests, are profitable, and can compete with the alternatives offered elsewhere. In addition to calculating the appropriate prices to charge for menu items, the kitchen equipment, staff skills, and seasonal ingredients must be taken into account when planning profitable menus.

The kitchen must be properly equipped to efficiently prepare the menu. When the appropriate equipment is not available, the operation is less efficient. This inefficiency results in higher payroll expenses.

The techniques used to prepare the items on a menu should be in line with the skills of the kitchen staff. If the staff lack the required skills, food products are likely to be wasted due to errors in preparation, which causes expenses to increase.

Some menu items should only be offered during certain times of the year when key ingredients are readily available at a reasonable price. All items on a menu should be popular enough with guests that the ingredients required to prepare them are not wasted.

Purchasing Food. Purchasing involves the selection and ordering of products required to prepare and serve the items on a menu. A *purchase specification* is a written form listing the specific characteristics of a product that is to be purchased from a supplier. Depending on the product, the information on the purchase specification may include the product's quality or grade, variety or place of origin, size, or packaging requirements. It is also common for the purchase specification to include food safety requirements such as the temperature at which products must be transported and received.

A par stock checklist is used to help make sure that the appropriate quantity is ordered. *Par stock* is the amount of a particular product that should be kept in inventory to ensure an adequate supply is on hand for normal production. Par stock values should be set high enough to ensure that the operation does not run out of a product. **See Figure 10-9.** However, the values should be low enough to avoid disposing of products because they were left in storage too long.

Par Stock

Item	Par Stock	Stock On Hand	Estimated Use Prior to Delivery	Quantity to Order
Tomatoes	5 cases	2 cases	1 case	4 cases
Lettuce	10 cases	5 cases	5 cases	10 cases
Cucumbers	2 cases	1 case	1 case	2 cases
Onions	1 case	15 cases	5 cases	0 cases

Figure 10-9. Par stock values affect the amount of product ordered.

Receiving and Storing Food. All food items need to be checked upon arrival to ensure that the product received actually matches what was on the purchase specification. **See Figure 10-10.** Products should be weighed or counted to make sure that the amount or quantity received is the same as specified on the invoice.

Receiving Products

Purchase Specification	
Item Name	baking potatoes
Variety	Idaho Russet
Grade	US #1
Count per Case	80
Net Weight per Case	50 lb
Packaging	heavy-duty cardboard box

Figure 10-10. Products need to be checked against the purchase specification upon arrival.

Special attention should be paid to perishable foods to make sure they are fresh, free of damage, of the appropriate weight or size, and at the proper temperature when received. Perishable foods include poultry, fish and shellfish, meats, dairy products, and produce. Thermostats on refrigerators and freezers need to be checked frequently to ensure that the units are operating at the temperatures required to keep perishable foods safe.

Nonperishable items have a much longer shelf life and generally can be kept for six months to a year. These items are normally stored at room temperature in their original packaging. The packaging should be carefully checked upon arrival to avoid accepting dented cans, crushed boxes, and torn or damaged packages. Food and beverage items in dry and cold storage areas must be kept 6 inches off the floor.

Storing foods properly is essential to control costs. Immediately after items have been received and checked-in they should be placed into inventory. Many foodservice operations use an issuing procedure, which allows items that are stored to be moved into production only after a requisition is issued. A *requisition* is an internally generated invoice that is used to aid in tracking inventory as it moves from storage to production.

Preparing and Serving Food. Many foods are trimmed or broken down before being served or used in a recipe. Sometimes the parts trimmed from products such as meats and vegetables can be used in other recipes, such as stocks or soups, instead of being disposed of as waste. Unnecessary waste and costs can also be prevented by minimizing overproduction of products.

The FIFO method of stock rotation is the first step in limiting spoilage. Other ways to limit spoilage include ordering the correct amount of an ingredient based on par stock and avoiding obsolescence. *Obsolescence* is the removal of an item from the menu while ingredients for that particular item are still in stock.

One of the most important ways foodservice workers can help minimize expenses is to serve food using the proper portion control techniques. *Portion control* is the process of ensuring that a specific amount of food or beverage is served for a given price. This is accomplished by using portion control equipment such as ladles, portion-controlled scoops, scales, and slicers. **See Figure 10-11.** When the quantity of food served exceeds the amount the menu price was based upon, food expenses increase and profit is reduced.

Workplace Policies

Workplace policies are typically communicated to new staff during staff orientation. *Staff orientation* is a period during which new staff members are introduced to workplace policies and procedures. A staff orientation can last from a few hours to a few days, depending on the position and the foodservice operation.

Supervisors must follow and enforce workplace policies in order to protect the foodservice operation, staff, and guests. Negligence, or failure to perform legal responsibilities, may damage the reputation of the operation or result in lawsuits. Such occurrences could jeopardize the existence of the operation. Workplace policies regarding discrimination, harassment, substance abuse, and a hostile work environment should always be specified in a written document. Supervisors are responsible for reporting individuals who break a workplace policy.

Portion Control Equipment

Carlisle FoodService Products

Detecto, A Division of Cardinal Scale Manufacturing Co.

Carlisle FoodService Products

Figure 10-11. Portion control equipment is used to ensure that a consistent amount of food is served.

Discrimination. *Discrimination* is the unfair treatment of an individual based on characteristics such as gender, race, ethnicity, age, religion, appearance, or disability. Discrimination can be avoided by treating all people equally. No staff member wants to be bullied, mistreated, or viewed as inferior. Effective supervisors appreciate the differences among their staff, respect cultural differences, and promote diversity in the workplace.

Harassment. *Harassment* is behavior that is found to be threatening or that is not considered acceptable to the general public. Harassment can range from words or actions that make an individual feel uncomfortable or demeaned to situations that make an individual feel threatened or endangered. *Sexual harassment* is harassment that includes unwelcome verbal or physical conduct of a sexual nature. Harassment of any kind creates an unfriendly work environment and should never be tolerated.

Substance Abuse. Zero-tolerance drug and alcohol policies clearly prohibit the use of drugs and alcohol in the workplace. Working with knives, kitchen equipment, and open flames while under the influence of drugs or alcohol endangers everyone in the workplace.

Hostile Work Environment. A hostile work environment exists when staff feel verbally or physically threatened in the workplace. Violence, physical threats, and intimidation have no place in the foodservice environment. Supervisors should report all incidents. In the event of physical violence, the supervisor should also notify the police.

Staff Evaluations

Evaluations can help staff members improve skills and increase their value as part of the team. **See Figure 10-12.** A thorough evaluation includes both formal and informal evaluations. A *formal evaluation* is a scheduled and documented performance evaluation conducted on a regular basis, usually annually, by a supervisor that becomes part of an employee's permanent record. An *informal evaluation* is a performance evaluation conducted on an as-needed basis that provides an employee with constructive feedback on daily tasks or overall performance. If a staff member is not meeting expectations, the supervisor needs to take disciplinary action.

Workplace policies such as those regarding tardiness, requests for absence, emergency call-offs, and paid time off, must be clearly communicated to staff. Not following the rules can lead to suspension or termination. Common reasons for disciplinary action include stealing, excessive tardiness or absences, harassment, hostility toward coworkers or guests, and unethical behavior.

Effective Management

Disciplinary Actions

Most staff members do their jobs and work hard to benefit the operation. Unfortunately, some staff members may be unreliable, irresponsible, or belligerent. Supervisors often discipline staff in a constructive manner using oral and written reminders. When those reminders are not effective, a termination may result.

Staff Evaluations

Figure 10-12. Evaluations can help staff members improve skills and increase their value as part of the team.

Oral Reminders. Oral reminders are used by a supervisor to explain the offense to the employee and the reason a change in behavior is required. The employee is given the opportunity to explain his or her position and return to work. An oral warning is meant to allow everyone to approach the next shift with a positive attitude.

Written Reminders. Written reminders are used by a supervisor to address a repeated offense with an employee. The supervisor explains the seriousness of the situation and that it is necessary to conform to workplace policy. The supervisor follows up with a written memo that summarizes the meeting and the discussions therein. The supervisor and employee both sign the memo, and it is added to the employee's personnel file.

Suspensions and Terminations. The suspension or termination of an employee usually occurs if an employee fails to change their behavior after being given both oral and written reminders. Suspensions are less common than terminations in the foodservice industry. Immediate terminations also can occur when severe offenses such as violence or theft occur.

CHECKPOINT 10-2

1. Identify four ways in which supervisors exhibit professionalism.
2. Explain the role of a mentor.
3. Describe on-the-job training (OJT).
4. Explain the purpose of workplace policies.
5. Differentiate between discrimination and harassment.
6. Explain zero tolerance.
7. Describe a hostile work environment.
8. Identify two types of staff evaluations.
9. Explain the progression of disciplinary actions.

PLANNING MENUS

Foodservice supervisors are often involved in planning menus. A menu lists items that guests may order. Every purchasing and staffing decision relates directly back to the menu. The menu is the primary form of communication that a foodservice operation uses to inform guests of the types of foods and beverages offered and how much each item costs. **See Figure 10-13.** The menu also reflects the target market of the operation. In addition, it dictates the type and amount of ingredients to purchase, the equipment needed, and the level of skill required by the foodservice staff.

Menus

Figure 10-13. The menu informs guests of the types of foods and beverages offered.

Various individuals may be involved in the planning of menus. For example, an owner may plan the menu for a family-run restaurant, but a central office generally determines the menu for a chain operation. There are a variety of resources available to assist in the menu planning process, including websites, design firms, consultants, and dietitians. Regardless of who plans the menu, all menus must follow the truth-in-menu guidelines set by the federal government.

Truth-in-Menu Guidelines

The federal government enacted truth-in-menu guidelines, which require accuracy in statements made on menus. The truth-in-menu guidelines are designed to protect guests from fraudulent food and beverage claims. **See Figure 10-14.** Failure to comply with truth-in-menus guidelines can result in legal claims being made against a foodservice operation for misleading or endangering guests. Given the direct link between food and health, guests rely on menus for accurate information.

Nutrition Labeling

Menu Classifications

Planning menus also involves determining the menu classification. The menu classification is determined by how often menu selections will change. Some menus remain fairly constant, while others change seasonally. Menu classifications include fixed menus, market menus, and cycle menus.

Fixed Menus. A *fixed menu* is a menu that stays the same or rarely changes. Fixed menus are commonly found in chain restaurants where the entire organization has the same menu. They can also be found at entertainment venues, such as ballparks and concert arenas, where the same menu is available throughout the season. Some foodservice operations add to a fixed menu by attaching a list of daily specials.

Market Menus. A *market menu* is a menu that changes frequently to coincide with changes in the availability of products. Market menus often revolve around seasonal items that are at their peak, which may only be for a short period of time. Farmers' markets are often a local source for seasonal ingredients. Buying locally promotes sustainability and provides guests with foods harvested at the height of freshness.

Cycle Menus. A *cycle menu* is a menu written for a specific period and is repeated once that period ends. **See Figure 10-15.** Cycle menus are common in institutional foodservice settings such as hospitals and cafeterias. For example, a cycle menu with predetermined items may repeat each month. Most cycle menus rotate items so that the same meal is not offered on the same day each week. For example, a menu might be written for an 11 day cycle. If lasagna is served on a Monday, the cycle menu may not list lasagna again until 11 days later on a Friday. This rotation helps keep the menu from becoming predictable.

Menu Types

Planning menus also involves determining the most appropriate menu type for the operation. Menu types are influenced by the menu classification as well as the nature of the foodservice operation. Some restaurants offer multiple courses for a set price, while other restaurants charge separately for each dish. A menu can be displayed in various ways, including on overhead signs, on a single sheet of paper, or in the style of a book. The physical form of the menu often reflects the menu type. Common menu types include à la carte, semi-à la carte, prix fixe, table d'hôte, California, and meal-specific.

Truth-In-Menu Guidelines	
Menu Label	*Example of Misrepresentation*
Portion size of an item	Advertising a 12 oz steak and serving a 10 oz steak
Quality or grade of an item	Listing USDA Prime beef and serving USDA Select beef
Preservation method	Advertising fish as fresh, but serving fish that was previously frozen
Preparation method	Claiming a food is housemade when it is a prepackaged item
Type of product served	Stating the use of extra virgin olive oil when vegetable oil was used
Certified foods	Claiming a food is organic that has not been certified as such
Point of origin	Stating "Florida" oranges when the oranges came from a different location
Nutrition information	Listing a product as low-fat when it does not meet the required criteria
Product brand	Serving a different brand than the one listed

Figure 10-14. The truth-in-menu guidelines are designed to protect guests from fraudulent food and beverage claims.

Eleven-Day Cycle Menu				
Monday	*Tuesday*	*Wednesday*	*Thursday*	*Friday*
Cream of Broccoli Soup Broiled Cod Sliced Roasted Pork Glazed Carrots Mashed Potatoes	Vegetable Soup Chicken Tetrazzini Italian Meatloaf Grilled Zucchini Pasta Marinara	Creamy Mushroom Soup Roast Turkey Baked Ham Bread Dressing Broccoli	Chicken Noodle Soup Grilled Chicken Blackened Catfish Rice Pilaf Cauliflower	Cream of Asparagus Soup Fettuccine Alfredo Teriyaki Chicken Fried Rice Carrots and Zucchini
Minestrone Soup Meatballs Baked Scrod Mostaccioli Italian Vegetables	Chicken and Rice Soup Pork Chop Chicken Tacos Spanish Rice Green Beans	Lentil Soup Roast Beef Veggie Lasagna Roasted Potatoes California Blend	Beef Barley Soup Oven-Fried Chicken Spaghetti & Meatballs Sweet Potatoes Peas	Corn Chowder BBQ Pork Stuffed Peppers Boiled Potatoes Crinkle-Cut Carrots
Navy Bean Soup Chicken à la King Pasta Primavera Steamed White Rice Brussels Sprouts	Cream of Broccoli Soup Broiled Cod Sliced Roasted Pork Glazed Carrots Mashed Potatoes	Vegetable Soup Chicken Tetrazzini Italian Meatloaf Grilled Zucchini Pasta Marinara	Creamy Mushroom Soup Roast Turkey Baked Ham Bread Dressing Broccoli	Chicken Noodle Soup Grilled Chicken Blackened Catfish Rice Pilaf Cauliflower
Cream of Asparagus Soup Fettuccine Alfredo Teriyaki Chicken Fried Rice Carrots and Zucchini	Minestrone Soup Meatballs Baked Scrod Mostaccioli Italian Vegetables	Chicken and Rice Soup Pork Chop Chicken Tacos Spanish Rice Green Beans	Lentil Soup Roast Beef Veggie Lasagna Roasted Potatoes California Blend	Beef Barley Soup Oven-Fried Chicken Spaghetti & Meatballs Sweet Potatoes Peas
Corn Chowder BBQ Pork Stuffed Peppers Boiled Potatoes Crinkle-Cut Carrots	Navy Bean Soup Chicken à la King Pasta Primavera Steamed White Rice Brussels Sprouts	Cream of Broccoli Soup Broiled Cod Sliced Roasted Pork Glazed Carrots Mashed Potatoes	Vegetable Soup Chicken Tetrazzini Italian Meatloaf Grilled Zucchini Pasta Marinara	Creamy Mushroom Soup Roast Turkey Baked Ham Bread Dressing Broccoli

Figure 10-15. A cycle menu is written for a specific period and is repeated once that period ends.

À la Carte Menus. An *à la carte menu* is a menu that has all food and beverage items priced separately. À la carte menus can be found in quick service venues where a hamburger is sold separately from a side of fries. À la carte menus are also found in restaurants where a side of grilled zucchini is sold separately from the entrée.

Semi-à la Carte Menus. A *semi-à la carte menu* is a menu that offers entrées along with additional menu items for a set price. The prices on a semi-à la carte menu often include an entrée and a choice of soup, salad, or side. However, appetizers and desserts are commonly sold separately. Some restaurants use semi-à la carte menus to offer value meal combinations that allow a guest to order a main dish, side dish, and beverage for a set price. An à la carte menu can be easily transitioned into a semi-à la carte menu by including side dishes in the price of the entrée instead of pricing each item separately. **See Figure 10-16.**

Semi-à la Carte Menus

CHEF'S TABLE

Caribbean Fiesta Salad – Bed of romaine and baby greens tossed in a peppercorn dressing with sliced tomatoes, mushrooms, sweet red peppers, shrimp, crabmeat, and chopped herbs $10.75

Blackened Chicken Wrap – Flour tortilla brushed with remoulade sauce and filled with shredded lettuce, diced tomatoes, green onions, Jack cheese, and slices of blackened chicken breast $9.50

Seared Duck and Grilled Radicchio Salad – Fresh spinach and a grilled wedge of radicchio topped with pan-seared duck breast, plum tomatoes, shiitake mushrooms, and goat cheese with a white balsamic vinaigrette $10.50

Each salad comes plated or tossed and wrapped in a large flour tortilla

Served with complimentary Soup du Jour

Italian Panini – Old world Panini bread brushed with olive oil and layered with sliced prosciutto, roasted peppers, onions, sun-dried tomatoes, artichoke hearts, olives, provolone, and topped with a tart vinaigrette $7.50

Individual Pizza of the Day – Homemade crust topped with the freshest ingredients layered with the finest cheeses. Ask about today's selection $7.75

Caprese Burger – Homemade ground beef patty cooked to order with fresh mozzarella, basil, plum tomatoes, and a balsamic drizzle $7.50

Complimentary Soup du Jour or House Salad served with each entrée, plus choose a complimentary side:

Vegetable du Jour
Hand-Cut French Fries
Garlic Bread

Figure 10-16. A semi-à la carte menu includes side dishes in the price of the entrée instead of each item priced separately.

Prix Fixe Menus. A *prix fixe menu* is a menu that offers limited choices within a collection of specific items for a multicourse meal at a set price. There are typically a few options within each course on a prix fix menu. For example, an appetizer, soup, salad, entrée, and dessert may be offered for a set price with three or four choices within each category. The guest is able to choose an item from each category. Prix fixe menus are often used in upscale restaurant settings. **See Figure 10-17.**

Prix Fixe Menus

Prix Fixe Four Course Menu
$75.00

First Course
(Select One)

Lightly smoked duck breast, fresh cheese, and carrot terrine
Shrimp and shaved baby carrot salad with honey coriander vinaigrette

Second Course
(Select One)

Silky cream of peppered butternut squash soup
Roasted baby beet salad
Frisée, green oak, and hazelnut salad

Third Course
(Select One)

Roasted chicken and au jus with ricotta dumplings
Roasted strip loin of beef with red wine sauce
Crispy shrimp, white wine butter sauce, and shellfish foam

Vegetable and Starch Selections
(Select Two)

Slow cooked fennel purée and warm tomato
Bacon, pearl onion, and fresh pea ragout
Leek and potato roulade
Shaved garlic broccolini
Glazed cipollini onions and mushrooms

Fourth Course

A tasting of frozen, warm, and fresh desserts
from our talented pastry department

Figure 10-17. Prix fixe menus offer limited choices at a set price and are often used in upscale restaurant settings.

Table d'Hôte Menus. A *table d'hôte menu* is a menu that identifies specific items that will be served for each course at a set price. A table d'hôte menu does not offer choices. A plated banquet at which all guests are served the same meal is an example of a table d'hôte menu.

California Menus. A *California menu* is a menu that offers all food and beverage items for breakfast, lunch, and dinner throughout the entire day. For example, a guest has the option of ordering breakfast items for dinner. California menus are commonly used by restaurants that are open 24 hours a day.

Meal-Specific Menus. A *meal-specific menu* is a menu that only offers a particular meal, such as breakfast, lunch, or dinner. Meal-specific menus feature items that are typical for that meal, such as eggs, pancakes, and waffles for breakfast or sandwiches and salads for lunch. All of the other types of menus, except the California menu, can be meal specific.

Menu Design

Planning menus also involves determining how to promote sales through an effective menu design. A well-designed menu clearly communicates in a manner that promotes sales and increases profitability. To achieve this goal, a menu design should reflect the atmosphere of the foodservice operation and leave a favorable impression on guests. For example, an outdoor café might present a handwritten menu to promote a casual, neighborhood atmosphere. In contrast, a fine dining restaurant may present a menu on linen paper with an elegant script. Even if two restaurants offer the same menu items, the menu design can help create the desired ambiance for each.

A successful menu design includes an effective menu mix. A *menu mix* is the assortment of items that may be ordered from a given menu. For example, a foodservice operation may have a menu mix that includes vegetarian entrées in addition to meat, poultry, and fish entrées. A well-balanced menu mix can promote guest satisfaction, attract more guests, and increase sales.

On average, guests spend approximately two minutes scanning a menu before selecting items. Within this brief time frame, a successful menu design can persuade guests to order items that will generate the greatest profits. A good menu design directs guests to the most profitable items by effectively using item descriptions, format, item placement, and graphic elements.

Item Descriptions. Menu item descriptions can entice guests by creating images of menu items in their minds. Some words have more selling power than others and can be used to help direct guests to higher-profit items. For example, "baked in our wood-fire oven" has more appeal than "baked."

In general, menu items that are common or familiar do not require descriptions. For example, New York strip steak, macaroni and cheese, and blueberry pie are menu items that present a clear image without further explanation. However, even if a menu item is familiar to guests, a powerful description can emphasize special components of the dish and elevate it from sounding good to sounding exceptional.

Items that are unclear or unfamiliar to most guests require further description. For example, "stuffed pork tenderloin" does not tell the guest enough about the dish, and some guests may be unfamiliar with an item such as gnocchi. Good menu item descriptions promote clarity, add appeal to the dish, and increase the likelihood that an item will be ordered. **See Figure 10-18.**

Menu Item Descriptions

Stuffed Pork Tenderloin

Hand-trimmed pork tenderloin filled with a flavorful blend of sweet Italian sausage and Granny Smith apples, pan-seared and glazed with a mandarin orange and honey BBQ sauce

Gnocchi

Delicate potato pasta dumplings simmered in a fresh tomato cream sauce, infused with basil and roasted garlic, and finished with freshly grated pecorino cheese

Figure 10-18. Good menu descriptions promote clarity, add appeal to the dish, and increase the likelihood that an item will be ordered.

Format. Effective menus require that careful consideration be given to the physical format of the menu and the placement of each menu item within that format. Menus are presented in four basic formats. Menu formats include single-page menus, bifold menus, trifold menus, and multipage menus. **See Figure 10-19.**

- Single-page menus present menu items on one page and may use both the front and back of the sheet.
- Bifold menus open like a book and present items on the left and right sides and possibly the front and back cover.
- Trifold menus have a right and a left panel that fold toward the center and overlap. Trifold menus may list items on the front and back of each panel.
- Multipage menus are similar to bifold menus, but include many more pages.

Regardless of the format chosen, menus come in a variety of sizes and shapes. They can be made from materials ranging from laminated card stock to leather-bound booklets. The menu format should reflect the style of the foodservice operation and be influenced by the menu classification and menu type. For example, a fixed, meal-specific menu that offers only breakfast may be formatted as a laminated bifold menu that has been designed for durability. In contrast, a single-page menu on heavy cardstock may be appropriate for a market menu that changes each day.

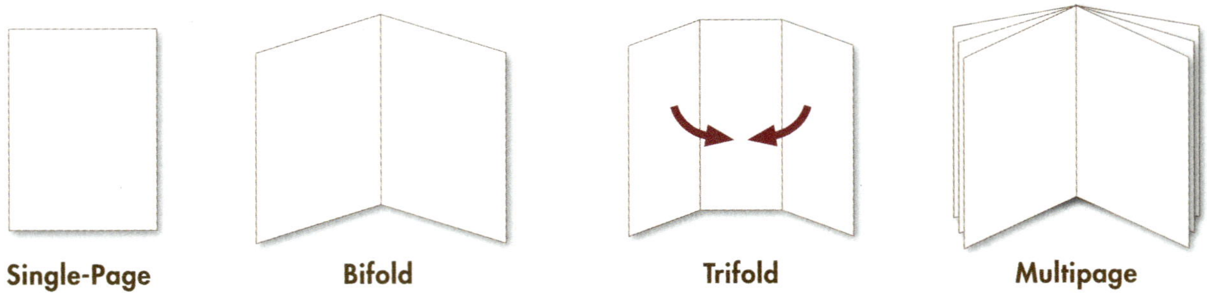

Figure 10-19. Menu formats include single-page, bifold, trifold, and multipage menus.

Item Placement. After the menu format is chosen, the placement of each menu item is the next consideration. Because most guests order food in the order it is eaten, items are typically grouped by type, such as appetizers, soups and salads, entrées, and desserts. The placement of items influences what guests choose because the eye follows a natural path when scanning a menu. For example, the eye naturally falls to the center of the page of a single-page menu. However, the eye naturally falls on the top left of a bifold menu, then moves to the top right, and finally rests in the center, making these three areas ideal positions for the most profitable items. **See Figure 10-20.**

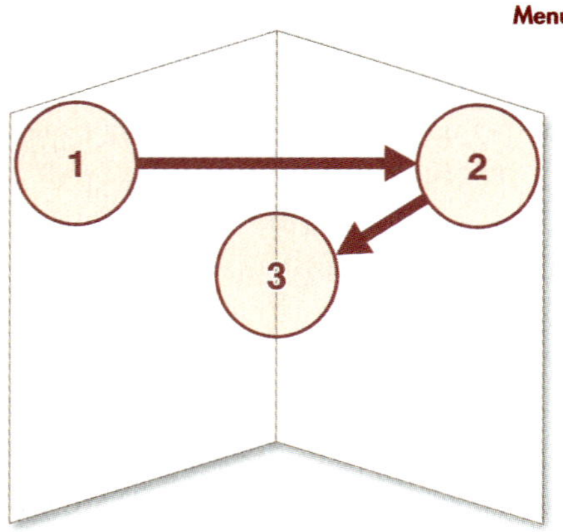

Figure 10-20. The eye naturally falls on the top left of a bifold menu, then moves to the top right, and finally rests in the center, making these three areas ideal positions for the most profitable items.

Graphic Elements. Menus should be legible and visually appealing. A legible menu starts with choosing lettering and a background color that make the text easy to read. For example, silver lettering on light-gray paper may strain the eyes and frustrate guests, but simple black letters on off-white paper can be easy to see.

The use of graphic elements, such as borders and images, should create a look that emphasizes high-profit items and reflects the theme of the foodservice operation. **See Figure 10-21.** For example, casual dining operations often use well-positioned photographs to direct guests to featured menu items. In contrast, an upscale restaurant may use an elegant border to highlight signature dishes.

Figure 10-21. The use of graphic elements, such as borders and images, should create a look that emphasizes high-profit items and reflects the theme of the foodservice operation.

Dietary Considerations

Planning menus also involves determining how to accommodate guests who choose meals based on food allergies or food intolerances, weight management issues, or plant-based diets. Dietary considerations often mean that guests have to limit or restrict the consumption of certain foods. Along with well-trained staff, menus that are clear and accurate can help guests make informed decisions.

Food Allergies. Most menus include food allergy warnings. A *food allergy* is a reaction by the immune system to a specific food. Allergic reactions can be minor or extremely severe. *Anaphylaxis* is a severe allergic reaction that causes the airway to narrow and prohibits breathing. An anaphylactic reaction can lead to unconsciousness, coma, or even death. Because of the potential severity involved with food allergies, some operations establish allergy-free work zones in the kitchen that prohibit the use of specific foods known to cause allergic reactions.

There are over 100 foods that can cause allergic reactions. However, eight specific foods are considered major allergens. According to the Food and Drug Administration (FDA), milk, eggs, peanuts, tree nuts, fish, shellfish, soy, and wheat account for 90% of all food allergies. **See Figure 10-22.** In addition to viewing the menu, some guests may request a list of ingredients to avoid having an adverse reaction. This essential information helps individuals with food allergies avoid an allergic reaction by avoiding the food that triggers it.

Sources of Common Food Allergens

Allergen	*Potential Food Sources*
Milk	• Many nondairy products contain casein (a milk derivative) • Some brands of tuna fish contain casein • Butter melted on top of grilled steaks (often not visible) or foods cooked in butter • Many baked products contain dry milk
Eggs	• Egg whites are often used to create the foam topping on specialty coffee drinks • Egg wash is sometimes used on pretzels before they are dipped in salt • Some brands of egg substitutes contain egg whites • Most processed cooked pastas (including those in prepared soups) contain eggs or are processed on equipment shared with egg-containing pastas
Peanuts	• Chili sauce, hot sauce, mole sauce, marinades, glazes, and salad dressings • Pudding, cookies, crackers, egg rolls, and potato pancakes • Some vegetarian food products advertised as meat substitutes • Foods that contain extruded, cold-pressed, or expelled peanut oil
Tree Nuts	• Salads and salad dressings, barbeque sauce, and honey • Meat-free burgers, fish dishes, pancakes, and pasta • Piecrust and the breading on meat, poultry, and fish
Fish	• Imitation fish or shellfish • Salad dressings, Worcestershire sauce, and sauces made with Worcestershire
Shellfish	• Many Asian dishes include fish sauce • Shellfish protein can become airborne in the steam released during cooking
Soy	• Baked goods, cereals, crackers, infant formula, soups, and sauces • Some brands of tuna and peanut butter • Soybean oil that has not been cold pressed, expeller pressed, or extruded
Wheat	• Some brands of hot dogs, imitation crabmeat, and ice cream contain wheat • Some Asian dishes contain wheat flour shaped to look like beef, pork, or shrimp

Figure 10-22. Milk, eggs, peanuts, tree nuts, fish, shellfish, soy, and wheat account for 90% of all food allergies.

Food Intolerances. A *food intolerance* is an adverse reaction to a food that does not involve the immune system. Symptoms of food intolerance usually include bloating, abdominal cramps, nausea, and diarrhea. When the offending food is eliminated by the body or removed from the diet, symptoms usually subside. Some people may be able to tolerate the problem food in small quantities, while other people may need to avoid the food. The most common food intolerances are gluten intolerance and lactose intolerance.

Gluten is a type of protein found in grains such as wheat, rye, barley, and some varieties of oats. Gluten is prevalent in flour and flour-based products such as breads, cereals, pastas, and baked goods. It can also be found in other foods such as bouillon cubes, soy sauce, and deli meats. **See Figure 10-23.** *Celiac disease* is a condition in which gluten damages the small intestine's ability to absorb nutrients. It is essential for a person with celiac disease to abstain from eating gluten. Gluten-free menu offerings can provide guests with safe and healthy alternatives.

Foods Containing Gluten

- Bouillon cubes
- Brown rice syrup
- Candies
- Cold cuts
- French fries
- Gravies
- Hot dogs
- Matzo
- Potato chips
- Rice mixes
- Salami
- Sauces
- Sausages
- Soups
- Surimi
- Tortilla chips

Figure 10-23. Gluten is found in flour-based products such as breads, cereals, pastas, and baked goods, as well as other types of foods.

Lactose is a sugar found in milk and dairy products. In addition to milk, yogurts, and cheeses, lactose can be found in many processed foods such as salad dressings, soups, and breads. Lactose intolerance is an inability to properly digest lactose. Alternatives such as lactose-free milk, almond milk, or products made from soy can be offered for those who are lactose intolerant.

Weight Management. A sedentary lifestyle and consistent consumption of foods that are high in fat and calories can contribute to weight gain. According to the U.S. Department of Health and Human Services, the number of individuals who are overweight or obese has risen from 13% to 35% over the past several decades. *Obesity* is a medical condition characterized by an excess of body fat. The rise in obesity correlates with an increase in obesity-related illnesses and diseases, such as high blood pressure, cardiovascular disease, type 2 diabetes, and some cancers. **See Figure 10-24.**

Weight-Related Illnesses and Diseases

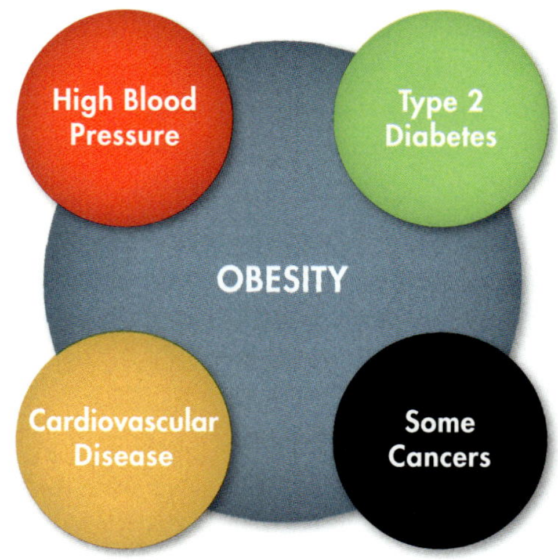

Figure 10-24. The rise in obesity correlates with an increase in obesity-related illnesses and diseases, such as high blood pressure, cardiovascular disease, type 2 diabetes, and some cancers.

To combat obesity, federal initiatives have been developed to promote healthy eating and physical activity. These initiatives encourage the foodservice industry to promote well-balanced and nutritious meals. Menus that include healthy options encourage guests to make more healthy choices. Restaurants that offer appropriate portion sizes can also help individuals manage weight and reduce obesity-related diseases.

Plant-Based Meals. Plant-based meals, also known as vegetarian meals, are being chosen by many people several days per week and every day by others. **See Figure 10-25.** There are many variations of plant-based meals. For example, some individuals exclude all animal-based foods while others may include dairy products or eggs. Menus that include vegetarian dishes, such as broccoli and cashew stir-fry or grilled vegetable risotto, fulfill the needs of this growing segment of the population. Studies have shown that eating plant-based meals can help reduce the risk of cardiovascular disease.

Plant-Based Meals

House Foods

Figure 10-25. Plant-based meals are chosen by many people on a regular basis.

CHECKPOINT 10-3

1. Explain the function of a menu.
2. Explain the purpose of the truth-in-menu guidelines.
3. Describe three classifications of menus.
4. Describe six menu types.
5. Explain three reasons why an effective menu mix is important.
6. Describe four elements of menu design.
7. Describe four types of dietary considerations that affect menu planning.
8. Identify eight major food allergens.

STANDARDIZED RECIPES

One of the most significant challenges for any foodservice operation is to produce food that is consistent. Consistency is important because guests expect food from a particular foodservice operation to look and taste the same every time it is ordered. Since the same cook does not always prepare the same items each day, foodservice operations use standardized recipes. A *standardized recipe* is a list of ingredients, ingredient amounts, and procedural steps for preparing a specific quantity of a food item. **See Figure 10-26.**

Standardized Recipe Elements

Meatloaf

Yield: 10 Servings | **Cooking Temperature:** 350°F

Portion Size: 6 oz | **Cooking Time:** 1 hour

Ingredients

1 tbsp	vegetable oil
3 oz	celery, small dice
4 oz	onion, small dice
2 fl oz	milk
2	eggs
1 c	breadcrumbs
1 tbsp	salt
1½ tsp	black pepper
1 tsp	thyme
3 lb	ground beef

Preparation

1. Heat oil in small sauté pan and sauté celery and onions until tender. Allow to cool.
2. Combine remaining ingredients and mix well.
3. Form mixture into a loaf and place in a greased bread pan.
4. Bake at 350°F for about 1 hour or until a thermometer inserted in the center of the loaf registers 160°F (to allow for carryover cooking).
5. Remove from oven, cover, and let rest for 15 minutes before slicing into 6 oz portions.

Nutrition information (per serving): 465.4 calories; 64% calories from fat; 32.7 g total fat; 149.2 mg cholesterol; 911.5 mg sodium; 503.2 mg potassium; 11.7 g carbohydrates; 0.9 g fiber; 4.0 g sugar; 10.8 g net carbohydrates; 28.6 g protein

Figure 10-26. Most standardized recipes include the same common elements.

Using standardized recipes helps ensure consistent quality, portion size, and cost. Standardized recipes usually contain the following common elements.

- The name of a recipe should be descriptive of the dish being prepared and should reflect the name used on the menu. For example, if a restaurant makes two kinds of chili sauce (a spicy version and a mild version) both recipes should not be named "Chili Sauce." Instead, more descriptive names, such as Spicy Red Chili Sauce and Mild Green Chili Sauce, should be used. Differences between the menu name and the standardized recipe name can lead to confusion in the preparation process, which may lead to unnecessary costs.
- A *yield* is the total quantity of a food or beverage item that is made from a standardized recipe. Yield may be given as portions (16 servings, 3 oz each), the size of the item produced (one 8 inch pie), or a measured amount of product (2 gal. of soup). A yield listed in ounces is a measurement of weight. A yield listed in fluid ounces is a measurement of volume.
- A *portion size* is the amount of a food or beverage item that is served to an individual person. Portion size is related to yield. For example, a soup recipe that yields 1 gal. can also be said to yield 16 portions of 8 fl oz each because 16 × 8 fl oz = 128 fl oz = 1 gal.
- The amount of each ingredient used in the recipe is listed next to the name of the ingredient. If an ingredient is to be prepared in a certain way prior to being measured, such as minced or chopped, that information is also provided. The ingredients are usually listed in the order that they are incorporated into the recipe to help ensure that no ingredient is left out.
- Preparation steps are listed in sequential order. Preparation steps direct when and how the ingredients are added to a recipe and are listed in sequential order. Some ingredients are cleaned or cut prior to being measured. For example, "½ cup sliced black olives" requires the olives be sliced before they are measured. In contrast, "½ cup black olives, sliced" requires the olives be measured before they are sliced. Therefore, more olives are required in the first example than in the second.
- The cooking temperature may be an exact temperature at which to set an oven, such as 400°F, or a more general indication of temperature such as "low heat" or "high heat." The temperature at which food is cooked greatly affects the outcome of the final product. For example, food that is cooked at too high of a temperature may burn on the outside before it is properly cooked in the center.
- Cooking times provided in standardized recipes are often treated as guidelines. However, professional cooks with experience rely more on the appearance or feel of an item than on a cooking time. They also often rely on exact measurements, such as an internal temperature checked with a thermometer, to determine when a food item is done.
- While nutrition information is not required to prepare a recipe, it is important information to have for menu planning and for guest inquiries. Due to health or dietary concerns, guests may ask how much fat, carbohydrates, or sodium there is in a menu item.

Scaling Recipes

A standardized recipe produces a specific yield. When a larger or smaller yield is needed, the recipe is scaled to produce the desired yield. *Scaling* is the process of calculating new amounts for each ingredient in a recipe when the total amount of food the recipe makes is changed. For example, a recipe that serves 4 people may be scaled for use in a restaurant that plans to make 50 servings of the recipe. Similarly, a recipe used by a banquet hall that normally serves 100 people may need to be scaled to make only 12 servings for a small party.

The scaling process starts by calculating a scaling factor. A *scaling factor* is the number that each ingredient amount in a recipe is multiplied by when the recipe yield is changed. **See Figure 10-27.** The formula for calculating a scaling factor is as follows:

$$SF = DY \div OY$$

where

SF = scaling factor

DY = desired yield

OY = original yield

Ingredients can be scaled by weight, volume, count, portion size, or product availability. Weight indicates the heaviness of a substance. Volume indicates the space a substance occupies. Count indicates the number of items being used. Scaling based on portion size or product availability may require additional calculations.

- **Scaling Based on Weight.** If a potato salad recipe that makes 4 lb of salad is scaled to make 80 lb of salad, the scaling factor is calculated by dividing the desired yield (80 lb) by the original yield (4 lb).

  $$SF = DY \div OY$$
  $$SF = 80 \text{ lb} \div 4 \text{ lb}$$
  $$SF = 20$$

- **Scaling Based on Volume.** If a soup recipe that makes 8 gal. of soup is scaled to make 3 gal. of soup, the scaling factor is calculated by dividing the desired yield (3 gal.) by the original yield (8 gal.).

  $$SF = DY \div OY$$
  $$SF = 3 \text{ gal.} \div 8 \text{ gal.}$$
  $$SF = 0.375$$

- **Scaling Based on Count.** If a cookie recipe that makes 24 cookies is scaled to make 84 cookies, the scaling factor is calculated by dividing the desired yield (84 cookies) by the original yield (24 cookies).

  SF* = *DY* ÷ *OY

  SF = 84 cookies ÷ 24 cookies

  SF = 3.5
- **Scaling Based on Portion Size.** If a fish recipe that makes 12 (8 oz) portions is scaled to make 30 (10 oz) portions, the yields must first be converted to a total number of ounces.

  OY = 12 × 8 oz = 96 oz

  DY = 30 × 10 oz = 300 oz

  Then the scaling factor formula can be applied.

  SF* = *DY* ÷ *OY

  SF = 300 oz ÷ 96 oz

  SF = 3.125
- **Scaling Based on Product Availability.** In some cases, a recipe is scaled because the amount of one ingredient needs to be changed based on availability. In situations like this, the scaling factor is calculated by dividing the available amount of the ingredient by the original amount of the ingredient. The formula for calculating a scaling factor based on product availability is as follows:

  SF* = *AA* ÷ *OA

  where

  SF = scaling factor

  AA = available amount

  OA = original amount

For example, if a beef stew recipe that calls for 15 lb of beef stew meat needs to be made with only 12 lb of beef stew meat, the scaling factor is calculated by dividing the available ingredient amount by the original ingredient amount.

SF* = *AA* ÷ *OA

SF = 12 lb beef stew meat ÷ 15 lb beef stew meat

SF = 0.8

On the other hand, if there are 18 lb of beef stew meat available and all of the meat is to be used, the scaling factor would be 1.2.

SF* = *AA* ÷ *OA

SF = 18 lb beef stew meat ÷ 15 lb beef stew meat

SF = 1.2

Calculating Scaling Factors

$$\text{Scaling Factor } (SF) = \frac{\text{Desired Yield } (DY)}{\text{Original Yield } (OY)}$$

Type of Yield	Original Yield	Desired Yield	Scaling Factor
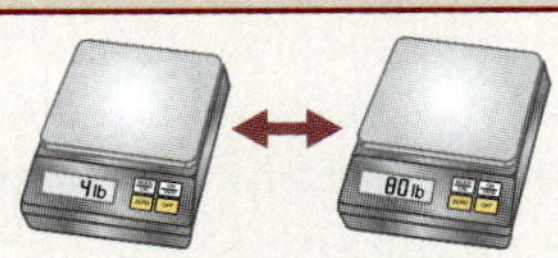 Total Weight Yield	4 lb	80 lb	80 ÷ 4 = 20
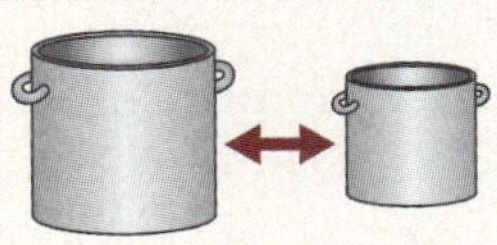 Total Volume Yield	8 gal.	3 gal.	3 ÷ 8 = 0.375
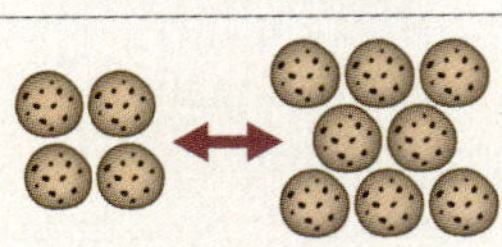 Count Yield	24 cookies	84 cookies	84 ÷ 24 = 3.5
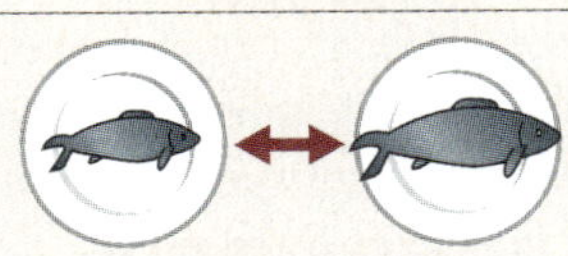 Portion Yield	12 (8 oz) portions (12 × 8 oz = 96 oz)	30 (10 oz) portions (30 × 10 oz = 300 oz)	300 ÷ 96 = 3.125

Figure 10-27. When a recipe is scaled based on yield, the scaling factor is calculated by dividing the desired yield by the original yield.

Scaling Considerations

When recipes are scaled, additional considerations must be taken into account before actually preparing the recipe. Addressing these considerations will help to ensure that the final product is of the same quality as the original recipe.

Measurements. If a new measurement is calculated that is not easily measured, such as 7.22 lb, the measurement will need to be adjusted slightly to make measuring more practical. **See Figure 10-28.** However, to avoid affecting the final result, the amount should not be adjusted drastically. For example, 7.22 lb should be adjusted to 7.25 lb.

Scaling Considerations

Edlund Co.

Figure 10-28. When scaling a recipe, minor adjustments may need to be made if the resulting measurement cannot be easily measured.

Equipment. Cooking equipment must properly accommodate the adjusted amount of food. It is important that cooking equipment is large enough to handle an increased yield.

Mixing Time. Instructions related to mixing times may need to be adjusted, especially when a recipe yield is significantly increased, to ensure that the appropriate amount of time is given to mix the ingredients adequately.

Cooking Temperature. A professional cook must determine whether the cooking temperature needs to be adjusted. Normally, cooking time is adjusted but in certain circumstances, such as when a recipe yield is significantly decreased, it may be necessary to reduce the cooking temperature to keep food from drying out.

Cooking Time. A professional cook must determine whether the cooking time for a recipe needs to be increased or decreased. If a larger roast is used for a recipe, a longer cooking time will normally be required. However, 10 steaks cooked on a grill will cook in the same amount of time as 2 steaks on a grill.

CHECKPOINT 10-4

1. Identify the common elements of standardized recipes.
2. Explain the difference between yield and portion size.
3. Differentiate between ounces and fluid ounces.
4. Explain the difference among weight, volume, and count.
5. Explain how to determine the scaling factor for a recipe.
6. Identify five factors that may have to be adjusted when scaling a recipe.

CALCULATING FOOD COSTS

A foodservice operation needs to continually monitor both the cost of food purchased and the cost of meals served to ensure the operation earns more in food sales than it pays in expenses. Calculating food costs involves much more than simply adding up the cost of ingredients. The terms "as purchased (AP)" and "edible portion (EP)" are commonly used to distinguish between a food item before and after it is trimmed of waste. An *as-purchased (AP) quantity* is the original amount of a food item as it is ordered and received. An *edible-portion (EP) quantity* is the amount of a food item that remains after trimming and is ready to be served or used in a recipe. **See Figure 10-29.**

As-Purchased Costs

An *as-purchased (AP) cost* is the amount paid for a product in the form it was ordered and received. Foodservice operations buy many food products in bulk quantities such as cases of hamburger patties, tubs of ice cream, crates of milk, or pails of pickles. Beverages are also typically purchased in bulk quantities such as 24-bottle cases of iced tea or half barrels (kegs) of beer. The AP costs of food and beverage products are equal to the prices listed on the invoice. An *invoice* is a document provided by a supplier that lists the items delivered to a foodservice operation and the prices of those items.

Figure 10-29. Carrot greens, tips, and peel are removed from the as-purchased (AP) form prior to use. The edible portion (EP) of the carrot remains.

When a product is purchased in bulk, it is often necessary to calculate a unit cost based on the AP cost. A *unit cost* is the cost of a product per unit of measure. Unit costs can be based on weight (ground turkey at $2.75 per lb), volume (milk at $3.00 per gal.), or count (bread at $2.00 per loaf). The *as-purchased (AP) unit cost* is the unit cost of a food item based on the form in which it is ordered and received. The AP unit cost of an item is calculated by applying the following formula:

$$APU = APC \div NU$$

where

APU = AP unit cost

APC = AP cost

NU = number of units

For example, to calculate the AP unit cost of an egg, divide the AP cost for a case of eggs by the number of eggs in the case. If the price paid for a case of eggs is $14.40 and there are 15 dozen eggs in the case, the AP unit cost would be $0.08 per egg.

$$APU = APC \div NU$$

APU = \$14.40 ÷ 15 dozen

APU = \$14.40 ÷ (15 dozen × 12 eggs/dozen)

APU = \$14.40 ÷ 180 eggs

APU = \$0.08/egg

Carlisle FoodService Products

The unit of measure used to calculate a unit cost should be based on how the product is used in recipes. Sometimes it may be helpful to calculate the unit cost of an item in more than one unit of measure. For example, some recipes may call for fluid ounces of heavy cream and other recipes may call for quarts of heavy cream. **See Figure 10-30.**

Calculating As-Purchased Unit Costs

AP Unit Cost = AP Cost ÷ Number of Units

1 Case Heavy Cream
AP Total Cost = $31.00
Number of Units = 12 qt

1 qt Heavy Cream
AP Unit Cost (per qt) = $2.58/qt
($31.00 ÷ 12 qt)

1 fl oz Heavy Cream
AP Unit Cost (per fl oz) = $0.08/fl oz
($2.58/qt ÷ 32 fl oz/qt)

Figure 10-30. As-purchased unit costs can be calculated based on more than one unit of measure.

Edible-Portion Costs

A similar calculation is performed to determine the edible-portion (EP) unit cost based on the AP unit cost. The *edible-portion (EP) unit cost* is the unit cost of a food or beverage item after taking into account the cost of the waste generated by trimming. Unless a food item has a yield percentage of 100%, the EP unit cost will always be higher than the AP unit cost. **See Figure 10-31.** Consider a mashed potato recipe that calls for 10 lb of peeled potatoes. The cost of the potatoes for the recipe will be more than the cost of 10 lb of whole potatoes because more than 10 lb of whole potatoes will need to be peeled in order to yield 10 lb of peeled potatoes.

Edible-Portion Unit Costs

Browne Foodservice

Figure 10-31. Unless the yield is 100%, the edible-portion (EP) unit cost will always be higher than the as-purchased (AP) unit cost due to the waste generated by trimming the product.

The EP unit cost of a food item is calculated by applying the following formula:

EPU* = *APU* ÷ *YP

where

EPU = EP unit cost

APU = AP unit cost

YP = yield percentage

For example, if a recipe calls for peeled potatoes and the AP unit cost of whole potatoes is $0.50 per lb and the yield percentage is 80%, what is the EP unit cost of the peeled potatoes?

EPU* = *APU* ÷ *YP

EPU = $0.50/lb ÷ 80%

EPU = $0.50/lb ÷ 0.80

EPU = $0.625/lb = $0.63/lb

The ingredient amounts provided in a recipe are EP quantities. The total cost of an ingredient in a recipe can be calculated by multiplying the EP quantities provided in the recipe by the EP unit cost for each ingredient. When calculating the EP unit cost of an ingredient, the unit of measure in the EP quantity and the AP unit cost must be the same. For example, if the EP quantity of sugar in a cake recipe is 12 oz and the AP unit cost of sugar is $0.64 per lb, the AP unit cost will need to be converted to a price per ounce.

This conversion is done by multiplying the original cost by the appropriate measurement equivalent. In this case, since 16 oz = 1 lb, $0.64 per lb is equivalent to $0.04 per oz. The total cost of the sugar in the recipe can then be calculated by multiplying the EP amount (12 oz) by the EP unit cost ($0.04 per oz), or 12 oz × $0.04/oz = $0.48.

Yield Percentages

Foodservice operations purchase many products in a form that is different from the way the product will be used. For example, meats may need to be trimmed of excess fat and bone, produce may need to be peeled and seeded, and seafood may need to be scaled, skinned, and have the heads and fins removed. In all of these cases, waste is generated. To account for waste, cost calculations are based on yield percentages.

A *yield percentage* is the edible-portion (EP) quantity of a food item divided by the as-purchased (AP) quantity and is expressed as a percentage. Yield percentages do not apply to food items that are served in the same form as they are purchased, such as premade pastries that are simply plated for service. Three different formulas involving yield percentages are used for the following reasons:

- An AP quantity is calculated when the amount of a food item that is required for a recipe (EP quantity) is known and the amount to be ordered needs to be determined.
- An EP quantity is calculated when a purchased amount of food (AP quantity) is already on hand and the edible or usable amount of the food needs to be calculated.
- A yield percentage is calculated when it cannot be found for a particular food item in any reference material.

A percentage circle can be used to help understand how calculations using yield percentages are performed. The circle is divided into three sections. The top section of the circle represents the EP quantity (EPQ), the part of the food item that is edible. The lower left section of the circle represents the AP quantity (APQ), the whole amount of the item. The lower right section contains the yield percentage (YP). When any two of the three variables are known, the third variable can be calculated. **See Figure 10-32.**

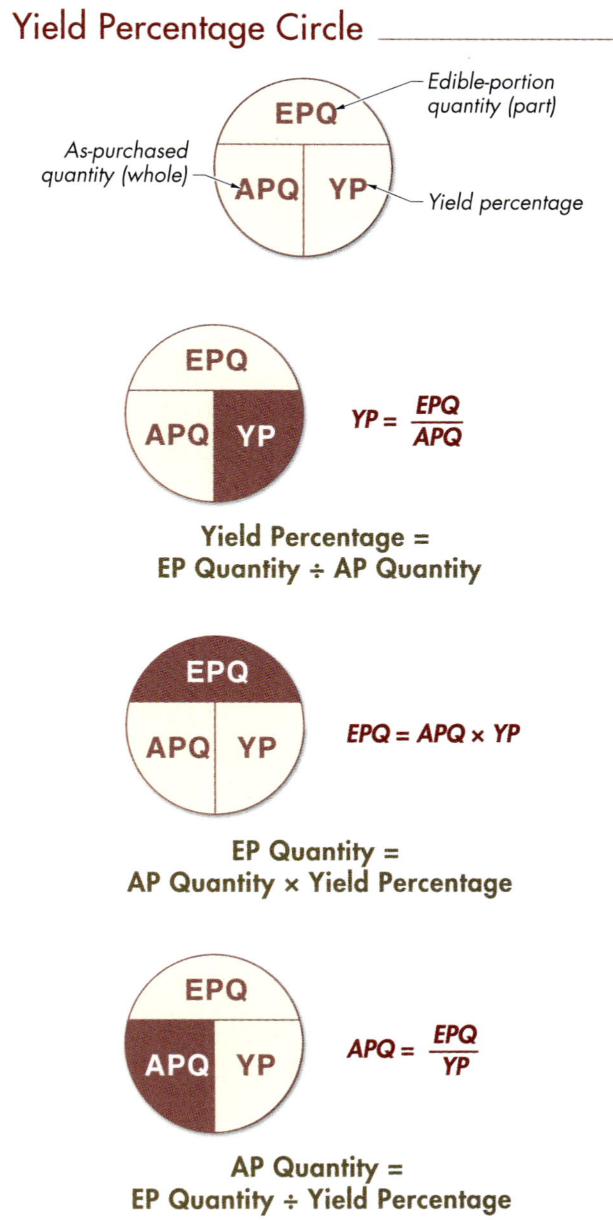

Figure 10-32. If any two variables in a percentage circle are known, the third variable can be calculated.

Calculating AP Quantities. Calculating an AP quantity is done to determine the proper amount of an ingredient to order to make a given quantity of food. The AP quantity is calculated by using the following formula:

$$APQ = EPQ \div YP$$

where

APQ = AP quantity

EPQ = EP quantity

YP = yield percentage

Consider a banquet for 100 people that includes grilled salmon on the menu. The recipe for the grilled salmon dish calls for a ½ lb fillet of salmon per serving. The total EP quantity of salmon can be calculated by multiplying the amount of salmon required per person by the number of people attending the banquet.

If whole salmon is ordered for the banquet, the heads, fins, skin, and bones of the salmon will need to be removed. If the yield percentage for cutting salmon fillets from whole salmon is 50%, the quantity of salmon to be ordered can be calculated by applying the formula for AP quantity as follows:

$$APQ = EPQ \div YP$$

$APQ = 50 \text{ lb} \div 50\%$

$APQ = 50 \text{ lb} \div 0.50$

$APQ = 100 \text{ lb}$

Calculating EP Quantities. Calculating an EP quantity is done to determine how much food can be prepared using an amount of an ingredient that is already on hand. The EP quantity can be calculated using the following formula:

EPQ = APQ × YP

where

EPQ = EP quantity

APQ = AP quantity

YP = yield percentage

For example, a supervisor may decide to take advantage of a special price being offered by a meat supplier and purchase 40 lb of whole beef tenderloin. The beef stroganoff recipe calls for 6 oz per serving of fully trimmed beef tenderloin cut into 1 inch pieces. The supervisor knows from experience that after trimming the fat and connective tissue from the beef tenderloin, only 75% of the tenderloin will be used. The number of servings of beef stroganoff that can be prepared is calculated using the following formula for EP quantity:

EPQ = APQ × YP

EPQ = 40 lb × 75%

EPQ = 40 lb × 0.75

EPQ = 30 lb

Common yield percentages are available in various reference tables. However, if a yield percentage for a particular food item is not known and cannot be found in reference material, it can be calculated by performing a raw yield test or a cooking-loss yield test.

Raw Yield Tests. A *raw yield test* is a procedure used to determine the yield percentage of a food item that is trimmed of waste prior to being used in a recipe. Yield percentage is calculated by using the following formula:

YP = EPQ ÷ APQ

where

YP = yield percentage

EPQ = EP quantity

APQ = AP quantity

To perform the raw yield test, a food item is purchased and weighed to determine the AP quantity. The food item is then trimmed and the edible portion is weighed to determine the EP quantity. For example, to calculate the yield percentage of carrots, the AP quantity of carrots is weighed. Then, the carrots are peeled and the greens and tips are removed as waste (unless the trimmings can be used in another recipe such as a stock). The EP quantity is determined by weighing the cleaned carrots. If 10 lb of carrots weigh 8.5 lb after being trimmed, the yield percentage can be calculated using the following formula:

YP = EPQ ÷ APQ

YP = 8.5 lb ÷ 10 lb

YP = 0.85

YP = 85%

Cooking-Loss Yield Tests. A *cooking-loss yield test* is a procedure used to determine the yield percentage of a food item that loses weight during the cooking process. For example, meat loses weight as fat is rendered and moisture is lost during cooking. The cooking-loss yield test is used when the EP quantity is based on the amount of cooked food to be served as opposed to the amount of raw food to be used in a recipe.

The yield percentage is calculated by dividing the EP quantity by the AP quantity. To perform the test, a food item is weighed to determine the AP quantity. The food item is then weighed again after being trimmed and/or cooked to determine the EP quantity. For example, a hamburger that weighs 8 oz prior to cooking might weigh 6 oz after cooking. The yield percentage in this case can be calculated by using the following formula:

YP = EPQ ÷ APQ

YP = 6 ÷ 8

YP = 0.75

YP = 75%

Calculations involving yield percentages are performed regularly in the professional kitchen. However, variables such as rounding, staff skill level, and product condition affect yield percentages.

Rounding. Special attention should be given to how the results for yield percentage calculations are rounded. **See Figure 10-33.** For example, if it is determined that 30.4 lb of fish should be ordered for a banquet, rounding the amount down to 30 lb using standard math rules would not result in enough fish for the banquet.

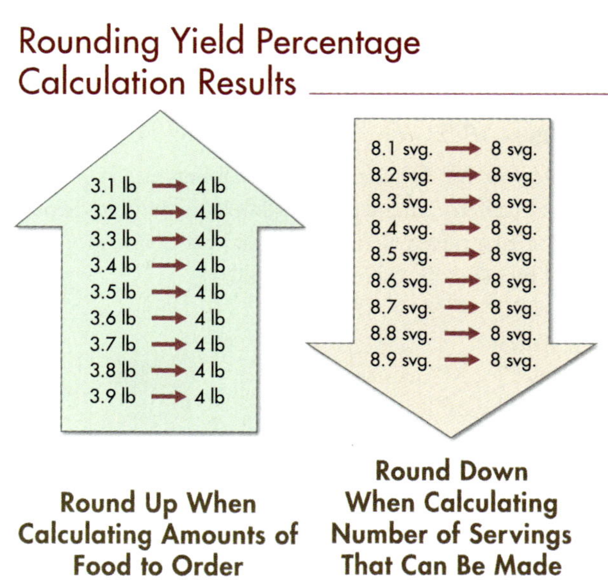

Figure 10-33. Special rules apply to how the results for yield percentage calculations are rounded depending on the quantity be calculated.

When ordering food, the yield percentage results should always be rounded up to ensure that enough food is purchased. However, when calculating the number of servings that can be made from a given recipe, such as 14.8, the yield percentage result should be rounded down to 14 because the recipe will not make 15 full servings.

As-Served Costs

Once the costs of the individual ingredients in a recipe are calculated, the costs are added to determine the total cost of the recipe. Then, the as-served costs for that recipe can be calculated. An *as-served (AS) cost* is the cost of a menu item as it is served to a guest. It is important to note that an AS cost is the total cost of the ingredients required to prepare one serving of a menu item.

Consider a chicken sandwich that is offered as a stand-alone item on a menu. If the sandwich consists of a grilled chicken breast served on a bun with a slice of tomato, a slice of cheese, lettuce, and mayonnaise, the costs of the ingredients are added to calculate the AS cost of the chicken sandwich. However, if the same chicken sandwich is offered on the menu as a meal that comes with fries and coleslaw, the AS cost of that menu item is equal to the total cost of the sandwich, fries, and coleslaw. On this menu, the chicken sandwich is treated as one ingredient of the chicken sandwich meal. **See Figure 10-34.**

To calculate the AS cost of a menu item that is prepared in large quantities and then portioned into servings, divide the total cost of the recipe by the number of portions the recipe yields. For example, if the total cost of the ingredients to prepare 2 gal. of tomato soup is \$15.00 and the recipe yields 32 (8 fl oz) servings, the AS cost of a serving of tomato soup would be \$0.47, or \$15.00 ÷ 32 = \$0.47/serving.

CHECKPOINT 10-5

1. Differentiate between as-purchased unit cost and edible portion unit cost.
2. Describe the relationship between edible-portion and trim loss.
3. Differentiate between calculating as-purchased (AP) quantities and calculating edible-portion (EP) quantities.
4. Differentiate between a raw yield test and a cooking-loss yield test.
5. Explain how to determine the total as-served cost of a menu item.

Calculating As-Served Costs

Chicken Sandwich						
Ingredients	*EP Quantity*	*EP Unit of Measure*	*AP Unit Cost*	*Yield Percentage*	*EP Unit Cost*	*Total Cost*
Chicken breast	4	oz	\$0.20 per oz	90%	\$0.22	\$0.88
Bun	1	each	\$0.40 per each	100%	\$0.40	\$0.40
Tomato slice	1	oz	\$0.18 per oz	90%	\$0.20	\$0.20
Cheese slice	1	each	\$0.20 per each	100%	\$0.20	\$0.20
Lettuce, shredded	0.25	oz	\$0.10 per oz	80%	\$0.13	\$0.03
Mayonnaise	0.5	oz	\$0.15 per oz	100%	\$0.15	\$0.08
					Total As-Served Cost:	**\$1.79**

Individual recipe cost becomes part of menu item cost

Chicken Sandwich Meal						
Ingredients	*EP Quantity*	*EP Unit of Measure*	*AP Unit Cost*	*Yield Percentage*	*EP Unit Cost*	*Total Cost*
Chicken sandwich	1	each	\$1.80 per each	100%	\$1.80	\$1.79
French fries	4	oz	\$0.20 per oz	100%	\$0.20	\$0.80
Coleslaw	3	oz	\$0.20 per oz	100%	\$0.20	\$0.60
					Total As-Served Cost:	**\$3.19**

Figure 10-34. An as-served (AS) cost is the total cost of the ingredients and the recipes that make up a menu item.

FOOD COST PERCENTAGES

Another way of examining food costs is through the use of food cost percentages. A *food cost percentage* is a percentage that indicates how the cost of food relates to the menu prices and food sales of a foodservice operation. The three types of food cost percentages used in the foodservice industry are menu-item food cost percentages, overall food cost percentages, and target food cost percentages.

- A *menu-item food cost percentage* is the AS cost of a menu item divided by the menu price, written as a percent.
- An *overall food cost percentage* is the total amount of money a foodservice operation spends on food divided by the total food sales over a defined period of time, written as a percent.
- A *target food cost percentage* is the percentage of food sales that a foodservice operation plans to spend on purchasing food.

Like a food cost percentage, a *beverage cost percentage* is a percentage that indicates how the cost of beverages relates to menu prices and beverage sales of a foodservice operation. The word beverage can be substituted for the word food in any of these calculations because the math is the same. If a menu item consists of both a food and beverage component that is going to be offered for a single price, the menu price must be based on a combined food and beverage cost percentage.

Menu-Item Food Cost Percentages

A menu-item food cost percentage is equal to the AS cost of a menu item divided by the menu price written as a percent. This percentage shows how the cost of the ingredients relates to the price charged for the menu item. To calculate a menu-item food cost percentage, apply the following formula:

$$IFC\% = ASC \div MP$$

where

IFC% = menu-item food cost percentage

ASC = AS cost

MP = menu price

A foodservice operation makes a profit by charging more for the items on a menu than it costs to prepare those items. Therefore, the menu price is always higher than the AS cost of that item. Since a menu-item food cost percentage is equal to the AS cost divided by the menu price, a menu-item food cost percentage will always be less than 100%.

For example, if the menu price for the chicken sandwich is \$6.95 and the AS cost is \$1.80, the menu-item food cost percentage for the chicken sandwich would be 25.9%.

$$IFC\% = ASC \div MP$$
$$IFC\% = \$1.80 \div \$6.95$$
$$IFC\% = 0.259$$
$$IFC\% = 25.9\%$$

Likewise, if the menu price for the chicken sandwich served with fries and coleslaw is \$9.95 and the AS cost is \$3.20, the menu-item food cost percentage for the meal would be 32.1%.

$$IFC\% = ASC \div MP$$
$$IFC\% = \$3.20 \div \$9.95$$
$$IFC\% = 0.321$$
$$IFC\% = 32.1\%$$

Overall Food Cost Percentages

In addition to calculating food cost percentages for individual menu items, foodservice operations also calculate overall food cost percentages based on total food sales. An overall food cost percentage is equal to the total amount of money a foodservice operation spends on food divided by its total food sales over a defined period of time. To calculate an overall food cost percentage, use the following formula:

$$OFC\% = FC \div FS$$

where

OFC% = overall food cost percentage

FC = total food costs

FS = total food sales

For example, if a restaurant spends \$9000 on food in a month when its total food sales were \$36,000 for that month, the overall food cost percentage for the restaurant would be 25%.

$$OFC\% = FC \div FS$$
$$OFC\% = \$9000 \div \$36{,}000$$
$$OFC\% = 0.25$$
$$OFC\% = 25\%$$

The other 75% of the sales would then be available for the foodservice operation to pay for expenses, such as payroll, rent, and utilities, and to contribute to the earnings of the owners of the operation. Lower food cost percentages translate into higher potential earnings for the owners of the operation.

Target Food Cost Percentages

A target food cost percentage is the percentage of total food sales that a foodservice operation plans to spend on purchasing food. Foodservice supervisors regularly compare overall food cost percentages with target food cost percentages to determine if food costs are higher or lower than planned. **See Figure 10-35.** When the overall food cost percentage of a foodservice operation is lower than the target food cost percentage, the operation has earned more money selling food than planned. Likewise, if the actual overall food cost percentage is higher than the target food cost percentage, the operation has earned less money selling food than planned. It is the responsibility of management to monitor food costs and make any necessary adjustments.

Food Cost Percentages

Daily Sales Report—Lunch Target Food Cost Percentage = 30%						
Menu Item	**AS Cost**	**Menu Price**	**Menu Item Food Cost % (AS Cost ÷ Menu Price)**	**Number Sold**	**Total Food Cost (AS Cost × Number Sold)**	**Total Food Sales (Menu Price × Number Sold)**
Appetizers						
Bruschetta	$1.75	$5.95	29.4%	48	$84.00	$285.60
BBQ shrimp	$4.25	$11.95	35.6%	27	$114.75	$322.65
Entrées						
Fried chicken	$2.65	$12.95	20.5%	43	$113.95	$556.85
Strip steak	$9.00	$25.95	34.7%	32	$288.00	$830.40
Desserts						
Strawberry shortcake	$2.25	$5.95	37.8%	40	$90.00	$238.00
Chocolate cake	$0.50	$3.95	12.7%	30	$15.00	$118.50
Total					**$705.70**	**$2352.00**
Overall Total Food Cost Percentage (Total Food Costs ÷ Total Food Sales)						**30.0%**

Figure 10-35. A target food cost percentage is the percentage of food sales that a foodservice operation plans to spend on purchasing food.

Managing Food Costs

CHECKPOINT 10-6

1. Define food cost percentage.
2. Identify three types of food cost percentages.
3. Explain how a foodservice operation makes a profit.
4. Explain the relationship between an overall food cost percentage and a target food cost percentage.

CALCULATING MENU PRICES

There are three primary methods used by foodservice operations to calculate menu prices. These methods include food cost percentage pricing, perceived value pricing, and contribution margin pricing. Regardless of the pricing method used, the AS costs of the menu items must be known first.

Beverage pricing is calculated the same way as food pricing. Simply substitute beverage costs for food costs in the original calculation for menu-item cost percentages.

Food Cost Percentage Pricing

Food cost percentage pricing begins by calculating a target price for each menu item. A *target price* is the price that a foodservice operation needs to charge for a menu item in order to meet its target food cost percentage. Target prices are calculated using the following formula:

$$TP = ASC \div TFC\%$$

where

TP = target price

ASC = AS cost of a menu item

$TFC\%$ = target food cost percentage

For example, if the AS cost of a sandwich is $1.45 and the target food cost percentage is 30%, the target price would be $4.83.

$$TP = ASC \div TFC\%$$
$$TP = \$1.45 \div 30\%$$
$$TP = \$1.45 \div 0.30$$
$$TP = \$4.83$$

Some foodservice operations will stop at this point and set the menu price equal to the target price. However, most foodservice operations tend to consider target prices as a starting point and then adjust the price up or down based on how management thinks guests will perceive the price.

Perceived Value Pricing

Perceived value pricing is the process of adjusting a target menu price based on how management thinks a guest will perceive the price. Many foodservice operations use a combination of food cost percentage pricing and perceived value pricing to determine final menu prices. **See Figure 10-36.**

There are two aspects to perceived value pricing. The first involves adjusting a target price to look like a price guests have become used to seeing on menus. For example, a target price for a menu item is calculated to be $7.03. Most foodservice operations will adjust the price down to $6.95 or $7.00.

Therefore, the actual food cost percentage for the menu item would be slightly higher or lower than the target food cost percentage. For example, if the target price of a bowl of soup is calculated to be $4.67 based on an AS cost of $1.40 and a 30% target food cost, the restaurant might increase the price to $4.95. At $4.95, the actual food cost percentage would be 28.3%, which is slightly lower than the target food cost percentage of 30%.

$$\boldsymbol{IFC\% = ASC \div MP}$$

$$IFC\% = \$1.40 \div \$4.95$$

$$IFC\% = 0.283$$

$$IFC\% = 28.3\%$$

Pricing Menu Items

Menu Pricing

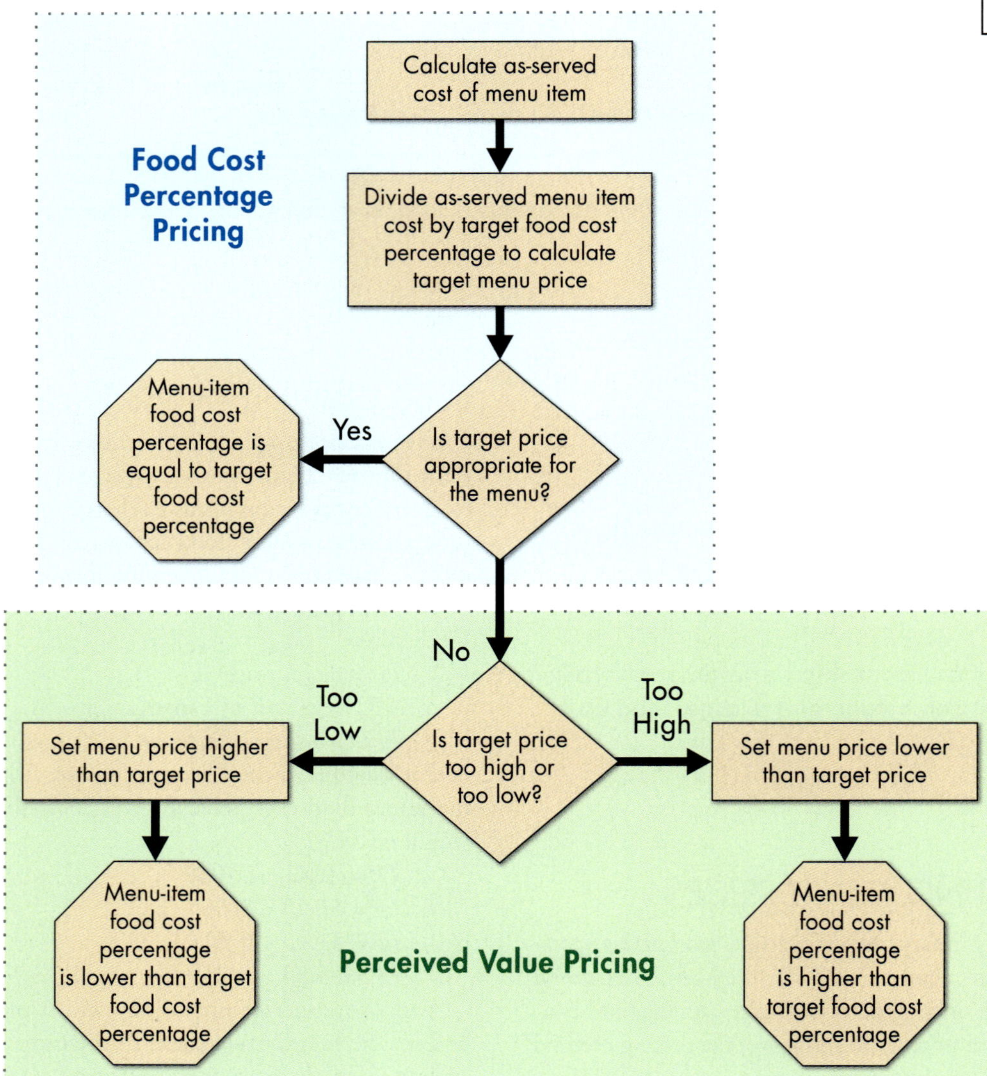

Figure 10-36. Many foodservice operations use a combination of food cost percentage pricing and perceived value pricing to calculate menu prices.

The second aspect of perceived value pricing involves adjusting a target price so that guests will view the price as reasonable relative to other items on the menu and similar items sold elsewhere. For example, a restaurant may offer an appetizer of nachos made with tortillas chips, taco meat, cheese, and tomatoes that has an AS cost of $1.05. If the target price is calculated based on a 28% target food cost percentage, the target price would be $3.75.

$TP = ASC \div TFC\%$

$TP = \$1.05 \div 28\%$

$TP = \$1.05 \div 0.28$

$TP = \$3.75$

If other appetizers on the menu include Buffalo wings for $8.95 and mozzarella sticks for $5.95, guests may be willing to pay more than $3.75 for the nachos because the perception is that nachos represent a substantial dish and should cost more than the mozzarella sticks but less than the Buffalo wings. In a case like this, the restaurant is likely to increase the price of the nachos to $6.95, which means the menu-item food cost percentage for the nachos would be 15.1%.

$IFC\% = ASC \div MP$

$IFC\% = \$1.05 \div \$6.95$

$IFC\% = 0.151$

$IFC\% = 15.1\%$

If the price was left at the target price of $3.75 the restaurant would earn $2.70 ($3.75 –$1.05 = $2.70) on each order of nachos. With the price adjusted to $6.95 the restaurant would earn $5.90 ($6.95 – $1.05 = $5.90) on each order of nachos. A lower menu-item food cost percentage means that the foodservice operation earns more on each order of nachos that are sold. However, just because lower percentages result in higher earnings per order, it does not mean that a foodservice operation can or should base all menu prices on a lower food cost percentage.

To determine the price of other items based on the same menu-item food cost percentage as the nachos, use the following formula:

$MP = ASC \div IFC\%$

where

MP = menu price

ASC = AS cost

$IFC\%$ = menu-item food cost percentage

For example, if the same restaurant selling the nachos prices a 24 oz rib-eye steak with an AS cost of $14.00 based on the same 15.1% food cost percentage, the menu price of the steak would be $92.71.

$MP = ASC \div IFC\%$

$MP = \$14.00 \div 15.1\%$

$MP = \$14.00 \div 0.151$

$MP = \$92.71$

It is unlikely that many guests would pay $92.71 for steak, even at a high-end restaurant. If the target food cost percentage is adjusted to 28%, the price of the steak would still be relatively high at $50.00 ($14.00 ÷ 0.28 = $50.00). If restaurant management thinks most guests will perceive a price of $50.00 as too high, the price may be reduced to be more in line with the perceived value of that restaurant's guests. If the final menu price set for the steak is $36.95, the food cost percentage will increase to 37.9%.

$IFC\% = ASC \div MP$

$IFC\% = \$14.00 \div \$36.95$

$IFC\% = 0.379$

$IFC\% = 37.9\%$

It is normal for items on a menu to have different food cost percentages. As long as there is a proper balance between items with lower food cost percentages and items with higher food cost percentages, the foodservice operation can maintain an overall food cost percentage that is close or equal to its target food cost percentage.

Contribution Margin Pricing

Another method used to calculate menu prices is referred to as contribution margin pricing. A *contribution margin* is the amount added to the AS cost of a menu item to determine a menu price. A foodservice operation calculates prices with a contribution margin to ensure that the amount of money made on each serving of food will cover expenses and still result in a profit. This type of pricing is not common in restaurants but is often used when calculating a price charged per person for a special event.

For example, a banquet is being prepared for 200 people. The menu includes a salad, an entrée, and a dessert that has a total AS cost of $20.00 per person. Therefore, the total AS cost for all of the food will be $4,000 (200 people × $20/person). The banquet manager calculates that all of the other expenses, such as labor and rental fees, total an additional $5000. It is banquet policy that a party of this size must generate earnings of $2000. Therefore, the contribution margin for this event would be $7000 ($5000 + $2000).

To calculate the contribution margin on a per person basis, the banquet manager simply divides the total contribution margin by the number of people attending the event ($7000 ÷ 200 people = $35/person).

This contribution margin is then added to the AS cost per person to calculate the final menu price for the party, which is $55 per person ($20.00/person + $35.00/person = $55/person).

Pricing Forms

A *pricing form* is a tool often used to help calculate the AS cost of a menu item and establish a menu price. **See Figure 10-37.** Pricing forms are updated when ingredient costs change to ensure that menu prices are adjusted as needed. The following steps are typically followed to complete a pricing form.

1. Enter information from the standardized recipe including the menu item name, number of portions, ingredients, and ingredient amounts (EP quantities).
2. Record the AP unit cost of each ingredient. If necessary, convert the unit cost to the same unit of measure as used in the EP quantity.
3. Record the yield percentage for each ingredient.
4. Calculate the EP unit costs by dividing the AP unit cost of each ingredient by the yield percentage of the ingredient.
5. Calculate the total cost of each ingredient by multiplying the EP quantity of each ingredient by the EP unit cost of each ingredient.
6. Calculate the total cost by adding the costs of each ingredient.
7. Calculate the cost per portion by dividing the total recipe cost by the number of portions.
8. Calculate the target price by dividing the total cost by the target food cost percentage.
9. Adjust the target price if desired and record the menu price.
10. Calculate the food cost percentage by dividing the cost per portion by the menu price.

CHECKPOINT 10-7

1. Identify three methods used to calculate menu prices.
2. Explain the purpose of a pricing form.

Pricing Forms

Total recipe cost ÷ number of portions

Total recipe cost ÷ target food cost

Cost per portion ÷ menu price

AP unit cost ÷ yield percentage

EP quantity × EP unit cost

Menu Item Name: Pot Roast
Number of Portions: 8
AS Cost per Portion: $3.95
Target Food Cost %: 30.0%
Target Price: $13.17

Menu Price: $13.95
Menu-Item Food Cost %: 28.3%

Ingredients	EP Quantity	EP Unit of Measure	AP Unit Cost (or recipe cost)	Yield Percentage	EP Unit Cost	Total Ingredient Cost
Eye of round, trimmed	5	lb	$3.80 per lb	85%	$4.47	$22.35
Vegetable oil	2	oz	$0.15 per oz	100%	$0.15	$0.30
Onions, diced	1	lb	$0.40 per lb	93%	$0.43	$0.43
Garlic, minced	0.5	oz	$0.25 per oz	85%	$0.29	$0.15
Tomato paste	6	oz	$0.20 per oz	100%	$0.20	$1.20
Beef stock	2	qt	$2.50 per qt	100%	$2.50	$5.00
Carrots, diced	2	lb	$0.40 per lb	85%	$0.47	$0.94
Celery, diced	1	lb	$0.50 per lb	70%	$0.71	$0.71
Cornstarch	3	oz	$0.15 per oz	100%	$0.15	$0.45
Salt	1	tbsp	$0.01 per tbsp	100%	$0.01	$0.01
Black pepper	2	tsp	$0.03 per tsp	100%	$0.03	$0.06
					Total Cost:	**$31.60**

EP quantity units and AP cost units must match

Sum of individual ingredient costs

Figure 10-37. A pricing form is used to help calculate the as-served (AS) cost of a menu item to establish a menu price.

PROFIT AND LOSS

Although every foodservice operation is somewhat unique, the methods for calculating and estimating profits and losses are fairly standard. Foodservice operations have revenues based on the sales of food and beverages. They also have expenses that can be categorized as either fixed or variable based on the amount of those sales. There are two categories of profit associated with any foodservice operation: gross profit and net profit.

Calculating Gross Profit

Gross profit is the calculated difference between total revenue and the cost of goods sold. The formula for calculating gross profit is as follows:

$$GP = TR - CGS$$

where

GP = gross profit

TR = total revenue

CGS = cost of goods sold

For example, if a restaurant had a total revenue of $31,000 in a month when the cost of goods sold was $10,500, the gross profit would be $20,500.

$$GP = TR - CGS$$
$$GP = \$31{,}000 - \$10{,}500$$
$$GP = \$20{,}500$$

Calculating Net Profit

Net profit is the calculated difference between the gross profit and operating expenses of a foodservice operation. The formula for calculating net profit is as follows:

$$NP = GP - OE$$

where

NP = net profit

GP = gross profit

OE = operating expenses

If the restaurant with a gross profit of $20,500 had $18,250 in operating expenses, there would be a net profit of $2250.

$$NP = GP - OE$$
$$NP = \$20{,}500 - \$18{,}250$$
$$NP = \$2250$$

However, if the same restaurant had $21,500 in operating expenses, the restaurant would have a net profit of –$1000, which is a loss of $1000.

$$NP = GP - OE$$
$$NP = \$20{,}500 - \$21{,}500$$
$$NP = -\$1000 \text{ or } (\$1000) \text{ or } <\$1000>$$

The relationship between gross profit and net profit is clearly shown on a standard profit and loss statement.

A *profit and loss statement* is a form that shows the revenue, expenses, and resulting gross and net profit (or loss) over a specific period of time. The purpose of a standard profit and loss statement is to provide management with an indication of the financial status of the foodservice operation. A profit and loss statement can be prepared at a detailed level or a summary level depending on the complexity and needs of the foodservice operation. **See Figure 10-38.** The net profit is the last entry on the statement and is referred to as the "bottom line."

Financial Data Analysis

Profit and Loss Statements

Summary Profit and Loss Statement

Total revenue	**$ 34,908.59**
Cost of goods sold	**$ 10,432.90**
Gross profit	**$ 24,475.69**
Operating expenses	**$ 20,768.25**
Net profit	**$ 3707.44**

Detailed Profit and Loss Statement

Total revenue	**$ 34,908.59**
Total food revenue	$ 27,801.23
Total beverage revenue	$ 7107.36
Cost of goods sold	**$ 10,432.90**
Total food costs	$ 7794.38
Total beverage costs	$ 2638.52
Gross profit	**$ 24,475.69**
Operating expenses	**$ 20,768.25**
Payroll expenses	$ 13,010.15
Social security taxes	$ 1050.75
Utilities	$ 1124.67
Rent	$ 2800.00
Interest	$ 325.19
Insurance and licenses	$ 356.80
Kitchen supplies	$ 803.21
Sales tax	$ 1047.26
Miscellaneous	$ 250.22
Net profit	**$ 3707.44**

Figure 10-38. A profit and loss statement can be prepared at a summary or a detailed level.

Payroll Expenses

Payroll expenses are frequently the most significant expense of a foodservice operation. **See Figure 10-39.** A *payroll expense* is any money paid to an employee who performs work for the operation. Employees are paid on either a salaried (fixed-cost) or hourly (variable-cost) basis. A salaried employee is paid based on a fixed amount of money. An hourly employee is paid based on an hourly wage multiplied by the number of hours worked.

Payroll Expenses

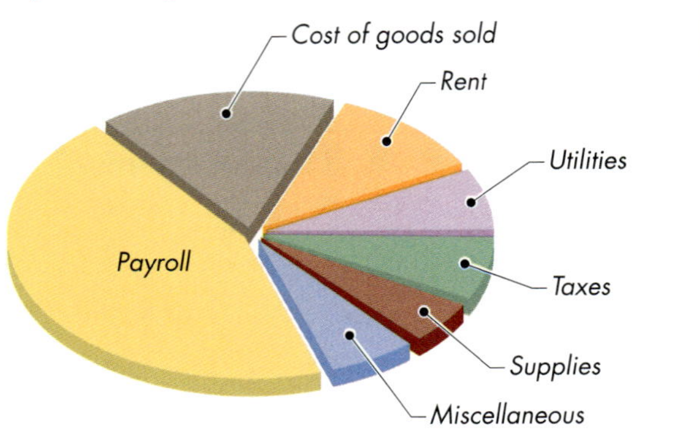

Figure 10-39. Payroll expenses are frequently the most significant expense of a foodservice operation.

Payroll expenses can be fixed or variable depending on how employees are paid. Generally, the payroll expenses of salaried employees are considered fixed expenses and the payroll expenses of hourly employees are considered variable expenses. Hourly employees are required to be paid at least 1.5 times their regular hourly wage for all hours over 40 worked in a week. For example, a cook earning $10 per hour is paid an overtime wage of $15.00 per hour ($10.00 × 1.5 = $15.00). This overtime wage is commonly referred to as "time-and-a-half."

Calculating Gross Pay. Payroll expenses for a foodservice operation are based on the gross pay of its employees. *Gross pay* is the total amount of an employee's pay before any deductions are made. Examples of deductions include federal, state, and local income taxes, Social Security taxes, employee-paid health insurance premiums, and contributions to retirement or savings plans. The gross pay for a salaried employee is the same for every pay period. The gross pay for an hourly employee may vary each pay period depending on the number of hours worked.

To calculate the gross pay of an hourly employee, regular pay is added to any overtime pay. Regular pay is equal to the number of hours worked up to 40 hours each week multiplied by the regular hourly wage. Overtime pay is equal to any hours worked over 40 hours in a week multiplied by the overtime hourly wage. For example, the gross pay for a line cook who earns $12 per hour and works 50 hours in one week would be calculated as follows:

Regular pay = 40 hours × $12.00/hour = $480.00
Overtime pay = 10 hours × ($12.00/hour × 1.5)
= 10 hours × $18.00
= $180.00
Total gross pay = $480.00 + $180.00 = $660.00

To calculate the total payroll expenses for a foodservice operation, the gross pay of each employee is added together. **See Figure 10-40.**

Calculating Net Pay. The net pay of each employee's paycheck will be less than the employee's gross pay. *Net pay* is the actual amount on an employee's paycheck and is equal to the gross pay minus payroll deductions. One of the most common categories of deductions is taxes. All employees are subject to pay Social Security and Medicare taxes. Most employees are required to pay federal income taxes. Depending on the city where the job is located, some employees are also required to pay state and/or local income taxes. Although taxes are not part of the net pay, they are part of the employee's gross pay and are considered foodservice operation expenses. The foodservice operation is required to withhold the amount of taxes owed by the employee and then pay those taxes directly to the government on behalf of the employee.

Additional Costs

Foodservice operations are required to pay additional taxes related to payroll that are above and beyond the taxes included in the gross pay. These taxes include the employer's share of Social Security and unemployment taxes. Foodservice operations are also required to carry various types of insurance such as liability insurance and workers' compensation insurance.

Workers' Compensation. *Workers' compensation* is a type of insurance that compensates an employee for medical expenses and the loss of wages due to work-related injuries and illnesses. Workers' compensation tax is collected by the individual states and each state has their own tax rates. Employers are required to provide workers' compensation insurance coverage to employees. They must also post information on workers' compensation, as specified by the state, in a prominent place in the operation to which all employees have access. Employees are required to report any work-related injuries, accidents, or illnesses to a supervisor within a specified time period.

Workers' Compensation

Calculating Payroll Expenses

Payroll Summary Report—Week of Jan 1–7, 20XX						
Employee	**Regular Hours**	**Overtime Hours**	**Pay Rate**	**Regular Pay** +	**Overtime Pay** =	**Gross Pay**
Prep cook	30	0	$9/hr	$270.00	$0.00	$270.00
Salad cook	40	0	$10/hr	$400.00	$0.00	$400.00
Line cook 1	40	10	$12/hr	$480.00	$180.00	$660.00
Line cook 2	40	5	$12/hr	$480.00	$90.00	$570.00
Server 1	40	0	$6/hr	$240.00	$0.00	$240.00
Server 2	40	6	$6/hr	$240.00	$54.00	$294.00
Chef	—	—	$800/wk	$800.00	—	$800.00
General manager	—	—	$1000/wk	$1000.00	—	$1000.00
					Total	**$4234.00**

Figure 10-40. The total payroll expenses of a foodservice operation are equal to the sum of the gross pay for all of the employees.

Unemployment Compensation. When employees lose their jobs for reasons beyond their control, they may be entitled to unemployment compensation benefits. *Unemployment compensation* is monetary payments made to workers who have lost their jobs through no fault of their own for a given period of time or until they find a new job. The amount is based on a percentage of the salary earned when the employee held the position within the operation. Unemployment compensation tax is collected by both the state and federal government.

Unemployment Insurance. The operation pays unemployment insurance when an employee files for this benefit after losing a job. The amount of unemployment insurance an operation pays is directly related to the number of employees it has terminated or laid off over a specific period of time. The greater the number of layoffs and applicable terminations, the higher the amount paid for unemployment insurance.

Benefit Packages. Employees receive payment for the work performed during a pay period. However, some employers also offer employees benefit packages to attract and retain them. These benefits are typically partially subsidized by the employer, and the employee contributes to pay for the premiums. Larger organizations may also offer stock purchase programs for their employees. Common benefits offered may include the following:

- paid holidays, vacations, and sick leave
- medical, dental, vision, and life insurance
- retirement and legal assistance plans
- discounts for hotels, theme parks, and health clubs

Canada Beef Inc.

CHECKPOINT 10-8

1. Differentiate between gross profit and net profit.
2. Name the type of statement used to indicate the financial status of a foodservice operation.
3. Identify the most significant expense of a foodservice operation.
4. Differentiate between gross pay and net pay.
5. Identify two additional taxes that foodservice operations are required to pay.
6. Differentiate between workers' compensation and unemployment insurance.

Refer to DVD for **Quick Quiz®** questions

SUSTAINABILITY CORNER

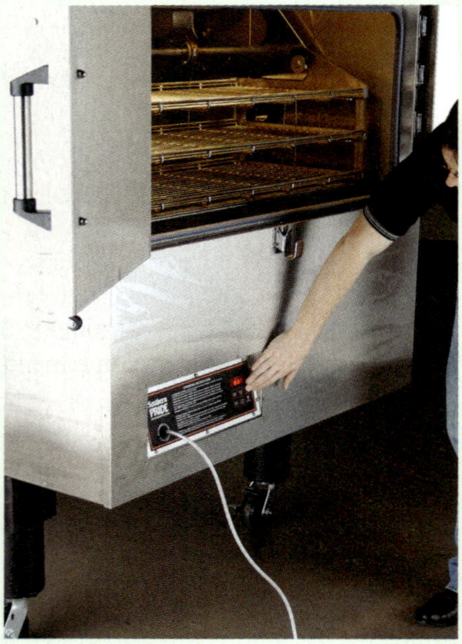

Southern Pride

There are millions of restaurant meals served daily in the United States. This provides the restaurant industry millions of opportunities to make sustainable choices. The restaurant and food industry has the power to become the environmental stewards of the planet. Not only are sustainable practices a way to reduce operating costs, but for an industry that is the largest user of energy and producer of food waste, the solutions make every member of staff a leader in sustainability.

As a supervisor, you can help make an establishment sustainable through leading by example. Examples can be as simple as always turning off a light when leaving a room or finding a leaky faucet and fixing it. Having an understanding of the impact that sustainable practices have on the business, the community, and the world will help you lead and motivate others to change their behavior.

Electrical use in a commercial kitchen is continual. A common practice is for the first staff member who arrives to power up all of the equipment in the kitchen. However, the equipment may not be used for hours and therefore both energy and profits are being wasted. As a supervisor, you can create a power up/power down schedule, which will save costs and energy and provide an opportunity to get everyone involved.

Researching who will pick up a facility's used fats, oils, and grease (FOG) versus throwing them down the drain is another sustainable practice that is hugely beneficial to the environment. FOG causes clogging problems in a drainage system, causes environmental problems, and can cost a restaurant money to fix. As a supervisor, you can turn the FOG into a profit for the restaurant. There are now businesses that will recycle the FOG to create biofuel for use in cars and trucks. Your research to find a company that will pay for FOG will benefit everyone.

There are many ways for a supervisor to teach staff how to run a sustainable facility, but the supervisor must lead by example and encourage best practices. Then, every staff member in the establishment can make a difference.

Recycling FOG

Certification Exam Preparation Questions

1. A(n) ___ is a person who oversees the work done by others and who has the responsibility of directing, training, and evaluating staff.
   A. owner
   B. supervisor
   C. leader
   D. scheduler

2. Effective communication requires ___, speaking, observing, and writing.
   A. prioritizing
   B. listening
   C. organizing
   D. scheduling

3. ___ is used to develop effective staff schedules in spite of fluctuating demand.
   A. Workplace policy
   B. Prioritizing
   C. Historical data
   D. Time management

4. ___ is the act of assigning a task to another person while remaining responsible for the results.
   A. Scheduling
   B. Supervising
   C. Delegating
   D. Prioritizing

5. ___ is the unfair treatment of an individual based on certain characteristics such as gender, race, ethnicity, age, religion, appearance, or disability.
   A. Obsolescence
   B. Zero-tolerance
   C. Discrimination
   D. Termination

6. A(n) ___ evaluation is a scheduled and documented performance evaluation conducted on a regular basis that becomes part of an employee's permanent record.
   A. verbal
   B. written
   C. formal
   D. informal

7. During OJT, new staff members are shown what needs to be done and how to do it and are then expected to be able to execute tasks ___.
   A. carefully
   B. independently
   C. slowly
   D. quickly

8. ___ food involves the selection and ordering of products required to prepare and serve the items on a menu.
   A. Scheduling
   B. Receiving
   C. Purchasing
   D. Storing

9. Staff ___ is a period during which new staff members are introduced to the policies and procedures of a workplace.
   A. orientation
   B. probation
   C. evaluation
   D. termination

10. A(n) ___ is an experienced person who coaches and guides a less experienced person.
    A. manager
    B. supervisor
    C. mentor
    D. owner

11. Disciplinary actions include oral reminders and ___ reminders.
    A. intermittent
    B. regular
    C. visual
    D. written

12. ___ is behavior that is found to be threatening or unacceptable to the general public.
    A. Harassment
    B. Zero-tolerance
    C. Discrimination
    D. Suspension

13. To accept gifts from a vendor is ___.
   A. encouraged
   B. forbidden
   C. ethical
   D. unethical

14. Every purchasing and staffing decision relates directly back to the ___.
   A. schedule
   B. par stock
   C. menu
   D. requisition

15. ___ is the removal of an item from the menu while ingredients for that particular item are still in stock.
   A. Obsolescence
   B. Par stock
   C. FIFO
   D. Scaling

16. A(n) ___ menu is a menu that changes frequently to coincide with changes in the availability of products.
   A. cycle
   B. market
   C. California
   D. à la carte

17. ___ is a condition in which gluten damages the small intestine's ability to absorb nutrients.
   A. Anaphylaxis
   B. Celiac disease
   C. Lactose intolerance
   D. none of the above

18. A ___ is an assortment of items that may be ordered from a given menu.
   A. cycle menu
   B. market menu
   C. menu mix
   D. pricing form

19. The ___ is the number that each ingredient amount in a recipe is multiplied by when the recipe yield is changed.
   A. edible yield
   B. edible portion
   C. yield percentage
   D. scaling factor

20. ___ is monetary payments made to workers who have lost their job through no fault of their own for a given period of time or until they find a new job.
   A. Social Security
   B. Workers' compensation
   C. Unemployment compensation
   D. Unemployment tax

21. List four key interpersonal skills of effective foodservice supervisors.

____

____

____

____

22. List the four workplace policies that should always be documented in writing.

____

____

____

____

23. Scale a recipe based on portion size.

____

____

____

____

____

____

____

24. Calculate the edible-portion cost of a food item.

____

____

____

____

____

____

SKETCHING EXERCISE

25. Sketch a fixed menu showcasing the menu mix, item placement, use of graphics, and price placement.

Notes

Glossary

A

abalone: A univalve contained in a brown, bowl-shaped shell with an iridescent, multicolored interior.

achiote seed: A seed spice made from a red, corn-kernel-shaped seed of the annatto tree. Also known as annatto seed.

acorn squash: A winter squash that looks like a large, dark-green acorn.

active dry yeast: A form of yeast that has been dehydrated and looks like small granules.

adductor muscle: A muscle that opens and closes the shell of a bivalve.

agave nectar: The nectar from an agave plant. Also known as agave syrup.

aggregate fruit: A cluster of very tiny fruits.

aging: The period of rest that occurs for a length of time after an animal has been slaughtered.

aioli: A thick emulsion made by slowly blending olive oil with raw garlic and egg yolks or cooked potatoes.

airline breast: A boneless skin-on chicken breast with the first wing section (bone-in) attached.

à la carte menu: A menu that has all food and beverage items priced separately.

albumen: The clear portion of the raw egg, which makes up two-thirds of the egg and consists mostly of ovalbumin protein. Also known as the egg white.

allemande sauce: A sauce made by adding fresh lemon juice and a yolk-and-cream liaison to velouté.

allspice: A berry spice made from the dried, unripe fruit of a small pimiento tree. Also known as Jamaican pepper.

almond: A teardrop-shaped fruit that grows on small almond trees.

ammonium bicarbonate: A chemical leavening agent that reacts as soon as it is heated without the help of an acid or base.

anaphylaxis: A severe allergic reaction that causes the airway to narrow and prohibits breathing.

anise: A seed spice made from a comma-shaped seed of the anise plant, which is a member of the parsley family.

Anjou pear: A plump, lopsided pear that is green in color.

appetizer: Food that is larger than a single bite and is typically served as the first course of a meal.

appetizer salad: A salad that is served as a starter to a meal.

apple: A hard, round pome that can range in flavor from sweet to tart and in color from pale yellow to dark red.

apricot: A drupe that has pale orange-yellow skin with a fine, downy texture and a sweet and aromatic flesh.

aram sandwich: A type of a cold wrap. Also known as a levant.

aromatic: An ingredient added to a food to enhance its natural flavors and aromas.

arrowroot: A thickening agent that is the edible starch from the rootstock of the arrowroot plant.

artichoke: The edible flower bud of a large, thistle-family plant that comes in many varieties.

Asiago: A grating cheese with a nutty, toastlike flavor.

Asian chili sauce: A red-colored Asian sauce made from puréed red chiles and garlic.

Asian pear: A round pear with the texture of an apple and yellow-colored skin. Also known as an apple-pear.

asparagus: A green, white, or purple edible stem vegetable that is referred to as a spear.

aspic: A savory jelly made from a consommé or clarified meat, fish, or vegetable stock that produces a clear finish when coating foods. Also known as aspic gelée.

as-purchased (AP) cost: The amount paid for a product in the form it was ordered and received.

as-purchased (AP) quantity: The original amount of a food item as it is ordered and received.

as-purchased (AP) unit cost: The unit cost of a food item based on the form in which it is ordered and received.

as-served (AS) cost: The cost of a menu item as it is served to a guest.

Atlantic oyster: A variety of oyster that has a fairly flat shell and a distinctive, salty-flavored, plump, and tender flesh. Also known as an Eastern oyster.

au jus: The natural, unthickened juices that flow from meat as it cooks.

avocado: A pear-shaped drupe with a rough green skin and a large pit surrounded by yellow-green flesh. Also known as an alligator pear.

avocado oil: A type of oil produced by pressing avocadoes.

B

baby back ribs: The meaty bones on the rib end of the pork loin.

baby T-bone: A 6–8 oz steak cut from the loin of veal.

bacon: Pork belly that has been cured and usually smoked.

bacteria: Pathogens that live in soil, water, or organic matter such as the bodies of plants and animals.

baked Alaska: A composed frozen dessert made with a sponge cake soaked in a liqueur-flavored simple syrup that is topped with ice cream and covered with an Italian meringue.

baker's cheese: A fresh cheese made from skim milk that is like cottage cheese but softer and finer grained.

baker's percentage: The weight of a particular ingredient expressed as a percentage based on the weight of the main ingredient in a formula.

baker's scale: A scale with two platforms that use a counterbalance system to measure weight. Also known as a balance scale.

baking: The dry-heat cooking technique in which food is cooked uncovered in an oven.

baking powder: A chemical leavening agent that is a combination of baking soda and cream of tartar or sodium aluminum sulfate.

baking soda: An alkaline chemical leavening agent that reacts to an acidic dough or batter by releasing carbon dioxide without the addition of heat. Also known as sodium bicarbonate.

ballotine: A boned poultry leg that has been stuffed with forcemeat.

balsamic vinegar: A vinegar made by aging trebbiano wine vinegar in a succession of casks made from a variety of woods for many years.

banana: A yellow, elongated tropical fruit that grows in hanging bunches on a banana plant.

barbeque sauce: A type of sauce that is used to baste proteins and that can be served as an accompaniment to a cooked protein.

barbequing: A dry-heat cooking technique in which food is slowly cooked over hot coals or smoldering hardwoods.

barding: The process of laying a piece of fatback across the surface of a lean cut of meat to add moisture and flavor.

bark spice: A type of spice derived from the bark portion of a plant.

barley: An ancient grain that resembles brown rice in shape, yet takes longer to cook.

Bartlett pear: A large, golden pear with a bell shape.

basil: A leaf herb with a pointy green leaf that is a member of the mint family.

basmati rice: A long-grain rice that only expands lengthwise when it is cooked.

basted egg: A fried egg with an unbroken yolk that is cooked like a sunny-side up egg, yet the tops are slightly cooked by tilting the pan and basting the top of the egg with hot butter.

basting: The process of using a brush or a ladle to place fat on or pour juices over an item during the cooking process to help retain moisture and enhance flavor.

batonnet cut: A stick cut that produces a stick-shaped item ¼ × ½ × 2 inches long.

battering: The process of dipping an item in a wet mixture of flour, liquid, and fat for frying.

Bavarian cream: A light dessert made with a flavored custard sauce that has been stabilized with gelatin while it is still warm.

bay leaf: A thick, aromatic leaf herb that comes from the evergreen bay laurel tree.

bay scallop: A fairly small scallop harvested from shallow saltwater.

bean: The oval or kidney-shaped edible seed of various plants in the legume family.

béchamel: A mother sauce that is made by thickening milk with a white roux and seasonings. Also known as a cream sauce or a white sauce.

beef: The flesh of domesticated cattle.

beef brisket: A thin section of beef that contains some of the ribs, the breastbone, and layers of lean muscle, fat, and connective tissue.

beef chuck: A shoulder primal cut of beef that contains the first five rib bones, some of the backbone, and a small amount of the arm and blade bones.

beef flank: A primal cut of beef that includes the thin, flat section of the hindquarters located beneath the loin.

beef rib: The primal cut of beef located between the chuck and short loin; contains seven rib bones, from the 6th to the 12th rib.

beef round: A primal beef cut of beef that includes a large grouping of muscles that represent the hind hip and thigh of the carcass.

beef rump roast: A roast cut from the primal round, above the back end of the hip bone.

beef shank: A bony section of beef that is surrounded by a small amount of very tough but flavorful meat.

beef short loin: A primal cut of beef located just to the rear of the primal rib; includes the 13th rib and a small section of the backbone.

beef short plate: A primal cut of beef that includes a thin portion of the beef forequarter located just beneath the rib cut.

beef sirloin: A primal cut of beef situated just behind the short loin; contains some of the backbone and hip bone.

beef strip loin: A short loin without a tenderloin.

beef tenderloin: An eye-shaped muscle running from the primal rib cut into the primal leg.

beer vinegar: A light-golden-colored vinegar made from beer with a strong malt taste.

beet: A round root vegetable that is a deep reddish purple or gold color.

beet green: The green, edible leaf that grows out of the top of the beet root vegetable.

Belgian endive: A leaf vegetable with a slender, tightly packed, elongated head that forms a point.

bell pepper: A fruit-vegetable with three or more lobes of crisp flesh that surround hundreds of seeds in an inner cavity.

belly: A primal cut of pork that is the lower portion of the hog between the shoulder and the leg.

bel paese: A lightly colored, semisoft dry-rind cheese with a buttery flavor that melts easily.

bench brush: A brush with long bristles set in vulcanized rubber attached to a handle.

berry: A type of fruit that is small and has many tiny, edible seeds.

berry spice: A type of spice made from the berry portion of a plant.

beurre blanc: A butter-based emulsified sauce made by whisking cold, softened butter into a wine, white-wine vinegar, shallot, and peppercorn reduction. Also known as white butter sauce.

beurre manié: A thickening agent made by kneading equal amounts, by weight, of pastry or cake flour and softened butter and can be whisked into a sauce just before service.

beverage cost percentage: A percentage that indicates how the cost of beverages relates to menu prices and beverage sales of a foodservice operation.

biological contaminant: A living microorganism that can be hazardous if it is inhaled, swallowed, or otherwise absorbed.

biscuit: A quick bread made by mixing solid fat, baking powder or baking soda, salt, and milk with flour.

biscuit mixing method: A mixing method used to produce a light and flaky baked product.

bisque: A form of cream soup that is typically made from shellfish.

bivalve: A mollusk with a top shell and a bottom shell connected by a central hinge that can close for protection.

blackberry: A sweet, dark-purple to black, aggregate fruit that grows on a bramble bush.

blanching: A moist-heat cooking technique in which food is briefly parcooked and then shocked by placing it in ice-cold water to stop the cooking process.

blast chiller: A specialized cooling unit that rapidly reduces the temperature of foods, rendering them safe for immediate storage.

blend: To mix two or more ingredients together until they are evenly distributed.

blender: A tall appliance with a slender canister that is used to chop, blend, purée, or liquefy food.

blind baking: A process in which a piecrust is baked for 10–15 minutes before a filling is added.

blintz: A filled crêpe that is only cooked on one side and not flipped over to cook the other side.

blood line: A dark-red, almost black muscle that runs along the lateral line of a fish.

blueberry: A small, dark-blue berry that grows on a shrub.

blue cheese: A blue-veined cheese made from cow milk and characterized by the presence of a blue-green mold. Also known as bleu cheese.

blue crab: A North American crab with blue claws and a dark blue-green, oval shell.

blue-veined cheese: A cheese produced by inserting harmless live mold spores into the center of ripening cheese with a needle.

boiling: A moist-heat cooking technique in which food is cooked by heating a liquid to its boiling point.

bok choy: A leaf vegetable that has tender white ribs, bright-green leaves, and a more subtle flavor than head cabbage.

bolster: A thick band of metal located where the blade of a knife joins the handle.

bombe: An ice cream dessert with at least two varieties of ice cream layered inside a spherical mold that has been frozen and then unmolded and decorated for service. Also known as a glacée.

boning knife: A thin knife with a pointed 6–8 inch blade used to separate meat from bones with minimal waste.

Bosc pear: A pear with a gourd-like shape and a brown-colored to bronze-colored peel.

boudin: A highly seasoned Creole link sausage made of pork, pork liver, and rice.

bouillon: The liquid that is strained off after cooking vegetables, poultry, meat, or seafood in water.

bound salad: A salad made by combining a main ingredient, often a protein, with a binding agent such as mayonnaise or yogurt and other flavoring ingredients.

bouquet garni: A bundle of herbs and vegetables tied together with twine that is used to flavor stocks and soups.

boysenberry: A deep-maroon, hybrid berry made by crossbreeding a raspberry, a blackberry, and a loganberry.

braising: A combination cooking technique in which food is browned in fat and then cooked, tightly covered, in a small amount of liquid for a long period of time.

bran: The tough outer layer of grain that covers the endosperm.

Brazil nut: A 1–2 inch long, white, richly flavored seed of a very large fruit that grows on a Brazilian nut tree.

breadfruit: An exotic fruit that is native to Polynesia and has bumpy green skin and a white starchy flesh.

breading: A three-step procedure used to coat and seal an item before it is fried.

bread knife: A knife with a serrated blade 8–12 inches long that is used to cut through the crusts of breads without crushing the soft interior.

bread pudding: A baked custard that is made by pouring a custard mixture over chunks of bread and baking it in the oven.

breast: The top front portion of the flesh above the rib cage of a bird.

breast quarter: Half of a breast, a wing, and a portion of the back of a bird.

brick cheese: A washed-rind semisoft cheese made from cow milk.

brie: A creamy, white soft cheese with a strong odor and a sharp taste.

brine: A salt solution that usually consists of 1 cup of salt per 1 gal. of water.

broccoli: An edible flower that is a member of the cabbage family and has tight clusters of dark-green florets on top of a pale-green stalk with dark-green leaves.

brochette: A food that is speared onto a wooden, metal, or natural skewer and then grilled or broiled.

broiler: A large piece of cooking equipment in which the heat source is located above or behind the food instead of below it.

broiler/fryer: A young male or female chicken that is less than five months old.

broiler/fryer duckling: A duck that is less than two months old.

broiling: A dry-heat cooking technique in which food is cooked directly under or over a heat source.

broken butter: A butter sauce made by heating butter until the fat, milk solids, and water separate or "break."

broth: A flavorful liquid made by simmering stock along with meat, poultry, seafood, or vegetables and seasonings.

brown rice: Rice that has had only the husk removed.

brown stock: A stock produced by simmering roasted meat, poultry, or game bones with mirepoix and an optional tomato product.

brunoise cut: A dice cut that produces a cube-shaped item with six equal sides measuring ⅛ inch each.

Brussels sprout: A very small round head of tightly packed leaves that looks like a tiny cabbage.

buckwheat: A dark, three-cornered seed of a plant unrelated to wheat that has a nutty, earthy flavor.

buckwheat groat: A crushed, coarse piece of whole-grain buckwheat that can be prepared like rice.

buffalo chopper: An appliance used to process larger amounts of a product into roughly equal-size pieces.

bulb vegetable: A strongly flavored vegetable that grows underground and consists of a short stem base with one or more buds that are enclosed in overlapping membranes or leaves.

butcher's knife: A heavy knife with a curved tip and a blade that is 7–14 inches in length.

butter: A fat made from cream that has a butterfat content of at least 80%.

butterflied fillet: Two single fillets from a dressed fish that are held together by the uncut back or belly of the fish.

butterhead lettuce: A lettuce with pale-green leaves and a sweet, buttery flavor, tender texture, and delicate structure.

butternut squash: A large, bottom-heavy, tan-colored winter squash.

button mushroom: A cultivated mushroom with a very smooth, rounded cap and completely closed gills atop a short stem. Also known as a white mushroom.

C

cake pan: A round, square, or specially shaped pan with short or tall sides used to bake cakes.

California menu: A menu that offers all food and beverage items for breakfast, lunch, and dinner throughout the entire day.

camembert: A soft cheese made from cow milk that has a yellow color and a waxy, creamy consistency.

Canadian bacon: A hamlike breakfast meat made from boneless, smoked, pressed pork loin.

canapé: An hors d'oeuvre that looks like a miniature open-faced sandwich.

canary melon: A fairly large, bright yellow melon with a smooth rind that is slightly waxy when ripe.

canned fuel: A flammable gel that provides several hours of heat once it is lit.

canola oil: A type of oil produced from rapeseeds. Also known as rapeseed oil.

cantaloupe: An orange-fleshed melon with a rough, deeply grooved rind.

caper: A flower spice from the unopened flower bud of a shrub; only used after being pickled in strongly salted white vinegar.

capon: A surgically castrated male chicken that is less than four months old.

capsaicin: A potent compound that gives chiles their hot flavor.

caramelization: A reaction that occurs when sugars are exposed to high heat and produce browning and a change in flavor.

caraway seed: A small, crescent-shaped brown seed of the caraway plant that is used as a spice.

cardamom: A berry spice made from the dried, immature fruit of a tropical bush in the ginger family.

carotenoid: An organic pigment found in orange or yellow vegetables.

carpaccio: Thin slices of meats or seafood that are served raw.

carrot: An elongated root vegetable that is rich in vitamin A.

carryover cooking: The rise in internal temperature of an item after it is removed from a heat source due to residual heat on the surface of the item.

cartilaginous fish: Any fish that has a skeleton composed of cartilage instead of bones.

casaba melon: A teardrop-shaped melon with a thick, bright-yellow, ridged rind and white flesh.

cashew: A nut from a kidney-shaped kernel of the fruit of the cashew tree that has a buttery flavor.

cassia: A bark spice made from the bark of a small evergreen tree. Also known as Chinese cinnamon.

cassoulet: A dish that consists of white beans stewed with duck fat, fresh sausage, and duck confit.

caul fat: A meshlike fatty membrane that surrounds sheep or pig intestines.

cauliflower: An edible flower that is a member of the cabbage family and has tightly packed white florets on a short, white-green stalk with large, pale-green leaves.

caviar: The harvested roe of sturgeon.

cayenne pepper: A berry spice made from dried, ground berries of certain varieties of hot peppers. Also known as red pepper.

celeriac: A knobby, brown root vegetable cultivated from a type of celery grown for its root rather than its stalk. Also known as celery root.

celery: A green stem vegetable that has multiple stalks measuring 12–20 inches in length.

celery seed: A tiny brown seed of a wild celery plant called lovage that is used as a spice.

celiac disease: A condition in which gluten damages the small intestine's ability to absorb nutrients.

cephalopod: Any of a variety of mollusks that do not have an external shell.

chafing dish: A hotel pan inside of a stand with a water reservoir and a portable heat source, such as canned fuel, underneath.

chalazae: The twisted cordlike strands of egg white that anchor the yolk to the center of the egg white.

Champagne vinegar: A vinegar made from Champagne grapes.

channel knife: A special cutting tool with a thin metal blade within a raised channel that is used to remove a large string from the surface of a food item.

chanterelle mushroom: A trumpet-shaped mushroom that ranges in color from bright yellow to orange and has a nutty flavor and a chewy texture.

charcuterie: The art of making sausages and other preserved items such as pâtés, terrines, galantines, and ballotines.

chard: A large, dark-green, leaf vegetable with white or reddish stalks.

chaud froid: A French term meaning "hot-cold."

Cheddar: A hard, aged cheese that is yellow or white and ranges in taste from mild to sharp.

cheesecake: A variety of baked custard that typically has a graham cracker or cookie crust.

cheese product: A processed food made of natural cheeses that may include additional ingredients such as emulsifiers.

chef's knife: A large and very versatile knife with a tapering blade used for slicing, dicing, and mincing. Also known as a French knife.

chemical contaminant: Any chemical substance that can be hazardous if it is inhaled, swallowed, or otherwise absorbed into the body.

chemical sanitizing: The process of using a chemical solution to reduce harmful pathogens to a safe level.

cherry: A small, smooth-skinned drupe that grows in a cluster on a cherry tree.

chervil: A leaf herb with dark-green, curly leaves with a flavor similar to parsley but more delicate and with a hint of licorice.

Cheshire: A hard cheese that is the oldest of the English-named cheeses.

chestnut: A nut from a rounded-off, triangular-shaped kernel found inside the burrlike fruit of the chestnut tree.

chèvre: A fresh cheese made from goat milk that varies in texture from soft to firm.

chicken cutlet: A boneless, skinless section of a chicken breast that has been tenderized.

chicory: A curly leaf vegetable with a slightly bitter-tasting flavor. Also known as escarole.

chiffon: An airy custard sauce most often used as a pie filling.

chiffonade cut: A slicing cut that produces thin shreds of leafy greens or herbs.

chiffon filling: A light, fluffy pie filling prepared by folding a meringue into a puréed fruit or a cream pie filling.

chile: A brightly colored fruit-vegetable pod with distinct mild to hot flavors.

chirashizushi: A type of sushi made by scattering toppings over or mixing toppings into vinegar-seasoned rice. Also known as scattered sushi.

chive: A stem herb with hollow, grass-shaped, green sprouts and a mild onion flavor.

chlorophyll: An organic pigment found in green vegetables.

chocolate: A food made from roasted and ground cacao beans that is often used to flavor other foods and beverages.

chopping: Rough-cutting an item so that there are relatively small pieces throughout, although there is no uniformity in shape or size.

chorizo: A Mexican pork sausage flavored with paprika that can be sweet or spicy.

chowder: A hearty soup that contains visibly large chunks of the main ingredients.

cider vinegar: Honey-colored vinegar made by fermenting unpasteurized apple juice or cider until the sugars are converted into alcohol.

cilantro: A leaf herb that comes from the stem and leaves of the coriander plant. Also known as Chinese parsley.

cinnamon: A bark spice made from the dried, thin, inner bark of a small evergreen tree.

citrus fruit: A type of fruit with a brightly colored, thick rind and pulpy, segmented flesh that grows on trees in warm climates.

clam: A bivalve found in both freshwater and saltwater.

clam knife: A small knife with a short, flat, round-tipped sharp blade that is used to open clams.

clarified butter: Butter fat that has been separated from the milk solids and water found in whole butter.

clarify: To remove impurities, sediment, cloudiness, and particles to leave a clear, pure liquid.

cleaning: The process of removing food and residue from a surface.

clearmeat: A cold, lean ground meat, fish, or poultry that is combined with an acid (such as wine, lemon juice, or a tomato product), ground mirepoix, egg whites, and an oignon brûlé.

clear plate: A rectangular slab of fat that contains a few strips of lean pork located just above the shoulder butt.

clear soup: A stock-based soup with a thin, watery consistency.

cleaver: A heavy, rectangular-bladed knife that is used to cut through bones and thick meat.

clementine: A citrus fruit that is similar to a tangerine, but with a rougher skin.

clove: A flower spice made from the dried, unopened bud of a tropical evergreen tree.

coagulation: The process of a protein changing from a liquid to a semisolid or solid state when heat or friction is applied.

coarse sugar: A white sugar composed of course-grain crystals that are about four times larger than granulated sugar. Also known as decorators' sugar.

cockle: A 1 inch wide bivalve with a shell that has deep ridges.

cocktail: A chilled appetizer served in a stemmed glass.

cocktail sauce: A thick tomato based product that usually contains vinegar, sugar, salt, spices, grated horseradish, hot sauce, Worcestershire sauce, and lemon juice.

cocoa powder: A reddish-brown, unsweetened powder extracted from ground cacao beans.

coconut: A tropical drupe with a white meat that is housed within a hard, fibrous brown husk.

cold closed sandwich: A variety of cold sandwich that consists of two pieces of bread, or the top and bottom of a bun or roll, coated with a spread and one or more fillings and garnishes.

cold open-faced sandwich: A variety of cold sandwich that consists of a single slice of bread that is often toasted or grilled and then coated with a spread and topped with thin slices of poultry, seafood, meat, partially cooked or raw vegetables, or a thin layer of a bound salad and a garnish.

cold-pack cheese: A creamy cheese product made by blending natural cheeses without the addition of heat.

cold wrap sandwich: A variety of cold sandwich in which a flat bread or tortilla is coated with a spread, topped with one or more fillings and garnishes, and rolled tightly.

collagen casing: A casing produced from collagen.

collard: A large, dark-green, edible leaf with a thick, white vein. Also known as a collard green.

combination cooking: Includes any cooking technique that uses both moist and dry heat.

combination curing: The process of combining either dry curing or wet curing with artery or stitch pumping to reduce processing time.

combi oven: An oven that has both convection and steaming capabilities. Also known as a combination oven.

Comice pear: A fairly large pear with a rotund body and a very short, well-defined neck.

composed salad: A salad that consists of a base, body, garnish, and dressing attractively arranged on a plate.

compound butter: A flavorful butter made by mixing cold, softened butter with flavoring ingredients such as fresh herbs, garlic, vegetable purées, dried fruits, preserves, or wine reductions.

compressed yeast: Yeast that has approximately 70% moisture and is available in 1 lb cakes or blocks. Also known as fresh yeast.

conch: A univalve that has a pinkish-orange shell with an interior that resembles a large snail.

Concord grape: A seeded grape with a deep black color.

conduction: A type of heat transfer in which heat passes from one object to another through direct contact.

confectioners' sugar: Granulated sugar that has been ground into a very fine powder. Also known as powdered sugar.

confit: A French term for meat that has been cooked and preserved in its own fat.

consommé: A very rich and flavorful broth that has been further clarified to remove any impurities or particles that could cloud the finished product.

contribution margin: The amount added to the AS cost of a menu item to determine a menu price.

convection: A type of heat transfer that occurs due to the circular movement of a fluid or a gas.

convection oven: A gas or electric oven with an interior fan that circulates dry, hot air throughout the cabinet.

convection steamer: A steamer that generates steam using an internal boiler, which circulates around the food to cook it rapidly.

cook-and-hold oven: An oven with two separately controlled compartments within one stainless steel cabinet that can be used to cook, roast, reheat, and hold a variety of foods. Also known as a retherm oven.

cooking: The process of heating foods in order to make them taste better, make them easier to digest, and kill harmful microorganisms that may be present in the food.

cooking-loss yield test: A procedure used to determine the yield percentage of a food item that loses weight during the cooking process.

coriander: A seed spice made from a light-brown, ridged seed of the coriander plant.

corn bread: A quick bread made from a batter containing cornmeal, eggs, milk, and oil.

Cornish game hen: Either a female or male chicken that is less than five weeks old. Also known as a Rock Cornish game hen.

corn oil: A type of oil produced from the germ of corn kernels.

cornstarch: The white, powdery, pure starch derived from corn.

corrective action: The point in a HACCP plan that identifies the steps that must be taken when food does not meet a critical limit.

cottage cheese: A pebble-shaped fresh cheese with a mildly sour taste.

cottage ham: The smoked, boneless meat extracted from the blade section of a pork shoulder butt.

coulis: A sauce typically made from either raw or cooked puréed fruits or vegetables.

country-style forcemeat: A coarsely ground forcemeat with a strong flavor. Also known as pâté de campagne.

court bouillon: A highly flavored, aromatic vegetable broth made from simmering vegetables with herbs and a small amount of an acidic liquid (usually vinegar or wine).

couscous: A tiny, round pellet made from durum wheat that has had both the bran and germ removed.

cracked grain: A whole kernel of grain that has been cracked by being placed between rollers.

cranberry: A small, red, round berry that has a tart flavor.

crayfish: A freshwater crustacean that resembles a tiny lobster. Also known as a crawfish or crawdad.

cream: (*v*) To combine fat and sugar in order to add air.

cream cheese: A soft, fresh cheese with a rich, mild flavor and smooth consistency that is made from cream or a mixture of cream and milk.

creaming mixing method: A mixing method used to produce a baked product with a fine crumb.

cream of tartar: A chemical leavening agent derived from tartaric acid, a by-product of wine production.

cream puff: A round pastry filled with sweetened whipped cream.

crème anglaise: A pourable vanilla custard sauce.

crème brûlée: A baked custard that has the texture of a stirred custard beneath a hardened sugar surface.

crème caramel: A baked custard served upside down and topped with hot, caramelized sugar. Also known as flan.

Crenshaw melon: A large, pear-shaped melon with a yellow-green, slightly ribbed rind and an orange or salmon-colored flesh.

crêpe: A French pancake that is light and very thin.

crisp cookie: A cookie prepared from dough that contains a high percentage of sugar.

crisphead lettuce: A lettuce with a large, round, tightly packed head of pale-green leaves.

critical control point (CCP): The point where a hazard can be prevented, eliminated, or reduced.

critical limit: The point in a HACCP plan where a minimum or maximum value is established for a CCP in order to prevent, eliminate, or reduce a hazard to a safe level.

crudité: A group of raw vegetables arranged on a platter and served with a dipping sauce.

crumb crust: A piecrust made from crumbled cookies or crackers that are held together with melted butter.

crushed red pepper: A berry spice made from a blend of dried, crushed berries from hot chili peppers. Also known as crushed chiles or chili flakes.

crustacean: A shellfish that has a hard, segmented shell protecting its soft flesh and that does not have an internal bone structure.

cucumber: A green, cylindrical fruit-vegetable that has an edible skin, edible seeds, and a moist flesh.

cumin: A seed spice made from a crescent-shaped amber, white, or black seed of the cumin plant, which is a member of the parsley family.

curd: The thick, casein-rich part of coagulated milk.

cured sausage: A type of raw sausage that contains sodium nitrite mixed into the forcemeat.

curing: The process of using salt and sodium nitrite alone or with flavorings or sugar to preserve a food item.

curing salt: A combination of table salt and sodium nitrite. Also known as TCM.

currant: A small red, black, or golden-white berry that grows in grape-like clusters.

curry leaf: A leaf herb with a small, shiny green leaf from the curry tree.

cut-in: To incorporate a solid fat into a dry ingredient.

cuttlefish: A translucent cephalopod with two tentacles, eight sucker-equipped arms, a hard internal cuttlebone, and large eyes at the base of its head.

cycle menu: A menu written for a specific period and is repeated once that period ends.

D

dandelion green: The dark-green edible leaf of the dandelion plant.

dark brown sugar: A moist sugar product that contains approximately 7% molasses.

dark chocolate: Chocolate that contains at least 35% chocolate liquor in addition to sugar, vanilla, and cocoa butter.

date: A plump, juicy, and meaty drupe that grows on a date palm tree.

deck oven: A drawer-like oven that is commonly stacked one on top of another, providing multiple-temperature baking shelves.

deep-frying: A dry-heat cooking technique in which food is completely submerged in very hot fat.

delegating: The act of assigning a task to another person while remaining responsible for the results.

demi-glace: A sauce made by adding equal parts of espagnole and brown stock together and reducing the mixture by half.

dessert salad: A sweet salad usually consisting of nuts, fruits, or sweeter vegetables such as carrots.

diagonal cut: A slicing cut that produces flat-sided, oval slices.

dice cuts: Precise cubes cut from uniform stick cuts.

Dijon mustard: A light-tan mustard that has a strong, tangy flavor and is made from brown or black mustard seeds, vinegar, white wine, sugar, and salt.

dill: A leaf herb with feathery, blue-green leaves that is a member of the parsley family. Also known as dill weed.

discrimination: The unfair treatment of an individual based on characteristics such as gender, race, ethnicity, age, religion, appearance, or disability.

distilled vinegar: Vinegar made by fermenting diluted, distilled grain alcohol. Also known as white vinegar.

docking: The process of making small holes in yeast dough before it is baked to allow steam to escape and to promote even baking.

double rib lamb chop: A rib chop cut to a thickness equal to two standard rib chops.

dough cutter: A flat, stainless steel blade attached to a sturdy handle. Also known as a bench knife.

dough docker: A rolling tool with small spikes that is used to perforate dough so that it will bake evenly without blistering in the oven heat.

dough sheeter: An electric appliance used to sheet, or roll out, doughs to the desired thickness.

dragon fruit: An exotic fruit of a cactus with an inedible pink skin, green scales, and white or red flesh speckled with small crunchy black seeds.

drawn fish: A fish that has had only the viscera removed.

dredging: The process of lightly dusting an item in seasoned flour or fine bread crumbs.

dressed fish: A fish that has been scaled and has had the viscera, gills, and fins removed.

dried fruit: Fruit that has had most of the moisture removed either naturally or through the use of a machine, such as a food dehydrator.

drummette: The innermost section of a wing located between the first wing joint and the shoulder of a bird.

drumstick: The lower portion of a leg of poultry, located below the hip and above the knee joint.

drupe: A type of fruit that contains one hard seed or pit. Also known as a stone fruit.

dry aging: The process of aging larger cuts of meat that are hung in a well-controlled environment.

dry curing: A curing method that involves the use of salt to dehydrate the protein in food.

dry-heat cooking: Any cooking technique that uses hot air, hot metal, a flame, or hot fat to conduct heat and brown food.

dry-rind cheese: A semisoft cheese that is allowed to ripen through exposure to air.

dry rub: A rub made by grinding herbs and spices into a fine powder and then rubbing the mixture into a protein prior to cooking.

dulse: A stringy, reddish-brown sea vegetable with a fishy odor that is rich in iron, iodine, potassium, and vitamin A.

Dungeness crab: A Pacific crab with a sweet-tasting flesh.

durian: An exotic fruit that contains several pods of sweet, yellow flesh and has a custard-like texture.

durum wheat: The hardest type of wheat.

Dutch-processed cocoa powder: A dark, unsweetened cocoa powder processed with an alkali to neutralize the natural acidity.

E

éclair: An oblong pastry filled with pastry cream and topped with a chocolate or fondant glaze.

Edam: A waxed-rind semisoft cheese made from cow milk that has a firm, crumbly texture.

edamame: Green soybeans housed within a fibrous, inedible pod.

edge: The sharpened part of the knife blade that extends from the heel to the tip.

edible flowers: The flowers of nonwoody plants that are prepared as vegetables.

edible-portion (EP) quantity: The amount of a food item that remains after trimming and is ready to be served or used in a recipe.

edible-portion (EP) unit cost: The unit cost of a food or beverage item after taking into account the cost of the waste generated by trimming.

eggless egg substitute: A yellow-colored liquid composed of soy, vegetable gums, and starches derived from corn or flour.

eggplant: A deep-purple, white, or variegated fruit-vegetable with edible skin and a yellow to white, spongy flesh that contains small, brown, edible seeds.

egg substitute: An egg product made from eggs that have been separated to remove the yolks.

egg yolk: The yellow portion of the egg.

emergency action plan: A written plan intended to organize employees during an emergency situation.

emulsification: The process of temporarily binding two liquids that do not combine easily, such as oil and vinegar.

emulsified forcemeat: A forcemeat made from a ratio of five parts protein, to four parts fat, to three parts crushed ice. Also known as the 5-4-3 style.

emulsion: A combination of two unlike liquids that have been forced to bond with each other.

endosperm: The largest component of a grain kernel and consists of carbohydrates and a small amount of protein.

English lamb chop: A 2 inch thick fabricated cut taken along the entire length of the unsplit loin.

English mustard: A type of yellow mustard that has a hot, spicy flavor and is made from ground yellow and brown or black mustard seeds, wheat flour, and turmeric.

enokitake mushroom: A crisp, delicate mushroom that has spaghetti-like stems topped with white caps. Also known as an enoki or a snow puff mushroom.

espagnole: A mother sauce made from a full-bodied brown stock, brown roux, tomato purée, and a hearty caramelized mirepoix.

espresso: An intensely flavored coffee made from beans that have been roasted to the very dark or espresso roasted stage.

essence: A concentrated fish stock that includes a large amount of aromatic ingredients such as celery, morels, a bouquet garni, a sachet, and fennel root.

European oyster: A variety of oyster with a relatively flat, cup-shaped shell and salty-sweet flavored, creamy-textured flesh.

exotic fruit: A type of fruit that comes from a hot, humid location but is not as readily available as a tropical fruit.

extra-virgin olive oil: A type of olive oil produced from the first pressing of the olives without the use of heat or chemicals that has an acid content of less than 1%.

extract: A flavorful oil that has been mixed with alcohol.

F

fabricated cut: A ready-to-cook cut that is packaged to certain size and weight specifications.

facility maintenance: The process of servicing a building, such as a foodservice operation.

farro: A hearty grain that tastes similar to wheat, yet resembles brown rice.

fatback: The layer of fat that runs along the back of the hog.

fennel: A celery-like stem vegetable with overlapping leaves that grow out of a large bulb at its base.

fennel seed: A seed spice made from an oval, light-brown, and green seed of the fennel plant.

fenugreek: A seed spice made from the pebble-shaped seed of the fenugreek plant in the pea family.

fermentation: The process by which yeast converts sugar into carbon dioxide and alcohol.

feta: A fresh cheese made from sheep milk or goat milk.

fiddlehead fern: The curled tip of an ostrich fern frond that has a nutty and slightly bitter flavor similar to asparagus and artichokes.

fig: The small pear-shaped fruit of the fig tree.

filé powder: A leaf herb that is made from the ground leaves of the sassafras tree.

fine brunoise cut: A dice cut that produces a cube-shaped item with six equal sides measuring 1/16 inch.

fine julienne cut: A stick cut that produces a stick-shaped item 1/16 × 1/16 × 2 inches long.

fingerling potato: A small, tapered, waxy potato with butter-colored flesh and tan, red, or purple skin.

fire-suppression system: An automatic fire extinguishing system that is activated by the intense heat generated by a fire.

first-in, first-out (FIFO): The process of dating new items as they are placed into inventory and placing them behind or below older items to ensure that older items are used first.

fish: Any of a classification of animal that has fins, gills, and an internal skeleton made of bones or cartilage.

fish fillet: The lengthwise piece of flesh cut away from the backbone.

fish sauce: A type of Asian sauce made from a liquid drained from fermented, salted fish.

fish steak: A cross section of a dressed fish.

fish stock: A basic stock prepared by adding fish bones or shellfish shells, vegetables, a sachet, and cold water to a stockpot and bringing it to a simmer over medium heat.

fixed menu: A menu that stays the same or rarely changes.

flaked grain: A refined grain that has been rolled to produce a flake. Also known as a rolled grain.

flaky piecrust: A piecrust prepared by cutting fat into flour until pea-size pieces of dough form.

flashbake oven: An oven that uses both infrared radiation and light waves to cook foods quickly and evenly from above and below.

flatfish: Any thin, wide fish with both eyes located on one side of the head and a backbone that runs from head to tail through the lateral line of the body.

flavonoid: An organic pigment found in purple, dark-red, and white vegetables.

flavored vinegar: Any vinegar in which other items such as herbs, spices, garlic, fruits, vegetables, or flowers are added.

flavoring: An item that alters or enhances the natural flavor of food.

flavoring emulsion: A flavored oil that has been mixed with water and an emulsifier.

flour: A fine powder that is created by grinding grains or nuts.

flower spice: A type of spice derived from the flower of a plant.

fluted cut: A specialty cut that leaves a spiral pattern on the surface of an item by removing only a sliver with each cut.

foie gras: The fattened liver of a duck or goose.

fold: To gently incorporate light ingredients into heavier ones using a smooth and gentle fold-over motion.

folded omelet: An omelet that is cooked until nearly done and then folded before serving.

fontina: A waxed-rind semisoft cheese made from cow milk.

food allergy: A reaction by the immune system to a specific food.

foodborne illness: An illness that is carried or transmitted to people through contact with or consumption of contaminated food.

food cost percentage: A percentage that indicates how the cost of food relates to the menu prices and food sales of a foodservice operation.

food intolerance: An adverse reaction to a food that does not involve the immune system.

food processor: An appliance with an S-shaped blade and a removable bowl and lid that can be used to quickly chop, purée, blend, or emulsify foods.

food spoilage indicator: A condition that signifies that food is deteriorating and is no longer safe for consumption.

food waste disposer: A food grinder mounted beneath a sink that is used to grind and eliminate solid food material.

forcemeat: A mixture of raw or cooked meat, poultry, seafood, or vegetables, fat, seasonings, and sometimes other ingredients.

formal evaluation: A scheduled and documented performance evaluation conducted on a regular basis, usually annually, by a supervisor that becomes part of an employee's permanent record.

freeze drying: The process of removing the water content from a food and replacing it with a gas.

French grill: A griddle with raised ridges that creates grill marks where the food touches the ridges.

frenching: A method of removing the meat and fat from the end of a bone; generally applied to chops.

fresh cheese: A cheese that is not aged or allowed to ripen.

fresh sausage: A type of sausage that is freshly made and has not been cured, smoked, dried, or further processed in any way.

fricassee: A combination cooking technique in which food is sautéed in butter, barely covered with a liquid, and simmered for a long period of time in a tightly covered pot.

fried ice cream: A dessert that is made by dipping rock solid frozen ice cream balls into a light batter and then deep-frying them in very hot oil.

frisée: A leaf vegetable with twisted, thin leaves that grow in a loose bunch. Also known as curly endive.

frittata: An omelet that is served open-faced after being browned in a broiler or hot oven.

fruit: The edible, ripened ovary of a flowering plant that usually contains one or more seeds.

fruit salad: A salad that is primarily made of fruits.

fruit-vegetable: A botanical fruit that is sold, prepared, and served as a vegetable.

fruit vinegar: A vinegar made from fruit juice that is first turned into fruit wine and then fermented by acetobacters (bacteria), which turns the wine into vinegar.

fryer: A cooking unit used to cook foods in hot fat.

fryer/roaster turkey: A male or female turkey that is less than three months old.

fumet: A concentrated stock made from fish bones or shellfish shells and vegetables.

G

galantine: Forcemeat that is wrapped in the skin of the animal the meat was taken from, shaped into a cylinder, trussed, and poached in a flavorful stock.

garlic: A bulb vegetable made up of several small cloves that are enclosed in a thin, husklike skin.

garlic chive: A stem herb with a solid, flat, grass-shaped sprout and a mild garlic flavor.

garnish: An edible item that complements the food presented on the plate while adding visual interest through the use of color, shape, and texture.

gastrique: A sugar syrup made by caramelizing a small amount of granulated sugar in a saucepan and deglazing the pan with a small amount of vinegar.

gazpacho: A cold soup made from stale cubes of bread, vinegar, water, olive oil, and diced tomatoes, cucumbers, and onions.

gelatin: A flavorless thickening agent made from animal protein.

gelatinization: The process of a heated starch absorbing moisture and swelling, which thickens the liquid.

gelatin salad: A salad made from flavored gelatin.

gelato: An Italian ice cream that has a creamier, denser texture than standard ice creams.

geoduck: A very large Pacific clam with a meaty siphon that protrudes from its shell.

germ: The smallest part of a grain kernel; contains a small amount of natural oils as well as vitamins and minerals.

giblets: The name for the grouping of the neck, heart, gizzard, and liver of a bird.

ginger: A root spice made from the large bumpy root of a tropical plant.

glace: A highly reduced stock that results in an intense flavor.

glazing: The process of covering an item with water to form a protective coating of ice before the item is frozen.

gluten: A type of protein found in grains such as wheat, rye, barley, and some varieties of oats.

gooseberry: A smooth-skinned berry that can be green, golden, red, purple, or white and has many tiny seeds on the inside.

Gorgonzola: A blue-veined cheese that is mottled with blue-green veins.

Gouda: A waxed-rind semisoft cheese that is similar to Edam but contains more fat.

grain: The edible fruit, in the form of a kernel or seed, of a grass.

grain-fed beef: Meat from cattle that were grain-fed in confined feeding operations for 90 days to 1 year.

granité: A frozen dessert made by frequently stirring a mixture of water, sugar, and flavorings, such as fruit juice or wine, as it is freezing. Also known as a granita.

granola: A baked mixture of rolled oats, nuts, dried fruit, and honey.

granulated sugar: A white sugar composed of small, uniformly sized crystals.

grape: An oval fruit that has a smooth skin and grows on woody vines in large clusters.

grapefruit: A round citrus fruit with a thick, yellow outer rind and tart flesh.

grapeseed oil: A type of oil produced from grape seeds.

grass-fed beef: Meat from cattle that were raised on grass or hay with little or no special feed.

gratin: Any dish prepared using the gratinée method.

gratinée: The process of topping a dish with a thick sauce, cheese, or bread crumbs and then browning it in a broiler or high-temperature oven.

gratin forcemeat: A forcemeat made from meat that is partially seared prior to grinding and a starchy binding agent.

grating cheese: A hard, crumbly, dry cheese grated or shaved onto food prior to service.

gravy: A sauce that has the same flavor as the food it accompanies when served.

griddle: A solid cooking surface made of metal on which foods are cooked.

griddling: A dry-heat cooking technique in which food is cooked on a solid metal cooking surface called a griddle.

grill: A cooking unit consisting of a large metal grate placed over a heat source.

grilled sandwich: A hot sandwich that is assembled by adding a precooked filling or cheese to bread and then heated on a griddle, in a sauté pan, or on a grill.

grilling: A dry-heat cooking technique in which food is cooked on open grates above a direct heat source.

grinder: A tool that consists of an auger, which pushes the forcemeat forward, and a cutting blade, which cuts the forcemeat as it is pushed through perforated stainless steel dies.

grits: A coarse type of meal made from ground corn or hominy.

gross pay: The total amount of an employee's pay before any deductions are made.

gross profit: The calculated difference between total revenue and the cost of goods sold.

ground poultry: Ground fabricated cuts of poultry.

Gruyère: A hard cheese that is similar in texture to Emmentaler but has a sharper flavor.

guava: A small oval-shaped tropical fruit, usually 2–3 inches in diameter, with thin edible skin that can be yellow, red, or green.

gumbo: A rich Creole soup made of broth, onions, celery, green peppers, okra, tomatoes, and rice.

gum paste: A very soft and extremely malleable white paste that dries to a porcelain-like finish and is used to make delicate decorative forms.

H

HACCP plan: A written document detailing what policies and procedures will be followed to help ensure the safety of food.

harassment: Behavior that is found to be threatening or that is not considered acceptable to the general public.

hard cheese: A firm, somewhat pliable and supple cheese with a slightly dry texture and buttery flavor.

hard-shell clam: An Atlantic clam with a blue-grey shell that contains a chewy flesh. Also known as a quahog.

hash: Shredded and chopped meat that has been mixed and cooked with diced potatoes, onions, and seasonings.

Havarti: A Danish, semisoft dry-rind cheese made from cow milk that has a buttery, somewhat sharp flavor.

hazard analysis: The process of assessing potential risks in the flow of food in order to establish what must be addressed in the HACCP plan.

Hazard Analysis and Critical Control Point (HACCP) system: A voluntary food safety management system that aims to identify, evaluate, and control contamination hazards throughout the flow of food.

hazardous material: A chemical present in the workplace that is capable of causing harm.

hazelnut: A grape-sized nut found in the fuzzy outer husks of the hazel tree. Also known as a filbert.

hazelnut oil: A type of oil produced from hazelnuts.

head cabbage: A tightly packed, round head of overlapping leaves that can be green, purple, red, or white in color.

headcheese: The spiced, pressed, and jellied meat from the head of a hog.

heart of palm: A slender, white, stem vegetable that is surrounded by a tough husk.

heat: Energy that is transferred between two objects or substances of different temperatures.

heat sanitizing: The process of using dry heat, very hot water, or steam to reduce harmful pathogens to a safe level.

heel: The rear portion of a knife blade; most often used to cut thick items where more force is required.

herb: A flavoring derived from the leaves or stem of a very aromatic plant.

hi-ratio shortening: An emulsified shortening that can absorb more sugar and moisture than hydrogenated shortenings.

hoisin sauce: A type of Asian sauce made from fermented soybean paste, garlic, vinegar, chiles, and sugar.

holding cabinet: A tall and narrow stainless steel box on wheels that accommodates standard sheet pans and contains temperature controls. Also known as a hot box.

hollandaise: A mother sauce made by thickening melted butter with egg yolks.

hominy: The hulled kernels of corn that have been stripped of their bran and germ and then dried.

honey: A sweet, thick fluid made by honeybees from flower nectar and approximately 1½ times as sweet as granulated sugar.

honeycomb tripe: The lining of the second stomach found in cattle.

honeydew melon: A melon with a smooth outer rind that changes from a pale-green color to a creamy-yellow color as it ripens.

honing: The process of aligning a blade's edge and removing any burrs or rough spots on the blade. Also known as truing.

hors d'oeuvre: An elegant, bite-size portion of food that is creatively presented and served apart from a meal.

horseradish: A root spice made from the large, brown-skinned root of a shrub related to the radish.

hot closed sandwich: A sandwich made by placing one or more hot fillings between two pieces of bread or a split roll or bun.

hot open-faced sandwich: A sandwich consisting of one or two slices of fresh, toasted, or grilled bread, topped with one or more hot fillings, and covered with a sauce, gravy, or a melted cheese topping.

hot wrap sandwich: A sandwich made by adding a spread and precooked fillings to a flatbread and then cooking it.

husk: The inedible, protective outer covering of grain. Also known as the hull.

hybrid fruit: A fruit that is the result of crossbreeding two or more fruits of different species to obtain a completely new fruit.

hydrogenation: A process that changes the molecular structure of oil into a solid.

I

ice cream: A frozen dessert made from cream, butterfat, sugar, and sometimes eggs.

icing: A sugar-based coating often spread on the outside of and between the layers of a baked product. Also known as frosting.

immersion blender: A narrow, handheld blender with a rotary blade that is used to purée a product in the container in which it is being prepared. Also known as a stick mixer.

impinger conveyor oven: An oven that directs heat from both above and below a food item as it moves along a conveyor belt.

inarizushi: A type of sushi made with vinegar-seasoned rice and toppings stuffed into a small purse of fried tofu.

individually quick-frozen (IQF): A designation for products preserved using a method in which each item is glazed with a thin layer of water and frozen individually.

induction cooktop: An electromagnetic unit that uses a magnetic coil below a flat surface to heat food rapidly.

informal evaluation: A performance evaluation conducted on an as-needed basis that provides an employee with constructive feedback on daily tasks or overall performance and is not documented.

infrared oven: An oven that uses infrared radiation to evenly and efficiently bake flat foods such as pizza.

infused oil: Any variety of oil with added herbs, spices, or additional ingredients that increase the flavor.

instant yeast: A form of yeast that does not need to be hydrated and may be directly added to a warm flour mixture.

insulated carrier: A container made of heavy polyurethane that is designed to hold pans of hot or cold foods at an appropriate temperature during transport.

invoice: A document provided by a supplier that lists the items delivered to a foodservice operation and the prices of those items.

irradiation: The process of exposing food to low doses of gamma rays in order to destroy deadly organisms such as E. coli O157:H7, campylobacter, and salmonella.

J

jackfruit: An enormous, spiny, oval exotic fruit with yellow flesh that tastes like a banana and has seeds that can be boiled or roasted and then eaten.

jicama: A large, brown root vegetable that ranges in size from 4 oz to 6 lb.

jowl: Meat from the cheek of a hog.

juice extractor: An electric machine that creates juice by liquefying raw vegetables and fruits and separating the fiber or pulp from the juice.

juicer: A device used to extract juice from fruits and vegetables.

julienne cut: A stick cut that produces a stick-shaped item ⅛ × ⅛ × 2 inches long.

juniper berry: A berry spice made from a small, purple berry of an evergreen bush.

jus lié: A sauce that is made by thickening a brown stock either by adding a cornstarch or arrowroot slurry or simply by a slow reduction. Also known as a fond lié.

K

kale: A large, frilly, leaf vegetable that varies in color from green and white to shades of purple.

kasha: Roasted buckwheat.

ketchup: A thick, tomato-based product that usually includes vinegar, sugar, salt, and spices.

key lime: A variety of lime that is smaller, more acidic, and more strongly flavored than other limes.

kielbasa: A Polish pork or beef sausage flavored with garlic, pimento, and cloves.

king crab: The largest-sized variety of crab, typically weighing between 6–20 lb and measuring as much as 10 feet from the tip of one leg to the tip of the opposite leg.

kiwano: An exotic fruit with jagged peaks rising from an orange and red-ringed rind that is native to Africa. Also known as horned melon.

kiwifruit: A small, barrel-shaped tropical fruit with a thin, brown, fuzzy skin.

kneading: The process of pushing and folding dough until it is smooth and elastic.

kohlrabi: A sweet, crisp, stem vegetable that has a pale-green or purple, bulbous stem and dark-green leaves.

kombu: A long, dark-brown to purple sea vegetable that is used to flavor dashi stock.

kombucha vinegar: Vinegar made with tea, acetobacters, yeast, and berries or other fruit.

kosher salt: A salt used to cure, season, and prepare kosher foods.

kumquat: A small, golden, oval-shaped exotic fruit with a thin, sweet peel and tart center.

L

lactose: A sugar found in milk and dairy products.

lamb: The meat from slaughtered sheep that are less than a year old.

lamb back: A lamb rack and loin that are still joined.

lamb bracelet: A hotel rack with the breast still attached.

lamb breast: A thin, flat, primal cut of lamb that contains the breastbone, the tips of the rib bones, and cartilage that is located under the shoulder and ribs.

lamb crown roast: A hotel rack containing sixteen ribs with the bones frenched, notched, and tied to create a circle that resembles a crown.

lamb foresaddle: The front half of the carcass consisting of the primal shoulder, rack, and breast cuts.

lamb hindsaddle: The rear half of the carcass consisting of the loin and leg.

lamb hotel rack: A lamb back that remains joined along the backbone.

lamb loin: A primal cut located between the rack and leg that includes the 13th rib, the loin eye muscle, the center section of the tenderloin, the strip loin, and some flank meat.

lamb rack: Eight rib bones located between the shoulder and loin of a lamb.

lamb riblet: A rectangular strip of meat cut from the lamb breast that contains part of a rib bone.

lamb shank: A cut of lamb that contains the upper foreshank bones.

lamb shoulder: The first four rib bones of each side and the arm and neck bones.

laminated dough: A dough that is layered with butter and used to make pastries, such as croissants, turnovers, and strudels.

langoustine: A small lobster that resembles a very large shrimp, except it has very long front arms with long, thin claws. Also known as a Norway lobster or a Dublin Bay prawn.

larding: The process of inserting thin strips of fatback into lean meat with a larding needle.

lavender: A leaf herb with pale-green leaves and purple flowering tops that is a member of the mint family.

leaf herb: A type of herb derived from the leaf portion of a plant.

leaf vegetables: Plant leaves that are often accompanied by edible leafstalks and shoots. Also known as greens.

lean dough: A yeast dough that is low in fat and sugar.

leavening agent: Any ingredient that causes a baked product to rise by the action of air, steam, chemicals, or yeast.

leek: A long, white bulb vegetable, with long, wide, flat leaves.

leg of lamb: A primal cut of lamb that contains the last portion of the backbone, hip bone, aitchbone, round bone, hindshank, and tailbone.

leg quarter: A thigh, a drumstick, and a portion of the back.

legume: The edible seed of a nonwoody plant that grows in multiples within a pod.

lemon: A tart yellow citrus fruit with high acidity levels.

lemongrass: A stem herb with long, thin, gray-green leaves, a white scallionlike base, and a lemony flavor. Also known as citronella.

lentil: A very small, dried pulse that has been split in half.

liaison: A thickening agent that is a mixture of egg yolks and heavy cream used to thicken sauces.

light brown sugar: A moist sugar product that contains approximately 3.5% molasses.

Limburger: A washed-rind semisoft cheese with a strong aroma and flavor.

lime: A small citrus fruit that can range in color from dark green to yellow-green.

liquor: A distilled liquid made from potatoes, corn, or other grains that has a high alcohol content.

loaf pan: A deep rectangular pan that is used to bake loaves of bread.

lobster: A saltwater crustacean with a brown to bluish-black external shell and two large claws.

loganberry: A red-purple hybrid berry made by crossbreeding a raspberry and a blackberry.

long-flake crust: A piecrust that absorbs a large amount of liquid because the flour and fat are rubbed together very lightly, leaving the fat in chunks about the size of a walnut.

long-grain rice: A rice that is long and slender and remains light and fluffy after cooking.

looseleaf lettuce: A mild flavored, rich-colored lettuce with a cascade of leaves held loosely together at the root.

lotus root: The underwater root vegetable of an Asian water lily that looks like a solid-link chain about 3 inches in diameter and up to 4 feet in length.

lowboy: A reach-in refrigerated unit located beneath a work surface.

lychee: An exotic drupe covered with a thin, red, inedible shell and has a light-pink to white flesh that is refreshing, juicy, and sweet.

M

macadamia nut: A tan, marble-sized nut with a hard shell that is the fruit of the macadamia tree.

mace: A seed spice made from the lacy red-orange covering of the nutmeg kernel.

mâche: A small, leafy green with a velvety texture. Also called lamb's lettuce.

Maillard reaction: A reaction that occurs when the proteins and sugars in a food are exposed to heat and merge together to form a brown exterior surface.

main course salad: A large salad containing an assortment of vegetables and a protein such as poultry, seafood, meat, eggs, or legumes that is served as an entrée.

Maine lobster: A large clawed lobster with a dark bluish-green shell, two heavy claws, and eight slender legs.

makizushi: A type of sushi made with vinegar-seasoned cooked rice layered with other ingredients and rolled in a dried seaweed paper called nori. Also known as a maki roll.

malt vinegar: Vinegar made from malted barley.

Manchego: A hard cheese with a tangy, slightly salty flavor that is produced from sheep milk.

mandarin: A small, intensely sweet citrus fruit that is closely related to the orange, but is more fragrant.

mandoline: A special cutting tool with adjustable steel blades used to cut food into consistently thin slices.

mango: An oval or kidney-shaped tropical drupe with orange to orange-yellow flesh.

mangosteen: A round, sweet and juicy exotic fruit about the size of an orange with a hard, thick, dark-purple rind that is inedible.

marinade: A flavorful liquid used to soak uncooked foods such as poultry, fish, and meats to impart flavor and sometimes to tenderize.

marjoram: A leaf herb with short, oval, pale-green leaves that is a member of the mint family.

marker: A round tool that has wire guides that leave marks indicating where to cut pies, round cakes, or pizzas into equal portions.

market menu: A menu that changes frequently to coincide with changes in the availability of products.

marzipan: A candy made from almond paste and sugar or corn syrup.

mascarpone: A fresh cheese that has a smooth texture, is white or pale yellow in color, and has a buttery, somewhat sweet flavor.

matignon: A uniformly cut mixture of onions, carrots, and celery and may also contain smoked bacon or ham.

mature duck: A duck that is more than six months old.

mature turkey: A turkey that is more than15 months old.

mayonnaise: A thick, uncooked emulsion formed by combining oil with egg yolks and vinegar or lemon juice.

meal-specific menu: A menu that only offers a particular meal, such as breakfast, lunch, or dinner.

mealy piecrust: A low-moisture piecrust prepared by rubbing fat into flour until the mixture resembles fine cornmeal.

mealy potato: A type of potato that is higher in starch and lower in moisture than other types of potatoes.

medium-grain rice: A rice that contains slightly less starch than short-grain rice but is still glossy and slightly sticky when cooked.

melon: A type of fruit that has a hard outer rind (skin) and a soft inner flesh that contains many seeds.

mentor: An experienced person who coaches and guides a less experienced person.

menu: A document that markets a foodservice operation to guests.

menu-item food cost percentage: The AS cost of a menu item divided by the menu price, written as a percent.

menu mix: The assortment of items that may be ordered from a given menu.

meringue: A mixture of egg whites and sugar.

mesclun greens: A mix of young greens that range in color, texture, and flavor.

Meyer lemon: A cross between a lemon and an orange.

microgreens: The first sprouting leaves of an edible plant.

microwave oven: A cooking unit that uses microwaves to heat the water molecules within foods.

milk: A nutritional beverage from the mammary glands of cows, goats, sheep, or water buffalo.

milk chocolate: A chocolate product that contains at least 12% milk solids and 10% chocolate liquor.

milled grain: A refined grain that has been ground into a fine meal or powder.

millet: A small, round, butter-colored grain that is gluten-free.

mincing: Finely chopping an item to yield a very small cut, yet not entirely uniform, product.

mint: A general term used to describe a family of similar leaf herbs, such as peppermint and spearmint.

mirepoix: A mixture of 50% onions, 25% celery, and 25% carrots, roughly cut into the appropriate size for the stock being produced.

mixer: A versatile electric appliance with U-shaped arms that secure a stainless steel mixing bowl under a rotating head that accommodates various attachments.

modified starch: A combination of starch and one or more vegetable gums.

modified straight-dough method: A method of mixing yeast dough in which the ingredients are added in sequential steps.

moist-heat cooking: Any cooking technique that uses liquid or steam as the cooking medium.

molasses: The dark syrup that is left after sugar cane has been processed.

mollusk: A shellfish with a soft, nonsegmented body.

Monterey Jack: A semisoft dry-rind cheese that has a smooth texture, a creamy white color, and a mild taste.

morel mushroom: An uncultivated mushroom with a cone-shaped cap that ranges in height from 2–4 inches and in color from tan to very dark brown.

mother sauce: One of five sauces from which the small classical sauces described by Escoffier are produced. Also known as a leading or grand sauce.

mousse: A rich and airy foam that has been lightened with whipped cream, whipped pasteurized egg whites, or both.

mousseline-style forcemeat: A forcemeat made from less-fibrous proteins such as poultry, seafood, veal, or pork.

mozzarella: A very tender fresh cheese with a soft, elastic-like curd.

mozzarella di bufala: A fresh cheese made from the milk of water buffalo or a combination of cow and water buffalo milk.

Muenster: A washed-rind semisoft cheese with a flavor between that of brick cheese and Limburger.

muesli: An unbaked mixture of rolled oats, wheat flakes, oat bran, raisins, dates, sunflower seeds, hazelnuts, and wheat germ.

muffin: A quick bread that is made with eggs, flour, and either oil, butter, or margarine.

muffin mixing method: A mixing method that uses liquid fats, such as vegetable oil or melted butter, to produce a rich and tender baked product.

muffin pan: A rectangular pan with cuplike wells used to bake teacakes, muffins, or cupcakes.

multidecker sandwich: A sandwich that consists of three pieces of bread, a spread, and at least two layers of garnishes and fillings.

mushroom: The fleshy, spore-bearing body of an edible fungus that grows above the ground.

muskmelon: A round, orange-fleshed melon with a beige or brown, netted rind.

mussel: A freshwater or saltwater bivalve with whisker-like threads that extend outside the shell to allow the animal to attach to items for protection.

mustard: A pungent powder or paste made from the seeds of the mustard plant.

mustard green: A large, dark-green leaf vegetable from the mustard plant that has a strong peppery flavor.

mustard seed: An extremely tiny seed from the mustard plant that is used as a spice.

N

nage: An aromatic court bouillon that is often used as a finishing sauce.

Napa cabbage: An elongated head of crinkly and overlapping edible leaves that are a pale yellow-green color with a white vein. Also known as celery cabbage.

nappe: The consistency of a liquid that thinly coats the back of a spoon and ensures that a sauce will cling lightly to another food.

natural casing: A casing produced from the intestines of sheep, hogs, or cattle.

nectarine: A sweet, slightly tart, orange to yellow drupe with a firm, yellow flesh and a large oval pit.

net pay: The actual amount on an employee's paycheck and is equal to the gross pay minus payroll deductions.

net profit: The calculated difference between the gross profit and operating expenses of a foodservice operation.

Neufchâtel: A soft, fresh cheese made from whole or skim milk or a mixture of milk and cream.

new potato: Any variety of potato that is harvested before the sugar is converted to starch. Also known as an early crop potato.

nigirizushi: A type of sushi made with small, hand-formed mounds of vinegar-seasoned cooked rice that is garnished with a topping.

noisette: A small, round, boneless medallion of meat.

nopal: The green, edible leaf of the prickly pear cactus.

nori: A thin, purple-black sea vegetable that turns green when it is toasted.

nougat: A candy made from caramelized sugar and often contains sliced nuts or fruit.

nut: A hard-shelled, dry fruit or seed that contains an inner kernel.

nutmeg: A seed spice made from an oval, gray-brown seed found in the yellow, nectarine-shaped fruit of a large tropical evergreen.

O

oat groat: An oat grain that only has the husk removed.

oats: Grains derived from the berries of oat grass.

obesity: A medical condition characterized by an excess of body fat.

oblique cut: A slicing cut that produces wedge-shaped pieces with two angled sides. Also known as a rolled cut.

obsolescence: The removal of an item from the menu while ingredients for that particular item are still in stock.

oca: A small, knobby tuber that has a potato-like flesh and ranges in flavor from very sweet to slightly acidic. Also known as a New Zealand yam.

octopus: A gray cephalopod with eight sucker-equipped arms, a birdlike beak, well-developed vision, and no internal or external shell.

offal: An edible part of an animal that is not part of a primal cut.

oignon brûlé: Half an onion that is charred on the cut side.

oil: A fat that remains in a liquid state at room temperature.

okra: A green fruit-vegetable pod that contains small, round, white seeds and a gelatinous liquid.

olive: A small, green or black drupe that is grown for both the fruit and its oil.

olive oil: A type of oil produced from olives.

olive-pomace oil: A type of olive oil produced using heat, and often chemicals, to extract additional oils from the olive pulp and olive pits after the first pressing.

omelet: An egg dish made with beaten eggs and cooked into a solid form.

onion: A bulb vegetable made up of many concentric layers of fleshy leaves.

onion piquet: Half of an onion studded with cloves and a bay leaf.

on-the-job training (OJT): A training method in which staff are trained to perform tasks while already on the job.

orange: A round, orange-colored citrus fruit.

oregano: A leaf herb with small, dark-green, slightly curled leaves that is a member of the mint family. Also known as wild marjoram.

oven spring: The rapid expansion of yeast dough in the oven, resulting from the expansion of gases within the dough.

overall food cost percentage: The total amount of money a foodservice operation spends on food divided by the total food sales over a defined period of time, written as a percent.

over-easy egg: A fried egg with an unbroken yolk that is cooked until the egg white has gone from translucent to white on the bottom and then flipped over and cooked on the other side just until the white is no longer translucent.

over-hard egg: A fried egg with a bright, firm white that is not rubbery and a pale yellow yolk that looks almost fluffy.

overhead warmer: A heat source located above a prepared food that keeps the food hot for service. Also known as a heat lamp.

over-medium egg: A fried egg with a completely cooked white and a yolk that is cooked nearly all the way through.

overrun: An increase in the volume of a frozen product as a result of the incorporation of air during churning and freezing.

oxtail: The tail from a cattle carcass.

oyster: A saltwater bivalve with a very rough shell that is coated with calcium deposits.

oyster knife: A small knife with a short, dull-edged blade with a tapered point that is used to open oysters.

oyster mushroom: A broad, fanlike or oyster-shaped mushroom that varies in color from white to gray or tan to dark brown.

P

Pacific oyster: A variety of large oyster that has fragile, curvy shells and a briny, sweet, and mild-tasting flesh. Also known as a Japanese oyster.

palette knife: A flat or offset narrow knife with a rounded, 3½–12 inch blade that varies in flexibility. Also known as a cake spatula.

pancetta: Unsmoked pork belly that has been cured in salt and spices, such as nutmeg and pepper, and then dried for a few months. Also known as Italian bacon.

pan-dressed fish: A dressed fish that has had its head removed. Also known as a headed and gutted (H&G) fish.

pan-frying: A dry-heat cooking technique in which food is cooked in a pan of hot fat.

Panini grill: An Italian clamshell-style grill made specifically to cook grilled sandwiches.

panning: The process of placing pieces of dough into pans.

papaya: A pear- or cylinder-shaped tropical fruit weighing 1–2 pounds with flesh that ranges in color from orange to red-yellow.

paprika: A berry spice made from a dried, ground sweet red pepper berry that is often used as a colorful garnish.

parasite: A pathogen that relies on a host for survival in a way that benefits the organism and harms the host.

paring knife: A short knife with a stiff 2–4 inch blade used to trim and peel fruits and vegetables.

parisienne scoop: A special cutting tool that has a half-ball cup with a blade edge attached to a handle and that is used to cut fruits and vegetables into uniform spheres.

Parmesan: A grating cheese that is extremely hard, has a granular texture, and can keep indefinitely when produced properly.

parsley: A leaf herb with curly or flat dark-green leaves that is used as both a flavoring and a garnish.

parsnip: An off-white root vegetable 5–10 inches in length.

par stock: The amount of a particular product that should be kept in inventory to ensure an adequate supply is on hand for normal production.

partial tang: A shorter tail of a knife blade that has fewer rivets than a full tang.

passion fruit: A small, oval-shaped exotic fruit that typically weighs 2–3 ounces and has firm, inedible skin that can be either yellow or purple.

pasta: A term for rolled or extruded products made from a dough composed of flour, water, salt, oil, and sometimes eggs.

pasteurized egg: An egg that has been heated to a specific temperature for a specific period of time to kill bacteria that can cause foodborne illnesses.

pastillage: A malleable white paste that dries to a matte finish and is used to make decorative forms. Also known as sugar paste.

pastry bag: A cone-shaped paper, canvas, or plastic bag that is fitted with a pastry tip.

pastry brush: A small, narrow brush that is used to apply liquids, such as egg wash or butter, onto baked products.

pastry cream: A thick cornstarch- or flour-based egg custard made with egg yolks, sugar, and milk. Also known as crème pâtissière.

pastry tip: A cone-shaped tip that is fitted into the narrow end of a pastry bag.

pastry wheel: A dough-cutting tool with a rotating disk attached to a handle.

pâté: A forcemeat that is typically layered with garnishes and baked in a mold.

pâte à choux: A pastry made by beating eggs into a paste of boiled water, butter, and flour. Also known as éclair paste.

pathogen: A microorganism that can cause disease.

pattypan: A round, shallow summer squash with scalloped edges that is harvested when it is no larger than 2–3 inches in diameter.

payroll expense: Money paid to an employee who performs work for the operation.

pea: The round edible seed of various plants in the legume family.

peach: A sweet, orange to yellow drupe with downy skin.

peanut: A legume that is contained in a thin, netted, tan-colored pod that grows underground.

pear: A bell-shaped pome with a thin peel and sweet flesh.

pearled grain: A refined grain with a pearl-like appearance that results from having been scrubbed and tumbled to remove the bran.

pecan: A nut with a light-brown kernel from the pecan tree.

pectin: A chemical present in all fruits that acts as a thickening agent when it is cooked in the presence of sugar and an acid.

peel: The thick outer rind of a citrus fruit.

peeler: A special cutting tool with a swiveling, double-edged blade that is attached to a handle and is used to remove the skin or peel from fruits and vegetables.

peppercorn: The dried berry of a climbing vine known as the Piper nigrum; used whole, ground, or crushed.

pepper sauce: A type of hot and spicy sauce made from various types of chili peppers; typically includes vinegar.

perceived value pricing: The process of adjusting a target menu price based on how management thinks a guest will perceive the price.

Persian melon: A green muskmelon with finely textured net on the rind. Also known as a patelquat.

persimmon: A bright-orange tropical fruit that grows on trees and is similar in shape to a tomato.

personal hygiene: The physical care maintained by an individual.

pesto: A sauce made from fresh ingredients that have been crushed with a mortar and pestle or finely chopped in a food processor before being mixed with olive oil.

phyllo: Paper-thin sheets of dough that are brushed with butter and used to make pastries.

physical contaminant: Any object that can be hazardous if it is inhaled, swallowed, or otherwise absorbed into the body.

pickling salt: A pure form of salt that contains no residual dust, iodine, or other additives. Also known as canning salt.

picnic: A cut of pork fabricated from the upper part of the foreleg that includes a portion of the shoulder.

picnic shoulder: A primal cut of pork that is the lower half of the shoulder of a hog.

pie pan: A round, shallow pan with sloped sides used for baking pies.

pineapple: A sweet, acidic tropical fruit with a prickly, pinecone-like exterior and juicy, yellow flesh.

pine nut: An ivory-colored, torpedo-shaped nut from the pine cone of various types of pine trees.

pistachio: A pale-green, bean-shaped nut from the pistachio tree.

pith: The white layer just beneath the peel of a citrus fruit.

plantain: A tropical fruit that is a close relative of the banana, but is larger and has a dark brown skin when ripening.

plum: An oval-shaped drupe that grows on trees in warm climates and comes in a variety of colors such as blue-purple, red, yellow, or green.

poaching: A moist-heat cooking technique in which food is cooked in a liquid that is held between 160°F and 180°F.

pome: A fleshy fruit that contains a core of seeds and has an edible skin.

pomegranate: A round, bright-red tropical fruit with a hard, thick outer skin.

poppy seed: A very small, blue-gray seed of the poppy plant.

porcini mushroom: An uncultivated, pale-brown mushroom with a smooth, meaty texture and a pungent flavor. Also known as a cèpe.

pork belly: A fatty slab of meat and skin from the side and belly of a hog.

pork leg: A primal cut of pork that is composed of the hind thigh and buttock of a hog. Also known as a ham.

pork loin: A primal cut that extends along the backbone, from about the second rib, through the rib and loin area of a hog.

pork shoulder butt: A square, compact area of the shoulder located just above the front legs of a hog. Also known as Boston butt.

pork spareribs: The long, narrow ribs and breastbone of a hog.

pork tenderloin: A fairly long, tapered strip of lean meat taken from the underside of the loin.

portion control: The process of ensuring that a specific amount of food or beverage is served for a given price.

portion size: The amount of a food or beverage item that is served to an individual person.

portobello mushroom: A very large and mature, brown cremini mushroom that has a flat cap measuring up to 6 inches in diameter.

Port Salut: A washed-rind semisoft cheese that has a soft, smooth, orange-colored rind and a glossy, ivory-colored interior.

potato: A round, oval, or elongated tuber that is the only edible part of the potato plant.

potentially hazardous food: A food that requires temperature control in order to keep it safe for consumption.

poultry: The collective term for various kinds of birds that are raised for human consumption.

poultry half: A full half-length of a bird split down the breast and spine.

poultry leg: A drumstick and thigh.

poultry tenderloin: The inner pectoral muscle that runs alongside the breastbone of a bird.

prepared yellow mustard: A type of bright-yellow mustard that is mild in flavor and made from ground yellow mustard seeds, vinegar, and turmeric.

prep sink: A sink that is used in the preparation of food, such as rinsing whole produce.

pressure steamer: A steamer that uses water heated within a pressure-controlled, sealed cabinet to cook foods much quicker than a convection steamer.

preventive maintenance: The systematic servicing of a facility in order to maintain safe operating conditions.

pricing form: A tool often used to help calculate the AS cost of a menu item and establish a menu price.

prickly pear: A pear-shaped tropical fruit with protruding prickly fibers that is a member of the cactus family.

primal cut: A large cut from a whole or a partial carcass.

prix fixe menu: A menu that offers limited choices within a collection of specific items for a multicourse meal at a set price.

processed cheese: A blend of fresh and aged natural cheeses that are heated and melted together.

profit and loss statement: A form that shows the revenue, expenses, and resulting gross and net profit (or loss) over a specific period of time.

profiterole: A round pastry filled with ice cream or savory fillings.

proofing: The process of letting yeast dough rise in a warm (85°F) and moist (80% humidity) environment until the dough doubles in size.

proofing cabinet: A holding cabinet that contains both temperature and humidity controls. Also known as a proofer.

prosciutto: A type of dry-cured Italian ham.

provolone: A hard cheese with an elastic texture and a mild to sharp taste depending on the age.

puff pastry dough: A laminated unleavened dough that rises as a result of the steam created when moisture in the layers of dough is released during baking. Also known as pâte feuilletée.

pulse: The dried seed of a legume.

pumpkin: A round fruit-vegetable with a hard orange skin and a firm flesh that surrounds a cavity filled with seeds.

punching: The process of deflating risen yeast dough to allow carbon dioxide to escape.

purchase specification: A written form listing the specific characteristics of a product that is to be purchased from a supplier.

pure olive oil: A type of olive oil produced using heat, and often chemicals, to extract additional oils from the olive pulp after the first pressing.

purple potato: A mealy potato with smooth, thin, bluishpurple skin and purple flesh. Also known as a blue potato.

Q

quarter of beef: A side of beef that has been divided into two parts between the 12th and 13th ribs.

quiche: A baked egg dish composed of a savory custard baked in a precooked piecrust.

quick bread: A baked product that is made with a quick-acting leavening agent such as baking powder, baking soda, or steam.

quick bread loaf: A loaf of quick bread made from a batter similar to muffin batter that has been flavored with fruits, vegetables, nuts, and/or spices.

quince: A hard yellow pome that grows in warm climates.

quinoa: A small, round, gluten-free grain that is classified as a complete protein.

R

radiation: A type of heat transfer that uses electromagnetic waves to transfer energy.

radicchio: A small, compact head of red leaves, similar to a small head of red cabbage.

radio frequency identification (RFID): A form of technology that uses electronic tags to store data that can be monitored from remote distances.

radish: A root vegetable with a white flesh and a peppery taste that comes in many colors and shapes.

raft: The clearmeat that has risen to the surface.

rambutan: Fragrant and sweet exotic fruit covered on the outside with soft, hair-like spikes.

ramp: A wild leek with a flavor similar to scallions, yet with more zing.

range: A large appliance with surface burners.

raspberry: A slightly tart, red aggregate fruit that grows in clusters.

ratio: A mathematical way to represent the relationship between two or more numbers or quantities.

rat-tail tang: A narrow rod of metal that runs the length of a knife handle but is not as wide as the handle.

raw bar: A presentation of a variety of raw and steamed seafood presented and served on a bed of ice.

raw yield test: A procedure used to determine the yield percentage of a food item that is trimmed of waste prior to being used in a recipe.

reach-in unit: A temperature-controlled cabinet for storing cold or frozen food items.

receiving area: The area of the professional kitchen where all delivered items are checked for freshness, proper amount, correct temperature, and accurate price.

recovery time: The time it takes for the fat to return to the set temperature.

red flame grape: A seedless grape that ranges from a light purple-red color to a dark-purple color.

red potato: A round, waxy, red-skinned potato with white flesh.

reduction: The process of gently simmering a liquid until it reduces in volume and results in a thicker liquid with a more concentrated flavor.

refined grain: A grain that has been processed to remove the germ, bran, or both.

relish: An assortment of uncooked vegetables that are served raw, marinated, or pickled.

remouillage: A stock made from using bones that have already been used once to make a stock.

rennet: A substance that contains acid-producing enzymes or an acid.

requisition: An internally generated invoice that is used to aid in tracking inventory as it moves from storage to production.

resinous herb: An herb that requires time to release its fragrant aroma.

revolving tray oven: A large oven with shelves that rotate like a ferris wheel to allow a large number of items to be baked at one time.

rhubarb: A tart stem vegetable that ranges in color from pink to red and is usually prepared like a fruit.

ribbon pasta: A thin, round strand or flat, ribbonlike strand of pasta.

rib eye: A large, eye-shaped muscle within the rib that is a continuation of the sirloin muscle.

rice bran oil: A type of oil produced from the bran (outer layer) of rice kernels.

rice vinegar: Vinegar made from rice wine.

rich dough: A yeast dough that incorporates a lot of fat, sugar, and eggs into a heavy, soft structure.

ricotta: A creamy fresh cheese that looks similar to cottage cheese but has a bland, yet slightly sweet flavor.

rind: An exterior layer of a food.

rivet: A metal fastener used to attach the tang of a knife to the handle.

roaster: A young male or female chicken that is 2–3 months old and has a ready-to-cook carcass weight of 5 lb or more.

roaster duckling: A duck that is less than four months old.

roasting: A dry-heat cooking technique in which food that contains fat or that has fat added to it is cooked uncovered at a high temperature in an oven or on a revolving spit over an open flame.

rolled-in dough: A yeast dough with a flaky texture that results from the incorporation of fat through a rolling and folding procedure.

rolled oats: Oats that have been steamed and flattened into small flakes. Also known as old-fashioned oats.

rolled omelet: An omelet that is cooked and then rolled onto a plate and cooked filling ingredients are added through a slit cut into the top. Also known as a French omelet.

rolling pin: A slim cylinder that is used to flatten pastry dough, bread crumbs, or other foods.

roll-in unit: A refrigeration unit that allows speed racks to be rolled in and out of the unit through a door opening that is just above floor height.

romaine lettuce: A lettuce with long, green leaves that grow in a loosely packed, elongated head on crisp center ribs.

Romano: A grating cheese that is similar to Parmesan but softer in texture and made from cow or goat milk.

rondelle cut: A slicing cut that produces disks. Also known as a round cut.

root spice: A type of spice derived from the root portion of a plant that grows underground.

root vegetable: An earthy-flavored vegetable that grows underground and has leaves that extend above ground.

Roquefort: A blue-veined cheese made from sheep milk and characterized by a sharp, tangy flavor.

rosemary: A leaf herb with needlelike leaves that is a member of the evergreen family.

rotating rack oven: A large oven that rotates 10–80 pans of food as it cooks. Also known as a reel oven.

rotisserie: A sideways broiler in which foods are placed on a steel rod or spit that revolves past the heat source to ensure even heating.

roundfish: Any fish with a cylindrical body, an eye located on each side of the head, and a backbone that runs from head to tail in the center of the body.

rounding: The process of shaping scaled dough into smooth balls.

roux: A thickening agent made by cooking a mixture of equal amounts, by weight, of flour and fat and is used to thicken sauces and soups.

rub: A blend of ingredients that is pressed onto the surface of uncooked foods such as poultry, fish, and meat to impart flavor and sometimes to tenderize.

russet potato: A mealy potato with thin brown skin, an elongated shape, and shallow eyes.

rutabaga: A round root vegetable with a yellow-tinted flesh that is the result of a cross between a Savoy cabbage and a turnip.

rye: A hearty grain with dark-brown kernels that are longer and thinner than wheat.

S

sabayon: A light and foamy custard prepared by whisking sugar and egg yolks over a double boiler into a silky foam. Also known as a zabaglione.

sachet d'épices: A mixture of spices and herbs placed in a piece of cheesecloth and tied with butcher's twine. Also known as a sachet.

safety data sheet (SDS): A document that provides detailed information describing a chemical, instructions for its safe use, the potential hazards, and appropriate first-aid measures. Formerly known as a material safety data sheet.

saffron: A flower spice made from the dried, bright-red stigmas of the purple crocus flower.

sage: A leaf herb with narrow, velvety, green-gray leaves that is a member of the mint family.

salad dressing: A sauce for a salad.

salad green: An edible leaf used in raw salads or as a garnish.

salamander: A small overhead broiler that is usually attached to a range.

salsa: A sauce made by mixing or puréeing diced vegetables or fruits, herbs, and spices together.

salsify: A white or black root vegetable that can grow up to 12 inches in length.

salt: A crystalline solid composed mainly of sodium chloride and used as a seasoning and a preservative.

sandwich base: The edible packaging that holds the contents of a sandwich.

sandwich filling: The main ingredient in a sandwich that is stacked, layered, or folded on top of the base to form a sandwich.

sandwich garnish: A complementary food item that is served on or with a sandwich.

sandwich spread: A slightly moist, flavorful substance that seals the pores of the bread and creates a thin moisture barrier.

sanitizing: The process of using chemicals or heat to destroy or reduce harmful microorganisms to a safe level.

Santa Claus melon: A large, mottled yellow and green variety of muskmelon that has a slightly waxy skin and soft stem end when ripe.

santoku knife: A knife with a razor-sharp edge and a heel that is perpendicular to the spine.

satsuma: A small, seedless variety of mandarin.

sauce: An accompaniment that is served with a food to complete or enhance the flavor and/or moistness of a dish.

sausage: A forcemeat that is typically stuffed into a casing or shaped into a patty.

sausage stuffer: A manual or electric piece of equipment that uses a piston to pump forcemeat through a nozzle into casings.

sautéing: A dry-heat cooking technique in which food is cooked quickly in a sauté pan over direct heat using a small amount of fat.

savory: A leaf herb with smooth, slightly narrow leaves that is a member of the mint family.

Savoy cabbage: A conical-shaped head of tender, crinkly, edible leaves that are blue-green on the exterior and pale green on the interior.

scaling (ingredients): The process of weighing an ingredient using a scale.

scaling (recipes): The process of calculating new amounts for each ingredient in a recipe when the total amount of food the recipe makes is changed.

scaling factor: The number that each ingredient amount in a recipe is multiplied by when the recipe yield is changed.

scallion: A small bulb vegetable with a slightly swollen base and long, slender, green leaves that are hollow. Also known as a green onion.

scallop: A bivalve with a fan-shaped shell and a cream-colored adductor muscle with a sweet, delicate flavor.

scallopini: A small, ¼ inch thick slice of veal (generally leg meat) that is 2–3 inches in diameter.

scimitar: A long knife with an upward curved tip that is used to cut steaks and primal cuts of meat.

scoring: The process of making shallow, angled cuts across the top of unbaked yeast bread to allow carbon dioxide to escape during baking.

searing: The process of using high heat to quickly brown the surface of a food.

sea salt: Salt produced through the evaporation of seawater.

sea scallop: A large scallop with a coarse texture that is harvested from deep saltwater.

seasoning: An item that intensifies or improves the flavor of food.

sea vegetables: Edible saltwater plants that contain high amounts of dietary fiber, vitamins, and minerals.

Seckel pear: A small pear that is sometimes called a honey pear or sugar pear because of its syrupy, fine-grained flesh and complex sweetness.

seed: The fruit or unripened ovule of a nonwoody plant.

seed spice: A type of spice derived from the seed portion of a plant.

semi-à la carte menu: A menu that offers entrées along with additional menu items for a set price.

semolina: The granular product that results from milling the endosperm of durum wheat.

separate course salad: A salad served as a course of its own, typically following the main course.

sesame oil: A type of oil produced from sesame seeds.

sesame seed: A small, flat, white or black seed found inside the pod on a sesame plant and used as a spice.

sexual harassment: Harassment that includes unwelcome verbal or physical conduct of a sexual nature.

shallot: A very small bulb vegetable that is similar in shape to garlic and has two or three cloves inside.

shaped pasta: A pasta that has been extruded into a complex shape such as a corkscrew, bowtie, shell, flower, or star.

sheet pan: A flat aluminum pan with very low sides.

sheet pan extender: An adjustable frame that extends the height of a sheet pan to accommodate items such as sheet cakes.

shellfish: The classification for aquatic invertebrates that may or may not have a hard, external shell.

shell membrane: The thin, skinlike material located directly under the eggshell.

sherbet: A frozen dessert made by combining puréed fresh fruit, simple syrup, and milk.

sherry vinegar: Vinegar made from sherry wine.

shiitake mushroom: An amber, tan, brown, or dark-brown mushroom with an umbrella shape and curled edges. Also known as a forest mushroom.

shortening: A flavorless fat made by chemically changing a vegetable oil into a solid form.

short-flake crust: A piecrust that absorbs slightly more liquid than mealy piecrust.

short-grain rice: Rice that is almost round in shape and has moist grains that stick together when cooked.

shucking: The process of opening a bivalve.

side of beef: A half of a carcass split along the backbone.

side salad: A small salad served to accompany a main course.

sift: To pass dry ingredients through a sieve in order to remove lumps and incorporate air.

silicone mat: A woven, nonstick mat that can withstand temperatures between −40°F and 580°F.

simmering: A moist-heat cooking technique in which food is gently cooked in a liquid that is between 185°F and 205°F.

siphon: A tubular organ that is used to draw in or eject fluids.

sleeper: A lobster that is dying.

slicer (equipment): An appliance that is used to uniformly slice foods such as meat and cheese.

slicer (knife): A knife with a narrow blade 10–14 inches long that is used to slice roasted meats. Also known as a carving knife.

slurry: A mixture of equal parts of cool liquid and a starch that is used to thicken other liquids.

smoked sausage: A type of cured sausage that is smoked during the curing process.

smoke point: The temperature at which an oil begins to smoke and give off an odor.

smoker oven: A gas or electric oven that generates wood smoke and is most often used to smoke or barbeque meats and poultry.

smoking: A dry-heat cooking technique in which food is slowly cooked over smoldering hardwoods in a vented enclosure.

smoothie: A blended drink made of fruit, such as berries or bananas, yogurt, and/or milklike beverages.

snow crab: A crab that is similar to a king crab but smaller and available in greater supply.

soft cheese: A cheese that has been sprayed with a harmless live mold to produce a thin skin or rind. Also known as a rind-ripened cheese.

soft cookie: A cookie prepared from dough that contains a lot of moisture.

soft filling: An uncooked pie filling that is baked in an unbaked piecrust.

soft-shell clam: An Atlantic clam with a thin, brittle shell that breaks easily. Also known as a long-neck or steamer clam.

soft-shell crab: A blue crab that has been harvested within 6 hours of molting, or shedding its shell in order to grow a larger shell.

solid pack: A canned product with little or no water added.

sorbet: A frozen dessert made by combining puréed fresh fruit and simple syrup.

sorrel: A large, green, leaf vegetable that ranges in color from pale green to dark green and from 2–12 inches in length.

soufflé: An airy and fragile custard made from a thick egg yolk sauce that has been lightened by stiffly beaten egg whites and then baked.

soufflé omelet: A lighter variation of a folded omelet. Also known as a fluffy omelet.

soup: A food that is made with a stock and pieces of meat, poultry, seafood, and/or vegetables.

soybean oil: A type of oil produced from soy beans.

soy nut: A legume from the soybean pod that has been soaked in water, drained, and then roasted or baked.

soy sauce: A type of Asian sauce made from mashed soybeans, wheat, salt, and water.

spaghetti squash: A dark-yellow winter squash with pale-yellow flesh that can be separated into spaghetti-like strands after it is cooked.

speed rack: A tall cart on wheels with rails that hold entire sheet pans of food. Also known as a tallboy.

spelt: An ancient grain with a nutty flavor and high protein content that is also a good source of riboflavin, zinc, and dietary fiber.

spice: A flavoring derived from the bark, seeds, roots, flowers, berries, or beans of aromatic plants.

spinach: A dark-green leaf vegetable with a slightly bitter flavor that may have flat or curly leaves, depending on the variety.

spine: The unsharpened top part of the knife blade that is opposite the edge.

spiny lobster: A lobster that has spines covering its body and five slender legs on each side. Also known as a rock lobster.

spit-roasting: The process of cooking food by skewering it and suspending and rotating it above or next to a heat source.

sponge method: A method of mixing yeast dough in two steps.

springform pan: A round pan with a metal clamp on the side that allows the bottom of the pan to be separated from the sides.

squash blossom: The edible flower of a summer or a winter squash.

squid: A translucent, head-footed cephalopod that has two tentacles, eight sucker-equipped arms, two lateral fins, and a flat, internal cuttlebone.

staff orientation: A period during which new staff members are introduced to workplace policies and procedures.

standardized recipe: A list of ingredients, ingredient amounts, and procedural steps for preparing a specific quantity of a food item.

star anise: A star-shaped seed fruit from the Chinese magnolia tree. Also known as Chinese anise.

star fruit: An exotic fruit that is shaped like a star when cut. Also known as a carambola.

steaming: A moist-heat cooking technique in which food is placed in a container that prevents steam from escaping.

steam injection: The process of adding water directly in the hot cavity of an oven so that steam is created.

steam-jacketed kettle: A large cooking kettle that has a hollow lining, known as a jacket, into which steam is injected to rapidly and uniformly cook foods. Also known as a steam kettle.

steamship round roast: The beef round with the shank and rump removed.

steam table: An open-top table with heated wells that are filled with water.

steel: A steel rod approximately 18 inches long attached to a handle and used to align the edge of knife blades. Also known as a butcher's steel.

steel-cut oats: Oat groats that have been toasted and cut into small pieces.

steer: A male calf that has been castrated prior to reaching sexual maturity.

stem herb: A type of herb that comes from the stem portion of a plant.

stem vegetable: The main trunk of a plant that develops buds and shoots instead of roots.

stewer: A female chicken that is more than 10 months old. Also known as a stewing hen.

stewing: A combination cooking technique in which bite-sized pieces of food are browned in fat, barely covered with a liquid, and simmered for a long period of time in a tightly covered pot.

Stilton: A blue-veined cheese made from cow milk with a flavor that is milder than Roquefort or Gorgonzola.

stir: To gently mix two or more ingredients together until they are evenly combined.

stir-frying: The process of quickly cooking items in a heated wok with a very small amount of fat while constantly stirring the items.

stock: An unthickened liquid that is flavored by simmering seasonings with vegetables, and often, the bones of meat, poultry, or fish.

stone crab: An Atlantic crab with a brownish-red shell and large claws of unequal size.

straight-dough method: A method of mixing yeast dough that combines yeast with warm water and then adds it to the rest of the ingredients, mixing them together all at once.

straight forcemeat: A forcemeat that consists of seasoned ground meat emulsified with ground fat.

strawberry: A bright-red, heart-shaped berry covered with tiny black seeds.

stuffed pasta: A pasta that has been formed by hand or machine to hold fillings.

suckling pig: A pig 4–6 weeks old that weighs 20–35 lb dressed.

summer squash: A fruit-vegetable that grows on a vine and has edible skin, flesh, and seeds.

sunchoke: A tuber with thin, brown, knobby-looking skin. Also known as a Jerusalem artichoke.

sunny-side up egg: A lightly fried egg with an unbroken yolk that is not flipped over to cook the other side.

supervisor: A person who oversees the work done by others and who has the responsibility of directing, training, and evaluating staff.

suprême sauce: A chicken velouté with the addition of cream.

surf clam: A large species of Atlantic clam that can grow to 8 inches in size.

surimi: A fish product made from a mixture of fish and/or shellfish and other ingredients.

sushi: A vinegar-seasoned rice dish garnished with raw fish, cooked seafood, eggs, or vegetables.

sweating: The process of slowly cooking food to soften its texture.

sweetbreads: The thymus glands of a calf, located in the neck.

sweet corn: A fruit-vegetable that has edible seeds called kernels that grow in rows on a spongy cob encased by thin leaves (husks), forming what is referred to as an ear of corn.

sweet potato: A tuber that has a paper-thin skin and flesh that ranges in color from ivory to dark orange.

Swiss cheese: The name used to describe varieties of hard cheeses with large holes, also known as eyes.

synthetic fibrous casing: A casing produced from a plastic-like synthetic material.

Szechuan pepper: A berry spice made from the dried berries of an ash tree.

T

table d'hôte menu: A menu that identifies specific items that will be served for each course at a set price.

tamis: A flat, round sieve with an aluminum or wood frame and a mesh screen bottom. Also known as a drum sieve.

tang: The unsharpened tail of a knife blade that extends into the handle.

tangelo: A hybrid of a tangerine and a grapefruit.

tangerine: A small citrus fruit with a slightly red-orange peel that can be easily removed without a knife.

tangor: A hybrid of a tangerine and a sweet orange.

tapioca: A thickening agent derived from the cassava plant.

target food cost percentage: The percentage of food sales that a foodservice operation plans to spend on purchasing food.

target price: The price that a foodservice operation needs to charge for a menu item in order to meet its target food cost percentage.

tarragon: A leaf herb with smooth, slightly elongated leaves; best known as the flavoring in béarnaise sauce.

tartare: Freshly ground or chopped raw meat or seafood.

tartar sauce: A mayonnaise-based sauce that contains finely chopped relish or pickles, capers, grated onions or minced chives, and fresh parsley.

tart pan: A round, shallow baking pan with sloped sides that are smooth or fluted and may have a removable bottom.

tatsoi: A spoon-shaped, emerald-colored leaf vegetable native to Japan.

tea: A leaf herb with jagged leathery leaves that comes from an evergreen bush in the magnolia family.

tea sandwich: A petite and delicate cold sandwich with a trimmed crust and a soft filling.

temperature danger zone: A range of temperature, between 41°F and 135°F, in which bacteria thrive.

tempura: A very light batter that is used on vegetables, poultry, seafood, and meats served in Asian cuisine.

tender: A small strip of a poultry breast.

terrine: A forcemeat that is baked in a mold without a crust.

thickening agent: A substance that adds body to a hot liquid.

thick soup: A soup having a thick texture and consistency.

thigh: The upper section of a leg of poultry located below the hip and above the knee joint.

Thompson grape: A seedless grape that is pale to light green in color.

thyme: A leaf herb with very small gray-green leaves that is a member of the mint family.

tilt skillet: A versatile piece of cooking equipment with a large-capacity pan, a thermostat, a tilting mechanism, and a cover. Also known as an electric braiser.

tip: The front quarter of a knife blade.

tisane: An herbal beverage created by steeping herbs, spices, flowers, dried fruits, or roots in boiling water.

tomatillo: A small fruit-vegetable with a thin, papery husk covering a pale-green skin that encases a pale-green flesh. Also known as a Mexican husk tomato.

tomato: A juicy fruit-vegetable that contains edible seeds.

tomato sauce: A mother sauce made by sautéing mirepoix and tomatoes, adding white stock, and thickening with a roux.

tossed salad: A mixture of leafy greens, such as lettuce, spinach, chicory, or fresh herbs, and other ingredients, such as fruits, vegetables, nuts, cheese, meats, and croutons, served with a dressing.

tourné cut: A carved, football-shaped cut with seven sides and flat ends.

tourné knife: A short knife with a curved blade that is primarily used to carve vegetables into a specific shape called a tourné, which is a seven-sided football shape with flat ends. Also known as a bird's beak knife.

tripe: The muscular inner lining of a stomach of an animal, such as cattle or sheep.

tropical fruit: A type of fruit that comes from a hot, humid location but is readily available.

trunnion kettle: A small steam-jacketed kettle that is tilted by pulling a lever or turning a wheel to empty the kettle.

trussing: The process of tying the legs and wings of a bird tightly to the body to keep a compact shape.

tube pasta: A pasta that has been pushed through an extruder and then fed through a cutter that cuts the tubes to desired length.

tuber: A short, fleshy vegetable that grows underground and bears buds capable of producing new plants.

turbinado sugar: A blond sugar that is purified and cleaned by steam before being dried into coarse-grain crystals.

turmeric: A root spice made from the root of a lily-like plant in the ginger family.

turnip: A round, fleshy root vegetable that is purple and white in color.

turnip green: A dark-green leaf vegetable that grows out of the top of the turnip root vegetable.

U

ugli fruit: A large, teardrop-shaped, seedless citrus fruit. Also known as a uniq fruit.

unemployment compensation: Monetary payments made to workers who have lost their jobs through no fault of their own for a given period of time or until they find a new job.

unit cost: The cost of a product per unit of measure.

univalve: A mollusk that has a single solid shell and a single foot. Also known as a gastropod.

unsweetened chocolate: Pure chocolate liquor that contains 50% to 58% cocoa butter.

utility knife: A multipurpose knife with a stiff 6–10 inch blade that is similar in shape to a chef's knife but much narrower at the heel.

V

vanilla bean: The dark-brown pod of a tropical orchid.

variety fruit: A fruit that is the result of breeding two or more fruits of the same species that have different characteristics.

variety spread: Any other food mixture of a spreadable consistency that can be added to a sandwich to complement or increase flavor and moisture.

veal: The flesh of calves, which are young cattle.

veal breast: A thin, flat cut of meat located under the shoulder and ribs; contains the breastbone, tips of the rib bones, and cartilage.

veal cutlet: A thin slice of veal.

veal foresaddle: The front half of a calf carcass consisting of the primal shoulder, rack, breast, and shank cuts.

veal foreshank: The upper portion of the front leg of a calf.

veal hindsaddle: The rear half of a calf carcass consisting of the loin and leg.

veal leg: A primal cut from the hind leg that contains the leg, sirloin, last portion of the backbone, pelvis (hip bone and aitchbone), round bone, hindshank, and tailbone.

veal loin: A primal cut of a calf located between the primal rack and leg; includes the 12th and 13th rib, the loin eye muscle, the center section of the tenderloin, the strip loin, and flank meat.

veal rack: A primal cut located between the shoulder and loin of a calf and containing seven rib bones.

veal shoulder: A primal cut of a calf that contains the first four rib bones, some of the backbone, and a small amount of the arm and blade bones.

vegetable: An edible root, bulb, tuber, stem, leaf, flower, or seed of a nonwoody plant.

vegetable salad: A salad that is primarily made of vegetables.

vegetable stock: A clear, light-colored stock produced by gently simmering vegetables with a white mirepoix and a sachet.

velouté: A mother sauce made from a flavorful white stock (either veal, chicken, or fish stock) and a blonde roux.

ventilation system: A large exhaust system that draws heat, smoke, and fumes out of the kitchen and into the outside air.

vertical cutter/mixer (VCM): An appliance used to cut and mix foods simultaneously using high-speed blades and a mixing baffle, which is used to manually move the product into the blades.

vichyssoise: A puréed cold soup made with milk, potatoes, leeks, onions, and chicken stock and finished with heavy cream.

vinaigrette: A temporary emulsion of oil and vinegar that may also include additional flavorings and seasonings.

vinegar: A sour, acidic liquid made from fermented alcohol.

virgin olive oil: A type of olive oil produced from the first pressing of the olives without the use of heat or chemicals that has an acid content of as much as 3%.

virus: A pathogen that grows inside the cells of a host.

W

wakame: A long, tender, grayish-green sea vegetable that expands to seven times its size when soaked in water.

walk-in unit: A room-sized insulated storage unit used to store bulk quantities of cold or frozen food.

walnut: A nut from the fruit of the walnut tree.

walnut oil: A type of oil produced from walnuts.

warewashing: The process of cleaning and sanitizing all items used to prepare and serve food.

wasabi: A root spice made from the light-green root of an Asian plant.

wash: A liquid that is brushed on the surface of dough prior to baking.

washed-rind cheese: A semisoft cheese with an exterior rind that is washed with a brine, wine, olive oil, nut oil, or fruit juice.

wastewater: Water that has been used to prepare food or beverages or for cleaning.

water chestnut: A small tuber with brownish-black skin and white flesh. Also known as a water caltrop.

watercress: A small, crisp, dark-green, leaf vegetable with a pungent, yet slightly peppery flavor.

watermelon: A sweet, extremely juicy melon that is round or oblong in shape, with pink, red, or golden flesh and green skin.

waxed-rind cheese: A semisoft cheese produced by dipping a wheel of freshly made cheese into a liquid wax and allowing the wax to harden.

waxy potato: A type of potato with a thin skin and slightly waxy flesh that is lower in starch and higher in moisture than mealy potatoes.

wet aging: The process of aging meat in vacuum-sealed plastic.

wet curing: A curing method in which foods are submerged in a brine. Also known as immersion curing.

wet rub: A rub made by incorporating wet ingredients, such as Dijon mustard, flavored oils, puréed garlic, or honey, into a dry rub mixture and then rubbing it into a protein prior to cooking.

wheat: A light yellow cereal grain cultivated from an annual grass that yields the flour used in many pastas and baked goods.

wheel: The round center cut of a large fish from which steaks are cut.

whetstone: A stone used to grind the edge of a blade to the proper angle for sharpness.

whey: The watery part of milk.

whip: To agitate ingredients in order to incorporate air.

white chocolate: A confectionary product made from cocoa butter, sugar, milk solids, and flavorings.

white mirepoix: A mirepoix made of onions, celery, and leeks or parsnips instead of carrots.

white potato: An oblong mealy potato with a thin, white or light-brown skin and tender, white flesh.

white stock: A light-colored stock produced by gently simmering poultry, veal, or fish bones in water with vegetables and herbs.

whole fish: The market form of a fish that is taken from the water and sold as is.

whole grain: A grain that only has the husk removed.

wine: The fermented juice of grapes or other fruit.

wine vinegar: Vinegar made from red or white wine.

wing: The tip, paddle, and drummette of a bird.

wing paddle: The second section of a wing located between the two wing joints of a bird. Also known as a wing flat.

wing tip: The outermost section of a wing of a bird.

winter squash: A fruit-vegetable that grows on a vine and has a thick, hard, inedible skin and firm flesh surrounding a cavity filled with seeds.

wood ear mushroom: A brownish-black, ear-shaped mushroom that has a slightly crunchy texture. Also known as a cloud ear or a tree ear mushroom.

workers' compensation: A type of insurance that compensates an employee for medical expenses and the loss of wages due to work-related injuries and illnesses.

work section: An area within the professional kitchen that is dedicated to a specific set of tasks.

work station: An area within a work section that is relatively self-sufficient and provides the necessary tools and equipment to perform specific tasks within the section.

Y

yam: A large tuber that has thick, barklike skin and flesh that varies in color from ivory to purple.

yearling turkey: A mature turkey that is less than 15 months old.

yeast: A microscopic, living, single-celled fungus that releases carbon dioxide and alcohol through a process called fermentation when provided with food (sugar) in a warm, moist environment.

yeast starter: A mixture of flour, yeast, sugar, and water.

yellow potato: An oval, waxy potato with thin, yellowish skin and flesh and pink eyes.

yellow squash: A summer squash that resembles a bowling pin with either a straight or crooked neck.

yield: The total quantity of a food or beverage item that is made from a standardized recipe.

yield percentage: The edible-portion (EP) quantity of a food item divided by the as-purchased (AP) quantity and is expressed as a percentage.

yogurt: A semisolid food made from milk that has fermented due to the addition of bacteria.

young goose: A goose that is usually less than six months of age and weighs approximately 4–10 lb.

young turkey: A male or female turkey that is less than eight months old.

Z

zest: The colored, outermost layer of the peel of a citrus fruit, which contains a high concentration of oil.

zester: A special cutting tool with tiny blades inside of five or six sharpened holes that are attached to a handle.

zucchini: An elongated summer squash that resembles a cucumber and is available in green or yellow varieties.

Index

Page numbers in italic refer to figures.

A

B

C

D

F

G

H

N

O

Q

R

S

V

W

Y

Z

USING THE *NATIONAL APPRENTICESHIP TRAINING PROGRAM FOR COOKS* INTERACTIVE DVD

Before removing the Interactive DVD from the protective sleeve, please note that the book cannot be returned for refund or credit if the DVD sleeve seal is broken.

WINDOWS SYSTEM REQUIREMENTS

To use this DVD on a Windows® system, your computer must meet the following minimum system requirements:

- Microsoft® Windows® 7, Windows Vista®, or Windows® XP operating system
- Intel® 1.3 GHz processor (or equivalent)
- 128 MB of available RAM (256 MB recommended)
- 335 MB of available hard disk space
- 1024 × 768 monitor resolution
- DVD drive (or equivalent optical drive)
- Sound output capability and speakers
- Microsoft® Internet Explorer® 6.0 or Firefox® 2.0 web browser
- Active Internet connection required for Internet links

MACINTOSH SYSTEM REQUIREMENTS

To use this DVD on a Macintosh® system, your computer must meet the following minimum system requirements:

- Mac OS X 10.5 (Leopard) or 10.6 (Snow Leopard)
- PowerPC® G4, G5, or Intel® processor
- 128 MB of available RAM (256 MB recommended)
- 335 MB of available hard disk space
- 1024 × 768 monitor resolution
- DVD drive (or equivalent optical drive)
- Sound output capability and speakers
- Apple® Safari® 2.0 web browser or later
- Active Internet connection required for Internet links

OPENING FILES

Insert the Interactive DVD into the computer DVD drive. Within a few seconds, the home screen will be displayed allowing access to all features of the DVD. Information about the usage of the DVD can be accessed by clicking on Using This Interactive DVD. The Quick Quizzes®, Illustrated Glossary, Flash Cards, Media Clips, Checkpoints, Culinary Math Applications, Certification Exam Preparation, Apprenticeship Online Portal, and ATPeResources.com can be accessed by clicking on the appropriate button on the home screen. Clicking on the ATP logo (www.atplearning.com) accesses information on related educational products. Unauthorized reproduction of the material on this DVD is strictly prohibited.

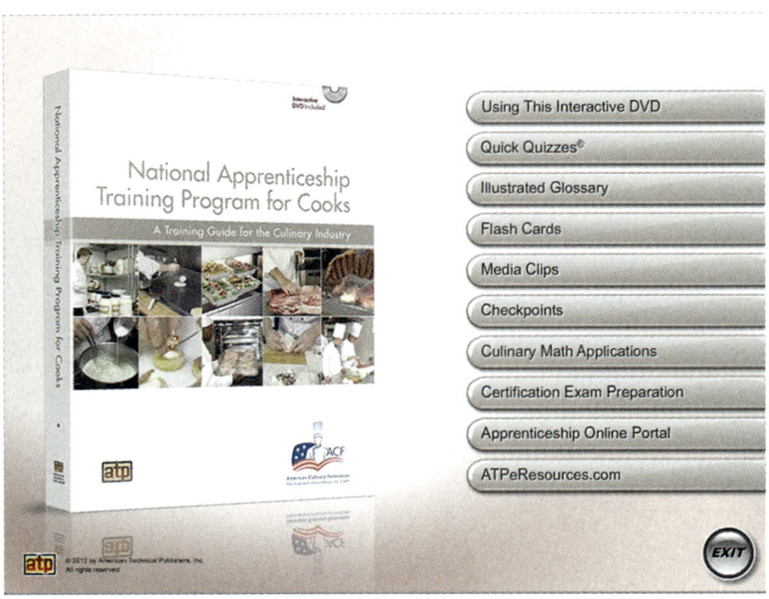